Evolution and Fossil Record of African Proboscidea

This book details the long, diverse, and complex phylogenetic history of elephants and their fossil relatives (the Proboscidea), whose origin is deeply embedded some 60 million years ago in Africa. Most of the major evolutionary events of proboscideans occurred in Africa, and these are examined in their geological, paleoecological, geographic, and faunal contexts. Updated information about feeding adaptations, taxonomy and systematics, genetics, and site occurrences is included and summarized in tables, figures, and charts. This is the first comprehensive review of African proboscideans and illustrates the need to more actively protect elephants and ensure their survival in modern ecosystems.

Key Features:

- Provides a comprehensive systematic review of the African proboscidean fossil record
- Includes a summary of taxonomy, geochronology, biogeography, and morphology
- Documents major faunal events including those associated with hominin origins
- Synthesizes new data from genomic, isotopic, and microwear analyses
- Emphasizes the role of elephants in ecosystems and the importance of conservation

Evolution and Fossil Record of African Proboscidea

William J. Sanders

CRC Press
Taylor & Francis Group
Boca Raton London New York

CRC Press is an imprint of the
Taylor & Francis Group, an **informa** business

Cover artwork by Karen Laurence-Rowe

First edition published 2024
by CRC Press
6000 Broken Sound Parkway NW, Suite 300, Boca Raton, FL 33487-2742

and by CRC Press
4 Park Square, Milton Park, Abingdon, Oxon, OX14 4RN

CRC Press is an imprint of Taylor & Francis Group, LLC

ISBN: 9781482254754 (hbk)
ISBN: 9781032474090 (pbk)
ISBN: 9781315118918 (ebk)

DOI: 10.1201/b20016

Typeset in Times
by codeMantra

"There is mystery behind that masked gray visage, an ancient life force, delicate and mighty, awesome and enchanted, commanding the silence ordinarily reserved for mountain peaks, great fires, and the sea."

—Peter Matthiessen, The Tree Where Man Was Born, 1972

"Ex Africa Semper Aliquid Novi"

—Pliny the Elder, Naturalis Historia, 77 A.D.

Dedicated to Mayra, who has lovingly and patiently sustained my passion for the study of elephants and their ancient forebears.

Contents

Foreword

There is a famous declaration by Pliny the Elder which maintains *Ex Africa Semper Aliquid Novi*. This statement avers that there is always something new out of Africa. The legitimacy of this historical declaration has survived over time and is as appropriate today as it was when first advanced. From the paleontological perspective, this proposition is strengthened by the occurrence of numerous paleontological sites and fossil remains within the continent revealed by many years of research. The richness of the fossil record demonstrates that Africa gave rise over deep time to innumerable, successive groups of mammals, of which many Pleistocene survivors still roam the continent today.

Africa occupies about a fifth of the world's landmass and is home to a disproportionally large percentage of all living mammals, inhabiting perhaps a greater diversity of habitats than any other continent. These species are and have been nurtured by the vast size of the continent, great contrasts in geological settings and landscapes, and tremendous ecological variability within the continent. Among these fossil and contemporary African mammalian assemblages are the majestic proboscideans, living and past. These animals arose in the continent or, more accurately, in what was once Afro-Arabia, with the great majority of their most significant evolutionary episodes happening within Africa. It is surprising to realize that for the almost first half of their 60-million-year existence, proboscideans were residents solely of the Afro-Arabian supercontinent, and their later significant migrations out of Africa to Eurasia and beyond began only some 35 million years after the origin of the Order. Modern proboscideans comprise only three species, the African savanna (*Loxodonta africana*) and forest (*Loxodonta cyclotis*) elephants, along with the Asian elephant, *Elephas maximus*. Modern elephants are the most prominent contemporary terrestrial animals and are fundamentally significant and vital members of their ecosystems because of their size. Many of their ancestral groups are preserved as fossils recovered from sites on the continent and are curated in African museums as widespread geographically as the Cairo Geological Museum, the National Museums of Kenya, and the Iziko South African Museum in Cape Town. New discoveries and a burgeoning fossil record, as well as new understandings about older collections, warrant a comprehensive review of these specimens.

Due to ongoing and intensified paleontological fieldwork efforts and an increased pace of discovery, since 1978 the fossil record of proboscideans has been extended farther back in time by nearly an additional 30 million years, and the number of new taxa has concurrently soared, accompanied by innovative studies of proboscidean morphology, novel phylogenetic analyses, and an impressive breadth of contextual studies refining the dating and paleoecology of the taxa. This is, then, a good moment to take stock of these advances and summarize the state of knowledge about the African representatives of the Proboscidea.

This book by Dr. William Sanders focuses on the documentation of new proboscidean taxa reported since the publication of the *Cenozoic Mammals of Africa* by Werdelin and Sanders in 2010, and updates prior work on previously known taxa, providing thorough descriptions of fossil material while refining the taxonomy and phylogeny of African proboscideans from the early Paleogene until the Holocene. Also, vitally, the author draws attention to concerns that modern elephants face today such as rapid extinction, and highlights the current threats facing the two species of African elephants, adding an essential voice of alarm from a paleontological point of view.

From an educational perspective in the continent, the African Union in its Charter for African Cultural Renaissance has reiterated the unity of Africa that is founded first and foremost on its history. For this reason, the African Union has sought to integrate the General History of Africa into educational programs within institutions in the continent. Such integration would form an essential component of decolonizing educational programs in Africa and worldwide. This book edition thus provides an excellent resource material in paleontological studies and research to enhance projects in African paleontology, particularly for a new generation of indigenous African paleontologists.

On the conservation standpoint, elephants may have been responsible for altering environments in a manner that favored the success of early hominins, our ancestors, the very lineage that threatens their survival today. Although the evolutionary history of proboscideans occurred over a long period of time, the modern elephants today are in danger of being decimated not through natural competitive factors but rather by human agency and the effects of human activities, including poaching, land conflicts, and anthropogenic-driven climate change.

At regional levels, the loss of elephants would be devastating for local ecosystems and fauna, perhaps catalyzing the loss of biotic diversity and entire habitats. At a global level, the failure to protect and save elephants would demonstrate an incapacity to maintain the planet with sound stewardship. As the long journey documented in the chapters of the book shows, the evolution of elephants has involved tremendous phyletic complexity and winnowing of lineages and morphology—this journey is irreproducible and elephants cannot be replaced.

Dr. Emma Mbua
Department of Earth Sciences,
National Museums of Kenya

Foreword Author's Biography

Emma Mbua is a paleoanthropologist in the Department of Earth Sciences at the National Museums of Kenya. She earned her Ph.D. from the University of Hamburg, Germany, in 2001 and was later appointed head of the Department of Earth Sciences, at the National Museums of Kenya. Dr. Mbua's research interests are focused on Plio-Pleistocene (ca. 4.0–0.5 Ma) hominin evolution and adaptations. She is currently undertaking research at the Kantis Fossil Site, an early hominin occurrence on the foothills of the Ngong Hills near Nairobi. Since the inception of her field research in this area, Dr. Mbua has recovered rich paleontological remains, including *Australopithecus afarensis* fossils dated to 3.5 million years old. In addition, she has published numerous peer-reviewed scientific papers in paleoanthropology individually and in collaboration with other scholars. Dr. Mbua is a recipient of the First Mary Leakey Award, presented by the Leakey Foundation in 1998, for her dedication to research in the field of paleoanthropology. In 2005, Dr. Mbua founded the East African Association of Paleoanthropology and Paleontology, whose main objective is to unite scholars through biennial scientific conferences and seminars. The association also undertakes mentorship of upcoming young scientists in eastern Africa in the field of prehistory.

Acknowledgments

The motivation for this book project was the belief of my editor, Chuck Crumly, that much more could be written about African proboscidean evolution and their fossil record than the thumbnail sketches of individual taxa in the *Cenozoic Mammals of Africa*, which I co-edited with his guidance a little over a decade ago, and his confidence that I could provide a more thorough account. He was correct about that—there are volumes that could be published about elephants and their ancestors—but he probably did not envision the amount of time it would take to obsessively organize the content of the current book, or the amount of research and field travel that were still needed between inception and completion of the project to make it truly relevant and current. Nonetheless, he kept the light burning for the delivery of the manuscript and has my deepest gratitude for his patience and perseverance, and for helping to make possible publications of this nature. A number of exceptional recent books on paleontology have Chuck's imprint on them, so I am honored to have had the opportunity to work with him again. I am also grateful to Kara Roberts and Iris Fahrer of Taylor & Francis, and Saranya Narayanan of CodeMantra for their professional assistance in organizing the diverse aspects of production of the book to bring it to publishable form.

I am extremely fortunate to have the artwork of Karen Laurence-Rowe grace the cover of the book. Her wildlife conservation paintings are among the most impressive of this or any other generation. The cover illustration is titled *Tsavo Land of the Ancient* and expertly captures some of the qualities of elephants that are so like us, as social animals that provide great care and protection of their young. The use of extant African elephants for the cover art corresponds with an inspiration for the book, the desire to demonstrate from a paleontological perspective the protracted and diverse evolutionary path that had to occur in order for animals such as elephants to come into existence, and therefore, how monumental a loss it would be for them to be extirpated—it is unbearable to contemplate a time when it would no longer be possible for artists like Karen to draw these magnificent animals from life.

My proboscidean studies and this project have benefited greatly from, and found inspiration in, discoveries made by many others living and past, and the paleobiological, taxonomic, and systematic foundations they have set, including as exemplars (but by no means limited to), the early 1900s work of C. W. Andrews on barytheres, moeritheres, and palaeomastodonts; recovery and study of Plio-Pleistocene proboscideans from North Africa and the Omo in Ethiopia by Camille Arambourg; Michel Beden's 1970s–1980s studies of elephants from the Omo, Ethiopia, Laetoli, Tanzania, and Koobi Fora, Kenya and his efforts to organize and partition *Elephas recki*; Thure Cerling's pioneering isotopic research that opened a window into the feeding behavior of ancient proboscideans; the research works of Basil Cooke on southern African proboscideans that spanned six decades; Yves Coppen's comprehensive efforts to describe and identify Central African proboscideans; Wilhelm Dietrich's descriptions and identifications of eastern African proboscideans, particularly from Olduvai and Laetoli, Tanzania, over a period from the 1910s to the 1950s; Emmanuel Gheerbrant's astute analyses recognizing the earliest proboscideans; the vital contributions of John M. Harris to our understanding of the adaptations and evolution of deinotheres and barytheres, as well as his careful documentation of the proboscideans of West Turkana, Kenya; recovery, description, and identification of proboscideans from the Middle Awash, Ethiopia by Jon Kalb (with Assefa Mebrate); laudable dissertations by Hassan Mackaye and Steven (Hanwen) Zhang, who bravely tackled enormous subject matters (a immense assemblage of Mio-Pliocene proboscideans from Chad, and novel reclassification of elephants based on cranial morphology, respectively); Vincent Maglio's magisterial taxonomic organization of the Elephantidae and novel descriptions of eastern African proboscideans in the 1960s–1970s, including erecting *Stegotetrabelodon orbus*, *Primelephas gomphotheroides*, *Loxodonta adaurora*, and *Elephas ekorensis* in a single historic publication; Hikoshichiro Matsumoto's careful, detailed investigations of moeritheres and palaeomastodonts in the 1920s that anticipated cladistic methodology, and his important accounts of Asian *Palaeoloxodon*; the explorations of the Sahara in the 1920s and 1930s and recovery at Sahabi, Libya of primitive elephants, stegodonts, and anancine gomphotheres by Carlo Petrocchi; the brilliant delineation of stegodontid phylogeny and taxonomy by Haruo Saegusa, and his admirable work on anancine gomphotheres, early elephants, and African palaeoloxodonts that embody elegant study of fossil taxa; Hezy Shoshani's enthusiastic embrace of all things proboscidean and especially his selfless fieldwork in Eritrea that led to the recovery of one of the oldest elephantimorphs; Pascal Tassy's unrivaled expertise on proboscidean phylogenetic systematics, and specifically his studies of proboscideans from the Miocene of eastern Africa, Lothagam, Kenya, the Western Rift in Uganda, and from the Baynunah Fm. in Abu Dhabi (United Arab Emirates); the sherlockholmesian deductions and precision of Adrian Lister's studies of mammoth evolution; and Heinz Tobien's encyclopedic accounting of craniodental morphological characteristics of an impossibly wide diversity of proboscidean taxa. There are many other researchers involved directly or indirectly with the collection, description, and identification of African proboscideans and studies of their geological, paleoecological, behavioral, biogeographical, and functional contexts whose work has taught me a great many things, but those included here and their accomplishments are sufficient to illuminate the tremendous breadth of prior and current discovery and knowledge that comprises the foundation necessary for contemporary review of the fossil record and evolution of these magnificent mammals.

One of the greatest pleasures of the research process that formed the background to this volume was the opportunity to study original fossil and modern specimens, in some of the true temples of science. The journey has taken me the length and breadth of Africa, from Cape Town to Cairo and Kinshasa to Dar es-Salaam, and beyond. For this, I am grateful to the following persons and institutions for their generous permissions and facilitating collections work: Meave Leakey, Emma Mbua, Fredrick Manthi, Mary Muungu, Pauline Mbatha, and Rose Nyaboke (National Museums of Kenya, Nairobi), Meave Leakey, Louise Leakey, the late Isaiah Nengo, and Martin Kirinya (Turkana Basin Institute, Turkwel and Ileret, Kenya), Thalassa Matthews, Kerwin van Willingh, and Graham and Margaret Avery (Iziko South African Museum), Muluneh Mariam, Alemu Admasu, and Zelalem Assefa (National Museums of Ethiopia, Addis Ababa), Ato Jara (Ethiopian Center for Research and Conservation of Cultural Heritage, Addis Ababa), the late Michael Mbago, Paul Msemwa, Amandus Kwekason, Avelin Malyango, Charles Msuya, and the late Christine Kiyembe (Tanzanian National Museums, Dar es-Salaam), the late Jean de Heinzelin (Museé Royal de l'Afrique Centrale, Tervuren), Ezra Musiime (Uganda Museum, Kampala), Mohammed Korany Abdel Gawad and Ahmed El-Barkooky (Cairo University), the late Mohammed el-Bedawi, the late Fathi Ibrahim Imbabi, and the late Yusry Attia (Geological Museum, Cairo), Mark Beech (Coastal Heritage and Palaeontology Section of the Historic Environment Department, Abu Dhabi Tourism and Culture Authority, Abu Dhabi, United Arab Emirates), Mohammed Arif (Geological Survey of Pakistan, Islamabad), Pascal Tassy and the late Leonard Ginsburg (Muséum national d'Histoire naturelle, Paris), Kay Behrensmeyer (National Museum of Natural History, Smithsonian Institution, Washington, DC), the late Malcolm McKenna (American Museum of Natural History), Adrian Lister, Jerry Hooker, Alan Gentry, Andy Currant, and Pip Brewer (Natural History Museum, London), Faysal Bibi (Museum für Naturkunde, Berlin), Philip Gingerich and the late Gregg Gunnell (University of Michigan Museum of Paleontology, Ann Arbor), and the late Elwyn Simons (Division of Fossil Primates, Duke University Lemur Center, Durham, NC).

I am also deeply grateful to those who provided access and in many instances generous financial and logistical support to study fossil proboscideans in their projects, either in museums or at paleontological field sites or both, notably Terry Harrison (Manonga and Laetoli, Tanzania), Laura MacLatchy (Moroto, Napak, and Bukwa, Uganda), Meave and Louise Leakey (Ileret, Turkwel, and Koobi Fora, Kenya), Ellen Miller (Wadi Moghara, Egypt and Buluk, Kenya), the late Isaiah Nengo (Buluk, Kenya), Xueping Ji and Nina Jablonski (Shuitangba, China), John Fleagle (Fejej, Ethiopia), Kieran McNulty (Rusinga and Mfwangano, Kenya), Noel Boaz (Lusso Beds, Democratic Republic of Congo and Sahabi, Libya), the late Hezy Shoshani (Dogali, Eritrea), John Kappelman (Chilga, Ethiopia and Sinap Fm., Turkey), Mikael Fortelius (Sinap Fm., Turkey), Mercedes Gutíerrez (Losodok and Nakwai, Kenya), Fredrick Manthi, Carol Ward, and Michael Plavcan (Kanapoi and Natodomeri, Kenya), Scott Simpson (Gona, Ethiopia), Iyad Zalmout (Shumaysi Fm., Saudi Arabia), the late Andrew Hill, Faysal Bibi, Mark Beech, and Brian Kraatz (Baynunah Fm., Abu Dhabi, United Arab Emirates), Susanne Cote (Moruorot and Kalodirr, Kenya), the late Bill Kimbel (Hadar Fm., Ethiopia), Yohannes Haile-Selassie (Woranso-Mille, Ethiopia), and Eric Seiffert (Birket Qarun, Egypt).

I am greatly thankful for logistical assistance and kind hospitality extended to me during my museum research and fieldwork travels, particularly by Ron Mininger (Democratic Republic of Congo), Christine Steininger and Margaret Avery (South Africa), Amandus Kwekason (Tanzania), Jerry Hooker and Alan Gentry (London, United Kingdom), Rose Nyaboke, Emma Mbua, and Meave Leakey (Kenya), Ellen Miller (Durham, North Carolina), Mark Beech (Abu Dhabi), the late Jean de Heinzelin (Belgium), and Samson Tsegaye (Ethiopia). Assistance with significant aspects of the book project, for which I am very appreciative, was provided by Gerhard (Sol) Mundinger (*Primelephas* data collation, analysis, and interpretation), Carol Abraczinskas, John Klausmeyer (particularly for teaching me the magic of Photoshop and Adobe Illustrator), and the late Bonnie Miljour for their assistance with constructing figures, Jacob Lusk (reference editing), and Melissa Wood (help with Meshlab and Blender and creating *Loxodonta adaurora* images). I am also very thankful for the encouragement of my Director in the University of Michigan Museum of Paleontology, Matt Friedman.

Special mention and highest appreciation are owed to Carol Abraczinskas for her dedicated and skilled expertise creating and organizing figures, and the numberless hours spent researching map localities and correcting tables. Her keen eye for detail caught many errors on my part and led to improvements in many aspects of the volume. Carol created Figures 2.1–2.7, 3.1–3.4, 3.6–3.9, 3.11–3.14, 3.17–3.21, 3.23–3.25, 4.1, 4.10, 5.1, 5.8, 6.1, and 7.1, including all of the maps. Her contributions were vital to the production of this volume, as is true of her work on numerous publications out of the University of Michigan Museum of Paleontology.

Many friends and colleagues have generously shared their ideas about fossil proboscideans and results of their research efforts, which have enriched my scope of knowledge and were vital to my work, including Emmanuel Gheerbrant, Gina Semprabon, Juha Saarinen, Steven Zhang, Adrian Lister, Wighard von Koenigswald, the late Hezy Shoshani, Ursula Göhlich, the late Jon Kalb, Thure Cerling, the late Haruo Saegusa, John Kingston, Pat Holroyd, Cyrille Delmer, the late Andrew Hill, Pascal Tassy, John Harris, the late Frank Brown, Rodolphe Tabuce, Lionel Hautier, the late Wendy Dirks, the late Basil Cooke, Kati Huttunen,

David Lambert, Georgi Markov, Fredrick Manthi, Pauline Mbatha, Nichole Lohrke, the late Richard Leakey, the late Peter Robinson, Yohannes Haile-Selassie, Alan Gentry, Martin Pickford, Al Roca, Jean-Jacques Jaeger, Adam Rountrey, Irisa Arney, Nancy Todd, and ShiQi Wang.

My deepest, heartfelt gratitude is extended to those who have had the most profound effects on my scholastic trajectory, fired my passion for this work, helped me at critical moments of my training, studies, and career, taught me invaluable lessons about how to do this work, and offered unconditional friendship and love that continues to sustain me: my late parents, Joan Eimer and Karl Sanders, who understood the value of zoos, museums, and libraries for a young imagination, encouraged me to follow my boyhood dreams of searching for fossils in Africa, and despite limited means sacrificed substantially to give me the opportunity of a higher education; my advisor, Terry Harrison, who mentored me with patience and kindness, took time to introduce me to senior colleagues, taught me the taxonomic value of properly collecting and organizing fossil data, encouraged my initial study of proboscidean morphology, and generously entrusted to me the study of proboscidean fossils in his projects at Manonga and Laetoli; the late Basil Cooke, who encouraged me to pursue the study of fossil proboscideans and provided valuable feedback and advice on my first efforts in the study of Western Rift fossils; Ray Bernor, who provided timely advice and support during my graduate student years; Alan Gentry, who took the time to answer my research questions when I was just starting out; John Kappelman, a stalwart friend who gave me wonderful opportunities to study proboscideans from his projects in the Sinap Formation in Turkey and Chilga, Ethiopia; my sisters, Jean Parsons and Christine Kesling Hankins, whose work as educators inspires me and whose reminders about my youthful imperfections keep me humble; the late Gregg Gunnell, my friend, co-worker, and colleague for over a quarter of a century and my in-house sounding board and mentor during his time at the University of Michigan; Ellen Miller, my great good friend from deep time in graduate school, whose approbation about my work means so much to me—some of the greatest joys in my career, after all these years, have been to work alongside Ellen at her field site of Buluk and to share with her the simple pleasure of watching the sun set over Lake Turkana from Ileret Camp; and above all, my wife and love of my life, Mayra Rodríguez, who has endured all my travels away and time spent on research projects, sustained me emotionally, materially supported my projects, encouraged me at all turns, "adopted" orphaned elephants with me, and who, through this long passage together across our lives, understood my enchantment with this work and Africa better than anyone else. I could not have achieved anything without her. She allows me to once again be that little boy who dreams of finding fossils in Africa.

Author's Biography

William J. Sanders, PhD, ("Bill") attended college at the University of Chicago and earned his PhD in Paleoanthropology from New York University in 1995. His doctoral dissertation on the australopithecine vertebral column was recognized for distinction by a Dean's Best Science Dissertation Award. At the University of Michigan, Bill holds the position of Senior Research Laboratory Specialist and is an Associate Research Scientist in the Museum of Paleontology and Department of Anthropology. His scholarly interests include the taxonomy, systematics, evolution, paleoecology, and morphological adaptations of Old World fossil mammals, particularly of Afro-Arabian proboscideans and catarrhine primates. Bill's research is field- and specimen-based, which has led to paleontological and museum work in China, Pakistan, Turkey, the United Arab Emirates, the United Kingdom, France, Belgium, Germany, and across Africa from "Cape to Cairo" (South Africa, Tanzania, Kenya, Uganda, Ethiopia, Democratic Republic of Congo, and Egypt). Results of his 40 years of research investigations have been widely published in academic journals and presented at professional venues. In 2010, he co-edited (with Lars Werdelin) the *Cenozoic Mammals of Africa* (University of California Press), which received a PROSE Award from the American Publishers Awards for Professional and Scholarly Excellence, Single Volume Reference in Science. In 2017, Bill was awarded a Lifetime Achievement Award by the Association for Materials and Methods in Paleontology for high standards of professionalism and mentoring of younger colleagues.

Abbreviations

MORPHOLOGICAL ABBREVIATIONS

ac pretrite anterior accessory conule (e.g., ac1 is an accessory conule anterior to the first loph(id) or plate on the pretrite side of the crown)
C or **c** upper or lower canine
DI or **di** upper or lower deciduous incisor
dP or **dp** deciduous premolar (e.g., dP3 is an upper third premolar and dp3 is a lower third deciduous premolar)
ET enamel thickness
H crown height
HI hypsodonty index, H×100/W
I or **i** upper or lower incisor (e.g., I2 is an upper second incisor and i2 is a lower second incisor)
L length
l. left
LF lamellar frequency, number of loph(id)s or plates per 100 mm
M or **m** upper or lower molar (e.g., M1 is an upper first molar, and m1 is a lower first molar)
MC metacarpal (e.g., MC III is the third or middle metacarpal)
MT metatarsal (e.g., MT III is the third or middle metatarsal)
P or **p** upper or lower premolar (e.g., P3 is an upper third premolar, and p3 is a lower third premolar)
pc pretrite posterior accessory conule (e.g., pc1 is an accessory conule posterior to the first loph(id) or plate on the pretrite side of the crown)
poa posttrite anterior accessory conule (e.g., poa1 is an accessory conule anterior to the first loph(id) or plate on the posttrite side of the crown)
pop posttrite posterior accessory conule (e.g., pop1 is an accessory conule posterior to the first loph(id) or plate on the posttrite side of the crown)
r. right
W width
x refers to a pre- or postcingula(id), as in x3x, denoting a tooth comprised of three loph(id)s with pre- and postcingulae(ids)
X denotes a large pre- or postcingulum(id) or nascent additional loph(id) or plate; **+**, indicates a missing portion of a tooth, and that the original dimension was greater (e.g., L=+125.0 mm indicates a missing morphology at the anterior end of a tooth); **1, 2, 3, …**, cheek tooth loph(id) or plate number counted from the anterior of the crown; **I, II, III, …**, cheek tooth loph(id) or plate number counted from the posterior of the crown.

GEOLOGICAL AND TEMPORAL ABBREVIATIONS

BP before present
Fm. formation
Ka thousand years
Kyr thousand years (passage of time)
Ma mega-annum, millions of years
Mb. member
Myr million years (passage of time)
y years.

INSTITUTIONAL ABBREVIATIONS

A Virginia, South Africa
AD Arrisdrift, Namibia
A.L. Afar locality, Hadar, Ethiopia
AM Auchas, Namibia
AMSE AMSE pit at Auchas, Namibia
-AT Aterir
BAR Cheparawa, Kenya
-BC Chemeron
BMNH followed by a number series, The Natural History Museum, London (formerly the British Museum [Natural History])
BOU-VP Bouri, Ethiopia
BQ Birket Qarun, Egypt
C followed by a series of numbers, Cairo Geological Museum
CGM Cairo Geological Museum
CH Chilga specimens in the National Museums of Ethiopia
CPSGM Collections Paléontologiques du Service Géologique du Maroc
DPC Duke University Primate Center
-EK Ekora, Kenya
EP Tanzanian National Museums (Eyasi Plateau)
-ER East Rudolf (East Turkana; e.g., Koobi Fora, Ileret)
FSC-SK Skoura specimen, Ain Chock Faculty of Science, Casablanca, Morocco;
FSL Kabylie, Algeria
GTS Dakhla area, "Garitas" locality, Laboratory of Paleontology, Institut des Sciences de l'Evolution, Montpellier
IPUB Laetoli, Tanzania

IE	Omo, Ethiopian National Museums
-KA	Karungu, Kenya
KBA	Kanam or Maboko
KI	Uganda Museum, Kampala (Kisegi-Nyabusosi)
KK	field designation, Kakesio, Tanzania
KL	Kollé, Chad
KNM	National Museums of Kenya
-KP	Kanapoi
KT	Koro-Toro, Chad
L	followed by a number series (e.g., L 124-1), Ethiopian National Museums (Middle Awash), or Gebel Zelten, Libya
LAET	Laetoli, Tanzania
-LS	Losodok
-LT	Lothagam
M	followed by a row of numbers, Natural History Museum, London
MB.Ma	Museum für Naturkunde, Berlin;
MCZR	Museo Civico, di Zoologia, Rome
-ME	Meswa Bridge
-MI	Mwiti (Kajong)
MKM-VP	Makah Mera, Ethiopia
MMK	McGregor Museum, Kimberley, South Africa
MNHN	Muséum national d'Histoire naturelle, Paris
MOR	Moroto, Uganda
-MP	Mpesida Beds
MSD-VP	Megsid Dora, Woranso-Mille, Ethiopia
NAP	Napak, Uganda
-NK	Lemudong'o
NAP	Napak
NK	Uganda Museum, Kampala (Nkondo-Kaiso area, Nyawiega)
-NW	Nakwai, Kenya
ONHM	Oman Natural History Museum, Muscat
PQAD	Arrisdrift, Namibia
PQ-L	followed by a number series (e.g., SAM-PQ-L 2562), Langebaanweg
PU	Windhoek Spring Deposit, Namibia
-RU	Rusinga
SAM	Iziko South African Museum
-SB	Kubi Algi, Kenya
SGS	Saudi Geological Survey
-SH	Samburu Hills, Kenya
SHUM	Shumaysi Fm., Saudi Arabia
-SO	Songhor, Kenya
SQU	Sultan Qaboos University, Dhofar Governorate, Oman
T	Djebel Nara, Tunisia
-TH	Tugen Hills
UM(MP)	University of Michigan Museum of Paleontology
UON	Bir el Ater, Algeria
WM	Tanzanian National Museums (Wembere-Manonga)
-WS	Buluk (West Stephanie)
-WT	West Turkana
Y. or **Y.S.**,	Mursi Fm., Ethiopia
Z	Gebel Zelten, Libya.

OTHER ABBREVIATIONS

Alt.	alternative, referring to competing calculations of geological ages of site units or taxonomic hypotheses
cm	centimeter
EQ	encephalization quotient, relative brain size, calculated as the ratio of observed to expected brain mass for an animal of a given size
g	gram
h	hectare
Hz	hertz, unit of frequency, number of cycles per second
kg	kilogram
km	kilometer
m	meter
mm	millimeter
sp.	species
spp.	multiple species
t	ton
?	occurrence uncertain.

Glossary

DENTAL DEFINITIONS

Abaxial conelet, the outer, main conelet in each half-loph(id) (Tassy, 1996a)

accessory central conules, enamel-covered pillars situated at the anterior and/or posterior faces of the loph(id)s or plates, or in the transverse valleys, partially blocking them centrally (Tobien, 1973b)

adaxial conelet(s), the inner, or meso-, conelet(s) in each half-loph(id) (Tassy, 1996a)

anancoidy, alternation of paired half-loph(id)s, in which lingual half-loph(id)s are anterior to buccal half-loph(id)s (Tobien, 1973b)

brachyodonty, condition of low crowns in cheek teeth, in proboscideans measured as crown height x 100/width, ≤75

bunodonty, cheek tooth crowns comprised of independent rounded or bulbous cusps that may be accompanied by mesoconelet(s) (e.g., *Phiomia* molars)

bunolophodonty, cheek tooth crowns comprised of rounded or bulbous outer cusps or conelets incompletely connected by transverse crests or conelets, usually divided by a median sulcus

cementum, cementodonty, hardened, calcified dental tissue invested around or covering cheek tooth loph(id)s or plates, sometimes infilling transverse valleys, and when abundant on a cheek tooth crown comprises the condition of "cementodonty" (Tobien et al., 1986)

chevroning, the arrangement of half-loph(id)s to occlusally form an anteriorly pointing V, or chevron (Tobien, 1975)

choerodonty, occurrence of accessory tubercles within transverse valleys (Osborn, 1942)

crescentoids, enamel crests running from the apices of abaxial conelets of pretrite half-loph(id)s to the bottom of transverse valleys, and ending near the middle axis of the crown (Tobien, 1975)

ectoflexus, an indentation on the outer side of a cheek tooth, usually between loph(id)s

endoflexus, an indentation on the inner side of a cheek tooth, usually between loph(id)s

hypsodonty, condition of high-crowned cheek teeth, in proboscideans measured as crown height × 100/width, ≥100

intermediate molars, dP4/dp4, M1/m1, and M2/m2

lophodonty, cheek tooth crowns comprised of uninterrupted transverse crests comprising the loph(id)s, which may be laterally anchored by incorporated cusps (e.g., deinothere molars)

mamelon(s), small tubercle(s) along a ridge or crest on a tooth

mesodonty, condition of intermediate relative height of cheek teeth, in proboscideans measured as crown height × 100/width, =76–99

posttrite, refers to the less worn half of each loph(id), which is lingual in lower and buccal in upper molars (Vacek, 1877)

pretrite, refers to the more worn half of each loph(id), which is buccal in lower and lingual in upper molars (Vacek, 1877)

ptychodonty, plication or infolding of enamel borders with grooving of the sides of the molars (Osborn, 1942)

trefoil, a tripartite enamel wear figure of a half-loph(id) formed by conelets and associated anterior and posterior accessory central conules

tusk, an outsized, projecting, or protuberant anterior tooth, which in proboscideans is an incisor, either I/i1 or I/i2

zygodont crests, enamel crests running from the apices of abaxial conelets of posttrite half-loph(id)s to the bottom of the transverse valleys, and ending near the middle axis of the crown (Tobien, 1975).

Introduction

DESIDERATA

> Morning's dew coalescing on the tent tarps and streaming off. Dissonant melody of Maasai cowbells as the herd passes. Giraffes silhouetted on the horizon across from camp, in the early morning mist, fifteen in number, loping along.
>
> **—W. J. Sanders' field camp notebook entry, 20 July 2004, Laetoli, Tanzania**

Africa holds extraordinary prospects for the paleontological investigation of mammalian evolution. Largely still inhabited by relic Plio-Pleistocene mammalian faunas, it is populated by animals such as giraffes, rhinos, lions, gazelles, zebras, okapis, and giant forest hogs, as well as descendants of endemics of longer residence such as hyraxes and catarrhine primates, beneficiaries of having been sheltered from the most extreme effects of climate and ecological change due to the position of the landmass astride the equator, the only continent to span both northern and southern temperate zones (Sanders and Werdelin, 2010). Occupying 20% of the world's landmass, Africa is home to approximately a quarter of all living mammals (Kingdon, 1997; Medellín and Soberón, 1999), encompassing a tremendous range of species and adaptations, fostered by the size and physical and ecological heterogeneity of the continent (Kingdon, 1989; O'Brien and Peters, 1999).

Elephants are the exemplars of long-term endemic mammal survivorship on the continent. Notably, Africa, or more accurately Afro-Arabia, was the deep-time landmass of proboscidean origins and of nearly all of their most significant evolutionary episodes. No eutherian mammals can trace their unbroken presence in Africa as distant in time as proboscideans. Remarkably, for approximately the first half of their 60-million-year existence, proboscideans were denizens solely of the island continent, and their meaningful forays out of Africa to Eurasia and beyond began only some 40 million years after their origin, first in the form of early Miocene emigrations of mammutids, gomphotheres, and deinotheres, and later by geographic expansion of elephants during the Plio-Pleistocene. Some Eurasian proboscideans, with ancient precursors originally from Africa, returned to the continent in the late Miocene, including stegodonts, anancine gomphotheres, and most importantly, tetralophodonts, which gave rise to elephants. The African history of Proboscidea is rich with taxonomic and adaptive diversification and instances of competitive winners and losers that constantly winnowed and reshaped the composition and morphological impulse of the Order, leading in intermittent fashion to the immense body size, graviportal postcranial structure, hypertrophy of incisors into tusks, development of trunks, infrasonic hearing and sound production, horizontal tooth displacement, and fore-aft proal mastication embodied in their totality by extant elephants. Understanding the breadth and depth of the phylogenetic journey that led to this impressive accumulation of morphological adaptations compels an encompassing review equivalent to the scope of African proboscidean evolution.

It is possible to more comprehensively document key phylogenetic events in the evolution of African proboscideans and other vertebrate groups now because their fossil records have expanded markedly over the last 40 years as a consequence of the increasingly insatiable hunt for human ancestors. These burgeoning fossil assemblages require large-scale revisions of their taxonomic, phylogenetic, and contextual frameworks. For example, the temporal range of fossil proboscideans is now understood to be twice the geological duration known in 1978 (contrast Coppens and Beden, 1978, Coppens et al., 1978, and Harris, 1978a with Sanders et al., 2010a). Advances in radiometric dating, chemostratigraphy, and paleomagnetic sequence stratigraphy have also greatly refined the chronostratigraphy of sites, improving calibration of fossil proboscidean temporal contexts (for instance, see Tamrat et al., 1995; Renne et al., 1999; Feibel, 2003; Seiffert et al., 2005a; WoldeGabriel et al., 2005; Deino et al., 2010; Brown and McDougall, 2011; McDougall et al., 2012; Coster et al., 2012; Kidane et al., 2014; Kocsis et al., 2014; Yans et al., 2014). Genomic innovations in the analysis of ancient DNA and large-scale genetic studies of extant mammals have contributed valuable and unexpected insights about proboscidean systematics, particularly about the relationships of other mammals to proboscideans and among elephants (e.g., Springer et al., 1997; Roca et al., 2001; Asher et al., 2003; Roca and O'Brien, 2005; Rohland et al., 2010; Meyer et al., 2017; Palkopoulou et al., 2018). Dietary preferences of proboscideans are more directly understood because of new analytical approaches involving dental isotopes, microwear, and mesowear (e.g., Koch, 1991; Cerling et al., 1999; Zazzo et al., 2000; Passey et al., 2005; Clementz et al., 2008; Levin et al., 2008; Liu et al., 2008; Kingston, 2011; Kingston et al., 2011; Uno et al., 2011; Saarinen et al., 2015; Lohrke, 2017; Manthi et al., 2017; Semprebon et al., 2017, 2018; Uno and Bibi, 2022). In addition, great emphasis has been placed on investigating the relationship between global climate, local environmental conditions, and tectonic activity and their effects on African faunal evolution (e.g., Williamson and Savage, 1986; Janis, 1993; Alemseged, 2003; Warny et al., 2003; Bobe and Behrensmeyer, 2004; Bonnefille et al., 2004; Sepulchre et al., 2006; Harzhauser et al., 2007; Pik et al., 2008; Edwards et al., 2010; Wichura et al., 2010; Kingston et al., 2011; Levin et al., 2011; Strömberg, 2011; Blumenthal et al., 2017; Linder, 2017). Because of these advances, phylogenetic patterns and adaptive drivers of proboscidean evolution can be investigated with greater accuracy and confidence.

Extant Proboscidea is composed of only three species, the African savanna and forest elephants, *Loxodonta africana* and *Loxodonta cyclotis*, and the Asian elephant, *Elephas maximus*. Elephants are the largest contemporary terrestrial animals, and because of their size are fundamentally significant and vital members of their ecosystems (Eltringham, 1992; Shoshani and Tassy, 1996). Since their inception in the late Miocene, there have been a great many species of elephants, distributed vastly more widely than today—indeed, proboscidean diversity, particularly in the Miocene, was once considerably broader. It is difficult to imagine the fecundity and biological opulence, the carrying capacity, of ecosystems that could have simultaneously accommodated multiple, massive-sized proboscidean species, yet the African fossil record reveals that in the past this was common. This informs us that there are no modern analogs of ancient African ecosystems, and that we are only beginning to understand the interplay between extinct proboscidean taxa, climate, and paleoecology.

The purpose of the present account of the fossil record and evolution of this impressive diversity of proboscideans in Africa, or Afro-Arabia to be precise for the Paleogene and the early Neogene, is two-fold: First, the fossil record of African proboscideans has greatly expanded and requires a more comprehensive, current reckoning. Delineating the interplay of competition, origination, and extinction in the evolution of African proboscideans requires a deep-time perspective of the taxa. Prior attempts to recount the entirety of the African proboscidean fossil record include the Proboscidea volumes of Osborn (1936, 1942), and chapters in Maglio and Cooke's (1978) landmark volume *Evolution of African Mammals* (Coppens and Beden, 1978; Harris, 1978a; Coppens et al., 1978). African fossil elephants were synopsized around the same time in authoritative works by Arambourg (1970), Maglio (1973), and Beden (1980), and soon thereafter Tassy (1986) provided a splendid, methodical review of the African Miocene proboscidean fossil record, based on the evidence from Kenyan sites. Much earlier, magnificent descriptive and taxonomic treatments of Paleogene proboscidean groups such as barytheres, moeritheres, and palaeomastodonts had been carried out by Andrews (1906) and Matsumoto (1922, 1923, 1924). These summaries were accomplished during heady intervals of novel expeditionary work throughout the continent. Since 1978, following renewed paleontological fieldwork efforts and an increased pace of discovery, the fossil record of proboscideans was extended farther back in time by nearly an additional 30 million years, and the number of new taxa has concomitantly soared, accompanied by innovative studies of proboscidean morphology and novel phylogenetic analyses (e.g., Tassy, 1981, 1982, 1988, 1990; Mahboubi et al., 1986; Roth, 1989; Court, 1994a, b; Shoshani, 1996; Mackaye, 2001; Pickford, 2003a; Sanders, 2004, 2007, 2017; Delmer, 2005, 2009; Gheerbrant et al., 2005; Tabuce et al., 2007; Saegusa and Gilbert, 2008; Saegusa and Haile-Selassie, 2009; Ferretti and Debruyne, 2011; von Koenigswald, 2014, 2016a; Larramendi, 2016; Larramendi et al., 2020; Saegusa, 2020; Zhang, 2020), as well as by improved contextual studies of stratigraphy, chronology, paleoecology, and biogeography (e.g., Tassy, 1990a; Feibel, 1999; Ludwig and Renne, 2000; Van der Made and Mazo, 2003; Levin et al., 2011; Uno et al., 2011; McDougall et al., 2012). As a result, it became possible to produce a more contemporary, updated review of African proboscideans (Sanders et al., 2010a) as part of the successor volume to the *Evolution of African Mammals* (Maglio and Cooke, 1978), the *Cenozoic Mammals of Africa* (*CMA*; Werdelin and Sanders, 2010). The present book offers documentation of new taxa reported since the publication of the *CMA*, provides more thorough descriptions of fossil material, greater contextual information, and hones the taxonomy and phylogeny of African proboscideans.

Second, a more urgent reason to describe the entirety of African Proboscidea is to draw attention to the pending extinction of the extant members of the Order—both species of African elephants are now considered endangered to critically endangered in the Red List of the International Union for Conservation of Nature (Thouless et al., 2016). It is essential to add the voice of alarm from a paleontological perspective about the ecological and evolutionary magnitude of such a tragedy. Documenting the phylogenetic journey of African proboscideans underscores the depth of biological tragedy that would occur by the extinction of these animals. The evolutionary path that led to African elephants has been tens of millions of years in the making, involved multiple, richly varied branchings, withstood tectonic, climatic, ecological, and faunal upheavals, and involved impressive morphological transformations that are equaled by few other mammalian Orders. Elephants and the evolutionary sequence that produced them are grand, singular phenomena and cannot be remade, replaced, or replicated.

Proboscideans have been keystone taxa in African ecosystems throughout their existence, sustaining those ecosystems and imposing outsized effects on community structure and habitat accessibility for other organisms. Ironically, elephants may well have been responsible for altering environments in a manner that favored the success of early hominins and subsequent human evolution, the very organisms that now threaten their survival (Sanders, 2020). Their intelligence, behavior, adaptations, and social structure rival and largely parallel our own, which makes their destruction at the hands of poachers and hunters particularly egregious and heart-rending. Although the deep evolutionary history of proboscideans is a long series of outcomes of winners and losers, modern elephants are in danger of being extirpated not through natural competitive factors, but rather by greedy, unnecessary, and avoidable human agency—conflicts over land, increased pace and amplitude of anthropogenic climate change, pursuit of bush meat, and ivory poaching. The paleontological record provides a powerful argument for urgent protection of the modern inheritors of a complex and irreproducible evolutionary trek of 60 million years, which involved the natural sifting of tremendous phyletic diversity. As intelligent beings, we should have a greater comprehension of the appalling magnitude of impending elephant extinction, given how arduous and intricate it was for evolution to produce elephants. The chapters of the book record the temporal and phylogenetic spans of that evolutionary journey. At regional levels, the loss of elephants would have profound impacts on ecosystems and faunas, triggering loss of biotic diversity and entire habitats. At a global scale, the looming extirpation of these grand animals should frighten us because it would slash a significant, irreparable hole in the fabric of organismal interdependence that protects our own existence.

Taxonomic chapters (2–7) that comprise the heart of the book are temporally ordered and arbitrarily partitioned around major phylogenetic events or branchings, following the provision of contextual information in Chapter 1 about the classification, higher order relationships, and temporal extent of African proboscideans. They proceed in the following order:

Chapter 2, early-middle Paleogene plesielephantiform origins of Proboscidea, succession of the earliest taxa, and possible earliest appearance of elephantiforms;

Chapter 3, late Paleogene appearances of barytherioids, moeritheres, the first substantial record of elephantiform proboscideans (palaeomastodonts), and the origin of deinotheres and elephantimorphs (mammutids and gomphotheres);

Chapter 4, early and middle Miocene phylogenetic diversification of elephantimorphs, their replacement of the old cohort of barytherioids, moeritheres, and palaeomastodonts, and the first significant emigration events of gomphotheriids, mammutids, and deinotheres out of Africa to Eurasia;

Chapter 5, late Miocene sequential immigration events of Eurasian proboscideans into Africa, notably tetralophodontines, anancine gomphotheres, and stegodonts, the disappearance of older gomphotheriid and mammutid lineages, and, most notably, evolution and initial diversification of the first elephants;

Chapter 6, early Pliocene replacement of archaic elephants by more derived species, some of which belong to crown elephant clades, and appearance of more advanced forms of stegodonts and anancine gomphotheres;

Chapter 7, mid-Pliocene to late Pleistocene dominance of elephants in the "*Elephas recki* complex"; competing hypotheses about evolutionary modes, phylogeny, and classification of elephants in the "*Elephas recki* complex"; persistence and ultimate success of loxodont elephants, northward sequestration and loss of mammoths from the continent, and extinction of the last African palaeoloxodonts, anancine gomphotheres, stegodonts, and deinotheres.

Chapters are accompanied by tables summarizing site occurrences of species and provide key geological and taxonomic references, and by maps showing site locations. The numbering system of site occurrences in the maps bears some explanation: In some instances, a dot on the map represents more than one site, accommodated by using the same number multiple times with different letters (for example, in Map 7.1, "4a. Oued Akrech, Morocco" and "4b. Fouarat, Morocco"); alternatively, authors discussing the same site might have used different iterations of names for the site, represented by a single number with multiple letters (e.g., in Map 7.1, "59a. Ileret, Kenya," and "b. Areas 1A, 6, 6A, 8A, 11, Ileret, Kenya"), or there might be individual localities within larger site areas, depicted with the same numbering scheme (e.g., in Map 7.1, "84a. Elandsfontein, South Africa" and "b. Bone Circle assemblage, Elandsfontein, South Africa"). Although in this system site names may appear redundant in places in the figure titles for maps, the goal was to account for the numerous iterations of site names common in the literature that might be familiar to the reader.

Figures of key fossils useful for encapsulating the morphology of species are included where possible, as are tables summarizing tooth formulae and comparative morphometric information. The majority of specimens discussed in the chapters have been examined first-hand over a 40-year interval in a comparative morphological framework (measured, drawn, annotated, photographed), although for a variety of reasons—embargoes, geopolitical hazards, specimens in study by other researchers, missing items—some fossil materials were inaccessible for this research. Nonetheless, fieldwork and museum studies are supplemented by thorough survey of published scholarly accounts of Afro-Arabian proboscideans and extensive communications with other researchers.

An effort has been made to accurately and fairly represent differing interpretations and hypotheses about the taxonomy and phylogeny of the constituent taxa of the book. Because of the temporal and geographic unevenness of the fossil record and the limits this places on interpretation, the inferences reached in the book represent ongoing conversations about taxonomy, systematic relationships, phylogenetic history, and evolutionary modes. Anatomical details are viewed through the lens of synapomorphies and symplesiomorphies, and distinctive or diagnostic characteristics are recounted for each taxon. However, production of large-scale cladistic analyses of African proboscideans is beyond the scope of the book. Instead, the goals of the volume are to illuminate the African proboscidean fossil record as fully as possible, achieve a reasonable classification and phylogeny, and to present the current state of knowledge about morphology and contexts—faunal, biogeographic, ecological, temporal—of the many proboscidean taxa that have existed in Africa.

Homoplasy is rife within Proboscidea—for instance, a number of different proboscidean lineages independently lost lower tusks and premolars, elephant lineages separately achieved hypsodonty, and amebelodonts, choerolophodonts, and anancine gomphotheres all convergently evolved transverse offset of their pre- and posttrite half-loph(id)s. Failure to recognize or properly account for parallelisms and convergences poses a hazard for the retrodiction of proboscidean relationships, potentially overwhelming synapomorphies with homoplasies. Nonetheless, based on the robustness of paleontological evidence revealed by the comparative morphological study of African proboscideans, the fossil record demonstrably provides a powerful instrument for systematic analysis and recovery of phylogenies, and therefore, much of what is inferred using this evidence constitutes well-supported hypotheses.

1 Context of African Proboscidean Evolution

It is impossible to speak of the objects of any study, or to think lucidly about them, unless they are named. It is impossible to examine their relationships to each other and their places among the vast, incredibly complex phenomena of the universe, in short to treat them scientifically, without putting them into some sort of formal arrangement.

—George G. Simpson (1945)

INTRODUCTION

In his review of taxonomy and classification, Simpson (1945: p. 11) advocated that the classificatory system of zoologists and paleontologists should constitute a method for researchers to communicate to one another—a "practical means by which [they] may know what they are talking about and others may find out"—and emphasized that its principal objective is transmission of phylogenetic information. Organizing fossil organisms into species permits the study of their higher order relationships, which, in turn, within the contexts of time, tectonics, biogeography, and paleoecology may provide insights into the evolutionary drivers of adaptive innovation, diversification, competition, and extinction. The phylogenetic systematic aspect of classification requires decisions about genetic and morphological homology, but consideration of phylogeny in its fullness mandates relationships to be studied within those contexts. The consanguinity of proboscideans with an odd and surprising set of other mammals reveals their deep roots in a distinctively African branch of taxa. The winnowing of proboscidean taxa and adaptations over time in Afro-Arabia shows the complex and asynchronous evolutionary path of these adaptations that produced extant elephants, including the savanna and forest elephants of the continent. From this perspective, the current classification is a road map to African proboscidean phylogeny as well as an accounting of their vast fossil record, requisite for retracing their magnificent evolutionary journey.

AFROTHERIAN SYSTEMATICS

Proboscidea is a eutherian mammalian Order, comprised of three extant species of elephants—the African savanna and forest elephants, *Loxodonta africana* and *L. cyclotis*, respectively, and the Asian elephant, *Elephas maximus*, along with a large, diverse group of extinct relatives such as phosphatheres, moeritheres, barytheres, deinotheres, palaeomastodonts, gomphotheres, mastodonts, and mammoths that are more closely related to elephants and one another than to other mammalian taxa. Proboscideans are part of a larger Supercohort, Afrotheria ("African beasts"), that improbably includes such morphologically dissonant animals as golden moles (Chrysochloridae), tenrecs (Tenrecidae), elephant shrews or sengis (Macroscelidea), aardvarks (Tubulidentata), hyraxes (Hyracoidea), and dugongs and manatees (Sirenia) (Figure 1.1; Asher et al., 2003; Robinson and Seiffert, 2004; Roca and O'Brien, 2005; Tabuce et al., 2007a, 2008; Asher and Seiffert, 2010). The tenrecs of Madagascar and otter shrews of West and Central Africa are now considered sister taxa and placed in the Order Afrosoricida within Afrotheria (Carter and Mess, 2007; Kingdon et al., 2013). Sieffert (2007b) placed golden moles in Afrosoricida, as well. Among eutherian mammals, afrotheres are possibly a primitive sister taxon to xenarthrans+[Euchontoglires (which includes primates and rodents)+Laurasiatheria (which includes perissodactyls, carnivores, and artiodactyls)] (Figure 1.1; Robinson and Seiffert, 2004; Roca and O'Brien, 2005).

As Asher (2001) has pointed out, it is unlikely on anatomical grounds alone that a clade of the composition of Afrotheria would be recognized without genetic evidence to support it (e.g., Springer et al., 1997, 2003; Stanhope et al., 1998; Murphy et al., 2001; Scally et al., 2001). Depending on the optimization model used in parsimony analysis of the group and the manner in which multistate characters are treated, there are few or no unambiguous morphological characters supporting Afrotheria: Presence of a naviculocalcaneal facet; internal carotid placement lateral to the anterior pole of pars cochlearis; scattered vomeronasal blood vessels; four allantoic vessel chambers; small P3 protocone; well-developed buccal cingulae rather than stylar shelves; relatively late eruption of permanent dentition; increase in numbers of thoracolumbar vertebrae; and testicondy (Sánchez-Villagra et al., 2007; Seiffert, 2007b; Asher and Lehmann, 2008). In addition, some afrotheres share low core body temperatures, daily heterothermy, and poor thermoregulation (Robinson and Seiffert, 2004).

Molecular evidence suggests a Cretaceous origin for Afrotheria consistent with the break-up of South America and Africa, followed by divergence of crown afrotherian Orders in the early Paleocene or possibly latest Cretaceous (e.g., Eizirik et al., 2001; Springer et al., 2003; Robinson and Seiffert, 2004; Seiffert 2007b; see Chapter 2). It has been suggested that the high morphological disparity of afrotheres is a result of explosive diversification into specialized niches in the later Cretaceous or early Paleogene that "erased" early afrotherian synapomorphies (Robinson and

DOI: 10.1201/b20016-1

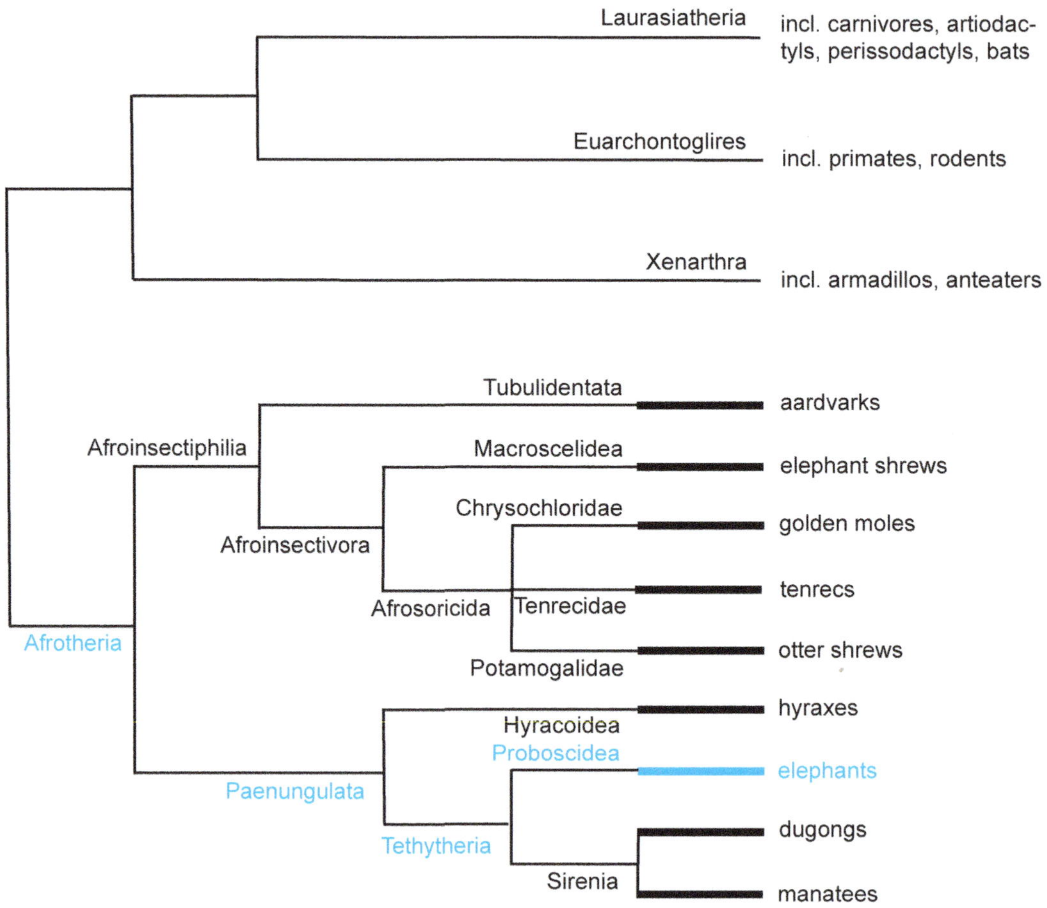

FIGURE 1.1 Cladistic tree depicting the composition and intra-group relationships of Afrotheria, and the relationship of Afrotheria to other major mammalian clades. Abbreviations: incl., including (but not comprehensive). The mammalian taxa to which elephants belong are highlighted in blue. (The figure is based on results of morphological and genetic analyses in Asher et al. (2003); Robinson and Seiffert (2004); Nishihara et al. (2005); Roca and O'Brien (2005); Seiffert (2007b, 2013a); and Kingdon et al. (2013)).

Seiffert, 2004). As indicated by the early fossil records of its constituent taxa, Afrotheria has deep roots in Africa at least as old as the Paleocene (Tabuce et al., 2007a; Gheerbrant et al., 2012, 2014, 2018). All afrothere crown taxa have their oldest fossil records in Africa except for sirenians, the most basal species of which occurs in middle Eocene Jamaica (Savage et al., 1994). If the hypothesis about afrotheres (and tethytheres; see below) originating and diversifying in Africa is correct, there must be even older, African, sirenians than Jamaican *Prorastomus* (Nishihara et al., 2005).

Within Afrotheria, there is good molecular and modest morphological evidence that sirenians, hyraxes, and proboscideans, along with the extinct and morphologically strange desmostylians and embrithopods, form a sub-clade "Paenungulata" (Figure 1.1), first recognized by Gregory (1910) and later formalized by Simpson (1945). However, a number of extinct mammals included in Simpson's (1945) conception of Paenungulata are no longer accepted as members of the group (e.g., Pantodonta, Dinocerata, Pyrotheria) (Seiffert, 2013a). Recently, Paleocene-aged *Ocepeia* from the Ouled Abdoun Basin of Morocco, represented by a partial cranium and upper dentition, was identified as a basal afrothere and suggested to be a possible paenungulate, as well (Gheerbrant et al., 2014). Morphological features of Paenungulata were reviewed by Seiffert (2013a) and include: Loss of the clavicle and metacromion; amastoidy; loss of lacrimal-palatine contact; presence of an alisphenoid canal; and a vertical or anterior orientation of the mandibular ramus. Additionally, a cladistic analysis placed tubulidentates as a sister taxon to Paenungulata in "Pseudoungulata" (and, correspondingly, allied elephant shrews with tenrecs and golden moles in "Afroinsectivora") (Barrow et al., 2010), whereas mtDNA analyses grouped tubulidentates in a monophyletic taxon with afroinsectivorans (Nishihara et al., 2005 and references therein).

There is morphological encouragement for a clade of proboscideans+sirenians (and desmostylians) (unfortunately beset by evident convergence in a few middle ear features; see Fischer, 1996; Benoit et al., 2013) within Paenungulata, "Tethytheria" (Figure 1.1; McKenna, 1975; Seiffert, 2007b). Summaries of proposed tethythere morphological synapomorphies were provided by Tassy and Shoshani (1980),

Domning et al. (1986), Shoshani (1992b, 1993), Fischer and Tassy (1993), and Asher et al. (2003). Some of these shared derived features are: Bilophodont molars with a tendency to develop an additional cusp on the posterior cingulum; anterior border of the orbit is anterior to the first molar; salient lateral flare of the squamosal portion of the zygomatic arch; the external auditory meatus and posterior part of the zygomatic arch are elevated; the external auditory meatus is ventrally partially closed; inflated tegmen tympani; bifid apex of the heart; digastricus muscle originates from the stylohyal; and sternohyoid muscle absent or fused with the sternothyroid muscle. Molecular evidence of the relationships of these taxa vis-à-vis hyraxes, however, did not produce a consensus about Tethytheria (Amrine and Springer, 1999; Nishihara et al., 2005; Roca and O'Brien, 2005; Kellogg et al., 2007; Pardini et al., 2007; Seiffert, 2007b). Moreover, study of orbital features of afrotheres indicated a closer relationship between proboscideans and hyraxes than either has with sirenians (Cox, 2006), supported by the shared features of: Loss of the lacrimal-jugal contact; presence of a lacrimal tubercle; presence of an alisphenoid canal; and separation of the foramen rotundum and the sphenorbital fissure. Tethytheria may also include embrithopods (Seiffert, 2013b; Gheerbrant et al., 2018). Within Domning et al.'s (1986: fig. 22) conception of Tethytheria, proboscideans and desmostylians are more closely related than either is to sirenians.

By comparison, Proboscidea is an uncontroversial taxon, although recognition of its earliest species is challenging due to their primitive morphologies with few shared derived features of the Order (see Chapter 2). Characters thought to be synapomorphic for Proboscidea include: Enlarged second incisors; radius positioned or fixed in pronation, tooth enamel with a "keyhole"-shaped prism in cross-section; a well-developed zygomatic process of the maxillary bone contributing significantly to the zygomatic arch and floor of the orbit; relatively large size of the pars mastoidea compared with the pars cochlearis internally in the periotic; buccal position of the hypoconulid; and retracted optical foramen in the orbitotemporal fossa, toward the posterior of the cranium (Shoshani et al., 1998; Gheerbrant et al., 2005; Shoshani and Tassy, 2013). Many of the features associated with more recent proboscideans such as elephants and mastodons, including ponderous trunks, tusks, skeletal adaptations for graviportal posture, gigantism, loss of most incisors, canines, and premolars, and development of horizontal tooth displacement, are absent in proboscideans prior to the late Paleogene (see Chapters 2 and 3).

AFRICAN PROBOSCIDEAN CLASSIFICATION

Taxonomic Considerations

The classification of African (or Afro-Arabian) proboscideans devised in Table 1.1 is based on the detailed taxonomic work of Chapters 2–7 and extensive incorporation of published studies of African fossil proboscideans. Most of it is conventional, with no compelling reasons to depart from custom; nonetheless, a number of systematic and taxonomic issues regarding African proboscideans remain to be resolved.

The most challenging problems arise from the bold attempt by Shoshani and colleagues (2001a) to create a taxon of the same hierarchical level as Elephantiformes, the Plesielephantiformes, to accommodate most of the remaining, more primitive fossil proboscideans. However, Plesielephantiformes is almost certainly paraphyletic (Gheerbrant et al., 2005), it is not obvious which taxa should be assigned to it, and some of their relationships are unresolved or contentious.

Indeed, even a brief survey of published phylogenetic systematic studies reveals widely disparate hypotheses of plesielephantiform clade trees (see Figure 1.2). Among these trees, the positions of Paleogene moeritheres, daouitheres, arcanotheres, barytheres, and chilgatheres are all unstable, and in the majority of them there is very little clade resolution (Gheerbrant et al., 2005; Delmer, 2009; Seiffert et al., 2012; Hautier et al., 2021). In addition, although supporting evidence is slender, deinotheres are commonly placed as a close sister group to elephantiformes (Figure 1.2B–D). In Gheerbrant et al.'s (2005) study, for instance, the proposed synapomorphies favoring this arrangement are few (e.g., large size of i2; m1 with a well-developed postentoconulid), redundant (i1 < i2), or vulnerable to homoplasy (e.g., P4 metacone well-expressed and separated from the protocone; perilymphatic foramen present; cochlear canal and round fenestra of the periotic absent). By comparison, however, elephantiforms and deinotheres exhibit considerable differences in cranial structure and loph(id) construction, whereas deinotheres share a great many craniodental features with various taxa in plesielephantiforms, such as lophodonty (Harris, 1978a) and Schmelzmuster enamel patterns (Tabuce et al., 2007b). This is reflected in the current classification. Nonetheless, alternative hypotheses of deinothere phylogenetic relationships are well supported by putative synapomorphies (see below) and therefore require deeper scrutiny. Other potentially contentious parts of the current plesielephantiform classification are placement of chilgatheres in Deinotheriidae (contra Delmer, 2009), *Arcanotherium* grouped together with *Barytherium* in Barytheriidae (contra Tabuce et al., 2007b; Delmer, 2009), and inclusion of *Numidotherium* in Barytherioidea (contra Mahboubi et al., 1986; Gheerbrant, 2009; Gheerbrant et al., 2012).

Moeritheres are also phylogenetically problematic. Primitively, they retain three incisors and a canine in each upper dental quadrant, but their molars resemble those of bunolophodont elephantiform proboscideans, while their skull is highly autapomorphic as a result of adaptation to semi-aquatic existence. Thus, placing them in their own Superfamily with uncertain Subordinal affinity is a reasonable classificatory judgment. Although the number of proposed moerithere species seems inflated by typological interpretation of morphological variation (Andrews, 1902,

TABLE 1.1
Classification of Afro-Arabian Proboscideans

Proboscidea Illiger, 1811
Plesielephantiformes Shoshani et al., 2001a
Family Incertae Sedis
***Eritherium* Gheerbrant, 2009**
***E. azzouzorum* Gheerbrant, 2009**
Phosphatheriidae Gheerbrant et al., 2005
Phosphatherium Gheerbrant et al., 1996
P. escuillei Gheerbrant et al., 1996
?Phosphatheriidae Gheerbrant et al., 2005
Khamsaconus Jaeger et al. in Sudre et al., 1993
K. bulbosus Jaeger et al. in Sudre et al., 1993
Family Incertae Sedis probably nov. Gheerbrant et al., 2005
Daouitherium Gheerbrant and Sudre in Gheerbrant et al., 2002
D. rebouli Gheerbant and Sudre in Gheerbrant et al., 2002
Barytherioidea Andrews, 1906
Numidotheriidae Shoshani and Tassy, 1992
Numidotherium Jaeger in Mahboubi et al., 1986
N. koholense Jaeger in Mahboubi et al., 1986
Barytheriidae Andrews, 1906
Barytherium Andrews, 1901b
B. grave (Andrews, 1901d)
Arcanotherium Delmer, 2009
A. savagei (Court, 1995)
***Omanitherium* Seiffert et al., 2012**
***O. dhofarense* (Seiffert et al., 2012)**
Gen. et sp. nov. [Nakwai] this volume, Chapter 4
Deinotherioidea Bonaparte, 1845
Deinotheriidae Bonaparte, 1845
Chilgatheriinae Sanders et al., 2004
Chilgatherium Sanders et al., 2004
C. harrisi Sanders et al., 2004
Deinotheriinae Bonaparte, 1845
Prodeinotherium Éhik, 1930
P. hobleyi (Andrews, 1911)
Deinotherium Kaup, 1829
D. bozasi Arambourg, 1934a
Suborder Incertae Sedis
Superfamily Incertae Sedis
Family Incertae Sedis
***Saloumia* Tabuce et al., 2019**
***S. gorodiskii* Tabuce et al., 2019**
Moeritherioidea Andrews, 1906
Moeritheriidae Andrews, 1906
Moeritherium Andrews, 1901d
M. chehbeurameuri Delmer et al., 2006
M. lyonsi Andrews, 1901d
M. trigodon Andrews, 1904a
Elephantiformes Tassy, 1988
Family Incertae Sedis
***Dagbatitherium* Hautier et al., 2021**
***D. tassyi* Hautier et al., 2021**
Palaeomastodontidae Andrews, 1906
Palaeomastodon Andrews, 1901d
P. beadnelli Andrews, 1901d
P. sp. nov. A [Chilga] (Sanders et al., 2004)

(Continued)

TABLE 1.1 (*Continued*)
Classification of Afro-Arabian Proboscideans

P. sp. nov. B [Chilga] (Sanders et al., 2004)
***P.* sp. nov. [Nakwai] this volume, Chapter 4**
Phiomia Andrews and Beadnell, 1902
P. serridens Andrews and Beadnell, 1902
P. major Sanders et al., 2004
Elephantimorpha Tassy and Shoshani, 1997 in Shoshani et al., 1998
Mammutida Tassy and Shoshani, 1997 in Shoshani et al., 1998
Mammutidae Hay, 1922
Losodokodon Rasmussen and Gutiérrez, 2009
L. losodokius Rasmussen and Gutiérrez, 2009
Eozygodon Tassy and Pickford, 1983
E. morotoensis (Pickford and Tassy, 1980)
Zygolophodon Vacek, 1877
Z. aegyptensis Sanders and Miller, 2002
***Z.* sp. nov. [small form; Buluk]**
Z. turicensis (Schinz, 1824)
Elephantida Tassy and Shoshani, 1997 in Shoshani et al., 1998
Gomphotheriidae Hay, 1922
Gomphotheriinae Hay, 1922
Eritreum Shoshani et al., 2007
E. melakeghebrekristosi Shoshani et al., 2007
Gomphotherium Burmeister, 1837
cf. *Gomphotherium* sp. nov. Kappelman et al., 2003
G. sp. [Kajong] Tassy, 1979a
G. libycum (Fourtau, 1918)
cf. *G.* sp. nov. [archaic form] Sanders, 2008a
G. spp. indet. [dwarfed forms]
Amebelodontinae Barbour, 1927
Progomphotherium Pickford, 2003a
P. maraisi Pickford, 2003
Archaeobelodon Tassy, 1984
A. sp. [primitive morph; Kajong]
***A.* sp. nov. [advanced morph; Buluk]**
Protanancus Arambourg, 1945
***P.* sp. nov.? [small form; Buluk]**
P. macinnesi Arambourg, 1945
Afromastodon Pickford, 2003a
A. coppensi Pickford 2003a
Platybelodon Borissiak, 1929
P. sp. Maglio, 1969a
Gen et sp. indet. [Samburu Hills]
Konobelodon Lambert, 1990
***K. cyrenaicus* (Gaziry, 1987b)**
Choerolophodontinae Gaziry, 1976
Afrochoerodon Pickford, 2001
A. kisumuensis (MacInnes, 1942)
Choerolophodon Schlesinger, 1917
C. zaltaniensis Gaziry, 1987a
C. sp. indet. [dwarfed species]
C. ngorora (Maglio, 1974)
Tetralophodontinae Van der Maarel, 1932
Tetralophodon Falconer, 1857
T. sp. [Bir el Ater 3]
T. sp. nov. [Samburu Hills] Tsujikawa, 2005a
T. sp. [Skoura]

(*Continued*)

TABLE 1.1 (*Continued*)
Classification of Afro-Arabian Proboscideans

T. sp. indet. [Kisegi]
T. sp. indet. [Chorora]
T. sp. indet. [Gebel Krechem el Artsouma]
***T.* sp. indet. [Hamra]**
Anancinae Hay, 1922
Anancus Aymard in Dorlhac, 1855
A. kenyensis (MacInnes, 1942)
A. osiris Arambourg, 1945
A. petrocchii Coppens, 1965
A. capensis Sanders, 2007
***A. ultimus* Sanders, 2011**
Stegodontidae Osborn, 1918
Stegodon Falconer and Cautley, 1847
***S.* sp. nov. Mackaye, 2001**
S. kaisensis Hopwood, 1939
Elephantidae Gray, 1821
Stegotetrabelodontinae Petrocchi, 1954
Stegotetrabelodon Petrocchi, 1941
***S. emiratus* Sanders, 2022**
S. syrticus Petrocchi, 1941
S. orbus Maglio, 1970b
Elephantinae Gray, 1821
Stegodibelodon Coppens, 1972
S. schneideri Coppens, 1972
Selenetherium Mackaye et al., 2005
S. kolleensis Mackaye et al., 2005
Primelephas Maglio, 1970b
P. korotorensis (Maglio, 1970b)
Mammuthus Brookes, 1828
M. subplanifrons (Osborn, 1928)
?*M.* sp. indet. probably nov. [Toros Menalla] Mackaye, 2001
M. sp. "Hadar type"
M. africanavus (Arambourg, 1952)
M. meridionalis (Nesti, 1825)
Elephas Linnaeus, 1758
E. ekorensis Maglio, 1970b
***E. brumpti* (Beden, 1980)**
***E. shungurensis* (Beden, 1980)**
***E. atavus* (Arambourg, 1947)**
Palaeoloxodon Matsumoto, 1924b
***P. recki* (Dietrich, 1915)**
***P. jolensis* (Pomel, 1895)**
Loxodonta F. Cuvier in F. Cuvier and Geoffry Saint-Hilaire, 1825 (anonymous emendation, 1827)
L. cookei Sanders, 2007
L. adaurora Maglio, 1970b
L. exoptata (Dietrich, 1941)
L. atlantica (Pomel, 1879)
L. africana (Blumenbach, 1797)
L. cyclotis (Matschie, 1900)

Bold text indicates recognition of a new taxon or novel application of a name to an existing taxon (since the last major overview of Afro-Arabian proboscideans in Sanders et al. [2010a]); ?=uncertain.

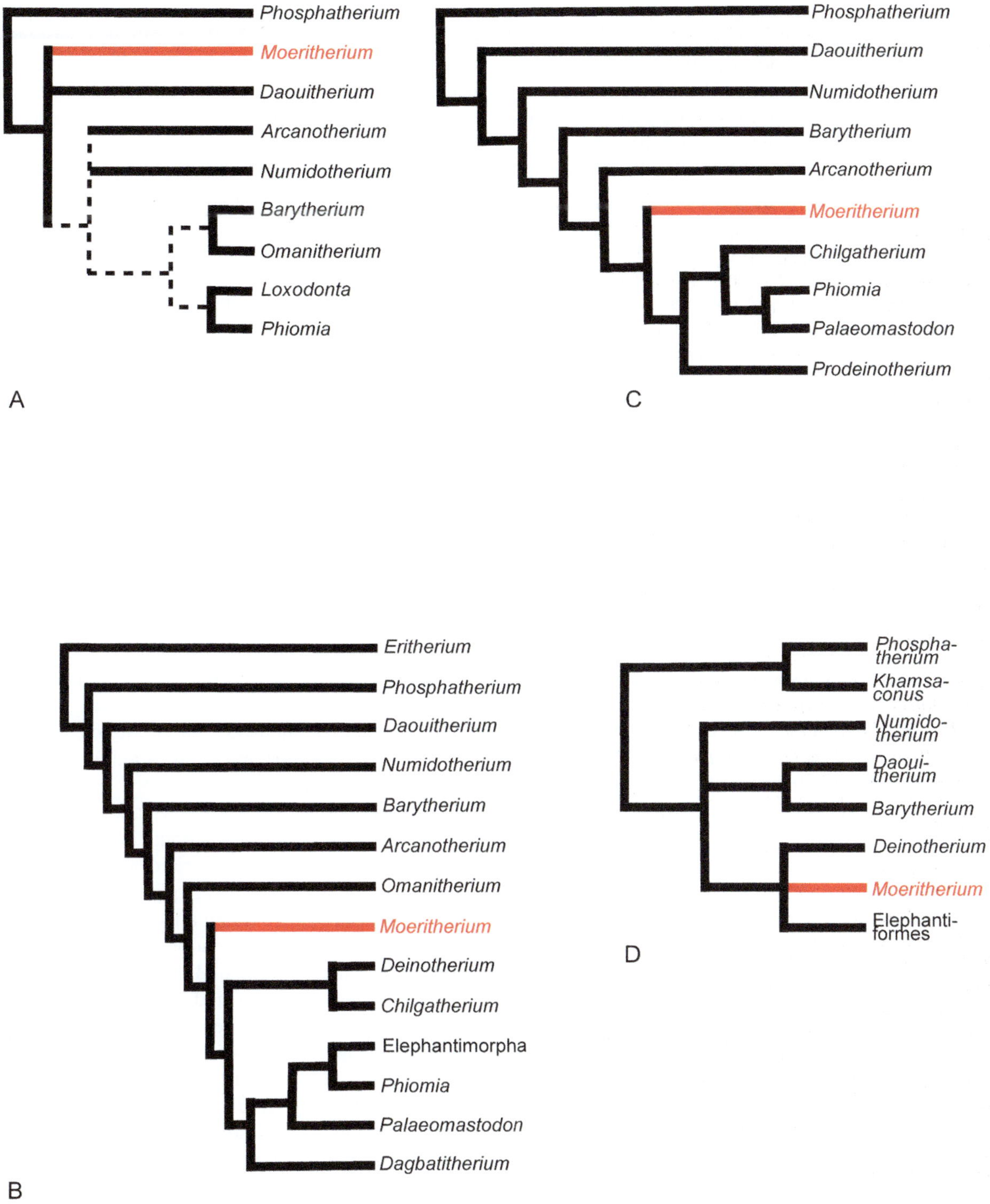

FIGURE 1.2 Comparison of selected cladistic trees of early proboscideans based on different analyses. Trees are adapted from published results; length of lines are arbitrary and do not reflect time or phylogenetic distance. The phylogenetic position of *Moeritherium* (in red) illustrates the unstable resolution of proboscidean relationships in these cladograms. (A) Based on Adams consensus tree of Seiffert et al. (2012). (B) Based on strict consensus tree of Hautier et al. (2021). (C) Based on most parsimonious tree of Delmer (2009). (D) Based on consensus tree of Gheerbrant et al. (2005).

1906; Schlosser, 1911; Matsumoto, 1922, 1923; Petronievics, 1923; Deraniyagala, 1955; Delmer et al., 2006), most of this craniodental variability is resolvable as sexual dimorphism and interindividual (rather than interspecific) differences (Sanders et al., 2010a). Therefore, the current classification restricts moeritheres to three time-successive species.

Elephantiformes is less taxonomically controversial. Because palaeomastodonts lack the mechanism of horizontal tooth displacement, they are excluded from Elephantimorpha. As with moeritheres, historically the Fayumian palaeomastodont material has been assigned to an excess of species (e.g., Andrews, 1901d, 1904a, 1905; Andrews and Beadnell, 1902; Pontier, 1907, Matsumoto, 1922, 1923, 1924), based on trivial morphological features or levels of variation that are explicable by sexual dimorphism and interindividual differences (Sanders et al., 2010a). As a result, *Palaeomastodon* and *Phiomia* from the Fayum are constrained in the current classification to one (sexually dimorphic) species each, with new species from Chilga, Ethiopia and Nakwai, Kenya clearly exceeding the morphometric variation of Fayumian species and therefore supportable (Sanders et al., 2004; this volume Chapter 4).

The separation of elephantimorphs into Mammutida and Elephantida is strongly defensible morphologically and points to a fundamental division of the group that may extend to the base of the late Paleogene and contradicts the monophyly of Palaeomastodontidae (Sanders et al., 2004) and Elephantimorpha (Tassy, 1994b). Elephantimorpha is monophyletic if its taxa comprise a sister group to the primitive elephantiform *Phiomia*; alternatively, if *Phiomia* is ancestral to gomphotheres and *Palaeomastodon* is the sister taxon to mammutids (Matsumoto, 1924; Tobien, 1971, 1978a; Sanders et al., 2004, 2010a), Elephantimorpha is diphyletic (Tassy, 1994b). In the current classification, Elephantimorpha is maintained as a useful concept, but strong consideration is given in Chapter 3 to the diphyletic hypothesis, indicating that Elephantimorpha and Palaeomastodontidae may not be monophyletic. If correct, a new classificatory scheme will have to be devised to reflect an earlier phylogenetic division of mammutids from gomphotheres.

The taxonomy of African mammutids in the classification is based on progressive changes in molar morphology (increasingly more prominent expression of zygodont crests and crescentoids) (see Tassy and Pickford, 1983; Sanders and Miller, 2002; Rasmussen and Gutíerrez, 2009) and the primitiveness of the cranium in *Eozygodon* (Tassy, 2018)—it is unlikely to cause debate. Future study of North African material and recovery of more informative fossil elements from eastern and southern Africa, however, may yield additional species of *Zygolophodon*.

Similarly, subdivision of gomphotheres in Elephantida into Gomphotheriinae, Amebelodontinae, and Choerolophodontinae is incontrovertible, although there are alternative preferences for these groups to be raised to Family level (e.g., Tassy, 1986). *Progomphotherium* was initially recognized as a gomphotheriine because of its unexpanded, rounded lower tusks (Pickford, 2003a), but the distribution of accessory conules on the posttrite side of its molars crowns is alternatively suggestive of inclusion in Amebelodontinae (Sanders et al., 2010a). New information about the dating (Watkins, 1989; Brown et al., 2016) and morphology of molars and tusks of fossils of the amebelodont *Archaeobelodon* from Kajong (Mwiti) and Buluk, Kenya (see Chapter 4) indicate that they are divisible into separate, successive species. Although the amebelodontine from late Miocene Sahabi, Libya was initially placed in *Amebelodon* (Gaziry, 1982), the tetralophodonty of second molars, six lophids in m3, and the great degree of lower tusk flattening and development dentinal tubules and rods all accord with identification as *Konobelodon* (Lambert, 2016).

The current classification follows Pickford (2001) in dividing choerolophodonts into plesiomorphic *Afrochoerodon* and more advanced *Choerolophodon*. Differences in cranial features and expression of molar features such as ptychodonty, cementodonty, chevroning, and number of loph(id)s are starkly contrasted between these choerolophodont genera. The hypothesis that the choerolophodont from Ngenyin, Kenya is a novel, dwarfed species (Pickford, 2004) is strongly supported by the diminutive size of its molars and postcrania. However, it cannot be named "*C. pygmaeus*" nor can fossil material from early middle Miocene Bosluis Pan, South Africa (Pickford, 2005a), as the type specimen to which that nomen was applied was shown to be abnormal or a supernumerary molar and probably not from a choerolophodont (Tassy et al., 2013). Additionally, Pickford (2001) hypothesized that *C. zaltaniensis* and *C. ngorora* are synonymous, which could be true for the early phase of *C. ngorora* from the late middle Miocene, but the type series from the early late Miocene phase is distinctive (Sanders et al., 2010a), and thus the current classification maintains both species.

The tetralophodont taxa *Tetralophodon* and *Anancus* are included in the current classification as gomphotheres, but some researchers prefer Gomphotheriidae to be limited to trilophodont taxa (e.g., Shoshani and Tassy, 2005), in which case these tetralophodont lineages would either be encompassed by Gomphotherioidea or placed in a novel Superfamily within their own Families. Several species of *Anancus* that are not individualized by others are maintained in the current classification: *A. petrocchii* was considered an advanced morph of *A. kenyensis* by Tassy (1986), although the evidence presented in Chapter 5 appears strong enough to consider them as separate species; *A. ultimus* was erected to accommodate pentalophodont, advanced stages of *A. kenyensis* (Sanders, 2011), not including *A. petrocchii*, that were detailed by Kalb and Mebrate (1993). If *A. kenyensis-A. ultimus* was an anagenetically evolving lineage, the subdivision of the lineage into different species is arbitrary, based primarily on intermediate molar loph(id) number; certainly, there is considerable overlap in molar crown morphology between temporally successive populations of these anancine gomphotheres (Kalb and Mebrate, 1993).

Elephants are generally recognizable and well-constrained phylogenetically by the structure of lamellae in their molars and raised, anteroposteriorly compressed skulls (Maglio, 1973; Saegusa, 2020). However, the plesiomorphic condition of the earliest elephants, the stegotetrabelodonts, which retained tetrabelodonty, permanent premolars, and had very brachyodont molars of only a few, nascent plates, illustrates the blurred morphological boundary between their tetralophodontine ancestors and basal Elephantidae. Because of this, *Stegotetrabelodon* was initially classified in Gomphotheriidae (e.g., Petrocchi, 1941). Similarly, there are problems regarding the subdivision and definition of Stegotetrabelodontinae versus Elephantinae (e.g., Mackaye, 2001). If lower tusks were lost only once in elephantids, then the taxonomic composition of these Subfamilies in the current classification is sound; if not, the infrafamilial relationships of early elephants will need to be reassessed, particularly for *Selenitherium* and *Stegodibelodon*.

Other taxonomic issues regarding elephants include the probable misidentification of the type specimen of *Mammuthus subplanifrons* (it more closely resembles molars of *Loxodonta cookei*; Sanders, 2007; contra Osborn, 1928), which if confirmed would necessitate making "*cookei*" a junior synonym of "*subplanifrons*" within *Loxodonta* and designating a new species nomen for the earliest mammoths.

The most critical taxonomic question for African elephants concerns the subdivision and classification of the "*Elephas recki* complex," the dominant Plio-Pleistocene proboscidean group on the continent. This is truly the "elephant in the room" for proboscidean classification! The archetypal approach was to arbitrarily partition the group into a series of anagenetically-evolved, successively progressive subspecies or stages of the species *E. recki* (e.g., Maglio, 1973; Beden, 1980; Sanders and Haile-Selassie, 2012), but this scheme is clearly outdated and overly simplistic. At the very least, the younger "subspecies" of the group belong to *Palaeoloxodon* and are specifically distinct (Saegusa and Gilbert, 2008; Larramendi et al., 2020; Zhang, 2020); the older "subspecies" are raised to specific status and in the current classification are retained in *Elephas*, but other hypotheses of their affinities have been offered (e.g., Saegusa and Gilbert, 2008; Zhang, 2020). By logical extension, "*E.*" *jolensis* is now classified in *Palaeoloxodon*.

Diagnostic Considerations

Reviewing databases in order to understand the diagnostic characteristics, character state polarities, and phylogenetic systematics of proboscidean taxa is complicated by the great morphological and taxonomic diversity of the group, and rampant homoplasy. Important large-scale morphological systematic investigations of proboscidean intraordinal relationships were carried out by the late Jeheskel Shoshani and Pascal Tassy (e.g., Tassy, 1982, 1990b, 1994b; Shoshani et al., 1996, 2001a; Shoshani and Tassy, 2005), supplemented in part by the researches of Emmanuel Gheerbrant and colleagues (e.g., 1998a), Rodolphe Tabuce and collaborators (2007b), and Cyrille Delmer (2009), among others. Although relatively contemporary, the results of these studies in part have been made obsolete by new fossil discoveries and ancient DNA analyses. Diagnostic characteristics of major African proboscidean taxa are provided as an accompaniment to the current classification and as an update of the state of knowledge of these taxa, relevant to discussions of taxonomy and phylogenetic relationships.

It is challenging to establish synapomorphies that might encompass all plesielephantiform proboscideans. For instance, most taxa in the Suborder are lophodont, but the earliest known plesielephantiform, *Eritherium*, has molars that are bunodont and only incipiently lophodont. It is primitive by its full eutherian lower dentition, including c and p1, small third molars, short mandibular symphysis, lack of a protoloph and weak metacone in its upper premolars, distinct mesostyle and ectocingulum in its upper molars, lower molar postmetacristid, absence of a submaxillary fossa, and position of the orbits above P4-M1 (Gheerbrant, 2009). Unambiguous proboscidean synapomorphies include enlargement of i1 > i2, i1–2 styliform, procumbent, and labiolingually compressed, i3 reduced, c reduced, p1 small and morphologically simple, hypoconulids in molars buccally placed, and enlarged coronoid retromolar fossa (Gheerbrant, 2009). Its retention of c and molar centrocristae and mesostyles, relative unimportance of p1, size of the retromolar fossa, and placement of hypoconulids are shared with *Phosphatherium* (Gheerbrant, 2009), providing a tenuous link with other plesielephantiforms.

True lophodonty of early Paleogene proboscideans first appeared in *Phosphatherium*, although its molar cusps are crowded inward and their connecting transverse crests are low. Although apomorphic in this regard in comparison to *Eritherium*, nonetheless *Phosphatherium* is also primitive in many aspects of its craniodental anatomy (Gheerbrant et al., 2005). For example, its cranium is relatively low, the nasal aperture is anteriorly located, postorbital constriction is pronounced, a sagittal crest is present, the zygomatic arches flare widely laterally, there is no premaxillary-frontal contact, canines are present, the full eutherian complement of upper teeth is retained, the mandibular symphysis is unfused, there is no inferior diastema and the upper diastema is very short, and premolars are of simple construction (a more extensive list of primitive features is provided by Gheerbrant et al. [2005]). Despite the lack of separation of the anterior dentition from the cheek teeth, the largest lower incisor of *Phosphatherium* is outsized and procumbent, a condition shared by other subsequent plesielephantiforms. Identification of anterior lower teeth to proper position is difficult (Gheerbrant et al., 1996), although it seems most parsimonious to interpret them as i1–2, c, p2–4. This formula is more advanced than that of the geologically younger *Daouitherium* (Yans et al., 2014), which is alternatively i1–3, c, p2–4 or i1–2, c, p1–4 or i1–3, p1–4 (see Gheerbrant et al., 2002), complicating the relationship between these successive plesielephantiforms.

Although cladistic analysis placed *Khamsaconus* as the close sister taxon to *Phosphatherium* (Gheerbrant et al., 2005), this study did not include *Eritherium*. Subsequent analysis suggests that it is more likely that *Khamsaconus* is closer to *Eritherium* (Gheerbrant et al., 2012), lacking true molar lophodonty (Gheerbrant et al., 1998b), of very small size, with a postentoconule in dP4, and having only radial enamel (Tabuce et al., 2007b). The presence of a posterior accessory conule in a partial lower molar of *Khamsaconus* is a feature more commonly identified in moeritheres and elephantiform proboscideans, but much more evidence is required to meaningfully test phylogenetic implications.

Aside from the apparent reversal to a more primitive tooth formula, *Daouitherium* is craniodentally more advanced than *Phosphatherium*: It was relatively large for its time, p3–4 are molarized, the corpus of the dentary is relatively deep and the ramus is high, indicating a more raised basicranium and higher skull (Gheerbrant et al., 2002), and dental enamel is primarily composed of Hunter-Schreger bands (Tabuce et al., 2007b). The molarization of its lower premolars has been invoked to argue that *Daouitherium* should be excluded from the ancestry of *Numidotherium*, with simpler premolars (Gheerbrant et al., 2002).

Numidotherium represents a marked morphological transformation of proboscideans, compared with earlier plesielephantiforms. Its craniodental anatomy anticipates features that characterize later, more advanced proboscideans. For example, the basicranium is raised well above the occlusal plane; the skull is profusely pneumatized; the nasal aperture is high (but not retracted); c and p1 are lost; its bilophodont molars are perfect smaller models of barythere homologs; body size is estimated to have been very large; the median lower incisor is hypertrophied, spatulate, and procumbent; the ramus is anteroposteriorly immense and high; and there is a substantial diastema isolating the incisors from the cheek teeth (Jaeger in Mahboubi et al., 1986). Primitive features of this proboscidean include cochlear structure consistent with a limited, high-frequency hearing range (Court, 1992), prominent zygomatic apophyses similar to those of *Eritherium* and *Phosphatherium* (Gheerbrant et al., 2005, 2012), low orbit position, sagittal crest formation, and development of an entepicondylar foramen in the humerus. Its autapomorphic postcranium is well-documented, unexpectedly more consistent structurally with an ambulatory gait and sprawling posture than with graviportal support of its massive body (Jaeger in Mahboubi et al., 1986; Court, 1994a). The scenario that emerges in the consideration of these early Paleogene plesielephantiforms is one of a succession (rather than radiation) of taxa that does not epitomize an unambiguous ancestral-descendant sequence.

"Barytheres" probably constitutes a paraphyletic group that are united by general resemblance of bilophodont molars with m3 possessing an enlarged distal heel, relatively simple, foreshortened upper premolars, molars usually rectangular in occlusal view and longer than wide, horizontally straight symphyses with somewhat hippo-tusk-like, projecting incisors (usually set well below the level of the cheek tooth row), and an enlarged set of downward-curved upper incisors.

However, cranial form differs wildly between *Arcanotherium* and *Barytherium*. In *Barytherium*, the cranium is relatively low in lateral view but has a very raised basicranium with occipital condyles far above the palatal plane, a profile resembling the crania of deinotheres, and the nasal aperture is moderately retracted, consistent with the development of a trunk (Shoshani et al., 1996; Delmer, 2005; Jaeger et al., 2012). By contrast, *Arcanotherium* has a higher cranial profile but lower basicranium and occipital condyles, and its nasal aperture is barely retracted, inconsistent with the presence of a trunk (Delmer, 2009; contra Jaeger et al., 2012). The height of the *Arcanotherium* cranium and anterior position of its nasal opening recall the morphology of *Numidotherium* crania. As well, the position of the orbits in *Barytherium* is probably an adaptation to aquatic existence, as they are anteriorly placed in line with the nasal aperture, whereas the orbits of *Arcanotherium* are located more posteriorly over the anterior molars.

Lower dental formulae of barytheres are similar, with all having two incisors, three premolars, and three molars (a late surviving barythere that is possibly Neogene in age has only one set of lower incisors); however, i2 is probably larger than i1 in *Arcanotherium* and inferred to be larger in *Omanitherium* (Pickford, 2015b), but not in *Barytherium*. Additionally, the large lower central incisors in *Arcanotherium* are labiolingually compressed and clad in enamel. Moreover, *Arcanotherium* retains I1–3 and C in its upper dentition, distinct from the more reduced condition in *Barytherium*.

Despite craniodental evidence for a closer relationship of *Arcanotherium* to *Numidotherium* and *Omanitherium* (e.g., skull form, incisor dominance, enamel covering of lower incisors; Pickford (2015b)), Delmer (2009) and Jaeger et al. (2012) formed an alternate hypothesis that a host of post-cranial and craniodental features (e.g., astragalar anatomy, enamel microstructure, degree of molarization of premolars, lower molar cristid proportions and development, upper molar conule development, and orbital position) supports a closer sister taxon relationship to elephantiforms and deinotheres. The contrasting phylogenetic implications of different sets of traits among barytheres indicates that their relationships to one another and to other proboscideans (numidotheres, deinotheres, elephantiforms), along with familial and superfamilial affiliation, are unresolved.

Deinotheriidae is morphologically distinctive and has been a defined group for nearly 200 years (Bonaparte, 1845). Unlike elephantimorphs, all of its adult dentition was deployed simultaneously once emerged. There does not appear to be any delay in emergence of molars beyond normal ontogenetic appearance of teeth, in contrast with primitive elephantiforms. In common with archaic plesielephantiforms and barytheres, molars are lophodont and loph(id)s wear to chisel-like, semi-vertical surfaces (Harris, 1978a), but autapomorphically first molars and deciduous fourth premolars are trilophed. Also autapomorphic are the loss of the upper tusks and downturned orientation of the lower tusks, which are *presumed* to be i2s. If so, the dominance of i2 is shared with moeritheres and elephantiform proboscideans that retain this tooth position, but it is also a condition of *Arcanotherium* (Delmer, 2009) and apparently *Omanitherium* (Pickford, 2015b; Al-Kindi et al., 2017).

Chilgatherium shares many molar characteristics of deinotheriines, but is autapomorphic in incipient development of tritolophodonty in m2 and M3/m3 (Sanders et al., 2004). The inferred different modes in which these teeth developed third loph(id)s in *Chilgatherium* (postmetaloph ornamentation versus inflation of the distal cingulid) suggest that care must be taken in assuming homology of tritoloph or hypolophid development between taxa as disparate as deinotheres and elephantiforms.

The deinothere upper and lower premolar formula of two in each quadrant differs from that of barytheres and moeritheres, which have an additional premolar in each quadrant, and from that of palaeomastodonts, which have three premolars on each side in the upper jaw, a condition retained in some primitive elephantimorphs. Crania of deinotheres have a braincase of low-to-moderate height, but the basicranium is raised well above the palate, and exhibit a short, downturned rostrum and moderately retracted, small nasal aperture consistent with the presence of a compact trunk. The overall similarity of their crania is closer to skulls of some barytheres than of primitive elephantiforms. The large skeletons of these immense proboscideans converge on the graviportal adaptations of elephantiforms, with only subtle differences. Tassy (1994b) amassed an impressive list of shared characters to support the sister-group relationship of deinotheres to elephantiform proboscideans, including lack of enamel in adult lower tusks, P3–4 with a hypocone, trilophodont first molars and fourth deciduous premolars, reduction of the paroccipital processes, dP2 with a protocone, nuchal fossa present in the occipital planum, and astragalus with an enlarged ectal facet, reduced fibular facet, and short neck. Many of these features are unknown for chilgatheres, but it will be important to learn their morphological state in the earliest deinothere in order to assess competing hypotheses of synapomorphy versus homoplasy of these features.

Moeritheres are difficult to assess phylogenetically because of the strong overprint of semi-aquatic existence on their skeleton, retention of numerous primitive features, but general resemblance of their bunolophodont cheek teeth to those of primitive elephantiforms. Plesiomorphic features include a low braincase with a sagittal crest, retention of I3 and C, and anteriorly placed nares (Matsumoto, 1923; Tassy and Shoshani, 1988; Shoshani et al., 1996; Tassy, 1996a). For a time, they enjoyed status as the most primitive of proboscideans, but the discovery of older, lophodont forms altered their phylogenetic position (Gheerbrant et al., 2005). Now, with the discovery of bunolophodont proboscidean taxa from earlier in the Paleogene, it is worth entertaining the hypothesis that moeritheres represent an ancient lineage of bunolophodont proboscideans that paralleled the archaic sequence of lophodont species. Features that characterize moeritheres include very anterior position of orbits (anterior to P2), tubular elongation of the cranium, reduction of the postorbital process of the frontal, loss of the lacrimal, presence of a metacone on P2–3, and paraconid and protostylid on p3–4 (Tassy, 1982), as well as elongation of the body by large numbers of thoracolumbar vertebrae, interpreted as an amphibious feature (Tobien, 1971). The large diameter of the nasal aperture and extreme anterior position of the orbit may also be related to a semi-aquatic existence. Despite their strange appearance, moeritheres have been placed close to primitive elephantiforms in the current classification, largely on the basis of similarity of their molars; the molars of moeritheres-*Phiomia-Gomphotherium* comprise a progressive series that evolved by iterative addition of loph(id)s. The features that could resolve the phylogenetic position of moeritheres remain to be found in more complete specimens of taxa such as mid-Eocene *Saloumia* and *Dagbatitherium* (Tabuce et al., 2019; Hautier et al., 2021), or their predecessors in even older Paleogene contexts.

Features uniting Elephantiformes include elongation of the face, presence of a nasal fossa, posterior shift of the orbits above M1 and retraction of the nasal aperture (associated with the development of a prehensile trunk), substantial enlargement and projection of ever-growing tusks, elongation of the symphysis, expansion of cranial diploë, absence of paroccipital processes, development of three-layered Schmelzmuster enamel structure, mesoconid expansion, transversely enlarged buccal molar cusps, individualization of third loph(id)s in intermediate molars, presence of a prominent hypoconulid and postentoconulid in m1–2, presence of accessory conules in m1–2, lower molars with an enlarged protoconulid, and M3 with an enlarged postentoconule (Tassy, 1994b, 1996a; Tabuce et al., 2007b; Hautier et al., 2021). Relevant features from this list lend support for inclusion of the new middle Eocene taxon *Dagbatitherium* in Elephantiformes (Hautier et al., 2021), alongside palaeomastodonts, mammutids, gomphotheres, stegodonts, and elephants.

Elephantimorpha has largely replaced "Elephantoidea" in referring to those proboscideans united by the mechanism of horizontal tooth displacement (Tassy, 1990b; Tassy and Shoshani, 1997 in Shoshani et al., 1998), surely one of the most important adaptations that proboscideans evolved, accounting in no small way for the extensive taxonomic diversification and expansive geographic distribution of more advanced elephantiform proboscideans to (Sanders, 2017). It was hypothesized by Tassy (1988) to have arisen once. Monophyly of the Elephantoidea is also supported by a number of other synapomorphies, such as elongation of upper tusks, M3 completely trilophodont and relatively narrow, m3 with four lophids and an accessory conule in the third transverse valley, loss of the ventral foramen for the temporal canal, enlargement of the ecotympanic portion of the bulla, confluence of the foramen ovale and foramen lacerum medium, elongation of the external incisive fossa formed by premaxillae between tusk alveoli, higher sagittal process in the floor of the nasal fossa, and reduced anterolingual cingulid of lower molars (Tassy, 1990b, 1994a).

However, if mammutid and gomphothere elephantimorphs separately derive from *Palaeomastodon* and *Phiomia*, respectively, horizontal tooth displacement would constitute a parallelism between these two major groups rather than a synapomorphy, evolved independently at least twice. Gomphotheres share with *Phiomia* bunodonty and the development of accessory conules in molars as well as trefoil enamel wear figures (Sanders et al., 2004). *Palaeomastodon* anticipated the occlusal morphology of mammutid molars by the expression of zygodont crests, more open transverse valleys, and the development of crescentoids rather than accessory conules (Sanders et al., 2004). Conservatively, because the hypothesis of diphyly of Elephantimorpha is not otherwise well supported, the current classification retains both Palaeomastodontidae and Elephantimorpha.

Mammutidae, with the exception of early Miocene *Eozygodon*, is poorly represented in Africa. Nonetheless, familial features are distinctive, particularly in molars: Intermediate molars are trilophed; molars are non-bunodont and characterized by mesoconelets markedly lower than outer, main conelets (zygolophodonty); half-loph(id)s are apically compressed anteroposteriorly; there may be salient vertical posttrite zygodont crests; there are no accessory conules but crescentoids may descend from pretrite half-loph(id)s into the floor of transverse valleys (Tobien, 1975, 1996). Macrowear features on worn loph(id)s have a strong vertical component in moderate use of the crown. African mammutid species are separable by the degree of expression of these features, along with differences in molar size and proportions (e.g., Tassy and Pickford, 1983; Sanders and Miller, 2002). These molar traits may have been important for mastication of leafy vegetation (Tobien, 1996).

Gomphotheriidae encompasses a large number of different taxa that primitively share features such as bunodont-bunolophodont molar crowns, development of accessory conules that may contribute to trefoil enamel wear figures, presence of rounded to (in cross-section) piriform-shaped lower tusks set in elongate mandibular symphyses, trilophodont intermediate molars, third molars with three–four loph(id)s, raised basicranial regions in crania but low overall cranial height, a shorter braincase than lower face, upper tusks downturned, and relatively long palate (based on Tassy, 2013). Gomphotheriines are a wastebasket taxon in that their molar morphology is so basic that they may accommodate a greater diversity of species than realized by teeth alone. They develop anterior and posterior accessory conules that contribute to trefoil wear figures on the pretrite side of the crown, loph(id)s are basically oriented transverse to the long axis of molars, and lower tusks are oval in cross-section. *Gomphotherium* has been divided into primitive "*annectens*" and more advanced "*angustidens*" groups, based on cranial and molar morphological differences such as number of lophs in M3 and size and degree of retraction of the nasal aperture, enumerated in Chapter 4 (see also Tassy, 1994a). Choerolophodonts and amebelodonts evolved variations on this bauplan (Tobien, 1973a; Tassy, 1984, 1985, 1986; Pickford, 2001; Sanders et al., 2010a): Amebelodonts exhibit evolutionary trends for transverse broadening and dorsoventral flattening of lower tusks (the most derived forms possess dentinal tubules and rods inside of their i2s) (Lambert, 1990, 1992, 2016), development of third molars with more than four loph(id)s, and display pseudoanancoidy, in which enlarged pretrite accessory conules may reach well anterior or posterior to their associated half-loph(id)s and are obliquely angled toward the posttrite side of the crown, and main outer pretrite conelets are posteriorly set relative to their associated mesoconelets (pre- and posttrite mesoconelets are transversely aligned across the crown). Accessory conules are usually present in association with both pre- and posttrite half-loph(id)s. Permanent premolars are retained. By contrast, choerolophodonts lost their lower tusks and premolars, developed molar chevroning, in which pretrite anterior accessory conules and mesoconelets are displaced anterior to posttrite half-loph(id)s, and pre- and posttrite half-loph(id)s are angled on one another, had distally upcurved I2s without enamel bands and trough-like mandibular symphyses, and evolved choerodonty (multiplication of accessory conules in transverse valleys), ptychodonty, (rugosity of molar enamel), and cementodonty (thick molar crown coverings of cementum).

Tetralophodontinae is characterized by very large-bodied species with elongate mandibular symphyses and prominent lower tusks, maintenance of permanent premolars, and molars that may exhibit accessory conules that are incorporated into trefoil wear figures on both pre- and posttrite sides of the crown. Molars are bunodont-bunolophodont and brachyodont to subhypsodont. Intermediate molars are tetralophed and third molars may have as many as six loph(id)s, which overlaps with the condition in the most primitive elephants (see Coppens et al., 1978; Göhlich, 1999; Markov, 2008; Tibuleac et al., 2015). Postcranial elements are strikingly similar to those of stegotetrabelodonts (e.g., Alberdi, 1971).

Anancine gomphotheres are readily distinguishable by anancoidy, or lateral offset, of their half-loph(id)s into a "checkerboard" pattern (Mebrate and Kalb, 1985; Tassy, 1986; Kalb and Mebrate, 1993). Primitive anancines display tetralophed intermediate molars, unfolded enamel, modest development of accessory conules, weak anancoidy, and five–six third molar loph(id)s; more advanced forms may have strongly folded enamel, thicker cementum coating loph(id)s, more complex distribution of accessory conules on pre- and posttrite sides of the crown and throughout the length of the crown, pentalophodont intermediate molars, and six–seven third molar loph(id)s (Sanders, 2007, 2011). Anancines have brevirostrine mandibles without lower tusks, and most species lack permanent premolars (Coppens et al., 1978).

Elephants are convergent on Stegodontidae in loss of lower tusks, shortening of mandibular symphyses, development of molar lamellae, and adoption of fore-aft horizontal shearing mastication. However, *Stegodon* remained resolutely brachyodont, and molar plates of more advanced stegodontids in particular are readily discernable because they are superficially subdivided into numerous, bilaterally compressed and equal-sized apical digitations, and have a two-stage enamel stepping ("Stüfenbildung") (Saegusa, 1987, 1996b, 2020).

Elephantidae is well-defined by block-like molars constructed of parallel lamellae that are reinforced by cementum, third molars with at least six lamellae, or plates, main, outer conelets and mesoconelets generally of similar size and height, increased elevation and anteroposterior compression of crania, forward displacement of temporalis muscles, delayed serial appearance of molars, and proal or horizontal shearing mastication (Sikes, 1967; Aguirre, 1969a; Maglio, 1973; Roth, 1989; Saegusa, 2020). Stegotetrabelodontinae, the Subfamily comprising the most archaic elephants, primitively retained lower tusks, long mandibular symphyses, permanent premolars, and low-crowned molars with few plates and thick, unfolded enamel (Maglio, 1973; Coppens et al., 1978). By contrast, the Subfamily Elephantinae is composed of species that have lost lower tusks, and generally, with a few exceptions, that also dispensed with permanent premolars and had shorter mandibular symphyses and third molars with at least seven plates (Maglio, 1973; Tobien, 1978b; Kalb and Mebrate, 1993; Kalb et al., 1996a; Sanders et al., 2010a). However, *Selenitherium*, which lacks lower tusks, exhibits an elongate symphysis and third molars with only six lamellae (Mackaye et al., 2005). *Stegodibelodon*, which also lacks lower tusks and has third molars with seven plates, plesiomorphically has an immensely long symphysis (Coppens, 1972). Because lower tusks were lost or suppressed multiple times over the course of elephantimorph evolution (Sanders, 2017), it is possible that lower tusks were also lost independently within Elephantidae, and therefore is a weak character for defining Elephantinae.

MAJOR EPISODES OF AFRICAN PROBOSCIDEAN EVOLUTION

Because many African proboscidean species have large, heavy skeletons, pillar-like limbs, sturdy molars, robust dentaries, and immense tusks, the fossil record of the Order is profuse. In addition, proboscideans are obligate drinkers of standing water (Poole et al., 2013), and therefore it is probable that they frequented water courses advantageous to preservation at death and eventual fossilization. For these reasons, it is predictable that proboscideans are common constituents of faunal assemblages from sites on the continent across disparate landscapes and through many geological ages. It is likely that geographic distribution of African proboscideans in the fossil record is limited as much as or more by geological factors than artifacts of taphonomy related to biology or behavior. The African proboscidean record stretches back 60 Ma to the early Paleogene (Figure 1.3), the longest continuous presence of a eutherian mammal group on the continent (Gheerbrant, 2009). However, evidence of the taxonomic abundance of Proboscidea is uneven though time, delimited in no small way by tectonic volatility and associated accumulation of sediments. Thus, for example, the oldest proboscideans, from the Paleocene and early Eocene, primarily derive from a single basin in North Africa (Gheerbrant et al., 1992, 2017) and are taxonomically sparse, while there was a greater diversity of sites and explosion of taxa extending to other parts of the continent during the late Oligocene and Miocene (Figure 1.3), particularly in eastern Africa, due to onset of substantial rifting, volcanism, and consequent sedimentary accumulation during that interval (e.g., Abbate et al., 2014).

This paleontological evidence shows that nearly all of the major events of proboscidean evolution occurred in Africa (Sanders et al., 2010a). A comprehensive view of the extent and temporal ranges of genera comprising African (or Afro-Arabian) Proboscidea reveals the timing and taxonomic richness of these events, the most significant of which are labeled by number in Figure 1.3:

1. The *first event* marks the *origin of proboscideans*, as well as a modest *succession of early plesielephantiforms*, in the early Paleogene (Gheerbrant et al., 2012, 2017). Most of the finds are from a single basin in northern Africa, and for this interval nothing of proboscideans is known elsewhere on the continent. The earliest proboscideans were diminutive and are morphologically so plesiomorphic that it would be difficult to recognize the ordinal affinities of precursors if they were more primitive. These plesielephantiforms evolutionarily acquired molar lophodonty and exhibited a tendency for reduction or loss of canines, with progressive emphasis on specialization of the anterior dentition via hypertrophy of at least one incisor position and development of sizable diastemata. Within a few million years, they achieved large body size (Mahboubi et al., 1984; Gheerbrant et al., 2002), establishing a dominance in their ecosystems that continued throughout the rest of proboscidean evolution.
2. The *second event* is represented by such sparse fossil evidence that it has gone undiscovered until very recently, with the recovery in middle Eocene sites of *predecessors of the semi-aquatic moeritheres and elephantiforms*, the group that includes palaeomastodonts, mastodonts, gomphotheres, stegodonts, and elephants (Tabuce et al., 2019; Hautier et al., 2021). These discoveries extend the occurrence of early Paleogene proboscideans to western Africa. They also address immense "ghost lineage" interludes prior to the moeritheres and palaeomastodonts of the late Eocene-early Oligocene, back to an age when ancestral moeritheres and elephantiforms must have existed, and hint that plesielephantiforms might not have comprised the ancestry but rather a sister group of elephantiforms. These new finds are dentally bunodont-bunolophodont and contrast with the molar lophodonty of plesielephantiforms. Because

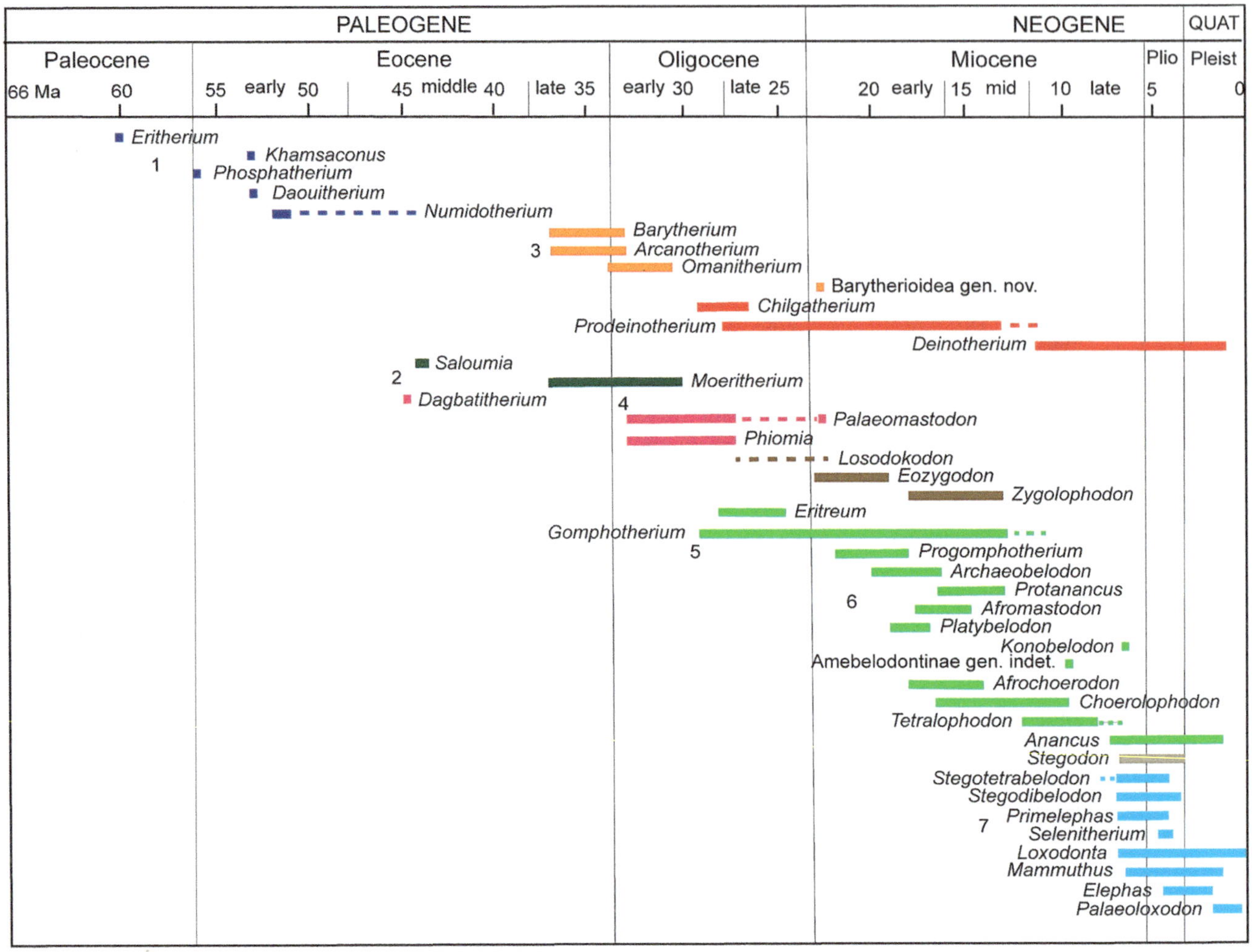

FIGURE 1.3 Temporal distribution of Afro-Arabian proboscidean genera throughout the Cenozoic. The scale of the chart does not permit representation of the Holocene, which dates from 11,700 y BP. *Loxodonta* is the only proboscidean genus represented in the African Holocene. Dotted lines indicate probable extensions of temporal ranges of taxa. Numbers represent arbitrarily selected major Afro-Arabian proboscidean evolutionary events. Colors are associated with different taxonomic groups: Dark purple, early plesielephantiforms; orange, barytheres; red, deinotheres; dark green, moeritheres and possible moerithere precursors; pink, early elephantiforms; dark brown, mammutids; light green, gomphotheres; light brown, stegodonts; blue, elephants.

there are no instances among other mammalian groups of lophodonty evolving into bunolophodonty (Hautier et al., 2021), it should not be surprising to find greater diversity of proboscideans and bunodont-bunolophodont ancestors of moeritheres and elephantiforms in the early Paleogene. It is also reasonable to assume that the ancestor of both these new bunolophodont proboscideans and lophodont plesielephantiforms of the early Paleogene was bunolophodont, and therefore could be even more ancient than *Eritherium*, with its nascent lophodonty.

3. *Event three* of African proboscidean evolution is marked by the *appearance of a variety of semi-aquatic barytheres and moeritheres* (Andrews, 1906; Matsumoto, 1923; Delmer, 2005; Shoshani et al., 2006; Clementz et al., 2008; Liu et al., 2008), with temporal ranges across the early-late Palaeogene boundary. These strange-looking proboscideans were hippo-like in their morphology (Osborn, 1909, 1936), and enjoyed success in their ecological milieu as the result of high sea-stand conditions of the time, diminished later by mid-to-late Oligocene drop in eustatic sea level and consequent loss of coastal habitats (Swezey, 2009). Their geographic range was primarily northern Africa and Arabia. Barytheres and moeritheres both exhibited impressive procumbent lower incisors, but diverged in molar morphology: Barytheres continued plesielephantiform lophodonty, whereas

moeritheres possessed bunolophodont cheek teeth. The molar morphology of moeritheres suggests a phylogenetic connection to elephantiforms, but the apomorphy of their postcrania and crania, dominated by aquatic features, prevents clear ascertainment of relationship to other proboscideans (Shoshani et al., 1996).

4. The *fourth event* encompassed the *appearance of definitive elephantiform proboscideans that exhibited evidence of trunks, upper and lower tusks, and reached immense body size, accompanied by evolution of graviportal postcranial adaptations* (Andrews, 1906). These elephantiforms, the palaeomastodonts, were late Paleogene inhabitants of northern and eastern Africa as well as Arabia, and may have persisted into the earliest Neogene in eastern Africa. In connection with the necessity of longer lifespans to reach enormous body sizes, the emergence of permanent premolars and molars was ontogenetically delayed in forming the full dentition, compared with the condition in plesielephantiform proboscideans.

   Palaeomastodonts were broadly ancestral to more advanced elephantimorph proboscideans—mastodonts and gomphotheres (and their descendants). Features of molar morphology suggest that mastodonts, with zygodont crests and accessory crescentoids, descended from the nominate palaeomastodont genus *Palaeomastodon*, and that the genus *Phiomia* gave rise to gomphotheres, with anterior and posterior accessory conules contributing to rounded trefoil enamel wear figures (Sanders et al., 2004). These terrestrial early elephantiforms were in the ascendancy as semiaquatic moeritheres and barytheres went extinct. Thick molar crown enamel and the lateral grinding motion of their mastication may have given palaeomastodonts an advantage over plesielephantiforms for feeding on a wider range of food items, including coarser plant materials. Although their nasal apertures are retracted, they are small and the area for muscle attachment around the aperture is modest, indicating that the palaeomastodont trunk may have been modest in size, perhaps more advantageous for manipulation of items in concert with the upper and/or lower tusks than for grasping food items close to the ground.

5. A *fifth event* involved the late early Oligocene *origin of deinotheres and elephantimorph proboscideans*, the two most highly successful groups of the Order in geological longevity and geographic distribution. Deinotheres rival gomphotheres for the longest proboscidean fossil record in Africa, spanning from the late Palaeogene until close to the end of the early Pleistocene (Vialli, 1966; Harris, 1978a; Sagri et al., 1998; Sanders et al., 2004; Rasmussen and Gutíerrez, 2009; Abbate et al., 2012, 2014). They are instantly recognizable by loss of upper tusks and presence of downturned lower tusks that may have been coordinated with the trunk for acquiring plant items. Although they are sometimes considered close sister taxa to elephantiform proboscideans (e.g., Delmer, 2009; Sanders et al., 2010a), deinothere lophodont molars and crania are more similar to barythere molars and skulls (Harris, 1973, 1976, 1978a). Therefore, it is possible that deinotheres derived from plesielephantiforms rather than as apomorphic offshoots of elephantiforms. Analysis of deinothere lophodonty indicates that most of their masticatory function involved Phase I vertical shearing of food items across the loph(id)s (von Koenigswald, 2014), particularly effective for leaf browsing. For most of their existence, deinotheres were evidently committed hyperbrowsers, feeding in closed settings, such as gallery forests (Harris, 1975; Cerling et al., 1999, 2015b; Uno et al., 2011; but see Lohrke, 2017 for exceptions). As a result of their dedicated browsing specializations and preference for feeding in closed canopy conditions, deinotheres remained morphologically conservative for the duration of their existence and may have avoided competition with many other large-bodied ungulates.

   Elephantimorph proboscideans are fundamentally separated phylogenetically into mastodonts (Mammutida) and gomphotheres and their descendants, including stegodonts and elephants (Elephantida). The initial Afro-Arabian advent of this group dates to the early Oligocene (Zalmout et al., 2010); representatives of both Mammutida and Elephantida were recovered from late Oligocene settings (Sanders et al., 2004; Shoshani et al., 2004; Rasmussen and Gutíerrez, 2009; Leakey et al., 2011; Abbate et al., 2012, 2014).

   The tremendous success of elephantimorphs, particularly gomphotheres and their descendants, is marked by their geological duration and diversification of taxa. In terms of winners and losers in proboscidean evolution, elephantimorphs can be viewed as the ultimate winners. The phylogenetic and ecological triumph of the group may be attributed to the *evolution of a key adaptation, horizontal tooth displacement* in which molars are delayed in emergence and "pushed out," or displaced more anteriorly, worn molars by anterior progression of newly emerged molars, by remodeling of the jaws and drift in a "conveyor belt" fashion (Shoshani, 2002; Tassy, 1996b; Ungar, 2010; Sanders, 2017). The advantages of this mechanism delaying the development and succession of molars cannot be overstated, and include the ability to survive the attrition of tooth wear from eating a wide range of harder or grittier plant items, longer ontogenetic growth intervals, and larger body size and increase in life-span and number of breeding cycles (Sanders, 2017).

Although early members of elephantimorph lineages possessed premolars that vertically replaced deciduous precursors, the evident disruption of the horizontal tooth displacement mechanism led to loss of permanent premolars among elephantimorphs, apparently independently in different lineages (e.g., mammutids, choerolophodonts, anancine gomphotheres, stegodonts, elephants—perhaps several times in different genera) (Sanders, 2017). Elephantimorphs that lost permanent premolars also dispensed with lower tusks and therefore had foreshortened dentaries in which premolars would have compromised the anterior rotation of molars (Sanders, 2017).

6. Following the origin of elephantimorphs in Afro-Arabia, this landmass continued as the locus of evolutionary innovation for proboscideans. The *sixth event*, starting in the early Neogene, was marked by *impressive diversification of gomphotheres* from basic trilophodont gomphotheriines to amebelodontines and choerolophodontines. In gomphotheriines, intermediate molars have three loph(id)s that are arranged transversely across their crowns; loph(id)s are pyramidal in shape in lateral view, and bunodont-bunolohodont; enamel loops on the pretrite side of the crown incorporate anterior and posterior accessory conules into trefoil wear patterns (Tobien, 1973a). The advantage of this pattern was to add surface area for Phase II lateral rotary grinding of the food bolus (von Koenigswald, 2014). The transverse orientation of loph(id)s and presence of accessory conules in the transverse valleys was altered and elaborated as "fabricational noise" in amebelodonts and choerolophodonts, which developed additional accessory conules and offset of half-loph(id)s, creating greater contact and surface area between enamel surfaces of lower against upper molars during mastication.

   In the course of amebelodont evolution, pretrite accessory conules progressively became larger, were angled obliquely across transverse valleys toward the posttrite side, and posttrite half-loph(id)s developed accessory conules as well. Offset, or pseudo-anancoidy, of half-loph(id)s was achieved by pretrite main outer conelets displaced slightly posterior relative to other conelets (MacInnes, 1942; Tassy, 1986). Moreover, lower tusks became larger and broader in association with specialized cropping of food items.

   Over the course of choerolophodont evolution, accessory conules were multiplied on both sides of molar crowns ("choerodonty"), and loph(id)s were chevroned by transverse offset of half-loph(id)s, in which the mesoconelet of each pretrite half-loph(id) projects anterior to the posttrite half-loph(id) and to its own paired abaxial conelet (Tassy, 1985, 1986). As in ambelodonts, this arrangement would have been effective at expanding contact and surface area of molar grinding surfaces. In addition, choerolophdonts evolved strong cementodonty (Tassy, 1985; Tassy et al., 2013), providing a liberal coating of cementum on molar crowns, helping to resist molar wear.

   Although there is a singular occurrence of a proboscidean incisor fragment from the late Oligocene Chitarwata Fm. in Pakistan (Antoine et al., 2003), significant records of elephantimorph and deinothere migration from Africa to Eurasia first occur during the early Miocene (Tassy, 1990a; van der Made and Mazo, 2003). In the late Miocene, proboscidean lineages that evolved in Eurasia from these early migrants returned to Africa in the form of tetralophodont gomphotheres (*Tetralophodon, Anancus*) and stegodonts (Sanders et al., 2010a).

7. The immigration into Africa of tetralophodont gomphotheres (*Tetralophodon*) as early as the late part of the middle Miocene provided an ancestor for the *seventh evolutionary event* in Afro-Arabian proboscidean history, the *origin and initial diversification of elephants*, during the second half of the late Miocene (Maglio, 1973). The key adaptation of this event and the rise of elephants was transformation of the bunolophodont gomphothere molar into *block-like teeth organized as a series of parallel transverse plates or lamellae, consolidated by a coating or embedding of the lamellae by cementum* (Roth, 1989; Saegusa, 2020). Furthermore, elephant crania and jaws were reorganized for deriving mastication across these molars by *fore-aft horizontal, or proal, shearing* (Maglio, 1972a). This transformation involved anteroposterior compression and heightening of the cranium, and in more progressive species also entailed loss of lower tusks and shortening of dentaries and symphyses, placing insertion of masticatory muscles closer to the center of mass of the lower jaw and making chewing action more effective (Maglio, 1972a; Sanders et al., 2021). Moreover, the most progressive elephant species evolved considerably *hypsodont molars.*

   Evolution and improvement of these key features has been associated with increasing availability of C_4 grasses in Africa starting in the late Miocene and concomitant ramping up of competition with other mammals, including stegodonts and anancine gomphotheres, for this resource (Maglio, 1973; Cerling et al., 1999, 2015b; Levin et al., 2011; Uno et al., 2011; Saegusa, 2020). The first elephants, some with very brachyodont molars, elongate jaws, and lower tusks (Petrocchi, 1954; Maglio, 1970b; Tassy, 1999, 2003; Sanders, 2008b, 2022), were morphologically "out-of-phase" with their behavior of eating substantial amounts of grasses (Lister, 2013), but crown elephant genera emerged very quickly and by the mid-Pliocene archaic elephants had been replaced by elephant species with molars constructed of a greater number of higher-crowned plates (e.g., Sanders, 2020; Sanders et al., 2021). These more advanced African elephants provided the phylogenetic sources for the subsequent evolution of Asian elephants, diversification of mammoths in Eurasia and North America, and radiation of palaeoloxodont elephants throughout Eurasia, by a series of migratory episodes in the late Pliocene-early Pleistocene and toward the end of the late Pleistocene (Maglio, 1973; Dudley, 1996; Lister, 1996; Lister and van Essen, 2003; Lister et al., 2005; Saegusa and Gilbert, 2008; Lister and Sher, 2015; Larramendi et al., 2020; Zhang, 2020).

By the Pleistocene, massive-sized elephants of the "*Elephas recki* complex" (now recognized as taxonomically more diverse than previously hypothesized [contrast Beden, 1980 and Sanders and Haile-Selassie 2012 with Saegusa, 2008, Larramendi et al., 2020, and Zhang, 2020]), with progressively more hypsodont molars incorporating additional plates, swept away other elephants to peripheral ranges, competitively pushed stegodonts and anancine gomphotheres to extinction, and became the dominant proboscideans across Afro-Arabia until the late Pleistocene (Manthi et al., 2019). Some of the elephants of this group were hypergrazers, superbly adapted to the expansion of C_4 grasses and grasslands in Africa and Arabia (Cerling, 1992; Ségalen et al., 2007; Cerling et al., 2011b, 2015a, b). Their catastrophic demise in Afro-Arabia during the late Pleistocene may have been driven by climatic factors leading to repeated mega-drought conditions that made grazing specialists vulnerable (Scholz et al., 2007; Faith, 2014; Manthi et al., 2019). The modern remnants of this seventh proboscidean evolutionary event are the forest and savanna African elephants, more eclectic in feeding preferences as browsers and mixed feeders, and perhaps with greater dietary plasticity (Koch et al., 1995; Cerling et al., 1999; Ziegler et al., 2016).

2 Early Paleogene
Origin and Evolution of the First Proboscideans

The primitive morphology of *Eritherium* suggests a recent and rapid paenungulate radiation after the Cretaceous-Tertiary boundary, probably favored by early endemic African paleoecosystems. At a broader scale, *Eritherium* provides a new old calibration point of the placental tree and supports an explosive placental radiation.

—Emmanuel Gheerbrant (2009)

INTRODUCTION

The most ancient record of proboscideans derives from early Paleogene sediments in Africa and constitutes the first of many important phylogenetic episodes in the evolutionary history of elephants and their relatives. Remarkably, nearly all origins of major proboscidean taxa are documented to have occurred on the continent, from the commencement of the Order to the first appearance of elephants (Sanders et al., 2010a). Moreover, until the end of the Paleogene, all proboscidean evolution was restricted to the island continent of Africa (Antoine et al., 2003), or more accurately for the period, Afro-Arabia. The split of Gondwana into Afro-Arabia and South America occurred during the Cretaceous, and for 80 Ma until the start of the Neogene Afro-Arabia remained largely isolated with only sporadic, tenuous contact with Laurasia/Eurasia (Rage and Gheerbrant, 2020). Ancestors of the oldest proboscideans and other afrotheres are themselves obscure, stemming either from archaic, unknown Gondwanan placental mammals (Gheerbrant et al., 2014) or more likely Laurasian immigrants to Africa, such as tribosphenidans, which upon their arrival would have found Mesozoic Afro-Arabia depauperate in native mammals (Luo et al., 2001; Kielan-Jaworowska et al., 2004; Gheerbrant and Rage, 2006; Gheerbrant, 2009; Gheerbrant et al., 2012; Rage and Gheerbrant, 2020; but see Sigogneau-Russell et al., 2001). Nonetheless, paleobiogeographic analysis of placental mammals firmly places afrotherian origins in Africa (Springer et al., 2011).

The very earliest record of fossil proboscideans is from a relatively warm time that preceded a brief global cooling event; subsequent episodes of first appearances of early Paleogene proboscidean taxa occur during a long-term worldwide warming trend and coincide with hypothermal events that punctuate that trend (Figure 2.1; Kocsis et al., 2014; Yans et al., 2014). In Holarctic settings coincident with the Paleocene-Eocene boundary, the first true primates, perissodactyls, artiodactyls, and rodents appeared (Gingerich, 2006; Rose et al., 2014), penecontemporaneous with early diversification of afrotheres in Africa. Afrotheres were free to exploit new niches and diversify without competition from these Holarctic taxa, in "island" Africa. During this time, the world was considerably warmer than it is today (Zachos et al., 2001), and the earliest proboscideans, inhabiting low latitudes, likely occupied tropical forest ecosystems, buffered from harsh effects of colder weather or seasonality. Although it is possible that climate change played a role in the first appearance of proboscideans, the sparseness of the African fossil record from the early Paleogene hinders attempts to establish causal connections between extrinsic factors and the early evolution of the Order.

The great antiquity of proboscideans establishes them as one of the longest surviving modern placental mammalian groups, and as the oldest remaining endemic mammals of Africa (Gheerbrant, 2009). Nonetheless, they derive from mammals that were probably part of a non-native afrothere stem group that originated in Laurasia; taxa identified as native Afro-Arabian mammals apparently did not survive across the Cretaceous-Paleocene boundary (Rage and Gheerbrant, 2020). At the start of the Paleogene, Africa (with Arabia) was separated from other Gondwanan landmasses and formed an island continent that enjoyed periodic faunal exchange with Laurasia, involving mostly southward migration of taxa of North American and European derivation via filter routes across the Tethys Sea (Gheerbrant, 1990; Gheerbrant and Rage, 2006; Rage and Gheerbrant, 2020). The original story of proboscidean evolution is exclusively North African (Table 2.1; Figure 2.2), but this may have more to do with the poor preservation or absence of early Paleogene sediments elsewhere on the continent (Gheerbrant, 1998) than with any special regional factors.

Competing hypotheses of the origin and diversification of placental mammals have been summarized alternatively by (a) the "Explosive Model," in which extant placental Orders and their extinct stem taxa arose and diversified in the early Tertiary; (b) the "Short-Fuse" or "Fast-Fuse Model," in which most or all of the extant placental Orders and the stem taxa of supraordinal groups uniting them originated in the Cretaceous; or (c) the "Long-Fuse Model," in which extant placental Orders originated in the early Tertiary but have stem taxa extending well back into the

DOI: 10.1201/b20016-2

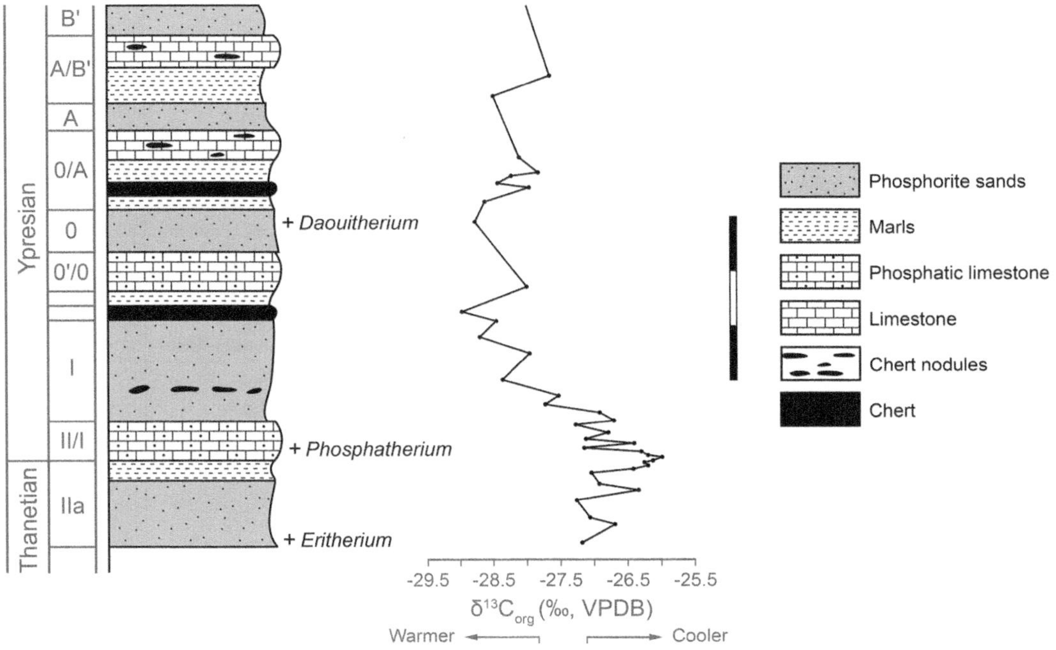

FIGURE 2.1 $\delta^{13}C_{org}$ chemostratigraphy and generalized stratigraphic section of the Ouled Abdoun Basin showing the relative position of proboscidean occurrences in the late Paleocene and early Eocene. (Adapted from Kocsis et al. [2014:fig. 1] and Yans et al. [2014:fig. 4], with permission of Elsevier, and from Gondwana Research through the Copyright Clearinghouse Center, respectively.

Cretaceous (Figure 2.3; Archibald and Deutschman, 2001; Yoder, 2013; Averianov and Lopatin, 2014; Davies et al., 2017). Paleontological and combined phenomic-molecular phylogenetic analyses have supported the first hypothesis (e.g., Foote et al., 1999; Archibald and Deutschman, 2001; O'Leary et al., 2013), whereas molecular phylogenetic analyses and some biogeographic scenarios (summarized in Gheerbrant and Rage, 2006) depict the origin of major placental mammalian clades and the diversification of crown mammal Orders, including Proboscidea, as having taken place well within the Cretaceous (hypothesis 2; e.g., Springer et al., 1997; Eizirik et al., 2001; Murphy et al., 2001; Bininda-Emonds et al., 2007).

Two other hypotheses of placental mammalian cladogenesis have been proposed, the "Soft Explosive Model" and "Trans-KPg Model" (Springer et al., 2019). The "Trans-KPg Model" is similar to the "Short-Fuse Model" in that it proposes most diversification of placental mammals occurred after the Cretaceous-Paleogene extinction event but differs in suggesting that this interordinal cladogenesis of placentals continued uninterrupted across the late Cretaceous-early Paleogene, coincident with a parallel radiation of multituberculate mammals in response to the availability of angiosperm plants (Liu et al., 2017). The "Soft Explosive Model" posits cladogenesis of the major mammalian groups (e.g., Afrotheria, Euarchontoglires, Xenarthra, Laurasiatheria) in the late Cretaceous but hypothesizes that the balance of placental interordinal diversification occurred around or after the Cretaceous-Paleogene boundary (Phillips, 2016). In this model, most crown placentals (but not all) arose after this boundary. As with the "Short-Fuse Model" and "Long-Fuse Model," these hypotheses are also hampered by the prevalence of ghost lineages.

At issue is the fidelity of the paleontological record for investigating the time of origin of placental mammalian taxa—do fossils give us meaningful approximations of the age of appearance of major taxa? Although O'Leary et al. (2013) claimed that there is no direct fossil evidence supporting hypothesis 2 or 3, Averianov and Lopatin (2014:p. 804) countered that these authors "deliberately ignored all the findings that were not consistent with their hypothesis." Furthermore, morphological divergence of sympatric stem paenungulates from the oldest known proboscideans indicates that North Africa was already a center of afrotherian radiation by the mid-Paleocene, raising questions about whether endemic evolution of placental mammals in Africa, including afrotheres, began after the end of the Cretaceous (Gheerbrant et al., 2014). Recent recovery of gnathodental fossils from the Ouled Abdoun Basin of early Ypresian age that exhibit well-defined embrithopod features (Gheerbrant et al., 2018) suggests a lengthy prior time of diversification of afrotheres. Furthermore, results of a recent large-scale, morphology-based cladistic analysis primarily of Paleocene mammalian taxa indicates that the origin of placental mammals could have predated the Cenozoic while reaffirming that no definitive crown placental mammals

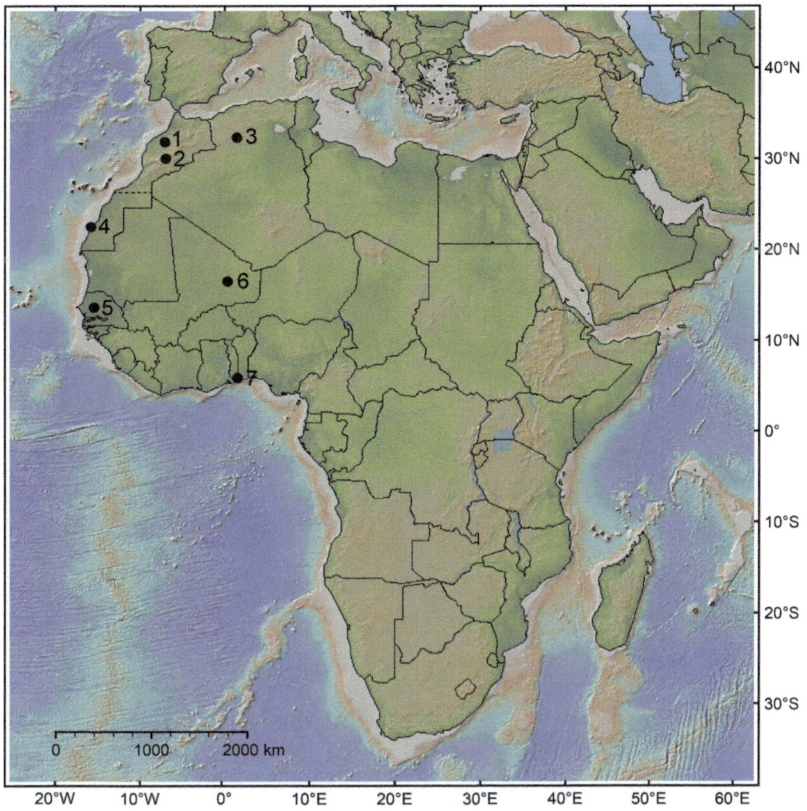

FIGURE 2.2 Map showing location of early Paleogene proboscidean sites. 1a, Quarry A4, Sidi Chennane quarries, Ouled Abdoun Basin, Morocco; 1b, Grand Daoui quarries, Ouled Abdoun Basin, Morocco; 2, N'Tagourt 2, Ouarzazate Basin, Morocco; 3, El Kohol, Brezina, Algeria; 4, Dakhla, Morocco; 5, M'Bodione Dadere, Senegal; 6, Tamaguélelt, Mali; 7, Dagbati Quarry, Togo.

can be presently identified from the Cretaceous (Halliday et al., 2015). Analysis of the completeness of the eutherian mammalian fossil record across the Cretaceous/Paleogene mass extinction event, however, demonstrates a genuine absence of placental mammals in the Cretaceous (Davies et al., 2017). Conversely, recent results of genomic investigation of chitinase genes provocatively suggest that interordinal diversification of placental mammals began in the Cretaceous, but genetic change associated with a shift from insectivory to more varied diets in placental mammals occurred after the Cretaceous/Paleogene boundary, meaning that morphological features permitting identification of different placental Orders would have lagged behind their actual origination (Emerling et al., 2018). Evidence for early Paleocene diversification of afrotheres suggests that stem afrotheres, at least, could have originated prior to the Cenozoic. Nonetheless, despite molecular indications of proboscidean origins in the late Cretaceous (e.g., Bininda-Emonds et al., 2007), the morphology of the first proboscidean, *Eritherium azzouzorum*, reminiscent in many features of primitive "condylarths" and exhibiting very few proboscidean synapomorphies, corresponds best with hypothesis 1 or 3 and a younger, Paleocene debut of the Order (Gheerbrant, 2009; Gheerbrant et al., 2012). Indeed, stochastic methods applied to the dating of clade origins indicate that it is likely that placental mammals first appeared in the late Cretaceous (much younger, however, than posited by most molecular-based hypotheses), but unlikely that afrotheres diverged until the Paleocene (Halliday et al., 2016).

Subsequent to the first appearance of proboscideans in the mid-to-late Paleocene, the geographical and stratigraphic limitations of the Afro-Arabian fossil record reveal a limited succession, rather than radiation (e.g., Gheerbrant and Tassy, 2009), of taxa through the early Eocene, but one that clearly anticipates adaptive trends of later taxa and that precociously establishes proboscideans as keystone species in their faunas early in their history. Most of the features characteristic of elephants, however, such as a freely mobile, pendulous trunk, outsized tusks, horizontal displacement of molars, gigantism, and associated graviportal postcranial adaptations (Gheerbrant and Tassy, 2009), are not present in proboscideans until much later in the Paleogene. On the contrary, the differences between these early taxa and

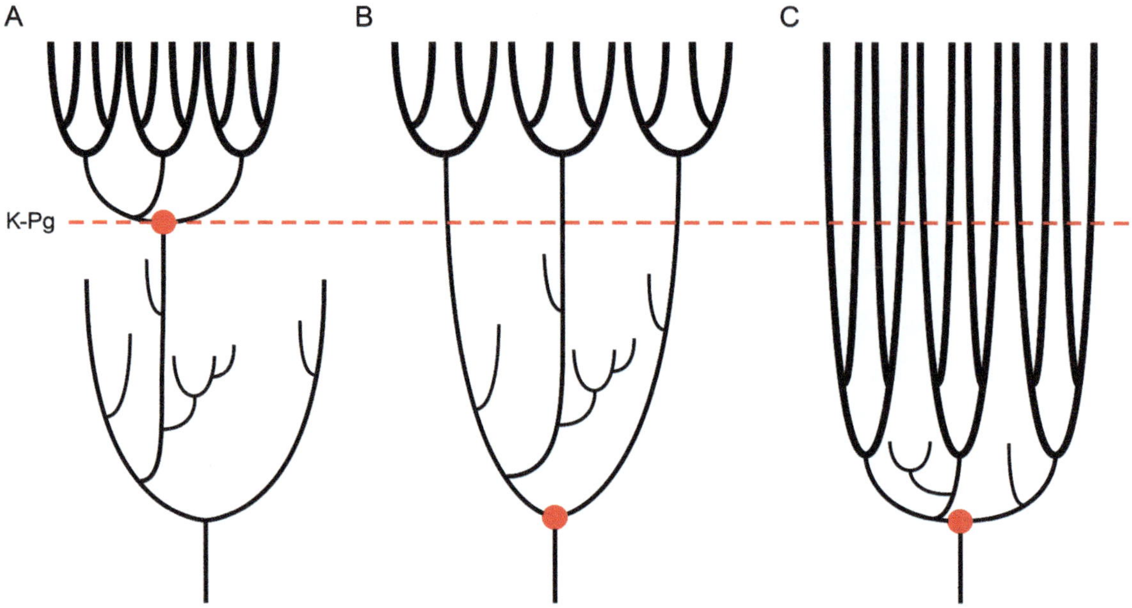

FIGURE 2.3 Alternate models of early placental evolution. Red circles indicate the point of origination of the clade and the thicker black lines indicate the timing of subsequent interordinal diversification. Abbreviation: K/Pg, Cretaceous-Paleogene boundary. (A) Explosive model. (B) Long-fuse model. (C) Short-fuse model. (Adapted from Davies et al. [2017:fig. 1], with permission of the authors and through open access under the Creative Commons CC BY by license.)

extant elephants point to the extraordinary adaptive plasticity of proboscideans through time and to morphological transformations as profound as any witnessed in the fossil record of other mammalian Orders.

SYSTEMATIC PALEONTOLOGY

PROBOSCIDEA Illiger, 1811
PLESIELEPHANTIFORMES Shoshani, Sanders, and Tassy, 2001a
Family Incertae Sedis
***ERITHERIUM* Gheerbrant, 2009**
***ERITHERIUM AZZOUZORUM* Gheerbrant, 2009**
(Tables 2.1–2.3; Figures 2.1, 2.2, 2.4, and 2.6)

Rather than starting with a grand entrance befitting their extant physical and ecological magnitude, the long evolutionary journey of elephants began inauspiciously, with animals of such small size and primitive anatomy that they are barely recognizable as proboscideans. The fossil remains of these elephant ancestors are dated to the late middle or late Paleocene epoch (attributed to the Thanetian because the local selachian assemblages do not facilitate recognition of the Selandian, but probably late Selandian in age) at ~60 Ma by stratigraphic placement, biochronological zonation of associated elasmobranchs, and clever application of local carbon and oxygen isotope chemostratigraphy in comparison with records of global isotopic excursions (Figure 2.1; Gheerbrant, 2009; Kocsis et al., 2014; Yans et al., 2014). They belong to rare assemblages of early Paleogene vertebrates from a time interval otherwise virtually undocumented in Africa, their recovery the result of assiduous field efforts and tenacious detective work by French-Moroccan teams to track down the proveniences of fossils from the Ouarzazate and Ouled Abdoun Basins of Morocco, many of the latter held in private collections or found in the marketplace (Cappetta et al., 1978, 1987; Sudre et al., 1993; Gheerbrant et al., 1998a, b, 2003, 2014; Augé and Rage, 2006; Gheerbrant, 2009; Solé et al., 2009). *Eritherium azzouzorum*, the first proboscidean, derives from a lower bone-bed horizon in phosphate quarries of the Ouled Abdoun Basin, the most ancient known placental mammal site in Africa, discovered by local collectors; its name denotes an "early beast" and honors its discoverers from the Ouled Azzouz Village (Gheerbrant, 2009; Gheerbrant et al., 2012). As is frequently the case for basal members of phylogenetic groups, virtually nothing of the morphology of *Eritherium* anticipates the structure of its later, elephant-like relatives, necessitating exceedingly careful systematic analysis to identify its relationships. It should be noted that not all cladistic treatments return results finding *Eritherium* to cluster with Proboscidea (Gheerbrant et al., 2018).

Comparative morphology and phylogenetic relationships of *Eritherium* have been comprehensively detailed by Gheerbrant (2009) and Gheerbrant et al. (2012). Based on tooth size-body mass regression and volumetric estimates of a body scaled to fit the skull, *Eritherium* is calculated to have weighed only between three and eight kg (Gheerbrant, 2009; Larramendi, 2016), overlapping in size with modern hyraxes (Shoshani et al., 2013), making it not only the oldest but also the smallest known proboscidean, except for the geologically younger, poorly documented *Khamsaconus*. Aside from its diminutive body size, other aspects of *Eritherium* that are plesiomorphic include a primitive

TABLE 2.1
Major Occurrences and Ages of Early to Mid-Paleogene African Proboscidea

Taxon	Occurrence (Site, Locality)	Stratigraphic Unit	Age	Key References
Proboscidea Illiger, 1811				
Plesielephantiformes Shoshani, Sanders, and Tassy, 2001				
Family incertae sedis				
Eritherium azzouzorum	Quarry A4, Sidi Chennane quarries, Ouled Abdoun Basin, Morocco (type)	Lower bone-bed horizon, phosphate bed IIa	Late middle or early late Paleocene, probably late Selandian, ~60 Ma	Gheerbrant, (2009), Gheerbrant et al. (2012), Kocsis et al. (2014), and Yans et al. (2014)
Phosphatheriidae Gheerbrant et al., 2005				
Phosphatherium escuilliei	Grand Daoui quarries, Ouled Abdoun Basin, Morocco (type)	Intercalary Beds II/I	Earliest Eocene, early Ypresian, ~56 Ma	Gheerbrant et al. (1996, 1998, 2003), Gheerbrant (1998), Kocsis et al. (2014), and Yans et al. (2014)
Family incertae sedis, probably nov. Gheerbrant et al., 2005				
Daouitherium rebouli	Grand Daoui quarries, Ouled Abdoun Basin, Morocco (type)	Probably phosphorite level 0	Early Eocene, early/middle Ypresian, ~53 Ma	Gheerbrant et al. (2002), Kocsis et al. (2014), and Yans et al. (2014)
?Phosphatheriidae Gheerbrant et al., 2005				
Khamsaconus bulbosus	N'Tagourt 2, Ouarzazate Basin, Morocco (type)	Ait Ourithane or Jbel Ta'louit Fm.	Early Eocene (middle Ypresian), ~52.9 Ma	Sudre et al. (1993), and Gheerbrant et al. (1998b)
Suborder incertae sedis				
Family incertae sedis				
Gen. et sp. indet. probably nov.	Phosphate beds Tamaguélelt, Mali	Tamaguélelt Fm.	Early Eocene, Ypresian	Patterson and Longbottom (1989), Moody and Sutcliffe (1993), and O'Leary et al. (2006, 2019)
Numidotheriidae Shoshani and Tassy, 1992				
Numidotherium koholense	El Kohol, Brezina, Algeria (type)	El Kohol Fm.	Early Eocene, late Ypresian, 52–51 Ma	Mahboubi et al. (1984, 1986), Noubhani et al. (2008), Coster et al. (2012), and Mahboubi et al. (2014)
?*Numidotherium* sp.	Dakhla, Morocco	Vertebrate level B1, Unit 2, ?Gerran Mb., ?Samlat Fm.	Middle to late Eocene	Adnet et al. (2010)
Family incertae sedis				
Saloumia gorodiskii ("cf. *Moeritherium*")	M'Bodione Dadere, Senegal (type)	Lam Lam Fm.	Middle Eocene, mid-Lutetian, ~44 Ma	Gorodiski and Lavocat (1953), and Tabuce et al. (2019)
Elephantiformes Tassy, 1988				
Family incertae sedis				
Dagbatitherium tassyi	Dagbati Quarry, Togo (type)	Phospharenite Mb., Habotoé-Kpogamé Phosphate Complex	Eocene, Lutetian, ~46.5 Ma <>~44.0 Ma	Hautier et al. (2021)

? = attribution or occurrence uncertain.

eutherian dental formula (the lower dental composition appears to be i1–3, c, p1–4, m1–3 [or 3-1-4-3], and the upper dental quadrant is reconstructed as I?, C, P1–4, M1–3 [or?-1–4-3], with all teeth simultaneously in occlusion in adults; Table 2.2), no diastema, small third molars, bunodont incipiently bunolophodont molars, orbit above the level of P4-M1, and maxillary bone not strongly developed on the orbit. The dentary also primitively exhibits a small coronoid foramen at the same level as the cheek tooth alveoli. There are no tusks, no evidence for development of a trunk, and no obvious aquatic adaptations in the skull. CT scan analysis indicates that the structure of the petrosal of *Eritherium* is more primitive than that of geologically younger *Phosphatherium* and more similar to that of the basal paenungulate *Ocepeia* in having an inflated tegmen tympani, very deep subarcuate fossa, and ossified canal for the superior ramus of the stapedial artery (Schmitt and Gheerbrant, 2016). Unambiguous proboscidean synapomorphies are few: the lower canine is

TABLE 2.2
Adult Dental Formulae and Incisor Dominance in Early Paleogene African Proboscidea

Taxon	Upper Permanent Dentition	Largest Upper Incisor	Lower Permanent Dentition	Largest Lower Incisor
Eritherium azzouzorum	?I, C, P1–4, M1–3	?	i1–3, c, p1–4, m1–3	i1
Phosphatherium escuilliei	?I1–3, ?C, P1–4, M1–3	?I2	i1–2, c, p2–4, m1–3 or i1–2, p1–4, m1–3 or i2–3, p1–4, m1–3 or i2–3, c, p2–4, m1–3	i1 or i2
Daouitherium rebouli	—	—	i1–3, c, p2–4, m1–3 or i1–2, c, p1–4, m1–3 or i1–3, p1–4, m1–3	i1
Numidotherium koholense	I1–3, C, P2–4, M1–3	I2	i1–2, p2–4, m1–3	i1

C/c, canine; I/i, incisor; M/m, molar; P/p, premolar; ?, presence unknown.

reduced in size, as is i3; (d)p1 is small and simple in structure; i1 is enlarged and bigger than i2; i1–2 have styliform, procumbent crowns; molar hypoconulids are shifted buccally; the coronoid retromolar fossa is enlarged.

While no postcranial elements have been described for *Eritherium*, its sympatry with hyaenodontids, ungulate-like stem afrotheres, and primitive condylarthran ungulates (Gheerbrant et al., 2014) suggests derivation from a terrestrial habitus, though Seiffert (2013) reasoned that the recovery of the early proboscideans *Eritherium*, *Phosphatherium*, and *Daouitherium* from marine phosphate sediments in the Ouled Abdoun Basin indicates a semi-aquatic existence, and based on elephant embryology others have argued for an aquatic origin for the Proboscidea (Gaeth et al., 1999). Nonetheless, the co-occurrence of terrestrial taxa and an abundant marine fauna can be explained by a "peculiar taphonomy" of floated carcasses, presumably disgorged by rivers into littoral settings (Gheerbrant et al., 2002:p. 493; 2003).

Conversely, the site sample provides a reasonably respectable accounting of its cranial and mandibular morphology (Gheerbrant et al., 2012), from which more direct interpretations can be made. The resulting overall image is one of a basic, primitive eutherian mammalian skull. The nasal aperture is small and anteriorly placed, with long nasal bones behind that are narrow anteriorly and broader posteriorly near the frontal. The infraorbital foramen is large and situated above the anterior end of P3. Orbital size is small and the orbits are located low above the tooth rows. The zygomatic arch is a thin, high bony blade with a distinctive ventral process, and does not flare widely from the cranium. The cranium exhibits pronounced postorbital constriction, and the parietals are inflated and dorsally form a high and sharp sagittal crest that is most pronounced posteriorly. The mandibular symphyses are short, terminate posteriorly below p2, and probably were not fused. The anterior mental foramen is located below p1, and the posterior mental foramen mostly likely was below the posterior edge of p4.

Eritherium azzouzorum is best represented by its dental sample (Gheerbrant et al., 2012), which unfortunately lacks the upper incisors. The upper canine alveolus suggests that this was a very small tooth and not procumbent. P3–4 are submolariform and bicuspid, but are not lophodont, and have poor metacone development (Figure 2.4A). Upper molars are only incipiently bilophodont, with inflated cusps and poor development of proto- and metalophs. M2 is enlarged relative to M1 and M3; upper molar size progression is M1 < M2 > M3 (Figure 2.4A). Lower molars have inflated, bunodont crowns; they exhibit two pairs of transversely aligned cusps which each contact via their internal edges to form rudimentary proto- and hypolophids (Figure 2.4B). The second lower molar is enlarged, especially compared to m1, and all three lower molars have strong distocingulids. Lower molar size progression is m1 < m2 ≥ m3 (Figure 2.4B). Mean lengths and widths of M3 and m3 are only 6.23 × 6.56 mm and 7.65 × 5.50 mm, respectively (Gheerbrant et al., 2012), markedly more diminutive than the third molars of elephants, which reach hundreds of millimeters in length (Sanders et al., 2010a). The first two lower premolars are small and simple, whereas p3–4 are more molarized, though dominated by the protoconid. Compared with other early Paleogene proboscideans, the cheek teeth of *Eritherium* are very small (Table 2.3). While i3 is the smallest tooth in the series, and c is also very small, the height and anterior projection of i1–2 forms a notable toothcomb-like structure (Gheerbrant et al., 2012).

Parsimony analysis of *Eritherium azzouzorum* places it in the proboscidean clade as a close sister taxon to the

TABLE 2.3
Cheek Tooth Dimensions of Early Paleogene Proboscideans and Selected Plesielephantiforms

Taxon/Specimen/Tooth	Loph(id) Formula	L	W	H
	Eritherium azzouzorum[g] (late middle or late Paleocene)			
P1 (N=2)	—	Mean=2.7 s.d.=0.0	Mean=2.4 s.d.=0.07	—
P2 (N=2)	—	Mean=4.7 s.d.=0.21	Mean=3.6 s.d.=0.14	—
P3 (N=6)	x1	Mean=5.3 s.d.=0.26	Mean=6.3 s.d.=0.60	—
P4 (N=7)	x1	Mean=5.3 s.d.=0.32	Mean=7.0 s.d.=0.53	—
M1 (N=7)	x2	Mean=6.5 s.d.=0.21	Mean=7.9 s.d.=0.40	—
M2 (N=7)	x2	Mean=6.6 s.d.=0.28	Mean=8.1 s.d.=0.57	—
M3 (N=7)	x2	Mean=6.2 s.d.=0.30	Mean=6.6 s.d.=0.43	—
p1 (N=1)	—	3.7	2.0	—
p2 (N=2)	—	Mean=4.0 s.d.=0.07	Mean=2.9 s.d.=0.71	—
p3 (N=8)	—	Mean=5.6 s.d.=0.46	Mean=3.8 s.d.=0.38	—
p4 (N=8)	—	Mean=5.9 s.d.=0.31	Mean=4.2 s.d.=0.39	—
m1 (N=7)	x2x	Mean=6.7 s.d.=0.36	Mean=4.8 s.d.=0.22	—
m2 (N=13)	x2x	Mean=7.5 s.d.=0.31	Mean=5.9 s.d.=0.32	—
m3 (N=12)	x2x	Mean=7.7 s.d.=0.45	Mean=5.5 s.d.=0.29	—
	Phosphatherium escuilliei[d] (early Eocene)			
P3 (N=3)	x1x	Mean=7.9 s.d.=0.56	Mean=8.8 s.d.=0.61	—
P4 (N=-4)	x1x	Mean=7.7 s.d.=0.36	Mean=9.7 s.d.=0.34	—
M1 (N=6)	x2x	Mean=8.8 s.d.=0.31	Mean=9.1 s.d.=0.32	—
M2 (N=5)	x2x	Mean=11.2 s.d.=0.44	Mean=11.1 s.d.=0.47	—
M3 (N=3)	x2x	Mean=12.2 s.d.=1.16	—	—
p3 (N=4)	—	Mean=8.0 s.d.=0.52	Mean=5.9 s.d.=0.53	—
p4 (N=5)	x1x	Mean=8.5 s.d.=0.38	Mean=5.4 s.d.=0.40	—
m1 (N=6)	2x	Mean=9.3 s.d.=0.40	Mean=6.5 s.d.=0.30	—
m2 (N=7)	x2x	Mean=10.9 s.d.=0.63	Mean=8.2 s.d.=0.44	—
m3 (N=4)	x2X	Mean=13.3 s.d.=0.98	Mean=8.4 s.d.=0.67	—
	Daouitherium rebouli[b] (early Eocene)			
?P4 CPSGM MA6	2	10.0	11.5	8.0
p2 CPSGM MA4	—	17.0	9.0	17.0
p3 CPSGM MA4	x1x	17.5	11.0	17.0
p3 MNHN PM3	x1x	18.0	12.0	16.0

(Continued)

TABLE 2.3 (*Continued*)
Cheek Tooth Dimensions of Early Paleogene Proboscideans and Selected Plesielephantiforms

Taxon/Specimen/Tooth	Loph(id) Formula	L	W	H
p4 CPSGM MA4	x2x	18.0	13.0	16.0
m1 CPSGM MA5	x2x	16.5	12.0	—
m1 CPSGM MA4	x2x	19.0	14.0	12.5
m1 MNHN PM3	x2x	21.0	15.0	12.5
m2 CPSGM MA4	x2x	28.0	18.0	15.0
m2 MNHN PM3	x2x	28.0	18.0	16.0
m3 CPSGM MA4	x2X	34.0	21.0	15.5
m3 MNHN PM3	x2X	35.0	21.0	16.0
	***Numidotherium koholense*[e] (early Eocene)**			
P2 (N=8)	—	Mean=21.5 s.d.=1.33	Mean=18.9 s.d.=0.70	—
P3 (N=11)	x1x	Mean=19.8 s.d.=0.94	Mean=23.8 s.d.=1.4	—
P4 (N=13)	x1x	Mean=19.2 s.d.=0.75	Mean=25.5 s.d.=1.19	—
M1 (N=13)	x2x	Mean=23.8 s.d.=1.40	Mean=24.7 s.d.=1.25	—
M2 (N=16)	x2x	Mean=32.3 s.d.=1.41	Mean=32.3 s.d.=1.57	—
M3 (N=14)	x2x	Mean=37.1 s.d.=1.49	Mean=35.8 s.d.=0.81	—
r. p2 (N=11)	—	Mean=21.5 s.d.=1.39	Mean=12.5 s.d.=0.91	—
l. p2 (N=11)	—	Mean=21.3 s.d.=1.62	Mean=12.1 s.d.=0.76	—
r. p3 (N=18)	—	Mean=20.0 s.d.=0.81	Mean=13.9 s.d.=0.98	—
l. p3 (N=11)	—	Mean=19.9 s.d.=1.19	Mean=13.5 s.d.=0.74	—
r. p4 (N=25)	x2	Mean=20.3 s.d.=0.94	Mean=15.6 s.d.=1.05	—
l. p4 (N=22)	x2	Mean=20.1 s.d.=1.15	Mean=15.5 s.d.=1.07	—
r. m1 (N=15)	x2x	Mean=24.2 s.d.=1.33	Mean=17.9 s.d.=1.02	—
l. m1 (N=10)	x2x	Mean=23.8 s.d.=1.23	Mean=17.9 s.d.=1.06	—
r. m2 (N=24)	x2x	Mean=31.7 s.d.=1.83	Mean=23.1 s.d.=1.29	—
l. m2 (N=19)	x2x	Mean=32.8 s.d.=2.23	Mean=23.4 s.d.=1.63	—
r. m3 (N=25)	x2X or x3	Mean=41.7 s.d.=2.15	Mean=26.3 s.d.=1.46	—
l. m3 (N=21)	x2X or x3	Mean=42.0 s.d.=2.68	Mean=26.2 s.d.=1.71	—
	?*Numidotherium* sp. (?early Eocene)			
GTS.1 (Dakhla, Morocco)[f]				
l. M1 or M2	2x	41.0	36.0	—
	***Saloumia gorodiskii* (middle Eocene)**			
MNHN.F.MCA 1 (M'Bodione Dadere, Senegal)[k]				
r. upper molar	x2	>10.7	12.3	—
	***Dagbatitherium tassyi* (middle Eocene)**			
		ULDG-DAG1		
(Dagbati, Togo)[l]	3x	37.5	29.2	Very brachyodont
r. m1 or m2				

(*Continued*)

TABLE 2.3 (*Continued*)
Cheek Tooth Dimensions of Early Paleogene Proboscideans and Selected Plesielephantiforms

Taxon/Specimen/Tooth	Loph(id) Formula	L	W	H
	***Barytherium grave* (late Eocene-early Oligocene)**			
C 100012 (Qasr el Sagha Fm., Fayum, Egypt)[a]				
l. P2		57.0	57.0	—
l. P3		39.0	65.0	—
l. P4		50.0	80.0	—
l. M1	2x	59.0	84.0	—
l. M2	2x	75.0	86.0	—
l. M3	2x	83.0	87.0	—
r. p2	—	51.0	e.35.0	—
r. p3	—	43.0	47.0	—
r. p4	—	57.0	55.0	—
r. m1	2x	57.0	—	—
r. m2	2x	86.0	64.0	—
r. m3	x2X	105.0	70.0	—
UM 2233 (Qasr el Sagha Fm., Fayum, Egypt)				
l. ?m2	2x	67.9	55.6 (1, 2)	Heavily worn
UM 2246 (Qasr el Sagha Fm., Fayum, Egypt)				
r. m2	x2x	78.7	59.1 (1)	—
UM 2237 (Qasr el Sagha Fm., Fayum, Egypt)				
l. M1	x2x	56.2	51.4 (1)	35.3 (1)
UM 2465				
l. M3	x2x	78.7	64.0 (1)	—
DPC 2917 (Qasr el Sagha Fm., Fayum, Egypt)				
l. M2	2x	70.0	71.0 (1)	38.5 (1)
DPC 20844 (Birket Qarun Fm., Fayum, Egypt)				
l. M3	x2x	76.2	66.0 (1)	36.8 (1)
DPC 20848 (Birket Qarun Fm., Fayum, Egypt)				
l. P2	x1 1/2x	44.0	45.2 (1)	22.8 (1)
DPC 20857 (Birket Qarun Fm., Fayum, Egypt)				
Two isolated lower molar lophids	—	—	52.5	—
	—	—	44.5	—
6819/16 (Idam Unit, Dur at Talha, Libya)				
r. M3	x2x	75.0	69.0	40.0
689/1 (Idam Unit, Dur at Talha, Libya)				
l. P2	x1x	41.1	41.4	—
l. P3	x1	33.1	43.6	—
l. P4	x1	32.9	50.0	—
l. M1	x2x	50.0	44.9	—
l. M2	x2x	68.6	59.3	—
l. M3	x2x	66.0+	63.1	—
6958/1 (M 92403) (Idam Unit, Dur at Talha, Libya)				
l. p2	—	35.1	25.1	25.9
r. p2	—	32.5	24.5	—
l. p3	1 1/2	31.4	32.2	33.0
l. p4	1 1/2	33.0	38.2	33.1
r. p4	1 1/2	36.4	37.5	35.6
l. m1	2x	48.5	37.8	27.2

(Continued)

TABLE 2.3 (*Continued*)
Cheek Tooth Dimensions of Early Paleogene Proboscideans and Selected Plesielephantiforms

Taxon/Specimen/Tooth	Loph(id) Formula	L	W	H
r. m1	2x	47.5	39.1	—
l. m2	2x	62.4	51.8	33.7
r. m2	2x	65.7	50.3	—
l. m3	2X	78.6	50.7	33.5
r. m3	2X	77.7	46.0+	—
689/1 (Idam Unit, Dur at Talha, Libya)				
l. M3	—	—	65.5	—
r. M3	—	—	63.3	—
l. m3	2X	82.8	+50.8	—
r. m3	2X	76.5	56.5	—
686/1 (Idam Unit, Dur at Talha, Libya)				
l. m3	2X	93.3	+53.6	—
6811/11 (Idam Unit, Dur at Talha, Libya)				
r. M3	x2x	79.3	73.5	—
6956/1				
r. M3	x2x	73.4	62.2	—
6819/40 (Idam Unit, Dur at Talha, Libya)				
l. M3	x2x	81.0	75.1	—
r. M3	x2x	82.3	74.1	—
6817 (Idam Unit, Dur at Talha, Libya)				
?l. M3	x2x	73.0	71.5	—
?r. M3	x2x	73.5	71.5	—
?r. m3	2X	90.2	59.6	—
	Arcanotherium savagei **(late Eocene-early Oligocene)**			
6821/10 (type) (junction, Evaporite and Idam Units, Dur at Talah, Libya)				
r. p2	1X	24.0	18.3	21.8+
r. p3	1X	24.3	21.6	—
r. p4	1X	24.1	24.6	—
r. m1	x2x	33.6	27.5	—
r. m2	x2x	45.0	30.0	—
r. m3	x2X	56.8	35.8	—
6955/1				
r. m3	x2X	62.5	30.5	24.5
686/2				
l. m3	x2X	59.9	38.5	—
	Omanitherium dhofarense **(early Oligocene)**			
SQU-290 (type) (Shizar Mb., Ashawq Fm., Oman)[h]				
r. p3	1X	22.4	20.9	—
r. p4	1X	25.6	24.9	—
l. p4	1X	26.4	24.8	—
r. m1	x2x	39.9	31.9	—
l. m1	x2x	39.5	32.2	—
r. m2	x2x	50.8	39.7	—
l. m2	x2x	48.6	39.3	—
l. m3	x2x	61.7	43.6	—
	"*Omanitherium* 17," Shizar Mb., Ashawq Fm., Oman[i]			
l. P3 ONHM TN 2017-16	—	22.5	24.0	—
r. P4 ONHM TN 2017-17	x1x	22.3	29.0	—
r. P4 ONHM TN 2017-44	x1x	24.4	30.7	—

(*Continued*)

TABLE 2.3 (*Continued*)
Cheek Tooth Dimensions of Early Paleogene Proboscideans and Selected Plesielephantiforms

Taxon/Specimen/Tooth	Loph(id) Formula	L	W	H
r. M1 ONHM TN 2017-18	x2x	31.2	—	—
l. M3 ONHM TN 2017-79	x2x	43.7	38.3	—
l. m2 ONHM TN 2017-50	—	—	37.3	—
	Thaytiniti, Oman, Shizar Mb., Ashawq Fm.[i]			
r. P2 ONHM 1478-TH4	—	22.0	23.0	—
r. P3 ONHM 1478-TH3	—	26.0	29.0	—
r. p4 SQU sans nº	—	—	23.0	—
r. m3 SQU-290	—	—	44.3	—
	Taqah, Oman, Shizar Mb., Ashawq Fm.[i]			
l. p4 ONHM TQ 15		25.0	25.6	—
	Chilgatherium harrisi, Chilga, Ethiopia (late Oligocene)[c]			
l. P3 CH37-V-17	x2x	34.0	27.5	20.2
r. P3 CH9-22	x2x	37.0	28.1	22.8
l. ?P4 CH9-7	x2x	38.2	29.9	23.8
r. M3 CH35-1	x3x	58.7	38.3	24.0
Partial upper molar CH12-4	x1+	33.5+	41.0	28.8
Partial upper molar CH12-3	x1+	38.0+	e. 34.5	31.4
l. p4 CH4-2a	+1x	+23.5	24.4	21.4
l. m1 CH4-2b	x2+	31.4+	26.5	—
r. m1 CH35-3d	+1x	+19.2	25.0	18.3
r. m2 CH35-3a	x3	50.5	33.7	25.0
l. m2 CH35-3b	+2	+22.5	32.0	—
l. m3 CH35-3c	x3	60.6	40.9	28.4
r. m3 CH35-3e	x2+	+42.4+	39.3	31.0
	Chilgatherium sp., Shumaysi Fm., Saudi Arabia (early Oligocene)			
r. m3 SGS-SHUM 2020-01	x3	60.0	41.0	33.0
	?*Chilgatherium* sp. (M2) or ?Barytherioidea cf. *Barytherium* (M1), Shumaysi Fm. Saudi Arabia (early Oligocene)			
l. M1 or M2 SGS-SHU 2018-063	x2x	48.1	39.0	21.5
	?Barytherioidea gen. et sp. nov., Nakwai, Kenya (early Miocene)			
r. m1 or m2	2x	e. 50.0	39.7 (2)	—
l. m1 or m2	2x	e. 50.4	+38.4 (2)	—
r. m2 or m3	x2X	54.2	+42.0 (1)	—
l. m2 or m3	x2X	55.0	42.5 (1)	—

All dimensions in mm. Abbreviations: e., estimated; H, height; l., left; L, length; M/m, upper or lower molar; N, number of specimens; P/p, upper or lower premolar; r., right; s.d., standard deviation; W, width; x, indicates an anterior or posterior cingulum(id); X, indicates a very large anterior or posterior cingulum(id); +, indicates missing morphology; (1, 2, . . .), number in parentheses indicates loph(id) at which the dimension was measured.

[a] Andrews (1906);
[b] Gheerbrant et al. (2002);
[c] Sanders et al. (2004);
[d] Gheerbrant et al. (2005);
[e] Noubhani et al. (2008);
[f] Adnet et al. (2010);
[g] Gheerbrant et al. (2012);
[h] Seiffert et al. (1912);
[i] Pickford (2015);
[j] Al-Kindi et al. (2017);
[k] Tabuce et al. (2019);
[l] Hautier et al. (2021).

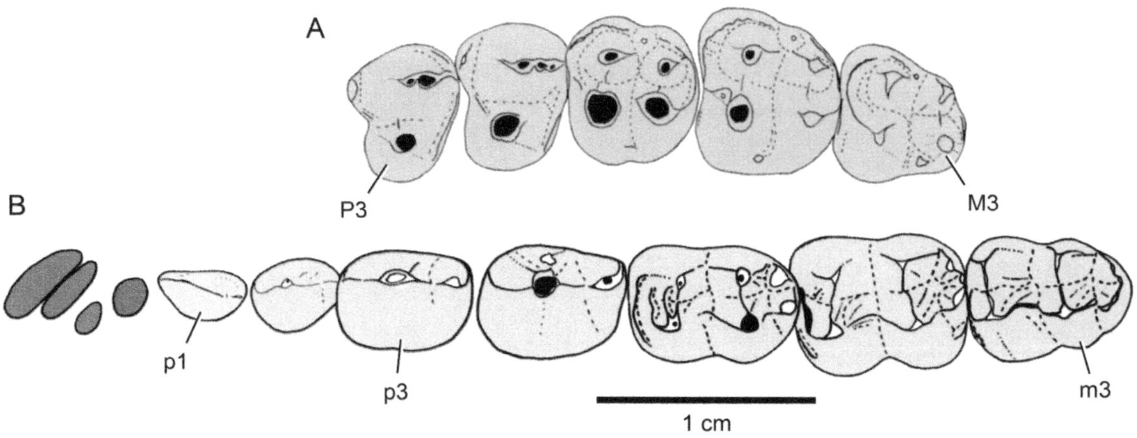

FIGURE 2.4 Dentition of *Eritherium azzouzorum*, probably Selandian, Ouled Abdoun Basin, Morocco. Abbreviations: M/m, upper/lower molar; P/p, upper, lower premolar. (A, B) are to the same scale. Anterior is to the left. (A) P3-M3, MHNT PAL 2006.0.18, occlusal view. (B) i1-m3, reconstruction, occlusal view. (Images adapted from Gheerbrant et al. [2012:fig. 18A and fig. 15D], courtesy of E. Gheerbrant and with permission of Schweizerbart Science Publishers [www.schweizerbart.de/journals/pala].)

early Eocene species *Phosphatherium escuillei* from the same basin in Morocco (Gheerbrant, 2009; Gheerbrant et al., 2012). Similarity of the molars of *Eritherium* to those of late Paleocene-early Eocene louisinids and of the earliest known macroscelideans *Chambius* and *Herodotius* (Gheerbrant et al., 2012) are especially critical for assessment of competing hypotheses about the identity of the common ancestor of crown afrotheres and their biogeographic origins (Robinson and Seiffert, 2004). The occurrence of *Eritherium* provides a fortunate opportunity to "observe" the emergence of a modern placental mammalian Order in the fossil record and contributes critical evidence for the investigation of the timing of emergence of endemic placentals in Africa and better understanding the stem-group relationships of afrotheres (Gheerbrant et al., 2012, 2014; O'Leary et al., 2013). The plesiomorphic status of its craniodental anatomy makes it difficult to envision that an older, more primitive proboscidean could be identified in the fossil record.

PHOSPHATHERIIDAE Gheerbrant, Sudre, Tassy, Amaghzaz, Bouya, and Iarochène, 2005
PHOSPHATHERIUM Gheerbrant, Sudre, and Cappetta, 1996
PHOSPHATHERIUM ESCUILLIEI Gheerbrant, Sudre, and Cappetta, 1996

(Tables 2.1–2.3; Figures 2.1, 2.2, 2.5, and 2.6)

Originally described from two fragmentary gnathodental specimens associated with selachian teeth as late Paleocene (Thanetian) in age (Gheerbrant et al., 1996, 1998a, b; Gheerbrant, 1998), *Phosphatherium escuilliei* is now more accurately recognized as earliest Eocene (Ypresian) in age and firmly provenienced to intercalary phosphorite Beds II/I, stratigraphically located above phosphorite Bed IIA of *Eritherium* and probably subjacent to *Daouitherium* from phosphorite level 0 of the Ouled Abdoun Basin, Morocco (Table 2.1, Figure 2.2); (Gheerbrant et al., 2002, 2003; Kocis et al., 2014; Yans et al., 2014). Gheerbrant et al. (1998b) have collated an impressive list of mammals recovered from the Ouled Abdoun Basin, including palaeoryctids, todralestids, adapisoriculids, possible creodonts, condylarths, and, notably, plesiadapiform primates along with the enigmatic primate *Altiatlasius koulchii*, which is possibly a very early euprimate or even an anthropoid (Tabuce et al., 2009; Williams et al., 2010). This provides the oldest evidence for penecontemporaneous co-occurrence of primates and proboscideans in the same basin. The occurrence of *Phosphatherium* is very close to the Paleocene-Eocene Thermal Maximum (PETM) at 56 Ma, which involved massive release of carbon into the ocean-atmosphere system and increase of global temperature between 5°C and 8°C (McInerney and Wing, 2011). In this geological setting, the morphology of *Phosphatherium* more securely reinforces the antiquity of African proboscideans in the early Paleogene, and, together with its relations in the Ouled Abdoun Basin and younger North African occurrence of *Numidotherium koholense* (Table 2.1), shows the beginnings of phylogenetic trends in Proboscidea for body size increase, progressive loss of anterior teeth, and acquisition of true molar lophodonty.

Named for its geological setting and the individual who obtained the type specimen from a local fossil dealer (Gheerbrant et al., 1996), *Phosphatherium escuilliei* is estimated to have weighed 10–15 kg based on tooth size allometry—length and widths of M1–2 of the type maxilla are 8.65 × 8.80 mm and 11.15 × 10.70 mm, respectively—and from cranial length (≈170 mm) to have weighed up to 17 kg with a conjectured shoulder height of 30 cm, as there are virtually no postcranials known for the species (Gheerbrant et al., 1996; Gheerbrant, 1998; Gheerbrant and Tassy, 2009; Larramendi, 2016). This makes it a much larger animal than the diminutive *Eritherium*, though compared with later proboscideans *Phosphatherium* was still quite small (see Table 2.3). The dental formula of the upper teeth is the same as can be inferred for *Eritherium*, 3-1-4-3, but there is a loss of several lower teeth, relative to the condition in *Eritherium*, and the lower dental formula is alternatively

2-1-3-3 or 2-0-4-3 (with either i2–3 or more likely i1–2 preserved; Table 2.2). Given that the most prominent lower incisor of other early proboscideans is i1, the loss of i1 is unlikely, and probably the third incisor and either the canine or first premolar are lost in the lower dentition (Table 2.2, Figures 2.5A and 2.6B). Once erupted, all adult teeth functioned together. Except for a single middle phalanx, the postcranium of *Phosphatherium* is unknown. The phalanx is robust, short, wide, and dorsoventrally flattened. Morphology of its distal extremity suggests that there was little capacity for much extension or flexion of the distal phalanx. Dimensions of the middle phalanx are: L = 15.2 mm; W = 11.1 mm; and height (at the base) = 7.1 mm.

The comparative morphology of *Phosphatherium* is discussed in detail in Gheerbrant et al. (2005), showing that its molars have a closer approximation of true (bi)lophodonty than in *Eritherium* (Figure 2.6A and B) anticipating the condition in *Daouitherium* and *Numidotherium* (Figure 2.6C and D). Molar enamel microstructure ("Schmelzmuster") is two-layered, composed of radial enamel in the outer portion and Hunter-Schreger bands in the inner portion of the enamel, comprising about 85% of enamel thickness, but lacks the 3D enamel prisms that are an apomorphy of Proboscidea (Tabuce et al., 2007b). Size progression of upper molars is M1 < M2 < M3, with M2 much larger than M1 and less of a jump in magnitude between M2 and 3 (Figure 2.5D). In the

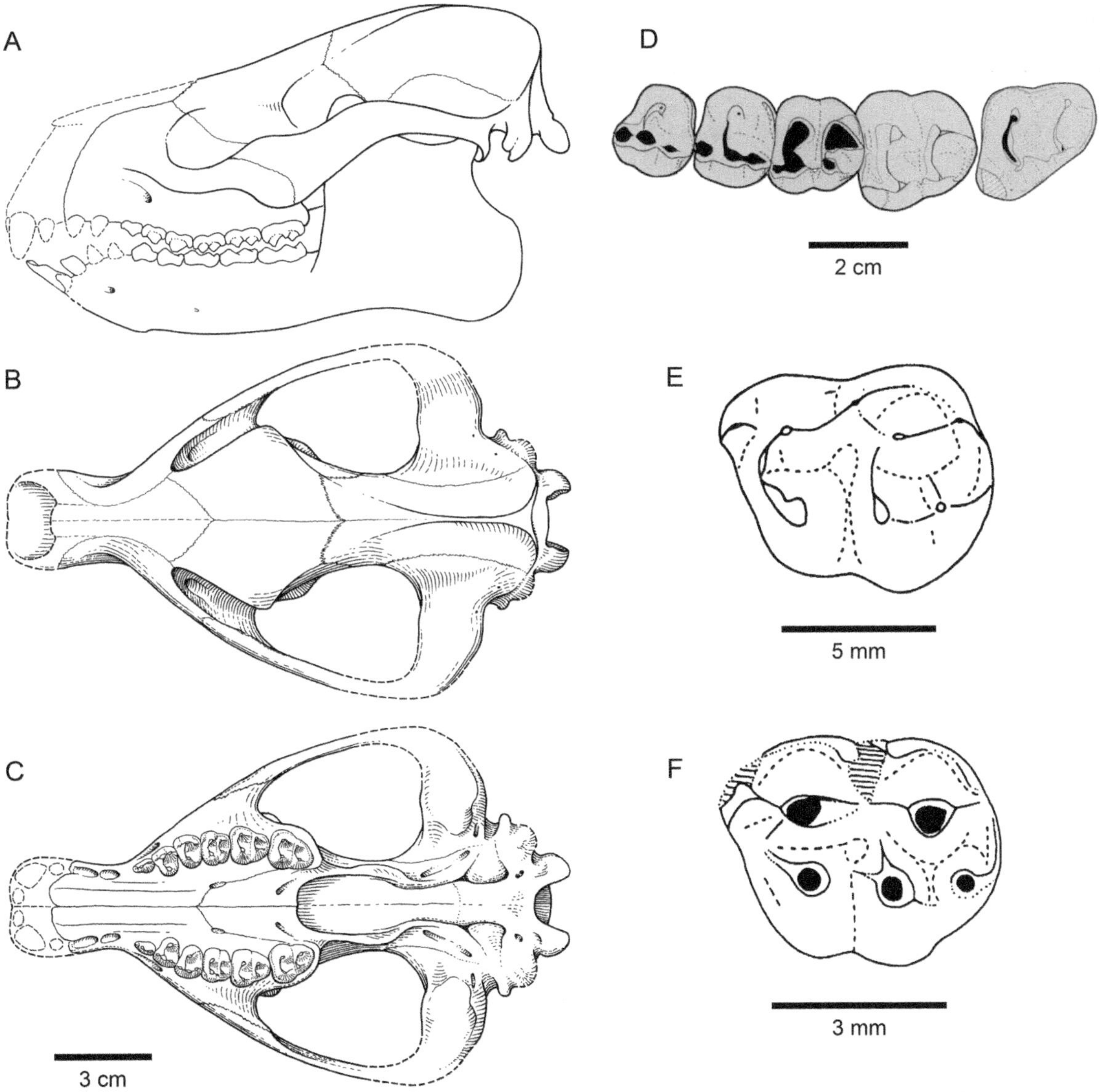

FIGURE 2.5 Skull and dentition of *Phosphatherium escuilliei*, early Ypresian, and *Khamsaconus bulbosus*, middle Ypresian, Ouled Abdoun Basin, Morocco. (A–C) are to the same scale. Anterior is to the left. (A) Reconstructed skull, left lateral view, *P. escuilliei*. (B) Reconstructed cranium, dorsal view, *P. escuilliei*, based on specimen OCP DEK/GE 305. (C) Reconstructed cranium, ventral view, *P. escuilliei*. (D) Left P3–M3 (rotated), composite specimen based on MNHN.F PM2 and MNHN.F PM17, occlusal view, *P. escuilliei*. (E) Right dP4, MNHN.F PM1, occlusal view, *P. escuilliei*. (F) Right dP4 (reversed), NTG 2-50, occlusal view, *Khamsaconus bulbosus*, holotype. ([A–C] are from Gheerbrant et al. [2005:fig. 6A, B and fig. 32A], [D] is from Gheerbrant et al. [2012, fig. 14D], and [E, F] are from Gheerbrant et al. [1998:fig. b]. Images courtesy of E. Gheerbrant and with permission of Schweizerbart Science Publishers [www.schweizerbart.de/journals/pala].)

upper molars, a continuous crest connects the cusps of the proto- and metalophs; the transverse crests are lower than the cusps that anchor them, which are crowded toward the center of the crown. This constitutes an intermediate condition between the nascent lophodonty of *Eritherium* and the true lophodonty of *Daouitherium* and *Numidotherium*. The proto- and metalophs are equivalent in importance, though the molar crowns are broadest anteriorly. A strong parastyle contacts the mesial cingulum on the buccal side and a distinct posthypocrista runs posteriorly to the distal cingulum (or distocrista), which is more developed in M2 than in M1. DP4 is similar to M1 but more trapezoidal in occlusal outline (Figure 2.5D and E), and additionally differs from M1 by having a relatively anteroposteriorly broader transverse valley between lophs, more curved protoloph, lower crown height, and is smaller overall. M3 is more asymmetric than M1–2 in occlusal view, with a transversely much broader proto- than metaloph. The protoloph has a strong mesially convex curvature, and the protocone is considerably more voluminous than the hypocone. The parastyle, postparacrista, and postmetacrista of M3 are all well developed.

C and P1 are documented by alveolar sockets that are separated from P2 by a short diastema. P2 is simple in structure, long and narrow in occlusal view, with a single cusp toward the front of the crown and a long crest descending from this cusp and running distally. There are tiny cuspules at the very mesial and distal ends of the P2 crown. P4 is generally larger than P3, and both premolars are wider than long in occlusal outline and more squared in shape than P2, especially P4. These premolars are semi-molariform, with a protoloph but no posterior loph. The parastyle forms a large cuspule that is greater in size than the metacone. With wear, the paracone and metacone connect and essentially comprise an ectoloph. Little can be said about the upper anterior dentition because these teeth are not preserved for *Phosphatherium*. Based on phylogenetic evidence—the preservation of I1–3 and C in *Numidotherium* and *Moeritherium*—the anterior dental arcade for *Phosphatherium* is reconstructed as I1–3 and C, with I2 presumably the dominant upper incisor (Table 2.2; Gheerbrant et al., 2005).

Lower molars are well represented in the type series and are all distinctly bilophodont, with crestiform cusp pairs (protoconid-metaconid; hypoconid-entoconid) connected by transverse ridges. The transverse ridges are posteriorly convex and have strong cristids descending mesially from the hypoconid and protoconid, respectively. The latter cristid connects with the anterior cingulid. The distal cingulid of m3 forms a prominent heel (but not third lophid), with a large, centrally located hypoconulid. The m1 and m2 distal cingulids are more modest and may sport smaller-sized, more buccally shifted hypoconulids, as well as small, lingually situated postentoconulids. Size progression of the lower molars is m1<m2<m3 (Figure 2.6B). The lower molars are longer than wide and are generally rectangular in shape in occlusal view. The protolophid is slightly wider than the hypolophid in each molar, and in lateral view, the protolophid is taller than the hypolophid. The lophids each have a chisel-like appearance. This configuration, with planar wear facets on the distal side of lophids and the mesial side of lophs, shows the beginning of what has been termed the "deinothere pattern of mastication (POM)," in which the lower jaw moves primarily in a steeply-inclined, mesial direction and both shear-cutting and compression of the food bolus is achieved in Phase I occlusion. *Daouitherium*, *Numidotherium*, and *Barytherium*, as well as deinotheres, also exhibit this POM mechanism (von Koenigswald, 2014, 2016a).

Several specimens permit accurate reconstruction of the cranium, particularly the nearly complete OCP DEK/GE 305 (Figure 2.5A–C; Gheerbrant et al., 2005). Notable features of the cranium of *Phosphatherium* include orbits with a strong maxillary contribution, posteriorly open and positioned with their anterior margins above P4; large infraorbital foramina and associated sulci, low and located above the dorsal margin of P2 on each side; dorsoventrally stout zygomatic arches broadly flared laterally, ascending toward their squamosal end, where they are robust and nearly at right angles to the long axis of the cranium, and with a large ventral apophysis at their malar end; pronounced postorbital constriction and salient posterior sagittal crest which connects with laterally strong nuchal crests; basicranium raised moderately above the level of the palate (to a lesser degree than in *Numidotherium* and more similar to the condition in *Moeritherium*); conformation of the maxillary permits envisioning a large, anteriorly situated nasal opening, large nasal fossae (though not as capacious as those in *Moeritherium* and *Numidotherium*) and a long contact between the nasal and maxillary bones; the external auditory meatus is open ventrally and bordered anteriorly by a postglenoid process and posteriorly by a posttympanic process; the pterygoid processes are well developed, elongate, and slightly divergent; the glenoid fossa is massively built and flattened; and there is a postglenoid foramen, as in *Moeritherium* and elephantiform proboscideans. While there is no evidence for a trunk in the cranial morphology of *Phosphatherium*, the large size of the infraorbital foramina and associated infraorbital fossae indicate powerful musculature to operate mobile, prehensile lips, foreshadowing later development of trunks in proboscideans.

The mandible has a gracile corpus with a moderately high ramus (Figure 2.5A). The ramus has a straight anterior margin set posterior to the m3, and the coronoid process is higher than the condyle. Primitively, it exhibits a small mandibular foramen at the level of the cheek tooth alveoli, a mandibular canal below (and not offset from) the cheek tooth alveoli, and doubled mental foramina (Ferretti and Debruyne, 2011). Posteriorly, the masseteric fossa is massive, and the angle of the lower jaw is squared. The symphysis is comparatively small, obliquely angled, and extends to the posterior end of the p3. Compared with *Daouitherium* and *Numidotherium*, the lower jaw of *Phosphatherium* is more primitive because of its lower ramus, more posterior position of the anterior margin of its ramus, and less posterior extension of its symphysis.

Although the skull of *Phosphatherium*, in overall aspect, is primitive and "condylarth"-like, nonetheless it preserves a number of proboscidean synapomorphies that confirm its inclusion in the Order: contribution of the maxillary bone to part of the zygomatic arch at the lower border of the orbit; strong development of the pars mastoidea of the periotic; buccal shift of molar hypoconulids; presence of a fronto-squamosal suture; configuration of the glenoid region of the cranium; and prominence of i1 (Gheerbrant and Tassy, 2009). Before the discovery of early Paleogene proboscideans, the dental anatomy of *Moeritherium* from the Fayum suggested that the ancestral condition for proboscideans included bunolophodont molars, but with the discovery of *Numidotherium*, followed by *Phosphatherium* and *Daouitherium*, it was then thought that proboscideans arose from an ancestor with lophodont molars (e.g., Gheerbrant and Tassy, 2009). The subsequent recovery of *Eritherium*, with only incipient molar lophodonty and bunodont cusps, has moved the pendulum back to consideration of a bunolophodont or bunodont ancestor. In this scenario, it is possible that bunolophodonty is secondarily evolved in *Moeritherium*. Skull and dental morphology in *Phosphatherium* leave no doubt about its proboscidean affinities, and cladistic analysis demonstrates its phylogenetic position firmly between *Eritherium* and *Numidotherium* (Gheerbrant, 2009).

?PHOSPHATHERIIDAE Gheerbrant, Sudre, Tassy, Amaghzaz, Bouya, and Iarochène, 2005
***KHAMSACONUS* Jaeger, Sige, and Vianey-Liaud in Sudre et al., 1993**
***KHAMSACONUS BULBOSUS* Jaeger, Sige, and Vianey-Liaud in Sudre et al., 1993**
(Table 2.1; Figures 2.1 and 2.5)

Because it is known from only two cheek teeth and does not fit the evolutionary narrative evoked by the succession of species in the Ouled Abdoun Basin and at El Kohol, *Khamsaconus bulbosus* is phylogenetically and taxonomically the least-understood early proboscidean. The smallest of all proboscideans (two times smaller than *Eritherium* [Gheerbrant et al., 2012]), and named after the presence of five inflated cusps in the type specimen, *Khamsaconus* has only been recovered from the site of N'Tagourt 2 in the Ouarzazate Basin of Morocco (Sudre et al., 1993; Gheerbrant et al., 1998b). Faunas from the Ouarzazate Basin are derived from a sedimentary series that ranges from about 57 Ma to 44-43 Ma in age, comprised of over 130 species of bony fish, sharks, amphibians, reptiles, and mammals, notably featuring the most ancient creodont and the possible earliest anthropoid, *Altiatlasius* (Gheerbrant et al., 1993, 2017). Although the Paleocene-Eocene Thermal Maximum, which is associated with the largest faunal turnover of the early Paleogene, occurred at about 56 Ma (McInerney and Wing, 2011) coincident with interval of deposition of beds in the Ouarzazate Basin, there is no sign of significant faunal changes in the basin compared with the effects on Laurasian faunas (Gheerbrant et al., 2017). This reflects the ecological buffering effect of being located close to equatorial latitudes.

The associated mammals from N'Tagourt 2 include several species of proteutherian didelphodonines, a condylarth, several species of ?liptophylan adapisoriculids, and a number of distinct but indeterminate eutherians. They derive from near-shore marine deposits and were initially dated by selachian, gastropod, and lamellibranch bivalves to the Ypresian (early Eocene), which was later more precisely calibrated by paleomagnetic reversal stratigraphy to the mid-Ypresian at 52.9 Ma (Cappetta et al., 1987; Gheerbrant et al., 1992, 1998b), contemporaneous with *Daouitherium* of the Ouled Abdoun Basin (Table 2.1). Specimens from the basin are all isolated elements, and most of the taxa they represent show a strong Laurasiatic influence, except for the evident endemicity of the creodonts and *Khamsaconus* (Gheerbrant et al., 1998b, 2017).

The type specimen is an isolated right molariform dP4, 4.66 mm long and 3.78 mm wide, in reasonably good condition. In occlusal view, the crown forms a rounded trapezoid, with five bunodont, bulbous cusps, as the species name indicates (Figure 2.5F). Anteriorly, it has a low cingular shelf that projects mesially more on the buccal side. There is a low, ribbon-like buccal cingulum that continues around to the distal end of the crown, though the distal cingulum is unimportant. There are two cusp pairs that do not form transverse lophs: protocone and paracone anteriorly, and hypocone and metacone posteriorly. The transverse valley, or interloph, is distinct and connects with a deeply notched entoflexus and ectoflexus. The protocone and hypocone are subequal in size. The paracone has pre- and postparacristae, and the metacone also has pre- and postmetacristae. Notably, there is a prominent fifth cusp posterior to the hypocone that is identified by Gheerbrant et al. (1998b) as a "postentoconule." The postentoconule is linked to the metacone via a well-developed distocrista connecting with the postmetacrista. The hypocone has cristae descending mesially and distally. The occlusal outline and composition of the dP4 of *Khamsaconus* resembles the dP4 of *Phosphatherium* (Figure 2.5E and F), though the latter deciduous premolar is distinguished by its larger size (L=8.85 mm, W=7.10 mm), formation of a protoloph, and more complex arrangement of cristae (Gheerbrant et al., 1998b). Despite the overall similarity to the dP4 of *Phosphatherium*, the layout of the main cusps, absence of a paraconule, and presence of a fifth, posterolingual cusp also recalls the morphology of this deciduous premolar in the hyopsodontid condylarthran *Monshyus*, which resulted in a tenuous initial assignment of *Khamsaconus* to the family ?Hypsodontidae and ?Louisininae (Sudre et al., 1993).

To the contrary, assignment of a fragment of a lower left molar to *Khamsaconus* led Gheerbrant and colleagues (1998b) to have greater confidence in the identification of the taxon as a proboscidean. This specimen was found at the same locality (N'Tagourt 2) as the type deciduous premolar. As with the type, it is very small in size, only 4.20 mm wide. The anterior (trigonid) portion is preserved and is composed of a low, rounded protoconid separated by a V-shaped median sulcus from a smaller metaconid. There is a low, diminutive posterior central accessory conule connected to the protoconid by a curved postprotocristid. Mesial to the main cusps is a small, centrally located paracristid that is connected to the protocone by a narrow ridge. The paracristid is preceded anteriorly by a very low cingulid that continues to the buccal side of the crown.

Khamsaconus exhibits cheek tooth morphology inconsistent with the trend over time from bunolophodont, incipient lophodonty in *Eritherium* to increasingly greater lophodonty in the proboscidean series in the Ouled Abdoun Basin and *Numidotherium*. The bunodonty or bunolophodonty of its cheek teeth, absence of true lophodonty, and presence of an accessory conule in the lower molar are more similar to elephantiform occlusal morphology, though it is questionable that it had any direct phylogenetic connection to elephantiform proboscideans. Gheerbrant and colleagues (2012) felt that the dental morphology of *Khamsaconus* suggests a closer relationship to *Eritherium*, which is a more reasonable view. Nonetheless, conspicuous differences in cheek tooth morphology among proboscideans in the same temporal interval suggest that greater diversification may have occurred in the early evolution of the group than we are able to perceive in the limited fossil record of the time.

Clearly, aspects of *Khamsaconus* mark it as a very primitive proboscidean: diminutive size; presence of a postentoconule in dP4; extreme brachyodonty of cheek tooth crowns; presence of a posterior accessory conule in the lower molar; and microstructure of the enamel. It has been suggested that features of enamel microstructure may have phylogenetic implications, and the presence of (only) radial enamel in *Khamsaconus* (shared with the outgroups Sirenia and Hyracoidea) indicates that it is more plesiomorphic in this regard than *Phosphatherium* and *Daouitherium*, which exhibit enamel structure composed of Hunter-Schreger bands, and the even more advanced *Numidotherium*, which displays 3D enamel structure (Tabuce et al., 2007b). *Khamsaconus* may be as little derived as *Eritherium*, though there are not yet enough anatomical elements in common for a comprehensive comparison (but see Gheerbrant et al., 2012), and it remains to be established whether *Khamsaconus* had any role in the subsequent evolution of African Proboscidea.

Phylogenetically, *Khamsaconus bulbosus* has been situated as the sister species to *Phosphatherium escuilliei*, an arrangement that is poorly sustained by small size and presence in upper cheek teeth of a parastyle and centrocrista, which led to the inclusion of this enigmatic proboscidean in the Family Phosphatheriidae (Gheerbrant et al., 2005; Gheerbrant and Tassy, 2009). However, because the analysis that established this clade did not include *Eritherium* (*Khamsaconus* is more similar in size dentally to *Eritherium* than to *Phosphatherium*, for example) and the characters that support the clade are vulnerable to homoplasy, these phylogenetic hypotheses are unpersuasive.

Family Incertae Sedis Probably Nov. Gheerbrant, Sudre, Tassy, Amaghzaz, Bouya, and Iarochène, 2005
DAOUITHERIUM Gheerbrant and Sudre in Gheerbrant et al., 2002
DAOUITHERIUM REBOULI Gheerbrant and Sudre in Gheerbrant et al., 2002

(Tables 2.1–2.3; Figures 2.1, 2.2, and 2.6)

From the same area that yielded *Eritherium* and *Phosphatherium*, *Daouitherium rebouli* is the youngest of this trio (Table 2.1) and punctuates the critical importance of the Ouled Abdoun Basin for documenting the earliest phases of proboscidean evolution. The species is named for the quarrying site from which it derives and for the generous donor of several of its specimens for study. It was originally believed to be from the same stratigraphic level as *Phosphatherium*, and therefore to represent the first evidence for proboscidean evolutionary diversification (Gheerbrant et al., 2002). Since then, revised interpretation of selachian fossils associated with *Daouitherium* suggests a placement in phosphorite Bed 0, of early-middle Ypresian age, close to the significant early Eocene ETM2 and ETM3 warming events (Yans et al., 2014). This chronology makes sense in terms of directionality of morphological transformation from *Eritherium* to late Ypresian *Numidotherium*, with trends for increase in body size, loss of anterior teeth, more complete lophodonty of molars, and greater separation and distinction of function between incisors and cheek teeth. Thus, rather than an adaptive radiation, *Eritherium-Phosphatherium-Daouitherium* likely comprise a phylogenetic succession covering the early part of the Ypresian. Evidence for greater taxonomic diversity in this interval derives from Ypresian beds at Tamaguélelt, Mali (Table 2.1), where dentaries lacking teeth appear to represent either one (sexually dimorphic) or two novel proboscidean species (O'Leary et al., 2006). Although placed in Numidotheriidae by the authors, because of the extension of the symphysis as posterior as p4, the inferred small size of the teeth (close to dimensions for *Phosphatherium*) and possible posterior inclination of the ramus suggest that these specimens are more primitive than that.

Daouitherium is particularly notable as the earliest known large mammal from Africa, with molars only about 10–20% smaller than those of the substantial-sized *Numidotherium* (Table 2.3; Gheerbrant et al., 2002). It is estimated from tooth size regression to have weighed between 80 and 170 kg and from volumetric analysis to have weighed about 200 kg, about as large as a living tapir (Gheerbrant et al., 2002; Gheerbrant and Tassy, 2009; Larramendi, 2016). Coinciding so closely as it now appears to be with significant Eocene warming events (Yans, 2014), it is difficult to comprehend its body size increase in the context of climate change; body size decrease rather than increase is expected with higher temperatures, as a mechanism of increasing surface area relative to mass to assist cooling the body ("Allen's Rule"). Aside from greater body size, *Daouitherium* is distinguished from its predecessors by its more complete molar lophodonty, greater molarization and lophodonty of premolars, and considerably higher position of the condyles on the mandibular ramus (Gheerbrant and Tassy, 2009), an advanced configuration shared with *Numidotherium* and *Barytherium*, indicating a dorsoventrally deeper cranium with a more raised glenoid and basicranium than in *Phosphatherium* and presumably *Eritherium*. Complicating the tidy polarity of the rest of its morphology, compared with its predecessors, *Daouitherium* appears to preserve one more tooth anterior to p2 than is present in the dentary of *Phosphatherium*, probably either an incisor or premolar (Table 2.2). However, the authors of the type paper (Gheerbrant et al., 2002) qualify this observation by admitting that better material is required to confirm these putative differences.

Nearly the entire fossil sample of *Daouitherium* is composed of lower teeth and dentary remains, well described in detail by Gheerbrant et al. (2002). The molars are true bilophodont teeth with crestiform cusps, the lophids transversely uninterrupted by midline sulci or notches. Crown size increases markedly through the molar series, m1<m2<m3. Well-developed cingulids are present on m1 and m2, whereas m3 exhibits a substantial distal heel or incipient tritolophid. In m1, the hypolophid is wider than the protolophid. Molars are longer than they are wide, a primitive feature compared with later Paleogene barytheres and arcanotheres, and crown height is also primitively modest (dimensions of the m3 of the type specimen are: length=34.0 mm, width=21.0 mm, and height=15.5 mm). The cristid obliqua is well developed. In occlusal morphology, the molar series is strikingly similar to that of *Numidotherium koholense*, which led to initial classification of *Daouitherium* in Numidotheriidae (Gheerbrant et al., 2002; Shoshani and Tassy, 2005). All adult teeth are simultaneously in occlusion.

The lower dentition forward of p2 is absent, but their alveoli reveal clearly that i1 was the largest anterior tooth (apparently not as enlarged as in *Numidotherium*), and their placement also show that there was no significant diastema between the anterior dentition and cheek teeth, unlike the condition in *Numidotherium*, primitively conserving the pattern in *Eritherium* and *Phosphatherium*. The lower premolars are increasingly molarized from p2 to p4, with p4 a smaller bilophodont version of m1. Although p2 is simpler, lacks transverse crests, and is triangular in occlusal outline, it possesses some molariform traits such as a large talonid, large hypoconid, and distinct cingulids. A large, buccally placed hypoconid is also observed in p3. Enamel is extended onto the buccal roots of the lower premolars. The lone upper premolar, a P4, exhibits a single (proto)loph with a strong distal cingulum and marked distal crest of the paracone. The metacone is virtually indiscernable. In terms of occlusal details, there is a remarkable continuity of form in P4 between *Phosphatherium*, *Daouitherium*, and *Numidotherium* (see Gheerbrant et al., 2002:fig. 11).

Other aspects of the dentary besides alveolar placement include a relatively deep corpus, mental foramen below the tooth anterior to p2, and a vertically high ramus with its anterior margin at the level of the posterior of m3 and only slightly angled forwardly. Primitively, *Daouitherium* shares with *Phosphatherium* a small, low coronoid foramen and mandibular canal below the cheek tooth alveoli (Ferretti and Debruyne, 2011). Unfortunately, the symphysis is not preserved. These gnathodental features contain important systematic information for interpretation of the phylogenetic position of *Daouitherium*, which cladistically is intermediate between *Phosphatherium* and *Numidotherium* (Gheerbrant et al., 2002). In addition, enamel microstructure in the molars of *Daouitherium* lacks the primitive radial enamel retained in *Phosphatherium* and is nearly completely composed of Hunter-Schreger bands, except for a small section of prisms that anticipate the 3D enamel morphology observed in *Numidotherium* (Tabuce et al., 2007). To summarize, traits primitively shared with *Phosphatherium* include the absence of a mandibular diastema, more than two teeth anterior to p2, posterior location of the origin of the ramus, presence of a paraconid on p3–4, and presumably a non-tusk-like lower central incisor; more advanced traits anticipating *Numidotherium* include the great height of the ramus, fully realized molar (bi)lophodonty, and overall larger size, though *Daouitherium* has less bunodont molar cusps and more molarized p3–4, which Gheerbrant and colleagues (2002) argue excludes *Daouitherium* from the ancestry of *Numidotherium*.

BARYTHERIOIDEA Andrews, 1906
NUMIDOTHERIIDAE Shoshani and Tassy, 1992
NUMIDOTHERIUM Jaeger in Mahboubi et al., 1986
NUMIDOTHERIUM KOHOLENSE Jaeger in Mahboubi et al., 1986

(Tables 2.1–2.3; Figures 2.2, 2.6, and 2.7)

At the time its discovery was first reported by Mahboubi and colleagues in 1984, *Numidotherium koholense* became the oldest proboscidean then known and added more than 10 Ma to the recorded antiquity of elephant precursors. The genus nomen is based on the Latin for "wild beast" and for the region, and the species nomen denotes the discovery site and formation name (El Kohol). The species was a valuable find because it more deeply embedded the ancestry of proboscideans in Africa and its hypodigm includes an impressive abundance of cranial, gnathodental, and postcranial remains, making it possible to thoroughly investigate its form and behavior. It is skeletally the most completely represented proboscidean from the early Paleogene (Benoit et al., 2013). Moreover, *Numidotherium* constitutes the last significant evidence of early Paleogene plesielephantiforms prior to a long middle Eocene hiatus.

Numidotherium koholense is part of a diverse vertebrate assemblage, associated with invertebrates and plants, that a French-Algerian team recovered from the marly, calcareous middle member of the El Kohol Fm. near Brezina, on the southern flank of the Saharan Atlas Range in Algeria (Mahboubi et al., 1984). Its age was tenuously inferred to

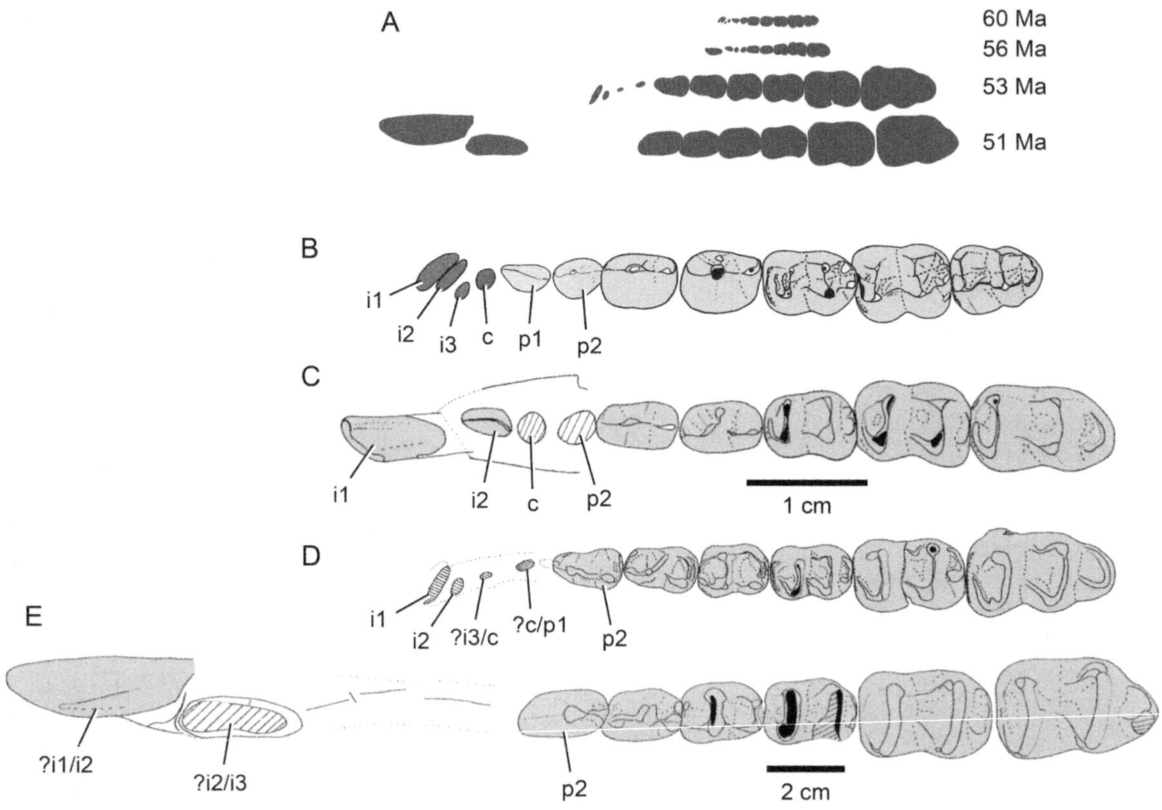

FIGURE 2.6 Comparison of lower tooth rows of early Paleogene proboscideans, in occlusal view. Abbreviations: c, canine; i, incisor; p, premolar. (B, C) are to the same scale; (D, E) are to the same scale. Anterior is to the left. (A) Occlusal silhouettes of lower tooth rows of *Eritherium* (top), *Phosphatherium*, *Daouitherium*, and *Numidotherium* (bottom), to the same scale, showing size increase over time. (B) *Eritherium azzouzorum*, reconstruction based on specimens OCP DEK/GE 307, MNHN.F PM28, and MNHN.F PM84, probably Selandian, Ouled Abdoun Basin, Morocco. The dental formula is 3-1-4-3. (C) *Phosphatherium escuilliei*, OCP-CPSGM MA8, early Ypresian, Ouled Abdoun Basin, Morocco. The dental formula is 2-1-3-3 or 2-0-4-3. (D) *Daouitherium rebouli*, reconstruction, middle Ypresian, Ouled Abdoun Basin, Morocco. The dental formula is alternatively 3-1-3-3, 2-1-4-3, or 3-0-4-3. (E) *Numidotherium koholense,* KAX-11, late Ypresian, El Kohol, Algeria. The dental formula is 2-0-3-3. ([B] is from Gheerbrant et al. [2012:fig. 15D], [C] is from Gheerbrant et al. [2005:fig. 27C], [D] is from Gheerbrant et al. [2005:fig. 27A], and [E] is from Gheerbrant et al. [2005:fig. 27B]. Images courtesy of E. Gheerbrant and with permission of Schweizerbart Science Publishers [www.schweizerbart.de/journals/pala].)

the end of the early Eocene (late Ypresian), primarily by reference to charophytes (algae) from the middle member and generally supported by stages of evolutionary development of faunal elements, including a very primitive hyrax, *Seggeurius amourensis* (Mahboubi et al., 1986; Court and Mahboubi, 1993; Rasmussen and Gutiérrez, 2010). Magnetostratigraphy of the deposits supports this age estimate (Coster et al., 2012). The fauna is composed of a mix of endemic vertebrates (proboscideans, hyraxes, crocodilians, lungfish) and immigrant taxa (didelphids, creodonts, "lipotyphlan insectivores") suggesting transatlantic and transtethysian connections. Of these, the proboscidean remains in particular have been intensively studied.

Survey of the *Numidotherium* sample from El Kohol, described in comprehensive detail by Jaeger in Mahboubi et al. (1986), reveals a striking and substantial morphological transformation from the incremental advances observed in the older Ouled Abdoun proboscidean succession. The most notable departure from the anatomy of earlier proboscideans, and comprising a novel morphological pattern within Proboscidea, is the structure of the cranium of *Numidotherium*. In the type specimen (KA1–18), it is raised high above the occlusal plane, with a basicranium and occipital condyles situated well superior to the palate (Figure 2.7A), and is richly pneumatized around the braincase. Other salient features include a transversely wide, thickened frontal (at least in adult males), a tall, vertical face with a short, narrow snout that encloses a high, narrow, non-retracted nasal aperture (that likely would not have supported a trunk), and deep submaxillary fossae lateral to the snout. The zygomatic arches are anteroposteriorly short but flare widely laterally, sweep steeply upwardly toward the back of the cranium, and sport large, downwardly projecting apophyses below their origins (Figure 2.7A and B). These apophyses recall the distinctive ventral processes on the zygomatic arches of *Eritherium* (Gheerbrant et al., 2012) and *Phosphatherium* (Gheerbrant et al., 2005). On each side of the cranium, the submaxillary fossa continues posteriorly below the orbit to the root of the zygomatic arch, producing a very unusual configuration. The orbit is situated low, above the premolars, and is unbounded dorsally and posteriorly (Figure 2.7A). The infraorbital foramen also opens low, above the anterior margin of P2. Temporal lines meet in the midline to form a sagittal crest that joins an impressively raised nuchal crest posteriorly. A lengthy diastema separates the upper canine from the anterior premolar.

Structure of the basicranium and ear region in *Numidotherium koholense*, including the semicircular canals, have been described in detail by Benoit et al. (2013). A few features are shared in common with elephants, such as the relatively small size of the oval and round windows of the petrosal, whereas semicircular canal dimensions appear to be autapomorphic, and the presence of a transpromontory sulcus on the petrosal suggests a primitive pathway for the internal carotid artery. Cochlear anatomy of *Numidotherium* indicates that it was unlikely to have had an elephant-like ability to hear infrasonic sounds; instead, the small number of coils and mode of coiling in this Paleogene proboscidean are consistent with a limited, high frequency range of auditory acuity (Court, 1992). Cladistic analysis of ear region morphology weakly links *Numidotherium* and *Phosphatherium* as a sister group to elephantimorph proboscideans and *Moeritherium* (Benoit et al., 2013; but see Schmitt and Gheerbrant, 2016).

Considerable variation occurs in the cranial sample of this species, attributable to age-grade development and

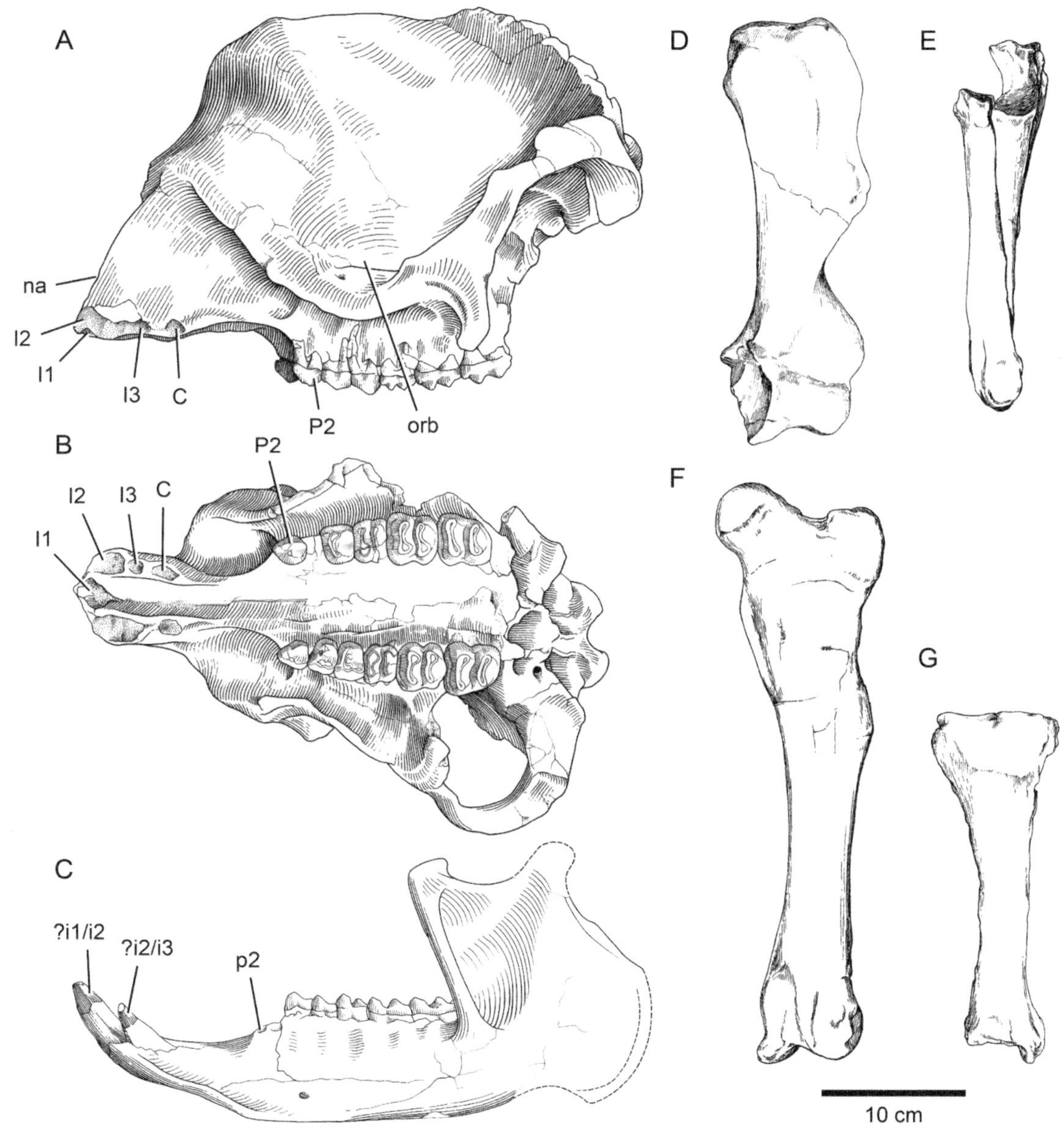

FIGURE 2.7 Skeletal elements of *Numidotherium koholense*, late Ypresian, El Kohol, Algeria. Abbreviations: C, upper canine; I/i, upper/lower incisor; na, nasal aperture; orb, orbit; P/p, upper/lower premolar. (A–C) anterior is to the left and the images are at the same scale. (D–G) proximal is to the top and the images are at the same scale. (A) Cranium, KA 1-18, holotype, right lateral view (reversed). (B) Cranium, KA 1-18, holotype, ventral view. (C) Right dentary, KAX-11, right lateral view (reversed). (D) Left humerus, KA 2-16, anterior view. (E) Right radius and ulna, KA 2-19, anterior view. (F) Right femur, KA 1-1, anterior view. (G) Right tibia, KA 2-18, anterior view. (From Mahboubi et al. [1986:figs. 4, 7, 8, 10–13], with permission of Schweizerbart Science Publishers [www.schweizerbart.de/journals/pala].)

sexual dimorphism (Noubhani et al., 2008). For example, in some individuals, the temporal lines do not meet in a sagittal crest, the position of the infraorbital foramen varies, there is no thickened bar of bone across the frontal, and the submaxillary fossae are shallower and less concave. These authors have also authoritatively documented the range of variation in dentary and tooth dimensions, as well as in occlusal morphology of cheek teeth. Dentary size appears to increase as a function of age, in parallel to differences observed in the cranium. Analyses of coefficients of variance for cheek teeth indicate an acceptable range of crown dimensions for a single species.

Compared with earlier proboscideans, the mandible of *Numidotherium* is distinctive in exhibiting a great spatial separation between incisors and the cheek teeth, characteristic of many of the later members of the Order, including early elephants (Figure 2.7C). The median incisor (i1?) is relatively large and procumbent, spatulate, laterally serrated, and is thinly covered by enamel. The lateral incisor (i2?) is also procumbent, but smaller, and has a distal ridge with many enamel serrations. The symphysis extends as posteriorly as p4. A mental foramen opens low on the corpus, below p3. The greatest depth of the corpus is under p2-p3; the ramus is anteroposteriorly immense, high and rooted on the corpus at the posterior of m3 (Figure 2.7C). The overall impression gained of this configuration is of an animal using its anterior dentition in a significant manner to procure food, for defense, and for social display.

The composition of the dentition is also advanced from that of earlier proboscideans. For upper teeth, the dental formula of *Numidotherium* is 3-1-3-3, but for lower teeth it is 2-0-3-3, signaling a loss of presumably the lower lateral incisor and canine, as well as both upper and lower first premolars (Table 2.2), in association with the extension of the diastema and emphasis on the tusk-like hypertrophy of i1. The permanent cheek teeth were preceded by three deciduous premolars in each dental quadrant. DP2 resembles P2, but DP3–4 are more molarized than P3–4; DP4 is a smaller version of M1 (Noubhani et al., 2008). Upper molar size progression is M1<M2<M3 (Table 2.3), and these cheek teeth are truly bilophodont and tapir-like. Molar cusps are approximately equal in height and the transverse crests uniting them are anteriorly convex. In M2 and M3, the protoloph is wider than the metaloph, whereas in M1 the lophs appear more equal in width. Postparacrista and postmetacrista are associated with the paracone and metacone, respectively. The anterior cingulum is low and closely appressed to the protoloph; the posterior cingulum forms a low, modest shelf. P2 is a triangular tooth that is slightly longer than wide, and has a simple crown morphology with a larger anterior buccal cusp and small protocone, no hypocone, and a small distal fovea. There is no transverse loph. P4 is a larger version of P3. Both of these premolars are wider than long. The anterior cusps form a transverse loph, there is no hypocone, and the buccal cusps essentially comprise an ectoloph. I1 is a narrow incisor with thin, smooth enamel described as low-crowned by Jaeger (Mahboubi et al., 1986), but this is not the case for I2 (Noubhani et al., 2008); its median ridge may be finely serrated. I2 is the largest upper incisor and is stout, high-crowned, canine-like, oval at the base, and is covered with thin enamel. As with the other upper incisors, it is downturned in orientation. The third upper incisor is much smaller than I1 and appears button-like and simple; contrary to Jaeger's opinion (Mahboubi et al., 1986), it is readily distinguishable from the canine (Noubhani et al., 2008). The upper canine is also small and low-crowned but differentiated from I3 by being slightly larger, more pointed apically, and having a distal crest.

Jaeger (Mahboubi et al., 1986) speculates that the large, procumbent, and spatulate lower central incisor is an i2. Based on presumed homology with the lower central incisors in the older proboscidean succession from the Ouled Abdoun Basin, and the likelihood that tooth loss occurred at the distal edge of the anterior tooth series in concert with increased emphasis on anterior dental function and development of a diastema to offset anterior teeth from cheek teeth, it seems more plausible that the lower central incisor in *Numidotherium* is i1 (see also Delmer, 2009). The smaller i2 resembles i1 but has a more asymmetric crown and has many serrations on its convex distal edge of its crown. p2 is relatively elongate and dominated by centrally placed protoconid; it is broadest posteriorly. This anteriormost lower premolar exhibits an odd wear pattern, a smooth excavation of the root at the mesial end of the tooth, just below the cervix of the crown, thought to be the result of browsing at or above the height of the head by stripping leaves from slender branches (Court, 1993). p3 is similar in occlusal morphology to p2 but is slightly wider, whereas p4 has some development of a transverse crest between the protoconid and the metaconid, which is offset distally from the protoconid, though none of the lower premolars is as molarized as in *Daouitherium*. Lower molar size progression is m1<m2<m3, and each molar is bilophodont, although m3 has a variable-sized but generally large talonid (this does not constitute a third lophid; see Noubhani et al., 2008). The lower molar transverse lophids are anteriorly concave and each has two crests that descend forward from the outer edges of the lophid. In each lower molar, the anterior cingulid is feeble and the posterior cingulid variably has a small hypoconulid.

The recovery of abundant postcrania of *Numidotherium koholense* (Mahboubi et al., 1986) makes it possible to make more accurate body size estimates and investigate the positional mechanics and habitus of the species (Court, 1994a). Significantly, it is the earliest assemblage in the fossil record for which we have evidence about how proboscideans moved across the landscape. The long bones of the appendicular skeleton are particularly well represented and are relatively robust. For example, the proximal and distal ends of the humerus are approximately 100–110 mm in width, compared to an overall length of about 300 mm. Moreover, the humerus exhibits a powerful deltoid ridge that extends at least halfway down the shaft and an equally impressive, flaring epicondylar crest (Figure 2.7D). The proximal tuberosity does not project much above the height of the head, which faces primarily posteriorly. Distally, the olecranon fossa is deeply excavated. It is notable that the humerus primitively preserves an entepicondylar foramen and small lesser trochanter. The ulna is similarly heavily constructed, with a large, posteriorly reflected and short olecranon process. It is fused distally and sometimes proximally with the smaller radius (Figure 2.7E), and thus the radius is fixed in pronation on the ulna. Together, these elements form a deep sigmoid notch proximally for the distal humeral joint, and a broad distal surface area for articulation with the carpus. Although no measurements have been published for the radioulnar complex, they are about 80% the length of the humerus.

As with the bones of the forelimb, the hindlimb elements are also relatively robust. The femur is longer than the humerus, with a length greater than 400 mm, and is especially heavy and wide proximally, in part due to flange-like lesser and third trochanters (Figure 2.7F). The shaft bows markedly outward laterally and is anteroposteriorly flattened. The greater trochanter is lower than the head but stout. Orientation of the head is medioproximal. Contra Mahboubi et al. (1986), the patellar facet is broad, and it faces primarily anteroventrally (Court, 1994a). The medial malleolus is much larger than the corresponding lateral malleolus. In overall aspect, the femur more resembles the femora of palaeomastodonts than the slender femur of *Moeritherium*. The tibia is similarly sturdy, much shorter than the femur (~260 mm), and possesses a very broad proximal end (~100 mm) (Figure 2.7G). It preserves a prominent medial malleolus distally, a relatively broad distal articulation for the astragalus that angles about 20–22° from the vertical, and a projecting, broad interarticular eminence proximally to accommodate the deep intercondylar notch of the distal femur. Unfortunately, the fibula is poorly preserved, but reveals evidence of powerful peroneal muscles.

Other postcranial elements known for *Numidotherium* include the axis (C1) and atlas vertebrae (C2), a fragmentary scapula and pelvis, astragalus and calcaneum, a few metapodials, isolated phalanges, and most of the carpus, including evidence of an os centrale. The first cervical vertebra resembles those of more advanced proboscideans (elephantiforms and deinotheres) by its craniocaudal shortness, flattened transverse processes, and absence of a hypopophysis. The second cervical is also relatively short and high, with a strong odontoid process; unlike C1, it is distinguished from the axis in elephantiforms and deinotheres by its proportions and flattened postzygapophyses, and in these features more resembles the C2 of moeritheres. The scapular remains of *Numidotherium* are very fragmentary, but preserve a bifid acromial process that overhangs the scapular neck, a robust coracoid process similar to that observed in *Barytherium*, and has a transversely narrow glenoid. Due to the incongruity of the scapular and humeral articular surfaces, contact between these joint surfaces is limited to a single point and it is thought that this configuration relates to a very mobile shoulder joint (Mahboubi et al., 1986; Court, 1994a).

There is some controversy about astragalar morphology in *Numidotherium*. Court (1994a) noted that the astragalus described by Mahboubi et al. (1986) exhibits features in common with *Palaeomastodon*, as was observed by those authors, but that it does not correspond to the calcanei or tibiae of the taxon and therefore might represent a second, rarer proboscidean taxon at the site. Its ectal facet is slightly larger than its figure-eight-shaped sustentacular facet (differing from the astragalus of *Barytherium*, which has a much larger sustentacular than ectal facet), and it is rectangular in shape with a prominent medial process, all resembling the astragalus of *Palaeomastodon* except for its smaller overall size (Mahboubi et al., 1986). Using a different astragalar specimen that articulates better with the distal tibia, Court (1994a) observed that it is characterized by having a head that articulates only with the navicular (as in other proboscideans), ectal and figure-eight-shaped sustentacular articular surfaces nearly equal in size, sustentacular facet that is anterolaterally oriented, navicular facet oddly angled to the lateral side, fibular facet extended anteroposteriorly, and naviculo-astragalar articulation lacking a stop facet. Most of these features are autapomorphic. No other evidence for the mystery taxon has been forthcoming, and the differences between these two astragali appear to have been exaggerated by Court (1994a). Overall, the calcaneum is proboscidean-like, but it has a very diminished plantar tubercle, its cuboid facet is small, and the ectal facet is more anteroposteriorly oriented than in other proboscideans (Court, 1994a).

Court (1994a) wrote in careful detail about the morphology of the wrist and manus elements. The site assemblage includes a cuneiform, lunate, unciform, magnum, and trapezoid. Although this review cannot adequately summarize Court's (1994a) anatomical study of these elements, several observations are critical for understanding the positional capabilities of *Numidotherium*. First, the proximal and distal rows of carpals are aligned serially, such that, for example, the cuneiform bone sits proximally directly above the unciform bone, and similarly, the lunate sits directly in line above the magnum. Second, joint excursions of the wrist bones relative to one another and to the forelimb elements and metapodials help to "maintain the integrity of substrate contact as the body weight passes over the hand" (Court, 1994a:p. 317) during the propulsive phase of locomotion, at which time the forelimb retracts, adducts, and rotates, bringing the body over the supporting manus. Third, the digits are splayed out in a nearly horizontal position from the wrist, making *Numidotherium* a plantigrade animal—very different from the specialized digitigrady of modern elephants. Additionally, the middle phalanges are relatively short and dorsoventrally flattened, and the terminal phalanges are very short, rounded elements that probably bore nails rather than hooves, also differing markedly from modern elephants, which possess hoof-bearing terminal phalanges (Court, 1994a).

The morphological complexes in the postcranium of *Numidotherium* are inconsistent with graviportal positional adaptations, despite impressive body size: between 1.0 and 1.5 m tall at the shoulders and estimated at between 250 and 300 kg in body mass (Gheerbrant and Tassy, 2009; Larramendi, 2016). Rather, locomotion has been reconstructed as consistent with a semi-sprawling or ambulatory gait on abducted limbs (Court, 1994a). Although joint surface configuration and posture in *Numidotherium* share much with the condition in some semi-aquatic mammals, the results of comparative oxygen isotope analysis indicate that it lived in terrestrial habitats, and these results are confirmed by the study of cross-sectional long bone features in *Numidotherium*, which exhibit thickened cortices and considerable medullary cavity space, also consistent with adaptation to terrestrial existence (Mahboubi et al., 2014).

As was pointed out by Jaeger in Mahboubi et al. (1986), the proboscidean synapomorphies of *Numidotherium* are clear and abundant, among them the relative height of the external auditory meatus and basicranium well above the level of the palate, the absence of mastoid processes, the prominence of I2 and i1, and the blocky construction and large medial process of the astragalus. Although the retention of an entepicondylar humeral foramen and inferred presence of a free os centrale in the wrist are very primitive traits, *Numidotherium* shares a number of features in common with the more advanced elephantoids and deinotheres, including the height and pneumatization of the cranium, the larger size of the ectal facet versus the sustentacular facet, the anteroposterior flattening and lengthening of the femur, and the smaller size of the radius than the ulna. Cladistically, the position of *Numidotherium* to other proboscidean taxa varies widely, including (a) between *Moeritherium* and deinotheres (Mahboubi et al., 1986), (b) sister-group relationship with deinotheres (Mahboubi et al., 1986), (c) sister-group relationship to *Moeritherium* (Gheerbrant, 2009; Gheerbrant et al., 2012), (d) sister-group relationship to *Daouitherium*+*Barytherium* (Gheerbrant et al., 2005), and (e) sister taxon of *Barytherium* (Gheerbrant et al., 2005). Negotiating a solution to this phylogenetic morass will require greater representation of missing elements in taxa (for instance, postcrania for proboscidean taxa older than *Numidotherium*; the cranium of *Daouitherium*), and more careful delineation of character states (for example, the relatively short faces of *Moeritherium* and *Numidotherium* should not be coded equivalently [see Gheerbrant et al., 2005], as they are morphologically dissimilar and short in different ways). Considered together, the skeletal features of *Numidotherium* suggest that it is a considerably more advanced proboscidean than earlier members of the Order, representing a novel and unexpected grade of morphological organization that is difficult to reconcile with its inferred terrestrial existence. Shoshani and Tassy (2005) placed *Numidotherium* in the same family as *Phosphatherium* and *Daouitherium*, but the degree of morphological apomorphy and advances in *Numidotherium* compared to these genera argue for their taxonomic separation at a higher level. With its uniquely derived cranial and postcranial anatomy, it is equally difficult to envision *Numidotherium* as a good ancestral model for the phylogenetic group out of which elephantiform proboscideans eventually evolved. Nonetheless, it seems reasonable based on molar similarities and degree of reduction of dentition to anticipate the origin of more derived barytherioids in the later Paleogene from a form similar to *Numidotherium koholense* (see Gheerbrant et al., 2005).

SUBORDER INCERTAE SEDIS
SUPERFAMILY INCERTAE SEDIS
FAMILY INCERTAE SEDIS
SALOUMIA Tabuce et al., 2019
SALOUMIA GORODISKII Tabuce et al., 2019

(Tables 2.1 and 2.3; Figures 2.2 and 2.8)

Synonymy—*Moeritherium* Gorodiski and Lavocat, 1953

A small broken molar, MNHN. F. MCA 1, from marine sediments at the village of M'Bodione Dadere, Senegal, was originally described and tentatively referred to *Moeritherium* by Gorodiski and Lavocat in 1953. It is an important specimen because it is Lutetian in age (Table 2.1) and therefore one of the very few proboscidean specimens known from the middle Eocene. A recent more detailed, comparative study of the molar found that in size and morphology it does not belong in *Moeritherium*, and therefore the authors placed it in its own genus and species (Tabuce et al., 2019).

The molar is broken distally but appears to have been nearly square in shape and extremely brachyodont. The apices of the four main cusps appear worn away, but Tabuce et al. (2019), with firsthand observation of the specimen, concluded that they were broken away prior to fossilization. The circular dentin exposures of each cusp reveal that it was bunolophodont. This aspect of the tooth differs from molars of other early Paleogene proboscideans except for *Khamsaconus* and *Dagbatitherium* (see below); however, it is too large for the former, too small for the latter, and closer to the size range of lophodont *Phosphatherium* (see Table 2.3). MNHN. F. MCA 1 has a low cingular ribbon on its three preserved sides, flaring out to form an entostyle lingually and a tiny ectostyle buccally. The protocone and paracone are subequal in size and the wear figure of the protocone reveals an anterior extension into a paraconule. The mesial half of the hypocone shows that it was likely the biggest cusp, set transversely across from the remnant of the smaller metacone. Although Tabuce and colleagues (2019) claim that the hypocone is centrally located on the crown, this may be an artifact of the flare of the lingual cingulum and not an accurate portrayal of crown morphology. The molar has vertical wrinkling of the enamel and lacks a continuous centrocrista and well-defined ectoflexus. Although the authors (Tabuce et al., 2019) claim that the enamel is thick, by proboscidean standards, even at the small size of the tooth, it instead appears to be rather thin.

Bunolophodonty of the M'Bodione Dadere molar, as well as the small size of the parastyle, quadrituberculy, and bluntness of the paraconule are features shared with *Moeritherium*, but as Tabuce et al. (2019) aptly point out, these features either differ in their expression from the condition of moerithere molars and/or are primitive features and therefore do not support any special relationship between *Saloumia* and *Moeritherium*. The development of the lingual cingulum, lack of a centrocrista, and wrinkled enamel are thought by the authors to be shared derived features between these genera, but the lingual cingulum seems different in *Saloumia* by its flare. It is alternatively possible that *Saloumia* documents "an early experiment in dental diversity among Paleocene-Eocene proboscideans, with no direct relationships with later Paleogene proboscideans" (Tabuce et al., 2019:p. 7). In the paucity of proboscidean-bearing sites from the middle Eocene, the presence of *Saloumia* at M'Bodione Dadere is intriguing and hints at the greater diversification of proboscideans that must have occurred during this interval. As such, Tabuce and colleagues (2019) deserve ample credit for their detailed and careful re-analysis of this important specimen.

SUBORDER ELEPHANTIFORMES TASSY, 1988
Family Incertae Sedis
DAGBATITHERIUM Hautier, Tabuce, Mourlam, Kassegne, Amoundkji, Orliac, Quillévéré, Charruault, Johnson, and Guinoit, 2021
DAGBATITHERIUM TASSYI Hautier, Tabuce, Mourlam, Kassegne, Amoundkji, Orliac, Quillévéré, Charruault, Johnson, and Guinoit, 2021
(Tables 2.1 and 2.3; Figures 2.2 and 2.8)

An intriguing right m1 or m2 (ULDG-DAG1), recently recovered from the phospharenite member of the Dagbati Quarry in Togo, is another of the very rare occurrences of Proboscidea in the middle Eocene (Table 2.1). The genus name refers to the "wild animal of Dagbati (quarry)," and the species name honors the great Pascal Tassy, who erected the category "Elephantiformes" and advanced the study of proboscideans more than any other

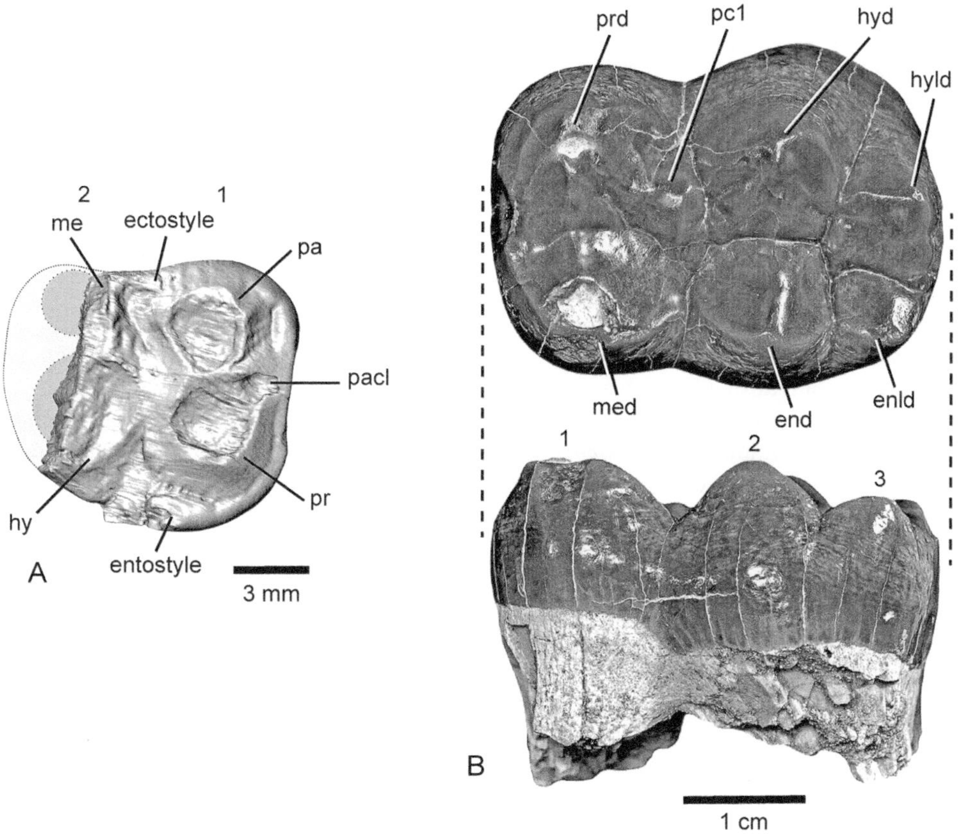

FIGURE 2.8 Isolated bunolophodont molars of middle Eocene African proboscideans. Abbreviations: end, entoconid; enld, entoconulid; hy, hypocone; hyd, hypoconid; hyld, hypoconulid; me, metacone; med, metaconid; pa, paracone; pacl, paraconule; pc1, accessory conule posterior to lophid 1; pr, protocone; prd, proconid; 1, 2, 3, …, loph(id) number counted from the anterior of the crown. (A) Incomplete upper molar, MNHN.F. MCA 1 from M'Bodione Dadere, Senegal, occlusal view, *Saloumia gorodiskii* (Family incertae sedis). Anterior is to the right. (Image courtesy of R. Tabuce [Tabuce et al., 2019].) (B) Right m1 or m2, ULDG-DAG1 from Dagbati quarry, Togo, occlusal and lateral views, *Dagbatitherium tassyi* (Elephantiformes). Anterior is to the left. (Image courtesy of L. Hautier [Hautier et al., 2021].)

researcher of his generation. What makes it so intriguing is that, unlike the lophodonty of proboscidean molars of the early Eocene, this specimen is bunolophodont with a clear resemblance to palaeomastodont molars. Cladistic treatment of ULDG-DAG1 places *Dagbatitherium* as the sister taxon to [*Palaeomastodon*+*Phiomia*]+Elephantimorpha, and accordingly its describers have referred it to Elephantiformes (Hautier et al., 2021). Prior to the discovery of *Dagbatitherium*, the oldest known elephantiform proboscideans were palaeomastodonts of the early Oligocene (Table 3.1). Its import for tracing the origins of elephantiforms is extraordinary.

ULDG-DAG1 is a rectangular-shaped molar, longer than wide, of modest dimensions (Table 2.3). Its overall size is between that of intermediate molars of moeritheres and palaeomastodonts; it is smaller than molars of all other later Paleogene proboscideans but bigger than molars of earlier Paleogene proboscidean taxa (see Table 2.3). It has three lophids and an anteroposteriorly very narrow, low strip of enamel comprising a postcingulid. The protolophid is composed of nearly equivalent-sized protocone and metacone, separated superficially by a narrow midline sulcus. A distinct accessory conule occupies the transverse valley posterior to the protoconid. The hypolophid is composed of similar-sized hypoconid and entoconid, and the tritolophid has dominant hypoconulid on the buccal side and smaller entoconulid on the lingal side. The tritolophid is transversely narrower than the first two lophids but distinct and separated from the hypolophid by a transverse valley. Both the hypolophid and tritolophid have median sulci. The main cusps on the pretrite, or buccal, side of lophids 1 and 2 exhibit subtle anterior and posterior cristids and the hypoconulid has an anterior cristid, as well. In posterior view, the buccal side of the crown is canted inward. In lateral view, the lophids are bulbous and very brachyodont. The apices of the lophids are slightly anteroposteriorly "pinched," resembling the condition of *Palaeomastodon* molars more than those of *Phiomia*. There are no cingulids or cementum.

Analysis of the Schmelzmuster pattern of enamel in *Dagbatitherium* resembles the complex, three-layered arrangement found in most elephantiform proboscideans, but is more derived than the condition in palaeomastodont molars (Hautier et al., 2021). It either means that palaeomastodonts had a reversal to two-layered Schmelzmuster, or that *Dagbatitherium*, like the plesielephantiform *Arcanotherium*, is convergent on derived elephantiform proboscideans in this feature. The trilophodonty (with individualized hypoconulid and [post]entoconulid), lateral expansion of the crown, lack of accessory conules in the second transverse valley, and bunolophodonty of ULDG-DAG1 resemble the condition of primitive elephantiform proboscideans and support its placement in the taxon (Hautier et al., 2021). Other features diagnostic of Elephantiformes, such as elongation of the mandibular symphysis, anteroposterior extension of the diastema between the lower tusks and cheek teeth, presence of evergrowing tusks, orbital retraction above M1, absence of paroccipital processes, and presence of a nasal fossa (Tassy, 1990b), are not possible to document from an isolated molar. Nonetheless, based on its apparent dental synapomorphies, placement of ULDG-DAG1 in Elephantiformes is a sound working hypothesis.

SUMMARY

1. Proboscideans originated in Africa, the first of many important phylogenetic events in the history of the Order that occurred on the continent. The oldest known taxa have fossil records limited to North Africa and are dated reliably as old the late middle or late Paleocene, ~60 Ma.
2. Evolutionary derivation of proboscideans is obscure, and it is not established whether they are descended from archaic Gondwanan mammals or more probably Laurasiatic immigrants that arrived during the Cretaceous. Early Paleogene basins in North Africa which have produced proboscideans provide evidence for the presence of a mix of endemic and Laurasiatic taxa, demonstrating that Africa was not completely isolated during this time.
3. There is no fossil evidence for a Cretaceous start of the Order, and contentious debate continues about whether crown placentals (including proboscideans), their stem lineages, or both first arose prior to the Cenozoic. Paleontological analysis of the completeness of the fossil record of eutherian mammals across the Cretaceous/Paleogene boundary indicates a real absence of placental mammals in the Cretaceous.
4. The first phase of proboscidean evolution covers an interval of about nine million years and begins with animals that exhibit no resemblance to extant elephants. They are very small in body size, lack large, protuberant tusks, have no sign of a mobile, prehensile proboscis (trunk), have full or nearly full dental batteries, and their molars are bunolophodont or incipiently lophodont, rather than each being composed of a series of plates.
5. By the end of this first phase, proboscideans became much larger, making them the dominant ungulates in their faunas, but they lacked the graviportal adaptations of later, elephantiform proboscideans and deinotheres. Enamel microstructure in molars appears to have evolved from a condition preserving all or some radial enamel (with accompanying zones of Hunter-Schreger bands) to a condition dominated by 3D enamel. Starting with *Phosphatherium* and continuing in *Daouitherium* and *Numidotherium*, the chewing mechanism has been characterized as a vertical shear-cutting, deinothere-like pattern of mastication. Crania became significantly raised and substantially pneumatized, and upper and lower

jaws each developed a pair of outsized incisors (the lower procumbent). The importance of the front teeth for display, food acquisition, and fighting was emphasized by extension of that part of the tooth row anteriorly, forming an extensive diastema between the anterior dentition and premolars. All adult teeth continued to be in occlusion simultaneously, and molars had achieved full lophodonty. By the end of the early Eocene, there was still no sign of nasal retraction that would have been consistent with development of a trunk.

6. Of the proboscideans comprising this first phase, substantial postcrania has only been recovered for *Numidotherium*. Although reconstruction of posture and joint excursions of skeletal elements appear to show some similarities with features of semi-aquatic mammals, dental isotope analysis indicates that these proboscideans were terrestrial. There is no strong evidence of an aquatic ancestry for proboscideans, and support for (semi-) aquatic existence in the earliest proboscideans derives primarily from their preservation in near-shore marine sediments.
7. The fossil record for the African early Paleogene is very restricted, with occurrences limited to a few basins in North Africa. Consequently, the diversity and phylogeny of the earliest proboscideans are likely poorly represented and very imperfectly understood. At the close of the first phase of proboscidean evolution during this time interval, *Numidotherium* appears to some researchers to be too autapomorphic cranially and postcranially to represent the ancestral condition for later proboscideans, particularly elephantiforms, though dentally shares many features with later barytheres.
8. By the end of the early Paleogene, proboscideans had become the dominant taxa in their ecosystems, but we know little of their subsequent evolution until much later in the Eocene, because of a long middle Eocene near-hiatus in mammalian site occurrences in Africa. Isolated molars of proboscideans from the middle Eocene diverged from the lophodonty of early Eocene plesielephantiforms by the bunodont-bunolophodont morphology. The more primitive of these molars has been placed in a new genus (*Saloumia*) and is very small but reminiscent of moerithere teeth; the more advanced molar, placed in a second new genus (*Dagbatitherium*), exhibits features suggesting that it is an early elephantiform proboscidean. Nonetheless, because of their stark morphological contrast with the molars of plesielephantiforms, the ancestry of these new genera is obscure.

3 Late Paleogene
First Major Diversification and Adaptive Radiation of Proboscideans

In 1900 the author predicted that the ancestors of the proboscideans ... would be discovered in Africa. Two years later the members of the British Geological Survey of Egypt discovered in the Oligocene of the Fayum remains of *Palaeomastodon* and of *Moeritherium*, which were at once regarded as the solution of the ancestry of the proboscideans.

—Henry Fairfield Osborn (1919)

INTRODUCTION

Following a sparse, lengthy middle Eocene (Lutetian, Bartonian) interlude in the proboscidean fossil record, late Eocene and Oligocene Afro-Arabian deposits yielded a substantial number of new taxa (Table 3.1). However, the phylogenetic connections of these taxa to early Paleogene proboscideans and to each other remain poorly delineated. Nonetheless, the assorted barytherioids, moeritheres, chilgatheres, deinotheres, palaeomastodonts, gomphotheres, and mammutids of the late Paleogene, which followed the phosphatheres, daouitheres, numidotheres, and tantalizing potential precusors of moeritheres and elephantiforms of the early Paleogene, offer clues about the origins of proboscideans that populated the African Neogene, the persistent deinotheres and a rich succession of elephantimorphs, including elephants. In addition, the late Paleogene interval provided the first evidence of co-occurrence of multiple proboscidean taxa at sites, evoking consideration of adaptive radiations and competitive exclusion. For example, at the late Oligocene site of Chilga, Ethiopia, four genera of proboscideans with at least five species—chilgatheres, gomphotheres, and palaeomastodonts—were sympatric (Sanders et al., 2004). Thus, this interval marked the beginning of the grand diversification of proboscideans that characterized faunas in Africa until well into the Pleistocene.

Proboscidea is basically divisible into the paraphyletic plesielephantiforms and elephantiforms, with the phylogenetic position of deinotheres and moeritheres vis-à-vis these taxa less certain. Toward the end of the Paleogene, plesielephantiforms and moeritheres waned and elephantiforms were in the ascendency (Sanders et al., 2010a). Equally importantly, it is now established that the fundamental split of elephantimorphs (derived elephantiforms) into Mammutida (mastodons) and Elephantida (gomphotheres, stegodonts, and elephants) transpired at least by the mid-Oligocene (Sanders et al., 2004; Shoshani et al., 2004; Rasmussen and Gutiérrez, 2009; Zalmout et al., 2010). Moreover, the key adaptation driving their success, horizontal tooth displacement (Sanders, 2017), can be traced back in time to basal members of Elephantimorpha (e.g., Shoshani et al., 2001b, 2004). The delay of emergence and successive "conveyor belt" replacement of molars, and their consequent enlargement, in elephantimorph horizontal tooth displacement provided evident adaptive advantages over the primitive vertical tooth emergence arrangement employed by plesielephantiforms, deinotheres, and primitive elephantiforms, in which all adult teeth were in simultaneous occlusal function, permitting exploitation of a wider and coarser range of food items and increased lifespans and number of offspring (Sanders, 2017). The evolution of this dental development mechanism, with its consequential impact on life history parameters, was central to the great phylogenetic radiation of elephantimorphs, particularly gomphotheres, that ensued in the African Neogene.

Afro-Arabia of the late Paleogene remained largely an island continent with proboscideans isolated from other landmasses until nearly the Paleogene-Neogene boundary. Only a single proboscidean specimen, a fragment of an elephantoid upper incisor deriving from the latest Oligocene Bugti Mb. of the Chitarwata Fm. in Balochistan (Pakistan), is known from outside Afro-Arabia during the Paleogene (Antoine et al., 2003, 2013). Although some mammals such as anthracotheres, anomaluroid and hystricognathous rodents, and possibly anthropoids made it into Africa from Eurasia during the late Paleogene (see Jaeger et al., 2010a) via sweepstakes dispersal or filter routes, the Afro-Arabian fauna was largely endemic and best characterized by Fayumian assemblages in Egypt, comprised of creodonts, ptolemaiids, proboscideans, a rich diversity of hyraxes, arsinoitheres, and parapithecoid and stem catarrhine primates (Simons, 1968; Bown et al., 1982; Gagnon, 1997). Nonetheless, *within* Afro-Arabia, the late Paleogene distribution of proboscideans became much more widespread than in the early Paleogene, with occurrences in what is now the Arabian Peninsula and East Africa, for example (compare Figures 2.2 and 3.1). An important factor contributing to the more prevalent incidence of late Paleogene mammalian sites across more of the continent was the extensive, periodic accumulation of terrestrial sediment driven by weathering of massive, tectonically-driven late Eocene-Oligocene Trap basalts in eastern Africa and western Arabia, resulting in a

DOI: 10.1201/b20016-3

TABLE 3.1
Major Occurences of Late Paleogene Afro-Arabian Proboscidea

Taxon	Occurrence (Site, Locality)	Stratigraphic Unit	Age	Key References
Proboscidea Illiger, 1811				
Plesielephantiformes Shoshani, Sanders, and Tassy, 2001a				
Barytherioidea Andrews, 1906				
Barytheriidae Andrews, 1906				
Barytherium grave	Fayum, Egypt (type)	Qasr el Sagha Fm.	Late Eocene	Andrews (1901a, b, 1904b, 1906), Simons (1968), Harris (1978a), and Sanders et al. (2010a)
		Birket Qarun Fm.	Late Eocene, 37 Ma	Seiffert (2006), Seiffert et al. (2008), Sanders et al. (2010a), and Sanders and Seiffert (2014)
	Dur at Talah, Libya (=Gebel Coquin; Dur el Talha; Dur at Talhah)	Idam Unit (alt. New Idam Unit)	Early Oligocene (alt. Bartonian, middle Eocene)	Arambourg and Magnier (1961), Wight (1971, 1980), Rasmussen et al. (2008), Delmer (2009), Jaeger et al. (2012), and Abouessa et al. (2012)
Arcanotherium savagei (Court, 1995)	Dur at Talah, Libya (type)	Evaporite and Idam Units (alt. New Idam Unit)	Late Eocene-early Oligocene (alt. Bartonian, middle Eocene)	Arambourg and Magnier (1961), Savage (1969, 1971), Harris (1978), Wight (1980), Court (1995), Shoshani et al. (1996), LeBlanc (2000), Delmer (2009), Jaeger et al. (2010a), and Abouessa et al. (2012)
Omanitherium dhofarense	Locality DPP-2010-1, Oman (type)	Shizar Mb., Ashawq Fm.	Early Oligocene	Seiffert (2006), Seiffert et al. (2012)
	"*Omanitherium* 17", Oman	Shizar Mb., Ashawq Fm.	33.7–33.3 Ma	Al-Kindi et al. (2017)
	Thaytiniti, Oman			Seiffert (2006) and Pickford (2015b)
	Taqah, Oman	Shizar Mb., Ashawq Fm.	31.5–31.0 Ma	Seiffert (2006) and Pickford (2015b)
?*Barytherium* sp. indet. ("cf. Barytherioidea")	Thaytiniti, Oman	Shizar Mb., Ashawq Fm.	Early Oligocene, 33.7–33.3 Ma	Thomas et al. (1989, 1999) and Seiffert (2006)
Suborder incertae sedis				
Moeritherioidea Andrews, 1906				
Moeritheriidae Andrews, 1906				
Moeritherium chehbeurameuri	Bir el Ater, Algeria (type)	Sandy, fluviodeltaic member	Late Eocene, Priabonian (alt. Bartonian-Priabonian)	Delmer et al. (2006)
	?Khenchela, Algeria		Late Eocene, ?Priabonian	Pickford and Tassy (1980), and Delmer et al. (2006)
	Fayum, Egypt	Birket Qarun Fm.	Late Eocene, ~37 Ma, early Priabonian	Seiffert (2006), Seiffert et al. (2008), and Sanders and Seiffert (2014)
Moeritherium lyonsi (including "*M. gracile*," "*M. ancestrale*," "*M. pharaonensis*," and "*M. latidens*")	Fayum, Egypt (type)	Qasr el Sagha Fm.	Late Eocene	Andrews (1901a, 1901b, 1902, 1906), Matsumoto (1923), Petronievics (1923), Osborn (1936), Deraniyagala (1955), Coppens and Beden (1978), and Holroyd et al. (1996)
	Dur at Talah, Libya	Evaporite Unit (alt. New Idam Unit)	Latest Eocene or early Oligocene (alt. Bartonian, middle Eocene)	Savage (1971), Wight (1980), LeBlanc (2000), Jaeger et al. (2010), Sanders et al. (2010a), Abouessa et al. (2012), and Sanders and Seiffert (2014)
Moeritherium trigodon (including "*M. andrewsi*" and "*M. trigonodon*")	Fayum, Egypt (type)	Gebel el Qatrani Fm.	Early Oligocene, ca. 33–30 Ma	Andrews (1904a, 1906), Schlosser (1911), Matsumoto (1923), Osborn (1936), Coppens and Beden (1978), Seiffert (2006), and Sanders et al. (2010a)

(Continued)

TABLE 3.1 (*Continued*)
Major Occurences of Late Paleogene Afro-Arabian Proboscidea

Taxon	Occurrence (Site, Locality)	Stratigraphic Unit	Age	Key References
	Dur at Talah, Libya	Idam Unit (alt. New Idam Unit)	Early Oligocene (alt. Bartonian, middle Eocene)	Savage (1971), Wight (1980), LeBlanc (2000), Jaeger et al. (2010), Sanders et al. (2010a), Abouessa et al. (2012), and Sanders and Seiffert (2014)
Suborder incertae sedis				
Deinotherioidea Bonaparte, 1845				
Deinotheriidae Bonaparte, 1845				
Chilgatheriinae Sanders et al., 2004				
Chilgatherium harrisi	Harrat al Ujayfa, Al Hijaz Province, Saudi Arabia	Shumaysi Fm.	Early Oligocene. 29–28 Ma	Zalmout et al. (2010)
	Chilga, Ethiopia (type)	Chilga Fm.	Late Oligocene, 28–27 Ma	Kappelman et al. (2003) and Sanders et al. (2004)
Subfamily indet.				
Gen. et sp. indet. ("*Prodeinotherium* sp."; "Gen. et sp. nov.")	Losodok (Lothidok), Kenya	Eragaleit beds	Late Oligocene, 27.5–24.0 Ma	Boschetto et al. (1992), Rasmussen and Gutíerrez (2009), and Leakey et al. (2011)
	?Asmara-Massawa Road, Eritrea	Coarse-grained channel deposit, Dogali Fm.	Late Oligocene, ca. 28.0±2.5 Ma to 26.8±1.5 Ma	Sagri et al. (1998) and Abbate et al. (2014)
	?Mai Gobro 1, Eritrea	Lower channel deposit	Oligo-Miocene, 24.0–22.0 Ma (alt. between basalts dated to 29.04 and 23.55 Ma)	Abbate et al. (2012, 2014)
	?Adi Ugri, Eritrea	Mai Gobro section	24.6±0.25 Ma on rhyolite lenticular bed in section	Vialli (1966), Zanettin et al. (2006), and Abbate et al. (2012)
Elephantiformes Tassy, 1988				
Palaeomastodontidae Andrews, 1906				
Palaeomastodon beadnelli (including "*P. parvus*" and "*P. intermedius*")	Fayum, Egypt (type)	Gebel el Qatrani Fm.	Early Oligocene, ~33–30 Ma	Andrews (1901b, 1905, 1906), Matsumoto (1922, 1924), Simons (1968), and Seiffert (2006)
Palaeomastodon sp. indet.	Harrat al Ujayfa, Al Hijaz Province, Saudi Arabia	Shumaysi Fm.	Late early Oligocene, 29–28 Ma	Zalmout et al. (2010)
Palaeomastodon sp. indet.	Zella, Sirt Basin, Libya	Continental and Transitional Marine Deposits	Early Oligocene	Arambourg and Magnier (1961), Savage (1971), and Coster et al. (2015)
Palaeomastodon sp. nov. A	Chilga, Ethiopia	Chilga Fm.	Late Oligocene, 28–27 Ma	Kappelman et al. (2003) and Sanders et al. (2004)
Palaeomastodon sp. nov. B	Chilga, Ethiopia	Chilga Fm.	Late Oligocene, 28–27 Ma	Kappelman et al. (2003) and Sanders et al. (2004)
Phiomia serridens (including "*Ph. minor*," "*Ph. wintoni*," and "*Ph. osborni*")	Fayum, Egypt (type)	Gebel el Qatrani Fm.	Early Oligocene, ~33–30 Ma	Andrews and Beadnell (1902), Andrews (1904a, 1905, 1906), Matsumoto (1922, 1924), Simons (1968), Seiffert (2006), and Sanders et al. (2010a)
	Zella, Sirt Basin, Libya	Continental and Transitional Marine Deposits	Early Oligocene	Arambourg and Magnier (1961), Arnould-Saget and Magnier (1961), and Savage (1971)
("*Ph. minor*" and "*Ph. wintoni*")	Dur at Talah, Libya	Idam Unit (alt. New Idam Unit)	Early Oligocene (alt. Bartonian, middle Eocene)	Savage (1971), Wight (1980), LeBlanc (2000), Rasmussen et al. (2008), Jaeger et al. (2010), Sanders et al. (2010a), and Abouessa et al. (2012)

(*Continued*)

TABLE 3.1 (*Continued*)
Major Occurences of Late Paleogene Afro-Arabian Proboscidea

Taxon	Occurrence (Site, Locality)	Stratigraphic Unit	Age	Key References
("*Ph. osborni*")	Gebel Bon Gobrine, Tunisia		?Early Oligocene (alt. late Oligocene)	Arambourg and Burollet (1962)
Phiomia major	Chilga, Ethiopia (type)	Chilga Fm.	Late Oligocene, 28–27 Ma	Kappelman et al. (2003) and Sanders et al. (2004)
Aff. *Phioma* sp.	Lokone, Lokichar Basin, Kenya	Lokone Sandstone Fm.	Latest early Oligocene or earliest late Oligocene	Ducrocq et al. (2010) and Leakey et al. (2011)
?*Phiomia* sp. indet.	Malembe, Cabinda, Angola		Early Oligocene	Pickford (1986a, 1987b) and Tabuce et al. (2021)
Phiomia sp. indet.	Taqah, Oman	Shizar Mb., Ashawq Fm.	Early Oligocene, 31.5–31.0 Ma	Thomas et al. (1999), Whybrow and Clements (1999), Seiffert (2006),
cf. *Phiomia* sp.	Minqar Tibaghbagh, Qattara Depression, Western Desert, Egypt	BOTM level	Early Oligocene	Underwood (2017) and Van Vliet et al. (2017)
Elephantimorpha Tassy and Shoshani, 1997 in Shoshani et al., 1998				
Mammutida Tassy and Shoshani, 1997 in Shoshani et al., 1998				
Mammutidae Hay, 1922				
Losodokodon losodokius	Lothidok (Losodok), Kenya (type)	Eragaleit beds	Late Oligocene, 27.5–24.0 Ma (capping basalt)	Boschetto et al. (1992), Rasmussen and Gutiérrez (2009), and Gutiérrez, (2011)
	Nakwai, Kenya	Sedimentary sequence, Nakwai	Early Miocene, >22.85–14.39 Ma (youngest capping basalt)	Leakey et al. (2011) and Reynoso (2014)
Elephantida Tassy and Shoshani (1997) in Shoshani et al. (1998)				
Gomphotheriidae Hay (1922)				
Eritreum melakeghebrekristosi	Wedeg Melatse Farm, Eritrea (type)	Dogali Fm.	Late Oligocene, ~28.0±2.5 Ma to 26.8±1.5 Ma	Drury et al. (1994), Bunter et al. (1998), and Shoshani et al. (2004)
Gomphotheriidae gen. et sp. indet.	Harrat al Ujayfa, Al Hijaz Province, Saudi Arabia	Shumaysi Fm.	Late early Oligocene, 29–28 Ma	Zalmout et al. (2010)
cf. *Gomphotherium* sp.	Chilga, Ethiopia	Chilga Fm.	28–27 Ma	Kappelman et al. (2003) and Sanders et al. (2004)
Gomphotheriidae gen. et sp. indet.	Losodok, Kenya	Eragaleit Beds	Late Oligocene, 27.5–24.0 Ma	Boschetto et al. (1992) and Leakey et al. (2011)
?Gomphotheriidae gen. et sp. indet.	Mai Gobro 1, Eritrea	Lower channel deposit	Oligo-Miocene, 24.0–22.0 Ma (alt. between basalts dated to 29.04 and 23.55 Ma)	Abbate et al. (2012, 2014)
Gomphotheriidae gen. et sp. indet.	?Gebel Zelten, Libya	?Marada Fm.	?Late Oligocene or basal early Miocene	Pickford (2003a) and Sanders (2008a)

?=Attribution or occurrence uncertain.

series of intertrappean beds that preserved recurrent floras and faunas (Abbate et al., 2014).

The late Paleogene underwent two massive climate upheavals, the first an early Oligocene glacial maximum that brought about global cooling, rapid growth of the Antarctic ice sheet, world-wide drop in sea level, and extinction of many carbonate-secreting marine organisms, marking an end to greenhouse conditions that had predominated since the Mesozoic (Prothero, 1994; Zachos et al., 2001; Pearson et al., 2008; de Vries et al., 2021). Associated with these changes at the onset of the Oligocene is evidence for a decrease in the partial pressure of atmospheric carbon dioxide and increase in Antarctic ice volume. At this time, the Earth was subjected to its coolest temperatures of the

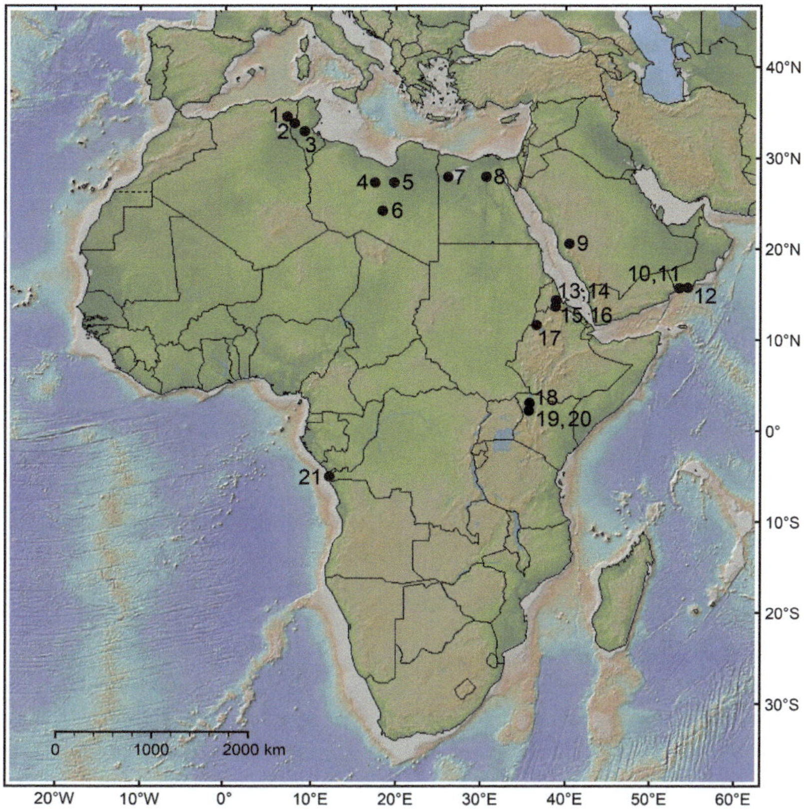

FIGURE 3.1 Map showing location of late Paleogene proboscidean sites. 1, Kenchela, Algeria; 2, Bir el Ater, Algeria; 3, Gebel Bon Gobrine, Tunisia; 4, Zella, Sirt Basin, Libya; 5, Gebel Zelten, Libya; 6a, Dur at Talah, Libya, b, Gebel Coquin, c, Dur el Talha, d, Dur at Talhah; 7, Minqar Tibaghbagh, Qattara Depression, Western Desert, Egypt; 8, Fayum, Egypt; 9, Harrat al Ujayfa, Al Hijaz Province, Saudi Arabia; 10, Locality DPP-2010-1, Oman; 11, Thaytiniti, Oman; 12, Taqah, Oman; 13, Wedeg Melatse Farm, Eritrea; 14, Asmara-Massawa Road, Eritrea; 15, Adi Ugri, Eritrea; 16, Mai Gobro 1, Eritrea; 17, Chilga, Ethiopia; 18a, Losodok, Kenya, b, Losodok (Lothidok), Kenya, c, Lothidok, Kenya, d, Lothidok (Losodok), Kenya; 19, Nakwai, Kenya; 20, Lokone, Lokichar Basin, Kenya; 21, Malembe, Cabinda, Angola.

Paleogene (DeConto and Pollard, 2003; Seiffert, 2007a). Due to its geographic position about the equator, however, Africa was buffered from the harsher effects of global cooling. Similar late Priabonian-early Rupelian distribution of shark and ray faunas along the coast of North Africa and substantial regional survival of elasmobranch taxa across the Eocene-Oligocene boundary indicate widespread comparable tropical environments and support the idea that Africa was protected from the consequences of global cooling at this time (Sweyden et al., 2019). Although many palm species went extinct and major floral shifts occurred in equatorial Africa during this event (Morley, 2005), nonetheless rain forests remained widespread throughout the continent into the late Oligocene (e.g., Jacobs et al., 2005). There is also evidence that many primate lineages in Africa (e.g., anthropoids, tarsiids, lorisids, lemurs) also survived, although the Fayum fossil sequence chronicles local extinction of adapids, plesiopithecids, galagids, and djebelemurines at the Eocene-Oligocene boundary (Seiffert, 2007a), and Afro-Arabia witnessed extinctions of anomaluroid and hystricognath rodents, some anthropoids, and carnivorous hyaenodonts, leading to a winnowing of peak late Eocene diversity in these groups by about 30 Ma (de Vries et al., 2021). Among proboscideans, barytheres (with an early Neogene exception) and moeritheres were nearly coterminous with these primates in their disappearance from the Afro-Arabian fossil record, while conversely elephantiforms in the guise of palaeomastodonts made their first appearance (Sanders et al., 2010a). Mid- and late Oligocene falls in eustatic sea level, related to development of the East Antarctic Ice Sheet (Swezey, 2009), may have played a role in diminishing coastal habitats critical to the survival of semi-aquatic moeritheres and barytheres, who disappeared coincident with these events.

It is possible that the great basaltic outpouring in Afro-Arabia during the late Oligocene contributed to the second cooling event of the Oligocene and concomitant mass extinctions on a global scale (White and McKenzie, 1989; Zhang et al., 2012; Abbate et al., 2014), although it has been argued that in the late Oligocene "run-up" to the episodic Eurasian faunal invasions of Africa in the early Neogene, African endemic faunas characteristic of the late

Eocene-early Oligocene were not in decline but were undergoing continued robust evolution (Sanders et al., 2004). Part of this evolutionary scenario included the Afro-Arabian emergence of new proboscidean taxa such as deinotheres, mammutids, and gomphotheres (e.g., Sanders et al., 2004; Shoshani et al., 2004; Rasmussen and Gutiérrez, 2009; Zalmout et al., 2010; Leakey et al., 2011), notably paralleled by the origin of more advanced anthropoids, including crown catarrhines (e.g., Zalmout et al., 2010; Leakey et al., 2011; Stevens et al., 2013). It is during the late Paleogene that primates began to regularly co-occur with proboscideans in African fossil sites. The fossil record is very persuasive that unlike many other mammalian groups that came to typify African faunas, the appearance of new proboscidean taxa and their ecological diversification was entirely an Afro-Arabian experience, completely independent of Eurasian immigration episodes in this interval.

SYSTEMATIC PALEONTOLOGY

BARYTHERIOIDEA Andrews, 1906
BARYTHERIIDAE Andrews, 1906
***BARYTHERIUM* Andrews, 1901b**
***BARYTHERIUM GRAVE* (Andrews, 1901d)**
(Tables 2.3, 3.1, and 3.2; Figures 3.1–3.4)

Synonymy—*Bradytherium grave* Andrews, 1901d; *Bradytherium grave* Andrews, 1901c; *Barytherium grave* Andrews, 1901b

Barytheres represent a continuation of the old plesielephantiform succession into the late Paleogene, characterized by only a modest reduction in tooth formulae (Tables 2.2 and 3.2), employment of all adult teeth simultaneously, and fully lophodont molars reminiscent of those in numidotheres. The molars of barytheres have been described as "bilophodont scissors" by von Koenigswald (2016b), in which shear-cutting of food occurs across two sharp blades (transverse loph(id)s) slicing against each other analogous to the action of a scissors, with the main function of the blades related to Phase I of the power stroke in mastication. This masticatory system is similar to the morphology and inferred chewing mechanisms of phosphatheres, daouitheres, numidotheres, and deinotheres. *Barytherium grave* was the first barytherioid taxon to be discovered, at the turn of the previous century (Andrews, 1901a–d, 1904b, 1906). The species is known with certainty only from two late Eocene-early Oligocene North African sites, the Fayum, Egypt and Dur at Talah, Libya (Table 3.1; Andrews, 1906; Arambourg and Magnier, 1961; Wight, 1980; Harris, 1978a; Delmer, 2005; Rasmussen et al., 2008; Seiffert et al., 2008; Sanders et al., 2010a; Jaeger et al., 2012; Sanders and Seiffert, 2014). *Barytherium* aptly means "heavy beast," doubly emphasized by Andrews' (1901d) use of *grave* for the species nomen, and it is distinguished as the oldest fossil proboscidean of truly elephantine size.

The first recovered remains of the species derive from the Qasr el Sagha Fm. in the Fayum (Andrews, 1906) and are rare elements in the fauna from that geological unit, co-occurring with more common remains of *Moeritherium lyonsi* (Figure 3.2A; Harris, 1978a). The late Eocene Qasr el Sagha Fm. conformably overlies the rich Birket Qarun Fm. and unconformably underlies the Jebel Qatrani Fm. of latest Eocene-early Oligocene age (Kusky et al., 2011). The Qasr el Sagha Fm. is divided into lithologically distinctive units; the lower 123 m comprise the Temple Mb. of glauconitic to limonitic arenaceous limestones, fine quartz sandstones, laminated sandy mudstones, and carbonaceous mudstones with abundant gypsum, while the upper 77 m comprise the Dir Abu Lifa Mb. made up of sandstones, mudstones, and limestone (Bown and Kraus, 1988). Paleomagnetic sequence stratigraphy suggests that the proboscidean-bearing Dir Abu Lifa Mb. of the Qasr el Sagha Fm. was deposited within the interval of 36–34 Ma (Kappelman, 1992; Kappelman et al., 1992; but see Seiffert et al., 2005b, Seiffert, 2006 for a revised interpretation of the initial paleomagnetic sequencing of the Fayum; Figure 3.2A). The depositional environment of the Dir Abu Lifa Mb. was a tide-dominated, deltaic system with offshore shallow marine sandstones, inshore sandstones (including alluvial lateral accretion sandstone) and mudstones, and mudstone channel deposits (Bown and Kraus, 1988; Kappelman, 1992; Legler et al., 2013). Notably co-occurring with the proboscideans in the formation are

TABLE 3.2
Adult Dental Formulae and Incisor Dominance in Late Paleogene African Proboscidea

Taxon	Upper Permanent Dentition	Largest Upper Incisor	Lower Permanent Dentition	Largest Lower Incisor
Barytherium grave	I1–2, P2–4, M1–3	I2	i1–2, p2–4, m1–3	i1
Arcanotherium savagei	I1–3, C, P2–4, M1–3	I2	i1–2, p2–4, m1–3	i2≥i1?
Omanitherium dhofarense	?I, ?C, P2–4, M1–3	Probably I2	i1–2, p2–4, m1–3	i2
Moeritherium spp.	I1–3, C, P2–4, M1–3	I2	i1–2, p2–4, m1–3	i2
Chilgatherium harrisi	?I, ?C, P3–4, M1–3	?	?i, ?c, ?p2, p3–4, m1–3	?
Palaeomastodon spp.	I2, P2–4, M1–3	I2	i2, p3–4, m1–3	i2
Phiomia spp.	I2, P2–4, M1–3	I2	i2, p3–4, m1–3	i2
Losodokodon losodokius	?I, ?C, ?P2, P3, P4, M1–2 (presumed M3)	?	?i, ?c, ?p2, p3–4, (presumed m1) m2–3	?
Eritreum melakeghebrekristosi	—	?	i2, ?p3–4, m1–3	i2

? = Presence unknown.

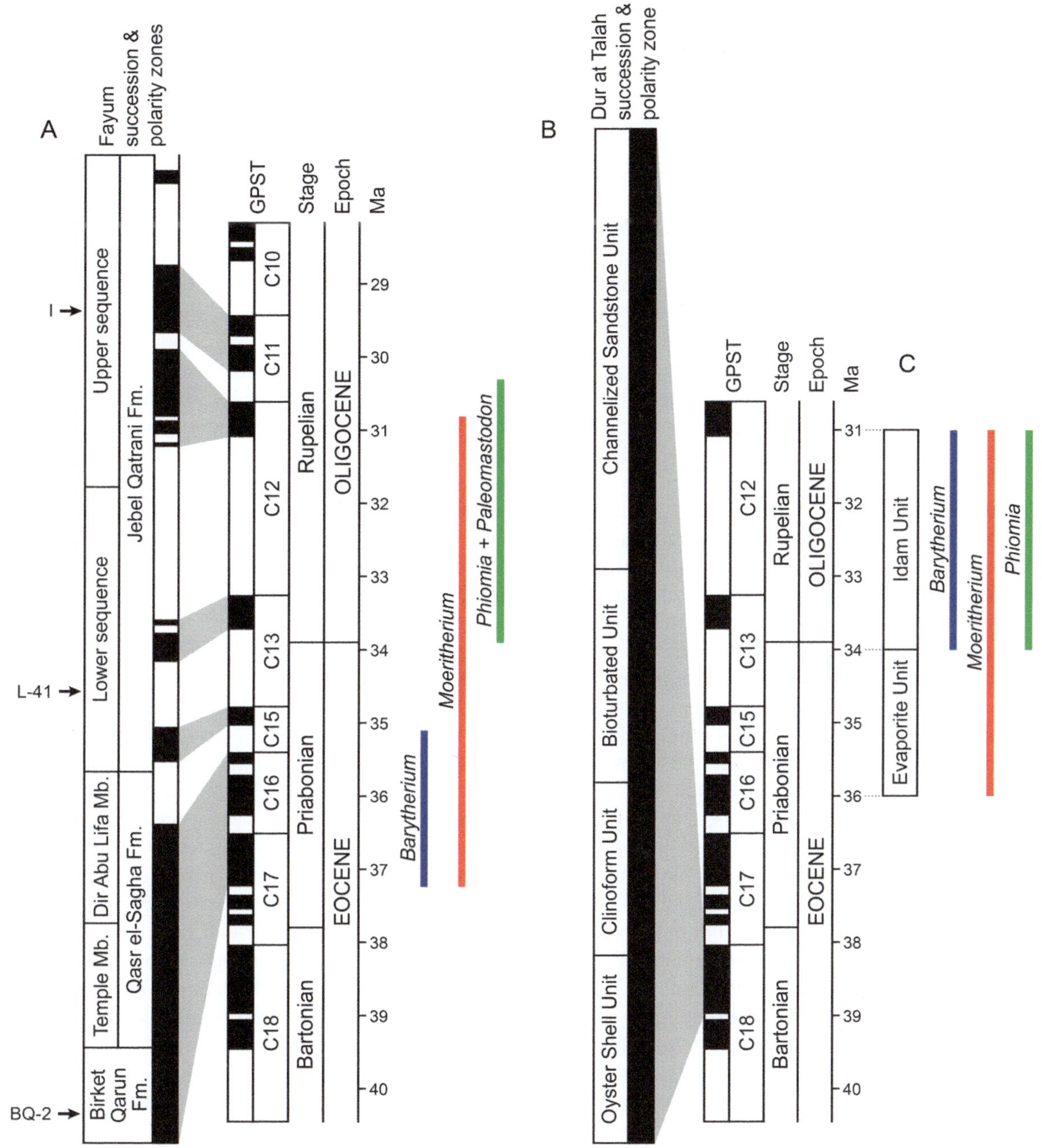

FIGURE 3.2 Comparative chronostratigraphic schema of the Fayum, Egypt and Dur at Talah, Libya. The ages of Fayum localities are based on robust sequence stratigraphy, magnetostratigraphy, and faunal calibration with Omani fossil sites Thaytiniti and Taqah, following Seiffert (2006). (A) Fayum sequence from Simmons et al. (2016:fig. 2) and distribution of proboscidean genera. (B) Dur at Talah sequence from Jaeger et al. (2010a:fig. 1). (C) Dur at Talah sequence and distribution of proboscideans according to Wight (1980). Note that the paleomagnetic sequence in (B) is a single black bar denoting normal paleomagnetism and does not constitute a "sequence"; it could be equally calibrated with GPST C13 or 15 as with C18, for example, and therefore could indicate Priabonian or Rupelian ages as well as Bartonian. The proboscidean taxa at Dur at Talah biochronologically support a younger age for the Evaporite and Idam (=Bioturbated) Units. (Image from Simmons et al. [2016] with permission of the American Museum of Natural History. Image from Jaeger et al. [2010] with permission of Springer Nature BV).

archaeocete whales (Gingerich, 2010), archaic sirenians in the genus *Eosiren* (Domning et al., 2010), pliohyracid hyraxes, creodonts, and the first recorded anthracotheres of the Fayum (Holroyd et al., 1996), as well as abundant remains of marine invertebrates (Kusky et al., 2011). Although Coppens et al. (1978) mentioned palaeomastodonts in the formation, Holroyd et al. (1996) did not list these proboscideans as being present in their review of the Qasr el Sagha non-marine fauna, nor did Simons (1968) in his earlier summation of Fayumian mammals. Specimens housed in the University of California Museum of Paleontology that are the basis of the report of Eocene-aged palaeomastodonts in the Qasr el Sagha Fm. are cataloged as deriving from Quarry B, which is in Gagnon's (1997) Fayum faunal zone FFZ 2, low in the Jebel Qatrani

Fm., not the Qasr el Sagha Fm., and are now dated to the earliest Oligocene (Figure 3.2A; Seiffert, 2006).

In contrast, barytheres are more common than moeritheres in the Idam Unit at Dur at Talah, Libya and their best record is from that site (see Jaeger et al., 2008, 2012; Delmer, 2005, 2009). Based on these occurrences, *Barytherium grave* can be considered a late Eocene-earliest Oligocene species (Sanders et al., 2010a). Fossil collecting at Dur at Talah commenced with oil exploration in the area of the site in the 1950s. From these collections, the first proboscideans from the site (moeritheres, barytheres) were identified, along with crocodiles, dyrosaurs, and a diverse sample of fish, snakes, and turtles (Arambourg and Magnier, 1961; Arambourg, 1963). A scientific expedition in the late 1960s from the University of Bristol (Savage, 1969) led to additional taxa being identified, including archaeocete whales, archaic sirenians, hyraxes, creodonts, rodents, and palaeomastodonts (Wight, 1971, 1980; LeBlanc, 2000). A brief American-led survey of the site in the mid-2000s also recovered palaeomastodonts, moeritheres, barytheres, arsinoitheres, probable hyraxes, long-nosed crocodiles, and abundant fish, particularly catfish, lungfish, sawfish, and gars (Rasmussen et al., 2008). The crocodilian taxa from the site, comprising gavialids, tomistomines, and a mesosuchian, share close affinity with crocodiles from the Qasr el Sagha and Jebel Qatrani Fms. of the Fayum (Agrasar, 2004a). More intensive efforts by a French-led expedition in the latter half of the 2000s led to the recovery of a considerable, significant new fauna, including a wealth of micromammal specimens from rodents, elephant shrews, bats, hyraxes, and, notably, strepsirrhine and anthropoid primates, along with a very abundant sample of barythere cranial and postcranial remains (Jaeger et al., 2008, 2010a; Grohe et al., 2012; Tabuce et al., 2012). For the African late Paleogene, the accumulated mammal assemblage from Dur at Talah is second in importance in its richness and diversity only to the fossil beds of the Fayum, Egypt.

The site comprises an escarpment approximately 120 km in length with about 100 m of relief, cut on its south face by wadis (Savage, 1971). Its stratigraphy was superbly detailed by Wight (1980), who divided the sediments, from base to top, into three units, Evaporite, Idam, and Serir (largely abiotic) and correlated the first two units closely with the Qasr el Sagha and Jebel Qatrani Fms. of the Fayum (see Figure 3.2A and C). To Wight (1980), the suite of marine pelecypods and gastropods, as well as the composition of the vertebrate assemblage, indicated a late Eocene age for the Evaporite Unit. The facies of the Evaporite Unit are very nearshore marine, or littoral marine, with rich oyster banks and large deposits of evaporites; the facies of the Idam Unit include intertidal to littoral environments as well as fluviodeltaic sediments, and particularly resemble the Jebel Qatrani Fm. in preserving distinctive limonitic casts of water-lily fruits; and the Serir Unit is composed of sandstones that furnish a hard cap to the subjacent sediments (Wight, 1980; Jaeger et al., 2008).

A somewhat different interpretation of the geology was made by Abouessa et al. (2012), who placed the Evaporite and Idam Units into a "New Idam Unit" while retaining the uppermost beds separately as the "Sarir Unit." These authors considered the New Idam Unit to represent shallow marine and tidal system deposits and the Sarir Unit to represent regressive, transitional marine into fluviodeltaic depositional environments. However, they concurred with Wight (1980) that these deposits are late Eocene to early Oligocene in age.

More controversial is Jaeger et al.'s (2010a, b) inference, based on their interpretation of the primate and rodent biochronology, that the Dur at Talah sequence (which they divided into Oyster Shell, Clinoform, Bioturbated, and Channelized Sandstone Units, from base to top, with the most important fossil-bearing levels in the Bioturbated Unit) is *middle* Eocene (late Bartonian) in age (Figure 3.2B), described as equivalent in age to Bir el Ater, Algeria and older than the Birket Qarun Fm., Egypt. Moreover, they interpreted their palaeomagnetic data to support this age and to indicate correlation with Chron 18n.1n, dated to 39–38 Ma. The Clinoform and Bioturbated Units have been equated with Wight's (1980) Evaporite and Idam Units, respectively (Jaeger et al., 2012). If these authors are correct, the oldest indisputable anthropoids in Africa are from Dur at Talah.

However, a number of lines of evidence challenge the interpretation of Jaeger and colleagues that the main fossil-bearing units of Dur at Talah are middle Eocene in age. To begin, the published paleomagnetic section of Jaeger et al. (2010a:fig. 1c) is a single black bar denoting normal polarity for the entirety of the sequence (Figure 3.2B), which does not constitute paleomagnetic sequence stratigraphy and therefore is not helpful in refining calibration of the chronostratigraphy of the site (Figure 3.2B). Because there are few late Paleogene African sites with limited numbers of overlapping mammalian species among them, it is difficult to biochronologically date site horizons (Sallam and Seiffert, 2016). Nevertheless, while the rodent fauna of Dur at Talah appears to be composed of more primitive species than terminal Eocene locality L-41 at the base of the Jebel Qatrani Fm. in the Fayum (Jaeger et al., 2012), they are clearly more derived than the hystricognathic rodents of the Birket Qarun Fm. (see Antoine et al., 2012), dated to about 37 Ma, and therefore, it may be more reasonably concluded from this evidence that Dur at Talah is intermediate in age between the Birket Qarun Fm. and L-41 (Sallam and Seiffert, 2016). The macroscelid *Nementchatherium* from Bir el Ater is more primitive than its congener at Dur at Talah (Tabuce et al., 2001, 2012), providing additional confirmation that Dur at Talah is younger than Bir el Ater. Furthermore, the proboscidean assemblage from the site corresponds more closely to Wight's (1980) age estimate for the Evaporite and Idam Units as late Eocene-early Oligocene (Figure 3.2C). Although Jaeger et al. (2012:p. 252) state that the proboscidean taxa "occur in all levels of the Clinoform and Bioturbated units and are not specific to any part of them," analysis of Savage's 1968–69 sample suggest that *Barytherium grave* and palaeomastodonts are only unquestionably found in the Idam (or Bioturbated) Unit, while the moeritheres belong to *Moeritherium lyonsi* and *M. trigodon* and are present successively in the Evaporite and Idam Units. Along with the presence of late Eocene-Oligocene arsinoitheres in the Idam Unit, the distribution of these taxa at Dur at Talah indicates that a middle Eocene age for the site is untenable (Figure 3.2; Sanders and Seiffert, 2014).

Cranial morphology is well documented (but previously poorly accounted) in *Barytherium grave*, primarily from Libyan specimens (Shoshani et al., 1996; Delmer, 2005; Jaeger et al., 2012). A fragmentary cranium from the Birket Qarun Fm. in Egypt (DPC 20844) exhibits a strongly raised basicranium, relative to the level of the palate, but more complete specimens from Dur at Talah show that the lateral outline of the cranium overall is low with a dorsally sloping profile descending anteriorly, downturned rostrum, and retracted nasal opening (Figure 3.3C). This profile recalls the outline of later *Prodeinotherium* crania (Figure 3.3D), a resemblance which might be meaningful phylogenetically. Cranial bones in *Barytherium* also exhibit extensive pneumatization, which is a hallmark of more advanced proboscideans. In ventral view, the cranium is broad with widely flared zygomatic arches and glenoids. It differs markedly from the condition in the cranium of the sympatric barytherioid *Arcanotherium* from Dur at Talah (Delmer, 2009), which is vertically higher but has a less raised basicranium and a shorter rostrum with more anterior position of the nasal aperture (Figure 3.3B). Although Jaeger et al. (2012) stated that retraction of the nasals in *Arcanotherium* indicates development of a trunk, the nasal aperture and nasal bones are not drawn back particularly far, and thus it is doubtful that this taxon had a prehensile trunk. Conversely, in *Barytherium* the nasal aperture is more retracted (Shoshani et al., 1996), consistent with development of at least a small, free proboscis. Additionally, in *B. grave*, the orbits are located anterior to the premolars in line with the nasal aperture (Delmer, 2005; very similar to the condition in *Moeritherium* and probably an aquatic adaptation), but in *Arcanotherium*, they are located more posteriorly above the anterior molars. The upper dental formula in *B. grave* is also reduced compared to that in *Arcanotherium*, specifically in the incisor-canine arcade (Table 3.2). In *B. grave*, there are only two upper incisors in each quadrant, both downturned, and no canine. Size of alveoli indicates that the upper central incisor is considerably smaller than the I2 (Shoshani et al., 1996; Delmer, 2005). I2 specimen 6817/21 from Dur at Talah is highly curved and large (290 mm straight across the curve, including crown and root), with labiolingual and mesiodistal crown dimensions of 67.5 mm and 44.6 mm, respectively (Figure 3.3E). Enamel is preserved on its labial face but does not appear to have been present on the lingual aspect of the crown (Figure 3.3F). Because of its large size, in addition to cropping plants, it is tempting to think of the *B. grave* I2s as functioning as defensive devices against predators and weapons for intraspecific competition, in concert with the large, projecting i1s.

The mandible of *B. grave* is massive and compact, with a very elongate diastema between the anterior premolars and lower incisors that internally accommodates incisor roots of considerable length (Figure 3.3J). Its fused symphysis internally extends as posteriorly as p4 or m1 (measuring 207.4 mm in Dur at Talah specimen 6958/1 (M 92403)) (Figure 3.3I). The corpus is deep, with a distinct blunt protuberance at its base near the posterior of the symphysis. A large mental foramen may be found above this protuberance. The presence of a single mental foramen is possibly a derived feature in Proboscidea shared by *Barytherium* with *Daouitherium* and *Numidotherium* (Ferretti and Debruyne, 2011). *Arcanotherium* and *Omanitherium* also appear to possess a single mental foramen in each dentary, though this is not certain due to missing bone in the anterior corpus of *Arcanotherium*. In the senescent type individual (C10012) from the Qasr el Sagha Fm., the protuberance is located anterior to the p2 position (Andrews, 1906:pl. 17, fig. 5), whereas in the young adult individual 6958/1 from Dur at Talah, the protuberance is positioned more posteriorly, below the premolar row. These differences may be attributed to severe tooth crowding and remodeling of the jaw in the older individual, which though part of the type specimen suggests that it is atypical for the species. Anteriorly, the incisors are set well below the level of the cheek tooth row (Figure 3.3J); this stepped anterior segment is characteristic of barytherioid species. The incisors are procumbent and the i1 is clearly larger than the i2 (Shoshani et al., 1996). The i1s are closely appressed to one another in the midline. In transverse cross-section, the i1 is pear-shaped with its greatest thickness, and enamel is only found on the labial and lateral aspects of the tooth, but not the lingual surface (Shoshani et al., 1996; Pickford 2015b). Labiolingual and mesiodistal measurements of the left and right i1s in specimen 6958/1 are 28.2 and 26.5 mm, and 31.9 and 26.6 mm, respectively. These incisors are relatively smaller than the largest lower incisors in *Omanitherium* and *Arcanotherium* (Pickford, 2015b). The ramus originates lateral to m2, and the coronoid process, which appears to be at the same height as the mandibular condyle, is situated well above the tooth row (110 mm above the molars in the type specimen). There is only a shallow coronoid depression between the process and condyle, and the anterior margin of the ramus is slighted canted forward.

As in most early Paleogene proboscideans and other barytherioid species, *B. grave* has bilophodont molars that increase in size posteriorly (Figure 3.3G–I; Table 2.3). Based on tooth size, it is evident that it is the largest known species of plesielephantiform proboscidean. Upper molars initially have chisel-like wear facets transversely across their proto- and metalophs, curved posteriorly at their lateral edges, indicating dominance of a deinothere-type Phase I chewing mechanism (von Koenigswald, 2014, 2016a, b). With greater dental use, wear patterns across the lophs open up dentine loops, most prominently on the lingual side in upper molars, suggesting addition of more Phase II chewing in older individuals (von Koenigswald, 2016b). In M1 and M2, a crescentoid descends medially from the protocone into the floor of the first transverse valley; the postcingulum is actually an elongate crescentoid attached to the hypocone, enclosing a diminutive talonid basin. Shorter crescentoids are present on the buccal side of

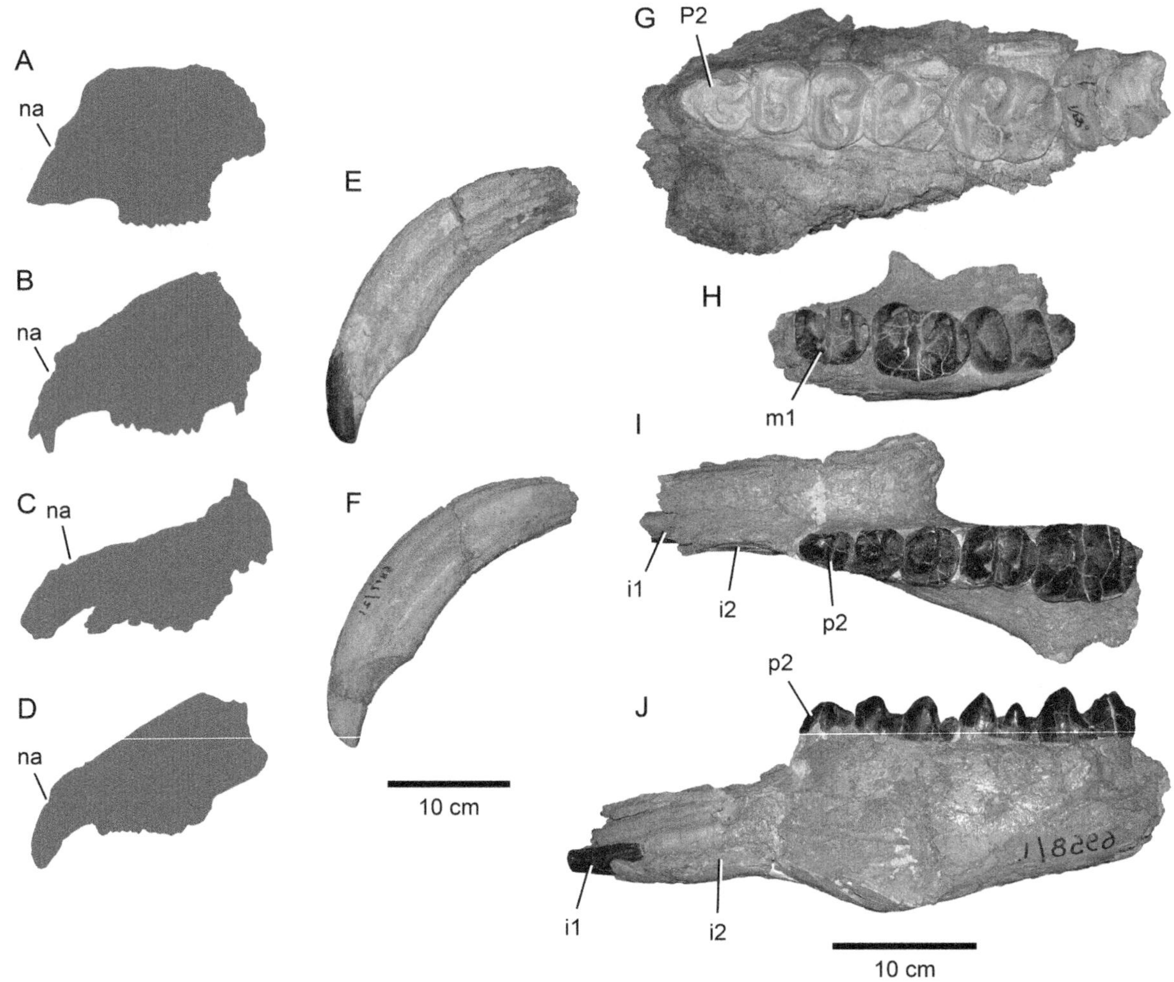

FIGURE 3.3 Comparative numidothere, barythere, and deinothere cranial silhouettes and gnathodental remains of *Barytherium grave* from the early Oligocene Idam Unit at Dur at Talah, Libya. Abbreviations: i, lower incisor; m, lower molar; na, nasal aperture; P/p, upper/lower permanent premolar. (A–D) Silhouettes represent crania of different sizes but are placed in the same scale to facilitate comparison. (E, F) to the same scale. (G–J) to the same scale. Anterior to the left. (A) Lateral view, cranial silhouette of *Numidotherium*. (B) Lateral view, cranial silhouette of *Arcanotherium*. (C) Lateral view, cranial silhouette of *Barytherium*. (D) Lateral view, cranial silhouette of *Prodeinotherium*. (E) Laterolabial view, I2, specimen 6817/1. (F) Laterolingual view, I2, specimen 6817/1. (G) Occlusal view, left maxillary fragment with P2-M3, specimen 689-1. (H), Occlusal view, right dentary fragment with m1–3, specimen 695-81. (I) Occlusal view, symphysis and left dentary fragment with p2–m2, specimen 695-81 (antimere of right dentary). (J) Lateral view, symphysis and left dentary fragment with p2–m2, specimen 695-81.

the lophs, as well. The M3 is similar to the first two upper molars but is narrower posteriorly and has less well developed or absent crescentoids.

P2–4 are strikingly short, especially P3–4, and exhibit only a single transverse loph in each (Figure 3.3G). P2 is distinguished by its triangular occlusal outline. In each of these premolars, the transverse wear-figure appears to be bent posteriorly on each side, particularly in P2 which has a metacone and possibly a nascent hypocone—for this reason, its occlusal wear-figure takes the shape of an anteriorly pointed chevron. These cheek teeth also progress in size from anterior to posterior position (Table 2.3).

The lower molars also have transverse lophids which initially wear as chisel-like cutting edges. In an unworn or lightly worn state, the lophids exhibit numerous mammillons, and their edges are curved anteriorly forward. The correspondence of anteriorly and posteriorly curved lophids and lophs in lower and upper molars of "bilophodont scissored" proboscideans such as *B. grave* probably helps hold the food bolus in place during Phase I shearing (von Koenigswald, 2016b). Strong crescentoids descend anteriorly into the transverse valleys from the buccal and lingual sides of the lophids. In m1–2, the postcingulid comprises a low shelf with an enlarged central conelet; the postcingulid of m3 is narrower but more posteriorly projecting and forms a strong heel. The second lower molar is far larger than m1, and as in the upper molar row, size increases from m1–3 (Figure 3.3H, I; Table 2.3).

The p2 is triangulate in occlusal shape with a tall, large cusp anteriorly and lower cusp posteriorly; p3 is more squared in shape, as is p4, and has a transverse anterior lophid, large hypoconid, and nascent buccal ectolophid, whereas p4 is more fully bilophodont. Crown size increases from p2–4 (Figure 3.3I; Table 2.3). Oxygen and carbon isotope values of Birket Qarun barythere teeth suggest that these proboscideans were using their scissor-like teeth to browse freshwater plants as amphibious denizens of aquatic habitats (Liu et al., 2008). However, while indicating diets of aquatic vegetation, $\delta^{13}C$ values for barytheres and moeritheres from Birket Qarun show little overlap (Liu et al., 2008:fig. 2A), suggesting competitive exclusion and exploitation of distinct aquatic habitats. Compared with living mammals, in size and habitus barytheres appear most like hippos.

Postcrania is poorly known for Fayumian *B. grave* and was originally described by Andrews (1906), based on a very few Qasr el Sagha Fm. specimens. This material is supplemented by a partial skeleton from Dor at Talah, BMNH 6817 (Savage, 1969; Delmer, 2005; see Figure 3.4). The Fayum scapula is large (+480 mm in length) and of general proboscidean configuration, marked by an exceptionally large, downturned supraglenoid process that anchors muscles inserting on the powerful deltoid ridge of the humerus (Shoshani et al., 1996). The width of the glenoid articular surface is 130 mm. The barythere scapula is also characterized by a narrow supraspinous fossa. The humerus is considerably longer than the radius, 604 mm versus 425 mm (570 mm versus 450 mm in the Libyan skeleton), and exhibits a salient bicipital groove, very prominent deltoid ridge, very broad, flat epicondyles distally, and a deep olecranon fossa. The head faces more medially than superiorly. The radius has two articular surfaces for the humerus, a large flat articulation for the ulna (indicating it was habitually held in pronated position), and a broad distal articular surface for the lunate and scaphoid bones of the carpus.

Although incomplete, the proximal features of the femur are elephant-like in proportion and upward facing, indicative of a graviportal posture, although distal femoral and proximal tibial configuration suggest a semi-flexed normal position of the bones (Figure 3.4). It also features a broad, anteroposteriorly flattened shaft, small lesser trochanter, and third trochanter. However, the distal condyles face more posteriorly than downward, indicating a bent knee posture. The femur from the Libyan skeleton is 775 mm in length, and the accompanying tibia is 480 mm long, producing a tibial-femoral index of 62, similar to that of elephants (Delmer, 2005). An unusual feature of this proboscidean is that the sustentacular facet is larger than the ectal facet in the astragalus (Shoshani et al., 1996). Astragalus DPC 20842 from the Birket Qarun Fm. is about twice the size of moerithere astragali, with a greatest length and width of 110 mm, and has a considerably larger sustentacular (L = 70.0 mm × W = 50.5 mm) than ectal (L = 50.0 mm × W = 26.5 mm) facet. The ectal facet projects posteriorly and is oval, while the sustentacular facet is more rectangular or trapezoidal in shape. The navicular articular surface is 94.5 mm wide and 62.3 mm high, and is accompanied by a secondary articular surface for the cuboid on the lateral side of the anterior face of the specimen. The medial tubercle is impressive in size and projects (Delmer, 2005). A partial calcaneum from the Fayum (UM 13973) has a large fibular facet, a prominent ventral tuberosity below the fibular facet, and a transversely compressed neck (Sanders et al., 2010a). It differs from the calcaneum of *Numidotherium* by the size of its more expanded, long tuber calcis. Reconstruction of the arrangement of podials and metapodials indicates that the manus was plantigrade, whereas the pes was more digitigrade in posture (Delmer, 2005).

Larramendi's (2016) body size calculation for *B. grave*, based on volumetric study of these postcrania, indicates that this was a very large animal, with a shoulder height of about 180–200 cm and mass of over two tons. Configuration of

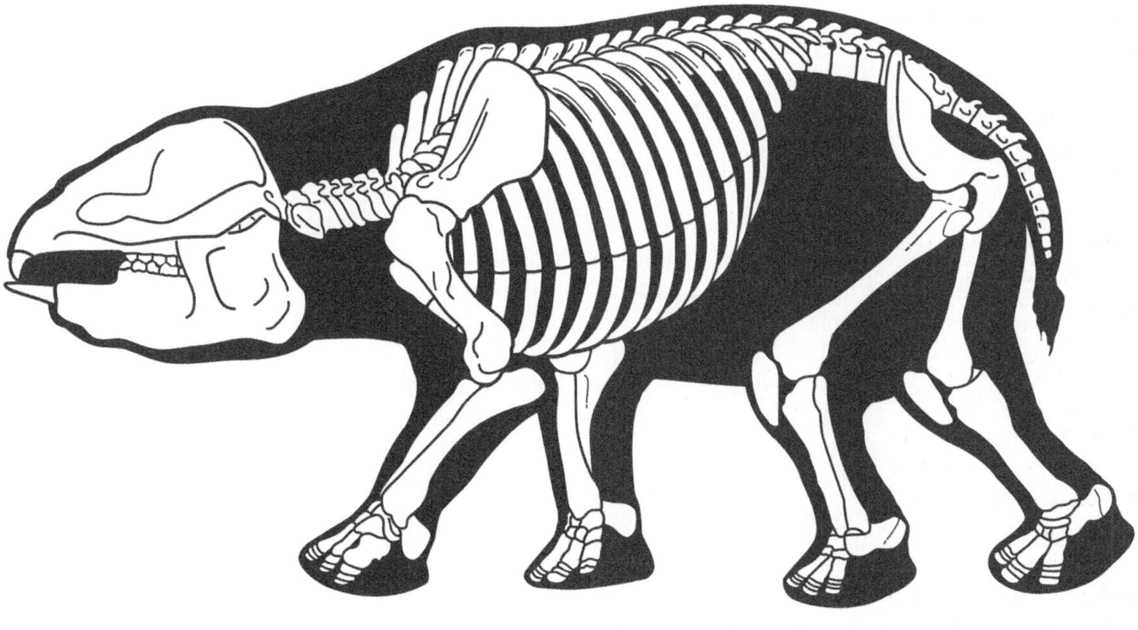

FIGURE 3.4 Body outline superimposed behind skeletal reconstruction of late Eocene-early Oligocene *Barytherium grave*. (Based on the reconstruction of M. Orliac and C. Delmer in Delmer [2005], with permission of C. Delmer).

the capitulum and tuberculum of ribs suggests that the thorax was barrel-shaped, concordant with a massive browser (Figure 3.4).

The relationships of *B. grave* are still not well established. Although Andrews (1901a) felt that the species might be a deinothere-like proboscidean, he placed it in Barypoda (Andrews, 1904b). He shortly thereafter changed his mind and formally included barytheres in the Proboscidea (Andrews, 1906), but he remained unsure of their affinities. Remarkably, influenced early on by Andrews' (1904b) placement of barytheres in the Barypoda, Osborn never changed his view on the subject and failed to even consider the issue in his two-volume compendium on proboscideans (1936, 1942; see discussion in Domning et al., 1986). Simpson (1945) firmly attributed barytheres to the Proboscidea in their own suborder, Barytherioidea. Alternatively, Derinyagala (1955) felt that species should not be placed in the Proboscidea if it could not be demonstrated that they had had probosces, and therefore removed *B. grave* from the order. At the time of Harris' (1978a) review of barytheres, they still seemed sufficiently strange compared to elephantoid proboscideans that he withheld judgment on their ordinal affinities. With an expanded fossil record of proboscideans into the early Paleogene and character-based phylogenetic analyses, it is now clear that *B. grave* and other barytheres belong in the Proboscidea.

Domning et al. (1986) published the first cladistic study that included barytheres, providing nine characters that they felt supported identification of *B. grave* as a proboscidean, including development of a proboscis, enlarged scapular coracoid process, graviportal stance, enlargement of i2, and frontopremaxillary contact. Other putative synapomorphies of the order listed by these authors are now shown to be incorrect (e.g., loss of the upper canine, loss of I3, loss of i1). Shoshani and Tassy (1996) and Shoshani et al. (1996) added to these proboscidean synapomorphies such features as an astragalar tuberculum mediale, m3s with two distal cusps, loss of p1, radius fixed in pronated position, pneumatization of the cranium, elevation of the nares (associated with a proboscis), and anteroposterior flattening of the femur. Loss of p1, i1, I3, and canines, are now understood as derived features within Proboscidea, and not synapomorphies uniting the entire order (see Tables 2.2 and 3.2).

Abel (1914, 1919, 1920) expressed the possibility of a close relationship of barytheres to deinotheres. However, because barytheres have simpler upper premolars, P2/p2s, and first molars with only two loph(id)s, compared with deinotheres, Harris (1978a) felt that they had very different masticatory specializations that precluded an ancestral-descendant bond between them. Whether deinotheres are derived from barytheres remains to be explored more rigorously, but the similarity of cranial profiles and shared presence of only 3D enamel prisms ("Schmelzmuster"; Tabuce et al., 2007b) are features that potentially support a hypothesis that they have a sister-taxon relationship. Conversely, differences in cranial configuration and tooth formula between *Barytherium* versus *Arcanotherium*+*Numidotherium* (Figure 3.3A and B) suggest that the latter two taxa are more closely allied phylogenetically to one another than either is to *Barytherium*. However, *Arcanotherium* appears more derived in Schmelzmuster pattern than does *Barytherium* (Delmer, 2009). There is less evidence available to assess the phylogenetic systematics of *Omanitherium* relative to these taxa, but based on the dominance of i2 in *Arcanotherium* and *Omanitherium*, and more complete covering of their lower incisors with enamel, Pickford (2015b) felt that it belonged in the *Arcanotherium*+*Numidotherium* clade. This hypothesis can be further tested by recovery of crania of *Omanitherium*, and Schmelzmuster analysis of its dental enamel.

ARCANOTHERIUM Delmer, 2009
ARCANOTHERIUM SAVAGEI (Court, 1995)

(Tables 2.3, 3.1, and 3.2; Figures 3.1, 3.3, and 3.5)

Synonymy—*Barytherium* sp. Harris, 1978a; *Barytherium* sp. Wight, 1980; *Barytherium* sp. nov. Shoshani et al., 1989; *Numidotherium savagei* Court, 1995; small *Barytherium* Shoshani et al., 1996

This species, with a somewhat complicated taxonomic history, is part of a modest diversification of barytherioids within a larger radiation of proboscidean taxa during the late Paleogene. *Arcanotherium savagei* is documented only at the site of Dur at Talah in Libya (Table 3.1), where it is found in both the late Eocene Evaporite Unit and early Oligocene Idam Unit, contemporaneous with moeritheres, other barytherioids, and palaeomastodonts (Delmer, 2009). It was originally identified as *Barytherium* sp. (e.g., Harris, 1978a; Wight, 1980). Court (1995) subsequently transferred it to the genus *Numidotherium*, based on skeletal resemblances to *Numidotherium koholense*, and named it in honor of R.J.G. Savage, who was instrumental in establishing the paleontological importance of Dur at Talah. More recently, Delmer (2009) erected a new genus for *N. savagei*, "*Arcanotherium*," placing greater emphasis on the differences than similarities between this species and *N. koholense*.

Court (1995) separated "*N.*" *savagei* from *B. grave* and allied it as the sister species to *N. koholense* based on a large number of features that differed from barytheres in the same manner for both species within his numidothere clade, including labiolingual compression of central lower incisors fully clad with enamel, more buccal location of cristid obliqua and lower trigonid than talonid cusps in lower molars, shallower mandibular corpus, more posterior origin of the ramus, scapular coracoid process medially-deflected rather than hook-like, presence of a fovea capitis femoris on the femoral head, very asymmetric distal femoral condyles, and ectal facet more equivalent in size to the sustentacular facet in the astragalus. These features are sufficient to conclude that "*N.*" *savagei* is not just a small variant of *Barytherium*.

Differences between "*N.*" *savagei* and *N. koholense* are primarily related to greater size and more derived gnathodental morphology of the former, including larger teeth

(Table 2.3), a nascent metaconid in p2 and metaconid well developed and paired with the protoconid in p3, anterior margin of the ramus located between m2 and m3, much wider lunate bone, and more robust ulna (Court, 1995). Delmer's (2009) erection of the new genus *Arcanotherium*, derived from the Latin word for enigmatic, is justified in his emended diagnosis of the species, which provided supplementary distinguishing features such as premolars more molarized than in barytheres and all early Paleogene proboscideans, upper molars with pronounced lingual and mesial cingula and a small paraconule anterior to the protoloph, and cheek tooth dentition less lophodont than in numidotheres and barytheres in having distinguishable cusp apices slightly higher than the crests between them.

The lower dentition of the species is particularly well preserved (Figure 3.5) and has a mandibular dental formula equivalent to that in barytheres, numidotheres, and moeritheres (Tables 2.2 and 3.2). Lower premolars are nearly the same in length but p4 is largest in overall crown dimensions. They are all double-rooted though the roots of p4 are more anteroposteriorly splayed than in p2–3. In each the protolophid is much higher than the shelf-like hypolophid. In p2, the protoconid and metaconid are fused to one another and rounded; the protoconid is much larger than the metaconid, and distally the hypoconid is considerably larger than the nascent entoconid. The p2 has a squat, semi-triangular crown shape. Lower p3–4 are more molarized but the protolophid remains dominant over the hypolophid in these premolars, and forms a true lophid, while the hypoconid and entoconid remain individualized. The protoconid and hypoconid are larger than the metaconid and entoconid in p3–4.

The lower molar series increases in size from m1 to m3, especially between m1 and m2. Although the cusp apices are rounded and distinguishable, they are incorporated into the lophids and the molars can be described as true *lophodont* teeth (and not *bunolophodont* as in palaeomastodonts or moeritheres, in which cusps on each side of the molar may be closely appressed on one another to form lophids with wear, but remain individualized with a midline sulcus between them when unworn). Cristids are well developed on the anterior pretrite sides of the lower molar crowns, angling mesiolingually into the transverse valleys; they are incorporated into occlusal dentinal wear-figures with wear. Unlike molars of numidotheres and barytheres, the *Arcanotherium* lower molars lack preentocristids or premetacristids (Court, 1995). Each lower molar is bilophodont, with the associated postcingulid increasing in size through the series. In m3, the postcingulid is transversely narrower than the lophids but constitutes a very extensive distal shelf that can almost be considered an incipient third lophid (the "hypoconulid lobe" of Court [1995]). In lateral view, the unworn protolophid is only slightly higher than the hypolophid.

Lower incisors of *Arcanotherium* are paired, with the alveoli for i1 and i2 similar in size but deeper for i2 (Pickford, 2015b). As in *Omanitherium*, in *Arcanotherium*, the i2 is probably larger than or similar in size to i1. The find of additional root fragments of the i1s of 68.21/10 demonstrates that these lower tusks had relatively short roots, probably

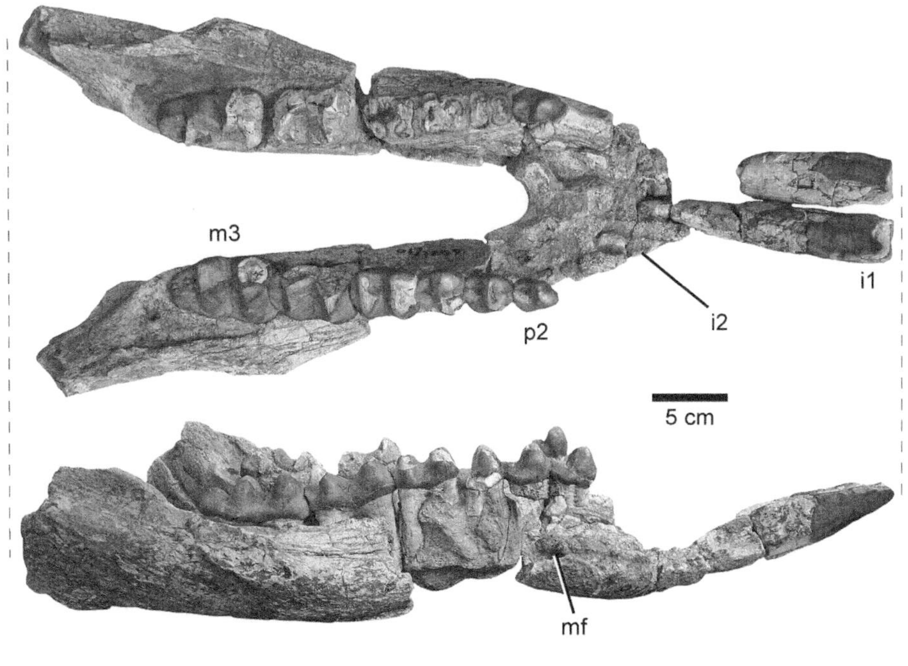

FIGURE 3.5 Dorsal and lateral views, mandible with right and left i1, right p2–m3, left p2 and partial left m2–3 of late Eocene *Arcanotherium savagei* from Dur at Talah, Libya (BMNH M82163) (type specimen). Abbreviations: i, lower incisor; m, lower molar; mf, mental foramen; p, premolar. Anterior is to the right. (Images courtesy of M. Pickford [Pickford, 2015b]).

did not project very far anterior to the symphysis, and projected anteriorly and slightly upward (Figure 3.5), as well as being angled on one another so that they are lower mesially than distally (Pickford, 2015b). Angulation of the incisors is evidenced by extensive interproximal facets on their mesial margins. The only way that the putative i1s of *Omanitherium* (see below) could be accommodated between the i2s in the lower jaw known for the species is if they were also angled against one another in a similar configuration. The i1s are wider than deep, with the right incisor having a mesiodistal length of 31.0mm, a labiolingual depth of 27.0mm, and a worn height of 52.0mm. Wear on the tips of the incisors suggests that they occluded against the I1s. A series of small, rounded crenulations are present along the distal margin of the i1s. A juvenile *Arcanotherium* symphyseal fragment, 68.6/9, from the Evaporite Mb., was erroneously attached to a palaeomastodont jaw by Wight (1971). This specimen preserves a beautiful i1 with many distal crenulations still partially in its crypt, exposed for a length of 48.4mm. Lateral and dorsal to the i1 is a small fragment of the i2. Immediately *lateral* to the i1 is a fragment of what should be identified as the di2 (Pickford, 2015b). Delmer (2009) identified the deciduous incisor as di1, based on mistakenly interpreting its position as *medial* to the i1.

Pieces of the mandible associated with the adult i1s and cheek teeth are informative. The symphysis is fused, broad, and relatively short, extending posteriorly only to the p2. There is a short diastema between p2 and the posterior aspect of the i2 alveolus. A small coronoid foramen is located posterolateral to the m3. Mandibular form corresponds more closely to *Numidotherium* than *Barytherium* in the relative shallowness of the corpus. Consequently, the lower incisors do not appear as stepped below the line of the cheek teeth as they are in *Barytherium* or *Omanitherium* (Figure 3.5). There is a mental foramen located below p2, but apparently not anterior to that in the corpus.

The account of Dur at Talah taxa by Jaeger and colleagues (2012) provides valuable information about the cranial morphology and upper dentition of *Arcanotherium*. Their work supports Delmer's (2009) and Court's (1995) differentiation of *Arcanotherium* from *Barytherium*. Compared with *Barytherium*, *Arcanotherium* has a dorsoventrally deeper cranium but less elevated basicranium, shorter rostrum, and more posteriorly positioned orbits (located above M1). It is reminiscent of the cranium of *Numidotherium* (Figure 3.3A and B), but its orbits are more posteriorly located, its infra-maxillary fossa is shallower, and its basicranium is not as high above the level of the palate (Jaeger et al., 2012). A salient difference from *Barytherium* is the retention in the upper dentition of *Arcanotherium* of a third incisor and canine, similar to the condition in *Numidotherium* and *Moeritherium*, with which arcanotheres share the same upper dental formula (Tables 2.2 and 3.2). The upper central incisor is downturned and spatulate, whereas the I2 is much more prominent though also downturned. The small size of the alveoli for I3 and C suggest that these teeth were as diminutive as in *Numidotherium* and *Moeritherium*.

There are few upper cheek teeth known for *Arcanotherium*. Delmer (2009) described the morphology of M1–3 as bilophodont with strong lingual cingulae; M1 exhibits a strong postprotocrista and smaller postparacrista, as well as a prominent distocrista extending posteriorly from the metaloph; M2 is larger and has a small paraconule extending anteriorly from the protoloph, a diminutive metaconule in the transverse valley, and well-defined postparacrista, postprotocrista, and posthypocrista; M3 is larger still and more triangular in shape (widest anteriorly), and exhibits a strong precingulum and prominent talon or distal shelf that accommodates a strong distocrista.

Delmer (2009) drew attention to postcranial features of *Arcanotherium* that differ from those of *Numidotherium*. In the ulna, for example, there is a lateral articular facet proximally (also different from *Barytherium* in this feature but resembling the condition in moeritheres and early elephantiform proboscideans), long olecranon process projecting posterosuperiorly (as in *Barytherium*, allowing full extension of the lower forelimb on the humerus), and greater extension of the lower than upper arc of the sigmoid articulation.

Court (1995) placed *Arcanotherium savagei* in *Numidotherium* with *N. koholense* as the most primitive known proboscidean outgroup to *Moeritherium*+(*Deinotherium*+(palaeomastodonts+derived elephantiforms)), but at that time, the sequence of more primitive phosphatheres and daouitheres had not yet been revealed. The cranial morphology and dental formula (Table 3.2) of *Arcanotherium* appear to support a closer relationship of this taxon with *Numidotherium* than with *Barytherium*. However, Delmer (2009) and Jaeger et al. (2012) argued that *Arcanotherium* is either a closer sister taxon to (moeritheres+elephantiforms)+deinotheres, or to elephantiforms directly. Features that Delmer (2009) and Jaeger et al. (2012) believed supported their hypotheses of arcanothere relationships with moeritheres and elephantiforms include the retracted position of the orbit, astragalar morphology, loss of the humeral entepicondylar foramen, derived enamel microstructure, lower molar preprotocristid>premetacristid, absence of m1–2 postmetacristid and preentocristid, upper molars with a paraconule, metaconule, strong lingual cingulum and reduced entoflexus, more molarized premolars than in barytheres and numidotheres, and more "individualized" cusps than in other lophodont taxa such as barytheres or deinotheres. In contrast, Pickford (2015b) hypothesized that *Arcanotherium* is the sister taxon to a clade of *Omanitherium*+*Numidotherium*, supported in part by similarities of dominance of the i2>i1 and full covering of enamel over lower incisor crowns in species of this clade, as well as by similarities in ulnar morphology. Pickford (2015b) identified an ulna from the Aydim Fm. of Dhofar as belonging to *Omanitherium* rather than to *Arsinoitherium*, as was originally reported by Al-Sayigh et al. (2008). The similarity between this ulna and ulnar specimen M 82176 of *Arcanotherium* is striking (see Al-Sayigh et al., 2008:fig. 4; Delmer, 2009:fig. 5A), suggesting that recovery of additional postcrania of *Omanitherium* might help unravel the relationships of these late Paleogene taxa.

In assessing these hypotheses, it is worth emphasizing again that the *bunolophodonty* of moeritheres and palaeomastodonts is very different from the molar *lophodonty* of arcanotheres and chilgatheres, which have true transverse crests that are anchored laterally by rounded cusp remnants which are not "individualized" (or free standing) as claimed but rather are fully integrated into the loph(id)s. Based on the same reasoning, *Chilgatherium* is more appropriately hypothesized as a primitive deinothere or derived barytherioid and its molar morphology does not support a unique sister-group relationship with primitive elephantiform proboscideans. As well, there is little about the dental morphology of moeritheres, including their enamel microstructure (Tabuce et al., 2007b), that suggests a close relationship with arcanotheres.

OMANITHERIUM Seiffert et al., 2012
OMANITHERIUM DHOFARENSE (Seiffert et al., 2012)

(Tables 2.3, 3.1, and 3.2; Figures 3.1, 3.6, and 3.7)

Synonymy—*Omanitherium dhofarensis* Seiffert et al., 2012; *Omanitherium dhofarense* Pickford, 2015b; *Omanitherium dhofarense* Al-Kindi et al., 2017; *Barytherium omansi* Al-Kindi, 2018

As with many other Paleogene proboscidean taxa, because of the vagaries of the geological preservation, the record of *Omanitherium dhofarense* is limited in time and space (Table 3.1) and comprises a modest fossil sample. The original fossil sample constituting this species when it was named was recovered from a single locality (DPP-2010-1) in the Shizar Mb. of the Ashawq Fm. in southwestern Oman (Seiffert et al., 2012). The Shizar Mb. proboscidean fossils derive from gypsiferous marls, which are vulnerable to damage by gypsum growth (Al-Kindi et al., 2017). Despite such limitations, these remains are sufficient to delineate a new barythere species, and have important implications for proboscidean phylogeny. Accompanying the *Omanitherium* remains in the Shizar Mb. are turtle, crocodile, lungfish, bat, rodent, primate, embrithopod, hyrax, and anthracothere fossils, derived from marls, silts, and sands representing alternating supratidal, lagoonal, and tidal flat depositional environments (Al-Kindi et al., 2017). The chronology of the Shizar Mb. is well constrained by magnetostratigraphy and biostratigraphic sampling of marine invertebrates, yielding a secure estimate of 33.7–33.3 Ma for the unit at Thaytiniti and establishing the occurrence of *Omanitherium* as earliest Oligocene in age (Seiffert, 2006; Seiffert et al., 2012).

A few strange, barythere-like teeth recovered previously from the locality of Thaytiniti, Oman in the same geological unit (Thomas et al., 1989, 1999) were thought by Seiffert et al. (2012) to potentially belong to this new species, as well, though they believed that a lower, possibly central incisor from Thaytiniti is more typical of other barythere species. This incisor differs from the i2s of *Omanitherium* by larger size, spatulate shape, and development of lateral serrations (Seiffert et al., 2012). Since the initial work by Seiffert's team, additional fossil collecting has been undertaken at and near the type site, as well as close by at Thaytiniti and Taqah (Pickford, 2015a–c; Al-Kindi et al., 2017), revealing much more about the dental morphology of the species.

The type specimen of *Omanitherium dhofarense* (SQU-290) is a partial mandible with most of the lower dentition of an older juvenile or subadult individual (Figure 3.6; Seiffert et al., 2012). Although p2 is lost, alveolar traces of it are retained, aiding the reconstruction of a lower cheek tooth formula of three premolars and three molars, with all adult teeth simultaneously in occlusion (Figure 3.6A–C). The corpora appear to have been relatively shallow, possibly because of the young age of the individual. Anterior to p2, the corpora make a precipitous drop down to the symphysis (Figure 3.6A). The preserved portion of the left ramus is rooted parallel to the posterior end of m2 and exhibits a tall coronoid process and deeply down-curved mandibular notch. Mental foramina are paired and set closely together below p2. The symphysis is fused and comprises a shallow, broad, and moderately long trough, extending as posterior as p3.

An interesting aspect of the entire specimen is the presence of two procumbent incisors, interpreted as i2s because of the gap between them, socketed at the lateral margins of the symphyseal segment of the jaw (Figure 3.6C and D; Seiffert et al., 2012). They were unerupted, as was m3, at the time of death, and though the symphysis is broken anteriorly in the midline, there is no trace of other incisors in the space between the i2s, or lateral to them. Loss of a lower incisor in each dentary is a derived feature among barytherioids. The i2s of SQU-290 are rounded (mesiodistal width = 24.0 mm; labiolingual thickness = 29.8 mm) and do not exhibit serrations. A second, larger set of incisors with the same morphology was found at the type locality, representing either a second individual or upper incisors of the type individual (Seiffert et al., 2012).

Since the discovery of the type jaw, however, incisors claimed to be i1 and I1 of the species have been found from near the type site and Thaytiniti (Figure 3.7; Pickford, 2015b; Al-Kindi et al., 2017). The i1, ONHM TN 2017-51, has a broken root but preserves a crown shape that is relatively compressed labiolingually, narrow mesiodistally, rectangular in outline and elongate in lingual view (Figure 3.7A). The more complete right i1 has dimensions of mesiodistal breadth (MD) = 31.0 mm, labiolingual thickness (LL) = 22.7 mm, and crown height (H) = 49.0+ mm (Al-Kindi et al., 2017). It would be a very tight squeeze for the type mandible to accommodate i1s of this size between its i2s. The suboval-shaped i2s accompanying the i1s are similar in size (Figure 3.7B), with dimensions of MD = 27.6 and 32.0 mm; LL = 20.7 and 23.0 mm; H = 50.0+ and 56.0 mm. A right I1 from Thaytiniti, ONHM TH 6, initially misidentified by Pickford (2015b) as a left i1, is spatulate with traces of mammillons along its distal margin (Figure 3.7C), and appears to have been downturned in alveolar position to articulate with the procumbent i1 in a pincer-like manner (Figure 3.7D; Al-Kindi et al., 2017:fig. 11). Its dimensions

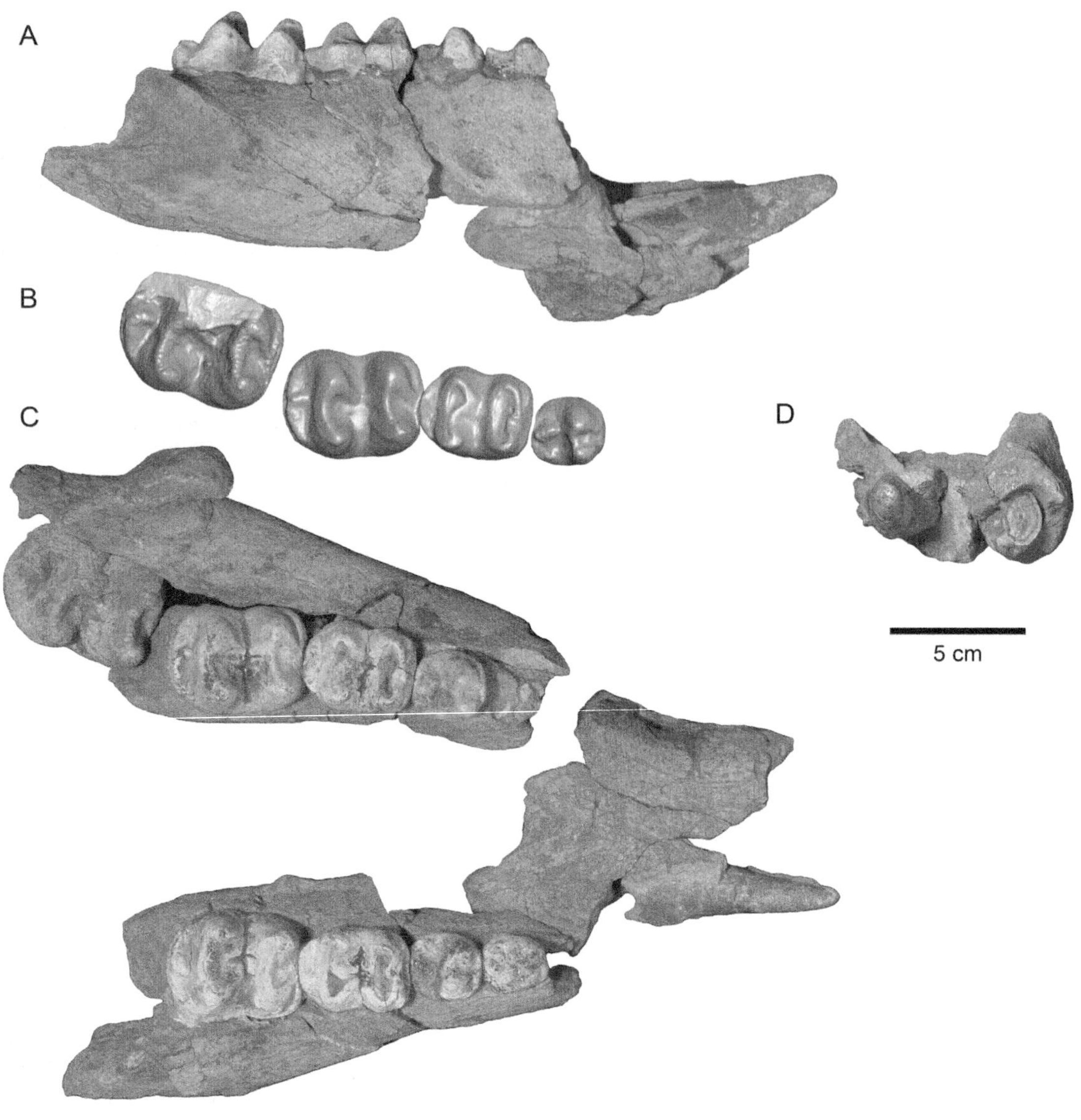

FIGURE 3.6 Partial mandible with right p3–m2, left p4–m3, and i2s emerging from their crypts, of early Oligocene *Omanitherium dhofarense* from Oman (type specimen). (A–D) to the same scale. (A–C) Anterior to the right. (A) Partial mandible, SQU-290, lateral view. (B) Scanned images of left p4-m3, SQU-290, occlusal view. (C) Partial mandible, SQU-290, occlusal view. (D) Right and broken left i2s emerging from their alveolar crypts, SQU-290, anterior view. (Images courtesy of E. Seiffert [Seiffert et al., 2012]).

are MD = 32.0 mm and LL = 22.2 mm (Pickford, 2015b). All incisors have or had an enamel covering. The I1 resembles the upper central incisor of *Arcanotherium*. Unfortunately, the lateral upper incisor is unknown as is the complete upper dental formula.

Molars are bilophodont (though cusp apices can be easily recognized at the edges of loph(id)s), with anteriorly concave proto- and hypolophids marked by numerous mammillons when unworn, strong oblique cristids (in m2–3), and at least m2–3 with centrally located hypoconulids. Molar size progression is m1<m2<m3 (Table 2.3). p3 is squared in occlusal outline and has equivalent sized meta- and protoconids that are closely appressed; the hypoconid is about half the size of the protoconid and the entoconid is much more diminutive. p4 is morphologically similar but slightly larger with taller cusps and a distinct protocristid connecting the proto- and meta-conids. There are no paraconids.

Collections made since 2012 have added critical information about the morphology of upper cheek teeth (Pickford, 2015b; Al-Kindi et al., 2017). Size of these teeth increases progressively from P2 to M3 (Table 2.3). The upper molars are bilophodont with low, diminutive pre- and postcingulae. Unworn transverse lophs may sport numerous mammillons. In P2, the protocone and paracone are bunodont and isolated from one another; there is also a small parastyle lateral to a small precingulum and posteriorly there is a very small, low postcingulum. P3 and P4 have large protocones lingually that are largely independent of buccally-situated ectolophs comprised of interconnected para- and metacones, with large basins between the two sides of the crown, particularly in P4 (Figure 3.7E and F). Pre- and postcingulae are low

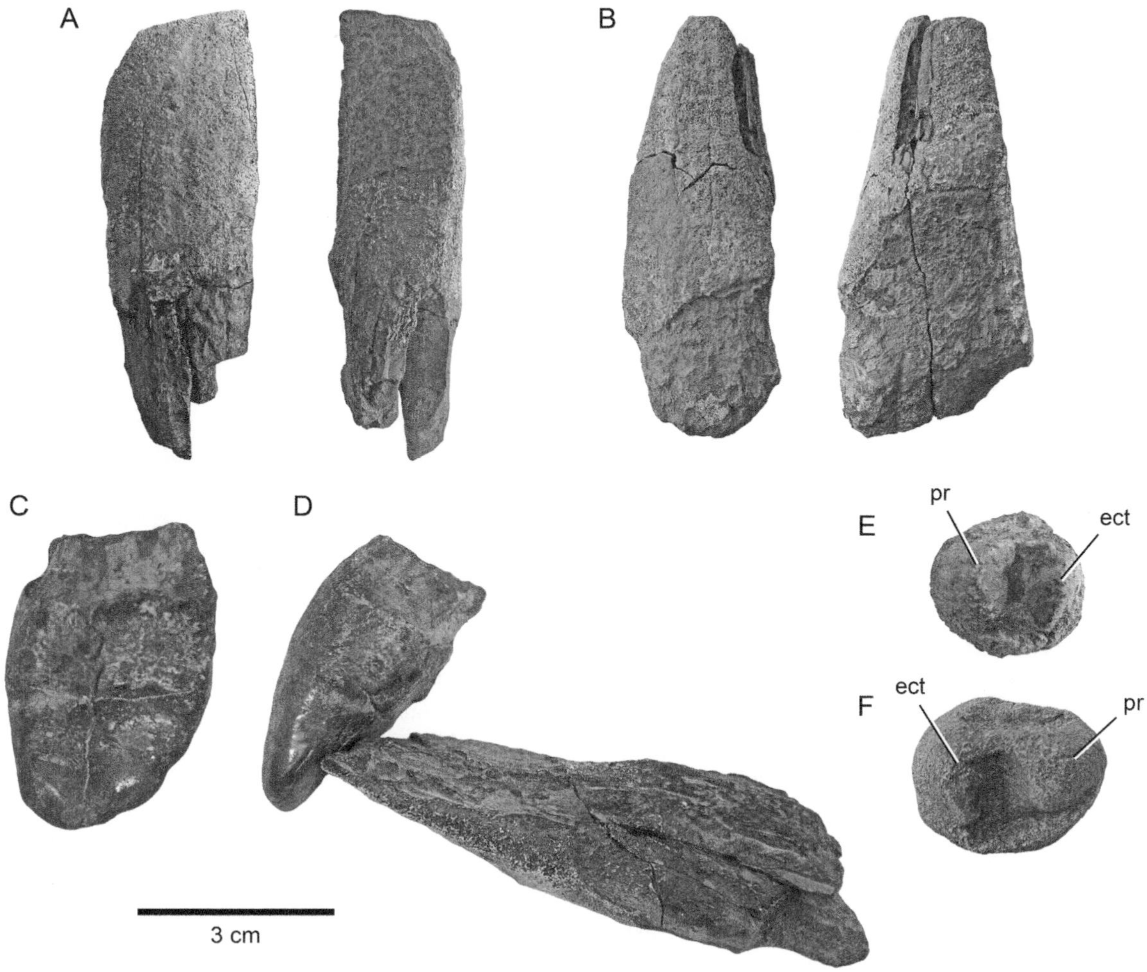

FIGURE 3.7 Incisors and premolars attributed to early Oligocene *Omanitherium dhofarense* from Oman. Abbreviations: ect, ectoloph; pr, protocone. (A–F) to the same scale. (A) Right i1, ONHM TN 2017-51, labial and lingual views. (B) Left i2, ONHM TN 2017-43, lateral and lingual views. (C) Right I1, ONHM TH 6, lingual view. (D) Right I1–i1 in pincer-like occlusion. (E) Right P3 (reversed), ONHM TH 3, occlusal view. (F) Right P4, ONHM TN 2017-44, occlusal view. (Images courtesy of M. Pickford [Pickford, 2015b; Al-Kindi et al., 2017]).

and unremarkable. In occlusal outline, these premolars are rounded to trapezoidal in shape.

Omanitherium is readily distinguished from other barytherioid taxa by the following features (Seiffert et al., 2012): from *Numidotherium koholense* by larger size, rounded, tusk-like i2s, relatively small inferred size of p2; large size of the metaconid in p3 and presence of entoconids in p3–4, more equal size and basal inflation of molar cusps, relatively low coronoid process and more anteriorly–rooted ramus, and long diastema between i2 and p2; from *Barytherium* by considerably smaller size, more spatulate I1s, incisors with enamel covering the entire crown, two-rooted p2-p4, no premolar hypocristids, and lower molars lacking ento- or premetacristids; and from *Arcanotherium* by a longer symphysis, more rounded i2s with no serrations, relatively small p2, no hypolophid in p4, more restricted postcingulid in m3, relatively broader m2–3s with more basally inflated cusps, and no molar ento- or premetacristids.

Based on similarities of molar and mandibular morphology, at least one parsimony analysis of phylogenetic relationships of *Omanitherium* places it as the sister taxon to *Barytherium*, supporting its inclusion in Barytheriidae (Seiffert et al., 2012). Monophyly of Barytherioidea, however, is not recovered in this study or earlier cladistic analyses of the relationships of *Numidotherium*, *Arcanotherium*, *Barytherium*, and now *Omanitherium* (Gheerbrant et al., 2005; Seiffert, 2007a; Delmer, 2009; Seiffert et al., 2012). Problems with reconstructing the relatedness of Paleogene proboscideans can be traced to difficulty identifying homoplasy, which appears to plague phylogenetic analyses of proboscideans. It is challenging, for instance, to embrace results of cladistic analyses that link lophodont taxa such as arcanotheres more closely than bunolophodont moeritheres with bunolophodont early elephantiform proboscideans (e.g., Delmer, 2009; Seiffert et al., 2012), without thinking that issues of convergence have yet to be worked out in coding character states in such analyses. Incisor morphology and enamel covering suggest a closer relationship of *Omanitherium* with *Arcanotherium* than with *Barytherium* (Pickford, 2015b). It is intriguing that the development of an ectoloph in P3–4 hints at a phylogenetic connection between *Omanitherium* and deinotheres (Pickford, 2015b), although cranially *Barytherium* seems to be a better sister-group candidate to pair with deinotheres than does an *Arcanotherium*+*Omanitherium* clade.

DEINOTHERIOIDEA Bonaparte, 1845
DEINOTHERIIDAE Bonaparte, 1845
CHILGATHERIINAE Sanders et al., 2004
CHILGATHERIUM Sanders et al., 2004
CHILGATHERIUM HARRISI Sanders et al., 2004

(Tables 2.3 and 3.1; Figures 3.1, 3.8, and 3.9)

Synonymy—Deinotheriidae gen. et sp. nov. Kappelman et al., 2003; "more derived . . . form" of *Arcanotherium savagei* Delmer, 2009

Chilgatherium harrisi derives primarily from the early late Oligocene site of Chilga in northeastern Ethiopia. It is named for its place of discovery and in honor of the considerable contributions to the field of African paleontology by John M. Harris, including significant advances in understanding the evolution and adaptations of deinotheres. It was the first non-elephantiform proboscidean recovered from the late Oligocene of Africa (Delmer, 2009). Chilga is among the most productive late Oligocene sites in Africa, with a diverse mammalian fauna reminiscent at the generic level of older Fayumian assemblages, including arsinoitheres, palaeomastodonts, hyraxes, and anthracotheres, and features one of the oldest occurrences of gomphotheres (Kappelman et al., 2003; Sanders et al., 2004; Jacobs et al., 2005). These mammals are accompanied by an abundant macroflora (Pan et al., 2006, 2010; Pan, 2010), a rare extravagance in the African fossil record. Unfortunately, it appears that a depositional taphonomic filter prevented the preservation of more diminutive taxa at the site. Consequently, Chilga has produced nothing in the size range of the copious assortment of primates and rodents found in the Fayum beds in Egypt.

Depositional settings for accumulation of fossils at Chilga include channel fills and floodplain deposits, overlain conformably on Oligocene flood basalts. The age of deposition is constrained by K-Ar dating of basalt below the sedimentary section to about 32 Ma, and refined by ^{40}Ar-^{39}Ar dates of 28–27 Ma on tuffs from within fossil-bearing beds in the sequence, supported by paleomagnetic sequence stratigraphy that places the sedimentary section in Chron C9n, between 27.972 and 27.072 Ma (Kappelman et al., 2003; Sanders et al., 2004). Thus, the Chilga proboscideans can be placed into a very precise early late Oligocene temporal context.

Chilgatherium harrisi is represented by only a small number of mostly isolated teeth, with no postcrania or skull material recovered for study. The exception is the type specimen CH35-3, which includes associated right partial m1, m2 (Figure 3.8A), and partial m3 and a left partial m2 and complete m3 in crypt (Figure 3.8B). The species is differentiated from other deinotheres by its smaller cheek teeth (Figure 3.9), upper premolars with more bunodont, independent cusps and weaker development of ectolophs (feeble connection of postparacrista and premetacrista), and incipient development of a tritoloph(id) in M3, m2, and m3 (Figure 3.8A–C); it contrasts with early Paleogene proboscideans and *Barytherium* by expansion of the m2 distocristid into an incipient third lophid, development of a postentoconulid in m3, greater expression of lingual cusps in upper premolars, and more bunolophodont condition of cheek teeth (Sanders et al., 2004). These morphological differences warrant at least subfamilial distinction of *Chilgatherium* from other deinotheres (Sanders et al., 2004). The upper premolars have four inflated cusps which fill the crown, instead of crested lophs (Figure 3.8D). The tritoloph of M3 is formed of a line of mammillons that retain a connection with the metaloph. Although the protoloph and metaloph of M3 have distinct cusps, they anchor transverse ridges with chisel-like wear along their anterior margins and are truly lophodont (Figure 3.8C). Strong posterior cristae on each side of the M3 crown extend into the transverse valleys. Lower molars exhibit inflated cusps that also secure continuous transverse crests. Small cristids extend anteriorly from the protolophid and hypolophid. Molar tritolophids of *Chilgatherium*, however, lack transverse crests (Figure 3.8A, B). Closely spaced last and penultimate lophids in the molar found immediately anterior to m2–3 in individual CH35-3 unmistakably indicate that m1 was trilophed (Delmer, 2009). A right mandibular fragment with a complete m3, SGS-SHUM 2020-01, recently recovered from the early Oligocene Shumaysi Fm. of Saudi Arabia, can be assigned to the genus (Table 3.1). The main cusps of each m3 lophid are bulbous and stand out from the lower transverse crests that they anchor. A substantial-sized crescentoid projects into the first transverse valley from the hypoconid. The tritolophid is large and independent, with a massive hypoconulid, but is lower overall that the protoconid and hypoconid, and its lingual cusp is not connected to the hypoconulid by a transverse crest. It differs from the Chilga m3 in having even heavier cusps, a more pronounced crescentoid, and a smaller, less projecting fovea anterior to the protolophid. Unfortunately, no symphyseal segment or lower incisor was recovered for the species. These would be exceptionally useful for testing the assignment of *Chilgatherium* to the Deinotheriidae.

A bilophed upper molar in a maxillary fragment, SGS-SHU 2018-063, was also recovered from the Shumaysi Fm. It is an appropriate size to fit as an M2 in the dentition of *Chilgatherium harrisi*, but the m2 of this species is trilophed. This new molar has sharp transverse, semi-vertical wear surfaces along its lophs; the metaloph is narrower than the protoloph; the distal cingulum sweeps upward to the lingual side to form a nascent cusp closely posterior to the hypocone. In size, proportions, and morphology, it is an even better fit as an M1 of *Barytherium grave*.

If *Chilgatherium* is a deinothere, as Sanders et al. (2004) thought based on upper premolar differences from barytherioids, the absence of a distal cingulid in m2 and presence of a posterior cingular ridge associated with the M3 tritoloph would support Harris' (1969, 1975, 1978a) hypothesis that deinotheres evolved m1 tritolophy via hypertrophy of the posterior cingulid, whereas the tritoloph of M1 evolved via postmetaloph ornamentation. It would constitute the oldest evidence of the Deinotheriidae yet recovered.

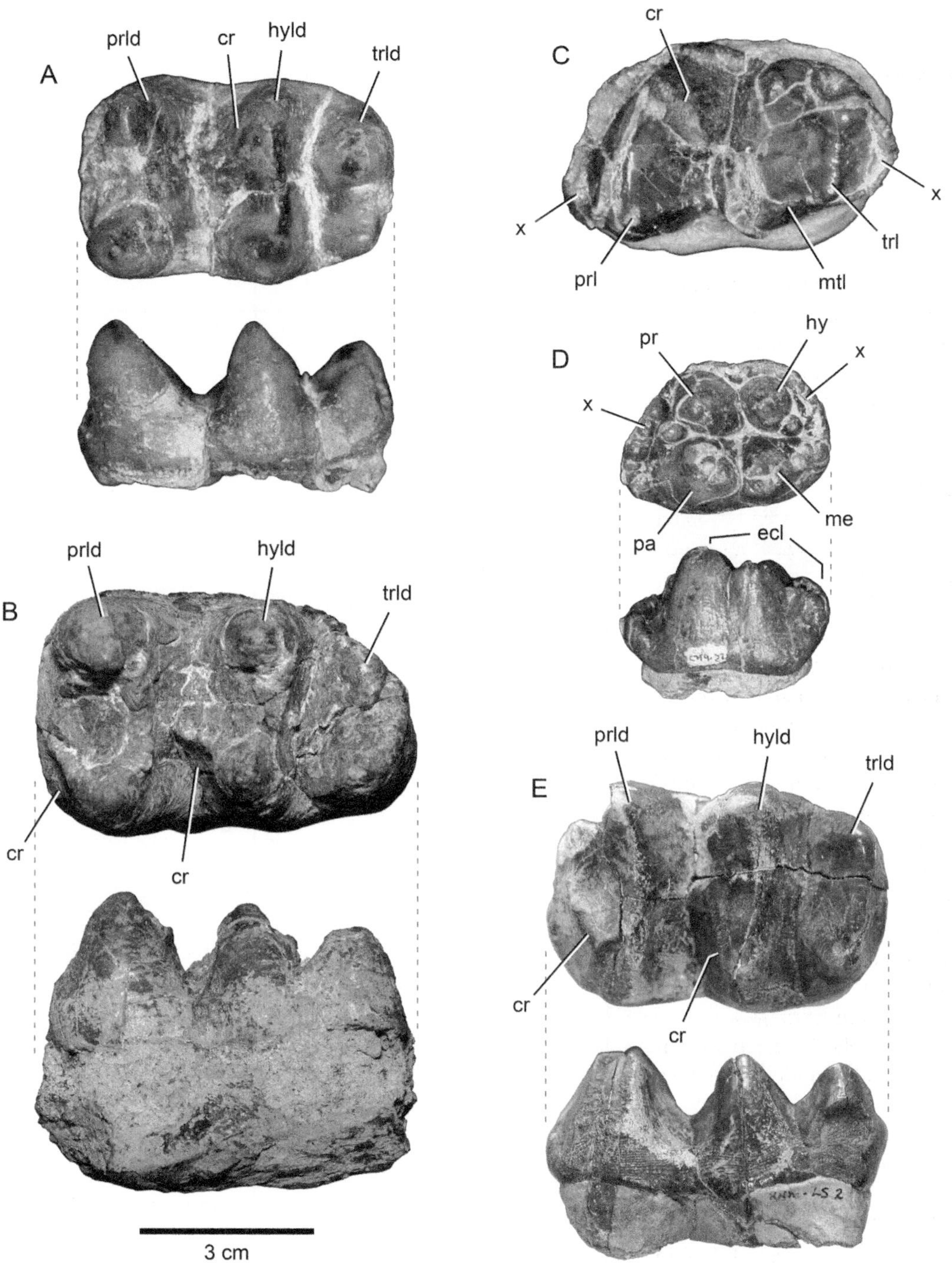

FIGURE 3.8 Cheek teeth of Oligocene African deinotheres. Abbreviations: cr, crescentoid; ecl, ectoloph; hy, hypocone; hyld, hypolophid; me, metacone; mtl, metaloph; pa, paracone; pr, protocone; prl, protoloph; prld, protolophid; trl, tritoloph; trld, tritolophid; x, pre- or postcingulum. (A–E) to the same scale. (A–D) *Chilgatherium harrisi*. (E) cf. *Prodeinotherium hobleyi*. Anterior is to the left. (A) Right m2 from Chilga, Ethiopia, CH35-3 (type), occlusal and lateral views. (B) Left m3 from Chilga, Ethiopia, CH 35-3 (type), occlusal and lateral views. (C) Right M3 from Chilga, Ethiopia, CH 35-1, occlusal view. (D) Right P3 from Chilga Ethiopia, CH 9-22, occlusal and lateral views. (E) Left m1 from Losodok, Kenya, KNM-LS 2, occlusal and lateral views.

An alternative phylogenetic interpretation is that *Chilgatherium* is a more derived relation of *Arcanotherium* (Delmer, 2009), which was thought to be a small variant of *Barytherium* (e.g., Shoshani et al., 1996) when *Chilgatherium* was described. Delmer (2009) stressed morphometric features he felt were shared by *Chilgatherium* and *Arcanotherium*: m2–3 dimensions very close, larger preprotocristid than premetacristid, weak lophodonty with cusps readily discernable, and absence of postmetacristid and preentocristid in m2. He also rejected the inclusion of the incipient deinothere-like upper premolars assigned to *Chilgatherium* by Sanders et al. (2004) as they were found at a different locality than its other cheek

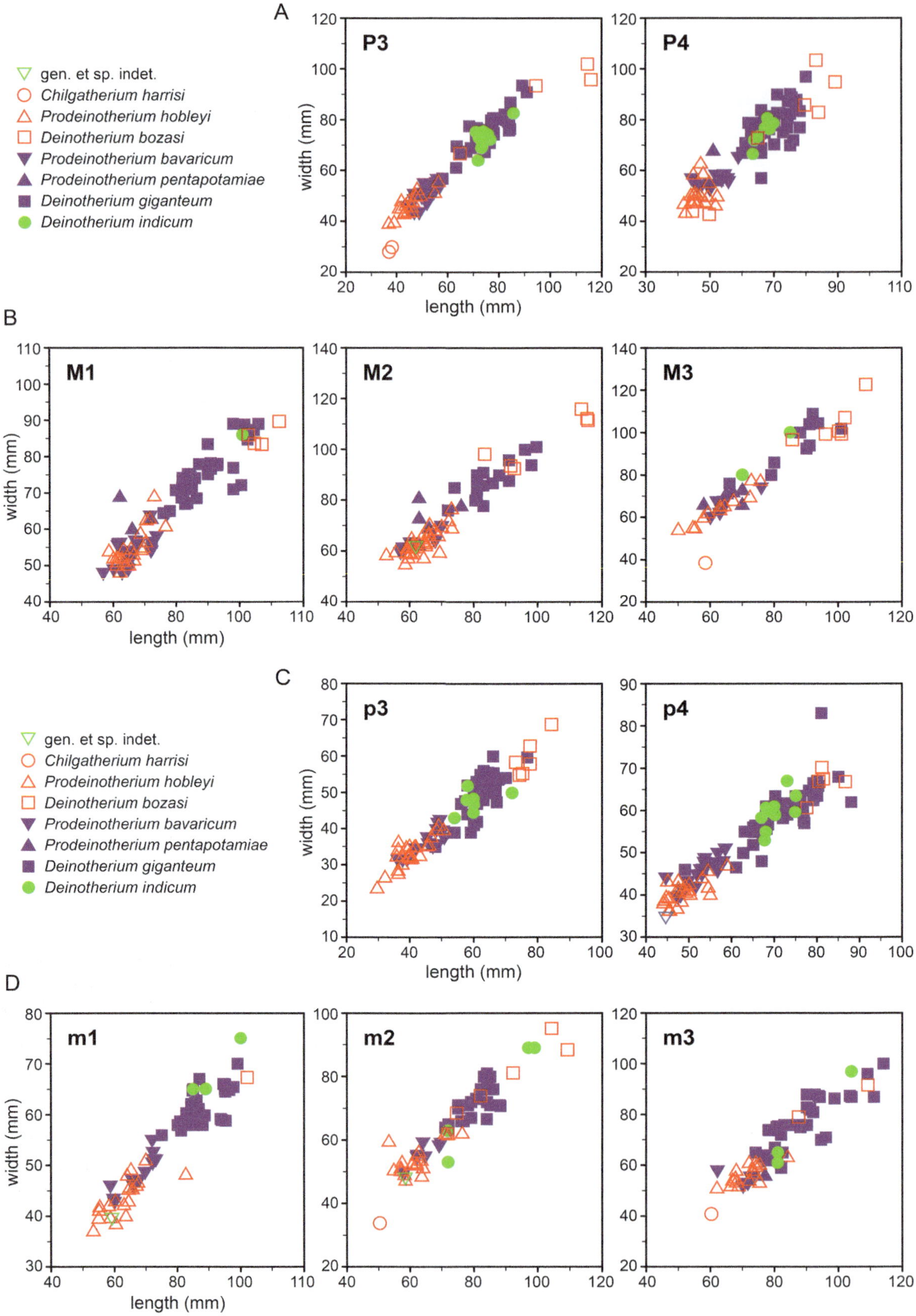

FIGURE 3.9 Bivariate plots of cheek tooth crown length versus width in deinotheres. Comparative dimensions supplementing original measurements are from Bachmann (1875), Weinsheimer (1883), Roger (1886), Andrews (1911), Forster Cooper (1922), Palmer (1924), Éhik (1930), MacInnes (1942), Gräf (1957), Sahni and Tripathi (1957), Symeonidis (1970), Harris (1973, 1977a, 1983b, 1987a), Gaziry (1976), Tobien (1988), Tsoukala and Melentis (1994), Huttunen (2000), Sach and Heizmann (2001), Sanders (2003), and Sanders et al. (2004). (A) P3–P4. (B) M1–M3. (C) p3–p4. (D) m1–m3.

teeth and because he felt that their bunodonty was inconsistent with the pattern of occlusion inferred for their lophodont molars. However, aside from comparable m2–3 dimensions in both taxa (Table 2.3), Delmer (2009) has overemphasized their morphological similarities. For example, m1 in *Arcanotherium* is bilophodont, consistent with its identity as a barythere, in contrast to the trilophodonty of m1 in *Chilgatherium*, a detail linking it with deinotheres. Postmetalophid ornamentations of m2–3 in both taxa are very different, m3 of *Arcanotherium* but not *Chilgatherium* exhibits a postmetacristid and preentocristid, and molar cusps are larger in *Chilgatherium*. As well, M3 in *Arcanotherium* (Delmer, 2009:fig. 3D) is bilophodont and much more like the same tooth in *Barytherium* than the trilophed M3 in *Chilgatherium*, which also sports considerably more prominent cristae emanating posteriorly from the proto- and metalophs (Sanders et al., 2004:fig. 4B). P3–4 of *Chilgatherium* does not fit with known homologs of other proboscideans from the site, but their incipient ectolophs and overall organization are what one might expect of a basal deinothere.

Additionally, molar enamel microstructure ("Schmelzmuster") in *Arcanotherium* is three-layered and grades from radial to Hunter-Schreger bands to 3D (Tabuce et al., 2007b), whereas in *Chilgatherium* the Schmelzmuster pattern is solely 3D (W. von Koenigswald, personal communication), identical to the condition in deinotheres, *Barytherium*, and *Numidotherium* (Tabuce et al., 2007b). It remains possible that *Chilgatherium* is part of a radiation of barytheres that developed convergences with deinotheres in enamel microstructure, some aspects of upper premolar morphology, and molar trilophodonty, but a hypothesis of sister-taxon relationship with *Arcanotherium* seems poorly supported. If it is a proto-deinothere, it would pose an alternative morphological transformation to the deinotheres of the Miocene-Pleistocene from what could be envisioned by derivation from a *Barytherium*-like ancestor (see Harris, 1978a). Finally, in Delmer's (2009:fig. 6) cladistic analysis, *Chilgatherium* is placed in a clade with palaeomastodonts, within a larger group that includes deinotheres. In the cladogram, *Moeritherium* and *Arcanotherium* are depicted as more distant sister taxa, respectively. This phylogenetic hypothesis is untenable and appears to rely on an overemphasis of cuspal bunodonty in *Chilgatherium*—there is nothing about molar morphology in this proboscidean that especially links it with palaeomastodonts (molar lophodonty versus bunolophodonty). Also, on dental grounds, it is difficult to accept a clade that includes palaeomastodonts and deinotheres to the exclusion of moeritheres.

DEINOTHERIINAE Bonaparte, 1845
CF. *PRODEINOTHERIUM* Éhik, 1930
CF. *PRODEINOTHERIUM HOBLEYI* (Andrews, 1911)

(Table 3.1; Figures 3.1 and 3.8)

Synonymy—*Deinotherium* cf. *hobleyi* Vialli, 1966; *Deinotherium* Sagri et al., 1998; *Prodeinotherium* sp. indet. Rasmussen and Gutiérrez, 2009; "*Deinotherium*" Abbate et al., 2012; cf. *Prodeinotherium* Abbate et al., 2014

The first indisputable deinotheres in the African fossil record are from the late Oligocene Eragaleit Beds at Losodok, Kenya (the geology and faunal composition of this site are described in the section on the mammutid *Losodokodon losodokius*), from a handful of dental specimens. These include KNM-LS 2, a lower left m1 (Figure 3.8E), KNM-LS 18240, a fragmentary left dentary with alveoli for dp2-dp3 and the dentin core of dp4, KNM-LS 18242, a left p4 missing much of its enamel cover but retaining its outline (not p3, contra Rasmussen and Gutíerrez, 2009), KNM-LS 5, a molar lophid, KNM-LS 3, a tusk fragment, KNM-LS 6, a fragment of molar enamel, and KNM-LS 18236, half of an anterior molar loph, either m2 or m3.

The molar dentition of the Losodok deinothere more closely resembles the cheek teeth of early to middle Miocene *Prodeinotherium hobleyi* than of slightly older *Chilgatherium harrisi*, in greater apical anteroposterior compression of loph(id)s, lesser inflation of cusp apices, straighter, more raised transverse loph(id)s between cusps, and larger size. Nonetheless, the complete m1 from Losodok, KNM-LS 2, is at the very low end of size for *P. hobleyi*.

The trilophodonty of KNM-LS 2 identifies it as an m1 and confirms these lophed teeth are those of a deinothere and not barythere (Figure 3.8E). Its protolophid and hypolophid are higher than its tritolophid, which is also narrower than the first two lophids. It is very small for a deinothere m1, with dimensions of L=57.5 and W=39.5, and is very brachyodont, with a crown height of only 24.2 mm (Rasmussen and Gutiérrez, 2009). In occlusal view, the protolophid and hypolophid are anteriorly slightly concave, with subvertical shearing surfaces present on their posterior faces. Each lophid has a smooth-edged crescentoid (cristid obliqua) that runs from the buccal cusp apex anteromedially downward, particularly well developed, though worn, on the hypolophid. Wear facets are sharp and vertical. The anterior cingulid is low and tightly appressed to the protolophid; the posterior cingulid is nearly nonexistent. The dp4 of KNM-LS 18240 is very small, with a length of 39.4 mm and greatest width of 25.1 mm at the second of three lophids. The p4 specimen KNM-LS 18242 has a length of 44.6 mm, greatest width of 34.9 mm, and height of 32.5 mm.

At Losodok, the deinothere specimens were recovered along with teeth of a primitive species of mammutid and dental remains of a possible gomphotheriid, represented by molar fragments composed of low, bulbous cusps, along with broken pieces of palaeomastodont teeth. These finds contribute to a growing fossil record of late Paleogene-earliest Neogene African proboscideans that reveals the step-wise replacement of late Eocene-early Oligocene barytheres, moeritheres, and palaeomastodonts by deinotheres and the first elephantimorph proboscideans (mammutids and gomphotheres) during the late Oligocene-earliest Miocene, well before early Miocene mammalian incursions from Eurasia began to impact African faunal evolution.

An m2 of small size (L=58.5 mm, W=48.0 mm) from the site of Adi Ugri, Eritrea was placed in "*Deinotherium* cf. *hobleyi*" and reported to be of early Miocene age by Vialli (1966). In size and development of its postcingulid the molar corresponds closely with the deinothere sample from Losodok. The m2 derives from a lignite-bearing intercalation between the Asmara/Aiba-Alaji basalts and Adi Ugri basalt in the Mai Gobro section. A rhyolite lenticular bed sampled in the middle of this intercalation was K-Ar dated to 24.6±0.25 Ma (Zanettin et al., 2006; Abbate et al., 2012). Thus, it may be contemporaneous with the Losodok deinotheres.

A deinothere M2 approximately 62 mm in length and width was recovered with a molar attributed to "Gomphotheriidae indet." from the same horizon as the Adi Ugri specimen, at the site of Mai Gobro 1, Eritrea (Abbate et al., 2012). This M2 resembles the morphology of early to middle Miocene *P. hobleyi* molars, but is smaller than M2s of that species. Unfortunately, there are no M2s in the Losodok sample for comparison. Abbate et al. (2014) allocated the molar to "cf. *Prodeinotherium*." Also, Sagri et al. (1998) reported but did not describe a lower jaw with teeth of "*Deinotherium*" from the Dogali Fm., along the Asmara-Massawa Road in Eritrea, that should be the same age as *Eritreum* (Table 3.1). However, possibly because deinotheres were not known from the Oligocene at the time of their discovery, Sagri and colleagues attributed the age of the occurrence to the early Miocene.

SUBORDER INCERTAE SEDIS
MOERITHERIOIDEA Andrews, 1906
MOERITHERIIDAE Andrews, 1906
MOERITHERIUM Andrews, 1901d

(Tables 3.1 and 3.2; Figures 3.1 and 3.10)

Perhaps the strangest of all proboscideans are moeritheres, a semi-aquatic, monogeneric group that is documented from a handful of late Eocene-early Oligocene mostly North African sites (Table 3.1), encompassing a lineage that persisted for about seven Myr. Inferences about their amphibious lifeways (Andrews, 1906, 1909a, b; Osborn, 1909, 1936; Coppens and Beden, 1978; contra Matsumoto, 1923) are bolstered by the discovery of their fossils from deltaic, estuarine, river channel, and nearshore marine sediments along what was the shoreline of the ancient Tethys Sea (Shoshani et al., 1996; Delmer et al., 2006; Clementz et al., 2008). Their anatomy led Osborn (1909, 1936) to speculate that moeritheres were sirenian or hippo ecomorphs. Ecological parallels to hippos in particular have now been corroborated by analyses of dental carbon and oxygen isotopes (Clementz et al., 2008; Liu et al., 2008) indicating that these proboscideans were semi-aquatic and fed on both freshwater and salt-stressed plants, including mangroves. Results of microwear analysis of moerithere teeth are consistent with consumption of a relatively soft diet and are congruent with prior reconstructions of dietary preference for aquatic vegetation, although there is no sign of the microwear features exhibited on hippo teeth that result from their nocturnal terrestrial grazing (Semprebon et al., 2017, 2018). Given their semi-aquatic lifeways, it is fitting that the name *Moeritherium* means "the wild beast of [Lake] Moeris (now Lake Birket Qarun)."

Morphologically, moerithere bodies are convergent on those of sirenians, desmostylians, and hippos (Osborn, 1936; Coppens and Beden, 1978; Tassy, 1981). The trunk is very long, with at least 19 thoracic vertebrae, four lumbars, and a sacrum of three fused vertebral elements, to which is attached a short tail (Figure 3.10A; Andrews, 1906). Relative to body length, limbs are short (Coppens and Beden, 1978), and were not held in graviportal position. Habitual flexed posture of the forearm on the humerus is indicated by the lack of posterior reflection of the high olecranon process of the ulna, and the posterior orientation of the distal condyles of the femur indicates that the knee was predominantly flexed. The innominate is very similar in form to that in early sirenians, with an expanded ischium and narrow ilium (Andrews, 1906), but with a much stronger attachment to the sacrum via the auricular surfaces (Andrews, 1909a), indicating greater terrestrial locomotor abilities. Postcranial features shared with other proboscideans include a (slightly) flattened femoral shaft, prominent medial tubercle on the astragalus, scapula with a strong, hooked coracoid process, and radius habitually pronated (Shoshani et al., 1996). Height of living moeritheres was estimated to have been approximately 0.5–0.7 m at the top of the shoulders (Coppens and Beden, 1978). Based in part on limb measurements of *Moeritherium lyonsi* from the late Eocene Qasr el Sagha Fm. of Egypt (Andrews, 1906), Larramendi (2016) estimated a body mass of 235 kg, about the size of extant pygmy hippos, which range in weight between 170 and 273 kg (Robinson, 2013). This should be considered a very generalized estimate, as moeritheres were highly sexually dimorphic.

The cranium is morphologically very distinct and its uniqueness is one of the reasons that moeritheres have not always been recognized as proboscideans (Coppens and Beden, 1978). The most striking features of the moerithere cranium are the long, straight tubular extension of the braincase to the face and the close anterior approximation of the orbits to the broad nasal aperture, retreated slightly from the anteriormost end of the skull (Figure 3.10B). It seems unlikely that this arrangement would have supported a trunk, but could indicate a prehensile upper lip (Matsumoto, 1923). Morphology of the cranium has been known since the first discoveries of moeritheres (e.g., Andrews, 1901c). Overall, it is low, with short nasals and longer frontal and parietal bones meeting in a straight line that ascends slightly from anterior to posterior. A low sagittal crest may be formed in larger, more mature individuals and joins the nuchal crest posteriorly. The orbits are small, situated very anteriorly, and are open above and behind. Zygomatic arches flare

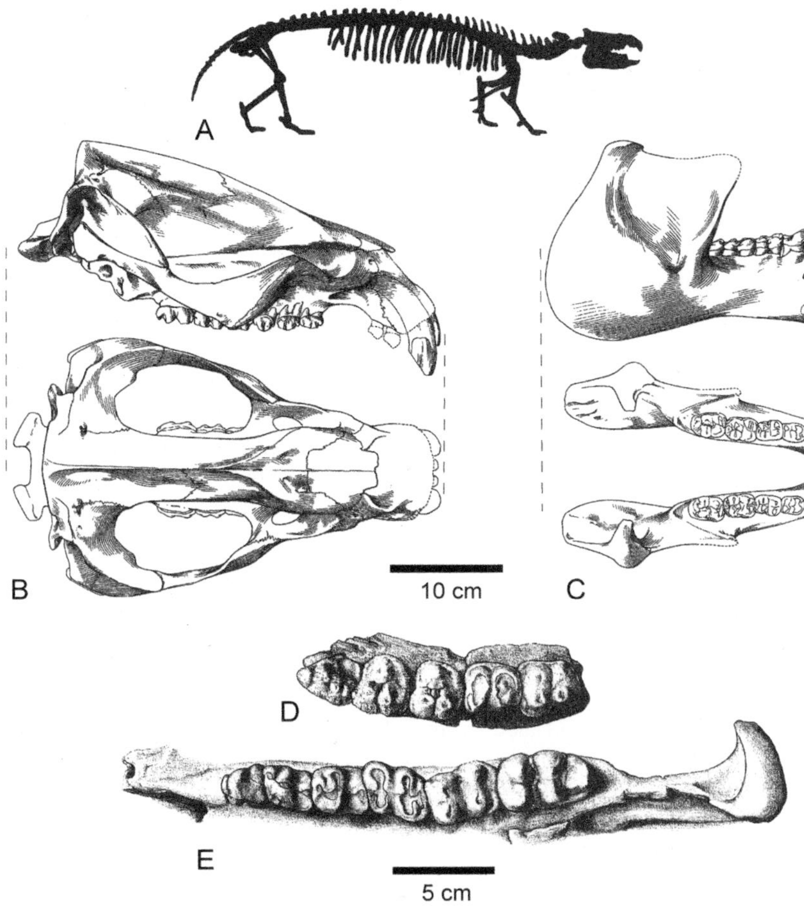

FIGURE 3.10 Skeletal remains of late Eocene-early Oligocene moeritheres from the Fayum, Egypt. (B, C) to the same scale. (D, E) to the same scale. (A–C) anterior to the right. (D, E) anterior to the left. (A) Silhouette, moerithere skeletal reconstruction, lateral view after Tobien (1971:fig. 7). (B) Cranial reconstruction, *Moeritherium trigodon*, lateral and dorsal views (Osborn, 1936:fig. 42A, A1). (C) Mandibular reconstruction, *M. trigodon*, lateral and occlusal views (Osborn, 1936:fig. 42B, B1). (D) Right P2–M2, *M. lyonsi*, occlusal view (Andrews, 1906:pl. X, fig. 2). (E) Left dentary with p2–m3, specimen C10001, *M. lyonsi*, occlusal view (Andrews, 1906:pl. X, fig. 1). (Images from Osborn [1936] with permission of the American Museum of Natural History. Figure in its entirety from Sanders et al. [2010a:fig. 15.4], by Bonnie Miljour).

laterally, angle forward, and are deep dorsoventrally. The external auditory meatus is nearly circular and bounded posteriorly by an outgrowth of the squamosal (Andrews, 1901c). Although the cranium is low with little angulation of the face on the neurocranium, the basicranium and external auditory meati are raised above the level of the palate. The palate is shallow and moderately broad. The glenoid processes are very broad and convex anteroposteriorly, and the occipital condyles are relatively large and indicate the ability to extensively move the cranium dorsoventrally on the neck. The braincase is modest in size, yielding an estimate cranial capacity of 240 cm^3 for an individual of *M. lyonsi* (Benoit, 2015). Benoit et al. (2019) recently calculated an encephalization quotient (EQ) for this individual of 0.20, which is very low compared with EQ values for elephantimorph proboscideans; the accuracy of their results, however, are called into question by their published body mass of 810 kg for *M. lyonsi*, attributed to Larramendi (2016), although he actually calculated a body mass of 235 kg for the species. Moerithere periotics lack fenestra and aqueductus cochleae, features shared with extant elephants and sirenians, and exhibit a secondarily undivided perilymphatic foramen, which is a synapomorphy shared by proboscideans and sirenians (Court, 1994b). These traits suggest that the ear region of moeritheres is more derived than that of numidotheres. Surprisingly, a cochlea from a moerithere cranium described by Tassy (1981) exhibits few differences from those of modern elephants and no obvious specializations expected for aquatic auditory function, and it is therefore possible that moeritheres could have discerned very low frequency sounds (Court, 1994b).

The upper dentition of moeritheres is primitive, retaining the canine and comprising a dental formula of 3-1-3-3 (Table 3.2). All adult teeth were in use simultaneously. The second incisor is prominently enlarged, chisel-like, and downward facing, while the third incisor and canine are relatively diminutive and difficult to differentiate as isolated specimens. Molars are bunolophodont and brachyodont with two lophs in each (Figure 3.10D) and proceed in a size series of M1 < M2 < M3. P3 and P4 are very similar in morphology and exhibit 1½ lophs. P2 is a more elongate tooth with an ectoloph but little development of transverse lophs.

The mandible is also distinctive, with a long, anteriorly projecting, spoutlike symphysis (Andrews, 1901c) that extends posteriorly to p4 and has an anteriorly procumbent incisor arcade, a very anteroposteriorly broad ramus that is rooted at the anterior end of m3, and coronoid process and transversely wide condyle nearly at the same level (Figure 3.10C). The presence of doubled mental foramina, position of the posterior mental foramen below p4, toothrow level of the mandibular foramen, and a large mandibular canal below the cheek teeth in moeritheres are primitive features (shared by *Phosphatherium* and an outgroup of hyracoids), but the loss of the coronoid foramen in moeritheres is autapomorphic (Ferretti and Debruyne, 2011). The second incisor is largest and spatulate (Table 3.2), while the central incisors are more triangular in shape and distally pointed in their unworn state. The tooth formula of the mandible is reduced compared with the upper dentition, 2-0-3-3 (Table 3.2). The lower molars are bilophodont (bunolophodont), each with a strong postcingulid heel, and there is greater development of bilophodonty of p3–4 in the lower dentition, though the cusps are much higher anteriorly than posteriorly (Figure 3.10E). Lower premolars usually exhibit a protostylid and anteriorly prominent paraconid; p4 may have stronger development of the metaconid and hypoconid than p3 (Delmer et al., 2006). Tobien (1978a) drew attention to the presence in moeritheres of a large cusp (hypoconulid) located in the center of the postcingulid in m1–2 or shifted to the pretrite side of the postcingulid in m3. Separation of the anterior from the posterior dentition in the mandible by a substantial diastema continues the trend for increasingly greater differentiation of anterior and cheek teeth observed in the successive series of early Paleogene proboscideans and suggests a specialized feeding function for the incisors of moeritheres, with the horizontally-oriented lower incisors working in concert with the vertically-oriented upper incisors for pruning vegetation.

A comprehensive overview of the discovery and taxonomy of species of *Moeritherium* is provided by Tassy (1981) and by Delmer and colleagues (2006). Moeritheres were the first proboscideans with non-elephant-like morphology to be discovered, at the turn of the previous century (Andrews, 1901a–d, 1906), but their odd morphology hindered systematic studies of their relationships and recognition of their inclusion in the Proboscidea (Andrews, 1909a; Osborn, 1909, 1921, 1936; Schlosser, 1911; Matsumoto, 1923; Deraniyagala, 1955; Tobien, 1971; Savage, 1976; Coppens and Beden, 1978). Andrews (1909a) felt that moeritheres might have been very near to the ancestry of both proboscideans and sirenians, as did Osborn (1936), though he believed that *Moeritherium* was a little closer to the former than the latter. A meticulous attempt to resolve the phylogeny of moeritheres was undertaken by Matsumoto (1923), who undertook a character-based comparative analysis of moeritheres, hyraxes, sirenians, and proboscideans that brilliantly foreshadowed later innovative cladistic methods. Matsumoto's (1923) analysis recovered 35 characters uniting moeritheres with other proboscideans, leading him to reason that the balance of evidence favored the inclusion of moeritheres in the Proboscidea, as possible antecedants to palaeomastodonts. Conversely, Deraniyagala (1955) rejected the proboscidean status of *Moeritherium* because of a presumed lack of a trunk. More recently, Coppens and Beden (1978) left the ordinal identity of moeritheres unresolved. Recognition of moeritheres as proboscideans was more rigorously established by careful descriptive analyses of the cranium and cladistic approaches by Domning et al. (1986), Tassy (1979b, 1981), and Shoshani et al. (1996). Ferretti and Debruyne (2011) hypothesized that moeritheres might be the sister taxon to a clade formed by Elephantiformes + Deinotherioidea. This hypothesis echoes the phylogeny of Tabuce et al. (2007b). Features culled from these analyses that unite moeritheres with other proboscideans, or with the more restricted group of elephantiforms + deinotheres, include hypertrophied second incisors, incipient bilophodonty, loss of first premolars and reduction or loss of canines, molar Schmelzmuster pattern primarily of Hunter-Schreger bands with some radial enamel microstructure (shared with palaeomastodonts), loss of the condyloid foramen, amastoidy, reduction of the anterior end of the jugal bone, laterally flared zygomatic arches, enlargement of the zygomatic process of the maxilla (laterally enclosing the orbit), raised external auditory meati enclosed ventrally by the posttympanic process of the squamosal, an enlarged scapular coracoid process, pronated position of the radius, and a well-developed medial tubercle on the astragalus (Matsumoto, 1923; Tassy, 1981; Domning et al., 1986; Shoshani et al., 1996; Tabuce et al., 2007b; Sanders et al., 2010a).

MOERITHERIUM CHEHBEURAMEURI Delmer et al., 2006

(Tables 3.1 and 3.2; Figures 3.1, 3.11–3.13, and 3.15)

Synonymy—*Moeritherium* sp., Mahboubi et al., 2003; primitive species of *Moeritherium*, Seiffert et al., 2008; *Moeritherium* sp., Sanders and Seiffert, 2014

The most primitive species of moerithere is *Moeritherium chehbeurameuri*, named for Dr. Arbeha Chehbeur-Ameur, who was a key member of the expedition that discovered the type series (Delmer et al., 2006). The type site, Bir el Ater, Algeria, produced a very modest handful of moerithere specimens, mostly isolated teeth, that can be assigned to a minimum of three individuals (Figure 3.11A, C, E, G, I, J; Delmer et al., 2006). Moeritheres from the site are part of a mixed marine-terrestrial vertebrate fauna derived from sandy fluviodeltaic sediments that includes hyraxes, elephant shrews, anomalurid and phiomyid rodents, insectivores, creodonts, and anthracotheres (Tabuce et al., 2000, 2001; Tabuce, 2002; Mahboubi et al., 2003; Jaeger et al., 2012; Solé et al., 2016). The presence of anthracotheres in this fauna is biochronologically critical, as the prior oldest record of anthracotheres in Africa was in the latest Eocene Dir Abu Lifa Mb. of the Qasr el Sagha Fm., Egypt (Holroyd et al., 1996). On faunal grounds, Bir el Ater seems temporally located equivalent to the lower part of the Qasr el Sagha Fm. (the entire formation is datable to ≈37–34 Ma) and close in age to, but slightly younger than, the early Priabonian Birket Qarun Fm., Egypt, which lacks anthracotheres (Seiffert et al., 2008). This narrows the temporal window for the anthracothere migration into Africa from Eurasia (Holroyd et al., 2010), and establishes a Priabonian age for Bir el Ater. Alternatively, but less convincingly, it has been suggested that Bir el Ater is as old as the middle Eocene (Tabuce et al., 2001; Mahboubi et al., 2003; Delmer et al., 2006; Solé et al., 2016). Even at the younger age estimate, Bir el Ater is one of the oldest sites in Africa to host both proboscideans and indisputable anthropoid primates (de Bonis et al., 1988; Mahboubi et al., 2003).

Moeritherium chehbeurameuri has been diagnosed as a small moerithere with nearly lophodont molars, diminutive I2, poorly molarized p3 with a narrow protolophid and poorly defined metaconid (Figure 3.11G), rectangular m1 crown lacking a buccal cingulid, and very wide P3 without a buccal cingulum (Figure 3.11; Delmer et al., 2006). A more accurate description is that the cheek teeth of this species

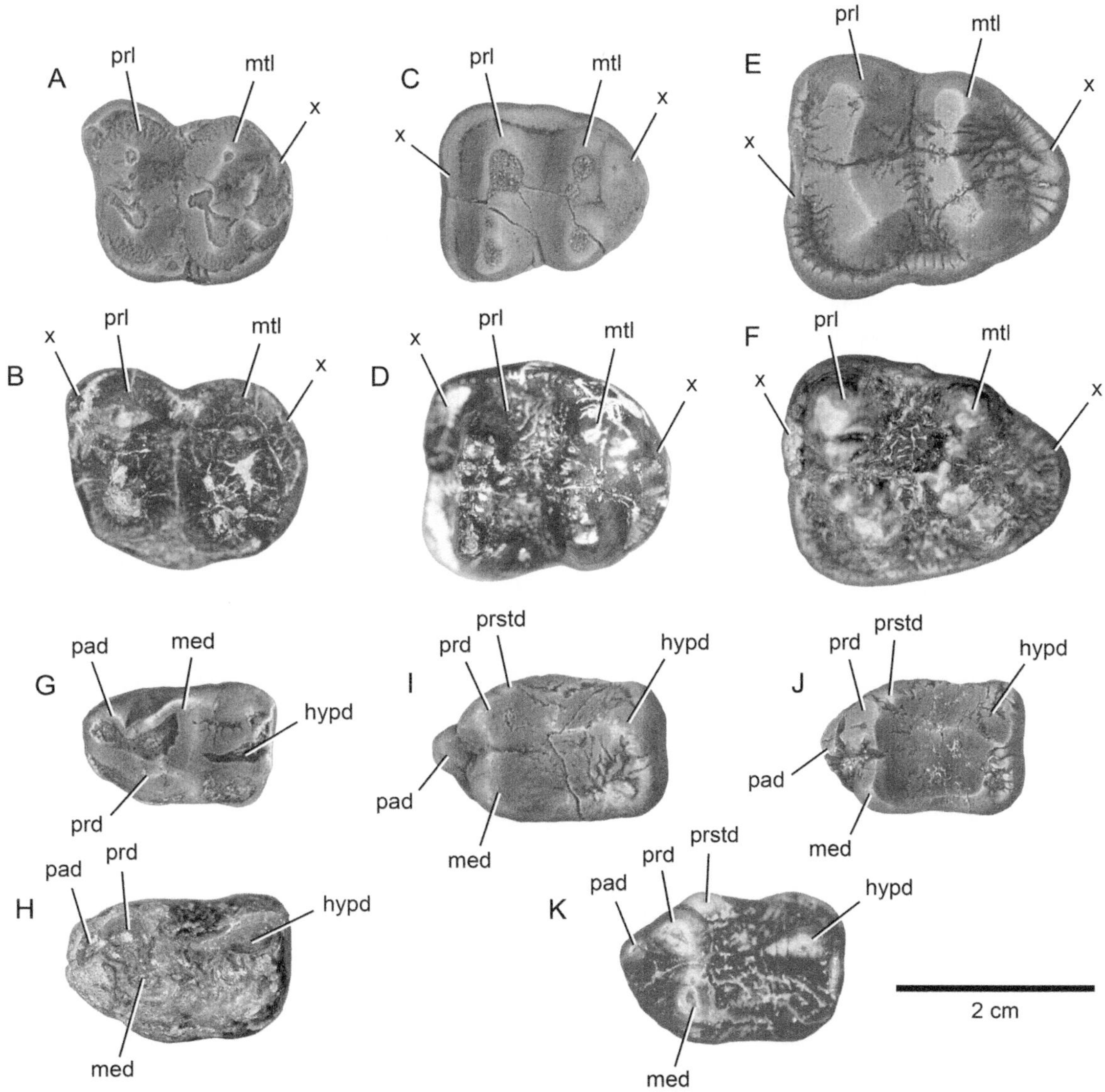

FIGURE 3.11 Selected molars of late Eocene *Moeritherium chehbeurameuri* from Bir el Ater, Algeria and Birket Qarun, Egypt. Abbreviations: hypd, hypoconid; med, metaconid; mtl, metaloph; pad, paraconid; prd, protoconid; prl, protoloph; prstd, protostylid; x, pre- or postcingulum(id). (A–K) to the same scale. Anterior is to the left. All views occlusal. (A) Left dP4, UON 84-375 from Bir el Ater, Algeria. (B) Left dP4, DPC 21453 from Birket Qarun, Egypt. (C) Right M1, UON 84-374 from Bir el Ater, Algeria. (D) Right M1, DPC 21829 from Birket Qarun, Egypt. (E) Left M3, UON 84-377 from Bir el Ater, Algeria. (F) Right M3 (reversed), DPC 21549 from Birket Qarun, Egypt. (G) Left p3, UON 84-379 from Bir el Ater, Algeria. (H) Left p3, DPC 21422 from Birket Qarun, Egypt. (I) Right p4, UON 84-378 from Bir el Ater, Algeria. (J) Right p4, UON 84-394 from Bir el Ater, Algeria; (K) Right p4, DPC 20851 from Birket Qarun, Egypt. Note the close morphometric similarity of teeth from the two sites. (Bir el Ater images from Delmer et al. [2006:pls. 1, 2], used with permission of John Wiley & Sons Books).

are bunolophodont with a tendency (seen to some degree in other species of moeritheres) to develop nearly continuous transverse ridges with moderate, semi-vertical wear of loph(id)s (e.g., Figure 3.11E). Nonetheless, molar crowns are simple, with weak or no development of accessory features between loph(id)s and small or absent mesoconelets in half-loph(id)s. Delmer et al. (2006) emphasized the transverse molar ridges of *M. chehbeurameuri* as a way of phylogenetically connecting moeritheres with older lophodont proboscideans of the earlier Paleogene and to contrast the species with what they characterize as more bunodont moeritheres of the younger Jebel Qatrani Fm. of the Fayum, Egypt.

The moerithere dentition from Bir el Ater seems to be taphonomically filtered to sample the small end of the size range of the species. A considerably more substantial assemblage of moerithere fossils that probably belongs to the same species derives from the Umm Rigl Mb. of the Birket Qarun Fm. in the Fayum, Egypt (Seiffert et al., 2008). The mammalian fauna from the locality of BQ-2 and associated nearby localities of the Birket Qarun Fm. is the most diverse of the entire Afro-Arabian Paleogene and includes the largest known assemblage of moerithere remains. Fossils recovered from Birket Qarun localities derive from fluvial depositional environments such as near-coastal shallow streams during a global sea-level lowstand, and reconstructions of the paleoenvironment indicate forested conditions at these localities (Seiffert et al., 2008). Faunal similarity is closest to Bir el Ater (with the notable absence of anthracothere artiodactyls), with the sites sharing the presence of the anthropoid *Biretia*, primitive anomaluroid rodents, hystricognathous rodents of *Protophiomys* grade, possibly the same or similar hyracoids, and a primitive species of *Moeritherium* (Seiffert et al., 2008). Other mammals present at Birket Qarun include barytheres, ptolemaiids, herodotiines, creodonts, chiropterans, insectivoran-grade placentals, adapid and djebelemurine stem strepsirrhine primates, and lorisiform crown strepsirrhines (Seiffert et al., 2008). Evidence from analyses of sequence stratigraphy, biochronology, and paleomagnetic reversal stratigraphy strongly correlate BQ-2 close to the base of Chron 17n.1n, between 37.24 and 36.51 Ma, essentially dated to 37 Ma (Figure 3.2; Seiffert et al., 2005b). Thus, the Birket Qarun Fm. provides the oldest known association of indisputable anthropoid primates and proboscideans.

Birket Qarun is a cornucopia of moerithere fossils, with abundant remains of crania, mandibles, gnathodental specimens, isolated teeth, and postcrania. These remains comprise a broader representation of interindividual variation in the tooth dimensions of *M. chehbeurameuri* and strong presumed sexual dimorphism in the size of crania and expression of anterior incisors. The I2 in male specimen BQ 02-341 is tusk-like (Figure 3.12A), measuring 23.6mm mesiodistally (MD), 32.8mm labiolingually (LL), and 53.6mm in height (H), compared with

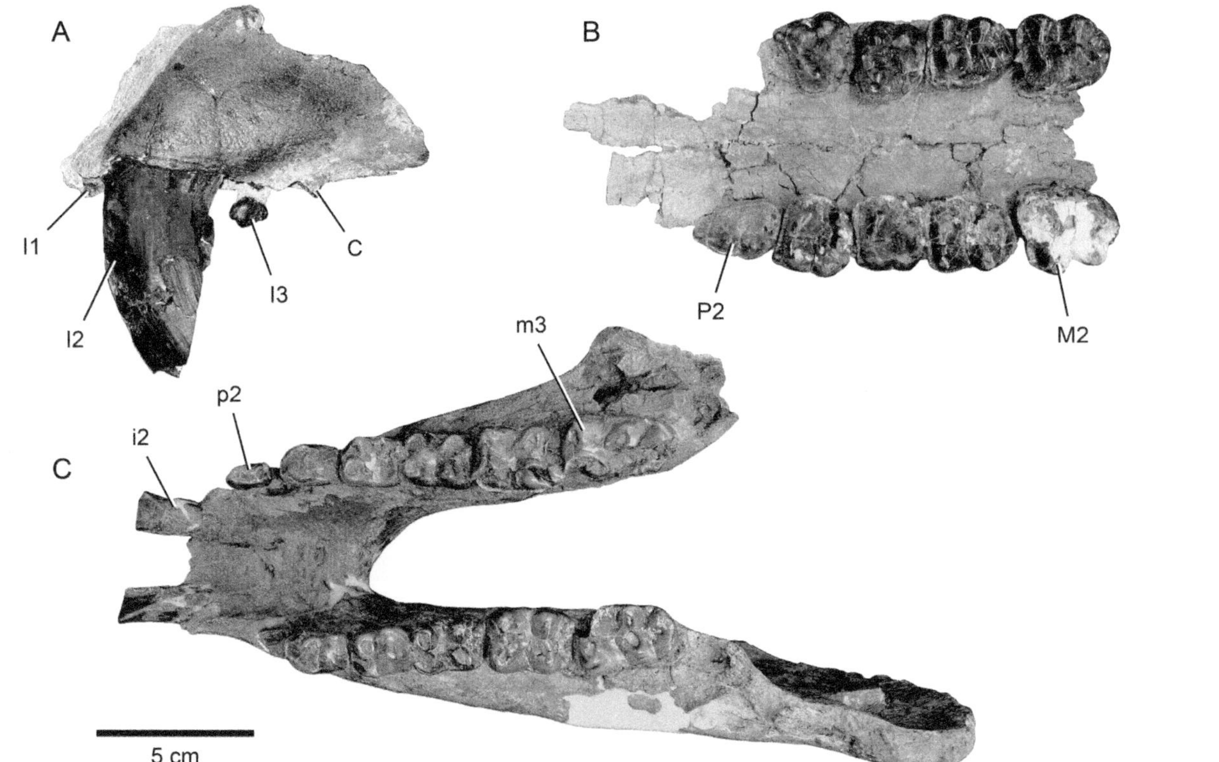

FIGURE 3.12 Palatal and mandibular remains of late Eocene *Moeritherium chehbeurameuri* from Birket Qarun, Egypt. Abbreviations: C, upper canine; I/i, incisor; M/m, molar; P/p, permanent premolar. All specimens to the same scale. Anterior to the left. (A) Premaxillary-maxillary fragment with broken I1, I2–3 and broken canine, BQ 32-341, lateral view. (B) Palate with right P–M2 and left P3–M2, BQ 10-137, occlusal view. (C) Mandible with right and left i2, right p2–m3, and left p3–m3, DPC 20834, occlusal view.

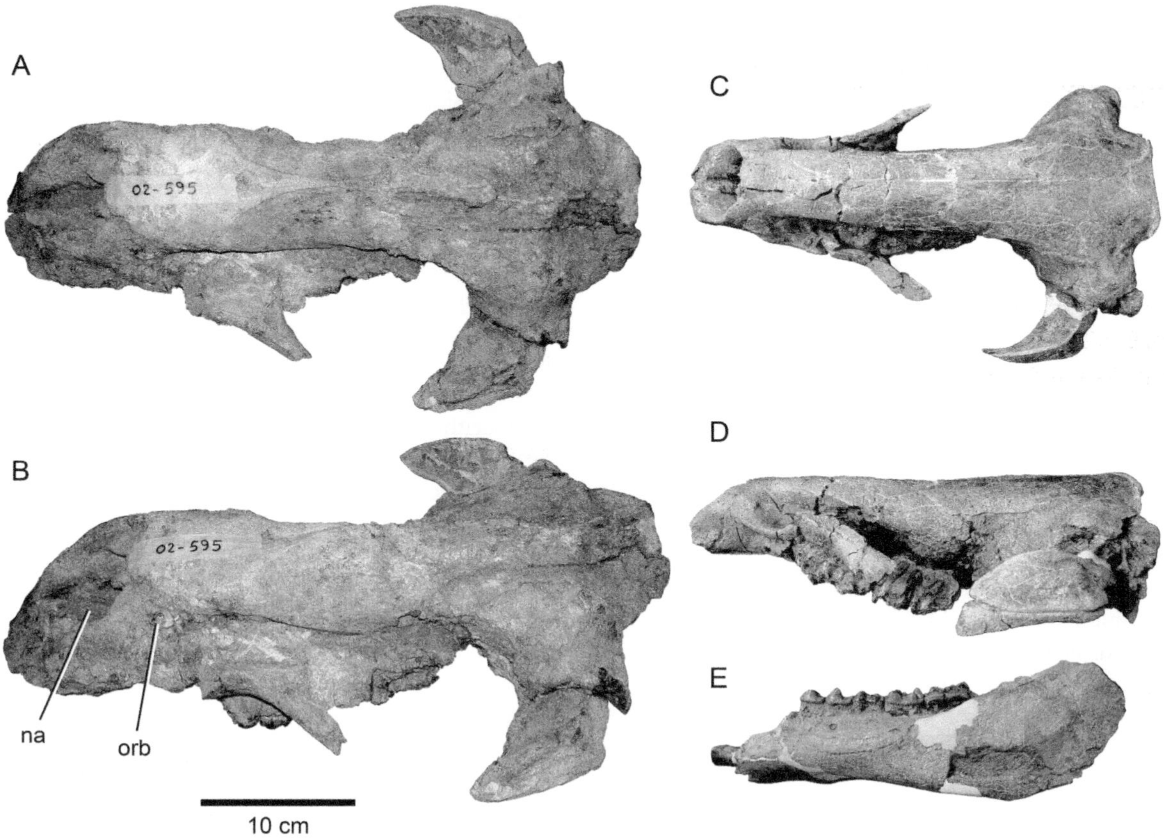

FIGURE 3.13 Skulls of late Eocene *Moeritherium chehbeurameuri* from Birket Qarun, Egypt. Abbreviations: na, nasal aperture; orb, orbit. All specimens to the same scale. Anterior is to the left. (A) Adult male cranium, BQ 02-595, dorsal view. (B) Adult male cranium, BQ 02-595, dorsolateral view. (C) Adult female cranium, DPC 20834, dorsal view. (D) Adult female cranium, DPC 20834, lateral view. (E) Adult female mandible associated with cranium DPC 20834, lateral view. Note the substantial size difference between crania BQ 02-595 and DPC 20834 due to sexual dimorphism.

dimensions of 18.8mm (MD)×22.5mm (LL) in female specimen DPC 21515. With estimated dimensions of 19.4mm (MD)×16.4mm (LL)×23.9mm (H), the small I2 from Bir el Ater is not extraordinarily tiny compared with I2s of other presumed female moeritheres. A large male cranium, BQ 02–595 (Figure 3.13A, B), measures 411mm in length and 292mm in greatest width (across the posterior zygomatic arches). Both the sagittal and nuchal crests are pronounced in this specimen, the zygomatic arches are dramatically flared enclosing capacious temporal foramina, the squamosals exhibit pronounced post-tympanic processes, and the articular glenoids are very broad. As in other moerithere crania, the face is very short, the large nasal aperture opens close to the front of the snout and the orbits are relatively small and located very anteriorly.

In contrast, a much more diminutive female cranium of the species missing its upper dentition, DPC 20854 (Figure 3.13C, D), is only 315mm long, an estimated 267mm in width, and lacks the sagittal crest and strong posteriorly overhanging development of the nuchal crest observed in the male specimen. It is much more gracile than the male cranium from Birket Qarun in all aspects of its morphology. DPC 20854 also includes a mandible with right and left p2-m3 (Figure 3.13E) that are the smallest cheek teeth known for moeritheres. The mandible including incisors is 276.6mm long but missing most of its rami (Figures 3.12C and 3.13E). Its symphysis extends posteriorly to a line across the anterior of the p4s, and measures 62mm anteroposteriorly. Its i2s are procumbent and modest in size, with left side mesiodistal (MD) dimensions of 14.8mm, 12.4mm labiolingually (LL), and a worn height of 13.0mm. An i2 of a presumed *M. trigodon* female, DPC 6036, is also relatively tiny, with dimensions of 14.6mm (MD)×14.4mm (LL)×30.0mm (H).

Comprehensive survey of cheek tooth dimensions shows that Bir el Ater molars and premolars are not unique in their diminutive size. Birket Qarun teeth range to even smaller dimensions, and *M. lyonsi* (including *M.* "*gracile*") and *M. trigodon* also overlap with Bir el Ater moeritheres in crown size at several tooth positions (though range to larger cheek tooth sizes) (Figure 3.14). Bigger samples show a greater degree of interindividual variation in tooth size within moerithere species, much of it driven by sexual dimorphism. Thus, molar size alone is not always sufficient for sorting isolated moerithere tooth specimens to species. Presumed males have considerably larger upper and lower second incisors than presumed females. Incisors may have been important in male-male competition for access to females, both for fighting and for display, in addition to their role in food acquisition (contra Andrews, 1909b; Osborn, 1936).

Morphologically, the dental sample from Birket Qarun is very similar to the more restricted sample from Bir el Ater (Figure 3.11B, D, F, H, K). P2 is elongate and sub-triangular in occlusal shape (Figure 3.12B), with a low, rounded parastyle protruding anteriorly. The largest cusp is the paracone, in line anteroposteriorly with the metacone which may have tiny mammillons on its mesial and distal flanks. Together, these cusps form a vertical cutting surface. The protocone is represented only by a slight swelling on the lingual cingulum. Posteromedially, the P2 features a low, flat basin enclosed by a low cingular ridge. P3 is a broader tooth with a large protocone, tiny parastyle set very laterally at the front of the crown, and paracone-metacone arrangement as in P2, also with a large distal basin. P4 is similar in form to P3 but may be slightly more squared in occlusal shape and is usually slightly smaller in size (Figure 3.12B); its parastyle is nearly indistinct. Each side of the protoloph in P3 and P4 may have small low mesoconelets closely appressed to the main cusps (paracone and protocone).

M1 is bilophodont (bunolophodont with a very narrow sulcus between half-lophs) with good development of the hypocone and a postentoconule posterior to the hypocone. M2 is similar to M1 but larger and with greater expression of the postentoconule complex. M3 is the largest molar and while lacking a distinct postentoconule has a salient distal talon shifted slightly to the lingual side of the crown (Figure 3.11F). In occlusal view, the half-lophs of the metaloph in M2 and particularly the M3 may be angled on one another forming a shallow chevron. Each molar has a low, ribbon-like precingulum that wraps around to the lingual side, where it remains low but may be substantial in size. The molars each also sport a very low postcingulum. Initial wear on the upper cheek teeth is subvertical. Chewing motion in moeritheres has been reconstructed as mesial with some inclination in Phase I and lateral in Phase II (von Koenigswald, 2016a). It seems obvious from the flattened wear patterns of teeth of older individuals that moeritheres incorporated more grinding transversely across the occlusal platform as they aged.

Moerithere lower cheek teeth are also abundant at Birket Qarun. The p2 is relatively small and narrow, with its cusps aligned anteroposteriorly. It is anteriorly pointed due to a low, narrow paraconid, posterior to which is the dominant cusp, the protoconid. A lower posterior cusp is separated from the paraconid by a bilaterally compressed "saddle." Occlusally, this premolar varies in shape from bean-shaped, to triangular, to rectangular. The p3 is also relatively narrow with a prominently projecting paraconid that is closely appressed to the protoconid. The protoconid is tightly associated with and slightly anterior to the smaller metaconid (Figure 3.11H). The low posterior talon is dominated by a

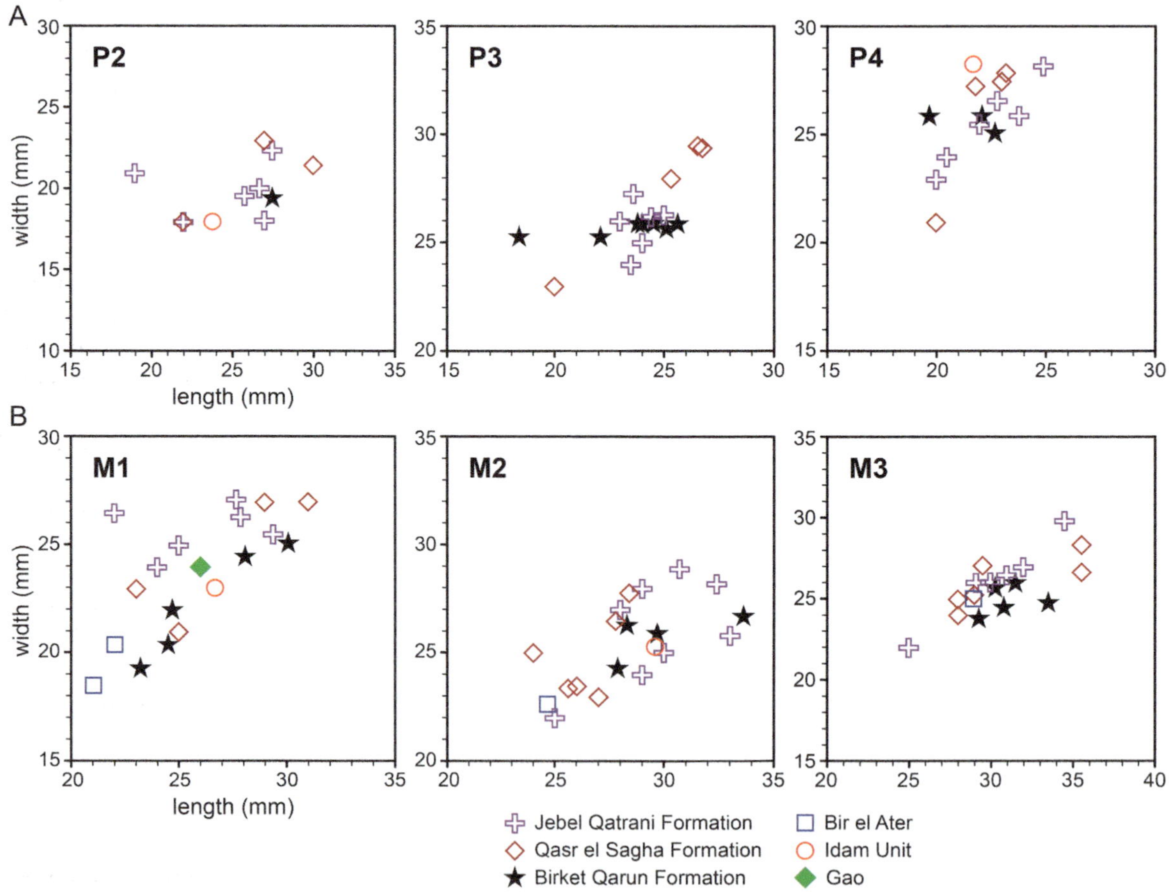

FIGURE 3.14 Bivariate plots of cheek tooth crown length versus width in moeritheres. Comparative dimensions supplementing original measurements are from Andrews (1906), Schlosser (1911), Matsumoto (1922, 1923), Derinyagala (1955), Tobien (1971, 1978a), Wight (1971), and Delmer et al. (2006). (A) P2–P4. (B) M1–M3. (C) p2–p4. (D) m1–m3.

(Continued)

buccally placed hypoconid. There is no true entoconid but rather its position is represented by a "dimple" on the low postmetacristid that descends posteriorly from the metaconid. In unworn crowns, a low, beaded precingulid descends buccally from the paraconid. Occlusal wear on the posterior of the protolophid is initially strongly vertical and chisel-like, while the heel wears more horizontally. The p4 is very similar to the p3 but the protolophid is formed of sub-equal protoconid and metaconid separated by a V-shaped median sulcus, and is anteriorly preceded by a smaller paraconid (Figure 3.11K). A protostylid in the form of a low cusp may be closely appressed laterally to the protoconid, but is sometimes obliterated by wear and is variable in expression. The entoconid in p4 is a low "dimple" or cuspule on the lingual rim of the crown, and may be slightly larger than the homologous tubercle in p3. Low crescentoids connect the proto- and hypoconids. Notably, the first moerithere specimen recovered from the Birket Qarun Fm., surface find p4 specimen DPC 5906 (Holroyd et al., 1996), is atypical for the sample. It is unique in the formation sample in having a prominent entoconid tubercle, accessory tubercle in the transverse valley, and signs of a well-developed crescentoid descending into the transverse valley from the hypoconid.

The moerithere lower molars from Birket Qarun are generally simple in occlusal detail. The lower first molar is bunolophodont with two transverse lophs and a small low heel with a well-developed hypoconulid offset toward the buccal side of the crown. Pretrite half-lophid 2 may have a small anterior crescentoid descending into the transverse valley. The lower second molar is similar to m1 but larger; expression of anterior crescentoid 2 is variable. Between m1 and m2, the increase in size is greater than between m2 and m3. Half-lophids in these molars may have tiny mesoconelets accompanying their main conelets. The lower third molar is also bilophed, but differs from m1 and m2 by its much more expansive heel that has a cingular ribbon running lingually from the hypoconulid. Crescentoid development is more pronounced in the partial m3 from Bir el Ater (UON 84-382) than in m3 specimens from Birket Qarun. Moerithere cheek teeth from Birket Qarun overlap in size with those of Bir el Ater, but with a much larger sample document greater dimensional variation, attributable to sexual dimorphism; the Bir el Ater sample appears to record specimens only of smaller females (Figure 3.14).

The postcranial sample from Birket Qarun concords with Andrews' (1906) interpretation of moerithere positional behavior. A femur (DPC 20845) lacking only its greater trochanter shows no graviportal adaptations, but instead is configured to be held in a habitually flexed posture, with its head facing laterally rather than proximally and distal condyles

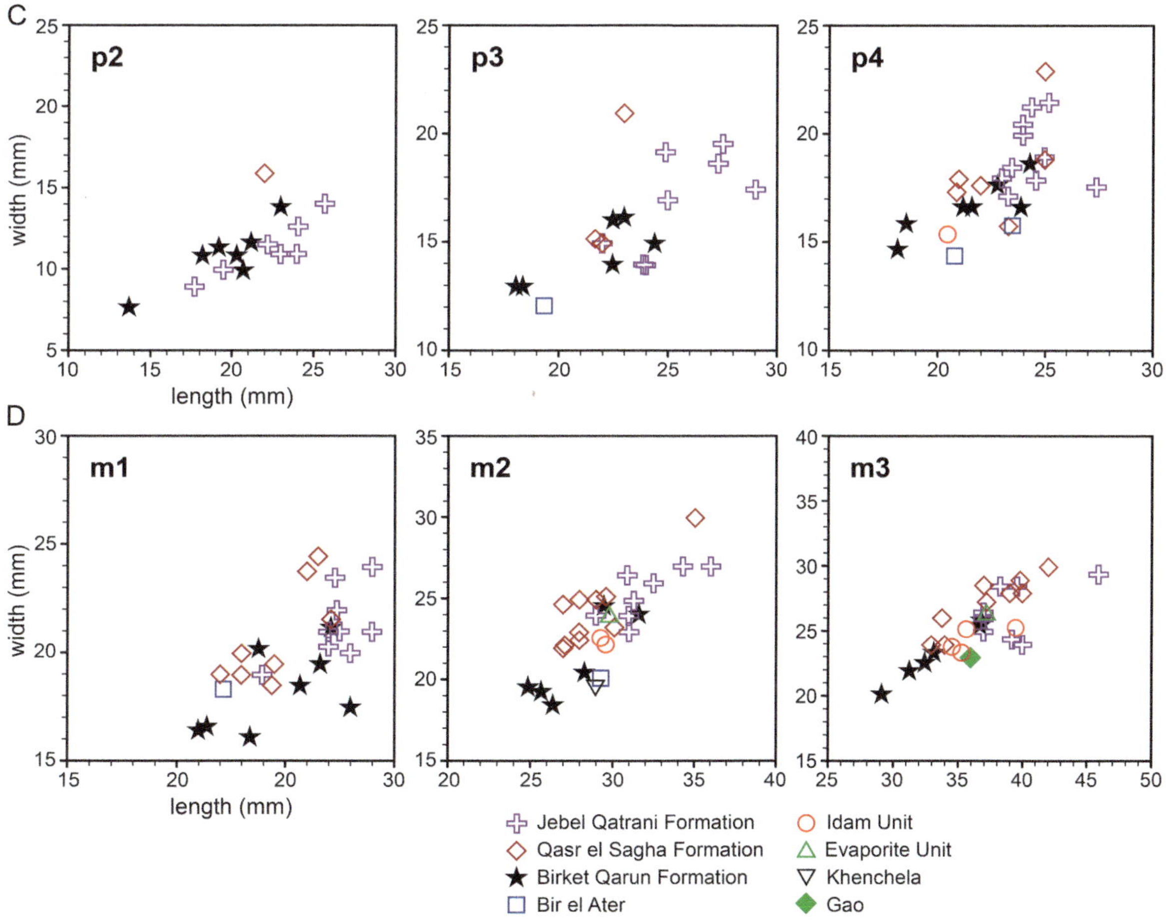

FIGURE 3.14 *(Continued)* Bivariate plots of cheek tooth crown length versus width in moeritheres. Comparative dimensions supplementing original measurements are from Andrews (1906), Schlosser (1911), Matsumoto (1922, 1923), Deraniyagala (1955), Tobien (1971, 1978a), Wight (1971), and Delmer et al. (2006). (A) P2–P4. (B) M1–M3. (C) p2–p4. (D) m1–m3.

primarily oriented posteriorly (Figure 3.15A). The greater length of the medial than the lateral condyle indicates that the femur was distally angled toward the midline of the body. It has a greatest length of ~285.0 mm, is slightly flattened with a midshaft width and depth of 28.0 and 25.8 mm, respectively, and broadens distally to a width of 53.3 mm. The patellar facet is relatively broad (26.0 mm) and shallow, suggesting movement of the lower leg on the femur was not limited to fore-aft movement in the manner of fast cursors. The inferoposteromedial placement of the fovea capitis femoris on the femoral head of other specimens confirms the presence of a strong ligamentum teres, habitual flexion of the femur on the hip, and outward rotation of the bone in neutral position. A prominent third trochanter is located on the posterolateral surface of the femoral diaphysis (Figure 3.15A), for gluteal musculature to act as strong lateral rotators, abductors, and powerful extensors of the leg. The femora of modern African elephants also display a third trochanter, more distally located along the shaft (Smuts and Bezuidenhout, 1994). Dimensional variation in the femoral sample from Birket Qarun is consistent with sexual dimorphism in body size—a femur from a possible female individual (DPC 21745) has midshaft width and depth dimensions of 21.4 mm and 21.3 mm, respectively, and is only 37.3 mm wide distally. Its distal condyles are approximately equal in size (medial width and height = 11.1 mm, 31.4 mm; lateral width and height = 13.0 mm, 29.2 mm), suggesting similar force transmission across both condyles.

A number of complete ulnae complement what is interpretable from the femur, that the limbs were habitually held in flexed position and not in graviportal alignment. The ulnae have very long, proximodistally straight olecranon processes topped by salient tuberosities for powerful extensor muscles (Figure 3.15B); the lack of flexure of the olecranon indicates a forelimb held in habitual flexed position with projection of the olecranon providing increased leverage for the extensors. Powerful forces must have been at play across this element, as the bifid sigmoid or trochanteric notch for articulation with the humerus is very broad (37.6 mm in CGM 02-343) compared with midshaft width of 25.7 mm and distal breadth of 28.1 mm, and a raised bar of bone runs the length of the specimen along its posterior aspect. Overall length of this element is 225.0 mm, to which the olecranon process and sigmoid articulation contribute 67.3 mm. The distal end of the ulna appears to have separate surfaces for articulating with the pisiform and cuneiform bones. A large flat surface on the shaft accommodates the radius in an unfused but probably regularly pronated position. A salient interosseus crest is located lateral to this surface (Figure 3.15B) and marks a strong connection between the radius on the ulna, presumably fixed in pronation. Greater projection of the upper than lower lip of the sigmoid notch must have accommodated a large degree of flexion of the lower forelimb and provided stability of the elbow during limb extension by locking into the deep olecranon fossa of the humerus. A proximally broken humerus from the Birket Qarun Fm. (DPC 20817; length = +200 mm minus the head and trochanter) has a strongly torqued shaft, much larger medial than lateral epicondyle, a deep olecranon fossa, and a marked bicipital groove anteriorly, but lacks an entepicondylar foramen. Midshaft width and depth are 17.0 and 40.3 mm, respectively, bicondylar width measures 69.3 mm, and transverse dimension of the trochlear articular surface is 38.5 mm. A more complete specimen from the Idam Unit at Dur el Talah has a relatively small head (W = 37.8 mm × H = 29.6 mm) that faces posterolaterally, a bicondylar width of 76.4 mm, and overall length of 232.2 mm. Andrews (1906) believed that the size of the medial epicondyle and salient development of the supinator crest on the distolateral aspect of the humerus indicated that the lower forelimb was capable of supination.

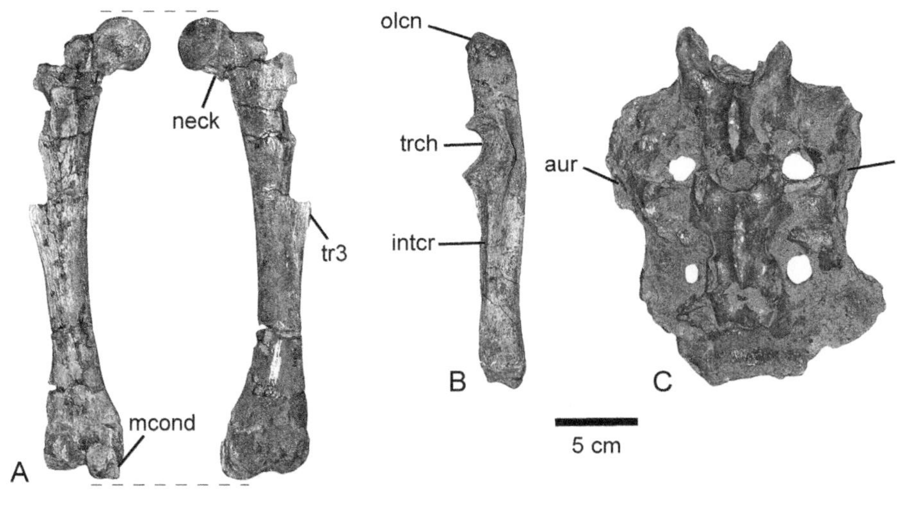

FIGURE 3.15 Postcranial elements of late Eocene *Moeritherium chehbeurameuri* from Birket Qarun, Egypt. Abbreviations: aur, auricular surface; intcr, interosseous crest; mcond, medial distal femoral condyle; olcn, olecranon process; trch, trochlear notch; tr3, third trochanter. All specimens to the same scale. Superior is to the top. (A) Partial left femur, DPC 20845, posterior and anterior views. (B) Left ulna, BQ 02-343, lateral view. (C) Sacrum, DPC 21771, posterior (dorsal) view. Note the distinct auricular surfaces for articulation with the innominates.

A scapular fragment (DPC 21074) preserves a prominent coracoid process overhanging the glenoid, and exhibits a constricted neck between the glenoid and scapular blade. A pelvic fragment (DPC 21415) exhibits typical moerithere morphology, with a nearly round acetabular articulation (width=45.8 mm, height=43.3 mm) and anteriorly angled base of the pubic ramus that with the base of the ischial ramus partially outlines what appears to have been a large obturator foramen. A more complete innominate from the Qasr el Sagha Fm. is approximately 390 mm in length, with a 195-mm-long ilium, nearly equivalent in length to the ischium (Schlosser, 1911).

A well-preserved sacrum formed of three fused vertebral elements (DPC 21771) provides important information relevant to interpretation of positional abilities in *Moeritherium*: the fusion of the vertebral elements into a coherent sacrum (Andrews, 1909a) makes it unlikely that this proboscidean was propelled in the water by undulation of its body, and the size of its auricular surfaces (Figure 3.15C) indicates a secure articulation with the rest of the pelvis, signaling that force transmission occurred through the pelvis during quadrupedal terrestrial progression. The size of the postzygapophyses at the caudal end of moerithere sacra generally indicates their articulation with a relatively large first caudal vertebra, which Andrews (1906) felt should be considered an unfused fourth sacral element. Overall shape of the sacrum from Birket Qarun is exactly like the sacrum from higher in the Fayum sequence, described by Andrews (1906). It is rectangular, longer than wide (estimated L=235 mm; W=145 mm), with the centrum articulation for the first caudal vertebra approximately as wide as the articular platform for the last lumbar vertebra. Lateral flanges at the caudolateral aspects of the sacrum (Figure 3.15C), homologous with transverse processes, may have acted as accessory attachment sites for gluteal musculature and anchored sacroiliac ligaments to stabilize the sacrum on the innominates. A larger sacrum from the Qasr el Sagha Fm. measures L=280 mm×W=180 mm (Schlosser, 1911).

MOERITHERIUM LYONSI Andrews, 1901d

(Tables 3.1 and 3.2; Figures 3.1, 3.10, 3.14, and 3.16)

Partial Synonymy—*Moeritherium gracile* Andrews, 1902, 1906; M. lyonsi Andrews 1903b, 1904a; *M. gracile* Matsumoto, 1922, 1923; *M. ancestrale* Petronievics, 1923; *M. gracile* Osborn, 1936; *M. pharoahensis* Deraniyagala, 1955; *M. latidens* Deraniyagala, 1955; *Moeritherium* cf. *lyonsi* Tassy, 1981; *M. gracile* Delmer et al., 2006

The nominal, or type, species of *Moeritherium* is *M. lyonsi*, based on a dentary with teeth associated with upper molars and a thoracic vertebra that derive from the late Eocene Qasr el Sagha Fm. of the Fayum, Egypt (Andrews, 1901d, 1906). As it was the first moerithere species described, there are few details from the original literature that are useful for differentiating it from other congenerics. Schlosser (1911) restricted specimens from the Qasr el Sagha Fm. to *M. lyonsi* and assigned specimens originally attributed to that species from the Jebel Qatrani Fm. to a new species, *M. andrewsi*. Coppens and Beden (1978) enumerated the following diagnostic features for *M. lyonsi*: very short lower premolars with p2 triangular in occlusal shape; lower molars very long; and all lower cheek teeth very broad. Tobien (1971) added that *M. lyonsi* is characterized by molars with four cusps lacking associated accessory conules and simple crown morphology. The differences between cheek teeth of *M. lyonsi* and *M. chehbeurameuri* are subtle, mostly involving greater inflation of cusps in *M. lyonsi* molars and premolars and some tooth positions ranging to larger size in the younger species (there is substantial overlap of cheek tooth dimensions between these species). *Moeritherium trigodon* (including "*M. andrewsi*"), from higher in the Fayum sequence, tends to have more complex cheek tooth crowns and its teeth reach a larger size range limit.

A number of names besides the nominal species of *Moeritherium* have been proposed for fossils from the Qasr el Sagha Fm. Andrews (1902) created a second species from the formation, "*M. gracile*," which he distinguished from *M. lyonsi* by its

> comparative lightness of structure, the narrowness of the palate, the smaller size of the upper molars and premolars, particularly of m3, the strong development of the cingulum in these teeth, and by the considerable inflation of the cranial region of the squamosal, which apparently contains air-sinuses.
>
> *(Andrews 1906: p. 127)*

Also, Petronievics (1923) recommended erection of a new species, "*Moeritherium ancestrale*," for a few specimens from the Qasr el Sagha Fm., on differences from *M. lyonsi* in the angulation of the occipital planum and size of the palatine vacuity between the maxillary inner borders of the alveolus for the I2s. Later, Deraniyagala (1955) proposed two additional moerithere species for the Qasr el Sagha Fm., "*M. latidens*" and "*M. pharaohensis*," based on shape and proportions of molars in two dentary fragments.

Subsequent researchers have tended to ignore *M. latidens* and *M. pharaohensis* as too poorly diagnostic to merit specific distinction and treated the features of *M. ancestrale* as exhibiting intraspecific variation that could be accommodated within *M. lyonsi* (Tobien, 1971; Delmer et al., 2006; Sanders et al., 2010a). Moreover, it has been widely considered that only one sexually dimorphic moerithere species is present in the Qasr el Sagha Fm., with "*M. gracile*" representing smaller individuals of *M. lyonsi* (Tobien, 1971; Coppens and Beden, 1978; Holroyd et al., 1996; Shoshani et al., 1996; Sanders et al., 2010a).

Alternatively, however, Matsumoto, in his 1923 landmark study of moeritheres, maintained *M. lyonsi* and "*M. gracile*" as good, separate species, based not only on size but also cheek tooth proportions, including relatively narrower, longer molars and shorter, broader premolars in *M. lyonsi*. Tassy (1981) followed this taxonomic scheme, recounting premolar-molar row proportional differences for each species, and Delmer et al. (2006) stridently defended the validity of "*M. gracile*" by delineating its diagnostic traits: square upper premolars, length of premolars shorter relative to molar series' length, smaller size, and presence of a labial cingulum in upper cheek teeth. However, as shown by Osborn (1936:p. 73), there is considerable inter- and intraspecific variation in moerithere cheek tooth dimensions and crown shape, including within "*M. gracile*" and *M. lyonsi*. Direct measurement of specimens from the Qasr el Sagha do not support Delmer et al.'s (2006) hypothesis; for example, the P3 shape index (L×100/W) for the small type of "*M. gracile*" is 87 but larger premolars presumably attributable to *M. lyonsi* have only slightly higher indices of 90–91 (relatively not much wider), and P4 indices are higher for "*M. gracile*" (95) than for *M. lyonsi* (80–84), indicating that the former has relatively broader upper fourth premolars than the latter! Tobien (1971:figs. 2, 3) showed that while *M. lyonsi* appears to have a relatively broad p3, its p4 has nearly the same proportions as p4 in "*M. gracile*." With a larger sample size, moerithere premolars from the Birket Qarun Fm. exhibit a greater a degree of metric and shape variation, with P3 occlusal shape indices ranging from 73 to 99. Moreover, if the cheek tooth proportional differences between the two purported species in the Qasr el Sagha Fm. cannot be sustained as consistent, it would be unusual to rely solely on the criterion of cingulum development to separate them. Thus, conservatively, "*M. gracile*" is synonymized here with *M. lyonsi*. It seems more likely that moeritheres of the Qasr el Sagha Fm. were sexually dimorphic and had a high level of interindividual variation.

A few gnathodental specimens of *M. lyonsi* warrant mention. Palatal specimen M 8900 preserving P2–4 is weathered but exhibits a strong zygomatic root that flares widely. The premolar row measures 73.8 mm, which would qualify as *M. lyonsi* by Tassy's (1981) criterion; P2 is elongate (L=26.7 mm, W=20.1 mm) and triangular in occlusal shape, P3 and P4 are relatively wide, with L=25.3 mm, W=28.0 mm and L=21.8 mm, W=27.3 mm, respectively, and are sub-triangular to square in occlusal shape. The upper premolars have lingual but not labial cingulae.

Palatal specimen M 8193 preserves M1–3 with simple crowns and slightly inflated cusps. Size progression is M1<M2<M3. Overall length of the molar row is 77.0 mm, within the upper range for "*M. gracile*" (Tassy, 1981), and labial cingulae are present as low, narrow ribbons alongside each metaloph. In another palatal fragment from the Qasr el Sagha Fm. that preserves M2–3 (M9114), the molars are large (M3 L=35.5, W=28.4) and appear to belong to the *M. lyonsi*-morph; nonetheless, there is a trace of a labial cingulum in each, very low, narrow, and discontinuous alongside the lophs.

Associated lower molars of specimen DPC 20613 from about 15 m below the base of the Dir Abu Lifa Mb. of the Qasr el Sagha Fm. fit the *M. lyonsi*-morph in size and shape (m1 L=26.0 mm, W=23.8 mm; m2 L=29.6 mm, W=25.2 mm; m3 L=37.0 mm, W=28.6 mm) and anticipate the more complex occlusal morphology of *M. trigodon*, with diminutive accessory conules in their transverse valleys, and m3 with a raised, not shelf-like, postcingulid, comprising an incipient third lophid.

MOERITHERIUM TRIGODON Andrews 1904a

(Tables 3.1 and 3.2; Figures 3.1, 3.10, 3.14, and 3.16)

Partial Synonymy—*Moeritherium trigonodon* Andrews, 1906; *M. lyonsi* (in part) Andrews, 1906; *M. trigonodum* Schlosser, 1911; *M. andrewsi*, Schlosser, 1911; *M. (trigodon) trigonodon* Matsumoto, 1922; *M. andrewsi* Matsumoto, 1923; *M. andrewsi* Osborn, 1936; *M. lyonsi* (in part) Osborn, 1936; *M. andrewsi* Lehmann, 1950; *M. andrewsi* Vaufrey, 1958

Erected originally by Andrews (1904a, 1906) to accommodate moerithere specimens from the "Fluvio-marine beds" of the Fayum (now the Jebel Qatrani Fm.), *Moeritherium trigodon* was based on few characters to distinguish it from *M. lyonsi*. These include a lower m3 talonid composed mainly of a large, single cuspule shifted to the buccal side of the crown; m3 occlusal profile tapered posteriorly; and striking size increase of cheek teeth toward the back of the tooth row. Lower molars are reported to be relatively narrower than those of *M. lyonsi* and to have some development of accessory conules between their lophids (Tobien, 1971; Coppens and Beden, 1978). After Schlosser (1911) created *M. andrewsi* to separate some specimens from the "Fluvio-marine beds" from moeritheres of the Qasr el Sagha Fm., Matsumoto (1922) subsumed the species into *M.* "*(trigodon) trigonodon*," but soon thereafter placed some material of larger individuals from the "Fluvio-marine beds" into *M. andrewsi*, treating it as a valid species separate from the smaller morph from the formation, *M. trigodon* (Matsumoto, 1923). Simons (1968) mentioned only *M. andrewsi* in his mammalian faunal list for the Jebel Qatrani Fm., but that was probably an oversight rather than a taxonomic decision.

Matsumoto (1923) provided diagnoses distinguishing the larger (*M. andrewsi*) and smaller (*M. trigodon*) moerithere morphs in the Jebel Qatrani Fm.: *M. andrewsi* has lower premolars not foreshortened compared with associated molars, relatively narrow lower cheek teeth, and a large and heavily built cranium that is posteriorly narrow, with a weak sagittal crest, and narrow zygomatic arches compared to overall length; *M. trigodon* is smaller (p2–4 L=63–70 mm, m1–3 L=83–85 mm versus p2–4 L=70–73 mm, m1–3 99–100 mm in *M. andrewsi*), with a small and lightly built cranium that has strong sagittal and nuchal crests and is relatively wide posteriorly with broad zygomatic arches, and a fusiform p2. Tobien (1971, 1978a) was evidently unimpressed by these differences, attributing them to interindividual variation, including sexual dimorphism, and consequently treated *M. andrewsi* as a junior synonym of *M. trigodon*. Other authors since then have tended to follow this interpretation (e.g., Coppens and Beden, 1978;

Shoshani et al., 1996; Sanders et al., 2010a). Given what is known from the large Birket Qarun Fm. sample about moerithere intraspecific variation, it is reasonable to have a working hypothesis of only a single moerithere species in the Jebel Qatrani Fm.

The most recent attempt to recalibrate the age of the Jebel Qatrani Fm., based primarily on sophisticated analysis of paleomagnetic sequence stratigraphy, suggests that the contact between the Qasr el Sagha and Jebel Qatrani Fms. is located about 48 m below the Eocene-Oligocene boundary (~33.9 Ma), making the lowermost part of the formation a little older than 34 Ma (Seiffert, 2006; Seiffert et al., 2008). Prior to this careful analysis, the lower part of the formation was considered to either be wholly Oligocene in age (e.g., Bown and Kraus, 1988 and references therein; Gingerich, 1993) or substantially within the late Eocene (e.g., Kappelman et al., 1992). Andrews (1906) believed the entire "Fluvio-marine" series to be late Eocene and the Qasr el Sagha Fm. to be middle Eocene in age. Though the Widan el-Faras basalt capping the formation has an $^{40}Ar/^{39}Ar$ date of 23.6 Ma (Kappelman et al., 1992), the highest vertebrate-bearing localities in the formation are about 30.2–29.5 Ma (Seiffert, 2006). These localities, which famously produced the well-known basal catarrhine primate *Aegyptopithecus zeuxis*, are in Gagnon's (1997) Fayum faunal zone 4, by which time moeritheres were no longer present in the sequence. Thus, the fossil record of moeritheres appears to span ~37–30 Myr.

The "Fluvio-marine beds" were renamed the Jebel Qatrani Fm. by Said (1962a), and are visually discernible from the subjacent Qasr el Sagha Fm. by their assorted brightly colored mudstones, sandstones, and gravelly sandstones, with little deposition of limestone, forming a sedimentary sequence reaching 340 m in thickness (Bown and Kraus, 1988). Geologically and paleontologically, it is the most intensely studied sequence of the African late Paleogene. The depositional environments represented in the Jebel Qatrani Fm. are dominated by alluvial accumulations, with facies including sediments laid down by braided streams, low-sinuosity streams, well-developed floodplain deposits, and ephemeral stream deposits (El-Araby and El-Barkooky, 2000). As at the younger Oligocene site of Chilga, Ethiopia, the formation preserves a rare and rich paleobotanical record (comprised of fruits, leaves, rhizomes and wood representing lianes, mangroves, tropical monsoonal forests, seasonally dry forests of tall trees, and possibly savanna woodlands) alongside substantial and diverse vertebrate assemblages, in a sub-tropical to tropical lowland coastal plain setting (Bown et al., 1982; Tiffney, 1991; Jacobs, 2004), as well as trace fossils that provide abundant evidence of invertebrates such as termites, annelids, and crustaceans (Bown, 1982).

The vertebrate fossil component of the formation embodies the classic endemic African fauna of the late Paleogene, providing a striking contrast with the succeeding African faunas of the early and middle Miocene that incorporated numerous immigrant taxa from Eurasia. In addition to palaeomastodonts and moeritheres, the formation has yielded the most diverse record of primitive anthropoids including the earliest basal catarrhine primates (Seiffert et al., 2010), strepsirrhine primates (Godinot, 2010; Seiffert et al., 2005b; Seiffert, 2006; Gunnell et al., 2018), and carnivorous creodonts (Lewis and Morlo, 2010), many different lineages representing at least eight genera of hyraxes (the dominant ungulate of the African late Paleogene; Rasmussen and Gutiérrez, 2010), the anthracothere genera *Bothriogenys* and the rarer *Qatraniodon* (Holroyd et al., 2010), immense-sized arsinoitheres (Tanner, 1978; Sanders et al., 2010b), the most diverse record of the small, enigmatic ptolemaiids (Gunnell et al., 2010), four new genera and four new species of bats (Gunnell et al., 2008; Gunnell, 2010), and a variety of rodents useful for biochronological correlation, including many hystricognathous taxa (Winkler et al., 2010; Sallam et al., 2011) (see also Simons, 1968; Gagnon, 1997).

Along with a tendency to have larger cheek teeth than earlier moeritheres (especially when "*M. andrewsi*" is incorporated into the species), *M. trigodon* molars and premolars are characterized by more swollen cusps, greater expression of accessory features (e.g., conules, crescentoids, protostylids), multiplication of mesoconelets, and tendency for molarization of p4. The m3 of the type molar series of *M. trigodon*, specimen M 8499, is archetypal in having as many as two mesoconelets per half-lophid, doubled anterior accessory conules, and doubled distal conelets on the heel or postcingulid, that is especially prominent in m2–3. The nearly unworn m3 has very salient enamel wrinkling, as well. In this molar, even though the talonid heel is very large and comprises an incipient third lophid, the hypoconulid is shifted to the buccal side and consequently the occlusal outline is distally tapered to that side. A huge p4, specimen DPC 2257, is 27.4 mm in length with a greatest width of 17.6 mm. It has a hypertrophied protostylid, a crescentoid descending anteriorly from the hypoconid, and nascent development of an entoconid. A large m2 (L=36.0 mm, W=27.0 mm), DPC 14066, from very low in the Jebel Qatrani Fm. sequence, has very rounded, bulbous cusps, strong development of accessory conules, and a very large hypoconulid. SMNS 47757, p2-m3, is probably a "*M. andrewsi*" morph of *M. trigodon*, with the premolar and molars series about 70 and 97 mm in length, respectively. The third premolar as well as p4 has a protostylid, enamel is strongly ptychodont, p4 is molarized, m2 has doubled conelets in its heel (=hypoconulid), accessory conules are well expressed, and mesoconelets are multiplied. In contrast, a smaller specimen from the same collection, p3-m3 SMNS 13092, appears to represent a "*M. trigodon*" morph of *M. trigodon*, with a smaller molar series of about 74 mm in length. The molars also exhibit strong development of accessory conules and large postcingulids, but the enamel is smoother and there are only single mesoconelets in each half-lophid.

Unquestionably, moeritheres of the Jebel Qatrani Fm. maintained the high degree of variation observed in other congenerics; for example, the type cranium of "*M. andrewsi*," subadult individual CGM 7867, has simpler teeth that look more like premolars and molars of *M. lyonsi*. P3–4 are particularly wide relative to length. Length of the cranium from nasion to the nuchal crest is 324.8 mm and greatest width across the zygomatic arches is 235.4 mm. A well-preserved subadult dentary from the upper part of the Jebel Qatrani Fm., DPC 6221, has a simple m1 crown with bulbous cusps and very minor development of a crescentoid anterior to the hypolophid, also more similar morphologically to *M. lyonsi*. Its premolar row is only about 66 mm in length, suggesting that it probably derives from a female individual (or "*M. trigodon*" morph). Its right i2 has modest dimensions of 21.5 mm (MD)×12.3 mm (LL). Its symphysis begins parallel to the anterior of p3, there is a short diastema between p2 and i2, and the anterior root of the ramus is located anterior to m2. The mandibular foramen is relatively small and opens at the same level as the tooth row, and the mental foramina are doubled and located below p2 and p4. These features are primitive among proboscideans (Ferretti and Debruyne, 2011). The coronoid process is only slightly lower than the glenoid process, and the ramus is anteroposteriorly very broad.

It is worth mentioning DPC 4211, a left astragalus cataloged as "*Moeritherium*" from low in the Jebel Qatrani Fm. (locality L43), as the moerithere astragalus has not been well described and is only figured in Gingerich et al. (1990). DPC 4211 has a greatest length of 74.1 mm, greatest width of 65.1 mm, and a tibial platform measuring L=53.0 mm×W=35.5 mm. Its navicular articular surface measures W=46.5 mm×H=31.3 mm and has medial and lateral facing segments. Ventrally, the laterally positioned ectal facet is clearly larger than the medially positioned sustentacular facet, and wraps around to the posteromedial side of the astragalus. The sustentacular facet does not contact the navicular facet. This astragalus also preserves a barely distinct medial tuberosity, not nearly as saliently projecting as depicted for *M.* "*gracile*" astragalus YPM 24186 from the Qasr el Sagha Fm. (Gingerich et al., 1990:fig. 2a, b), which it does not resemble. In YPM 24186, the ectal facet is shown as much larger than the sustentacular facet, and it also does not contact the navicular articular surface (Gingerich et al., 1990:fig. 2b). Dimensions given in Schlosser (1911) for a *M. lyonsi* astragalus (L=36 mm, W=28 mm) are impossibly too small for an animal the size of a moerithere. It is alternatively possible that DPC 4211 is instead a very squat example of a creodont astragalus, as a number of hyaenodontid creodonts occur in the Jebel Qatrani Fm. (Lewis and Morlo, 2010).

The moeritheres of the Fayum sequence, set in a reliable chronostratigraphic framework (Figure 3.2A), have potential for biochronological correlation with other sites. This is particularly true for dating the fossil beds of Dur at Talah, Libya. Arambourg (1963) was the first to report moeritheres from the site. Savage (1969, 1971) subsequently described the Evaporite and Idam Units together as lower and upper marls, and identified marine mammals in the lower marls and moeritheres throughout the middle and upper marls. Wight's (1971, 1980) more detailed work on the proboscideans from the site recognized *M. lyonsi* in the Evaporite Unit (his "Raqaba Mb.") and *M. trigodon* in the higher Idam Unit. In comparison with the Fayum sequence, this moerithere record supported his hypothesis that these units are late Eocene and early Oligocene, respectively. In his 1971 dissertation, Wight listed 14 localities at Dur at Talah yielding moerithere fossils, most of them in the Idam Unit; in his 1980 publication, he expanded the list to 18 localities. Nonetheless, moeritheres are relatively rare denizens at Dur at Talah. Later, Tassy (1981), in a classic anatomical study, described the cranium of a moerithere (MNHN specimen) from the Evaporite Unit and placed it in *M.* cf. *lyonsi*. Tassy (1981) noted that Matsumoto (1923) had defined *M. lyonsi* as having an upper premolar and molar row measuring 67–78 and 85 mm, respectively, compared with 62 mm for the upper premolar series and 75–79 mm for the molar row in "*M. gracile*," and observed that the reconstructed length for upper cheek teeth of the cranium from the Evaporite Unit falls between these ranges. However, the length of the cranium from the nuchal crest to anteriormost point on the premaxilla is 314 mm, suggesting it was a small, possibly female, individual. Morphologically, the preserved portions of the specimen do not differ from other moerithere crania and there is no reason to doubt its identification as *M. lyonsi*.

In contrast, Jaeger et al. (2012) claimed that new (undescribed and not figured) moerithere specimens from Dur at Talah document the presence of *M. chehbeurameuri* at the site. The new fossils include a cranium that is said to be very different from the MNHN cranium and other *M. lyonsi* crania. They erroneously claim that in *M. lyonsi*, the orbit is located anterior to the premolars, and that in the new material that they place in *M. chehbeurameuri*, the rostrum is longer than in other moeritheres with the orbits over the premolars. However, in the MNHN cranium and other specimens of *M. lyonsi*, as well as in the *M. chehbeurameuri* crania from the Birket Qarun Fm. and a *M. "andrewsi"* (*M. lyonsi* of Andrews (1906:pl. IX)) cranium, the orbits are located above P2 and the rostra are not elongated (Osborn, 1936; Tassy, 1981). One specimen where the orbits *are* situated anterior to the premolar row and thus appears to have an elongated rostrum is AMNH 13430, Osborn's reconstructed *M. trigodon* cranium (Matsumoto, 1923:fig. 7; Osborn, 1936:fig. 46A). Thus, rather than representing a *primitive* condition, it is likely that this constitutes a *derived* character state for moeritheres.

Examination of moerithere teeth from the Evaporite and Idam Units confirm Wight's (1980) identification of *M. lyonsi* and *M. trigodon* in these beds (Sanders and Seiffert, 2014). An incomplete mandible (68.15/5) with alveoli for incisors, roots of left and right p2-m1, and worn left and right m2–3 from the Evaporite Unit fits with the large morph of *M. lyonsi* (Figure 3.16A). The m2–3 have dimensions of L=29.8 mm, W=24.1 and 37.2 mm, 26.5 mm, respectively, simple occlusal morphology, and exhibit very large hypoconulids distally. An upper molar (68.4/14) in moderate wear from the Idam Unit was listed as an M3 (Wight, 1971) but is probably an M2. It has rounded cusps and a prominent posthypoconule, and seems to be a better fit morphometrically with *M. trigodon* than with *M. chehbeurameuri*. Similarly, other cheek teeth from the Idam Unit also can be accommodated readily in *M. trigodon*: 68.12/10, listed as a P3 (Wight, 1971), but is probably a P4, is relatively wide, has rounded cusps, and a distinct postmetaconule; a series of lower teeth (right p3–m3, left m2–3) in "exploded," fragmentary dentaries (69.50/6) are very derived, particularly the m3s, which are large and elongate (L=39.5 mm, W=25.3 mm), with distinct anterior crescentoids running into the transverse valleys and a substantial heel with a massive hypoconulid; an M1 with little wear (69.50/45) is much larger than M1s from Bir el Ater (L=26.7 mm, W=23.0 mm) and has both postmeta- and posthypocristae; and left and right m3s (68.3/1), triangular in occlusal shape with heavily wrinkled enamel, bulbous cusps, and strong expression of accessory conules (compare with Jebel Qatrani m3 M8499, *M. trigodon*; Figure 3.16B and C). The lower m3s in particular are more derived than moerithere m3s from the late Eocene Fayum sequence, supporting an early Oligocene age for the Idam Unit.

ELEPHANTIFORMES Tassy, 1988

The taxon Elephantiformes was erected to encompass palaeomastodonts and more advanced proboscideans now placed in the Elephantimorpha, that is, elephant-like proboscideans (Tassy, 1988). These taxa are contrasted with the lophodont proboscideans of the Paleogene (e.g., phosphatheres, numidotheres, barytheres), as well as with moeritheres and deinotheres. Elephantiformes is well-defined primarily by gnathodental characters and appears to be monophyletic (Tassy, 1988). Although the list of traits defining the taxon is extensive, the most important of these relate to development and functional isolation of true upper and lower tusks (long mandibular symphysis, widened premaxilla, mandibular diastema posterior to the distal symphyseal border, tusks evergrowing), development of a trunk (marked by facial retraction and increased space between upper incisors; orbits retracted above M1, nasal fossa present), and expansion in size and loph(id) number of molars (presence of molar hypoconulids and postentoconulids,

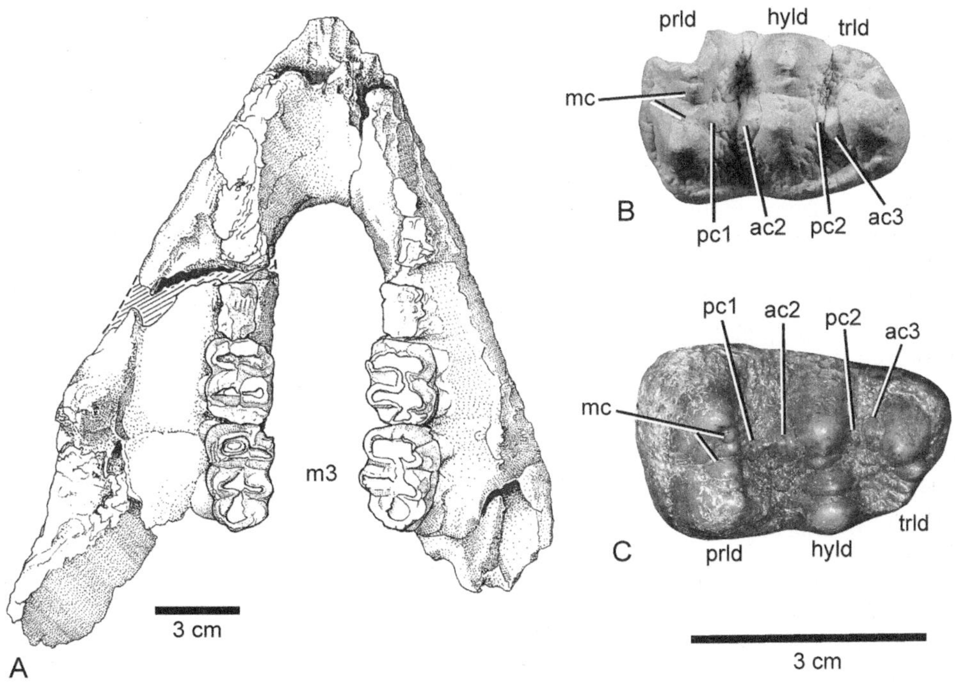

FIGURE 3.16 Mandibular and molar specimens of late Eocene-early Oligocene moeritheres from Dur at Talah, Libya, with a molar from early Oligocene Fayum, Egypt for comparison. Abbreviations: ac, anterior accessory conule; hyld, hypolophid; m, lower molar; mc, mesoconelet; pc, posterior accessory conule; prld, protolophid; trld, tritolophid. (A) anterior to the top. (B, C) anterior to the left and to the same scale. (A) Partial mandible with intact left and right m2–3, specimen 6815-5 from the late Eocene Evaporite Unit, Dur at Talah, Libya, occlusal view, *Moeritherium lyonsi* (Wight, 1971:fig. 3.2. (B) Left partial m3 (cast), specimen 683-1 from the early Oligocene Idam Unit, Dur at Talah, Libya, occlusal view, M. *trigodon* (image courtesy of A. Lister). (C) Right m3, M8499 from the early Oligocene Jebel Qatrani Fm. Fayum, Egypt, occlusal view, *M. trigodon*. Note the similar occlusal complexity of specimen 683-1 and M8499, typical of progressive *Moeritherium trigodon* third molars.

second interloph(id)s in molars, molars trilophodont and transversely expanded), features which in part can be linked to substantial increase in body size, also evidenced by graviportal skeletal morphology (Figure 3.17A and B; see Tassy, 1988). Although the mechanism of horizontal tooth displacement was not present in basal elephantiforms (Sanders, 2017), tooth wear patterns in cheek teeth of primitive elephantiforms such as palaeomastodonts indicate delay of sequential emergence of molars, such that first molars were heavily worn by the time that third molars were in occlusion, an important initial component of the mechanism and with implications for expanding the life cycles of these animals. Although elephantiforms began to become the dominant proboscideans during the late Paleogene, there is new evidence suggesting that their ancestry is more deeply rooted in bunolophodont forms from the earlier Paleogene (see Chapter 2; Hautier et al., 2021). Moreover, there are no good late Paleogene candidates to serve as potential phylogenetic predecessors to elephantiforms, and most previously documented proboscideans from the early Paleogene are lophodont species—it is easier to envision evolution of lophodonty from bunodonty, but not the reverse (Hautier et al., 2021).

PALAEOMASTODONTIDAE Andrews, 1906

(Tables 3.1, 3.2, 4.1, and 4.2; Figures 3.1, 3.17, and 4.1)

Palaeomastodonts have been known since the early 1900s (e.g., Andrews, 1901d, 1903a, b, 1905, 1906; Andrews and Beadnell, 1902), and until the 1980s were, with moeritheres and barytheres, the oldest known proboscideans. They are principally Oligocene taxa (Table 3.1; but see Table 4.1) that comprise the first of the elephant-like proboscideans in the Elephantiformes. All palaeomastodont fossils from the Fayum sequence derive from the Jebel Qatrani Fm., mostly from the "Lower Fossil Wood Zone" (Table 3.1; Andrews, 1906; Simons, 1968; Coppens et al., 1978; Gagnon, 1997). Although Coppens et al. (1978) allude to the presence of palaeomastodonts in the Qasr el Sagha Fm., they are conspicuously absent from authoritative faunal lists for this geological unit (e.g., Osborn, 1936; Simons, 1968; Holroyd et al., 1996; and see Andrews, 1903b).

Palaeomastodonts are primitive compared with other elephantiforms in their retention of a vertical tooth replacement mechanism and simultaneous use of their adult dentition (Figure 3.17C and D; Andrews, 1903a; Sanders, 2017). Thus, there is no reason to seriously consider Coppens et al.'s (1978) suggestion that palaeomastodonts belong in the Gomphotheriidae. These proboscideans are characterized by short nasal bones and retraction of the nasal aperture, suggesting presence of an elephantine-like trunk, which corresponds functionally with the presence of upper and lower tusks. Alternatively, however, Osborn (1909) proposed a functional scenario that envisioned the lower tusks of palaeomastodonts working in concert with prehension of the upper lip to seize food, obviating development of a true proboscis. Their body size was also elephantine: Allometric analyses of body mass to limb bone size indicate mean weights of 973–3,691 kg for *Palaeomastodon beadnelli* (Christiansen, 2004), and a volumetric approach to estimating body mass from limb bone dimensions yields a mass of >2.5 tons and a shoulder height of 220 cm for this species (Larramendi, 2016). Andrews (1906) estimated that palaeomastodonts varied considerably between moeritheres to Asian elephants in size.

The molars of palaeomastodonts are bunolophodont and brachyodont, with two and one-half to three loph(id)s in each. Their teeth are similar to moerithere molars, except that they exceed two loph(id)s in each. Because of the very specialized cranial morphology of moeritheres associated with their semi-aquatic existence, however, it is unlikely that moeritheres and palaeomastodonts had an ancestral-descendant relationship. Compared with moeritheres, the palaeomastodont dental formula is reduced, involving loss of some upper and lower incisors, upper canines, and a lower premolar (Table 3.2). The phylogenetic antecedants to palaeomastodonts remain elusive, as there are no suitable ancestral candidates among older, lophodont plesielephantiform proboscideans. Although taxonomy of the Palaeomastodontidae has been unstable since its inception (Moustafa, 1974a, b; El-Khashab, 1979), the Family is now widely accepted to include two genera, *Palaeomastodon* and *Phiomia*; systematic interpretation of these taxa has profound implications for monophyly of the group and conceptions of the origins of more derived elephantimorph proboscideans.

Andrews (1906) defined Palaeomastodontidae on traits of the skull and dentition: presence of enlarged upper and lower incisors, the lowers procumbent and projecting anteriorly beyond the limits of the cranium and the uppers downturned, with a lateral band of enamel; mandible with a long, spoutlike symphysis; and cranium with a retracted nasal aperture. At first, he split the Family between different forms of *Palaeomastodon*, based on position of the posterior of the symphysis and anterior mental foramen relative to the premolars, and degree of development of third molar loph(id)s and accessory conules (Andrews, 1905). Later, he recognized both *Palaeomastodon* and *Phiomia* (Andrews, 1906). By 1908, however, Andrews had changed his mind and subsumed all palaeomastodonts back into *Palaeomastodon*. Subsequently, Matsumoto (1922, 1924a) initiated taxonomic reorganization of the Palaeomastodontidae and re-allocated its species among both genera. Similar to the methods of his 1923 taxonomic revision of moeritheres, Matsumoto's (1924a) reconsideration of palaeomastodont taxonomy was character-driven and well supported by metric data. He distinguished *Phiomia* from *Palaeomastodon* primarily on differences in mandibular and cheek tooth features, including: more elongate symphysis, longer, narrower palate, more complete development of third loph(ids) in molars, and presence of molar accessory conules. These traits remain effective for differentiation of palaeomastodont genera (Sanders et al., 2010a). Osborn (1936) was even more effusive about differentiation of *Phiomia* from *Palaeomastodon*, and placed them in separate families, Bunomastodontidae and Mastodontidae (later, Palaeomastodontidae [Osborn, 1942]). Alternatively, Tobien (1971) and Coppens et al. (1978) preferred to divide these taxa only at the level of subgenera within the genus *Palaeomastodon*. A more extreme taxonomy is that of Lehmann (1950), who placed all palaeomastodonts into a single genus and species, *P. beadnelli*. However, differences in molar occlusal morphology among different palaeomastodont forms validate maintenance of two genera, and phylogenetic implications of these differences may require an even higher level of taxonomic partition (Sanders et al., 2004).

Postcranial remains of palaeomastodonts are rare and do not appear to differ significantly between the two genera. The specimens that have been recovered anticipate the skeletal morphological bauplans of elephants and mastodons, exhibiting adaptations to graviportal posture in support of heavy bodies (Andrews, 1906). The femur is pillar-like and its head faces superolateral (Figure 3.17A); the tibia is also pillared (Figure 3.17B); the humerus has a deep olecranon fossa to stabilize the forelimb in extension and its head is oriented proximally; and the ulna has a low olecranon process that is reflected posteriorly to permit habitual vertical

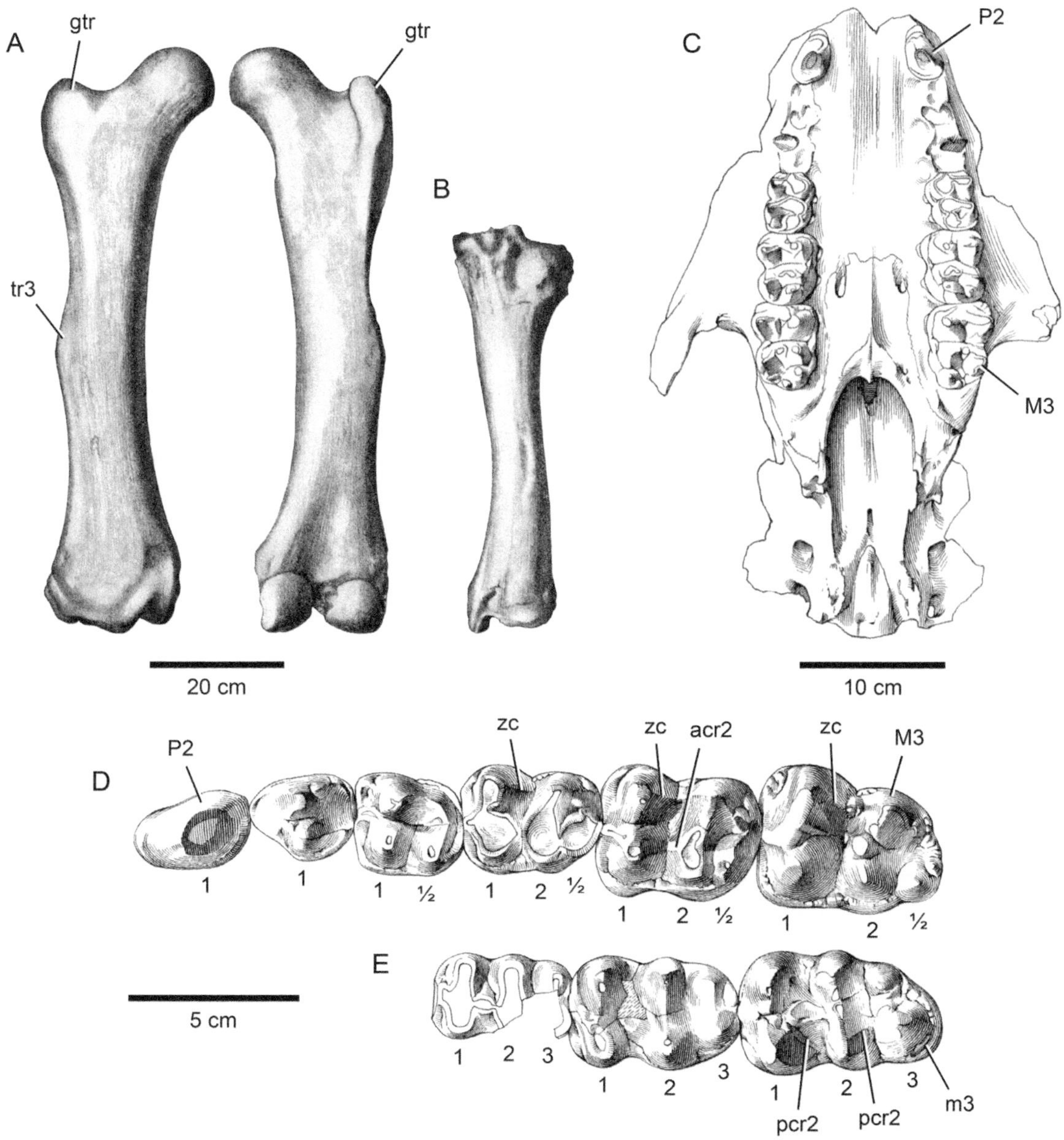

FIGURE 3.17 Cranial, postcranial, and dental remains of early Oligocene *Palaeomastodon beadnelli* from the Jebel Qatrani Fm., Fayum, Egypt. Abbreviations: acr, anterior crescentoid; gtr, greater trochanter; M/m, upper/lower molar; P, upper premolar; pcr, posterior crescentoid; tr3, third trochanter; zc, zygodont crest; 1, 2, 3, …, loph(id) numbers counted from the anterior of the crown. (A, B) to the same scale. (D, E) to the same scale and anterior to the left. (A) Femur, C10061, anterior and posterior views, from Andrews (1906:pl. XVI, fig. 3). (B) Tibia, C10015, anterior view, from Andrews (1906:pl. XVI, fig. 4). (C) Partial cranium, AMNH 13449, occlusal view, from Osborn (1936:fig. 91A1). (D) Upper tooth series P2–M3, AMNH 13449, occlusal view, from Osborn (1936:fig. 92A). (E) Lower molar series m1–3, AMNH 14547, occlusal view, from Osborn (1936:fig. 92B). (Images from Osborn [1936] with permission of the American Museum of Natural History).

alignment on the humerus (Andrews, 1906; Sanders et al., 2010a).

Femoral and tibial dimensions, measured on bones of different individuals (C10017 and C8883), suggest proportions of the lower to upper leg slightly greater than those of extant elephants (tibial L=635 mm, femoral L=875 mm). As in extant elephants, the internal distal condyle of the femur projects more than does the outer condyle, corresponding to asymmetry in height of the proximal tibial condyles (Smuts and Bezuidenhout, 1994), indicating that the femora were angled moderately inward toward lower legs set under the body in neutral stance. The large third trochanter located about midshaft on the femur (a little higher than in elephants) is evidence for an ability to powerfully extend the leg. The tibia has a prominent distal malleolus and deep distal articular surface for stability on the astragalus. The palaeomastodont astragalus has a very salient median process, a deep articulation for the tibial malleolus, much larger ectal than sustentacular facet, and its sustentaculum contacts the navicular articulation, which is strongly canted to the medial side.

The humeral shaft is torqued, with salient deltoid and supinator crests, a deep bicipital groove, and a broad distal articulation. This element also exhibits a proximally projecting greater trochanter but lacks an entepicondylar foramen. It differs from the humerus of *Moeritherium* in the orientation of its head and more modest flare of its epicondyles. Dimensions of a complete specimen (C8881) are L=452 mm, width at the distal end=138 mm (Andrews, 1906). An even larger humerus reported by Andrews (1908), that belongs to a partial skeleton associated with a skull and dentition of *Phiomia* ("*Palaeomastodon wintoni*"), measures 560 mm in length and 145 mm in breadth at its distal end. Length of a complete ulna (C8028) is 380 mm. The radius fits tightly against this element in habitual pronated position (Andrews, 1908). It differs from the ulna of *Moeritherium* especially in having a less projecting proximal than distal arc of the triradiate articulation for the humerus and a shorter, considerably more posteriorly reflected olecranon process. These differences illustrate the contrast between animals that habitually stand on flexed versus graviportal limbs.

PALAEOMASTODON Andrews, 1901d
PALAEOMASTODON BEADNELLI Andrews, 1901d

(Tables 3.1 and 3.2; Figures 3.1, 3.17 and 3.19)

Partial Synonymy—*Palaeomastodon parvus* Andrews, 1905; *P. barroisi* (in part) Pontier, 1907; *P. intermedius* Matsumoto, 1922; *P. parvus* Matsumoto, 1923; *P. intermedius* Matsumoto, 1923; *Palaeomastodon* (in part) Lehmann, 1950; *Palaeomastodon* (*Palaeomastodon*) *beadnelli* Coppens et al., 1978

Following re-allocation of palaeomastodont fossils among different genera, the few specimens remaining in *Palaeomastodon* indicate that this taxon was far less common in the Jebel Qatrani Fm. than *Phiomia* (Matsumoto, 1924a). *Palaeomastodon* is distinguished from *Phiomia* by the following features: palate broad and short (Figure 3.17C), mandibular symphysis shorter (anterior mental foramen located below p3 and back rim of the symphysis located far anterior to the premolar series), molars short and wide, greater anteroposterior compression of loph(id) apices, rudimentary or no development of accessory conules, no trefoil wear-figures of pretrite half-loph(id)s, and poor formation of molar third loph(id)s (molar loph(id) formulae usually 2½) (Figure 3.17D; Andrews, 1906; Matsumoto, 1924a).

Palaeomastodon beadnelli, the "ancient mastodon" named for Hugh Beadnell, who was instrumental in study of the geology of the Libyan desert and paleontology of the Fayum, is now used to encompass all the Fayum material assigned to *Palaeomastodon* (e.g., Coppens et al., 1978; Sanders et al., 2010a). Matsumoto (1924a) had divided the genus into three species on the basis of tooth size, with *P. "parvus"* having a p3-m3 series 197 mm in length, with m1–3 L=130 mm; *P. "intermedius"* with m1–3 L=159 mm, P2-M3 L=250 mm, and M1–3 L=150–152 mm; and *P. beadnelli* with p3-m3 L=285 mm and m1–3 L=194 mm. In the absence of more compelling features to allocate the genus among separate species, it is more reasonable to recognize only a single species of *Palaeomastodon* in the Fayum beds.

The cranium is less well represented than in *Phiomia serridens*, but preserved portions and dental formulae (Table 3.2) are similar, except for palatal proportions and occlusal features of the cheek teeth (Mastumoto, 1924a; Moustafa, 1974b; El-Khashab, 1979). Dental formulae are 1-0-3-3/1-0-2–3. Lophid formulae for P2-P4 are 1, 1½, and 1½, and for p3–4 are 1 and 2, respectively (Figure 3.17D). Molars tend to have incompletely developed third loph(-id)s, with poor expression of the posttrite tritoloph(id) (Figure 3.17D). For example, in AMNH 13449, the M3 paratype of *P. "intermedius"* (Osborn, 1936:figs. 94, 94), the third pretrite half-loph consists of an isolated conelet that is accompanied on the posttrite side by a tiny conelet, forming a very incomplete tritoloph. Where accessories are present in molars, they tend to take the form of crescentoid ridges rather than accessory conules, and early occlusal wear produces sharp ridges apically (Figure 3.17E). Zygodont crests may be present on the outer conelets of posttrite half-loph(id)s. The presence of zygodont crests and crescentoids (Figure 3.17D), along with antero-posteriorly compressed loph(id) apices, anticipates the morphology characteristic of mammutid molars.

An encephalization quotient (EQ) of 0.29 calculated by Benoit and colleagues (2019) for an individual of the species is very low compared with values for elephantimorph proboscideans. However, this result is unreliable as the cranial capacity of the individual (771 cm^3) was not directly measured but rather was estimated from drawings of an incomplete braincase, and the body mass that they used (2,500 kg) is cited from Larramendi (2016), though he does not provide this in his study. Christiansen (2004) calculated body masses for *P. beadnelli* ranging from 671 to 4,015 kg. Despite these problems, even coarse estimates of EQ suggest that *P. beadnelli* was far less encephalized than elephantimorphs of the Miocene to the present. Unfortunately, the lack of preserved braincases for the earliest, Oligocene examples of elephantimorphs makes it impossible to know the timing of significant increases in relative brain size among primitive elephantiform proboscideans (contra Benoit et al., 2019). Nonetheless, a casting of a brain cavity in a cranium of *M. lyonsi* produced an informative replica of a brain that Andrews (1906) felt was relatively very large for a mammal of its geological age, and interpreted the divisions of the lobes as similar to those observed in elephants.

PALAEOMASTODON SP. NOV. A (Sanders et al., 2004)

(Table 3.1; Figures 3.1 and 3.18)

Synonymy—aff. *Palaeomastodon* sp. nov. A Kappelman et al., 2003; aff. *Palaeomastodon* sp. nov. A Sanders et al., 2004

A small number of isolated molars comprise this species, which is documented only at early late Oligocene Chilga, Ethiopia, as part of a modest diversity of palaeomastodonts recovered from the site. It is differentiated from *Palaeomastodon beadnelli* by the larger size of some of its M3s, though both share incomplete or weak M3 trilophodonty.

The M3s (N=3) of *Palaeomastodon* sp. nov. A are composed of two and one-half-lophs (loph formula=x2½x) that do not exhibit accessory conules in the transverse valleys, but possess wrap-around cingular shelves on the lingual side of the crown (Figure 3.18A–C). Half-lophs may be superficially subdivided into two to three apical digitations. Zygodont crests are present posteriorly on the outer edge of the proto- and metalophs. The pretrite side of the tritoloph is composed of a stout conelet that may be subdivided into two apical digitations, and is accompanied on the posttrite side by a single diminutive conelet. Half-lophs are superficially divided in the midline by a shallow median sulcus. Very slight crescentoids may also be present on the posterior side of the first two pretrite half-lophs. Dimensions of the Chilga M3s range from L=67.2–80.2 mm, W=49.0–57.5 mm (Sanders et al., 2004). The molars are very brachyodont, with maximum height of only 30.5 mm. Arrangement of the bulbous cusps is bunolophodont. Wear on the occlusal surface of the lophs (for example, in M3 specimen CH 35-V-23) indicates that the chewing pattern is gomphotherian (von Koenigswald, 2016a), with orthal interdigitation of cusps of corresponding upper and lower molars on the buccal side during the Phase I power stroke and strong Phase II lingual or mesiolingual movement of these molars to grind across the pretrite side of the occlusal platform.

Features such as incomplete formation of the tritoloph in M3, lack of accessory conules, and some expression of crescentoids and zygodont crests differ from the condition typical of *Phiomia* molars and mark a resemblance of the Chilga specimens to molars of *Palaeomastodon beadnelli* (Matsumoto, 1924a; Tobien, 1978a; Tassy, 1994b). M3s from Chilga are particularly similar morphologically to the paratype M3 of *Palaeomastodon "intermedius"* from the Fayum, Egypt (AMNH 13449; see Osborn, 1936:figs. 93, 94), particularly in the obliquity of the second pretrite half-loph. The greater dimensions of Chilga *Palaeomastodon* molars compared with those from the Fayum indicates continued phyletic size increase over time, contributing to the emerging scenario that proboscideans were dynamically evolving through the Oligocene, prior to and independent of episodic faunal invasions from Eurasia in the early Neogene.

PALAEOMASTODON SP. NOV. B (Sanders et al., 2004)

(Table 3.1; Figures 3.1, 3.18, and 3.19)

Synonymy—aff. *Palaeomastodon* sp. nov. B Kappelman et al., 2003; aff. *Palaeomastodon* sp. nov. B Sanders et al., 2004

This species is represented by just a couple of molars from the early late Oligocene of Chilga, Ethiopia, most notably a severely step-fractured M3 (CH14-11) that was restored to its original shape by re-aligning cast sections across breaks in the crown (Figure 3.18D and E; Sanders et al., 2004). The M3 is distinguished from other palaeomastodont molars at Chilga and elsewhere in the sample by its large size (L=93.7 mm, W=66.4 mm; Figure 3.19), broad rectangular occlusal outline, salient expression of zygodont crests and pretrite crescentoids, more pronounced lingual cingulum, and greater enamel rugosity (Sanders et al., 2004). It is half again larger than M3s of *Palaeomastodon beadnelli* from the Fayum, Egypt (Andrews, 1906; Matsumoto, 1924a; Lehmann, 1950; Sanders et al., 2004). The crown is very low, with a height of only about 38.0 mm. Each half-loph is composed of a large outer conelet accompanied by a smaller, lower mesoconelet, except for the tritoloph, which exhibits the condition typical of *Palaeomastodon* in having only a relatively small, isolated conelet comprising its posttrite half-loph.

A very large left M2, CH25-16, appears to correspond in size and morphology to M3 specimen CH14-11 and is assigned to the same taxon. It exhibits a loph formula of x3x, with a narrow, low cingular ribbon continuing on both the lingual and buccal sides of the crown, and has dimensions of L=83.0 mm, W=59.2 mm, and H=+30.7 mm (very brachyodont); the enamel is very thick at 5.0 mm. These dimensions are quite large compared with those for *Palaeomastodon beadnelli* M2s. The specimen is fully in wear but enough remains of the crown to see that the half-lophs have one to two conelets and are low and bulbous. In lateral view, the transverse valleys are very open. There are no accessory conules on the occlusal surface of the crown, but the posttrite half-lophs have prominent zygodont crests and the proto- and meta- pretrite half-lophs have small

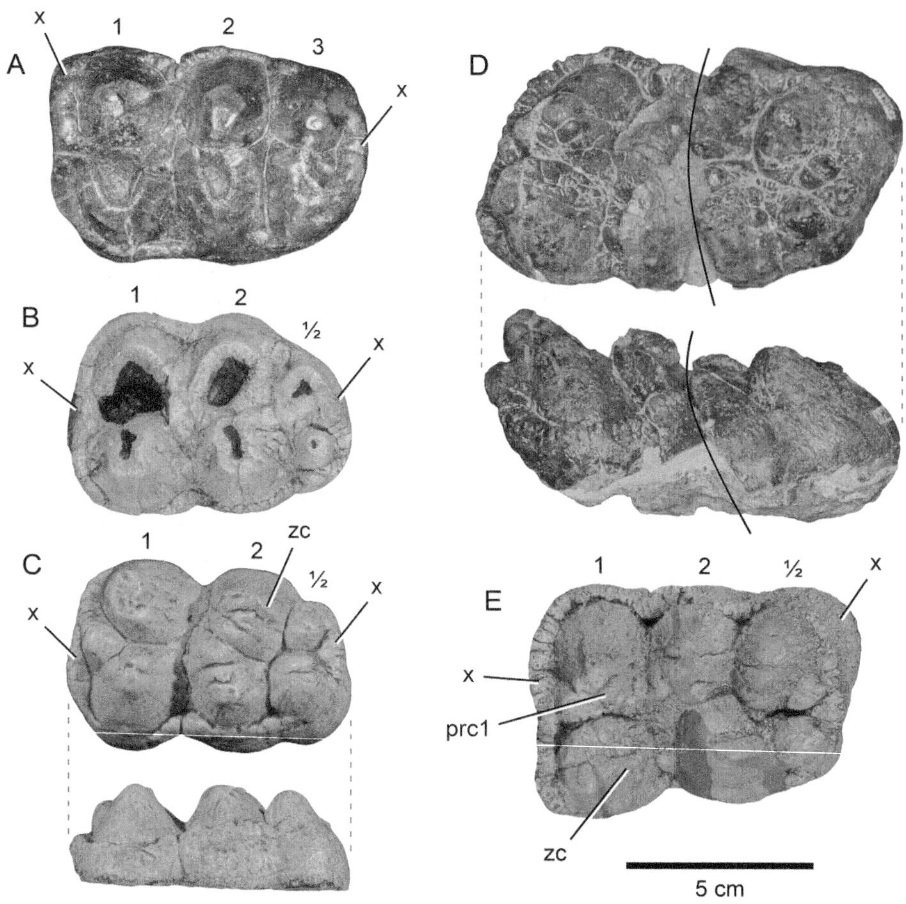

FIGURE 3.18 Palaeomastodont molars from early late Oligocene Chilga, Ethiopia. Abbreviations: prc, posterior crescentoid; x, pre- or postcingulum; zc, zygodont crest; 1, 2, 3, ..., loph number counted from the anterior of the crown. Curved line indicates a step fracture through the crown. All specimens to the same scale and anterior is to the left. (A) Left M3, CH71-20, occlusal view, *Palaeomastodon* sp. nov. A. (B) Right M3 (cast), CH35-V-23, occlusal view, *Palaeomastodon* sp. nov. A. (C) Left M3 (cast), CH14-V-12, occlusal and lateral views, *Palaeomastodon* sp. nov. A. (D) Right M3, CH14-11, occlusal and lateral views, *Palaeomastodon* sp. nov. B. (E) Reconstructed right M3 (cast), CH14-11, occlusal view, Palaeomastodon sp. nov. B.

posterior crescentoids that descend into the transverse valleys.

Features of CH14-11 and CH25-6 such as zygodont crests and crescentoids and the zygolophodonty of the crown anticipate morphological traits distinctive of mammutids, and support the hypothesis erected by Matsumoto (1924a) and resurrected by Tobien (1971, 1978a) and more recently by Sanders et al. (2004, 2010a) that *Palaeomastodon* is ancestral to mammutids and *Phiomia* is the precursor to gomphotheriids. In this scenario, Palaeomastodontidae is paraphyletic. Alternatively, however, *Phiomia* has been hypothesized as ancestral to amebelodonts (Osborn, 1919, 1936; Borissiak, 1929; Tobien, 1973a), or, based on cladistic studies of proboscidean phylogeny, is the sister taxon to all elephantimorphs with *Palaeomastodon* as their immediate outgroup (Tassy, 1994b, 1996c; Shoshani, 1996). Recovery of Oligocene-aged gomphotheres and mammutids from Afro-Arabia (Shoshani et al., 2001b; Gutiérrez and Rasmussen, 2007; Rasmussen and Gutiérrez, 2009; Zalmout et al., 2010) makes it likely that the common ancestor of elephantimorph proboscideans was at least as geologically old as early Oligocene in age.

PHIOMIA Andrews and Beadnell, 1902
PHIOMIA SERRIDENS Andrews and Beadnell, 1902

(Tables 3.1 and 3.2; Figures 3.1, 3.19, 3.20, and 3.21)

Partial Synonymy—*Palaeomastodon minor* Andrews, 1904a; *P. wintoni* Andrews, 1905; *P. minus* Andrews, 1905; *P. wintoni* Andrews, 1906; *P. beadnelli* (in part) Andrews, 1906; *Phiomia serridens* Andrews, 1906; *P. barroisi* (in part) Pontier, 1907; "probably" *P. minor* Andrews, 1908; *P. wintoni* Andrews, 1908; *Ph. osborni* Matsumoto, 1922; *Ph.* (*minus*) *minor* Matusmoto, 1922; *Ph. wintoni* Matsumoto, 1922; *Ph. minor* Matsumoto, 1924a; *Ph. wintoni* Matsumoto, 1924a; *Ph. osborni* Matsumoto, 1924a; *Palaeomastodon* (in part) Lehmann, 1950; *Palaeomastodon* (*Phiomia*) *serridens* Coppens et al., 1978; *Ph. serridens* Sanders et al., 2010a

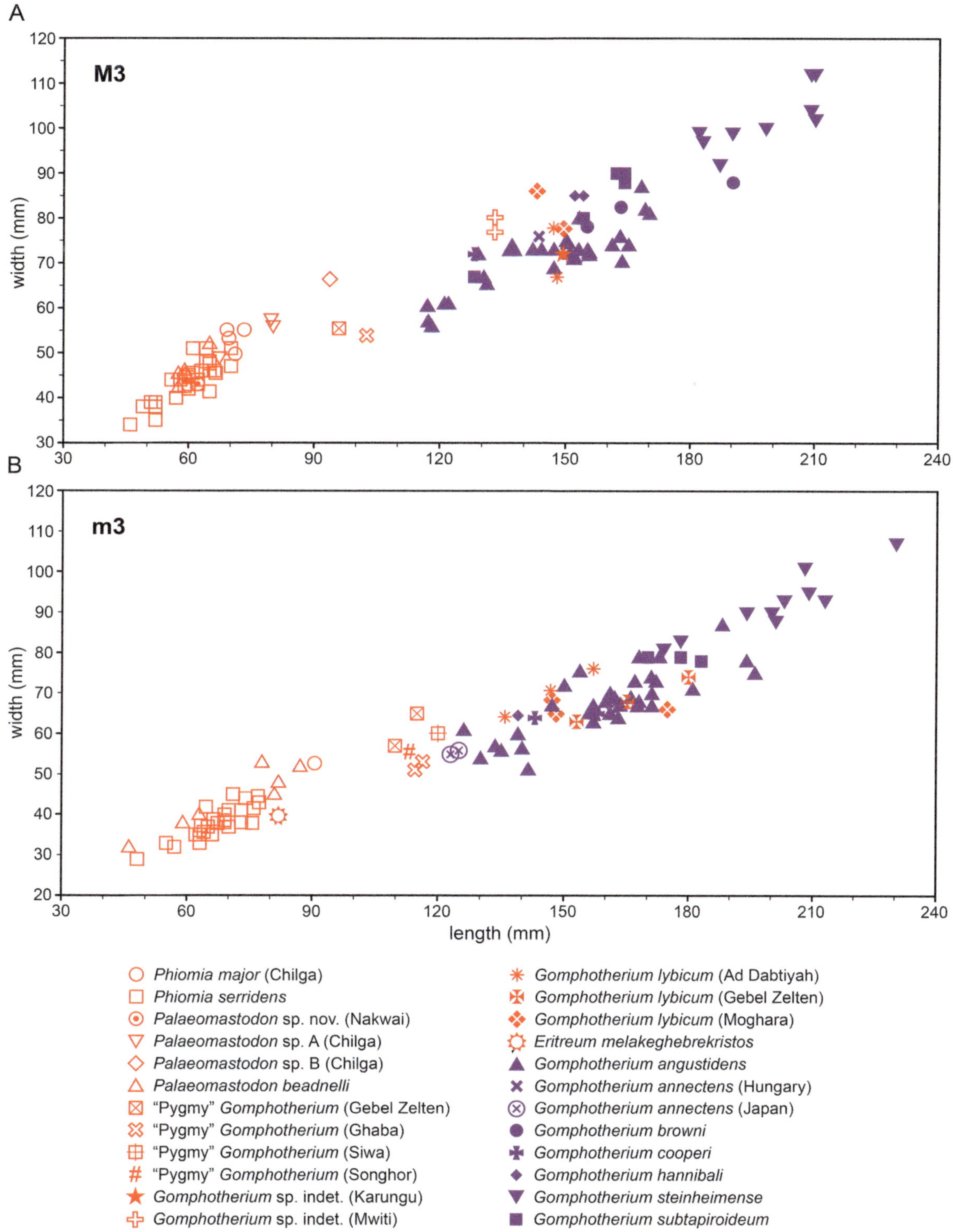

FIGURE 3.19 Bivariate plots of cheek tooth crown length versus width in palaeomastodonts and selected Afro-Arabian and Eurasian gomphotheriines. Comparative dimensions supplementing original measurements are from Andrews (1906), Fourtau (1918), Forster Cooper (1922), Matsumoto (1924a), Lehmann (1950), Bergounioux and Crouzel (1959), Arambourg (1961), Hamilton (1973), Gaziry (1976, 1987a), Tassy (1983b, 1985, 1986), Gentry (1987a), Roger et al. (1994), Göhlich (1998), Sanders and Miller (2002), Pickford (2003a, 2004, 2005a), Sanders (2003), and Sanders et al. (2004). (A) M3 dimensions. (B) m3 dimensions.

Phiomia serridens was first depicted as a probable creodont, based on a right dentary fragment preserving dp2–3 and d?i2 (Andrews and Beadnell, 1902). The name denotes the "saw-toothed animal of the Fayum," in reference to its dominant, procumbent incisor ornamented with crenulations along its distal margin, offset from the cheek teeth by a relatively substantial diastema (Figure 3.20A). Presumed carnivory of this animal was supported by the authors' supposition that the dp2 was a carnassial-like tooth, though it lacks a cutting blade and instead comprises an anterior paraconid behind which are a bunodont protoconid-metaconid pair, transverse to one another. The dp3 is bunolophodont and bilophodont with a very large paraconid anteriorly and a large accessory conule in its transverse valley. The paratype upper jaw fragment with a sharp tusk belongs to a hyrax (Matsumoto, 1924a). Andrews (1908) later considered that the dentary is that of a juvenile individual of "*Palaeomastodon*," possibly "*P. minor*." The bunolophodonty of the dp3 and its large accessory conule suggest that the specimen should be attributed to *Phiomia* rather than *Palaeomastodon*. Matsumoto (1924a) placed the specimen into *Phiomia wintoni*, but that nomen is a junior synonym and therefore the name "*Ph. serridens*" has priority. The similarity of the type dp3 and ?di2 to homologs in a nearly complete jaw of *Ph.* "*wintoni*" recovered from the Fayum in 1906 (Andrews, 1908) confirms the identity of the type. Dimensions of deciduous premolars in this specimen are: dp3, L=19.5 mm, W=14.5 mm; dp4, L=43.0 mm, W=21.0 mm. Measurements of upper deciduous premolars of *Ph. serridens* are: dP2, L=24.0 mm, W=11.0 mm; dP3, L=29.0 mm, W=20.0 mm; dP4, L=40.0 mm, W=23.0 mm (Andrews, 1908).

Matsumoto (1924a) asserted the validity of three species of *Phiomia*, and distinguished them as follows: *Ph. minor*, feeble development of the third lophid in m1–2, posterior end of the mandibular symphysis just anterior to p3, m3 short, length of p3–m3=180–202 mm, m1–3 L=126–147 mm, P2-M3 L=202–215 mm, M1–3 L=125–130 mm; *Ph. wintoni*, morphologically similar to *Ph. minor* but p3–m3 L=225–250 mm, m1–3 L=160–173 mm, P2–M3 L=255–280 mm, M1–3 L=145–170 mm; *Ph. osborni*, stronger development of m1–2 third lophids, posterior end of the mandibular symphysis at posterior end of p3, m3 very long and narrow with an incipient fourth lophid, p3–m3 L=250–255 mm, m1–3 L=177–180 mm. Other authors (e.g., Coppens et al., 1978; Sanders et al., 2010a) have combined these taxa into a single, variable (and presumably highly sexually dimorphic) species.

Cranial morphology of *Ph. serridens* is superbly known from the remains of several individuals (Figure 3.20B and D; Andrews, 1908; Osborn, 1936). In lateral view, the nasal opening is retracted to a position above M1 or P4 and the orbit is posteriorly located, above M2. The orbit is open posteriorly and communicates with the temporal fossa. Anteriorly, the lower face slopes downward to house the upper tusks, which are downturned and slightly everted. A relatively prominent infraorbital foramen opens below and in front of the orbit. A diastema of modest length separates the tusk from the cheek teeth. The external auditory meatus is situated very posteriorly and well above the level of the palate, and the occipital condyles are also raised to the same height and project posteriorly. The squamosal part of the zygomatic arch is short and underlain by an extension of the jugal portion of the arch. The braincase slopes upward posteriorly, in a straight line with the mid and lower face.

In dorsal view, a sagittal crest is formed and connects with the nuchal crest. The cranium is broadest across the zygomatic arches but they do not flare dramatically outward (*contra* Andrews, 1908). The nasal bones are quite short and slightly overhang the nasal aperture, which is relatively narrow. A shallow incisive fossa is present between the tusk alveoli in the anterior midline of the lower face.

In ventral view, the palatines project into the palate as anterior as the front of M2 (Figure 3.20D). The palate is shallow and does not deviate much in width between the toothrows. Behind the palate, the pterygoid plates are prominent and are oriented outward. The glenoid articulations are broad and transversely concave. In the basicranial area, the stylomastoid, carotid, and posterior lacerum foramina are all large, and basioccipital form resembles that of elephants, though is relatively longer. The postglenoid processes fold over the external auditory meati, nearly enclosing them as tubes. The temporal foramina are relatively spacious. As a whole, the cranium of *Phiomia* is low (220 mm) and only moderately wide (447 mm) relative to length (700 mm) (Andrews, 1908; Osborn, 1936), a primitive configuration compared with elephant crania. It is also poorly pneumatized (Andrews, 1903a, b).

Mandibular morphology is also well established for the genus. The mandible is very elongate, with a long diastema between the lower tusks and cheek teeth (Figure 3.20B and C). The symphysis and tusks either project straight forward or descend below the level of the cheek tooth alveoli at a low angle. With the exception of the hypertrophied type specimen of *Ph.* "*osborni*" (AMNH 13468), the posterior rim of the symphysis is usually located anterior to the premolar rows in the species. There are commonly two mental foramina, the most prominent one located well forward of the cheek teeth and posterior rim of the symphysis, just behind the socketed base of the i2, and the second one below the premolars. The ramus is anteroposteriorly broad and moderately high, with the coronoid process at about the same height as the condyle. The condyle is transversely broad and biconvex. The angle is rounded and large. Andrews (1908) provided dimensions for a nearly complete mandible of *Ph. serridens* (=his "*Palaeomastodon wintoni*"): Length not including the tusk, 765 mm, symphyseal length=273 mm, width across the condyles=344 mm, depth of the corpus below m3=100 mm. In addition, length of the incisors beyond their alveoli is 110 mm and the width of each i2 at the base is 83.0 mm.

Upper and lower cheek teeth in *Phiomia serridens* are bunodont to bunolophodont, very brachyodont, lack cementum, and in the uppers may have pronounced lingual cingulae. Molar size progression is usually M1/m1<M2/m2<M3/m3 (occasionally, M2 may be slightly longer than M3), with a marked size increase between the first and second molars (Figure 3.20C and D). Loph(id) formulae are as follows: P2, 1; P3/p3, 1½-2/1; P4/p4, 2/2; M1/m1, 3/3; M2/m2, 3/3; M3/m3, 3/3. They are distinct from

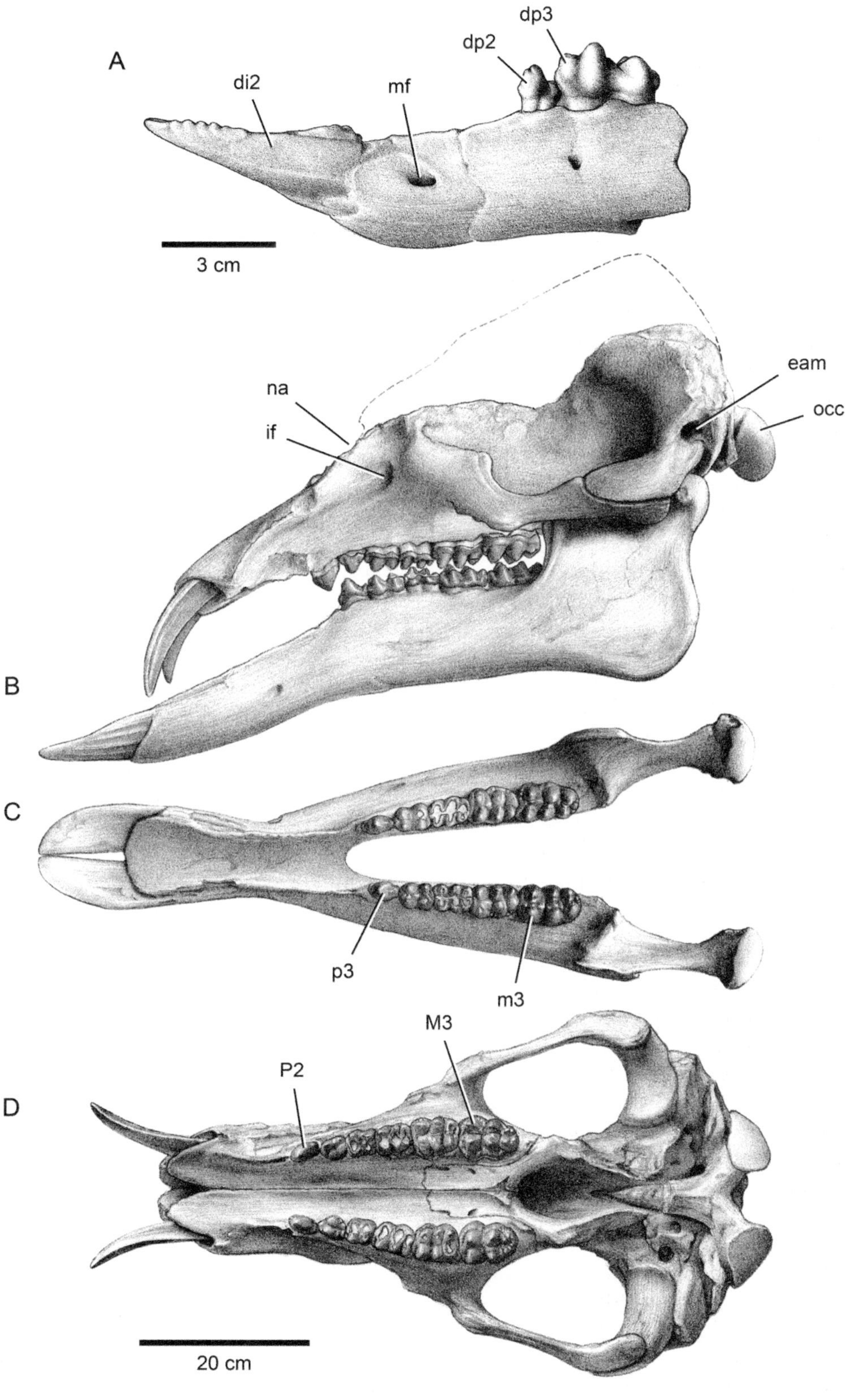

FIGURE 3.20 Cranial and mandibular specimens of *Phiomia serridens* from the early Oligocene Jebel Qatrani Fm., Fayum, Egypt. Abbreviations: di, lower deciduous incisor; dp, lower deciduous premolar; eam, external auditory meatus; if, infraorbital foramen; na, nasal aperture; occ, occipital condyle; M/m, upper/lower molar; mf, mental foramen; P/p, upper/lower premolar. (B–D) to the same scale. Anterior is to the left in all specimens. (A) Partial mandible with di2 and dp2–3 (type), C10007, lateral view, from Andrews (1906:pl. XVIII, fig. 4). (B) Reconstruction of partial skull, lateral view, from Andrews (1908:pl. 31, fig. 1). (C) Reconstruction of mandible with left and right i2 and p3–m3, dorsal view, from Andrews (1908:pl. 31, fig. 2). (D) Reconstruction of cranium with left and right I2 and P2–M3, ventral view, from Andrews (1908:pl. 31, fig. 3).

those of *Palaeomastodon beadnelli* in their development of accessory conules, possession of more inflated cusps, full trilophodonty of molars, and absence of crescentoids and zygodont crests.

Microwear analysis of Fayumian *Phiomia serridens* teeth indicates that they were browsers with more abrasion in their diet than moeritheres and were also substantial consumers of fruits and seeds (Semprebon et al., 2017, 2018). Dental carbon isotope values for *Palaeomastodon* and *Phiomia* are similar to those of moeritheres, but palaeomastodont oxygen isotope values are more consistent with terrestrial lifeways (Clementz et al., 2008). Masticatory pattern reconstructed from attritional and abrasive wear on palaeomastodont molars indicates that they chewed like gomphotheres, with a strong mesio-lingual grinding movement of teeth during Phase II occlusion, interpreted as an adaptation to browsing (von Koenigswald, 2016a).

P2 of *Ph. serridens* is a simple premolar composed of a single conical cusp that is sub-oval in occlusal view. P3 is triangulate in occlusal shape with subequal-sized proto- and paracones paired anteriorly. It has a small metacone more posteriorly located, peripheral to the talon basin. Development of the hypocone and distal cingular complex is variable in this premolar. P4 is square-shaped, with full bilophodonty, and may have a low, narrow cingulum surrounding much of the crown. M1 is fully trilophed, with greatest cingular development anteriorly and lingually. It consistently exhibits considerably more wear than M2–3, evidence of the delayed emergence of the posterior molars prior to simultaneous recruitment of the full dentition for mastication later in life. This phenomenon is observed in *Palaeomastodon*, as well, and may constitute a preadaptation to the horizontal tooth displacement mechanism of elephantimorph proboscideans, for slowed post-juvenile ontogeny and longer life spans (see Sanders, 2017). M2 is similar to M1 but considerably larger. M3 typically exhibits a greater tendency to develop mesial conelets, accessory conules, and has a more pronounced postcingulum. Unlike the rectangular crown shape of M1–2, the occlusal profile of M3 is usually narrower posteriorly.

The anteriormost premolar in the lower dental sequence, p3, resembles P2 but the anterior moiety of the crown may be superficially subdivided and posteriorly it has a low talonid shelf. p4 is bilophed and rectangular in outline, and may have an accessory conule in the transverse valley between the proto- and hypolophids. The lower molars are all trilophed, usually with some expression of large accessory conules on the pretrite side of the crown, particularly on the posterior side of lophids.

The upper incisors are sub-triangulate in cross-section (broadest dorsally), with an outer band of enamel that keeps the tusk sharp in wear, and have longitudinal striations. In their alveoli, they are downturned and slightly everted. Lengths of *Ph. serridens* I2s vary from 197 to 250 mm (Andrews, 1906; Mastomoto, 1924; Osborn, 1936). Lower incisors are rounded proximally and more flattened distally—it is easy to understand why *Phiomia serridens* was conceived of as a potential ancestor to the shovel-tusked amebelodonts (Osborn, 1919, 1936; Borissiak, 1929; Tobien, 1973a). Nonetheless, the earliest amebelodonts had more rounded lower tusks (Sanders et al., 2010a), and thus the similarity with more derived amebelodonts is homoplasic. The lower tusks of *Ph. serridens* are convex ventrally with strong longitudinal striations and concave dorsally. They are procumbent in the mandible, closely appressed to one another in the midline, and may be slightly angled downward toward each other medially. In the Berlin mandibular specimen MB.Ma. 28106, the i2s project 90 mm from their alveoli (Figure 3.21A and B), and if their socketed proximal portions extend nearly to the posterior of the symphysis, they are about 400 mm in total length, rivaling *Phiomia major* in lower tusk length (Sanders et al., 2004). This specimen is an outlier, however, and size of symphyses in other conspecifics suggests that they possessed much shorter i2s. Neither the upper nor lower tusks project markedly, compared with overall size of the skull; they may have had a more important function in food acquisition than for display, defense, or fighting.

Palaeomastodonts are rare elements of the fauna at Dur at Talah, Libya. In his dissertation and later publication about the site, Wight (1971, 1980) recognized the presence of *Phiomia "wintoni"* in the Idam Mb. (Table 3.1), based on a symphyseal fragment and partial upper tusk from a single locality. The posterior margin of the symphysis is close to the premolars and the anterior mental foramen is mesial to the premolars, confirming Wight's identification. Length of the symphysis is 235 mm. The tusk has a triangular cross-section (broadest dorsally) and large mesioventral wear facet; it preserves a length of 263.4 mm and is 75.1 mm wide proximally. Wight (1971) also described a dentary fragment preserving the roots of p3 and p4-m2 (68.7/1) from a different locality in the Idam Mb., which he compared with "*Palaeomastodon minor*" due to the small size of its cheek teeth. The posterior mental foramen is located below p4 but the symphysis is missing, hindering generic identification. Wight erroneously fit a symphysis onto this specimen that preserves a large central incisor and alveolus for a smaller lateral incisor. It probably belongs instead to *Barytherium grave*. A more fragmentary symphyseal fragment (68.6/9) from the Evaporite Mb. (= "Raqaba Mb.") that Wight (1971) assigned to "*Palaeomastodon* sp." certainly belongs instead in *Arcanotherium savagei*. Thus, despite reports of *Phiomia* and *Palaeomastodon* in the Idam and Evaporite Mbs. at Dur at Talah (Savage, 1969; Wight, 1971, 1980), it appears that only *Phiomia* was unequivocally sampled by Savage and Wight, limited to the Idam Mb. Therefore, there is no record of palaeomastodonts in the Evaporite Mb., further supporting the idea that it is late Eocene and the Idam Mb. is early Oligocene in age. More recent survey of the site by Rasmussen and colleagues (2008) produced dental fragments that they attributed to *Palaeomastodon*.

Palaeomastodonts have also been recovered from Zella in the Sirt Basin of central Libya. They were first reported by Arambourg and Magnier (1961) as "*Palaeomastodon*" and "*Phiomia wintoni*" in a vertebrate assemblage that includes hyraxes, anthracotheres, sharks, rays, crocodiles,

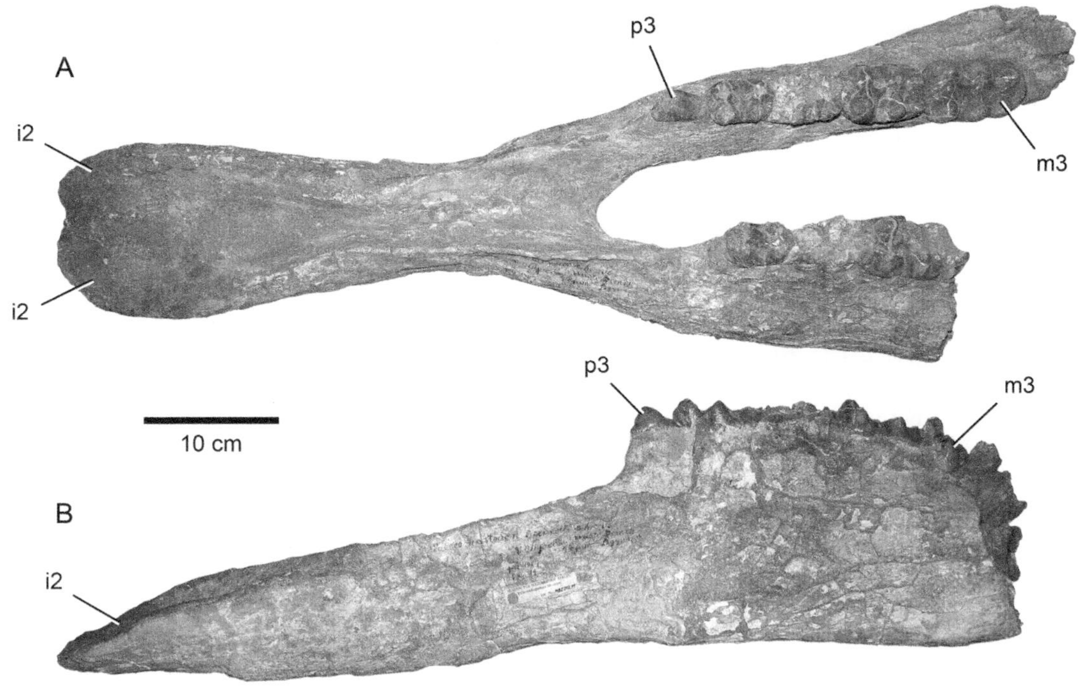

FIGURE 3.21 Mandibular specimen MB.Ma 28106 of *Phiomia serridens* from the early Oligocene Jebel Qatrani Fm., Fayum, Egypt. Abbreviations: i, lower incisor; m, lower molar; p, lower premolar. Anterior is to the left. (A) Dorsal view. Note distal broadening of lower incisors and elongate diastema. (B) Lateral view. Note distal flattening of lower incisors.

and tortoises (see also Arnould-Saget and Magnier, 1961; Arambourg, 1963; Savage, 1971). Arnould-Saget and Magnier (1961) placed the few palaeomastodont teeth (P4, M3, fragmentary m3) from Zella all in *Palaeomastodon* "*intermedius*," based on their study of the morphometric features of these specimens. More contemporary work at the site has added bony fish, sirenians, hystricognathous rodents, and anthropoid primates to this assemblage (Coster et al., 2015). An early Oligocene age was established for the site from biochronological correlation of foraminifera, selachians, and mammals (Coster et al., 2015). Combined with the other sites of palaeomastodont occurrences, Zella provides additional reinforcement for the observation that these proboscideans are largely Oligocene in temporal distribution (Table 3.1).

PHIOMIA MAJOR Sanders et al., 2004

(Tables 3.1 and 3.2; Figures 3.1, 3.19, and 3.22)

Partial Synonymy—*Phiomia* sp. nov. Kappelman et al., 2003; aff. *Phiomia* Ducrocq et al., 2010

Phiomia major has the best representation of any of the proboscidean species from the early late Oligocene site of Chilga, Ethiopia. Its sample comprises numerous isolated teeth, a few well-preserved tusks, and a very informative symphyseal portion of a mandible. The species nomen refers to the superior size of its teeth, compared with those of congenerics. It is readily distinguished from the sympatric species of *Palaeomastodon* present at Chilga by having third molars with full trilophodonty, development of accessory conules rather than crescentoids, and absence of zygodont crests. The presence of *Ph. major* alongside a primitive species of gomphothere at Chilga records one of the oldest instances of sympatric putative ancestral and descendant forms of proboscideans (Table 3.1).

The most striking remains of this species are the tusks and symphysis of the type specimen, CH17-1, which are outsized compared with the same elements of *Ph. serridens* from the Fayum (Figure 3.22A–C; Sanders et al., 2004). The upper incisor is robust and strongly curved, with an enamel band along its outer face (Figure 3.22A). From distal tip to preserved base, it measures 450 mm in a straight line. In cross-section, it has a piriform shape toward the base, and is more ovoid toward the tip. A prominent wear facet is present for 160 mm along the ventromedial surface of the distal end of the tusk, and medially it exhibits a sulcus for nearly its entire length. The I2s would have curved downward and slightly outward when in their alveoli.

Of the lower tusks, the right i2 is best preserved (Figure 3.22B) for a complete length of 460 mm, of which 214 mm projected anteriorly from its alveolus, more than twice the greatest projected length of an i2 in *Ph. serridens* (Sanders et al., 2004). Proximally, the i2 is dorsoventrally deep and transversely flattened, with a piriform shape in cross-section (Figure 3.22B), whereas it becomes very rounded toward its distal tip. In *Ph. serridens*, however, the lower tusks remain flattened distally and are relatively broader (Andrews, 1908: pl. 31). A shallow sulcus is present for nearly the whole medial surface of the i2. It is longitudinally straight with a 90 mm flat wear surface ventromedially toward the distal tip. In anterior view, the tusk is angled in its alveolus obliquely from ventromedially to dorsolaterally at about 30° from vertical. Similar to the upper tusks of this species, the lower tusks are ornamented with longitudinal striations. The lower tusks are set quite close to one another, as in the Fayum species.

Compared with *Ph. serridens*, the symphysis of *Ph. major* is more immense and proportionately exceeds the difference in molar size between these species (Andrews, 1906, 1908; Sanders et al., 2004). The length of the symphysis is 382 mm, and its width remains fairly constant along its alveolar portion, ranging from 89 to 93 mm (Figure 3.22C). The largest palaeomastodont symphysis known from the Fayum is the Berlin specimen MB.Ma. 28106, allocated to *P. beadnelli* but from its accompanying dentition it is clearly *Ph. serridens*—it has a length of 310 mm, with a greatest width of 121.3 mm and narrowest breadth of 63.7 mm (Figure 3.21). Unquestionably, the symphysis of *Ph. serridens* broadens more anteriorly than in *Ph. major*, though in other features the symphyses of these species are similar morphologically (Andrews, 1908:pl. 31). Breakage at the posterior end of the symphysis shows that the mandibular canal terminates in a rounded, closed antechamber that is positioned lateral to the i2 alveolus. This demonstrates that tusks are not developed in the mandibular canals or their antechambers, but rather in the more medially positioned alveoli. In the mandible of *Ph. serridens*, passage to the outside of the jaw for nerves and blood vessels from the antechamber occurs via the anteriormost mental foramen, which is located just posterior to the inserted base of the lower tusk and far anterior to both the cheek teeth and distal end of the symphysis (Andrews, 1908:pl. 31; see Matsumoto, 1924a:3). On the dorsal surface of the symphyseal segment, prominent longitudinal ridges demarcate a broad, shallow midline channel.

Except for their larger size, the cheek teeth of *Ph. major* are morphologically identical to those of *Ph. serridens*, with molars fully trilophodont, bunolophodont, and very low-crowned (Figure 3.22D–H). The molars are ornamented with a large posterior accessory conule at least behind the first loph(id), and have very low, narrow cingulae(ids) anteriorly and posteriorly that are closely appressed to the crown. The main conelets are dominant but there may be smaller mesoconelets accompanying them. There is no cementum. M3 narrows markedly posteriorly (Figure 3.22H), while the other molars are basically rectangular in occlusal shape. In size, these molars are intermediate between those of Fayum palaeomastodonts and elephantimorph proboscideans (except for the diminutive teeth of the basal gomphothere *Eritreum*). Phyletic tooth size increases over time in *Phiomia* may eventually prove useful for biochronological dating.

The tooth formulae of *Ph. major* and *Ph. serridens* appear to be the same (Table 3.2). Upper and lower fourth premolars are bilophodont; P4 is squared in occlusal outline, with a narrow cingular band running from anterior to posterior along the lingual margin of the crown, and p4 is rectangular in occlusal outline and exhibits a large pretrite accessory conule behind the first lophid. The lower third premolar is transversely narrow and much taller anteriorly than posteriorly, with equal-sized proto- and metaconids; a very small hypoconid is posterior to the protoconid. P3 has a projecting anterior cingular shelf, and is taller anteriorly than posteriorly, with loph 1 formed of equal-sized para- and protocones; the paracone has a distinct metacone behind it and posterior to the protocone is a small, incipient hypocone, making this premolar nearly bilophodont. P2 is worn but appears to be a simple premolar with a single main cusp anteriorly that is taller than the rest of the crown.

A palaeomastodont m2 reported from Lokone, Kenya is nearly identical morphometrically to m2s of *Ph. major* from Chilga (Ducrocq et al., 2010). Thus, there is little reason for its cautious assignment to "aff. *Phiomia*." Lokone is located in the Lokichar Basin, in the southwestern end of the greater Turkana Basin, and its diverse mammalian assemblage (including phiomyid rodents, bothriodontine anthracotheres, creodonts, and saghatheriine hyraxes) suggests a late early or late Oligocene age for the site (Table 3.1; Ducrocq et al., 2010; but see Werdelin, 2010). The Lokone *Phiomia* could be the last occurrence of palaeomastodonts in the Paleogene, but there is evidence (see Table 4.1) that these primitive elephantiform proboscideans persisted briefly into the Neogene. Nonetheless, they did not survive long in the face of competition with more advanced elephantimorph proboscideans that had the advantages of horizontal tooth displacement.

ELEPHANTIMORPHA Tassy and Shoshani, 1997 in Shoshani et al., 1998
MAMMUTIDA Tassy and Shoshani, 1997 in Shoshani et al., 1998
MAMMUTIDAE Hay, 1922
LOSODOKODON Rasmussen and Gutiérrez, 2009
LOSODOKODON LOSODOKIUS Rasmussen and Gutiérrez, 2009

(Tables 3.1 and 3.2; Figures 3.1, 3.23, and 3.24)

With the most mellifluous species appellation of any proboscidean, named for its site of origin and Greek for "tooth," *Losodokodon losodokius* was discovered by a 2007 Kenyan-American expedition (Rasmussen and Gutiérrez, 2009), firmly establishing the existence of mammutids in the Oligocene. Although mammutids enjoyed greater long-term success in more northerly latitudes in Eurasia and North America (Tassy, 1985; Mazo, 1996; Saunders, 1996; Tobien, 1996), the recovery of this taxon from Losodok (or Lothidok), Kenya points to Africa as their place of origin. It also shows, along with penecontemporaneous finds from Eritrea and Ethiopia, that more advanced elephantiform

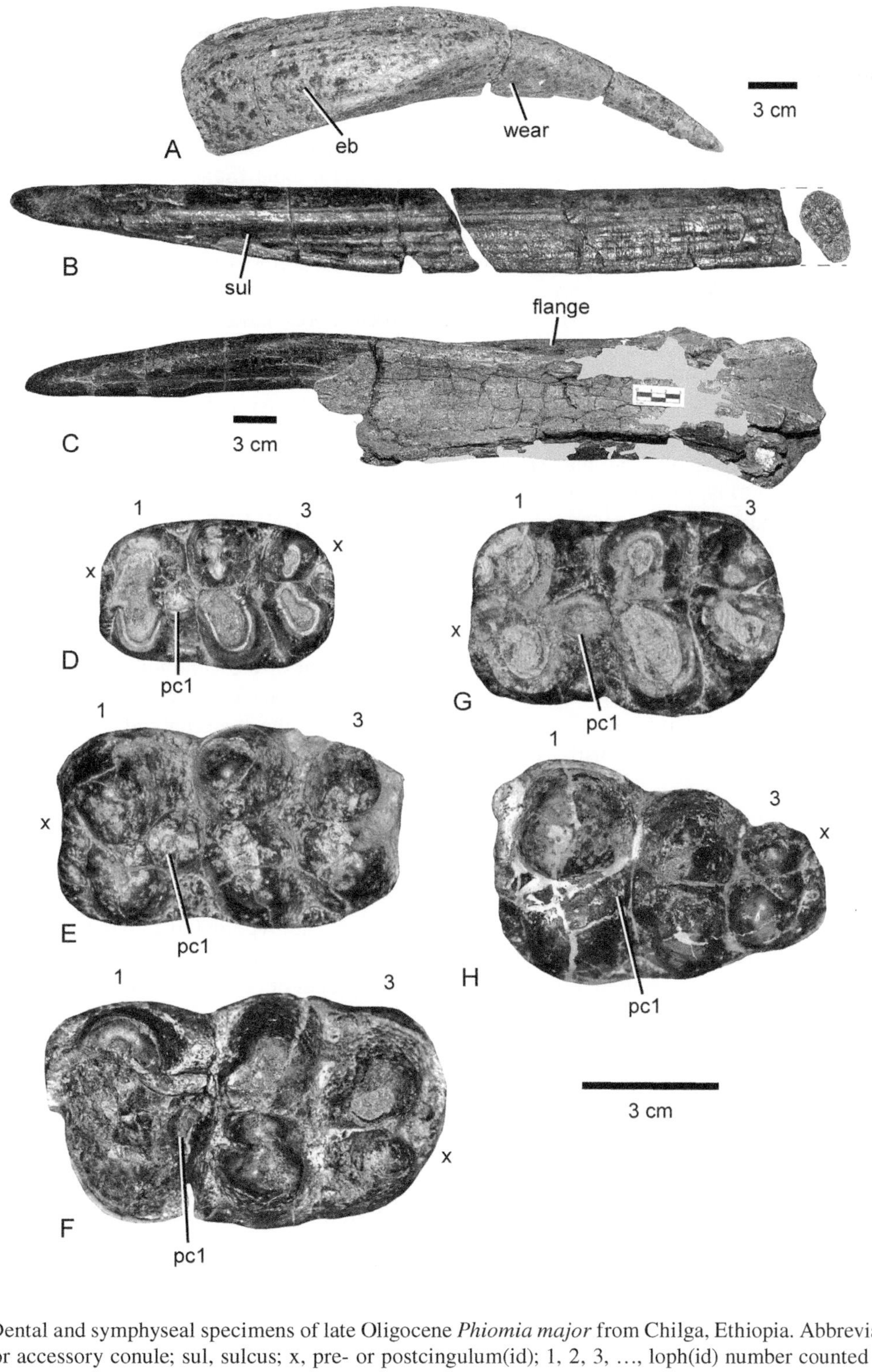

FIGURE 3.22 Dental and symphyseal specimens of late Oligocene *Phiomia major* from Chilga, Ethiopia. Abbreviations: eb, enamel band; pc, posterior accessory conule; sul, sulcus; x, pre- or postcingulum(id); 1, 2, 3, …, loph(id) number counted from the anterior of the crown. (A, B) to the same scale. (D–H) to the same scale. (A) anterior to the right. (B–H) anterior to the left. (A) Upper tusk, CH17-1 (type), lateral view. Note the salient wear and presence of an enamel band on the lateral surface of the tusk. (B) Lower right tusk, CH17-1 (type), dorsomedial view. Note the longitudinal sulcus and striations. (C) Symphyseal segment with right i2 in place, CH17-1 (type), dorsal view. Note the lateral flanges on each side. (D) m1, CH25-1, occlusal view. (E) m2, CHS4-1, occlusal view. (F) m3, CH9-2, occlusal view. (G) M2, CH9-1, occlusal view. (H) M3, CH22-1, occlusal view.

proboscideans were evolving from primitive palaeomastodonts in place in Africa prior to the dramatic, large-scale Eurasian migration of mammals into Africa that started at the beginning of the Miocene, setting the stage for the tremendous diversification of proboscideans observed in the early and middle Miocene of Africa (Sanders et al., 2004).

Losodok is located to the west of Lake Turkana in the Lothidok range in the Turkana Depression, and has been studied geologically in detail by Boschetto et al. (1992). *Losodokodon* derives from a conglomerate-sandstone-siltstone sequence in the Eragaleit beds, delimited in age to the late Oligocene between ~27 and 24 Ma by K/Ar dates on basalts above and below the beds (Boschetto et al., 1992). The accompanying mammalian fauna includes such typical African Oligocene denizens as arsinoitheres, primitive hyraxes (e.g., *Thyrohyrax kenyensis*), anthracotheres, a small propliopithecid catarrhine primate, and creodonts (Rasmussen and Gutiérrez, 2009; Gutiérrez, 2011; Leakey et al., 2011). However, remains of some mammalian taxa anticipating the composition of African Miocene faunas are also present, including a small species of deinothere, fragments of teeth of an indeterminate gomphothere, a cercopithecoid (?Victoriapithecidae gen. et sp. nov.) so primitive that it is almost defies identification, and a large, enigmatic catarrhine primate, *Kamoyapithecus hamiltoni* (Rasmussen and Gutiérrez, 2009; Gutiérrez, 2011; Leakey et al., 2011), interpreted by Zalmout et al. (2010) as a primitive hominoid.

Losodokodon is also present at Nakwai, believed on faunal evidence to be of similar age to the Eragaleit beds at Losodok (Rasmussen and Gutiérrez, 2009). Nakwai is also west of Lake Turkana in the Turkana Basin, with fossiliferous horizons dominated by finer sediments such as siltstones and mudstones. It does share faunal similarities to the Eragaleit beds, including the same mammutids and large catarrhine primate (Rasmussen and Gutierrez, 2009; Leakey et al., 2011); however, it is differentiated by having a true carnivore (*Mioprionodon hodopeus*, the earliest record of Carnivora in Africa [Leakey et al., 2011]) along with a cohort of other Neogene carnivoran and hyaenodont taxa (Friscia et al., 2020), hyraxes recorded at early Miocene sites elsewhere, and lacks arsinoitheres (Reynoso, 2014). In addition, Nakwai has produced an extremely primitive cercopithecoid monkey, *Alophia metios* (Rasmussen et al., 2019). New $^{40}Ar/^{39}Ar$ dating brackets the upper portions of the Nakwai section to the earliest Miocene, ≤23 Ma (Reynoso, 2014). Friscia et al. (2020) temporally situate the entire Nakwai sequence in the interval 24.0–21.5 Ma. *Losodokodon* has been recovered from locality section TOP 69 and TOP 73 at Nakwai, which have ages of ~22.0 and 22.2 Ma, established by radiometric dating and paleomagnetic sequence stratigraphy (Rasmussen et al., 2019). This somewhat unexpected result reveals that some constituents of the African Oligocene fauna were unaffected by initial contact of the continent with Eurasia, and offers a glimpse into a largely undocumented time interval in the evolution of African proboscideans and other mammals.

The sample of *Losodokodon* from Losodok is meager, comprising only a handful of teeth, but nonetheless it is consequential because it provides indisputable evidence of mammutids in the Paleogene, and with the early gomphotheres from Dogali, Eritrea (Shoshani et al., 2004), the Shumaysi Fm., Saudi Arabia (Zalmout et al., 2010), and Chilga, Ethiopia (Sanders et al., 2004), reveals the deep division in time of these two main branches of elephantimorph proboscideans. Included in the sample are several permanent premolars, which are lost in terminal mammutid species such as *Mammut borsoni* and *M. americanum* (Green, 2002; Green and Hulbert, 2005). Although there is evidence only for third and fourth permanent premolars in *Losodokodon*, the presence of P2 in palaeomastodonts (Andrews, 1906) and in *Zygolophodon turicensis* (Tassy, 1977a, 1985), which phylogenetically bracket this species, indicates that *Losodokodon* probably also possessed P2 in its upper dental quadrants.

The p3 (KNM-LS 18234) has a trigonid of two conelets separated by a median sulcus, and is accompanied by a lower talonid of two smaller conelets, also separated by a sulcus. It is a small tooth, approximately 22 mm in length and 15 mm in width. It has a paracristid but otherwise no development of crescentoids or accessory conules, and no presence of sharp crests or lophids. An interproximal facet is present on the posterior end of the crown.

The p4 (KNM-LS 18212) is more typical of mammutid teeth, with a bilophodont crown. It is trapezoidal in occlusal view, with each lophid formed of two conelets that connect across a very narrow median sulcus (Figure 3.23A). A narrow, low cingulid surrounds the crown on its mesial, buccal, and distal sides. Although the protolophid is heavily worn, the hypolophid is bunolophodont in aspect with some anteroposterior constriction of its apex. There is no development of crescentoids or accessory conules. Greatest length (47.21 mm) exceeds width (39.03 mm).

P4 (KNM-LS 18214) is more squared in occlusal outline, and like p4 is bilophodont (Figure 3.23B). The metaloph forms a true crest, while the protoloph is composed of two large conelets narrowly separated by a median sulcus. A low, narrow cingulum borders the crown on its mesial, lingual, and distal sides. The crown is slightly longer (44.55 mm) than wide (41.42 mm). It is possible that these premolars are all from the same individual.

Molars include a left M2 and right M3 (KNM-LS 18244), both trilophed (Figure 3.23C, D). They are small compared to all other mammutids (Figure 3.24) and at the low end of dimensions for homologous teeth in Miocene elephantimorphs, but considerably larger than palaeomastodont molars and molars in the penecontemporaneous gomphothere *Eritreum*, L × W = 85.50 × 52.02 mm and 92.26 × 53.66 mm for the holotype M2 and M3, respectively (Rasmussen and Gutiérrez, 2009). M2 is rectangular-shaped in occlusal view, with very low and narrow cingulae mesially and lingually. Lophs are transversely straight and orthogonal to the long axis of the crown. The first two lophs are subdivided by a very pinched mesial sulcus, and the third loph is continuous across the crown. There

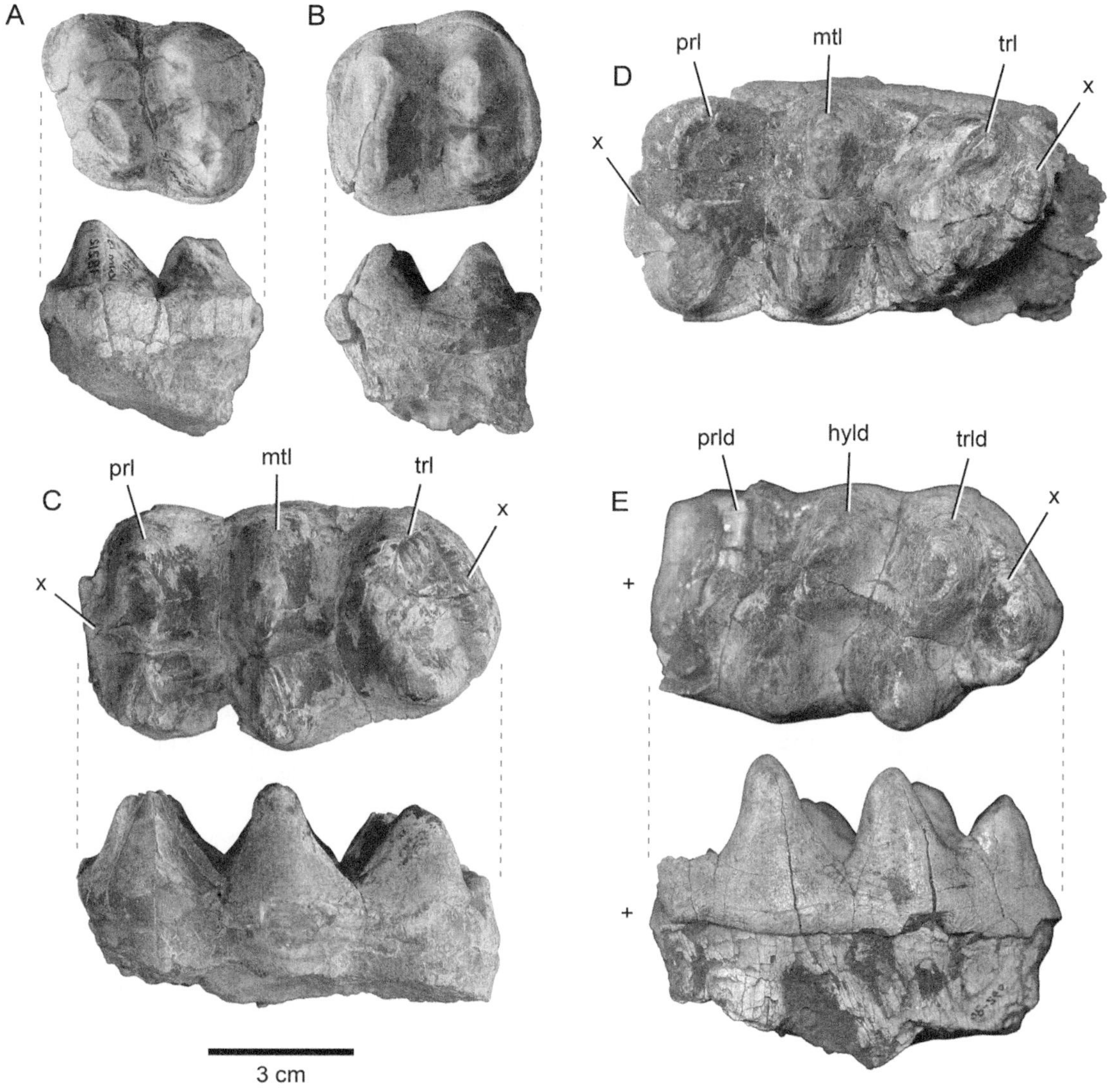

FIGURE 3.23 Cheek teeth of late Oligocene-early Miocene *Losodokodon losodokius*. Abbreviations: hyld, hypolophid; mtl, metaloph; prl, protoloph; prld, protolophid; trl, tritoloph; trld, tritolophid; x, pre- or postcingulum; +, indicates missing morphology. All specimens to the same scale and anterior is to the left. (A) Left p4, KNM-LS 18212 from Losodok, Kenya, occlusal and lateral views. (B) Left P4, KNM-LS 18214 from Losodok, Kenya, occlusal and lateral views. (C) Left M2, KNM-LS 18244 (type) from Losodok, Kenya, occlusal and lateral views. (D) Right M3, KNM-LS 18244 from Losodok, Kenya, occlusal view. (E) Left m3, KNM-NW 58580 from Nakwai, Kenya, occlusal and lateral views. Kenya.

is a low narrow posterior cingulum formed of numerous cuspules. Crown apices are mesiodistally compressed and transverse valleys between lophs are very broad and open. Crescentoids are imperceptible, present only as wear surfaces extending into the transverse valleys posteriorly from lophs 1 and 2, and a zygodont crest is present on loph 2. As in M3, wear is greatest on the pretrite side of the crown. This tooth is relatively narrower than its homolog in the early Miocene mammutid *Eozygodon*. Each half-loph is formed of a single conelet. M3 is also relatively narrower than in *Eozygodon*. It has low, narrow cingulae anteriorly and buccally, and a low distal heel composed of small cuspules that is closely appressed to the tritoloph. As in M2, crown apices are anteroposteriorly compressed and transverse valleys are broad and open, and height is very low (Rasmussen and Gutiérrez, 2009:table 6). Crescentoid development is weak (nearly imperceptible), although a salient zygodont crest is present on the posttrite side of the metaloph. In contrast to the M2, M3 tapers considerably posteriorly.

Contra Rasmussen and Gutiérrez (2009), the M3 of *Eozygodon* is not distinguished from that of *Losodokodon* by presence of an incipient fourth loph; only the lower m3 of *Eozygodon* has a nascent fourth lophid (Sanders et al., 2010a). The primary features that differentiate the M3 of *Losodokodon* from that of *Eozygodon* are smaller size (Figure 3.24A), relatively narrower crown, greater posterior tapering of the crown, weaker development of crescentoids, and fewer conelets per half-loph. The tritoloph of M3 in *Losodokodon* is transversely more constricted than lophs 1 and 2, but nonetheless it is a full loph (contra Rasmussen and Gutiérrez, 2009), and thus it is advanced over the condition

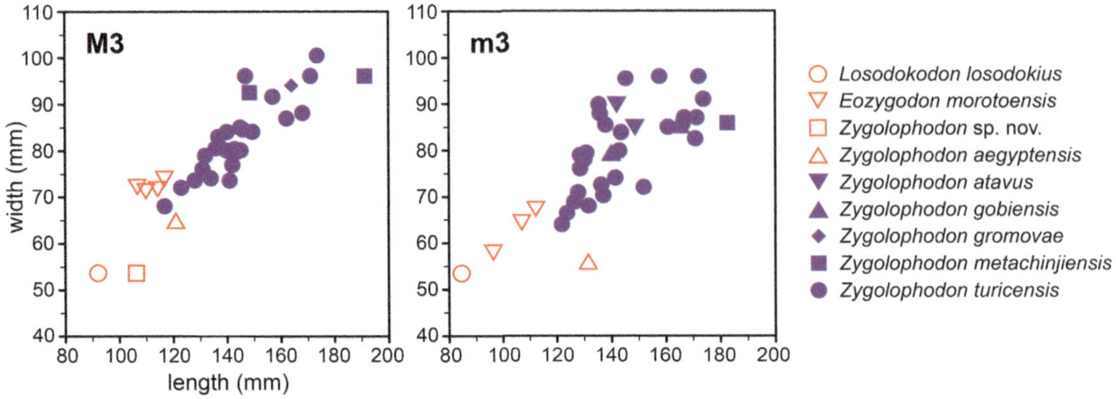

FIGURE 3.24 Bivariate plots of M3/m3 tooth crown length versus width in selected African and Eurasian late Paleogene-Neogene mammutids. Comparative dimensions supplementing original measurements are from Osborn (1936), Tassy (1977a, 1983a, 1985), Göhlich (1998), Sanders and Miller (2002), and Rasmussen and Gutíerrez (2009).

in *Palaeomastodon*, in which the tritoloph of M3 comprises only a half-loph.

An m3 from Nakwai, KNM-NW 58580, matches the morphology of *Losodokodon* molars from Losodok (Figure 3.23E). It is broken anteriorly but clearly had a lophid formula of 3x (very primitive for an elephantimorph m3), and is similar in size to the upper molars of Losodok, L=+86.2 mm, W=53.5 mm, and with a height of only 39.9 mm, is brachyodont (HI=75). Its lophids are obliquely oriented relative to the long axis of the crown, with very open transverse valleys between them. Zygodont crests and pretrite crescentoids are weakly developed. There is no cementum, and enamel thickness is moderate (ET=2.4 mm). The postcingulid is composed of two stout conelets that are lower than the lophids.

It is likely that chewing mechanics of *Losodokodon* were the same as in other mastodons, undergoing what von Koeingswald (2014, 2016a) referred to as the "mammutid pattern of mastication," in which the bolus of food is compressed orthally during Phase I jaw motion, followed by grinding in a lingual direction in Phase II jaw motion. Though the apices of loph(id)s are compressed, unlike the condition in deinotheres, they do not form a continuous planar wear surface across the crown and therefore probably did not function as a shear-cutting device (von Koenigswald, 2016b).

ELEPHANTIDA Tassy and Shoshani, 1997 in Shoshani et al., 1998
GOMPHOTHERIIDAE Hay, 1922
GOMPHOTHERIINAE Hay, 1922
CF. *GOMPHOTHERIUM* SP. NOV. Kappelman et al., 2003

(Table 3.1; Figures 3.1 and 3.25)

Gomphotherium serves as a collect-all taxon because "good" species included in the genus (e.g., middle Miocene *G. angustidens* from Europe [Tassy, 2013]) are dentally morphologically conservative, with trilophed intermediate molars, third molars with three to four loph(id)s, and pretrite half-loph(id)s typically with anterior and posterior accessory conules that form trefoil-patterned enamel loops with wear (simple molar crowns with no accessory conules on the posttrite side of the crown) (Sanders et al., 2010a). Thus, isolated gnathodental specimens that exhibit this combination of features are often placed in "*Gomphotherium*" as a gradistic sorting device. Proboscideans of this grade begin to appear at the end of the early Oligocene, such as in the Shumaysi Fm. fauna from Saudi Arabia around 29–28 Ma (Zalmout et al., 2010). They are rare in Oligocene faunal assemblages, and also have very small molars compared with those of Miocene gomphotheres. For example, in the Shumaysi Fm., a gomphothere partial M3 (SGS-UM 2009-063) is extremely brachyodont (H=41.5 mm) and only 50.1 mm wide, with very bulbous cusps but clear expression of anterior and posterior accessory conules. This specimen best resembles the gomphothere sample from Chilga, Ethiopia.

Associated right and left m1s or m2s from Chilga are of a morphological grade more advanced than sympatric palaeomastodont teeth (CH 25-V-12a, b). They are trilophed with distinct pre- and posttrite mesoconelets in each half-lophid, and with pretrite anterior and posterior accessory conules with each lophid (Figure 3.25A). In size, they appear as miniature versions of gomphothere molars, with dimensions of L=74.5–74.7 mm, W=50.0–50.4 mm, and heights of 35 and 37 mm, yielding HI=70 and 83. The crown widens noticeably at lophid 2, suggesting that these are m1s rather than m2s. If so, they are considerably larger than m1s of *Phiomia major* from the same assemblage (Sanders et al., 2004). There is no cementum, and traces of cingulids are primitively expressed on both sides of the crown in the lateral margins of the transverse valleys.

The presence of a basal gomphothere at Chilga is also attested to by the recovery of a partial right third molar (CH14-V-14) that exhibits anterior and posterior accessory conules in association with its pretrite half-lophs (Figure 3.25B). It is more advanced in the expression of these features and development of mesoconelets in its pre- and posttrite half-lophs than the molars of sympatric

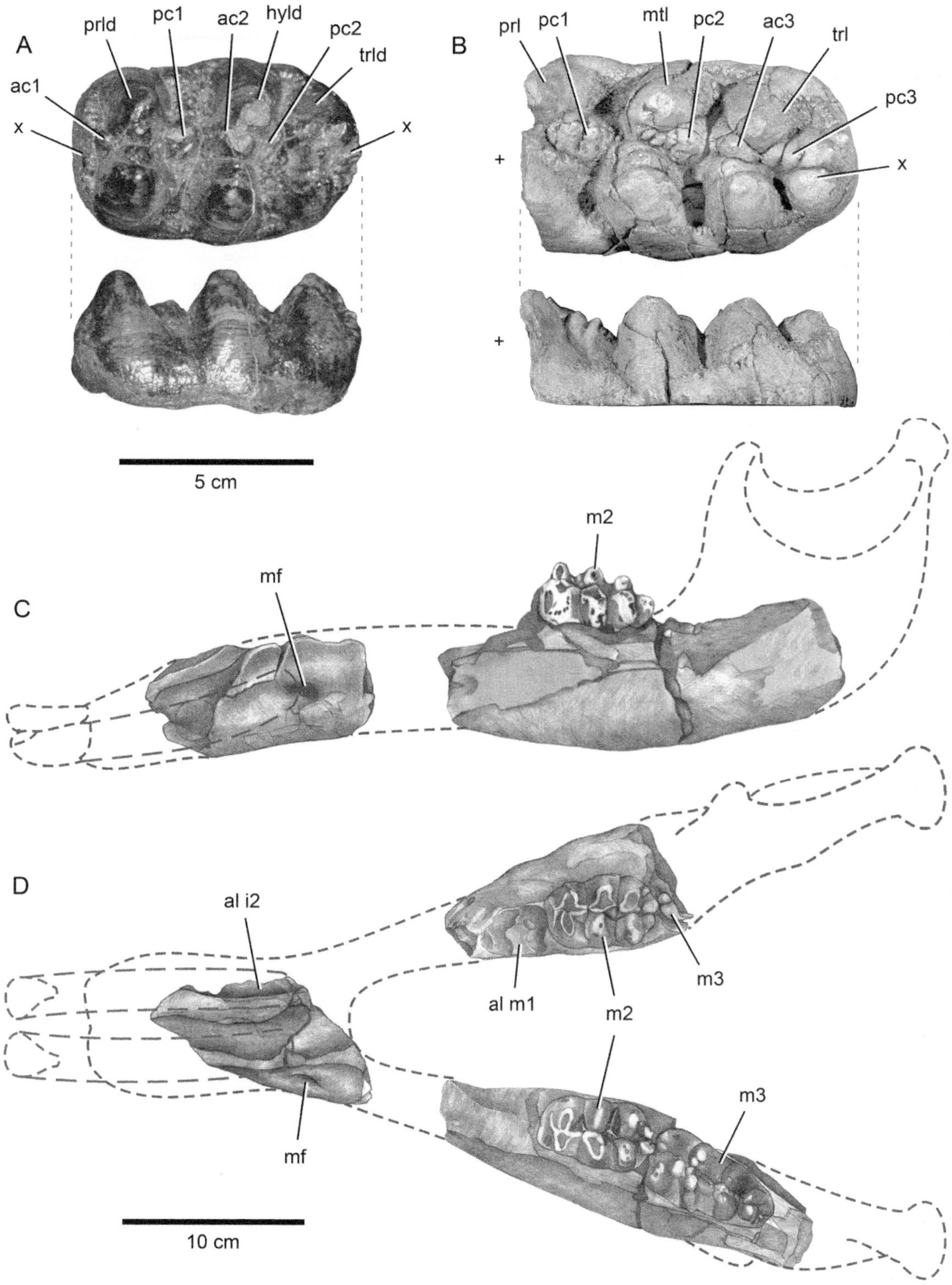

FIGURE 3.25 Late Oligocene gomphothere specimens. Abbreviations: ac, anterior accessory conule; al, alveolus; hyld, hypolophid; i, lower incisor; m, lower molar; mf, mental foramen; mtl, metaloph; pc, posterior accessory conule; prl, protoloph; prld, protolophid; trl, tritoloph; trld, tritolophid; x, pre- or postcingulum(id); +, indicates missing morphology. (C, D) to the same scale. Anterior to the left. (A) Right m2, CH25-V-12a from Chilga, Ethiopia, occlusal and lateral views, cf. *Gomphotherium* sp. nov. (B) Partial right m3 (cast), CH14-V-14 from Chilga, Ethiopia, occlusal and lateral views, cf. *Gomphotherium* sp. nov. (C, D) Partial mandible with alveoli for i2s, partially resorbed alveolus for right m1, right and left m2, and left m3, field number DOG87.1 from Dogali, Eritrea, lateral and dorsal views, *Eritreum melakeghebrekristosi*. (Images drawn by Gary Marchant [Shoshani et al., 2006:fig. 2A, B].)

palaeomastodonts from the site. While it is uncertain how much of the specimen is lost anteriorly, it may be nearly complete, with three full lophids and an incipient fourth lophid formed of two stout conelets posteriorly. It would be unlikely to find a gomphothere third molar with more than four loph(id)s at this early stage in the evolution of the group. The main conelets are torpedo-shaped in lateral view, rather than pyramidal as in palaeomastodonts, and very brachyodont (only 41.0 mm high in loph 1), enamel is very thick (4.7–5.4 mm), and the overall size of the specimen likely would have exceeded that of all of the Chilga palaeomastodont third molars (L=+87.5 mm, W=55.4 mm). A huge doubled posterior central conule is associated with the first loph; the second loph has a tiny anterior central conule and large posterior conule; and the third loph has equally small anterior and posterior accessory conules. A very low, tightly appressed postcingulid wraps around to the pretrite side of the crown. Other notable specimens attributed to this taxon include a fragmentary dentary with a p4 (CH25-3). The p4 has a lophid formula of x2x, and differs from the p4s of palaeomastodonts from the site in being larger (L=44.5 mm, W=33.2 mm, H=31.0 mm) with greater development of accessory conules and mesoconelets (Sanders et al., 2010a). Although larger than Chilga and Fayumian palaeomastodont homologs, as well m2–3 in the penecontemporaneous gomphothere *Eritreum*, nonetheless these cheek teeth are smaller than those of Miocene gomphotheres (Sanders et al., 2004; Shoshani et al., 2006).

A right P4 (CH25-V-12c) may also be from the same individual as the lower p4 and m2s. It is much larger (L=42.6 mm, W=39.9 mm, H=25.0 mm) and has better development of mesoconelets than the P4s of Chilga and Fayum palaeomastodonts. In addition, the anterior and posterior swellings on pretrite half-loph 2 resemble those present in *Gomphotherium* P4s (Sanders et al., 2004).

These gomphothere teeth from Chilga are not only morphologically, but also temporally intermediate between their palaeomastodont ancestors in the Fayum and earliest successors in the genus *Gomphotherium* in Africa and Eurasia between about 20 and 18 Ma (Tassy, 1985, 1986, 1988, 1994a, 1996b, Pickford, 1986a; Mazo, 1996, Sanders and Miller, 2002). The greater expression of accessory conules on the pretrite side of each cheek tooth (lingual in uppers and buccal in lowers) is related to the gomphotherian form of mastication, in which in Phase I the buccal sides of upper and lower molars are aligned, but in the longer Phase II, the mandible moves in a lingual direction, grinding the buccal cusps of the lower molars against the lingual cusps of the upper molars, creating the intense wear observed on the pretrite sides of teeth (von Koenigswald, 2016a). This mechanism was already occurring in palaeomastodonts, and may have been enhanced in effectiveness by the increased occlusal ornamentation observed in early gomphotheres such as those at Chilga.

An elephantimorph maxillary fragment with M2 and partial M3, from Mai Gobro 1, Eritrea was recovered in association with a deinothere M2 (see above) and dates to the late Oligocene (Abbate et al., 2012, 2014). The more complete M2 shows that this is from a trilophodont taxon, clearly much larger (M2 L=82 mm; W=48 mm) than penecontemporaneous *Eritreum* (see below). Though described as a gomphothere and placed in "Gomphotheriidae indet." (Abbate et al., 2012, 2014), it has morphological features that distinguish it from typical primitive gomphotheres, such as very prominent anterior crescentoids that sweep forward and medially from pretrite half-lophs 1 and 2, anteroposteriorly "pinched" pretrite half-loph apices, and semi-vertical, chisel-like wear facets. These features are convergent on a robust mammutid morph; however, the M2 does not resemble *Losodokodon* molars. It also does not resemble gomphothere teeth from the early Oligocene Shumaysi Fm., Saudi Arabia (Zalmout et al., 2010), early late Oligocene of Chilga, Ethiopia (Sanders et al., 2004), or molars of *Eritreum* from Dogali, Eritrea (Shoshani et al., 2004). The strong beaded lingual cingulum is primitive for Elephantimorpha. This specimen points to a late Oligocene diversification of elephantimorph proboscideans that is slowly being revealed from the handful of late Oligocene African paleontological sites.

ERITREUM Shoshani et al., 2007
ERITREUM MELAKEGHEBREKRISTOSI Shoshani et al., 2007

(Table 3.1; Figures 3.1 and 3.25)

The remains of this proboscidean were recovered serendipitously by a farmer for whom the species is named. Not only was it fortuitous that the type and only specimen of the taxon was found in its remote locale, but that its discoverer also recognized its importance and brought it to the attention of researchers. The genus is named after the country of origin, Eritrea. Despite its ungainly species name, *Eritreum melakeghebrekristosi* holds great importance for its age, systematic position, and ontogenetic implications of its gnathodental morphology (Shoshani et al., 2007).

The lone remains of *Eritreum* are comprised of left and right dentary posterior corpora, a partial symphysis with incisor alveoli, right and left m2s, a complete left m3 unemerged from its crypt, and the first loph of right m3 also in its crypt (Figure 3.25C and D). In addition, the right dentary preserves a mostly resorbed alveolus for m1. The remains derive from sandy-conglomerate-sediments in the Dogali Formation, exposed at the locality of Wedeg Melatse Farm at the base of the eastern escarpment in Eritrea, temporally constrained between basalts dated radiometrically at 28.0±2.5 and 26.8±1.5 Ma (Drury et al., 1994; Bunter et al., 1998; Shoshani, 2004; Shoshani et al., 2004; Abbate et al., 2014). Although the Dogali specimen was found in isolation, a partial deinothere tooth was found about 1,500 m to the northwest, in the same coarse-grained channel sediments, and thus is also attributable to the Dogali Fm. (Sagri et al., 1998; Abbate et al., 2014). As very few late Oligocene Afro-Arabian fossil vertebrate sites have been identified, proboscideans are consequently poorly known from this time interval, conferring great potential importance on the Dogali find.

Alveolar shape for the missing incisors is pyriform, based on preservation on the right side, higher (35 mm) than wide (23 mm). Estimated size of the mandible when complete is approximately 50 cm in length and 30 cm in width (Shoshani et al., 2004). The m2s are trilophed with anterior and posterior cingulids, yielding a lophid formula of x3x. Dimensions of the m2s are L=64.4 mm, 63.6 mm; W=40.6 mm, 39.5 mm, ET=3.0 mm. Each half-lophid is composed of two conelets, the outer abaxial conelets being larger than the inner mesoconelets. The first pretrite half-lophid in each m2 exhibits a trefoil wear pattern, due to incorporation of small anterior and posterior central accessory conules, which are also present in association with the 2nd and 3rd lophids. Crown height is very brachyodont (H=25.5 mm), and width is greatest posteriorly. The lower third molar is considerably larger than m2 (L=82.0 mm, W=39.5 mm), with four lophids and a diminutive precingulid (lophid formula x4). This molar is similar to the m2s except for the fourth lophid, which is composed of two bulbous conelets. Accessory conules are well developed in association with each lophid except the last. Greatest width is anterior and the crown is very low. Overall morphological affinity is with gomphotheriines, although m2 dimensions are well within the size range for Fayum palaeomastodonts (m3 is larger than all Fayumian *Phiomia* m3s, however) and smaller than corresponding molars in all Miocene gomphotheres and mammutids, as well as similar-aged palaeomastodonts and gomphotheres from nearby Chilga, Ethiopia (Figure 3.19). Scaling tooth size to palaeomastodont body dimensions yields body mass estimates of 446–521 kg (about the weight of a small African buffalo) and a shoulder height of between 122 and 132 cm (Shoshani et al., 2004).

Perhaps influenced by its small size, relative to gomphotheres, Shoshani et al. (2004) placed *Eritreum* in the Elephantimorpha as intermediate between palaeomastodonts and mammutids+Elephantida (gomphotheres, stegodonts, and elephants). In support of this phylogenetic assessment, they cite the absence of a postcingulid in m3 as a primitive feature for elephantimorphs. However, if the phylogenetic roots of mammutids and gomphotheres are present in the Palaeomastodontidae, then the occlusal morphology of *Eritreum* strongly indicates affinity with gomphotheres within Elephantida. Its molars exhibit none of the features of mammutids, such as zygodont crests, crescentoids, or "yoked" mesoconelets, but rather show the typical traits of gomphotheriines—bunolophodont lophids with anterior and posterior accessory conules contributing to trefoil enamel wear patterns on the pretrite side of molar crowns. Along with other Oligocene specimens from Chilga, Ethiopia, and the Shumaysi Formation, Saudi Arabia, *Eritreum* is one of the oldest recognizable gomphotheres in the fossil record.

More importantly, the delay in emergence of the m3—in palaeomastodonts, it would have already been in occlusion given the degree of wear on m2—and the very posterior position of m3 in its crypt in the dentary, as well as the resorbed alveolus for m1, are suggestive of horizontal tooth displacement in *Eritreum*. If this is the case, it is the oldest evidence of this tooth replacement mechanism, which is shared by all elephantimorph proboscideans (Mammutida+Elephantida) (Sanders, 2017).

SUMMARY

1. Subsequent to their origin and initial succession during the early Paleogene, there was a near-hiatus in the proboscidean fossil record during the middle Eocene Lutetian and Bartonian. However, the late Eocene-Oligocene of Afro-Arabia witnessed an eruption of diverse new taxa, including barytherioids with an apparent connection to early Paleogene plesielephantiforms, along with moeritheres, deinotheres, and the first substantial record of elephantiform proboscideans.
2. The sympatry of multiple proboscidean species that characterized African sites until well into the Pleistocene started to become common during the late Paleogene; for example, in the Idam Unit of Dur at Talah, Libya, moeritheres co-occur with several barytherioid species and palaeomastodonts. This indicates the presence of fertile tropical ecosystems capable of supporting a cohort of megaherbivores. As well, primates, including early anthropoids, began to be ecologically linked with proboscideans in late Paleogene African faunas, beginning with the sites of Birket Qarun, Egypt, Bir el Ater, Algeria, and Dur at Talah, Libya.
3. The earliest proboscidean occurrences are limited to northwest and western Africa, but by the late Paleogene extended across North-Central and northeastern Africa into the Arabian Peninsula and eastern Africa. Some of the expanded geographic record of proboscideans at this time may be an artifact of increased accumulation of terrestrial sediments resulting from the weathering of repeated episodes of massive outpouring of basalts in these regions, favoring recurrent preservation of floras and faunas.
4. Sedimentary sequences of the Fayum, Egypt (Birket Qarun, Qasr el Sagha, and Jebel Qatrani fms.) and Dur at Talah, Libya (Evaporite and Idam Units) preserve the richest and most important fossil successions of the African late Paleogene, providing a wealth of information about proboscidean diversity, morphology, habitats, and ecology. The Fayum beds, especially well dated by sophisticated palaeomagnetic sequence stratigraphy, are critical for biochronological correlation of other sites. This is especially pertinent for assessing hypotheses of sites with contentious ages, such as Bir el Ater and Dur at Talah.

5. An important proboscidean group from the late Eocene-early Oligocene is composed of the barytherioids *Arcanotherium*, *Omanitherium*, and *Barytherium*. Each of these barytherioids has limited distribution in time and space, and all resemble early Paleogene plesielephantiforms in having bilophodont molars, vertical tooth replacement, and simultaneous employment of their adult dentitions. The taxon may be paraphyletic. *Arcanotherium* shares features with early Eocene *Numidotherium*, such as a high cranial vault and anteriorly placed nasal aperture, relatively shallow mandibular corpus, retention of an upper dental formula of 3-1-3-3, and incisors with a complete enamel covering. *Omanithitherium* resembles *Arcanotherium* in having incisors fully clad with enamel and a dominant i2; however, it has a deep corpus like *Barytherium*. *Barytherium* is the largest of the barytherioids, with some graviportal adaptations in its postcranium (although probably with more flexed elbows and knees than in elephants), and is more derived than *Arcanotherium* in its upper dental formula (2-0-3-3) and loss of enamel on the lingual aspect of its incisors. Studies of dental isotopic composition reveal that barytheres were semi-aquatic. Despite its primitive features, *Arcanotherium* has been hypothesized as a sister taxon to elephantiforms based on bunodonty of molar cusps and enamel microstructure.
6. Another taxon of importance in the late Eocene-early Oligocene is the Moeritheriidae. Moeritheres are also revealed by isotope and microwear studies to have been amphibious with a diet of soft aquatic vegetation. Much of their strange skeletal morphology appears adapted to life in the water. They have been described as hippo- or sea cow-like. Their crania are characterized by a large, anteriorly-positioned nasal aperture in very close approximation to the orbits; the viscerocranium has a long, straight tubular connection to the neurocranium, and the skull appears low and elongate. The upper dental formula is primitive, 3-1-3-3. Lower incisors were procumbent and occluded with downward-oriented upper incisors to crop vegetation. The body shape is long and low, with a very long thorax, and flexed fore- and hindlimb joints. Moeritheres are monogeneric, with a species succession over about seven Myr from *Moeritherium chehbeurameuri-M. lyonsi-M. trigodon*. They remained morphologically conservative over that time, with minor differences in degree of bunodonty of cheek tooth cusps and expression of accessory crown features. Moerithere molars differ from those of plesielephantiforms in being bunolophodont, and "anticipate" the occlusal configuration of early elephantiform teeth. Each moerithere species appears to have been very sexually dimorphic in body size and tooth dimensions, particularly of I2/i2.
7. The Oligocene was marked by two major cooling events that involved accumulation of polar ice and lowering of sea level. These events may have played a role in loss of coastal habitats and the extinction of semi-aquatic barytherioids and moeritheres. Nonetheless, Afro-Arabian sites of this time were buffered from the severest effects of global cooling by their tropical position, and many endemic mammalian taxa on the continent continued to flourish through the Oligocene.
8. The early Oligocene witnessed the origin of elephantiform proboscideans, in the form of palaeomastodonts. These are the first substantial group of elephant-like proboscideans, represented by the genera *Palaeomastodon* and *Phiomia*. They are best known from the Fayum and Dor at Talah, but their range extends into the Arabian Peninsula and eastern Africa (where they had their last occurrences). Palaeomastodonts are characterized by elephantine-sized bodies supported by graviportal adaptations in the skeleton, single sets of procumbent upper and lower tusks accompanied by a trunk (perhaps smaller than in extant elephants), and bunolophodont, trilophed molars. They almost certainly gave rise to more derived elephantiforms (and eventually elephants) in the Elephantimorpha. The ancestry of palaeomastodonts is obscure, except for a single middle Eocene molar (see Chapter 2); although the cheek teeth of moeritheres are very similar, moerithere postcranial and cranial features are too autapomorphic, as a result of adaptation to semi-aquatic habitats, to constitute precursor conditions for elephantiform proboscideans. Dental evidence supports a hypothesis that the two main phylogenetic subdivisions of Elephantimorpha, mammutids and gomphotheres, evolved from *Palaeomastodon* and *Phiomia*, respectively. Thus, it is possible that the Palaeomastodontidae is paraphyletic.
9. The first evidence for gomphotheriids and mammutids, in the Arabian Peninsula and eastern Africa, shows that their initial divergence can be traced to at least the end of the early Oligocene, and that the occlusal morphological features that distinguish them were already well established with their first records. Correspondingly, the first sign of horizontal tooth displacement is documented in the late Oligocene Eritrean gomphotheriid genus *Eritreum*. This mechanism involves delayed emergence of cheek teeth, ejection of cheek teeth anteriorly from the alveolar row by more posteriorly emerging teeth, and successive, rather than simultaneous, use of cheek teeth. It provided elephantimorphs with a distinct advantage over more primitive elephantiform

proboscideans and plesielephantiforms by fostering slowed ontogenetic development and ability to eat a wider range of potentially tougher food items, expanding access to new ecosystems and life history parameters.

10. The first fossils of deinotheres are documented in the Oligocene, possibly with the presence of *Chilgatherium* in Ethiopia and Saudi Arabia, and certainly by the recovery of teeth resembling molars of early-middle Miocene *Prodeinotherium hobleyi*, in Eritrea and Kenya, from late Oligocene sediments. These occurrences marked the beginning of a highly successful lineage that managed to populate much of the Old World (primarily Africa, Europe, and South Asia) during the Miocene, surviving in Africa until the end of the early Pleistocene. Deinotheres appear to be an offshoot of Plesielephantiformes, possibly closely related to barytheres. Despite the disadvantage of not possessing the mechanism of horizontal tooth displacement, they appear to have avoided competing with elephantimorphs and enjoyed remarkable evolutionary success by adapting to specialized niches as hyperbrowsers in closed environments such as gallery forests. Their morphology, characterized by tapiroid molars, loss of upper tusks, and downward curve of a single set of lower tusks, remained remarkably conservative over nearly 26 Myr, except for massive body size increase in later species.
11. The fossil record of Afro-Arabia indicates that proboscideans (and other endemic mammalian taxa) were robustly evolving prior to early Miocene docking and faunal exchange with Eurasia. The loss of some taxa such as moeritheres and barytheres and rise of other taxa such as elephantiforms were probably driven by tectonic and climatic forces expressed at the continental and even local habitat levels, rather than competition with Eurasian immigrants. As well, new elephantimorph taxa—gomphotheres and mastodons—appear to have benefited from the evolution of a new mechanism for molar emergence, horizontal tooth displacement, which contributed to the competitive replacement of more primitive elephantiform proboscideans and most plesielephantiforms by the end of the Paleogene.

4 Early and Middle Miocene *Diversification of Proboscideans and Dominance of Elephantimorphs*

Palaeontology demonstrates in the Proboscidea a remarkably rich history, unsuspected from the extant relictual diversity, and one of the most spectacular morphological evolutions in the Mammalia.

—Emmanual Gheerbrant and Pascal Tassy (2009)

INTRODUCTION

Early Neogene proboscidean evolution was marked by the rapid diversification of elephantimorphs in Afro-Arabia, particularly of the Elephantida (ultimately comprised of gomphotheres, stegodonts, and elephants). Although elephantimorph gomphotheres and mammutids first diverged from palaeomastodont ancestors at least by the late Paleogene, the breadth of the Old World gomphothere radiation was not reached until the early to middle Miocene, with the appearance and phylogenetic flowering of choerolophodonts and amebelodonts. Trends for increasing body size and concomitant increase in tooth size, including greater hypertrophy of tusks, and development of more loph(id)s (particularly in third molars, which took on greater importance in horizontal tooth displacement) characterized the proboscideans of this interval. As well, increased occlusal complexity marked by greater numbers of conelets in loph(id)s and multiplication/hypertrophy of accessory conules and crescentoids occurred in parallel among the different elephantimorph lineages during the early Neogene.

While the late Eocene-Oligocene was marked by broad morphological disparity among barytheres, moeritheres, deinotheres, and palaeomastodonts postcranially and especially in craniodental features, influenced by contrasting adaptations to terrestrial versus aquatic lifeways, conversely in the early and middle Miocene proboscidean diversity was high but disparity consisted of subtler variations on skeletal and dental bauplans among elephantimorphs. This may be attributed to the relative success within the group of conservative graviportal postcranial adaptations for increased body size and the shared evolutionary novelty of horizontal tooth displacement (Sanders, 2017), as well as abandonment of semi-aquatic lifeways.

In horizontal tooth displacement, emergence of molars is delayed and adult cheek teeth are employed successively (Shoshani, 1992; Tassy, 1996a). In addition, cheek teeth at the anterior of the jaw are "ejected" by the emergence and forward movement of more posteriorly arising teeth. This appears to occur by mesial drift (Death and Coulson, 2016), whereby the horizontal progression of teeth along the jaw is effected by bony resorption and deposition along the alveolar row; teeth also undergo resorption of their roots as they migrate to anterior positions (Sikes, 1967) and can be pushed out because of overall net forward pressure by molars arising posteriorly. This mechanism in proboscideans parallels the polyphyodonty of manatees, in which molars continuously erupt and move forward by resorption and deposition of bone in the interalveolar septa, a phenomenon precisely documented and explicated by Domning and Hayek (1984). The benefits of horizontal tooth displacement include extended longevity, concomitant increase of number of breeding cycles, longer juvenile learning stages (important in animals such as proboscideans with complex social systems), and enhanced ability to exploit a wider, tougher range of food resources without exhausting the entire dental battery simultaneously (Sanders, 2017).

Such advantages were at least partially responsible for the replacement of plesielephantiforms and primitive elephantiforms by elephantimorph proboscideans that began in the Oligocene. In addition, the adaptive flexibility afforded by horizontal tooth displacement provided a template for rapid diversification of elephantimorphs, particularly gomphotheres, during the early Miocene, by supporting the exploitation of a wider range of resources. Presumably, differential selection among gomphothere populations for processing assorted food items led to morphological diversification of gnathodental structures. In basic gomphothere molars, crown morphology is characterized by half-loph(id)s composed of large outer main conelets accompanied by single, smaller mesoconelets, with modest-sized independent accessory conules on the anterior and posterior faces of pretrite half-loph(id)s. Accessory conules were incorporated into pretrite half-loph(id)s with wear, forming trefoil enamel loop patterns. Possibly in response to niche partitioning of ecosystems, evolutionary "experiments" at varying the manner of shearing, compressing, and grinding the food bolus in the different phases of chewing via novel occlusal features ("fabricational noise") were accommodated by the relatively open space between loph(id)s; accessory conules were hypertrophied, multiplied, and added to the posttrite side of the crown, mesoconelets were multiplied, and half-loph(id) pairs were dislocated transversely

DOI: 10.1201/b20016-4

and re-aligned. Thus, for example, amebelodont molars became distinguished by hypertrophy (and in some species oblique angulation) of pretrite accessory conules, development of posttrite accessory conules, and pseudoanancoidy of half-loph(id)s by bilateral offset of main conelets, while choerolophodont molars were differentiated by increase in size of mesoconelets, chevroning of half-loph(id) pairs by their oblique re-orientation, and mesial advancement of pretrite mesoconelets and their accompanying anterior accessory conules relative to their own abaxial main conelets and posttrite half-loph(id) elements, and, in more derived species, choerodonty (multiplication of accessory conules on both sides of the crown and in molar talon(id)s), as well as by cementodonty. It is apparent that these morphological departures from the basic gomphothere occlusal plan convergently achieved more effective comminution of food by increasing the area of molar grinding platforms.

Climatic and tectonic upheavals that characterized the Miocene must have also contributed to elephantimorph diversification, in part by vicariant fragmentation of populations and also by causing dramatic transformations of geomorphology and habitats. Global climatic changes that must have impacted on African ecosystems were as dynamic in the early-middle Miocene as taxonomic and morphological changes that proboscideans were undergoing. High-resolution deep-sea $\delta^{18}O$ isotope records calibrated closely to the geological time scale correlate with major climate shifts in the early and middle Miocene (Zachos et al., 2001), including an early Miocene cooling event followed by the Miocene Climatic Optimum between 17 and 15 Ma, representing the warmest temperatures of the Neogene (Kennett, 1995; Denton, 1999; Böhme, 2003). Global annual mean surface temperature at the maximum is estimated to have been 18.4°C, about 3°C higher than at present; this approximates expectations for the next century as a result of human-driven global warming (You et al., 2009). According to Van Zinderen Bakker and Mercer (1986), the Miocene rise in global temperature would have favored proliferation of woody vegetation across Africa, and spread of the tropical rain forest north and south of the Congo Basin. With the continent located in a more northerly position at that time, rainfall was greater in North Africa and consequently woodland and savanna ecosystems would have been present there, as well as in the eastern and southern sectors of the continent. The thermal maximum was followed by a major cooling episode between 14.8 and 14.1 Ma, linked with increased production of cold Antarctic deep waters and growth of the East Antarctic ice sheet (Flower and Kennett, 1994; Kennett, 1995; Denton, 1999; Böhme, 2003). As has been pointed out by Kender and colleagues (2009), however, climatic perturbations were more complex than that, with three episodes of polar cooling prior to the major cooling episode during the interval of the climatic optimum, each lasting as long as 100–300 Kyr.

The geodynamics of Africa were also volatile during the early and middle Miocene. Evidence has emerged of significant uplift of the East African plateau to a minimum of 1400 m by 13.5 Ma, even prior to rifting (Wichura et al., 2010). Uplift of the East African and Ethiopian plateaus prior to 13.5 and 20 Ma, respectively, was associated with the onset of more arid conditions in the region (Pik et al., 2008) and early establishment of a grass-dominated savanna biome in eastern Africa around 16–14 Ma (Kingston et al., 1994; Jacobs, 2004).

Significant tectonic events that affected early and middle Miocene Africa include southwest extension of the trans-African rift system, uplift of southern and eastern Africa, and maximum widening of the Red Sea (Partridge, 1997; Morley et al., 1999; Sepulchre et al., 2006; Potter and Szatmari, 2009). In addition, in North Africa the Atlas Mountains were uplifted, accommodating a substantial amount of western Mediterranean plate convergence between Africa and Iberia beginning in the early Miocene (Gomez et al., 2000; Potter and Szatmari, 2009). By the latter part of the early Miocene, internal massifs, such as the Rif and Kabylies, belonging to the Eurasian Plate, were compressed into North Africa, creating additional uplift along the coast and contributing to the rise of the Atlas Mountains (Mauffret, 2007). The closing of the Tethys epicontinental seaway was associated with these events, giving way to the formation of separate Mediterranean, Black, and Caspian Sea basins (Bernor, 1983; Janis, 1993). Outside of Africa, the opening of Drake's passage between Antarctica and South America produced a cold circumpolar current around Antarctica, promoting a steeper latitudinal thermal gradient (Janis, 1993). This could have driven a dominantly southward intercontinental exchange of mammals once geographic conditions accommodated it.

Low sea levels and docking of Afro-Arabia with Anatolia due to collisions of the Afro-Arabian and Eurasian plates (Rögl, 1999; Harzhauser et al., 2007; Sen, 2013) permitted early Miocene faunal movements between Eurasia and Africa—Africa was no longer an island continent. Much of the exchange involved immigration of Eurasian taxa into Africa, including squirrel-like rodents, insectivores, rabbits, rhinos, suids, chalicotheres, ruminant artiodactyls, and fissiped carnivores, supplementing endemics such as proboscideans and hyraxes, and older immigrant taxa such as creodonts and anthracotheres (Andrews and Van Couvering, 1975; Bernor et al., 1987; Janis, 1993; Agustí and Antón, 2002; Guerin and Pickford, 2003). The flow of mammals out of Africa during the early Miocene included gomphotheres, mammutids, and deinotheres, which emigrated from Africa to Eurasia penecontemporaneously, around 20–18 Ma (see Tassy, 1985, 1990a; Van der Made and Mazo, 2003). Docking of Afro-Arabia with Eurasia was multi-episodic (Harzhauser et al., 2007), and thus deinotheres made it into western Europe later than did gomphotheres (Göhlich, 2020)—the "Proboscidean Datum Event" involved several dispersal events (Tassy, 1990a; Antoine et al., 2013). By the late early to middle Miocene, African faunas increasingly featured woodland mammals such as horned bovids, giraffoids, listriodont suids, tragulids, rhinos, and monkeys, some older endemics vanished (archaic hyraxes, for instance), and taxonomic diversity increased (Pickford, 1981; Janis, 1993). Perhaps catalyzed by new competitors, more advanced amebelodont and choerolophodont species evolved to replace their confamilial predecessors. Recent work by Cote et al. (2018) suggests that this faunal transition was rapid and occurred around a narrow interval of 20–19 Ma, although proboscidean evolution was not necessarily in phase with the turnover of other taxa.

The highly variegated topographic profile of Africa that resulted from Miocene tectonic forces is well illustrated by East African uplift that accompanied extensive rifting regionally (Bernor, 1983). The rise of western and eastern rift highlands created a double rain shadow that disrupted moist tropical currents from the Atlantic and Indian Oceans and previously had supplied abundant year-round rainfall across equatorial Africa, leading to greater seasonality and latitudinal climate zonation, as well as the spread of seasonally adapted open woodlands in middle Miocene East Africa (Bernor, 1983). Elephantimorph proboscideans, with their adaptations to exploit a wide range of vegetation, not only survived this climatic and tectonic tempest but also mastered the transition from forest habitats to drier, more seasonal woodland and grassland ecotomes, with some African species becoming epicontinental (Table 4.1).

Unlike the Paleogene, the early and middle Miocene did not have substantial hiatuses of the fossil record, though the earliest Miocene and interval just prior to the start of the late Miocene are sparsely represented by mammal-bearing sites. Eastern Africa has a superior fossil record (in part because of rifting, in part due to geopolitical factors) compared with other African regions (Table 4.1; Figure 4.1). Although systematic work on North African early and middle Miocene proboscideans has lagged, nonetheless occurrences there contributed to a wider distribution of taxa in Africa than during the Paleogene.

SYSTEMATIC PALEONTOLOGY

BARYTHERIOIDEA Andrews, 1906
BARYTHERIIDAE Andrews, 1906
GEN. ET SP. NOV.
(Tables 2.3, 4.1, and 4.2; Figures 4.1 and 4.2)

A strikingly enigmatic mandible from locality TOP-85 at Nakwai, Kenya, KNM-NW 51755 (Figure 4.2), may be from the earliest Miocene (Table 4.1) and represents a surprisingly late-surviving barytherioid. Overall length of the specimen, including tusks, is 680 mm. The mandible is missing most of its left ramus, the coronoid process of the right ramus, and has a postmortem compound fracture of the corpus on its left side just anterior to the alveolus. In lateral view, the corpus and ramus are low; the symphysis curves downward gently on the corpus and is long (176.7 mm) and straight; and a single lower incisor on each side projects forward. There are three mental foramina, with the most anterior one the largest and recessed in the corpus. The angle is not large.

In dorsal view, there are two molars preserved on each side at the posterior of the alveolus, but the rest of the cheek teeth are broken away, hindering precise identification (Figure 4.2A). The symphysis is narrow, moderately deep and concave dorsally, and is bounded dorsally by sharp ridges. Distance between the ridges is about 66 mm. The tusks are not appressed in the midline and diverge slightly toward their distal ends. Relative to height, the corpus is relatively wide (at m1, H=71.3 mm; W=85.7 mm). The retromolar space forms a broad planum on each side. On the right side, the condyle is wide (61.0 mm) relative to anteroposterior length (33.5 mm) and is convex in both directions.

The tusks are straight and rounded, similar in shape to those of *Omanitherium*, but with no enamel covering. Incisor height and width are 25.5 mm×27.6 mm and 30.4 mm×25.9 mm on the right and left sides, respectively. Identification of the molars is difficult but appear to be m2–3. The m3s are bilophodont with a stout postcingulid dominated on the buccal side by a large conelet connected to a low series of smaller digitations that curve lingually (Figure 4.2B). The postcingulid is much narrower than the lophids. Each lophid comprises a transverse crest with a chisel-like, semi-vertical wear facet along its distal surface. The lophids are concave anteriorly and there are no independent cusps. The hypolophid in particular exhibits a very strong prehypocristid that angles forward and medially. The m2s are broken occlusally but preserve enough of their crowns to show that they each had a broad, low shelf-like postcingulid, are bilophodont, and were in wear to the extent that the lophids had dentine exposed apically (Figure 4.2B); they are slightly smaller than the m2s (Table 2.3).

It seems unlikely that this specimen is from a deinothere, as the space anterior to the m2 appears insufficient to have socketed p3–4 along with a trilophodont m1 longer than this molar. The deinothere m1 and p3–4 from Losodok are too large to fit in a jaw of this size. However, a trilophodont first molar from TOP-69, KNM-NW 66406, and a ?p3 from TOP-84, KNM-NW 66355, are small enough to have been accommodated in the -NW 51755 jaw, measuring L=41.5 mm, W=31.2 mm and L=21.9 mm, W=12.1 mm, respectively. By comparison, barytherioids have p2–4 and a bilophodont m1, but their premolars are relatively more anteroposteriorly compressed than those of deinotheres. The space for premolars in this jaw is short, particularly if accommodating p2–4, favoring identification of the specimen as a barythere. Also, it is evident that m1 was much shorter than m2 in KNM-NW 66406, which is not typical of deinotheres. The low position of the symphysis relative to the rest of the corpus and anterior projection of the incisors are typical of barytherioids rather than deinotheres. Phylogenetic alternatives to inclusion in the Barytherioidea include consideration of this specimen as a late-surviving, primitive proto-deinothere linking barytherioids with *Prodeinotherium*.

TABLE 4.1
Major Occurrences of Early to Middle Miocene Afro-Arabian Proboscidea

Taxon	Occurrence (Site, Locality)	Stratigraphic Unit	Age	Key References
Proboscidea Illiger, 1811				
Plesielephantiformes Shoshani, Sanders, and Tassy, 2001a				
?Barytherioidea Andrews, 1906				
?Barytheriidae Andrews, 1906				
Gen. et sp. nov.	Nakwai, Kenya (type)	Nakwai Fm.	<23.5 Ma (alt. 24.0–21.5 Ma)	Reynoso (2014), Rasmussen et al. (2019), and Friscia et al. (2020)
Suborder incertae sedis				
Deinotherioidea Bonaparte, 1845				
Deinotheriidae Bonaparte, 1845				
Deinotheriinae Bonaparte, 1845				
Prodeinotherium hobleyi	Sbeitla, Tunisia		Early Miocene	Vialli (1966), and Harris (1978a)
	Tébessa, Algeria		Early Miocene	Brives (1919), Vialli (1966), and Harris (1978a)
	Moroto II, Uganda		>21.6 Ma (alt. >20.6 Ma) (alt. ~17.5 Ma)	Bishop and Whyte (1962), Bishop (1967), Gebo et al. (1997), Pickford et al. (2003), and MacLatchy et al. (2019)
	Arrisdrift, Namibia	Arrisdrift Gravel Fm.	~20–19 Ma	Anonymous (1976), and Harris (1977a)
	Napak, Uganda		~20–19 Ma	Bishop (1964), Pickford (2003a), MacLatchy et al. (2006), Musalizi et al. (2009), and Cote et al. (2018)
	Koru, Kenya	Koru Fm.	19.5 Ma	Bishop (1967), Harris (1978a), and Pickford (1986a)
	Kajong (Mwiti), Kenya	Kajong Fm.	>19.2±1.2 Ma (probably within interval of 20.5–19.2 Ma) (alt. 23.0–19.0 Ma)	Savage and Williamson (1978) and Brown et al. (2016)
	Bukwa, Uganda		~19.0 Ma (alt. 17.5 Ma) (alt. 19.5–19.1 Ma) (alt. 22.0 Ma)	Bishop et al. (1969), Brock and Macdonald (1969), Walker (1969), Bishop (1971), Harris (1978a), Pickford (1981, 2002), Drake et al. (1988), MacLatchy et al. (2006), and Cote et al. (2018)
	Samburu Hills, Kenya	Nachola Fm.	19.2–15.0 Ma	Pickford et al. (1987)
	Wadi Moghara, Egypt	Moghara Fm.	~18.0–17.0 Ma (alt. ~21.0–16.0 Ma)	Fourtau (1918), Osborn (1936), Harris (1978a), Miller (1996, 1999), Sanders and Miller (2002), and Pickford et al. (2010)
	Loperot, Kenya		Early Miocene, > or = 18 Ma (minimum age = 17 Ma) (alt. ~19–18 Ma)	Bishop (1967), Maglio (1969a), Coppens et al. (1978), Savage and Williamson (1978), Pickford (1981), Leakey and Leakey (1986), Tassy (1986), Brown and McDougall (2011), Grossman et al. (2014), and Liutkus-Pierce et al. (2017, 2019)
	Rusinga, Kenya	Wayondo, Hiwegi, and Kulu Fms.	Slightly >17.8 Ma; 17.8 Ma; slightly<17.8 Ma	MacInnes (1942), Bishop (1967), Harris (1969, 1973), Pickford (1981, 1986a), and Tassy (1986)
	Mfwangano, Kenya	Hiwegi Fm.	17.8 Ma	Bishop (1967), Pickford (1986a), and Drake et al. (1988)
	Karungu, Kenya		>17.7 Ma (alt. 17.7–17.5 Ma)	Drake et al. (1988), Pickford (2003a), and Lukens et al. (2017)
	Lothidok, Kenya	Moruorot Mb., Lothidok Fm.	17.5–17.4 Ma (alt. 17.9–17.5 Ma)	Bishop (1967), Madden (1972), Pickford (1981), Tassy (1986), Boschetto et al. (1992), and Butts et al. (2018)

(Continued)

TABLE 4.1 (*Continued*)
Major Occurrences of Early to Middle Miocene Afro-Arabian Proboscidea

Taxon	Occurrence (Site, Locality)	Stratigraphic Unit	Age	Key References
	Moruorot, Kenya	Moruorot Mb., Lothidok Fm.	17.5–17.4 Ma (alt. 17.5–16.8 Ma)	Boschetto et al. (1992), and Butts et al. (2018)
	Kalodirr, Kenya	Moruorot Mb., Lothidok Fm.	17.5–17.4 Ma (alt. 17.5–16.8 Ma)	Boschetto et al. (1992), and Butts et al. (2018)
	As Sarrar, Saudi Arabia	Dam Fm.	Early middle Miocene, ~17.0–15.0 Ma (alt. ~19.0–17.0 Ma)	Thomas et al. (1982), and Whybrow and Clements (1999)
	Ghaba, Oman	Dam Fm.	Latest early Miocene or early-middle Miocene	Roger et al. (1994), and Pickford (2003a)
	Arongo Uyoma (Chianda), Kenya		Early Miocene, cf. Hiwegi and Kulu Fms.	Pickford (1986a)
	Gebel Zelten, Libya	Qaret Jahannam Mb., Marada Fm.	~16.5 Ma	Harris (1973), Pickford (1991a), and Sanders (2008a)
	Buluk (West Stephanie), Kenya	Buluk Mb., Bakate Fm.	16.3 Ma (alt. <17.3±1.4 Ma)	Harris and Watkins (1974), Pickford (1981), McDougall and Watkins (1985, 1988), Tassy (1986), Hurford and Watkins (1987), and Watkins (1989)
	Muruyur, Tugen Hills, Kenya	Mururyur Beds	15.5–15.0 Ma (alt. 14.3–13.2 Ma)	Bishop (1972), Harris (1978a), and Pickford et al. (2009)
	Mbagathi (Kirimun), Kenya	Kirimun Fm.	Middle Miocene	Pickford (1981), Pickford et al. (2009), Ishida and Ishida (1982), and Tassy (1986)
	Kaimogool, Kenya	Nyakach Fm.	15.0 Ma	Wynn and Retallack (2001)
	Maboko, Kenya (type)	Maboko Fm.	Beds 3 and 5, somewhat >14.7 Ma; between Bed 8 and the Maboko Phonolite, 14.7–13.8 Ma	MacInnes (1942), Bishop (1967), Tassy (1986), and Feibel and Brown (1991)
	Majiwa, Kenya	Maboko Fm.	Beds 3 and 5, somewhat >14.7 Ma; between Bed 8 and the Maboko Phonolite, 14.7–13.8 Ma (alt. ~16.0 Ma)	Andrews et al. (1981)
	Beni Mellal, Morocco		Middle Miocene, ~14.0 Ma	Lavocat (1961), and Remy (1976)
	Alengerr, Kenya	Alengerr Beds	~14.0–12.5 Ma	Bishop et al. (1971), Bishop (1972), and Harris (1978)
	Ft. Ternan, Kenya	Ft. Ternan Beds	13.7±0.3 Ma	Shipman et al. (1981), Tassy (1986), and Pickford et al. (2006)
	Ombo, Kenya		Middle Miocene	Bishop (1967), and Tassy (1986)
	?Sinda Area, Lower Semliki, Democratic Republic of Congo	Upper Mb., Sinda Beds	?middle Miocene	Hooijer (1963), Makinouchi et al. (1992), and Yasui et al. (1992)
	Gebel Cherichera, Tunisia	Beglia Fm.	~13.0–11.0 Ma	Harris (1978a)
	Ngorora, Tugen Hills, Kenya	Mbs. A–D, Ngorora Fm.	Within 13.0–8.5 Ma interval (alt. 13.5–11.5 Ma)	Bishop et al. (1971), Hill et al. (1985, 1986), Hill (2002), and Pickford et al. (2009)
Elephantiformes Tassy, 1988				
Palaeomastodontidae Andrews, 1906				
Palaeomastodon sp. nov.	Nakwai, Kenya (type)	Nakwai Fm.	<23.5 Ma (alt. 24.0–21.5 Ma)	Reynoso (2014), Rasmussen et al. (2019), and Friscia et al. (2020)

(*Continued*)

TABLE 4.1 (*Continued*)
Major Occurrences of Early to Middle Miocene Afro-Arabian Proboscidea

Taxon	Occurrence (Site, Locality)	Stratigraphic Unit	Age	Key References
Elephantimorpha Tassy and Shoshani, 1997 in Shoshani et al., 1998				
Mammutida Tassy and Shoshani, 1997 in Shoshani et al., 1998				
Mammutidae Hay, 1922				
Eozygodon morotoensis	Meswa Bridge, Kenya	Muhoroni Agglomerate	23.5–19.6 Ma (probably 23.0–22.0 Ma)	Bishop et al. (1969), Pickford and Tassy (1980), Tassy and Pickford (1983), and Tassy (1986)
	Elisabethfeld, Namibia		~21.0 Ma	Pickford (2008)
	Moroto I (type) and II, Uganda		>21.6 Ma (alt. >20.6 Ma) (alt. ~17.5 Ma)	Pickford and Tassy (1980), Tassy and Pickford (1983), Pickford et al. (1986, 2003), Gebo et al. (1997), Pickford (2003a, 2007b), Pickford and Mein (2006), and MacLatchy et al. (2019)
	Auchas, Namibia	Arrisdrift Gravel Fm.	~20.0–19.0 Ma	Pickford (2003a), and Tassy (2018)
Zygolophodon aegyptensis	Wadi Moghara, Egypt (type)	Moghara Fm.	~18.0–17.0 Ma (alt. ~21.0–16.0 Ma)	Miller (1996, 1999), Sanders and Miller (2002), and Pickford et al. (2010)
Zygolophodon sp. nov. [small]	Buluk (West Stephanie), Kenya (type)	Buluk Mb., Bakate Fm.	16.3 Ma (alt. <17.3±1.4 Ma)	Harris and Watkins (1974), McDougall and Watkins (1985, 1988), Tassy (1986), Hurford and Watkins (1987), and Watkins (1989)
Zygolophodon turicensis	Tugen Hills, Kenya	Mb. A, Ngorora Fm.	~13.0 Ma	Pickford (2007b)
	Gebel Cherichera, Tunisia	Beglia Fm. (Mbs. A-D)	(alt. ~13.0–11.0 Ma) (alt. ~12.0–11.0 Ma)	Gaudry (1891), Robinson (1974), Robinson and Black (1974), Pickford (2007b), and Pickford et al. (2009)
Zygolophodon sp. indet.	Rusinga, Kenya	Hiwegi Fm.	18.1–18.0 Ma (alt. 17.8 Ma)	Pickford and Tassy (1980), Pickford (1981, 1986b), and Drake et al. (1988)
	Lothidok 4 (Losodok), Kenya	Moruorot Mb.	17.9–17.5 Ma	Tassy (1986), Boschetto et al. (1992), Tobien (1996), and Pickford (2003a)
	Daberas Mine, Namibia		Early Miocene (alt. basal middle Miocene)	Pickford and Senut (2000), Pickford (2005b), and Pickford (2007b)
	Gebel Zelten, Libya	Qaret Jahannem Mb., Marada Fm.	~16.5 Ma	Savage and Hamilton (1973) and Sanders (2008a)
Elephantida Tassy and Shoshani, 1997 in Shoshani et al., 1998				
Gomphotheriidae Hay, 1922				
Gomphotheriinae Hay, 1922				
Gomphotherium sp.	Mwiti 5, Kajong (Mwiti), Kenya	Kajong Fm.	>19.2±1.2 Ma (probably within interval of 20.5–19.2 Ma) (alt. 23.0–19.0 Ma)	Savage and Williamson (1978), Tassy (1986), and Brown et al. (2016)
Gomphotherium libycum	Wadi Moghara, Egypt (type)	Moghara Fm.	~18.0–17.0 Ma (alt. ~21.0–16.0 Ma)	Fourtau (1918), Miller (1996, 1999), Sanders and Miller (2002), and Pickford et al. (2010)
	?Karungu, Kenya		>17.7 Ma (alt. 17.7–17.5 Ma)	Drake et al. (1988), Pickford (2003a), and Lukens et al. (2017)
	?Fejej, Ethiopia	Bakate Fm.	>16.18 Ma	Tiffney et al. (1994), and Richmond et al. (1998)
	Gebel Zelten, Libya	Qaret Jahanneam Mb., Marada Fm.	~16.5 Ma	Hormann (1963), Savage and Hamilton (1973), Coppens et al. (1978), Pickford (2003a), and Sanders (2008a)
("*Gomphotherium angustidens*")	?Djebel Nara, Tunisia	El Haouaria Fm.	?early Miocene	Robinson (2015)
	Ad Dabtiyah, Saudi Arabia	Dam Fm.	Early middle Miocene (alt. ~19–17 Ma)	Andrews et al. (1978), Gentry (1987a), Whybrow et al. (1987), Whybrow and Clements (1999), and Sanders and Miller (2002)
	?Al Jadidah, Saudi Arabia	Hofuf Fm.	Middle Miocene, ~14 Ma	Whybrow and Clements (1999)

(*Continued*)

TABLE 4.1 (*Continued*)
Major Occurrences of Early to Middle Miocene Afro-Arabian Proboscidea

Taxon	Occurrence (Site, Locality)	Stratigraphic Unit	Age	Key References
	Gebel Cherichera, Tunisia	Beglia Fm.	~13–11 Ma	Gaudry (1891), Robinson (1974), Robinson and Black (1974), and Pickford (2003a)
Gomphotherium spp. indet., "pygmy" or "dwarfed" gomphotheres	Songhor, Kenya		Early Miocene, 19.5 Ma	Bishop et al. (1969), Pickford and Andrews (1981), and Pickford (1986b)
	Siwa, Egypt		Early Miocene	Hamilton (1973), and Coppens et al. (1978)
	Gebel Zelten, Libya	Qaret Jahannem Mb., Marada Fm.	~16.5 Ma	Arambourg (1961), Coppens et al. (1978), Gaziry (1987a), and Sanders (2008a)
("*Gomphotherium* sp. indet.")	Buluk (West Stephanie), Kenya	Buluk Mb., Bakate Fm.	16.3 Ma (alt. <17.3±1.4 Ma)	Harris and Watkins (1974), McDougall and Watkins (1985, 1988), Tassy (1986), Hurford and Watkins (1987), Watkins (1989), and Sanders et al. (in prep.)
	Ghaba, Oman	Dam Fm.	Latest early Miocene or early-middle Miocene	Roger et al. (1994), and Pickford (2003a)
cf. *Gomphotherium* sp. nov. [archaic form]	Gebel Zelten, Libya	Qaret Jahannem Mb., Marada Fm.	~16.5 Ma	Harris (1969), and Sanders (2008a)
Gomphotherium sp. indet.	Locherangan, Kenya	Uppermost unit	17.7–16.8 Ma	Anyonge (1991)
Gomphotheriidae gen. indet.	Kaimogool, Kenya	Nyakach Fm.	15.0 Ma	Wynn and Retallack (2001)
Gomphotherium sp.	Gebel el Hendi Testour, Tunisia		14.0–13.0 Ma	Robinson and Black (1973), Salaj and Van Houten (1988), and Kowalski and Rzebik-Kowalska (1991)
Gomphotheriidae gen. et sp. indet.	Kabylie, Algeria		Miocene	Depéret (1897), Pickford (2005a), and Tassy et al. (2013)
Amebelodontinae Barbour, 1927				
Progomphotherium maraisi	Moroto I and II, Uganda		>21.6 Ma (alt. >20.6 Ma) (alt. ~17.5 Ma)	Gebo et al. (1997), Pickford (2003a), Pickford et al. (2003), and Pickford and Mein (2006)
	Auchas, Namibia (type)	Arrisdrift Gravel Formation, Pit AM 02	20–19 Ma	Pickford et al. (1995), Pickford and Senut (1997), and Pickford (2000, 2003a)
cf. *Progomphotherium maraisi* ("*Gomphotherium* sp. indet.")	Auchas, Namibia	Arrisdrift Gravel Fm., Pit AMSE	~18 Ma	Pickford et al. (1995) and Pickford (2003a)
Archaeobelodon sp. [primitive morph] (cf. *Archaeobelodon* sp.)	?Legetet, Kenya	Legetet Fm.	Early Miocene, ~20 Ma	Pickford (1986b), and Tassy (1986)
	Napak, Uganda		~20–19 Ma	Bishop (1964), Pickford (2003a), MacLatchy et al. (2006), Musalizi et al. (2009), Cote et al. (2018), and Pickford (2020)
	?Songhor, Kenya		Early Miocene, 19.5 Ma	Bishop et al. (1969), Pickford and Andrews (1981), and Pickford (1986b)
	Kajong (Mwiti), Kenya	Kajong Fm.	>19.2±1.2 Ma (probably within interval of 20.5–19.2 Ma) (alt. 23.0–19.0 Ma)	Savage and Williamson (1978), Tassy (1986), and Brown et al. (2016)
	Rusinga, Kenya	Hiwegi Fm. and Kulu Fm.	18.1–18.0 Ma; >17.25 Ma (alt. ~17.0–15.5 Ma)	Tassy (1986), Fox et al. (2009), and Peppe et al. (2009)

(*Continued*)

TABLE 4.1 (*Continued*)
Major Occurrences of Early to Middle Miocene Afro-Arabian Proboscidea

Taxon	Occurrence (Site, Locality)	Stratigraphic Unit	Age	Key References
	Wadi Moghara, Egypt	Moghara Fm.	~18–17 Ma (alt. 21–16 Ma)	Miller (1996, 1999), Sanders and Miller (2002), and Pickford et al. (2010)
	?Loperot, Kenya		Early Miocene, > or = 18 Ma (minimum age = 17 Ma) (alt. ~19–18 Ma)	Brown and McDougall (2011), Grossman et al. (2014), and Liutkus-Pierce et al. (2017, 2019)
("cf. *Progomphotherium*")	Karungu, Kenya		>17.7 Ma (alt. 17.7–17.5 Ma)	Drake et al. (1988), Pickford (2003a), and Lukens et al. (2017)
Archaeobelodon sp. nov. [advanced morph] ("*Protanancus macinnesi*")	Kalodirr, Kenya	Tiati Grits. Lothidok Fm.	17.5–17.4 Ma (alt. ~18.0–16.0 Ma) (alt. 17.7–16.6 Ma) (alt. 17.5–16.8 Ma)	Drake et al. (1988), Boschetto et al. (1992), and Butts et al. (2018)
	Nachola, Kenya	Nachola Fm.	~17–16 Ma	Pickford et al. (1987)
("*Archaeobelodon* cf. *filholi*")	Buluk (West Stephanie), Kenya (type)	Buluk Mb., Bakate Fm.	16.3 Ma (alt. <17.3±1.4 Ma)	Harris and Watkins (1974), McDougall and Watkins (1985, 1988), Tassy (1986), Hurford and Watkins (1987), and Watkins (1989)
?Amebelodontinae gen. et sp. indet., cf. *Archaeobelodon* ("*Platybelodon*")	Ad Dabtiyah, Saudi Arabia	Dam Fm.	Early-middle Miocene (alt. ~19–17 Ma)	Hamilton et al. (1978), Gentry (1987a), Whybrow et al. (1987), Whybrow and Clements (1999), and Sanders and Miller (2002)
?Amebelodontinae gen. et sp. indet.	As Sarrar, Saudi Arabia	Dam Fm.	Early middle Miocene, ~17.0–15.0 Ma (alt. ~19–17 Ma)	Thomas et al. (1982), and Whybrow and Clements (1999)
Protanancus sp. nov. [small]	Buluk (West Stephanie), Kenya	Buluk Mb., Bakate Fm.	16.3 Ma (alt. <17.3±1.4 Ma)	Harris and Watkins (1974), McDougall and Watkins (1985, 1988), Hurford and Watkins (1987), and Watkins (1989)
Protanancus macinnesi	Tugen Hills, Kenya	Muruyur Beds	15.5–15.0 Ma	Pickford et al. (2009)
(*Platybelodon*)	Nyakach, Kenya	Nyakach Fm.	15.0 Ma	Tassy (1986), Wynn and Retallack (2001), and Pickford (2003a)
	Kipsaraman, Tugen Hills, Kenya	Muruyur Fm.	~14.5 (alt. 15.5 Ma)	Hill (1999), Grundling and Hill (2000), Behrensmeyer et al. (2002), Pickford (2003a, 2007a), and Pickford and Kunimatsu (2005)
	Maboko, Kenya (type)	Maboko Fm.	Beds 3 and 5, somewhat >14.7 Ma; between Bed 8 and the Maboko Phonolite, 14.7–13.8 Ma	MacInnes (1942), Tassy (1986), and Feibel and Brown (1991)
	Kaloma, Kenya	Maboko Fm.	Middle Miocene	Pickford (1982)
	Majiwa, Kenya	Maboko Fm.	Middle Miocene	Pickford (1981, 1982) and Tassy (1986)
	Ft. Ternan, Kenya	Ft. Ternan Beds	13.7±0.3 Ma	Shipman et al. (1981), Tassy (1986), and Pickford et al. (2006)
	Sinda River area, near Ongoliba (Semliki 531), Democratic Republic of Congo	?Sinda Beds	?middle Miocene	Lepersonne (1959), Hooijer (1963), and Boaz (1994)
	?Ombo, Kenya		Middle Miocene	Van Couvering and Van Couvering (1976), Pickford (1981), and Tassy (1986)
	Alengerr, Kenya		<14.0 Ma, >12.4 Ma	Bishop (1972), and Tassy (1986)
Afromastodon coppensi	Arrisdrift, Namibia (type)	Arrisdrift Gravel Fm.	Early-middle Miocene, ~16.0 Ma (alt. ~17.5–17.0 Ma)	Corvinus and Hendey (1978), Hendey (1978a), and Pickford (1995, 2003a)
?Amebelodontinae cf. *Afromastodon coppensi*	BUR1, Burji, Ethiopia		~17–15 Ma	Suwa et al. (1991), and WoldeGabriel et al. (1991)

(*Continued*)

TABLE 4.1 (*Continued*)
Major Occurrences of Early to Middle Miocene Afro-Arabian Proboscidea

Taxon	Occurrence (Site, Locality)	Stratigraphic Unit	Age	Key References
Platybelodon sp.	Loperot, Kenya		Early Miocene, ≥18 Ma (minimum age=17 Ma) (alt. ~19–18 Ma)	Maglio (1969a), Coppens et al. (1978), Tassy (1986), Brown and McDougall (2011), Grossman et al. (2014), and Liutkus-Pierce et al. (2017, 2019)
Choerolophodontinae Gaziry, 1976				
Afrochoerodon kisumuensis	Wadi Moghara, Egypt	Moghara Fm.	~18–17 Ma (alt. 21–16 Ma)	Miller (1996, 1999), Sanders and Miller (2002), and Pickford et al. (2010)
	Buluk (West Stephanie), Kenya	Buluk Mb., Bakate Fm.	16.3 Ma (alt. <17.3±1.4 Ma)	Harris and Watkins (1974), McDougall and Watkins (1985, 1988), Tassy (1986), Hurford and Watkins (1987), and Watkins (1989)
	Bosluis Pan, South Africa		~16–15 Ma	Senut et al. (1996), and Pickford (2005b)
	Tugen Hills, Kenya	Muruyur Beds	15.5–15.0 Ma	Pickford et al. (2009)
	Maboko, Kenya (type)	Maboko Fm.	Beds 3 and 5, somewhat >14.7 Ma; between Bed 8 and the Maboko Phonolite, 14.7–13.8 Ma	MacInnes (1942), Tassy (1977b, 1979b, 1986), and Feibel and Brown (1991)
	Kaloma, Kenya	Maboko Fm.	Middle Miocene	Pickford (1982)
	Majiwa, Kenya	Maboko Fm.	Middle Miocene	Pickford (1982)
	Kipsaraman and Chepawara, Tugen Hills, Kenya	Muruyur Fm.	~14.5 (alt. 15.5 Ma)	Bishop (1972), Tassy (1986), Hill (1999), Grundling and Hill (2000), Pickford (2001, 2003a), Behrensmeyer et al. (2002), and Pickford and Kunimatsu (2005)
Choerolophodon zaltaniensis	Gebel Zelten, Libya (type)	Marada Fm.	16.5 Ma	Arambourg and Magnier (1961), Magnier (1962), Selley (1967, 1969), Savage and Hamilton (1973), Gaziry (1987a), Pickford (1991a), and Sanders (2008a)
Choerolophodon sp. indet. probably nov. [small] (=*C.* "*pygmaeus*") (?=*C. ngorora*)	Ngenyin, Tugen Hills, Kenya	Mb. A, Ngorora Fm.	~13.0 Ma	Pickford (2004) and Pickford et al. (2009)
Choerolophodon ngorora, primitive morph	Ft. Ternan, Kenya	Ft. Ternan Beds	13.7±0.3 Ma	Shipman et al. (1981), Tassy (1986), and Pickford et al. (2006)
	Tugen Hills, Kenya	Mbs. A-D, Ngorora Fm.	~13.0–11.0 Ma (alt. 13.5–11.5 Ma)	Maglio (1974), Tassy (1986), and Pickford et al. (2009)
Choerolophodon ngorora, advanced morph	Mbagathi (Kirimun), Kenya	Kirimun Fm.	Middle Miocene	Pickford (1981), Pickford et al. (2009), Ishida and Ishida (1982), and Tassy (1986)
Choerolophodon sp. indet.	Kaimogool, Kenya	Nyakach Fm.	15.0 Ma	Wynn and Retallack (2001)
	Majiwa, Kenya	Maboko Fm.	Middle Miocene	Pickford (1981) and Tassy (1986)
	Gebel Cherichera, Tunisia	Beglia Fm.	~13–11 Ma	Robinson (1974), and Robinson and Black (1974)
	Henchir Beglia, Tunisia	Beglia Fm.	~13–11 Ma	Robinson (1974), and Robinson and Black (1974)
	?Bled Douarah, Tunisia	Beglia Fm.	~13–11 Ma	Robinson and Black (1974), and Geraads (1989)

?=attribution, age, or occurrence uncertain; alt., alternative.

TABLE 4.2
Adult Dental Formulae and Incisor Dominance in Early-Middle Miocene African Proboscidea

Taxon	Upper Permanent Dentition	Largest Upper Incisor	Lower Permanent Dentition	Largest Lower Incisor
Barytherioid gen. et sp. nov.	—	?	i1 or i2, ?p (two or three premolars), m1–3	i1 or i2
Prodeinotherium hobleyi	P3–4, M1–3	—	i2, p3–4, m1–3	i2
Palaeomastodon sp. nov.	I2, ?P3–4, M1–3	I2	—	—
Eozygodon morotoensis	I2, ?P3–4, M1–3	I2	i2, ?p3–p4, m1–3	i2
Zygolophodon turicensis	I2, ?P3–4, M1–3	I2	i2, ?p3–4, m1–3	i2
Gomphotherium libycum	I2, P3–4, M1–3	I2	?i2, p3–4, m1–3	?i2
Progomphotherium maraisi	I2, P3–4, M1–3	I2	i2, p3–4, m1–3	i2
Archaeobelodon sp. nov.	I2, (P2 is unknown) P3–4, M1–3	I2	i2, p3–4, m1–3	i2
Protanancus macinnesi	I2, P3–4, M1–3	I2	i2, p3–4, m1–3	i2
Afromastodon coppensi	I2, ?P, M1–3	I2	i2, ?p, m1–3	i2
Afrochoerodon kisumuensis	I2, ?P, M1–3	I2	?i, ?P, m1–3	?
Choerolophodon ngorora (primitive morph)	I2, ?P, M1–3	I2	?i, ?P, m1–3	?
Choerolophodon zaltaniensis	I2, ?P, M1–3	I2	?i, ?P, m1–3	?

? = Presence unknown.

DEINOTHERIOIDEA Bonaparte, 1845
DEINOTHERIIDAE Bonparte, 1845
DEINOTHERIINAE Bonaparte, 1845
***PRODEINOTHERIUM* (Éhik, 1930)**
***PRODEINOTHERIUM HOBLEYI* (Andrews, 1911)**
(Tables 4.1 and 4.2; Figures 4.1 and 4.3)

Partial Synonymy—*Dinotherium hobleyi* Andrews, 1911; *Dinotherium cuvieri* Brives, 1919; *Deinotherium bavaricum* (in part) Gräf, 1957; *D. cuvieri* Savage, 1967; *Prodeinotherium hobleyi* Harris, 1973.

Following their first appearance in the late Oligocene, deinotheres began to flourish in Africa during the early Miocene. They are ubiquitous and sometimes abundant at African sites of that age (e.g., Rusinga and Buluk, Kenya; Gebel Zelten, Libya). They were a highly successful lineage of proboscideans that became widespread throughout the Old World, surviving in Africa until the end of the early Pleistocene (Harris, 1978; Beden, 1985; Behrensmeyer et al., 1995; Sanders et al., 2010a). Traditionally, deinotheres have been divided into older, smaller *Prodeinotherium* and its larger successors in *Deinotherium* (Harris, 1978; Huttunen, 2000, 2002; but see Pickford and Pourabrishami, 2013). Although the number of species names proposed for these genera has proliferated wildly (see, for example, the discussion of the number of deinothere species in South Asia in Sankhyan and Sharma [2014]), a conservative taxonomic scheme recognizes only *P. hobleyi* and *D. bozasi* in Africa, *P. pentapotamiae*, *D. indicum*, and sometimes *P. orlovii* in South Asia (*P. pentapotamiae* was also tentatively identified from middle Miocene horizons of the Tha Chang sand pits in Thailand [Thasod et al., 2012]), and *P. bavaricum* and *D. giganteum* in Europe (Harris, 1976, 1978a; Huttunen, 2000, 2002; Tiwari et al., 2006; Singh et al., 2020). European deinotheres with immense teeth are sometimes placed in *D. gigantissimum* (Garevski and Markov, 2011), but others (Pickford and Pourabrishami, 2013; Konidaris et al., 2017) claim that *D. proavum* has priority over "*gigantissimum*" and that *D. indicum* is synonymous with this species. The former assertion may be correct, but survey of metric data of *D. indicum* does not support the contention that it has immense molars or its synonymy with *D. proavum*. A more current taxonomic scheme for European deinotheres recognizes primitive *Prodeinotherium cuvieri* from early Miocene MN 3–4 times, *P. bavaricum* from MN 5–6 or 7, *Deinotherium levius* from MN 7–8, and *D. giganteum* from MN 9–10, overlapping with the immense-sized *D. proavum* (Göhlich, 2010; Konidaris et al., 2017).

Despite these issues, African *Prodeinotherium hobleyi* and *D. bozasi* are sound, well-defined species readily separable by time, size, and morphology. Harris (1978a) collated skeletal differences between *Prodeinotherium* and *Deinotherium*, which are applicable to the African species of these genera. For *Prodeinotherium*, these distinctions include a convex astragalar facet for the tibia, a more plantigrade manus, femur of similar length to that of *Elephas maximus*, scapula with a stout spine, metacromion, acromion, and well-developed supraspinous fossa, humerus with a proximally tapered lateral epicondyle, nearly vertically orientated lower tusks, M3s with well-developed postmetaloph ornamentation, P3–4s usually lacking mesostyles, nasal aperture about as high as wide, and cranial rostrum narrow, deep, and ventrally flexed. Postcranial differences (e.g., in deinotheres, ulnae stout with large olecranon processes; femora with lower medial epicondyles and more rounded condyles; tibiae with more concave femoral facets and quadrate astragalar facets) between deinotheres and elephantimorphs were listed by Harris (1978a) but generally are

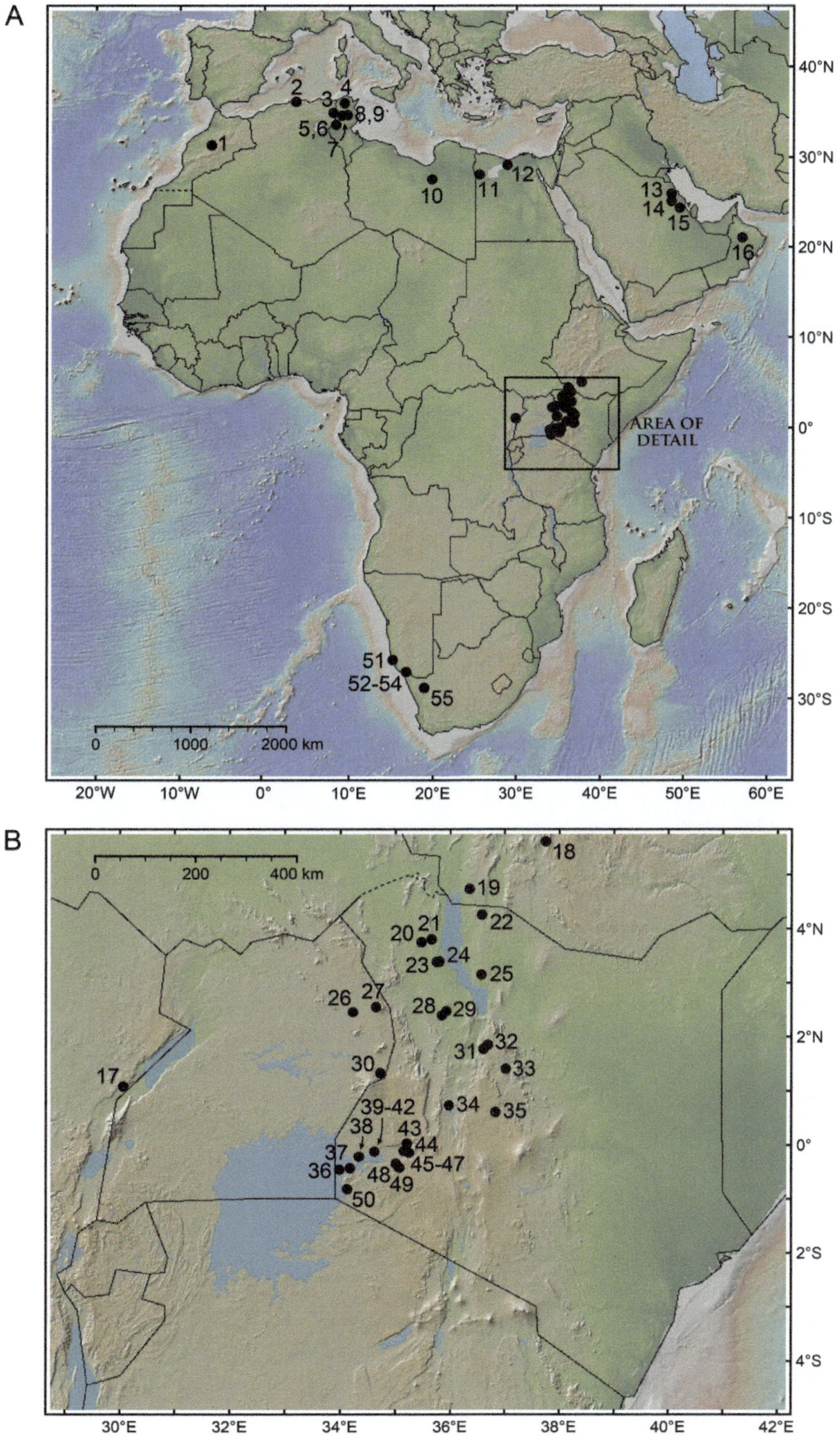

FIGURE 4.1 Maps showing location of early and middle Miocene proboscidean sites, (A, B). Area of detail, (B). 1, Beni Mellal, Morocco; 2, Kabylie, Algeria; 3, Tébessa, Algeria; 4, Gebel el Hendi Testour, Tunisia; 5, Henchir Beglia, Tunisia; 6, Bled Dourah, Tunisia; 7, Sbeitla, Tunisia; 8, Djebel Nara, Tunisia; 9, Gebel Cherichera, Tunisia; 10, Gebel Zelten, Libya; 11, Siwa, Egypt; 12, Wadi Moghara, Egypt; 13, As Sarrar, Saudi Arabia; 14, Ad Dabtiyah, Saudi Arabia; 15, Al Jadidah, Saudi Arabia; 16, Ghaba, Oman; 17a, Sinda Area, Lower Semliki, Democratic Republic of Congo; 17b, Sinda River area, near Ongoliba (Semliki 531), Democratic Republic of Congo; 18, BUR1, Burji, Ethiopia; 19, Fejej, Ethiopia; 20, Moruorot, Kenya; 21, Locherangan, Kenya; 22, Buluk (West Stephanie), Kenya; 23, Kalodirr, Kenya; 24a, Lothidok, Kenya, b, Lothidok 4 (Losodok), Kenya; 25a, Kajong (Mwiti), Kenya, b, Mwiti 5, Kajong (Mwiti), Kenya; 26, Napak, Uganda; 27a, Moroto I and II, Uganda; 27b, Moroto II, Uganda; 28, Loperot, Kenya; 29, Nakwai, Kenya; 30, Bukwa, Uganda; 31, Samburu Hills, Kenya; 32, Nachola, Kenya; 33, Alengerr, Kenya; 34a, Kipsaraman and Chepawara, Tugen Hills, Kenya; 34b, Kipsaraman, Tugen Hills, Kenya; 34c, Muruyur, Tugen Hills, Kenya; 34d, Ngenyin, Tugen Hills, Kenya; 34e, Ngorora, Tugen Hills, Kenya; 34f, Tugen Hills, Kenya; 35, Mbagathi (Kirimun), Kenya; 36, Mfwangano, Kenya; 37, Rusinga, Kenya; 38, Arongo Uyoma (Chianda), Kenya; 39, Maboko, Kenya;

(Continued)

Figure 4.1 (*Continued*) 40, Kaloma, Kenya; 41, Ombo, Kenya; 42, Majiwa, Kenya; 43, Songhor, Kenya; 44, Meswa Bridge, Kenya; 45, Koru, Kenya; 46, Ft. Ternan, Kenya; 47, Legetet, Kenya; 48, Kaimogool, Kenya; 49, Nyakach, Kenya; 50, Karungu, Kenya; 51, Elisabethfeld, Namibia; 52, Daberas Mine, Namibia; 53, Auchas, Namibia; 54, Arrisdrift, Namibia; 55, Bosluis Pan, South Africa.

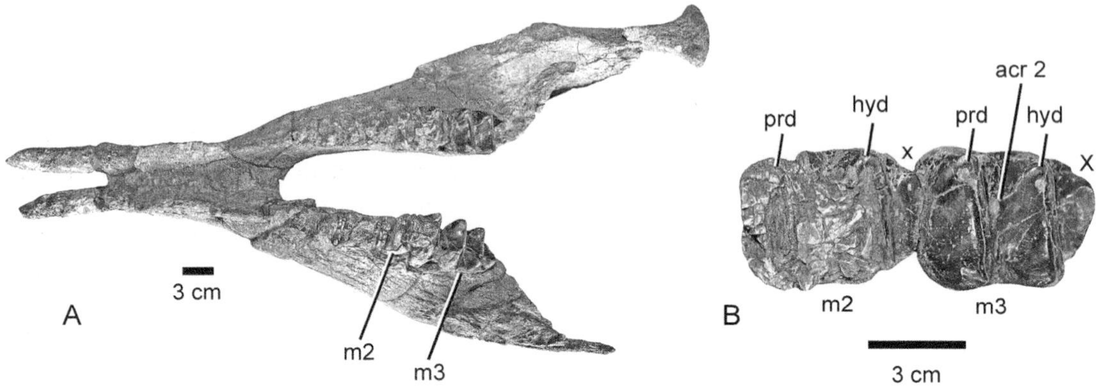

FIGURE 4.2 Mandible KNM-NW 51755 with right and left tusks, alveoli for right and left ?p2–m1, right and left partial m2s, and right and left m3s, ?Barytherioidea gen. et sp. nov. from ?early Miocene Nakwai, Kenya. Abbreviations: acr2, anterior pretrite crescentoid, pretrite half-lophid 2 (=prehypocristid); hyd, hypolophid; m, molar; prd, protolophid; x, postcingulid; X, large postcingulid. Anterior is to the left. (A) Mandible, occlusal view. (B) Enlarged right m2–3, occlusal view.

not marked. Volumetric body size estimates for deinotheres indicates that although both genera exhibit numerous skeletal adaptations to graviportal posture, *Prodeinotherium* (body mass 3.1–4.3 tons, shoulder height=247–278 cm) was considerably smaller than *Deinotherium* (body mass 8.8–13.2 tons, shoulder height=359–401 cm) (Larramendi, 2016). Based on allometric study of body mass to long bone dimensions, Christiansen (2004) estimated even heavier body masses for *Deinotherium*, perhaps up to 20,000 kg, which indicates that these deinotheres were among the largest of all proboscideans.

Prodeinotherium hobleyi is an early and middle Miocene species, though a few isolated teeth from the late Oligocene of eastern Africa (see Chapter 3) are indistinguishable morphologically from this species if at the low end of its range of cheek tooth dimensions. *Prodeinotherium* migrated out of Africa during the early Miocene and is present in Greece by 18.4 Ma, land mammal stage MN3, though it did not become common in western Europe until stage MN4 (Konidaris et al., 2017; Göhlich, 2020). African emigrant deinotheres gave rise to congeners in Europe (von Meyer, 1831), South Asia (Falconer, 1868; Forster Cooper, 1922), and possibly Asia (*P. sinense* [Qiu et al., 2007]). In the middle to late Miocene, *Prodeinotherium* was replaced by species of *Deinotherium*, which, outside of Africa, did not persist past the late Miocene (Harris, 1978; Barry and Flynn, 1989; Sanders, 2003; G. Markov, pers. comm.). However, the first record of *Deinotherium* in Africa is later than in Eurasia, and if its origin was parallel but independent on different continents, a new taxonomic scheme is warranted. Alternatively, Pickford and Pourabrishami (2013) felt that *D. bozasi* could be a tropical variant of *D. proavum*.

Cranial remains of *P. hobleyi* are rare and best represented at the terminal early Miocene site of Gebel Zelten, Libya. Description of the Gebel Zelten material, in particular subadult individual M 26665, was provided in detail by Harris (1973) in a classic comparative study. The cranium is characterized by having a short, downturned rostrum, low braincase that gently slopes downward anteriorly but with a basicranium raised well above the level of the palate, and moderately retracted, relatively narrow nasal aperture suggesting presence of at least a short proboscis. The greatest length of M 26665 is 940 mm. Posteriorly, the nuchal planum is very low and broad with a transversely broad, deep fossa for insertion of the nuchal ligament. The greatest width of the cranium is across the nuchal planum, 596 mm. The occipital condyles are not situated far below the ligamentous fossa and face mostly posteriorly. The nuchal crest is concave about the midline.

In ventral view, the palate is relatively narrow, the zygomatic arches do not appear to flare beyond the frame of the cranium, the glenoid articulations are anteroposteriorly convex and very broad, there is a deep postglenoid fossa behind each glenoid (but no postglenoid process), and the internal nares open posterior to the tooth rows (Figure 4.3A). The adult dental formula, 0-0-2-3/1-0-2-3, reflects the unusual absence (for proboscideans) of upper incisors, as well as loss of P2/p2 (Table 4.2).

In dorsal view, the nasals are very short and do not project much over the nasal aperture; the premaxillae

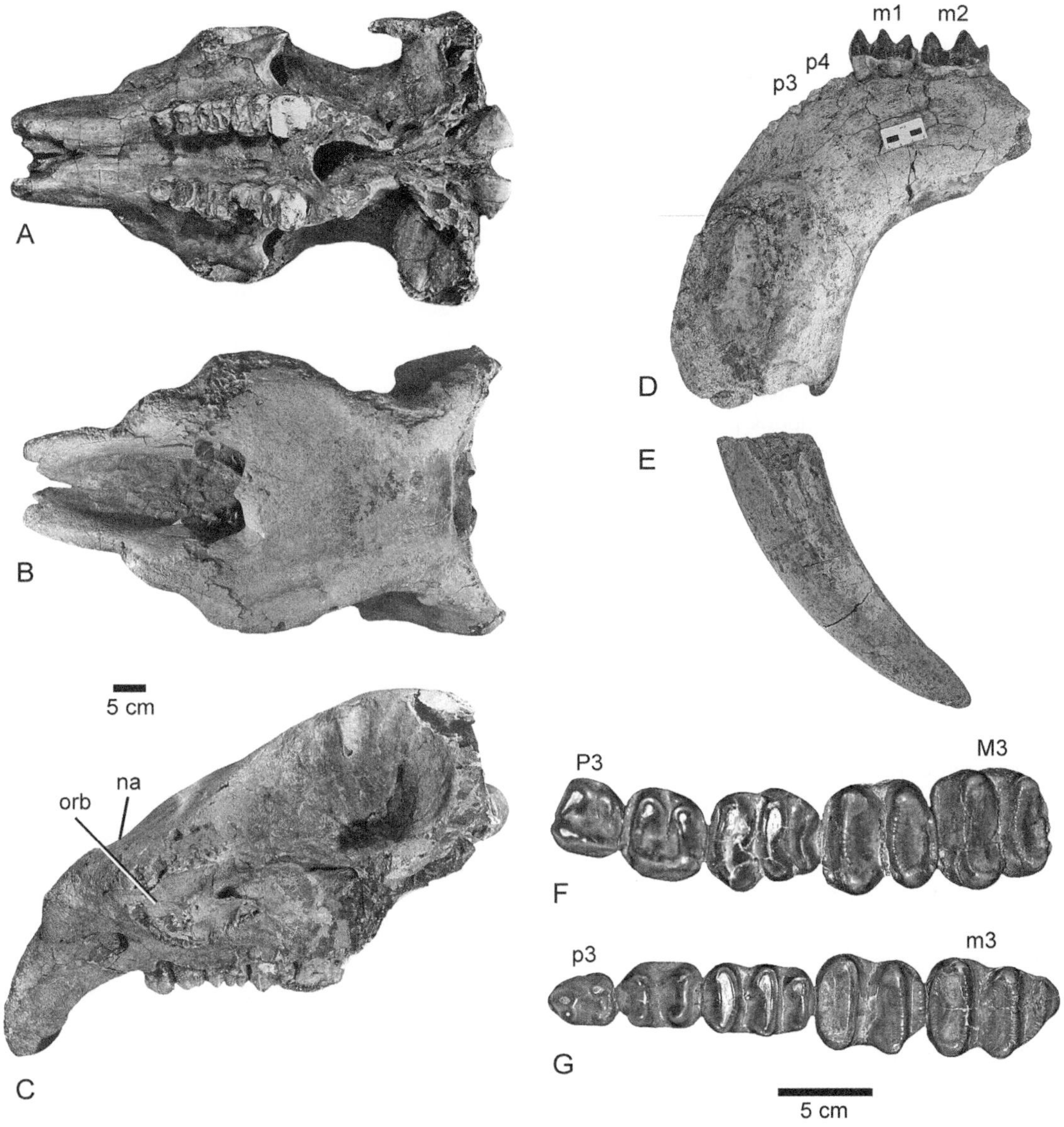

FIGURE 4.3 Skull and dental specimens of early and middle Miocene *Prodeinotherium hobleyi*. Abbreviations: M/m, upper/lower molar; na, nasal aperture; orb, orbit; P/p, upper/lower premolar. Anterior is to the left. (A) Cranium with cheek teeth, M 26665 from Gebel Zelten, Libya, ventral view. (B) Cranium, M 26665 from Gebel Zelten, Libya, dorsal view. (C) Cranium, M 26665 from Gebel Zelten, Libya, right lateral view (reversed). (D) Dentary with symphysis, m1–2, and alveoli for p3–4, KNM-WS 65638 from Buluk, Kenya, lateral view. (E) Lower tusk, field #533 from Buluk, Kenya, lateral view. Not necessarily from the same individual as KNM-WS 65638. (F) Upper cheek tooth row, M 15713 from Arongo Uyoma, Kenya, occlusal view. Same individual as the lower tooth row. (G) Lower cheek tooth row, M 15713 from Arongo Uyoma, Kenya, occlusal view. (Images A–C from Harris [1973:pl. 1, fig. 1b; pl. 2, fig. 2b; pl. 3, fig. 3b].)

are deeply inset between the maxillae in a narrow rostral trough; and the frontal and parietals are mostly flat and broad (the frontal may be convex anteriorly in the midline and the parietals are slightly concave transversely) (Figure 4.3B).

In lateral view, the orbit is small and low, situated as anteriorly as above P3 (Figure 4.3C). The occipital condyles do not project much posteriorly. Overall, the cranium is distinguished from that of *Deinotherium* by having a less retracted nasal opening, relatively wider and longer roof, more vertically inclined nuchal planum, shorter paroccipital processes, and lower position of occipital condyles.

The mandible of the species has a saliently downturned, wide symphysis that accommodates large, downwardly projecting lower tusks (Figure 4.3D and E). Posterior extension of the symphysis continues to the distal edge of p3. There are multiple mental foramina. Although corpora of deinothere mandibles are generally relatively slender, symphyses may be much more immense in depth. In *P. hobleyi*, the ramus is rooted at the posterior end of the last molar, has a vertical leading edge, and is not particularly high. The mandibular condyle is a little higher than the small coronoid process. In contrast, the angle of the ramus is large and rounded, suggesting that the masseter and pterygoid muscles had more powerful masticatory roles than the temporalis muscles (Harris, 1975).

The most striking feature of the skull in deinotheres are the downturned lower tusks (i2s?), which are often quite large. These incisors lack enamel and are oval in cross-section. They may have had enamel caps when first erupted and unworn. That they have a role in food acquisition (but likely not in digging) is indicated by distal wear on their mediolabial and lingual aspects. It is possible that some tusk attrition resulted from passive rubbing of the trunk, along with the effects of the trunk and tusks working synchronously to grasp vegetation. Deciduous precursors to these incisors are covered in thin enamel (Harris, 1973, 1978a).

In deinotheres, all erupted adult cheek teeth were used concurrently and even with aging premolars were not ejected anteriorly from the jaws (Figure 4.3F and G). Vertical tooth succession and molar lophodonty differentiate deinotheres from elephantimorphs and were primitively retained from the condition of more ancient plesielephantiform proboscideans. Second and third molars are bilophodont but first molars and deciduous fourth premolars are distinctively trilophed. Molar cusps are incorporated into transverse crests and unworn molars sport numerous small mammillons along the apices of the loph(id)s. Lophids are anteriorly concave and lophs are correspondingly anteriorly convex. With wear, lophids form sharply-crested transverse planar facets on their posterior sides that interdigitate with planar facets on the posterior aspects of lophs; these facets are primarily vertical in lightly moderately worn teeth.

It appears from study of tooth wear that nearly all deinothere mastication occurs during Phase I chewing, with motion of the lower teeth mostly mesial with a small component of mesiolingual movement to achieve both shear-cutting and compressive functions; there is no significant Phase II chewing activity in the deinothere pattern of mastication (von Koeingswald, 2014, 2016a). Harris (1975) reconstructed the masticatory cycle of deinotheres as follows: (a) The temporalis muscles contracted to elevate the mandible in pre-occlusion, with action of the pterygoid muscles to help guide the jaws into position; (b) further contraction of the temporalis muscles occluded the transverse loph(id)s of the molars, and contraction of the masseter muscles applied maximum shear force across the loph(id)s; (c) additional contraction of the masseter muscles separated the shearing facets of the molars and initiated the grinding phase of the premolars; and (d) the recovery phase of the cycle involved relaxation of the masseter, temporalis, and pterygoid muscles and contraction of the digastric muscles, allowing the lower jaw to open and retract. The prominent anterior cristids of lower molars (pre-proto-, -meta-, -hypo-, and -ento-cristids) and posterior cristae of upper molars (post-proto-, -para-, -hypo-, and -meta-cristae) that descend into transverse valleys medially from the lateral corners of the loph(id)s almost certainly functioned to secure the food bolus during the compressive part of mastication. First and second molars have low, transversely rectangular postcingular(id) shelves, and m3s exhibit a more prominent talonid that projects more on the pretrite side of the crown.

Upper premolars in *P. hobleyi* are square-shaped in occlusal view, with free protocones and formation of ectolophs by connection of para- and metacones. The hypocone is usually free-standing in P3 but may extend a transverse crest toward the metacone in P4. P3 is usually smaller than P4 and may have a more prominent, anteriorly projecting parastyle. In both premolars, the trigon basin is relatively large and open. Lower premolars are either triangulate (p3) or rectangular (p4) in occlusal shape. In p3, the protoconid+hypoconid form a prominent ectolophid, while the metaconid and entoconid are variably expressed and more marginally situated on the crown. By comparison, p4 is bilophodont and more molarized, with a protolophid and hypolophid; on its buccal side, both lophids may have anterior buttresses. Individual cusps are subsumed into these lophids. During occlusion, the buccal edge of the p3 ectolophid sheared against the lingual edge of the P3 ectoloph, limiting lateral movement of the jaws (Harris, 1975). Morphologically, upper and lower deciduous premolars (dP2/dp2-dP4/dp4) are smaller versions of the permanent premolars and first molars.

Harris (1975) described deinotheres as "shearing browsers," which is an apt term, given what is known about their masticatory mechanics and diets. Initial observations on wear across the chisel-like surfaces of the loph(id)s indicated a diet of soft vegetation and continual self-sharpening of upper and lower beveled edges of loph(id)s (Harris, 1975). However, although unpublished results of microwear analyses of *P. hobleyi* molars by Gina Semprebon and Juha Saarinen are consistent with an interpretation of deinotheres as browsers, they indicate consumption of coarser forage. Isotopic analysis of *P. hobleyi* molars by Cerling et al. (1999) revealed no trace of C_4 vegetation in their diets and promulgated the view that deinotheres were dedicated browsers. However, a recent large-scale study of isotopes in dental enamel by Lohrke (2017) discovered exceptions to dietary homogeneity in deinotheres, including at the middle Miocene site of Muruyur in the Tugen Hills of Kenya, where *P. hobleyi* was incorporating an unexpectedly substantial amount of C_4 vegetation in a mixed-feeding regimen. Overall, deinotheres were important constituents of early and middle Miocene faunas and their presence generally signals habitats with an abundance of leafy vegetation. The maneuverability of the *Prodeinotherium* skull, suggested by the high position of the occipital condyles and projection of the paroccipital processes, combined with brachydont, lophodont molars adapted to shearing leaves and evidence for a small trunk, indicated to Harris (1978a) a jungle or forest habitus, along densely vegetated river margins and swamps.

Since the late eighteenth century, deinotheres have had a colorful history of misidentification, as rhinos, tapirs, hippos, sea cows, whales, and marsupials (Harris, 1978a and references therein; Shoshani et al., 1996). Simpson (1945) placed them firmly in the Proboscidea, but up until fairly recently others remained skeptical of this assignment (e.g., Maglio, 1973; Harris, 1978a). More current research has firmly supported deinotheres as proboscideans (e.g., Tassy, 1982, 1985; Tassy and Shoshani, 1988; Sanders et al., 2010a). Derived proboscidean traits of deinotheres include a retracted nasal aperture (indicative of a trunk), loss of canines and most incisors (Table 4.2), tusk-like lower incisors, pneumatized cranium, radius fixed in a pronated position, and a well-developed astragalar medial tubercle (Shoshani et al., 1996).

Understanding their origins and relationships has been hindered by the fact that deinotheres were known for a very long time prior to the discovery of most of the plesielephantiform taxa of the Paleogene. Because of the nearly identical morphology of second and third molars of *Prodeinotherium* and *Barytherium*, Harris (1978a) explored their potential sister-taxon relationship, rejecting it because of differences in anterior tooth formulae, tritolophy of fourth deciduous premolars and first molars in deinotheres, and simpler construction of premolars in barytheres, suggesting differences in masticatory functions. The cladistic analysis of Delmer (2009) did not recover a special sister-group relationship of *Chilgatherium* and deinotheres because of *C. harrisi*'s M3 tritoloph, absence of a postmetacristid and presence of two cusps on the talonid of m3, instead placing *Chilgatherium* as a sister taxon to deinotheres+palaeomastodonts. These results seem improbable—deinotheres share no credible synapomorphies with palaeomastodonts. Deinotheres are also positioned phylogenetically as the sister taxon to palaeomastodonts by Gheerbrant and Tassy (2009), though they admit that the features supporting this arrangement are liable to convergence. Based on the consideration of ten mandibular characters relating to canals and foramina, Ferretti and Debruyne (2011) placed *P. hobleyi* as a sister taxon to palaeomastodonts rather than with plesielephantiforms. There is no credence to their phylogenetic systematic results: All of the features shared by deinotheres and palaeomastodonts are symplesiomorphies. Conversely, Shoshani and Tassy (2005) placed deinotheres in Plesielephantiformes, alongside barytheres in their classification. Schmelzmuster patterns in molar enamel support this taxonomy (Tabuce et al., 2007). It seems very unlikely that deinotheres are informative about the origins of elephantiforms. Instead, the main issue is whether deinotheres evolved from a lophodont taxon similar to *Barytherium*, or from predecessors resembling *Chilgatherium*, which in turn *might* have had a bunolophodont ancestry.

ELEPHANTIFORMES Tassy, 1988
PALAEOMASTODONTIDAE Andrews, 1906
PALAEOMASTODON Andrews, 1906
PALAEOMASTODON SP. NOV. [Nakwai]

(Tables 4.1 and 4.2; Figures 4.1 and 4.4)

A crushed and consequently heavily step-fractured and plastically-distorted cranium from Nakwai locality TOP 120, KNM-NW 58953 (Figure 4.4A–C), is labeled in the collections at the National Museums of Kenya as "*Losodokodon losodokius*." Despite taphonomic alterations, however, it is clearly not mammutid but rather is from a very late-surviving palaeomastodont. If the age of the locality (TOP 120) is similar to those of other nearby localities at the site (see Reynoso, 2014; Rasmussen et al., 2019; Friscia et al., 2020), this new specimen represents the only palaeomastodont known from the Neogene. As is the case for the probable barytherioid from the site, the temporal context of this specimen is a pleasant and unexpected surprise, bolstering the idea that African Paleogene-Neogene faunal change was episodic rather than a singular event, and that some taxa, particularly proboscideans, had their own evolutionary trajectories that were more influenced by local factors than contact with Eurasia (Sanders et al., 2004).

KNM-NW 58953 belongs to a small palaeomastodont species. Midline length of the cranium is 525 mm, width across the nuchal planum is estimated at 250 mm, and rostral width is 133 mm. The rostrum is relatively short, with a length of 225 mm. The specimen preserves a partial right I2, left P3-M3, and right P4-M3. The M3s were exposed but still in their alveolar crypts at time of death. The M1s are heavily worn in comparison with the premolars, revealing a tooth emergence pattern that can be reconstructed as dP2-dP3-dP4-M1-M2-P3-P4-M3, with dP2-4 lost and replaced by the permanent premolars.

In dorsal view, the temporal lines converge toward the midline over the braincase but do not meet to form a crest; the nasal aperture is vertical, retracted to the anterior of the orbital position, and relatively small, opening above the site of M2; and the rostrum is relatively moderately broad with a very shallow external incisive fossa separating the tusk alveoli, which are straight. The parieto-frontal midline area is transversely and anteroposteriorly concave, while the nasals are convex and only slightly overhang the nasal opening.

From the lateral perspective, the cranium is low and slopes gently forward, but the basicranium is moderately elevated above the level of the palate and not steeply angled upward. The orbit is not very high above the alveolar row, is small, and has a strong postorbital process. Its lateral margin hides the opening of the small infraorbital foramen. The incisor curves downward and it is evident that the tusks were only slightly laterally divergent.

In ventral view, the tooth rows converge anteriorly and posteriorly and are widest at M2 (Figure 4.4B). Prior to deformation, the palate was probably shallow. The basicranium is short and terminates posteriorly at large, downward-facing occipital condyles (each 50 mm wide × 30 mm in length). The preserved articular glenoid is transversely saddle-shaped. It is evident that the widest point of the cranium is low across the nuchal planum (W = 270 mm), and consequently the outwardly-flared squamosal root of the zygomatic arch is widely placed.

The remaining portion of tusk is 19.0 mm in width and 32.5 mm in height at its distal break, has a thin veneer of enamel laterally with longitudinal striations, and is very saber-like in cross-sectional shape. Although the molars are similar in dimensions to Fayumian palaeomastodont molars, the diminutive size of the premolars is remarkable

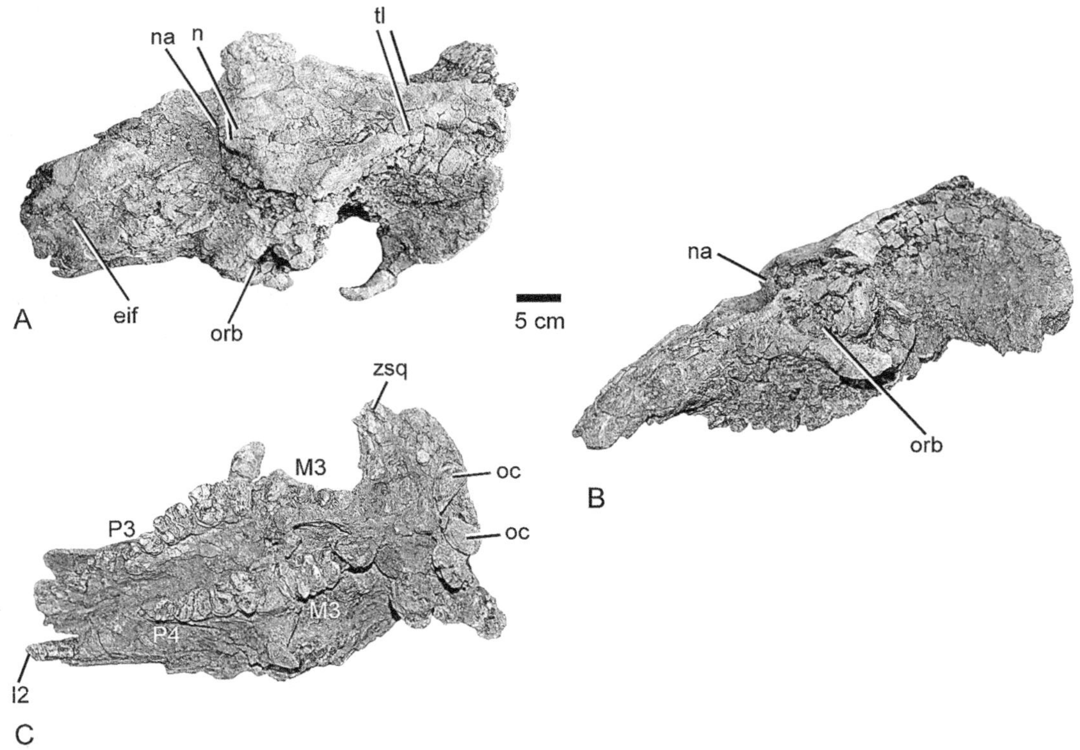

FIGURE 4.4 Partial cranium of *Palaeomastodon* sp. nov. from ?early Miocene Nakwai, Kenya. Abbreviations: eif, external incisive fossa; I, upper incisor; M, upper molar; n, nasal bone; na, nasal aperture; oc, occipital condyle; orb, orbit; P, upper premolar; tl, temporal line; zsq, squamosal portion of the zygomatic. Anterior is to the left. All images to the same scale. (A) Dorsolateral view, KNM-NW 58953. (B) Lateral view, KNM-NW 58953. Note the relatively low height of the vault and short rostrum. (C) Ventrolateral view, KNM-NW 58953. (Images courtesy of S. Cote.)

(dimensions of P3 are L = 15.7 mm, W = 14.8 mm, and of left and right P4 L = 21.5 mm and 25.0 mm, W = 23.3 mm and 25.0 mm, respectively). The P3 is bilophed and bunolophodont, with four distinct cusps and an anteriorly pointed cingulum. P4 is also bilophed and more trapezoidal in occlusal shape. Its anterior cusps are rounded and much larger than the posterior set of cusps. Its metacone is much larger than its hypocone. It has a small, low precingulum that wraps around to the lingual side of the crown. The molars are trilophed and it is possible to observe in M3 strong anterior and posterior crescentoids descending almost vertically into the transverse valleys. M3 lophs are anteroposteriorly divergent at their apices and except for the crescentoids the transverse valleys are very open. Tooth size increases from P3 to M1. The cheek tooth dentition differs from that of geologically older species of palaeomastodonts in lacking definitive signs of alveoli for P2s, and thus may have a reduced dental formula (Tables 3.2 and 4.2).

It is tempting to erect scenarios to explain the final extirpation of palaeomastodonts, so close to the Neogene boundary. Because they had not evolved horizontal tooth displacement, it is possible palaeomastodonts were at an adaptive disadvantage compared to the rapid diversifying elephantimorphs of the late Oligocene-early Miocene, which are defined by this apomorphy. However, the strong wear on the occlusal surface of M1 while M3 was not yet erupted and the premolars had just emerged suggests a palaeomastodont preadaptation to horizontal tooth displacement consistent with lengthening of life history stages and a phased use of cheek teeth; therefore, the adaptive advantage of elephantimorphs initially must have been muted. Nonetheless, the lack of other Neogene records of palaeomastodonts suggests that they had seriously declined by the late Oligocene, prior to major contact and faunal exchange with Eurasia.

ELEPHANTIMORPHA Tassy and Shoshani, 1997 in Shoshani et al., 1998
MAMMUTIDA Tassy and Shoshani, 1997 in Shoshani et al., 1998
MAMMUTIDAE Hay, 1922
***EOZYGODON* Tassy and Pickford, 1983**
***EOZYGODON MOROTOENSIS* (Pickford and Tassy, 1980)**
(Tables 4.1 and 4.2; Figures 4.1 and 4.5)

Synonymy—Mastodont nov. gen. Bishop, 1967; gen. and sp. indet Madden, 1972; Mammutid aff. *Zygolophodon* part. Van Couvering and Van Couvering, 1976; Mammutidae indet. Coppens et al., 1978; *Zygolophodon* sp. Tassy, 1979a; *Zygolophodon*

aff. *turicensis* Tassy, 1979a; *Zygolophodon morotoensis* Pickford and Tassy, 1980; *Zygolophodon* sp. Madden, 1980; *Z. morotoensis* Pickford, 1981; *Eozygodon morotoensis* Tassy and Pickford, 1983

Mammutids first appeared and diverged from gomphotheres in Africa during the Oligocene. Morphological evidence supports the hypothesis that they evolved from *Palaeomastodon* (Tobien, 1978; Pickford, 2003a; Sanders et al., 2004, 2010). Mammutids first emigrated from Africa to Eurasia during the early Miocene interval of the late MN3 or early MN4 (van der Made and Mazo, 2003) and subsequently spread more widely throughout Eurasia and to the New World by the middle Miocene (Tassy, 1986; Mazo, 1996; Saunders, 1996; Tobien 1996), where they ultimately had greater success than in Africa. In contrast, mammutids became marginalized in African faunas as the Miocene progressed, perhaps in response to the diversification and spread of gomphotheriids (Coppens et al., 1978; Tobien, 1996). Nevertheless, in the early to early middle Miocene mammutids were probably keystone taxa, though apparently not abundant, in some African faunas. Early mammutids are chronostratigraphically useful, since their morphologies and ages indicate a succession from Oligocene-earliest Miocene *Losodokodon* through ~23–20 Ma *Eozygodon* to *Zygolophodon*, <20 Ma (Tables 3.1 and 4.1).

Prior to the discovery of *Losodokodon* in late Oligocene sediments at Losodok, Kenya (Table 3.1; Rasmussen and Gutiérrez, 2009), early Miocene *Eozygodon morotoensis* (Table 4.1) was the oldest known mammutid, as its genus name implies. Originally placed in *Zygolophodon* (Pickford and Tassy, 1980), Tassy and Pickford (1983) transferred the species to *Eozygodon*, recognizing morphological differences that justified a new genus. The type material is from Moroto, Uganda, providing the species name. Gnathodental and postcranial specimens from Meswa Bridge, Kenya, were also referred to the species, but contra Tassy and Pickford (1983) a fragmentary P4 from Moruorot, Kenya (which Madden [1980] originally correctly placed in *Zygolophodon*) should not be included in the species. The purported occurrence of the species at Wadi Moghara, Egypt, based on an edentulous dentary (Pickford, 2003a), cannot be confirmed and seems unlikely given the presence of more advanced mammutids at the site. *Eozygodon* is strictly an African taxon and the range of its geological age constrains the timing of migration of mammutids to Eurasia to <20 Ma.

Comparative study of the dentition from the original species sample shows that *E. morotoensis* is primitive in having relatively small molars, trilophed intermediate molars, upper third molars also with only three lophs, weak development of accessory crescentoids on pretrite half-loph(id)s, and modest expression of zygodont crests on posttrite half-loph(id)s, fourth lophids of m3s transversely narrowed, very brachyodont crowns, and prominent lateral cingulae(ids) variably present primarily on upper molars. Development of zygodont crests and crescentoids is weaker and cingular expression is greater in molars from the older site of Meswa Bridge (Table 4.1) than in molars from Moroto. *Eozygodon* molars are relatively wide compared with those of other mammutids. Nonetheless, its molars are clearly zygolophodont, with anteroposterior compression of loph(id) apices, mostly open and spacious transverse valleys, main conelets laterally much higher than medially-placed mesoconelets ("yoked" transverse profile of loph(id)s), and deep median sulci. Cementum is absent and enamel is only moderately thick (1.5–3.6 mm in molars). Von Koenigswald (2016a) stressed that although Phase I occlusion was dominant and vertical wear facets are present in mammutids, they differed from deinotheres by employing more lateral grinding in Phase II mastication.

P4s from Moroto (MOR I 61E) have two lophs formed of four transversely elongated, anteroposteriorly compressed cusps that merge into transverse lophs with wear. Their crown dimensions are L = 37.8–39.2 mm and W = 40.2–40.5 mm, producing a square occlusal outline. A P4 from Auchas, Namibia is also square in occlusal view, with dimensions of L = 44.3 mm and W = 42.1 mm. Molar dimensions of the combined eastern and southern African samples are as follows: M1 L = 57.2–66.2 mm, W = 49.8–53.7 mm; M2 L = 77.9–90.2 mm, W = 62.4–68.6 mm; M3 L = 107.9–117.1 mm, W = 71.8–74.0 mm; m2 L = 82.4–83.0 mm, W = 58.5–59.6 mm; m3 L102.2–124.5 mm, W = 58.0–68.3 mm. Hypsodonty indices of third molars range from 55 to 60.

Eozygodon morotoensis tusks are primitive in a manner similar to those in palaeomastodonts: I2 is short (KNM-ME 7543 from Meswa Bridge L = 510 mm), with a downward and slightly lateral orientation, strong curvature, ventrolateral enamel bands, dorsal and ventral longitudinal sulci, and ovoid to piriform cross-sections (higher than wide; H = 64.5 mm, W = 51.0 mm); lower tusks are piriform in cross-section with a distinct dorsomedial longitudinal sulcus and are slightly flattened (KNM-ME 19 from Meswa Bridge L = 420 mm, H = 58.8 mm, W = 37.7 mm).

Cranial remains of *E. morotoensis* have been recovered from the early Miocene site of Auchas, Namibia (Pickford, 2003a). These specimens were cursorily described by Pickford (2003a) and later in a more comprehensive manner by Tassy (2018) in his classic review of the cranial anatomy of the species. The best preserved cranium from Auchas is AM 02. From the nuchal crest to the anteriormost tip of the premaxillaries, it measures 600 mm, and 730 mm in length in a straight line from the occipital condyles to the end of the premaxillaries (Pickford, 2003a).

AM 02 exhibits a mix of primitive retentions and apomorphies that clearly distinguish it from the cranium of *Progomphotherium maraisi* from the site and from crania of other mammutids. It is not known how many permanent premolars originally erupted in the specimen, but P4 is still in place on the right side with M3 emergent, indicating an extremely delayed horizontal displacement mechanism (see Sanders, 2017). As at Moroto and Meswa Bridge, the intermediate molars and M3 are trilophed. In contrast to the cranium of *P. maraisi*, in AM 02 the cranial roof is more raised and the basicranium is consequently

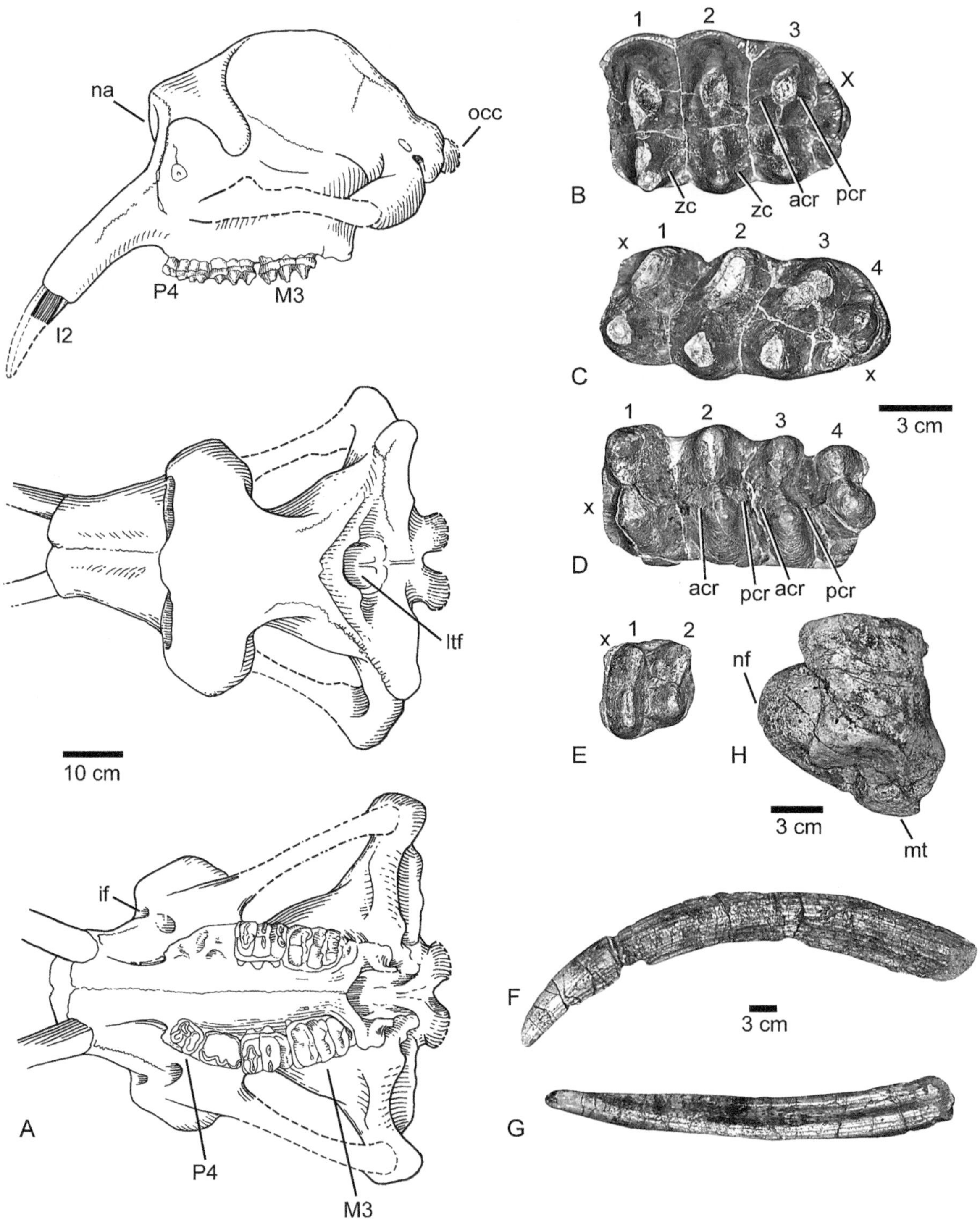

FIGURE 4.5 Skeletal elements of early Miocene *Eozygodon morotoensis*. Abbreviations: acr, anterior crescentoid; I, upper incisor; if, infraorbital foramen; ltf, fossa for insertion of the nuchal ligament; M, upper molar; mt, medial tubercle; na, nasal aperture; nf, navicular facet; occ, occipital condyles; P, upper premolar; pcr, posterior crescentoid; x, pre- or postcingulum(id); X, large postcingulum(id) or nascent loph(id); zc, zygodont crest; 1, 2, 3, …, loph(id) numbers counted from the anterior of the crown. (B–E) to the same scale. (F, G) to the same scale. Anterior is to the left. (A) Reconstructed cranium, AM 02 from Auchas, Namibia, lateral, dorsal and ventral views. (Artwork by D. Visset. Images courtesy of T. Tassy [Tassy, 2018:figs. 2A, B, 3A].) (B) Right M3, KNM-ME 17 from Meswa Bridge, Kenya, occlusal view. (C) Right m3, KNM-ME 7546 from Meswa Bridge, Kenya. (D) Left m3, MOR I 61E from Moroto, Uganda, occlusal view. (E) Left P4, MOR I 61E from Moroto, Uganda, occlusal view. (F) Upper tusk, KNM-ME 7543 from Meswa Bridge, Kenya, Lateral view. (G) Lower tusk, KNM-ME 19 from Meswa Bridge, Kenya, lateral view. (H) Right astragalus, ME 7554 from Meswa Bridge, Kenya, dorsal view.

positioned much higher than the level of the palate, though M2–3 wear gradients indicate that the angle of molar eruption was low.

The rostrum of AM 02 is relatively narrow, shallow across the rostral trough and short, the palate is deep and broadest at mid-toothrow, and the occipital planum is angled forward superiorly. In addition, the orbit is tall and posteriorly bounded by a very robust zygomatic process of the frontal. Primitively, the lacrimal has a lacrimal foramen. The nasal aperture is broad, situated above the level of the orbits, and appears anteroposteriorly narrow because of the overhang of the nasals and impingement of the premaxillary processes that separate it from the rostrum. The zygomatic arches flare widely at their squamosal ends; bizygomatic width is the broadest part of the cranium. In dorsal view, the cranium is "waisted" across the parietals, with the temporal lines convergent posteriorly toward the nuchal crest. The occipital condyles are high and project posterior to the occipital planum. In posterior view, the occipital planum is triangular in shape (similar to the condition in palaeomastodonts) with its broadest aspect low on the squamosals and exoccipitals more robustly developed than supraoccipitals; a large, ovoid fossa for insertion of the ligamentum nuchae dominates the planum and is subdivided by a midline crest.

A small postcranial sample, mostly from Meswa Bridge, can be attributed to *E. morotoensis* and was described by Tassy and Pickford (1983). Notably, the astragalus is high and has a prominent medial tuberosity and an extensive facet for the tibial malleolus; the downward orientation of the navicular surface and tall, narrow phalanges suggest a digitigrade posture of the hindfoot. Dimensions of astragali from Moroto II and Meswa Bridge are similar in length and width (128.2 mm × 136.5 mm versus 112.2 mm × 129.9 mm, respectively). They contrast with the morphology of the astragalus of *P. maraisi* from Moroto I, which is smaller (75.0 mm × 71.9 mm), with a less prominent medial tuberosity and navicular facet that faces forward rather than downward, suggesting a more pronograde hindfoot posture. Femoral length > 1,000 mm is consistent with a body size comparable to that of the American mastodon (Tassy and Pickford, 1983; Christiansen, 2004).

At Auchas and Moroto, *Eozygodon* is associated with *Prodeinotherium hobleyi* and *Progomphotherium maraisi*. At early Miocene Elisabethfeld, Namibia, *Eozygodon* co-occurs with a gomphotheriid that could be *Progomphotherium* (Pickford, 2008). It is not clear how these proboscideans partitioned their ecosystems, as their dental carbon isotope values overlap substantially and indicate browsing adaptations for each taxon (Kingston et al., 2011). Tassy and Pickford (1983) infer from the paleoecology of *Eozygodon* in eastern Africa that it inhabited dry, dense forests with open environments nearby. At the *Eozygodon* site of Auchas, paleobotanical evidence is consistent with woodland or savanna on sandy soils, with more annual rainfall than at present in the region (Pickford et al., 1995). Paleoclimate reconstruction of the early to early middle Miocene Namib indicates a very warm, humid interval (Pickford, 2000).

Mammalian faunal remains from channel deposits of the Muhoroni Agglomerates at Meswa Bridge (Pickford and Andrews, 1981) are not common but include the earliest record of tragulid ruminants in Africa (Geraads, 2010), tiny, primitive pedetid rodents (Pickford and Mein, 2011), an immense species of hyaenodont carnivore (Borths and Stevens, 2019), and the most primitive form of the fossil ape *Proconsul*, *P. meswae* (Harrison and Andrews, 2009). Modern pedetids are fossorial and favor semiarid to arid habitats (Winkler et al., 2010), whereas the presence of *Proconsul* indicates more forested environs.

Aside from the proboscideans at Moroto, mammals from the site include hyraxes, four genera of rodents, an anthracothere, a pecoran ruminant, an early hippopotamid (Orliac et al., 2010), a possible hyaenodont, and notably the fossil ape *Morotopithecus* alongside several other catarrhine primate species and a lorisoid (MacLatchy, 2004). A number of these species represent first appearance records. *Morotopithecus* exhibits a number of postcranial features that parallel modern ape adaptations for climbing and orthograde posture (e.g., Sanders and Bodenbender, 1994; Gebo et al., 1997; MacLatchy et al., 2000), suggesting the presence of forested conditions. The age of the site (Table 4.1) has been established by repeated radiometric dating of lavas in the sequence (e.g., Gebo et al., 1997); the proboscidean taxa present at Moroto are consistent with an age > 20 Ma.

Mammalian remains at Auchas are not numerous and include rhino bones, a *Prohyrax* specimen, a few specimens of the rodent *Diamantomys*, and a lower molar of *Myohyrax*, along with the aforementioned proboscidean fossils. Based primarily on the size of the *Prohyrax*, Pickford and colleagues (1995) infer an age for Auchas ca. 20–19 Ma, younger than Elisabethfeld. Biochronological correlation with East African sites (Table 4.1), however, is corroborative with an older age estimate for Auchas, >20 Ma.

A recent claim of an occurrence of *Eozygodon* in early-middle Miocene sediments at Yingbeihou Village, China (Zhang and Wang, 2020) can be dismissed. There is no good evidence that MN2-age-equivalent proboscideans such as *Eozygodon* (contra Zhang and Wang's [2020] claims that it is an early to middle Miocene taxon in Africa) made it out of Africa to Eurasia. Morphologically, the Yingbeihou mammutid is placed in *Eozygodon* largely because of the evidence for small lower tusks with a piriform cross-section. However, the Yingbeihou tusk is not identical with the Meswa Bridge i2; it is much smaller (W = 22.0 mm, H = 13.5 mm) and does not have both dorsal *and* ventral longitudinal sulci. The condition of lower tusks in African species of *Zygolophodon* is unknown because the genus has such a poor fossil record on the continent. dP2, dP3, dp2, dp4, and p4, recovered at Yingbeihou, are unknown for *E. morotoensis*. A dp3 of *E. morotoensis* from Moroto (MUZM 193 MOR I 1995) is smaller and narrower than dp3 specimen IVPP V2080 from Yingbeihou (L = 30.2 mm, W = 21.7 mm, versus L = 39.9 mm, W = 32.9 mm) and has much weaker development of posttrite conelets and pretrite crescentoids. The permanent molars from Yingbeihou are not preserved well enough for comparison with molars of

E. morotoensis. It is much more likely that the Yingbeihou mammutid is a congener of *Zygolophodon*.

ZYGOLOPHODON Vacek, 1877
ZYGOLOPHODON AEGYPTENSIS Sanders and Miller, 2002

(Tables 4.1 and 4.2; Figures 3.24, 4.1 and 4.6)

A more derived mammutid, *Zygolophodon aegyptensis*, was recovered from the extensive North African site of Wadi Moghara, embracing at least 40 terrestrial mammalian localities spread over 55 km in the northeastern part of the Qattara Depression, Egypt (Miller, 1999). It is a historically important paleontological area that was first prospected in the late 1800s (see Blanckenhorn, 1901; Fourtau, 1918). The fossil-bearing sequence is approximately 230 m thick, composed of sediments deposited by slow-moving fluviatile, estuarine, and lagoonal systems (Said, 1962a, b), manifested as sandstones, siltstones, and calcareous shales. The many silicified tree trunks at the site attest to the presence of early Miocene forests where there is now desert (Bown et al., 1982). Occurrence of numerous remains of birds such as storks, herons, and pelicans further attests to the presence of abundant fresh water at early Miocene Wadi Moghara (Smith, 2013). Investigations of other taxa from the site have emphasized its forested, well-watered setting (e.g., Pickford, 1991b).

Following renewed investigation of the site in the 1980s–1990s, Miller (1999) made detailed comparisons of its mammals with faunas from the Libyan site of Gebel Zelten and other early Miocene sites of East Africa. The results of the study, supplemented by subsequent investigations, enumerate a rich and diverse mammalian assemblage, including the proboscideans *Z. aegyptensis*, *Prodeinotherium hobleyi*, *Gomphotherium libycum*, *Afrochoerodon kisumuensis*, and cf. *Archaeobelodon* sp., abundant anthracotheres in four genera and six species (with a strong degree of narrowly localized species endemism), several creodont species alongside four carnivore species, several sanithere and suid species, tragulids, giraffids, rhinos, and the primitive victoriapithecid monkey *Prohylobates* (Miller, 1999; Sanders and Miller, 2002; Morlo et al., 2007; Pickford et al., 2010; Miller et al., 2014). Because the nature of the sediments is insufficient for standard radiometric dating or paleomagnetic sequence stratigraphy, the age of Wadi Moghara has been estimated by faunal coefficients of similar in comparison with other, absolutely dated sites. In this approach, an age of 18–17 Ma has been obtained (Miller, 1999). Alternatively, the long, alternating terrestrial-aquatic sequence at Moghara could indicate that the age estimate for the site is time-averaged and overly dependent on assumptions of species contemporaneity; Pickford et al. (2010) believe that an interval as long as 21–16 Ma and successive faunal sets could be represented at the site. Nonetheless, it is entirely possible that five proboscidean genera could have been present synchronously, since there are confirmed examples of the same level of proboscidean sympatry at other early Miocene sites (e.g., Buluk, Kenya and at localities of the Siwalik Series in the Potwar Plateau, Pakistan). Improving the temporal calibration of mammutids from Wadi Moghara is critical for understanding the origin of the *Zygolophodon* lineage, relationship to congeners from other parts of Africa, and timing of proboscidean "datum" events marking the spread of African deinotheres and elephantimorphs into Eurasia. The earliest record of *Zygolophodon* in Europe is in mammal zone MN3b (Tassy, 1985, 1990a; Van der Made and Mazo, 2003), pene-contemporaneous with the occurrence of the genus at Wadi Moghara and in eastern Africa.

Few specimens of *Z. aegyptensis* have been collected, though the existing sample is sufficient for definitive identification of the species (Sanders and Miller, 2002). Named for the country of its discovery, *Z. aegyptensis* is diagnosed as a small form of *Zygolophodon* (Figure 3.24) with relatively narrow molars, particularly m3 (L = 148.8 mm, W = 55.8 mm). Intermediate molars are trilophed. Lower m3 (DPC 9009) has four lophids with a low, moderately prominent postcingulid and anteroposteriorly narrow precingulid (Figure 4.6A). Pretrite cresentoids are well-developed posterior to lophids 1 and 2 and are smaller posterior to lophid 3 and anterior to lophids 2 and 3. Lophids 3 and 4 are extremely strongly convex anteriorly, which is an apomorphy of the species. Transverse valleys are spacious anteroposteriorly, corresponding with broad spacing of lophids. There is no cementum. The crown is supported by a small transverse root primarily below lophid 1, and a larger, posteriorly angled root under lophids 2–4.

M3 (DPC 12598) is advanced over the condition in *Eozygodon*, with four lophs, although loph 4 is transversely narrower than the other lophs (Figure 4.6B). Loph formula is x4x; the pre- and postcingulae have the form of low enamel ribbons closely appressed to lophs 1 and 4, respectively. A low cingular ribbon also runs along the pretrite side of the crown. As is typical of upper molars, M3 is relatively wider and shorter (L = 121.1 mm, W = 64.7 mm) than m3. Half-loph pairs are divided by a salient median sulcus, and outer main conelets are higher than mesoconelets, yielding the typical "yoked" transverse profile of mammutids. There are marked pretrite posterior crescentoids associated with lophs 1 and 2, and on the posttrite side there are strong zygodont crests that descend toward the midline from the mesoconelets. In lateral view, lophs are pyramid-shaped with correspondingly open, V-shaped transverse valleys, and crown apices are anteroposteriorly constricted. Crown height is brachyodont (hypsodonty index = 66). Enamel is unworn and rugose. There is no cementum.

A more robust, larger partial M3 (DPC 5920) has stronger development of anterior and posterior crescentoids and a striking zygodont crest on posttrite half-loph 3 (not preserved in broken lophs 1 and 2). Loph 4 is narrower than the other lophs but is wider than in DPC 12598. Enamel is considerably thicker than in *Eozygodon* (ET = 5.8–6.8). The presence of "gracile" and "robust" morphs is sampling interindividual variation in a single species that is probably due to sexual dimorphism. This is typical of the polymorphic nature of mammutid species (Tobien, 1975, 1996; Tassy, 1985).

Dental morphology of *Z. gromovae* from the middle Miocene of Ulan Tologoj, western Mongolian People's Republic (Dubrovo, 1974) most closely resembles that of

Z. aegyptensis. This hints at a possible mammutid connection between Afro-Arabia and Eurasia, starting in the early Miocene (Sanders et al., 2010a).

ZYGOLOPHODON SP. NOV. [dwarfed; Buluk]

(Table 4.1; Figures 3.24, 4.1, and 4.6)

Recent fieldwork at terminal early Miocene Buluk, Kenya (Table 4.1) by an American-Kenyan team led by Ellen Miller and the late Isaiah Nengo produced a wealth of new proboscidean fossils, curated in the Turkana Basin Institute at Ileret, Kenya, that augment extensive site collections housed in the National Museums of Kenya. Among the new specimens are several teeth that reveal the heretofore unrecorded presence of mammutids at the site, placed in a new species, *Zygolophodon* sp. nov. The best example of the taxon is a complete right M3 (KNM-WS 65209), with a loph formula of x4x, that will be designated as the type specimen of *Zygolophodon* sp. nov. (Figure 4.6C).

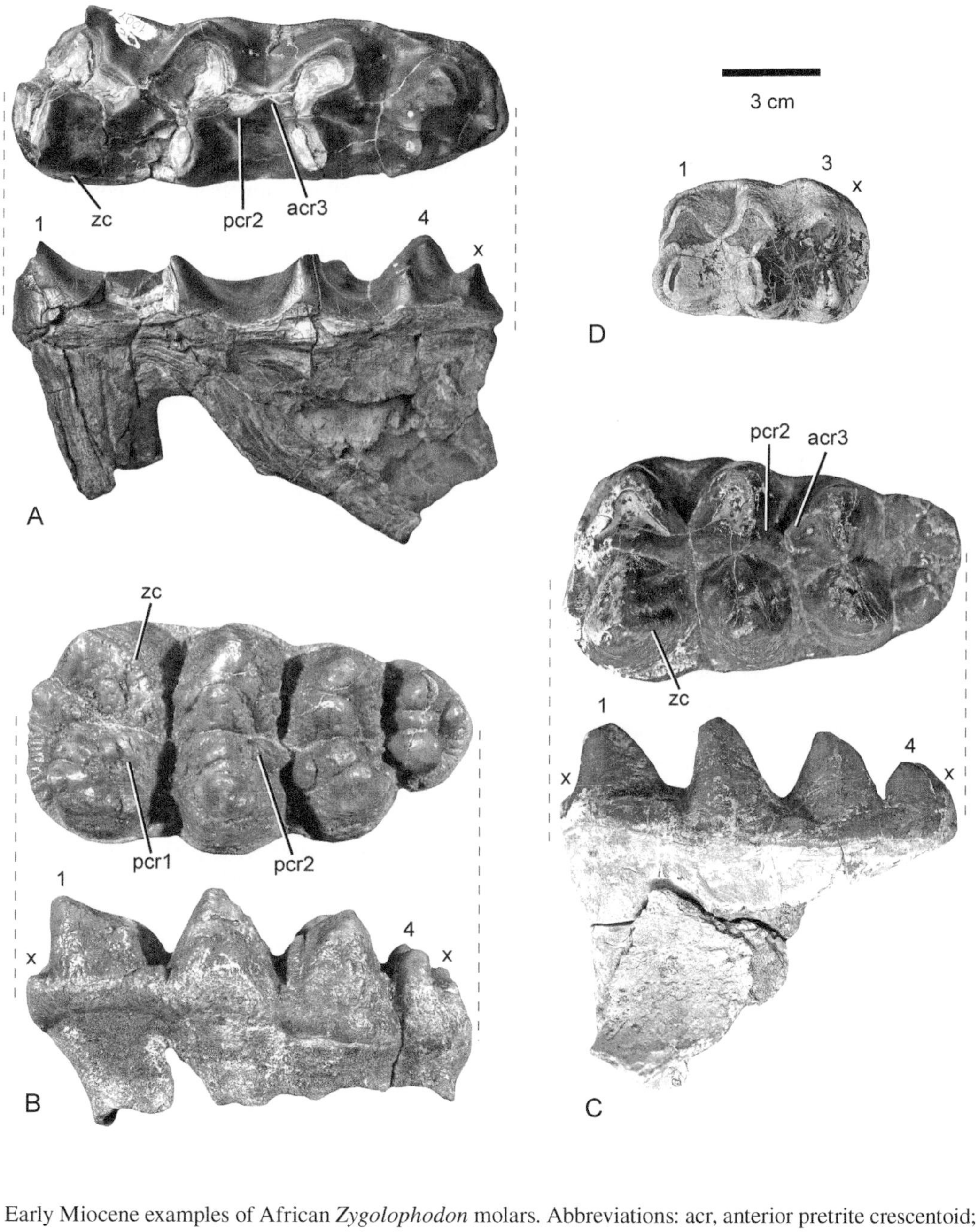

FIGURE 4.6 Early Miocene examples of African *Zygolophodon* molars. Abbreviations: acr, anterior pretrite crescentoid; pcr, posterior pretrite crescentoid; x, pre- or postcingulum(id); zc, zygodont crest; 1, 2, 3, …, loph(id) number counted from the anterior of the crown. All specimens are approximately to the same scale. Anterior is to the left. (A) Right m3, DPC 9009 (type) from Wadi Moghara, Egypt, occlusal and lateral views, *Zygolophodon aegyptensis*. Note the broad anteroposterior spacing between lophids and the anteroposterior compression of lophid apices. (B) Left M3, DPC 12598 from Wadi Mogahra, Egypt, occlusal and lateral views, *Z. aegyptensis*. (C) Right M3, KNM-WS 65209 (intended type) from Buluk, Kenya, occlusal and lateral views, *Zygolophodon* sp. nov. (dwarfed). (D) Right M1, KNM-WS 49459 from Buluk, Kenya, occlusal view, *Zygolophodon* sp. nov. (dwarfed).

Contrary to the general trend for Miocene African mammutids, the teeth of this species appear to be anomalously small for its geological age (Table 4.1; Figure 3.24). KNM-WS 65209 is only 106.4 mm in length and 53.6 mm wide, barely larger than the diminutive M3 of late Oligocene-earliest Miocene *Losodokodon*. Lophs 1 and 2 have posttrite zygodont crests, posttrite transverse valleys are very open and posttrite half-loph apices are anteroposteriorly constricted, and pretrite half-lophs 1 and 2 have prominent, vertically worn (chisel-like) crescentoids. Pretrite half-loph 3 has a strong anterior crescentoid. There is no cementum, and enamel is moderately thick (3.4–3.6 mm). Crown height is low. The fourth loph is narrower and simpler in construction than the other lophs. Pre- and postcingulae are inconsequential in extent, as are lateral cingulae on both sides of the anterior end of the crown. Root structure helps to identify this specimen as an upper molar: there is an anterior root below posttrite half-loph 1, a smaller anterior root below pretrite half-lophs 1 and 2, and a large posterior root under posttrite half-loph 2 and lophs 3 and 4. By itself, the diminutive size of the M3, combined with salient expression of crescentoids and zygodont crests, warrants erection of a new mammutid species to accommodate it. *Zygolophodon* sp. nov. is distinguished from *Losodokodon losodokius* by molars of slightly larger size with stronger expression of crescentoids and zygodont crests; from *Eozygodon morotoensis* by smaller molars that are relatively narrower; and from *Z. aegyptensis* and *Z. turicensis* by smaller size and simpler construction of loph(id)s.

A smaller complement to the type is KNM-WS 49459, a right M1 measuring 65.3 mm in length, 46.3 mm in width, and with a low crown height of 28.5 mm. Enamel thickness is modest, 2.1 mm, and it has no cementum. Loph formula is 3x; the postcingulum comprises a low, anteroposteriorly narrow shelf (Figure 4.6D). Mesoconelets are small and not well differentiated from the larger main conelets. Enamel wear figures reveal the distinctive mammutid pointed trefoils formed by wear on the pretrite half-lophs and their incorporated anterior (lophs 1–3) and posterior (lophs 1, 2) crescentoids. There is a zygodont crest on the posterior surface of posttrite half-loph 1. Despite the heavy wear on the pretrite side of the crown, the posttrite half-lophs remain anteroposteriorly compressed at their apices.

A partial m1 or m2, KNM-WS 65752, was originally identified as a deinothere molar on its accession card but belongs to this taxon. Each half-lophid has a larger outer main conelet accompanied by a smaller, lower mesoconelet, forming the classic "zygodont," or yoked, transverse crown profile. The lophids have an oblique orientation relative to the long axis of the crown. The lophid apices are strongly anteroposteriorly compressed, and there are modest anterior and posterior crescentoids descending from pretrite half-lophids into the floor of transverse valleys, which are much more open on the posttrite side. It is a narrow molar, with a width of 50 mm, and is low-crowned with a height of 29.6 mm. Enamel thickness is modest, 2.2–2.4 mm, and it has no cementum.

A small p4, KNM-WS 12685, is also labeled as a deinothere in the collections at the National Museums of Kenya, but belongs to this mammutid species, as well. It has a lophid formula of x2x, and dimensions of L = 37.6 mm, W = 31.0 mm. The posttrite half-lophids are anteroposteriorly compressed and it has stout crescentoids descending from pretrite half-lophids into transverse valleys or toward the postcingulid.

Buluk yields deinotheres, mammutids, gomphotheriines, multiple species of amebelodonts, and choerolophodonts. There is no mention of mammutids in earlier summaries of the Buluk fauna (e.g., Harris and Watkins, 1974; Leakey and Walker, 1985), perhaps because in comparison with the rich representation of *Archaeobelodon* and *Prodeinotherium* dental remains, mammutids, choerolophodonts, and gomphotheriines are rare in the Buluk assemblage. This diversity of sympatric proboscideans may be normal rather than exceptional for the early and middle Miocene Old World, based on information from site complexes such as Wadi Moghara, Egypt (Sanders and Miller, 2002) or the Siwalik sequence in the Potwar Plateau, Pakistan (Tassy, 1983a–c). Details of depositional circumstances at Buluk confirm the synchroneity of preservation of these taxa and provide a powerful picture of the extraordinary fecundity of this ancient African ecosystem.

The Buluk Mb. of the Bakate Fm. is composed of a lower bed of conglomerates, sandstones and siltstones overlain by tuffaceous sediments that have yielded a range of radiometric dates coalesced around an age just older than 16 Ma (e.g., McDougall and Watkins, 1985; Hurford and Watkins, 1987; Watkins, 1989). Dense concentrations of fossils are found in channel sandstones and conglomerates (Watkins, 1989), with no evidence for major transport; there is virtually no time averaging for the Buluk assemblage and the recovered taxa may be considered contemporaneous.

Besides the proboscideans, the vertebrate remains from the site include lungfish, turtles, crocodiles, and about 30 mammalian species, among them hyraxes, rhinos, suids, tragulids, giraffids, anthracotheres, and amphicyonid carnivores (Harris and Watkins, 1974; Leakey and Walker, 1985; Geraads and Miller, 2013). In addition, primitive cercopithecoid monkeys (*Noropithecus bulukensis*) are abundant at Buluk and are accompanied by a stem hominoid (*Afropithecus turkanensis*) and a small-bodied species of stem catarrhine (Leakey, 1985; Leakey and Walker, 1985; Miller et al., 2009).

Recent geological investigations of the site reveal evidence of strong seasonality in which the local floodplain is alternately well-drained and saturated. Preliminary isotopic investigations (Cerling et al., 1999; Arney, 2019) and microwear analysis indicate that the Buluk proboscideans are browsers (Sanders et al., 2020). The $\delta^{13}C$ values from proboscidean teeth indicate that the elephantimorph taxa, including the mammutid species, browsed in areas of broken canopy or woodland habitats, whereas the deinotheres from the site foraged in more covered areas (Arney, 2019). This information, coupled with an abundance of fossil wood and the presence of apes and water-dependent vertebrates, suggests that despite evidence for seasonal, subhumid woodland paleoenvironments at Buluk, forested areas and permanent water sources were proximate to the depositional environment.

ZYGOLOPHODON TURICENSIS (Schinz, 1824)
(Tables 4.1 and 4.2; Figure 4.1)

Several relatively large M2s from the Tugen Hills, Kenya, have been attributed to a robust form of *Zygolophodon turicensis* (Pickford, 2007b). These specimens show clear mammutid features: posttrite half-lophs are anteroposteriorly compressed, transverse valleys are very open except for salient anterior and posterior pretrite crescentoids that give the pretrite enamel wear figures a trefoil pattern, and posttrite half-lophs exhibit zygodont crests. A very narrow, low cingular ribbon runs along the lingual side of the crown, and the precingulum is also low and unobtrusive. The postcingulum forms a low, moderately-sized shelf. There is no cementum. Dimensions of the M2s are L=133.3–134.5 mm and W=81.3–82.8 mm (Pickford, 2007b). The crown is worn but appears to have been brachyodont.

A robust morph of *Zygolophodon* is present in the early Miocene Hiwegi Fm. at Rusinga, Kenya (Table 4.1), with trilophed intermediate molars, loph(id)s composed of multiple conelets on each side, and ornamented by salient crescentoids and zygodont crests. It is possible that these mammutids belong in *Z. turicensis*, and perhaps other early-middle Miocene African sites with fragmentary evidence of mammutids may also document this species. Given the long interval of time covered by these sites and the incomplete nature of the mammutid record, it is more likely that the species is serving as a wastebasket taxon until better fossils can be recovered. It is apparent that mammutids vanished from Africa prior to the late Miocene.

ELEPHANTIDA Tassy and Shoshani, 1997 in Shoshani et al., 1998
GOMPHOTHERIIDAE Hay, 1922
GOMPHOTHERIINAE Hay, 1922
GOMPHOTHERIUM Burmeister, 1837

As mentioned in Chapter 3, Gomphotheriinae and *Gomphotherium* also function as wastebasket taxa, lacking the derived dental traits that define other gomphotheriid groups such as amebelodonts and choerolophodonts. Instead, molar crown morphology of gomphotheriine species is generally simple, with accessory conules contributing to trefoil enamel wear patterns in pretrite half-loph(id)s but rarely present on the posttrite side of the crown; pre- and posttrite half-loph(id)s are transversely aligned without pseudoanancoidy or chevroning (unless loph(id)s are transversely crowded distally); intermediate molars are trilophodont and third molars usually do not exceed four loph(id)s in composition; and lower tusks are typically rounded to piriform-shaped in cross-section (Tassy, 2014). The lack of molar crown morphological specialization makes gomphotheriines good candidates for the deep ancestry of elephants (Sanders et al., 2010a). It is likely that any summary of Afro-Arabian gomphotheriines will underestimate species diversity, because in most instances occurrences consist of isolated teeth and the group is conservative dentally. It was established in Chapter 3 that the taxon arose in the Oligocene. Gomphotheriines spread across Afro-Arabia during the early to middle Miocene (Table 4.1) and rapidly throughout the Old World during the same interval. *Gomphotherium* is found in South Asia, Europe, China, and as distant as Japan by Neogene mammal zone MN3 (Takai, 1954; Bulot and Ginsburg, 1993; Qiu et al., 1999; van der Made and Mazo, 2003).

Old World gomphotheriines have been subdivided into a primitive *Gomphotherium* "*annectens* group" comprised of early Miocene *G. annectens* (Japan), *G. cooperi* (South Asia), *G. inopinatum* (China), and *G. sylvaticum*, *G. hannabali*, and *G. steinheimense* (Europe), (Gasparik and Markov, 2009) and a more advanced group extending into the middle and late Miocene *Gomphotherium* "*angustidens* group" typified by *G. connexum* (China), *G. browni* (South Asia), and *G. angustidens* and possibly *G. subtapiroideum* (Europe) (Tassy, 1983b, 1985, 1994a, 1996c; Welcomme, 1994; Wang, 2014; Wang et al., 2015, 2017a; but see Göhlich, 2010). This subdivision of the genus has been muddled by Wang et al. (2017a), who placed early-middle Miocene *G. browni* (South Asia) and middle-late Miocene *G. steinheimense* (Europe) into a third, separate group, "derived *Gomphotherium*." These groups represent grades but not necessarily clades, and the ensemble of trilophodont gomphotheriines are considered in their entirety to be paraphyletic by Tassy (1994a). There is morphological evidence for both the *G.* "*annectens* group" and *G.* "*angustidens* group" in the African fossil record.

African collections of *Gomphotherium* lack good examples of associated skeletons and well-preserved skulls. Nonetheless, European discoveries have provided abundant fossil material characteristic of the genus, particularly of middle Miocene *G. angustidens*, for which are preserved abundant dentition and an age-grade series of skulls and adult male and female specimens (Tassy, 1996b, 2013, 2014). Tassy (2013) described the skull of *G. angustidens* in detail. Compared with palaeomastodonts, the cranium of *G. angustidens* is derived in a number of morphological aspects, including having an enlarged facial region with a thickened maxilla, wide prenasal area, and more posterior placement of the nasal fossa. The nasal aperture is wider. The alisphenoid lateral wing is inflated and covers the maxillary capsule, a trait associated with adoption of the mechanism of horizontal tooth displacement. The premaxillae are long but parallel, housing (in males, at least) long, downward-oriented and slightly outward-curved tusks, with a shallow external incisive fossa between them. Because females either lack or have reduced upper tusks, their rostra are less robust than those of males. There is no sagittal crest. Zygomatic arches do not flare laterally and are very gracile. The occipital planum is low, not inflated, widest across its lower half, and angles forward moderately toward its apex. Its nuchal crest is rounded without signs of parieto-occipital bossing. The fossa for the nuchal ligament is deep, bifid, and located about halfway between the lowest and highest edges of the occipital planum. The glenoid articular surface is wide and narrow anteroposteriorly; behind it the postglenoid region is short and lacks a fossa. The occipital condyles are positioned low, project posteriorly, and are relatively wide. Primitively, *G. angustidens* retains a relatively low cranium, though the basicranial region is well raised with respect to the level of the palate, with a long facial area, and permanent premolars (Table 4.2). The braincase is shorter than the face, although relatively longer than in modern elephants. The orbits sit low above the alveolus for the last molar. The palate is long and narrow.

The mandible of adult males has a very long, dorsoventrally shallow symphysis that may be downturned; it accommodates lower tusks that are closely spaced and do not project much anteriorly from their alveoli. The symphysis is longer than the alveolar portion of the corpus. Distally, the lower tusks are rounded but may be piriform in cross-section more proximally. The symphysis is waisted toward its proximal end, flaring out more distally. Width of the symphysis can be contrasted with its greater breadth in *Archaeobelodon filholi* (Tassy, 2013). The ramus is high with the condyle elevated above the level of the coronoid process, which does not project much. An autapomorphy of the species is the absence or vestigial presence of the coronoid foramen (see Ferretti and Debruyne, 2011). The masseteric fossa is limited in extent but deep, and the posterior angle of the jaw is reduced. Primitively, there are two mental foramina on each side of the mandible, with the anterior one having larger size (see Ferretti and Debruyne, 2011). In *G. steinheimense*, the symphysis is also longer than the alveolus, and downturned, but the ramus appears to be lower and the coronoid process may be more elevated than the condyle (Göhlich, 1998).

Skull and dental traits of gomphotheres in the *G. "annectens* group" are more primitive (Tassy, 1994c; Welcomme, 1994). For example, the nasal aperture is not as retracted as in *G. angustidens* and is narrower, the zygomatic arch is much more robust and flaring at its maxillary root, the mandibular symphysis is shorter than the alveolar segment of the corpus and projects straight forward, upper molars are relatively wider, lack cementum, and M3 is trilophodont. In addition, accessory conules are more poorly developed in the molars.

Body mass must have varied widely among African gomphotheriines, given differences in molar dimensions, but in general these were very large, elephantine-sized animals. Based on allometric regressions of long bone dimensions, average body size for North American *G. productum* was estimated between 2,694 and 6,568 kg (Christiansen, 2004), and skeletal dimensions produced shoulder heights of 251–317 cm and volumetric estimates of 4.6–6.7 tons for *G. productum* and *G. steinheimense*, respectively (Larramendi, 2016). Accordingly, gomphotheriine postcrania closely resemble elephant skeletal elements, with numerous adaptations for graviportal positional behavior, though more robust in overall construction.

Gomphotheriine postcrania have been well documented and described by Tassy (1977) and Göhlich (1998). The skeleton typically has a long thorax with about 20 vertebrae. Neck vertebrae are short and anterior thoracic vertebrae have high neural spines, adaptations for effective muscular and ligamentous support of the skull on a stable cervical base. The lumbar vertebral region is short, bringing the ribcage into close approximation with broad ilia. The sacrum is short and blocky for resistance of bending stresses from the innominate-vertebral column articulations. Although it is composed of about 20 vertebrae, the tail is relatively short and likely did not provide balance during locomotion. Long bones are typically adapted to graviportal posture; for example, the femoral head and condyles predominantly are oriented vertically, as are the distal and proximal tibial articulations. The same conditions predominate in the forelimb; the humeral joints are vertically oriented, the fixed, pronated position of the radius on the ulna contributes proximally to an anteriorly extended trochlear notch for articulation with the distal humerus, and the low, posteriorly-oriented ulnar olecranon process permits full extension of the forelimb. Carpals and tarsals are blocky and arranged in taxeopodal alignment (see Rasmussen et al., 1990), with minimal interlocking of elements. As in elephants, the ectal facet of the astragalus is considerably larger than the sustentacular facet. Metapodials are robust and only modestly long, while phalanges are relatively wide and very short. Typical proboscidean traits present in the postcranium include the broad, squat dorsal articular surface of the astragalus for the tibia, a salient medial astragalar tuberosity, and fixed pronation of the radius on the ulna.

GOMPHOTHERIUM SP. [Kajong] Tassy, 1979
(Table 4.1; Figures 4.1 and 4.7)

A gomphotheriine from the early Miocene site of Kajong (Mwiti), Kenya, has features consistent with placement in the *G. "annectens* group." Specimen KNM-MI 1 consists of a palate with left and right M2–3 (Figure 4.7A; Tassy, 1986). Overall length of the palate in the midline is about 400 mm. The M2s are trilophodont with a loph formula of x3x and are heavily worn. The right M2 has dimensions of L = 97.0 mm and W = 62.3 mm. There is no sign of trefoil wear patterns in their enamel loops, though it is possible that anterior pretrite accessory conules are incorporated into them. The third molars each have three good lophs and incipient fourth lophs, which on the right side can be considered a hypertrophied postcingulum. On the left side, the nascent fourth loph is formed of four low conelets tightly appressed to the third loph, with a very low postcingulum composed of a row of tiny conelets tightly behind it. Loph formula for the right M3 can be expressed as x3X and for the left M3 as either x3Xx or x4x. Half-lophs are constructed of massive, low outer main conelets with poorly differentiated, tiny mesoconelets, and are separated by a very narrow, deep median sulcus. Crown height is very low and the hypsodonty index for the left M3 is 63. Overall size of these M3s is modest: L = 133.6, 133.2 and W = 77.4, 77.6 for the left and right M3, respectively. Lamellar frequency is low, 3.75. In occlusal view, the M3s are rectangular-shaped. Accessory conules are small and tightly appressed to their pretrite half-lophs. It is doubtful that they would have contributed to trefoil-shaped enamel wear figures. There is no cementum. Evidence for horizontal tooth displacement is provided by the evident initiation of resorption of the M1 alveolus.

The proboscidean assemblage from Kajong also includes scant evidence of *Prodeinotherium hobleyi*, and several important specimens of a primitive morph of *Archaeobelodon* sp. (see below). Other mammals from the site include sanitheres, the rhino *Brachypotherium*, rodents, hyraxes, anthracotheres, and tragulids (Pickford, 1984; Geraads, 2010; Geraads and Miller, 2013; Brown et al., 2016). The fossils derive primarily from fluvial deposits in the Kajong Fm., dated by $^{40}Ar/^{39}Ar$ to older than 19.2 Ma (Brown et al., 2016).

GOMPHOTHERIUM LIBYCUM (Fourtau, 1918)

(Tables 4.1 and 4.2; Figures 4.1 and 4.7)

Synonymy—*Mastodon angustidens* var. *libyca* Fourtau, 1918:84; *M. spenceri* Fourtau, 1918:89; *Trilophodon cooperi* Osborn, 1932:figs. 1–2; *Rhynchotherium spenceri* Osborn, 1936:485; *Trilophodon angustidens* var. *libycus* Osborn, 1936:260; *Gomphotherium angustidens* Hamilton, 1973:276; *G. angustidens* Tobien, 1973b:214–215; *G. angustidens* Coppens et al., 1978:342–343; "*G.*" *pygmaeus* Coppens et al., 1978:344; *G. cooperi* Gentry, 1987a; *G. spenceri* Savage, 1989:594; *G. angustidens* Savage, 1989:594; *G. angustidens libycum* Sanders and Miller, 2002; *G. angustidens libycum* Sanders et al., 2010a; *G. angustidens* Robinson, 2015

The species name derives from the Berber language for the region west of the Nile River, which is appropriate as the type series is from Wadi Moghara, Egypt (Fourtau, 1918, 1920; Sanders and Miller, 2002). It is diagnosed as a medium-sized gomphothere distinct from *Gomphotherium angustidens* by variable development of a fourth loph in M3 (Figure 4.7B), and from *G. steinheimense* and *G. browni* by smaller size and weaker subdivision of outer- and mesoconelets. It is also distinguished from more primitive "*annectens*"-grade species of *Gomphotherium* such as *G. cooperi* from the Bugti Beds, Pakistan, *G. annectens* from the Hiramaki Fm., Japan, and *G. sylvaticum* of early Miocene western Europe by stronger expression of accessory conules. Occlusal morphology of molars are more like those of *G.* "*angustidens* group" taxa. Moreover, it is differentiated from so-called "pygmy" gomphotheres of Siwa, Egypt, Ghaba, Oman, and Gebel Zelten, Libya by larger size (for comprehensive summary of these differences see Fourtau, 1920; Forster Cooper, 1922; Osborn, 1932; Arambourg, 1961; Tassy, 1983a, 1985, 1994a, 1996b; Gaziry, 1987a; Gentry, 1987a; Roger et al., 1994; Göhlich, 1998).

A small sample of isolated gnathodental specimens comprises the Wadi Moghara species sample. Intermediate molars are trilophodont (Figure 4.7C), and upper and lower third molars also may have as few as three loph(id)s. Not enough is known of the complete dentition to document the tooth formula, but it is assumed from comparison with other gomphotheriine species that it was likely 1-0-2-3/1-0-2-3. Molar mesoconelets are generally much smaller than outer, main conelets. Enamel is thick and unfolded, unworn cusps are low and bulbous, and cementum is light or absent. The accessory conules are generally large (blocking transverse valleys) and simple, combining with pretrite mesoconelets to form trefoil wear figures. There are no posttrite accessory conules. Size variation of molars is considerable; for example, m3s range in size from L = 147.1 to 175.0 mm and M3s from L = 121.1 to 149.5 mm (Sanders and Miller, 2002). This variation is unsurprising as gomphotheres exhibit a strong degree of sexual dimorphism (Tassy, 1985; Göhlich 1998).

The taxonomic history of the species is complex, starting with Fourtau's (1918) placement of specimens in *Mastodon angustidens* var. *libyca*, though he did not specify a type specimen for the variety. Osborn (1936:260–261), never shy about taxonomy, designated an unnumbered maxillary fragment with associated M2–3 that was figured in Fourtau (1918:84) to be the type. The third molar of the type has three complete lophs, an incipient fourth loph, and a tiny postcingulum; in addition, it has a low lingual cingulum, is widest anteriorly, has prominent accessory conules, and is brachyodont (Fourtau, 1918; Osborn, 1936). The material from Wadi Moghara was finally placed in *Gomphotherium* (*G. angustidens*) by Hamilton in 1973, along with molars from Siwa, Egypt. Sanders and Miller (2002) maintained the taxon as a subspecies (*G. angustidens libycum*). However, Tassy (2014) pointed out that *G. angustidens* has been over-used in naming gomphothere species and made a compelling argument that the nomen should be restricted to middle Miocene European bunodont gomphotheriines with trilophed intermediate molars, M3 with four lophs and m3 with four-five lophids. For that reason, the species is emended here to "*Gomphotherium libycum*."

Other fossil material from Wadi Mogahra (not specified) has been referred to "*Gomphotherium pygmaeus*" (Coppens et al., 1978:343), possibly justified by the similarity of faunas from Wadi Moghara and Gebel Zelten, Libya (Savage and Hamilton, 1973). Nonetheless, Savage (1989) listed *G. angustidens* at both sites but *G. pygmaeus* only in his faunal list for Gebel Zelten.

In addition, another species, *Mastodon spenceri*, was erected for gomphothere material from Wadi Moghara (Fourtau, 1918) based on downward deflection of the symphysis of a partial dentary with m2. Osborn (1936) subsequently interpreted this feature to re-allocate the specimen to the New World genus *Rhynchotherium*. In turn, Tassy (1985) speculated that its downturned rostrum and deep symphyseal gutter might indicate choerolophodont affinities.

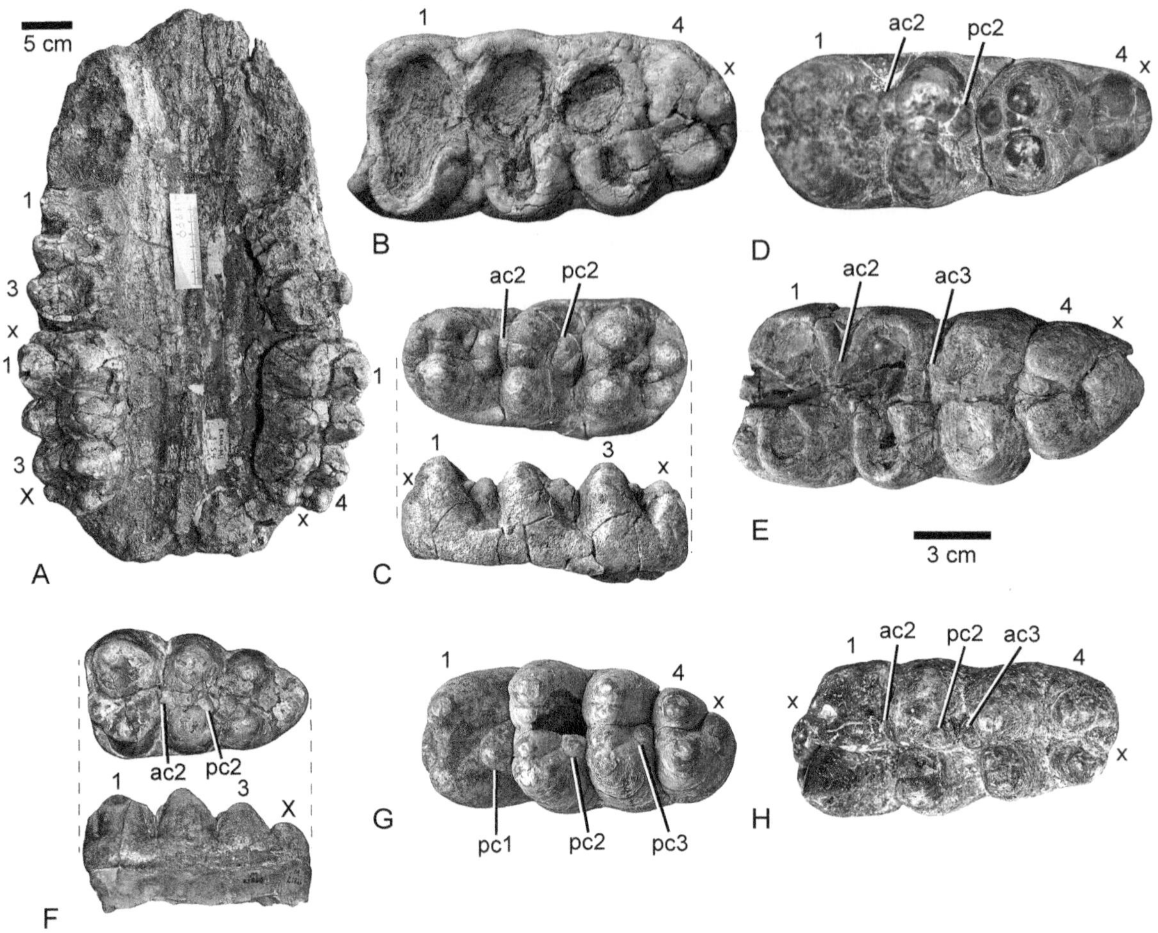

FIGURE 4.7 Early Miocene specimens of African *Gomphotherium* spp. Abbreviations: ac, pretrite anterior accessory conule; pc, pretrite posterior accessory conule; x, pre- or postcingulum(id); X, large pre- or postcingulum(id) equivalent to an incipient loph(id); 1, 2, 3, …, loph(id) number counted from the anterior of the crown. (B–H) anterior to the left and specimens are to the same approximate scale. (A) Palate with right and left M2-3, KNM-MI from Mwiti (Kajong), Kenya, ventral view, *Gomphotherium* sp. indet. Anterior is to the top. (B) Right M3, field number 97-656 from Wadi Moghara, Egypt, occlusal view, *Gomphotherium libycum*. (C) Right m2, DPC 12926 from Wadi Moghara, Egypt, occlusal and lateral views, *Gomphotherium libycum*. (D) Right m3, X7 from Gebel Zelten, Libya, occlusal view, *Gomphotherium libycum*. (E) Right M3, KNM-KA 57650 from Karungu, Kenya, occlusal view, *Gomphotherium* sp. (F) Left M3, M 21866 from Gebel Zelten, Libya, occlusal and lateral views, cf. *Gomphotherium* sp. nov. (archaic form). (G) Left m3, KNM-SO 37979 from Songhor, Kenya, occlusal view, *Gomphotherium* sp. indet. ("pygmy"). (H) Right M3, KNM-WS 65810 from Buluk, Kenya, occlusal view, *Gomphotherium* sp. indet. ("pygmy"). Probably not the same species as the Songhor molar.

However, there are no choerolophodont features, such as chevroning of lophids, present in the m2, and the ventral aspect of the symphysis is only gently angled downward, not inconsistent with identification as *Gomphotherium* (Sanders and Miller, 2002). Thus, there are no good reasons to justify recognition of these taxa in the Wadi Moghara proboscidean sample.

Some isolated molar specimens from the site of Ad Dabtiyah, Saudi Arabia (Table 4.1) are referable to *Gomphotherium* and may belong to the same species as the Wadi Moghara gomphotheriines. The occlusal morphology of these molars resembles gomphotheriines included in Tassy's (1985) "*Gomphotherium angustidens*" group based on stronger development of accessory conules than in the so-called "*Gomphotherium annectens*" group. The Ad Dabtiyah molars are more advanced than those of South Asian *G. cooperi* and therefore do not belong in that species. Because of the conservative nature of gomphotheriid molar morphology, it is difficult to be certain that the limited samples from Wadi Moghara and Ad Dabtiyah are synonymous (Sanders et al., 2010a). The Ad Dabtiyah proboscidean material was described in detail by Gentry (1987a). Site specimens include M 42940, left partial m2–m3. The third molar has a lophid formula of x4x; each half-lophid has a large outer, main conelet and a smaller mesoconelet, which are aligned transversely across the crown. Accessory conules are limited to the pretrite side of the crown and are present in association with the first three lophids; the posterior accessory conules are larger than the anterior conules and it is obvious that in older individuals, they would have contributed to trefoil wear patterns. Crown length is 136.0 mm and width is 64.2 mm. There is no cementum, crown height is low, and in lateral view lophids are pyramidal in shape. M 42945, another m3 with slightly less wear on the crown, is quite similar in morphology but even larger in size: estimated L=157.0 mm and W=76.1 mm.

Upper molars of this species exhibit the same morphology. M3 specimen M 42943 has a loph formula of x4, though its fourth loph is very narrow and resembles a hypertrophied postcingulum. Crown size is L=148.0 mm and W=66.8 mm. Specimen M 42946 has a trilophed M2 that is heavily worn with trefoil pretrite enamel figures and an M3 that if complete would have had a broader fourth loph than in M 42943. The trefoil enamel wear figures in this molar are especially well expressed. Crown size of the M3 is 147.0 and W=78.0 mm. As in the lower molars, there is no cementum and crown height appears low. There are no posttrite accessory conules and no development of pseudoanancoidy in these molars. Its provenience is less well established than other specimens from the site (Whybrow et al., 1987).

Two other specimens from Ad Dabtiyah may instead document the presence of an amebelodont at the site: M 42941, a lower tusk, and M 42942, a right dentary with m3. Although Gentry (1987a) did not feel that the tusk was transversely expanded enough to belong to an amebelodont, the cross-sectional shape is piriform with a dorsal sulcus, more resembling typical lower tusks of *Archaeobelodon* than *Gomphotherium*. It is broader than high (W=44.5 mm, H=26.3 mm), yielding a compression index of 59, which falls in the range of *Archaeobelodon* sp. nov. [Buluk], though far less compressed than lower tusks of more derived subfamilials. The m3 of M 42942 exhibits large posttrite anterior accessory conules in association with lophids 3–4, and the pretrite main conelets are offset posteriorly, producing pseudoanancoidy of the lophids. However, the pretrite accessory conules are not particularly massive and do not angle across the transverse valleys. There are no other specimens with which to explore the hypothesis of multiple gomphothere species at the site, but this would not be surprising given what is known about other coeval sites from Afro-Arabia.

Vertebrate-bearing sediments of the Dam Formation at Ad Dabtiyah (Table 4.1) were deposited in fluvial settings and are composed of sandy carbonate muds and conglomerates (Whybrow et al., 1987). Along with the proboscideans, the sediments produced turtles, fish, crocodiles, a hyrax, rhinos, tragulids, primitive bovids, listriodontine and kubanochoerine suids, indeterminate rodents, giraffids, and carnivores, and, notably, a hominoid (Andrews et al., 1978; Hamilton et al., 1978; Buffetaut, 1984; Andrews and Martin, 1987; Gentry, 1987b, c; Greenwood, 1987; Pickford, 1987; Whybrow et al., 1987). Hamilton et al. (1978) mentioned recovery of *Platybelodon* specimens from Ad Dabtiyah, but this identification was not corroborated by Gentry (1987a). The composition of the fauna suggests a well-watered and forested or wooded paleoenvironment.

The species may also be present at the latest early Miocene site of Gebel Zelten, Libya (Table 4.1; Sanders, 2008a). A small sample of molars exhibit features similar to those of molars from Wadi Moghara: trilophodont intermediate molars, third molars with as many as four loph(id)s, composed of heavy, bulbous outer conelets and diminutive mesoconelets that are transversely aligned across the crown. Pretrite anterior and posterior accessory conules contribute to trefoil wear figures. Molars are very brachyodont and have little or no cementum. Specimen X7, a right m3 (Figure 4.7D), is typical of the site assemblage. It has a lophid formula of x4x, low lamellar frequency of 3.25, and dimensions of L=153.0 mm, W=63.0 mm. It is low-crowned (H=46.0 mm; HI=73), lacks cementum, has small, single mesoconelets, and its lophids are transversely aligned straight across the crown. Anterior and posterior pretrite accessory conules are present in association with each lophid.

An age-grade series of dentaries and mandibles were also recovered for this species from Gebel Zelten, including specimen L113, with dp3–4 and an m1 developing in the crypt, specimen P4 with an alveolus for dp4 above p4 in its crypt, roots of m1, and m2 in its crypt, an unnumbered right dentary with dp4 (not m1 as reported by Hormann, 1963), p4 in its crypt, and alveoli for p3 and m1, specimen P2, with roots of m, and m2 in its crypt along with an alveolus for the left i2, specimen L32, a mandible with alveoli for p4, m1, and m2 erupting, along with alveoli for right and left i2s, and H2, a dentary with m2 and partial m3. Pickford (2003a) considered L32 to belong to a species of pygmy *Gomphotherium*, based on misidentification of an incomplete posterior molar with an estimated length of 111 mm (actual length is +96.0+ mm) as an m3; however, it is an m2 preceded by alveoli for p4 and m1. In size and morphology, it belongs in *G. libycum*. The morphology of these fragmentary mandibular remains is distinctive. Symphyseal gutters are wide and symphyseal angulation increases substantially with age; there are three mental foramina, the anterior one being the largest (described as "torpedo-tube-like"; Sanders, 2008a) and there are medial mental foramina communicating to the inner surface of the jaw from the mandibular canal. This latter feature is remarkable because medial mental foramina have been considered unique synapomorphies of the Elephantinae (Ferretti and Debruyne, 2011).

Gomphotheriine palatal remains with damaged remains of left M2–3 and right M2, and a well-preserved right M3 (KNM-KA 57650; Figure 4.7E) were recently recovered from early Miocene Karungu, Kenya. In size and morphology, the M3 is indistinguishable from M3 in Wadi Moghara *G. libycum*. The M3 is 149.3 mm in length and widest anteriorly, at 72.2 mm. Loph formula is x4x and lamellar frequency is a low 3.5, fitting with the bulbous appearance of the lophs. Enamel is very thick, 4.5–5.5 mm, and there is no cementum on the M3 crown. Crown height is evidently low but occlusal wear prevents calculation of a hypsodonty index. Mesoconelets, where present, are much smaller than the outer, main conelets. Lophs 1–3 have larger anterior pretrite accessory conules than small or absent posterior accessory conules; loph 4 has no accessory conules. There are no posttrite accessory conules and no sign of chevroning or pseudoanancoidy. The precingulum is low but not weak, and the postcingulum is formed of three small conelets in a triangular pattern.

The palate of this specimen is +555+ mm in length. The height of the root of the zygomatic arch above M3 is only 70 mm, very low. Width across the palate between the M3s anteriorly is 100 mm and posteriorly is 135 mm. The orbit is present above the posterior end of M3.

Fauna from Karungu derives from fluviatile and lacustrine sediments. Isotopic analysis of these sediments shows that as much as 21–57% of standing biomass was from C_4 vegetation (Lukens et al., 2017). This is consistent with an interpretation of open canopy, dry woodland to woody grassland based on prior faunal analysis (Andrews, 1974; Andrews et al., 1979). The fauna resembles modern woodland-bushland or floodplain faunas and locomotor categories are those of non-forest animals. The Karungu assemblage is diverse and includes artiodactyls, very few or no carnivores (depending on the locality), hyraxes, lagomorphs, lipotyphlans, macrosceliids, rodents, fish, crocs, turtles, gastropods, and, aside from the *Gomphotherium* specimen, amebelodonts and deinotheres (Table 4.1).

Gomphotherium dental remains from a number of North African sites, including early Miocene Djebel Nara in Tunisia, may belong to this species (Table 4.1). A partial left dentary, specimen T-4446, with alveoli for i2, p2(?)-4, m2, and an m1 in place, was recovered from a bed comprised of sandstones and conglomerates in the El Haouaria Fm. at the locality (Robinson, 2015). The m1 has a simple crown, with a lophid formula of x3x, and exhibits no ptychodonty or cementodonty. In occlusal view, its enamel loops on the pretrite side would have formed trefoil patterns in full wear; anterior accessory conules are small and posterior accessory conules are massive, except in the tritolophid where they are of equal size. Posttrite half-lophids are located straight across the crown from the pretrite side and have no accessory conules. Dimensions are L=77.9 mm, W=42.5 mm, and H=28.4 mm. The resulting hypsodonty index is 67, indicating that the crown is brachyodont. Size of the m1 is consistent with first molar size of the Moghara type series of *G. libycum* (Sanders and Miller, 2002). The cross-sectional outline of the incisor alveolus suggests that the lower tusk was not flattened, and the opening measures 30×21 mm.

CF. *GOMPHOTHERIUM* SP. NOV. Sanders, 2008a [ARCHAIC FORM]

(Table 4.1; Figures 4.1 and 4.7)

Synonymy—*Mastodon pygmaeus* Harris, 1969; *Gomphotherium pygmaeus* Savage, 1989

A single M3 from Gebel Zelten is incongruously archaic compared with other gomphothere molars from the site, given the age of the site (Table 4.1). It was presented to the Natural History Museum in London in 1962 by the Oasis Oil Company, placed close to the southwest face of the Gebel Zelten escarpment by its field coordinates. Its small size (L=96.0 mm, W=55.4 mm) is at the upper end of the range for palaeomastodonts and is even more diminutive than molars attributed to so-called "pygmy" gomphotheres of the African Miocene. However, its size is probably related to its status as a basal gomphotheriine rather than to dwarfing (Sanders, 2008a). It has a loph formula of x3X and is brachyodont (H=35.0 mm; HI=63). In occlusal view, it has the shape of an elongate triangle, with a substantial distal heel or postcingulum (Figure 4.7F). Each half-loph is formed of a massive, bulbous outer main conelet accompanied by a single tiny mesoconelet; these are aligned straight across the crown. Each loph has a small anterior and posterior accessory conule tightly appressed to the main pretrite conule. There is no cementum.

GOMPHOTHERIUM SPP. INDET. ["PYGMY"]

(Table 4.1; Figures 4.1 and 4.7)

A number of early Miocene African *Gomphotherium* have smaller molars than those of all other gomphotheriines, except for archaic species at Oligocene Chilga, Ethiopia and Dogali, Eritrea. The diminutive size of these molars seems to be due to a degree of dwarfing, rather than as a consequence of being primitive (as, for example, in M 21866 from Gebel Zelten), because they are reduced versions of molars of other Miocene gomphotheriines primarily in the *G. "annectens* group." These indeterminate species of *Gomphotherium* occur at Songhor and Buluk, Kenya, Siwa, Egypt, Gebel Zelten, Libya, and Ghaba, Oman (Table 4.1; Figure 4.1). Morphologically, they are dissimilar enough to indicate that they probably do not all belong in the same species and are only gradistically grouped together here.

Songhor, Kenya—A lone left m3 from this early Miocene site (Table 4.1), KNM-SO 37979, has a simple crown structure. The specimen has four good lophids yielding a lophid formula of x4x (there is an ecto- and entoflexus delineating the separation of lophids 3 and 4), with insignificant pre- and postcingulids (Figure 4.7G). Each half-lophid is composed of a dominant outer main conelet and subordinate mesoconelet. Mesoconelets are larger than those in the molars from Ghaba, Oman and are the same height as the main conelets. Lophids 1–3 have strong posterior accessory conules throughout the pretrite side but no anterior accessory conules. The lack of trefoil patterns on the pretrite side is consistent with *G. "annectens* group" morphology. There are no posttrite accessory conules. Overall crown size is diminutive (L=113.2 mm, W=55.6 mm). There is no cementum. Crown height is low (45.0 mm; HI=81).

Songhor has produced a rich vertebrate assemblage, which derive from clayey red-beds intercalated within deposits of volcaniclastic carbonatite deposits (Hay, 1986). Mammals from the site other than the "dwarfed" gomphotheriine include primitive amebelodonts, hyraxes, a large species of tenrec, barbourofeline, hyaenodontid, and viverrid carnivores, chalicotheres, suids, tragulids, paleomericids, a diversity of rodent taxa, galagos, and the fossil catarrhines *Rangwapithecus gordoni*, *Kalepithecus songhorensis*, *Limnopithecus evansi*, and *Proconsul major* (Jacobs et al., 1987; Nengo and Rae, 1992; Morales et al., 2001a; Tsujikawa and Pickford, 2006; Kunimatsu et al., 2017). Other vertebrates present include land tortoises, monitor lizards, touracos, chanting goshhawks, and owls (Harrison, 1980; Meylan and Auffenberg, 1986; Clos, 1995; Walker and Dyke, 2006).

Siwa, Egypt—In 1920 a small collection of fossil vertebrates from near Siwa Oasis was purchased from Lady Moon by the British Museum (Natural History) (Hamilton, 1973). The Siwa fauna is comprised of a crocodilian, *Tomistoma*, a fish, *Pristis*, and remains of a creodont, cetacean, the rhino *Brachypotherium*, anthracothere *Brachyodus*, and a lone right m3 of a small-sized gomphotheriine species. The molar, M 11964, has a lophid formula of 4x and exhibits a simple occlusal morphology. It has modest dimensions of L = 120 mm and W = 60 mm. Half-lophids are constructed of a large outer main conelet and smaller mesoconelet. At least lophids 2–3 have a large anterior accessory conule on the pretrite side; there are no posterior pretrite accessory conules or posttrite accessory conules. Consequently, the enamel wear figures do not form trefoils, which is consistent with the *G.* "*annectens* group." A low, ribbon-like cingulid is present on the lingual side of the crown. There is no cementum. Biochronologically, the age of the site is assessed as broadly similar to the early Miocene interval represented by nearby Wadi Moghara and Gebel Zelten (Hamilton, 1973).

Gebel Zelten, Libya—Gomphotheriine molars of relatively small size from late-early Miocene Gebel Zelten have been reported by Arambourg (1961) and Gaziry (1987a). Gaziry (1987a) placed these specimens in *Gomphotherium angustidens pasalarensis*, but they are separable on size to a different species (Sanders, 2008a; Sanders et al., 2010a). Intermediate molars are trilophodont and third molars have four loph(id)s. Half-loph(id)s are composed of large, bulbous outer main conelets accompanied by smaller mesoconelets, aligned straight across the crown. The molars have no crown cementum or posttrite accessory conules, and exhibit no sign of ptychodonty or choerodonty. Posterior pretrite accessory conules are much larger than barely discernable or small anterior pretrite accessory conules. Lower third molar Z13 has dimensions of L = 115.0 mm and W = 65.0 mm (Gaziry, 1987a); M3 specimen 1961-5-12 is even smaller (L = 110.0 mm and W = 57.0 mm; Arambourg, 1961). An anteriorly broken m3, specimen W3, exhibits the same morphology and also may belong in this group, preserving three lophids with a length of +84.5 mm and width of +50.2 mm (Sanders, 2008a).

Ghaba, Oman—Fossil vertebrates recovered from the latest early or earliest middle Miocene Dam Fm. at this site are modestly diverse and document fish (such as *Lates* cf. *niloticus* and *Clarias* sp.), crocodilians (including ?*Crocodylus* sp. and a species in the *Tomistoma-Euthecodon* group), several chelonian species, the anthracothere *Afromeryx zelteni*, the giraffe *Canthumeryx* sp., *Prodeinotherium* and *Gomphotherium*, and indeterminate rhinos, bovids, carnivores, and tragulids (Roget et al., 1994). The assemblage is most similar to the fauna from Gebel Zelten. Fossil vertebrates from Ghaba derive from fluviatile sandstone facies (Roger et al., 1994).

The Ghaba gomphotheriine sample has only four molars. An upper series has a trilophodont M2 and M3 with four lophs or three lophs and a hypertrophied talon that is narrower than the preceding three lophs. The M3 dimensions are L = 102.6 mm, W = 53.7 mm. Lophs are composed of large, bulbous outer main conelets and tiny mesoconelets and are associated with small, low posterior pretrite accessory conules. There are no posttrite accessory conules and no cementum on the crown. Lower third molars each have four good lophids of similar composition as those in M3. Their dimensions are L = 114.6 mm and 116.4 mm, and W = 51.0 mm and 53.0 mm. Anterior and posterior pretrite accessory conules are small in association with lophids 1–3 in one m3; in the other m3, the posterior pretrite accessory conule of lophid 1 is massive and large behind lophid 2, while anterior accessory conules of lophids 2–4 are smaller. Most of the pretrite half-loph(id)s of the Ghaba molars would not have formed complete trefoil enamel figures with wear. Morphologically, these molars have affinity with the *G.* "*annectens* group."

Buluk, Kenya—Buluk is a terminal early Miocene site featuring an impressive roster of proboscideans (Table 4.1), including a small species attributable to *Gomphotherium* (Sanders et al., in prep.). Intermediate molars are trilophodont and M3s have four lophs. M3 crown construction features pretrite half-lophs composed dominantly of a large, outer main conelet and posttrite half-lophs have nearly equal-sized meso- and main conelets. Crown height is very low, enamel is thick, and in lateral view lophs are bulbous and pyramidal. There are no posttrite accessory conules. On the pretrite side, lophs 1 and 2 incorporate anterior and posterior accessory conules into their enamel wear figures to form trefoils; loph 3 has a strong anterior accessory conule but no posterior accessory conule. Cementum is absent. Spacing between the centers of loph(id)s is broad, based on low lamellar frequencies. This morphological combination suggests a gomphothere of *G.* "*annectens* grade."

Two complete M3s in the Buluk gomphotheriine sample are small for an elephantimorph of this geological age, ranging in length from 118.7 to 126.8 mm and in width from 58.1 to 63.3 mm (Sanders et al., in prep.), including KNM-WS 65810 (Figure 4.7H). The modest size of these molars is characteristic of other proboscideans from the site, as well, and presents a unique aspect of the assemblage.

AMEBELODONTINAE Barbour, 1927
PROGOMPHOTHERIUM Pickford, 2003a
PROGOMPHOTHERIUM MARAISI Pickford, 2003a

(Table 4.1; Figures 4.1 and 4.8)

Synonymy—Longirostral mastodont Pickford et al., 1995

Best known from Auchas, Namibia, *Progomphotherium maraisi* was tenuously dated by biochronological correlation to ~20–19 Ma (Pickford et al., 1995; Pickford, 2000). The site documents a time of warmer, wetter conditions (Pickford and Senut 1997), represented by fossil beds comprised of fluviatile sediments ranging from coarse conglomerates to sands, silts, and clays. As mentioned earlier, the site assemblage is not abundant and its faunal composition is meager, including the proboscideans *P. maraisi*, *Eozygodon morotoensis*, *Prodeinotherium hobleyi*, hyraxes, rodents, rhinos, crocodiles, and turtles. Fossil wood from Auchas appears to derive from trees similar to those occurring in modern savannas and woodlands of the region. The association of proboscideans at Auchas is the same at the early Miocene site of Moroto, Uganda (Table 4.1).

The species is named for Daan Marais, the diamond mine pit superintendent at Auchas, and the genus nomen means "before gomphotheres," in reference to its archaic phylogenetic status within the Gomphotheriidae. Dental features indicate that *P. maraisi* is a basal amebelodont

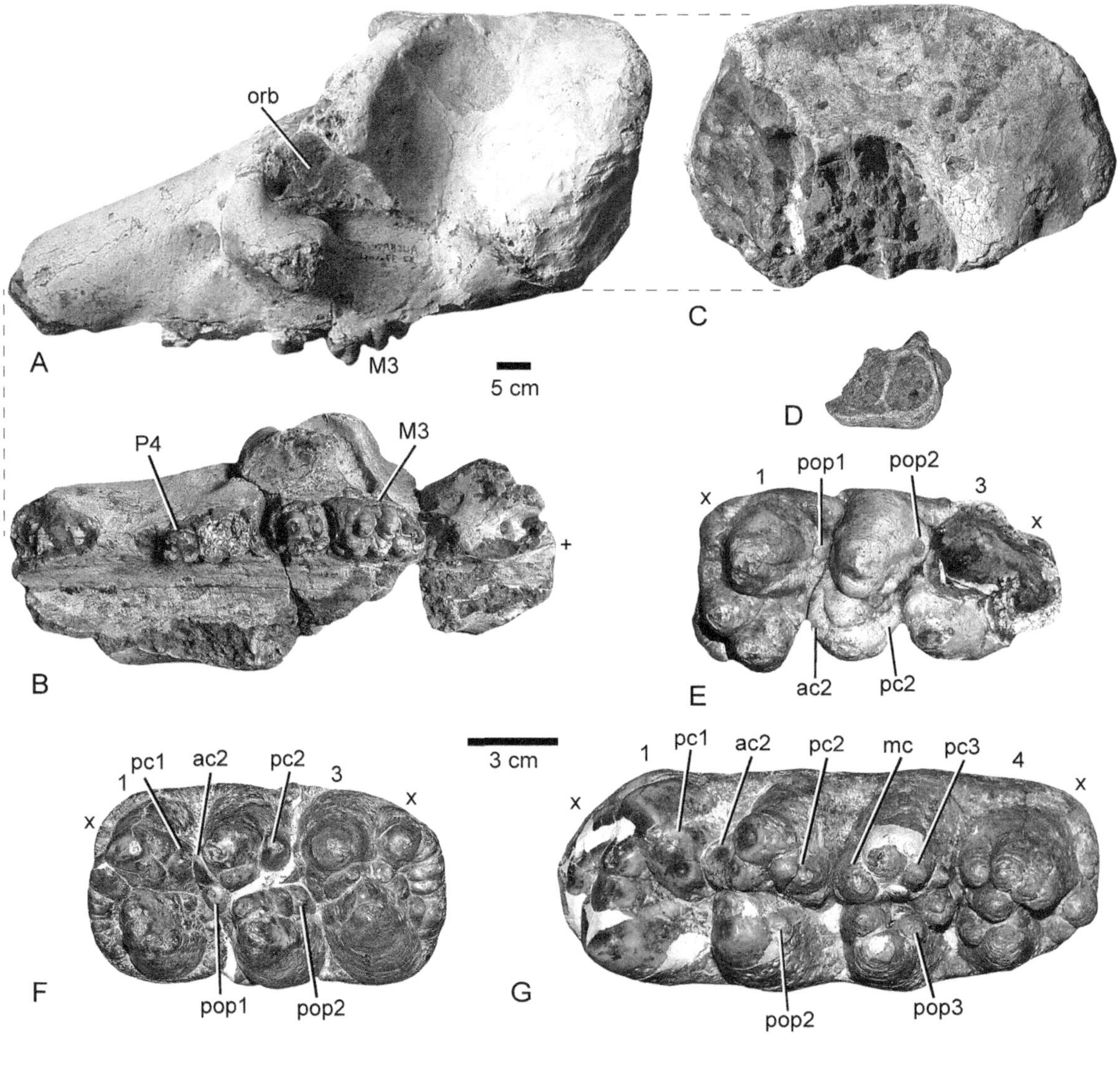

FIGURE 4.8 Cranial and dental specimens of early to early middle Miocene southern African amebelodonts. Abbreviations: ac, anterior pretrite accessory conule; M, upper molar; mc, mesoconelet; orb, orbit; P, upper premolar; pc, posterior pretrite accessory conule; pop, posterior posttrite accessory conule; x, pre- or postcingulum(id); +, indicates missing morphology; 1, 2, 3, ..., loph(id) number counted from the anterior of the crown. (A–D) to the same approximate scale; (E–G) to the same approximate scale. (A, B, E–G) anterior to the left. (A) Partial cranium, AM 1'95 (type) from Auchas, Namibia, right lateral view (reversed), *Progomphotherium maraisi.* (B) Anterior segment of partial cranium, AM 1'95 (type) from Auchas, Namibia, ventral view (reversed), *Progomphotherium maraisi.* (C) Partial cranium, AM 1'95 (type) from Auchas, Namibia, posterior view, *Progomphotherium maraisi.* (D) Lower tusk alveoli in symphyseal fragment, AM 5'99 from Auchas, Namibia, anterior view, *Progomphotherium maraisi.* Note the relatively tall oval shape of the complete left alveolus. (E) Right M3 (reversed), AM 1'95 (type) from Auchas, Namibia, occlusal view, *Progomphotherium maraisi.* (F) Right M2, PQAD 1065 from Arrisdrift, Namibia, occlusal view, *Afromastodon coppensi.* (G) Right m3, PQAD 257 from Arrisdrift, Namibia, occlusal view, *Afromastodon coppensi.* (Images courtesy of M. Pickford [Pickford, 2003a].)

with more primitive cranial and dental morphology than other subfamilials. Despite the paucity of fossils from the site, recovered mostly by mine workers, the species is firmly documented at Auchas, and includes cranial material useful for assessing its level of evolutionary development.

The holotype cranium (AM 1'95) is long and low and exhibits a number of other strikingly primitive features for an elephantimorph, including the low height of its basicranium relative to the level of the palate, posterior convergence but not confluence of the temporal lines, a low angle of eruption of M3, and low position of the orbits above the molar alveoli (Figure 4.8A). P4-M3 are present simultaneously with M3 emerging from its crypt, and the alveoli for P3 are still patent (Figure 4.8B). This is surprising, as horizontal tooth displacement is an elephantimorph synapomorphy (Sanders, 2017); in the case of *P. maraisi*, however, the permanent premolars were retained at least until M3 had completely emerged from its crypt (Pickford, 2003a). Because the vault is so low, it is offset at a steep angle from the viscerocranium, which is oriented downward. The rostrum is relatively short, about the same length as the neurocranium. Its incisor alveoli are parallel to one another and separated by a moderately deep but narrow external incisive fossa. In ventral view, the cranium has an hourglass shape, widest across the orbits and transversely constricted across the temporal fossae. The palate is moderately wide and deep. The orbit is posteriorly located relative to the dental row, above the posterior of M2 and anterior of M3. The nasal aperture is as broad as the rostrum but not high and is upturned at its lateral edges (bean-shaped). Tusk alveoli indicate that I2 was proximally ovoid in shape, higher than wide, and tusk remains show that it had a lateral enamel band. The nuchal planum is flat centrally and slightly swollen laterally, not very broad, and angles posteriorly at the nuchal crest (Figure 4.8C). The fossa for the nuchal ligament is located relatively low on the nuchal planum and is broad and deep. Internally, the cranium is extensively pneumatized. Pickford (2003a) felt that the cranium of *P. maraisi* is not very advanced morphologically over the condition of palaeomastodont crania. The primitive morphology of this species provides evidence for the African origin of amebelodontines.

A mandibular fragment (AM 2'95) preserves part of the symphysis, which is short (~200 mm), relatively broad (~200 mm), and not very deep (~125 mm) (Pickford, 2003a). The lower tusks appear to have been oval in shape and higher than wide at their alveolar insertions (Figure 4.8D; W = 46 mm × H = 65 mm; cross-sectional shape index = 141). The lack of dorsoventral compression of lower tusks in the earliest amebelodonts shows that the flattened lower tusks of *Phiomia serridens* from the Fayum are not representative of the basal condition for the shovel-tusked amebelodonts and weakens support for the hypothesis that they comprise an ancestor-descendant relationship (contra Osborn, 1919, 1936; Borissiak, 1929; Tobien, 1973a).

The Auchas molars are also morphologically primitive. Intermediate molars are trilophodont (Figure 4.8E). M3s have three lophs with a postcingulum or an incipient fourth loph. The lower third molars have four lophids without an appreciable postcingulid. Main conelets are massive and mesoconelets are poorly differentiated. Anterior and posterior pretrite accessory conules are weakly developed and pretrite trefoil enamel figures may not be formed with wear. Posttrite accessory conules may be multiplied and are small. In M3, anterior pretrite accessory conules are larger than posterior pretrite conules and lophs 2 and 3 are obliquely oriented to the long axis of the crown. The presence of posttrite accessory conules and strong transverse wear gradients in molars identifies *P. maraisi* as an amebelodont. In addition, the molars of the species are diminutive compared with those of other amebelodonts.

Isolated teeth of *P. maraisi* have also been reported from the locality of Moroto II in Uganda (Pickford and Mein, 2006), co-occurring with *Prodeinotherium hobleyi* and the mammutid *Eozygodon morotoensis*. The Moroto molars are even more primitive than those from Auchas. They are smaller, lack multiplication of accessory conules, and mesoconelets are not apparent in some specimens. A trilophodont left M2 from Moroto II, MOR IIb 1'98, is much more deeply worn on the pretrite than posttrite side, has distinct pretrite posterior accessory conules incorporated into enamel wear figures in lophs 1–3, swellings but not well-defined accessory conules on the anterior faces of pretrite half-lophs 1–3, and a low, stout accessory conule posterior to posttrite half-loph 1. Enamel is thick (3.9–4.6 mm), cementum is absent, the postcingulum is diminutive, and crown size is modest (L = 82.8 mm, W = 49.0 mm). The left M3 from the same individual has a loph formula of x3x, poorly distinguishable mesoconelets, bulbous, low (H = 28.4 mm; HI = 57) main conelets, very small pretrite and posttrite accessory conules, and only a wisp of cementum at the bases of the transverse valleys. It is also small for an amebelodont M3, with a length of 91.6 mm and the greatest width estimated at 51.0 mm. Its third loph is transversely narrower than lophs 1 and 2, a disparity emphasized by the salient ectoflexus between lophs 2 and 3. A fragmentary M3, MOR II 1'98+4'01, has very diminutive pre- and posttrite accessory conules anterior to loph 2. Its transverse valley between lophs 2 and 3 is coated in the lower half with cementum. The postcingulum of this M3 is composed of four low conelets enclosing a small basin posterior to the third loph, but this does not approach formation of an incipient fourth loph. The primitiveness of the Moroto amebelodont molars supports studies indicating that the site is >20.6 Ma in age (e.g., Gebo et al., 1997).

Little is known about the postcranium of this species, but an astragalus from Moroto I (1959 Moroto I) may belong to *P. maraisi*. It is distinguished from a second proboscidean astragalus from Moroto II (MOR IIc 5 Jan 62) by its smaller size and less ventrally oriented navicular facet, indicating weaker development of digitigrady in the pes. This is a primitive feature among elephantimorphs. The larger astragalus is attributed to *Eozygodon morotoensis*, based on similarities to an astragalus from Meswa Bridge assigned to that species. The smaller astragalus does not resemble deinothere astragali in morphology and thus is tentatively placed in *P. maraisi*. The greatest length of 1959 Moroto I is 75.0 mm, and the greatest width is 71.9 mm. The ectal facet (L = 52.9 mm, W = 47.9 mm) is considerably larger than the sustentacular facet (L = 38.5 mm, W = 26.4 mm), which is confluent with the navicular facet. The navicular facet faces anteriorly more than ventrally and has dimensions of W = 65.6 mm, H = 32.3 mm. Its medial tuberosity is modest in size, rugose, and non-projecting. The tibial platform is trochleated.

Several molars from pit AMSE at Auchas, sampling a level higher in the Arrisdrift Gravel Fm. than that of pit AM, which produced the type material for *P. maraisi*, were described by Pickford (2003) as being similar to *Archaeobelodon* aff. *filholi* in size while resembling *P. maraisi* in occlusal morphology. The AMSE pit is thought to be intermediate in age between pit AM at Auchas and the site of Arrisdrift, Namibia (Table 4.1). The M3 was erupting at a higher angle to the occlusal plane of M2 than the low eruption angle observed in *P. maraisi*, indicating that the cranium was more elevated. Posterior damage to the crown makes it impossible to know the full loph formula. This tooth seems too advanced to be considered *P. maraisi*: it has a very stout accessory conule posterior to posttrite half-loph 1, the anterior accessory conules of pretrite half-lophs 1 and 2 are very large, and there is some transverse dislocation between the pre- and posttrite half-lophids, reminiscent of the condition in *Protanancus macinnesi*. There are no obvious signs of mesoconelets, unless what Pickford (2003) refers to as pretrite accessory conules are actually enlarged mesoconelets, advanced anterior to the posttrite half-lophs, in which case third molar AMSE 1'95 would belong to a choerolophodont rather than an amebelodont species. An accompanying lower tusk is ovoid in shape. Pickford (2003) placed the AMSE molars and tusk in "*Gomphotherium* sp. indet.," but the molar occlusal morphology shows no sign of pretrite trefoil formation and is inconsistent with that identification.

Progomphotherium is also claimed to be present at Napak, Uganda, based on a fragment of a lower tusk that has been described as "peg-like" (Pickford, 2020). The tusk fragment is small, a little more than 15 cm in preserved length and 31.0 mm × 44.9 mm in width and height, respectively. *If* correctly oriented, these dimensions yield a cross-sectional index of 145, similar to the index calculated for *Progomphotherium* from Auchas. However, because lower tusks vary in cross-sectional shape throughout their lengths and the position of this fragment along the tusk cannot be ascertained, and as there is no accompanying molar evidence for the taxon at the site, this claim cannot be accepted unequivocally. If the tusk fragment is oriented differently, its cross-sectional index would be 69, within the range of *Archaeobelodon* sp., which is certainly present at Napak.

ARCHAEOBELODON Tassy, 1984
ARCHAEOBELODON SP. [PRIMITIVE MORPH]
(Table 4.1; Figures 4.1, 4.9, and 4.10)

Partial Synonymy—*Trilophodon angustidens kisumuensis* MacInnes, 1942; mammutid aff. *Zygolophodon* Van Couvering and Van Couvering, 1976; cf. *Gomphotherium angustidens* (in part) Coppens et al., 1978; cf. *Archaeobelodon* sp. Tassy, 1986; *Archaeobelodon* aff. *filholi* (in part) Tassy, 1986; *Protanancus macinnesi* Drake et al., 1988; *Archaeobelodon filholi* Brown et al., 2016

Tassy (1984) erected the genus *Archaeobelodon* to accommodate amebelodonts of primitive aspect. The genus nomen refers to the primitive cross-sectional shape of the lower tusks, presumably in comparison with those of more derived, shovel-tusked amebelodonts. His diagnosis of the taxon included: a relatively narrow mandibular symphysis (presumably compared with more derived amebelodonts), lower incisors composed of concentric dentine without dentinal tubules and with little bilateral expansion, powerful I2s directed downward, possessing a lateral enamel band but without torsion, trilophodont intermediate molars, molars very brachyodont and bunodont, viscerocranium elongated and more developed than the neurocranium, and persistence of P2.

Some African specimens are assignable to a basal, unnamed species of *Archaeobelodon*, barely more advanced than *Progomphotherium maraisi* in molar morphology. Cheek teeth of this species are small to moderate in size with modest expression of amebelodont features. For example, an incomplete third molar from Wadi Moghara, Egypt (CGM 30893) has an estimated size smaller than third molars of "*Archaeobelodon* aff. *filholi*" from the younger site of Buluk, Kenya (estimated L = 120.0 mm; W = 60.0 mm), four lophs and a low postcingulum closely appressed to loph 4, is subdivided into numerous small conelets in typical amebelodont fashion, has a relatively narrow crown, and shows barely a trace of cementum in the transverse valleys (Figure 4.9A; Sanders and Miller, 2004). It is primitive in lacking transverse offset of half-loph main conelets and has diminutive pre- and posttrite mesoconelets (better developed than in *Progomphotherium*), petite anterior accessory conules associated with each posttrite half-loph, and modest-sized anterior and posterior accessory conules associated with each pretrite half-loph (these do not extend obliquely across the transverse valleys as in "*Archaeobelodon* aff. *filholi*" or *Protanancus macinnesi*). This species includes gnathodental specimens from Kajong (Mwiti; see below), and also probably from Rusinga, Kalodirr, Legetet, Napak, and Songhor, Kenya (Table 4.1; Tassy, 1986; Pickford, 2020).

A palatal specimen from Rusinga (KNM-RU 4423) is from an adult, with very worn M2-M3 on both sides. The M3s are relatively narrow and exhibit four lophs. The fourth loph is narrower than lophs 1–3 but is fully formed and demarcated by an ectoflexus. Half-lophs on the pretrite side dwarf those of the posttrite side. This feature is distinctive of *Archaeobelodon*. There is no cementum. Tassy (1986) has described this palate in detail along with other specimens

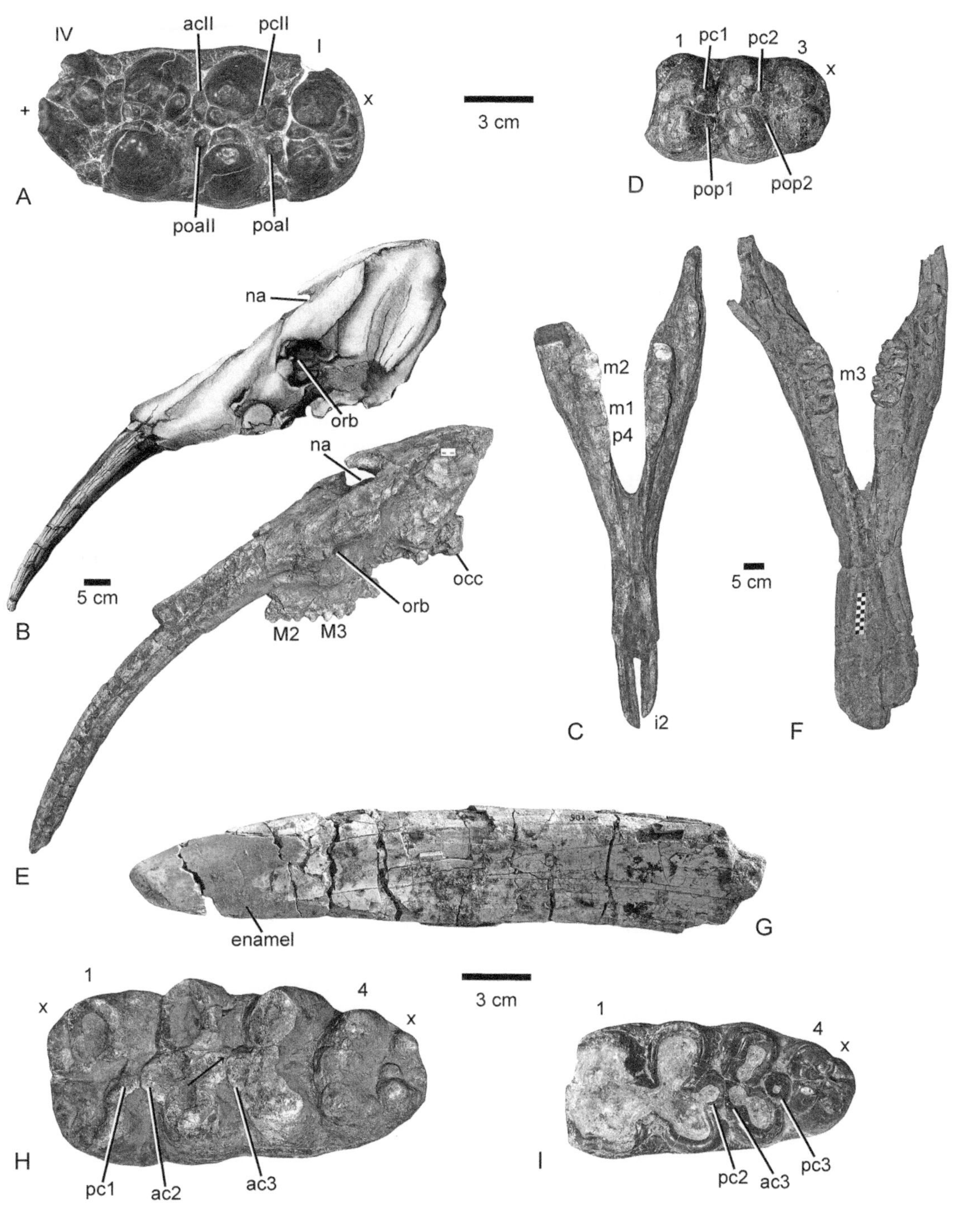

FIGURE 4.9 Skull and dental specimens of early Miocene African archaeobelodonts. Abbreviations: ac, anterior pretrite accessory conule; i, lower incisor; M/m, upper/lower molar; na, nasal aperture; occ, occipital condyles; orb, orbit; p, lower premolar; pc, posterior pretrite accessory conule; poa, anterior posttrite accessory conule; pop, posterior posttrite accessory conule; x, pre- or postcingulum(id); 1, 2, 3, …, loph(id) number counted from the anterior of the crown; I, II, III, …, loph number counted from the posterior of the crown; +, indicates missing morphology. (A, D, G–I) to the same scale and anterior is to the left. (B, E) to the same scale and anterior is to the left; (C, F) to the same scale and anterior is to the bottom. (A) Partial right M3, CGM 30893 from Wadi Moghara, Egypt, occlusal view, *Archaeobelodon* sp. [primitive morph]. (B) Cranial reconstruction, KNM-MI 7532 from Kajong (Mwiti), Kenya, lateral view, *Archaeobelodon* sp. [primitive morph]. Drawing by E. Damstra. (C) Mandible, KNM-MI 7532 from Kajong (Mwiti), Kenya, dorsal view, *Archaeobelodon* sp. [primitive morph]. (D) Right m1 or m2, KNM-MI 7588 from Kajong (Mwiti), Kenya, occlusal view, *Archaeobelodon* sp. [primitive morph]. (E) Cranium, KNM-WS 12660 from Buluk, Kenya, lateral view, *Archaeobelodon* sp. nov. [advanced morph]. (F) Mandible, KNM-WS 12660 from Buluk, Kenya, dorsal view, *Archaeobelodon* sp. nov. [advanced morph]. Note contrast between width of symphyses and incisors in (C and F). (G) Right i2, KNM-WS 65336 from Buluk, Kenya, dorsal view, *Archaeobelodon* sp. nov. [advanced morph]. Note distal preservation of enamel. (H) Left M3, KNM-WS 12660 from Buluk, Kenya, occlusal view, *Archaeobelodon* sp. nov. [advanced morph]. Note arrow indicating oblique angle of pc2. (I) Left m3, KNM-WS 49464 from Buluk, Kenya, occlusal view, *Archaeobelodon* sp. nov. [advanced morph]. Note size and orientation of pretrite accessory conules and pseudoanancoidy and chevroning of half-lophids.

of his "cf. *Archaeobelodon* sp.," including deciduous premolars, premolars, and most notably upper and lower tusk fragments from Rusinga. The upper tusk lacks longitudinal torsion and has an enamel band on its lateral aspect; the lower tusk is ovoid in cross-section, slightly wider than high (a more derived condition than i2 in *Progomphotherium*), but appears to lack the longitudinal sulci present in i2s of *Archaeobelodon* sp. nov. and *Protanancus macinnesi*.

Several dental specimens from Napak, Uganda confirm the presence of *Archaeobelodon* at the site, including a distal fragment of a lower tusk recovered in 1957 (Bishop, 1958). Pickford (2020) listed cross-sectional dimensions of W=26.4 mm, H=40.0 mm, but surely meant the converse, as the specimen is clearly slightly flattened horizontally and wider than high. These dimensions yield a cross-sectional index of 66, close to the indices of the lower tusks in the *Archaeobelodon* sp. mandible (KNM-MI 7532) from Kajong, Kenya. Specimen NAP XXXII 40'17 is a trilophodont right upper molar in wear with dimensions of L=92.3 mm×W=55.8 mm. Though listed as an indeterminate gomphothere by Pickford (2020), its pseudoanancoidy of loph 3, strong lateral wear gradient, and oblique angulation of small pretrite anterior accessory 2 indicate that this belongs in *Archaeobelodon*. Similarly, NAP I 1'98, a left lower molar heavily worn, also identifies with this genus by its strong lateral wear gradient and pseudoanancoidy.

ARCHAEOBELODON SP. NOV. [ADVANCED MORPH]

(Tables 4.1 and 4.2; Figures 4.1, 4.9, and 4.10)

Partial Synonymy—*Platybelodon kisumuensis* Harris and Watkins, 1974; *Gomphotherium kisumuensis* Savage and Williamson, 1978; *Platybelodon* sp. Pickford, 1981; *Archaeobelodon* aff. *filholi* (in part) Tassy, 1984; *Archaeobelodon* aff. *filholi* (in part) Tassy, 1986; *Archaeobelodon* cf. *filholi* Pickford et al., 1987; *Archaeobelodon filholi* Brown et al., 2016

The European species *Serridentinus filholi* was originally named in an aside and footnote by Frick (1933) and later transferred to a new genus, *Archaeobelodon*, by Tassy (1984). Its type mandible and a referred skull were figured by Tobien (1973a:pl. 25, figs. 16, 17) under the name "*Platybelodon danovi filholi*." Early Miocene African amebelodonts from Kajong (Mwiti) and Buluk, Kenya were identified by Tassy (1984) as having an affinity with this European species. Despite the less advanced condition of molars in these Kenyan amebelodonts, including simpler occlusal surfaces, more open transverse valleys, and smaller posttrite accessory conules, as well as a dissimilar lateral profile of a cranium from Kajong (dorsally straight in lateral view, rather than stepped), Tassy (1984) resisted providing the African species with a different name. Working with a larger sample from Buluk, including previously undescribed skull material, Sanders and colleagues (in prep.) felt it was important to recognize the distinctions of the African species from its European congeners and other African amebelodonts.

$^{40}Ar/^{39}Ar$ dating at Kajong recently provided an age of 19.2±1.2 Ma for a capping basalt and 20.3 Ma for a volcanic clast within the site sequence (Brown et al., 2016). Fossil mammals recovered from clastic sediments of the four subunits of the Kajong Fm. are limited in diversity and include archaeobelodonts, *Prodeinotherium hobleyi*, ?gomphotheriines, hyraxes, *Paraphiomys* sp., *Dorcatherium* cf. *pigotti*, poorly defined suids, *Brachypotherium* sp., and anthracotheres (Savage and Williamson, 1978; Williamson and Savage, 1986; Geraads and Miller, 2013; Brown et al., 2016). The sediments are mostly of fluvial derivation, with some lacustrine contribution to the lowermost unit of the formation (Brown et al., 2016).

Tassy (1984, 1986) relied heavily on a skull from Kajong (Mwiti), KNM-MI 7532, to differentiate a more derived form of *Archaeobelodon*, "*A.* aff. *filholi*," from archaeobelodonts at sites such as Rusinga ("cf. *Archaeobelodon* sp."). With a complete upper and two lower tusks, and traces of P4/p4-M2/m2, it represents a subadult individual. This superb specimen was recovered in 1973 through the efforts of the Koobi Fora Research Project, led by Richard Leakey (Brown et al., 2016). Tassy's (1986) description of the specimen is comprehensive and well figured. Much of the cranium and mandible are preserved. In lateral view, the cranium is low and has a relatively long rostrum that is nearly in a straight line with the braincase (Figure 4.9B). The orbito-temporal fossa is narrow and the orbit is situated low above the alveolar plane. The basicranium is not raised much above the level of the palate. Length of the cranium from the nuchal crest to the anteriormost point on the premaxillaries is 920 mm in dorsal view, and the cranium appears widest across the occipital planum (590 mm). The temporal lines converge posteriorly toward the nuchal crest but do not meet in the midline. The nasal aperture is much wider (400 mm) than high and bean shaped, with its lateral edges downturned, and the nasal bones overhang the aperture considerably. Lateral flare and scope of the temporal fossae are modest. The external incisive fossa is not very deep or wide and broadens a little anteriorly. The upper tusk is relatively extensive (110×102 mm in cross-section at its alveolar exit; 610 mm in exposed length), downturned, and slightly everted, with an enamel band on its lateral aspect. In posterior view, the nuchal planum is rounded with its widest point at about mid-height. The fossa for the insertion of the nuchal ligament is rounded, salient, and bifid, located at mid-height and bordered by slight swellings of the planum. The foramen magnum is ovoid and wider than high (88×53 mm). In ventral view, the cranium is severely waisted just anterior to the large, flattened glenoid fossae. Postglenoid fossae are absent. The palate is narrow and wider anteriorly than posteriorly. The zygomatic arches are missing.

The accompanying mandible is missing its rami but the corpora and symphysis are well preserved, with i2s in place (Figure 4.9C). In lateral view, the symphysis extends forward on the corpus in a straight line. The greatest length of the specimen is 920 mm, and the greatest depth of the corpus is below p4, 131 mm. Anteriorly, the symphysis is 114 mm wide, and its overall length is about 370 mm. The symphysis is relatively narrow and does not flare much anteriorly; the lower tusks are narrow, as well. Width of the symphysis is nearly constant throughout its length. The lower tusks are slightly flattened and pyriform in cross-section, with a dorsal longitudinal sulcus. They are wider (slightly >50 mm in each) than high (slightly >30 mm in each), with modest shoveling indices (H/W × 100) of 60–61.

It is unfortunate that the cheek teeth of KNM-MI 7532 are broken away and yield little information for comparison with amebelodonts from other sites. Thus, a small, well-preserved, isolated m2 or m1 from Kajong, with three lophids, KNM-MI 7588, is extremely valuable. It is simpler in occlusal details than homologues from Buluk. It measures L = 78.8 mm × W = 47.0 mm and is extremely brachyodont (H = 29.4 mm). Main conelets are bulbous and dominate poorly defined mesoconlets. There are accessory conules posterior to pretrite half-lophids 1 and 2 that do not reach obliquely across the transverse valley, and low, diminutive posterior conules tightly appressed to the posterior sides of posttrite half-lophids 1 and 2 (Figure 4.9D). The precingulid is largely worn away by interproximal contact, and the postcingulid is anteroposteriorly narrow. There is no cementum. The organization of the occlusal surface of the molar resembles that of the third molar of cf. *Archaeobelodon* sp. (here, "*Archaeobelodon* sp.") from Wadi Moghara (Sanders and Miller, 2002). Although there are many similarities in cranial morphology between the Kajong and Buluk archaeobelodonts, given the separation between them in geological age (Table 4.1) and contrasts in details of their molar and symphyseal morphology, they appear to represent different stages of an evolving lineage and are treated here as separate paleospecies.

The proposed type specimen of the new archaeobelodont species from Buluk (Sanders et al., in prep.) is KNM-WS 12660, comprised of a well-preserved cranium and associated mandible. The cranium is similar in size to the Kajong cranium, 950 mm in length from the nuchal crest to the anteriormost edge of the incisor alveoli and 825 mm in basal length. In lateral view, it is low, with a flat, transversely convex forehead, nuchal crest overhanging the occipital planum, a retracted orbital position (to the posterior of the tooth row), and a relatively long rostrum (L = 540 mm) gently sloping modestly downward nearly in a straight line from the braincase (Figure 4.9E). The basicranium is moderately raised above the level of the palate. Length of the braincase is considerably shorter than rostral length, measuring 300 mm from the posterior of the nasal aperture to the nuchal crest.

In anterior view, the external incisive fossa is moderately deep and broad, bordered by parallel tusk alveoli; the nasal aperture is broad, upturned at its lateral edges, located anterior to the orbits, and overhung by strongly projecting nasal bones. Width across the orbits is approximately 500 mm. The right I2 is preserved and oriented downward and outward. From the alveolus to its broken tip, its length is 675 mm, and its greatest height and width are 88.3 mm × 75.0 mm.

In posterior view, the occipital planum is distorted by step fracturing, but the exoccipitals appear to be well developed and the occipital condyles are moderately large (left and right dimensions are L = 85.1 mm, W = 80.0 mm and L = 84.3 mm, W = 76.8 mm). They project more ventrally than posteriorly. The fossa for the nuchal ligament is large and round and occupies the middle of the occipital planum, bordered by distinct swellings of the planum. Its superior margin is closer to the nuchal crest (140 mm) than its inferior margin is to the upper edge of the foramen magnum (300 mm).

The mandible is distinguished from that of the Kajong specimen by the much greater anterior bilateral splay of its symphysis (Figure 4.9C and F). Overall, the preserved length of the mandible of KNM-WS 12660 is 1,100 mm, with the symphysis contributing 550 mm to that. The mandible is relatively narrow, having a width across the coronoid processes of only 360 mm. The tusk alveoli are very long and distally broad. Between them, the symphysis is shallowly concave and is only slightly downturned. A very large mental foramen is present low and anteriorly facing just posterior to the proximal-most insertion of the lower tusk; there are four mental foramina on the right side.

Lower tusk remains attributable to *A.* sp. nov. from Buluk include KNM-WS 65325, preserving +330 mm in length to the distal tip. In its middle section, its height and width are 28.4 mm and 44.8 mm, yielding a flattening index (H/Wx100) of 63, less compressed than lower tusks in *Protanancus* (see below). In cross-section, it has a piriform shape, rounder medially and flatter laterally. There is no enamel band. Wear along the medial aspect makes it clear that the lower tusks were in contact with one another. There is a narrow longitudinal sulcus on the ventral side. The tip of the tusk is heavily worn on both ventral and dorsal aspects, with wear striations oriented in longitudinal direction. In lateral view, the tusk is convex ventrally and concave dorsally.

Another *Archaeobelodon* sp. nov. i2 from Buluk, KNM-WS 65336, preserves +275 mm of length to the tip, with a flattening index of 50. This tusk exhibits both ventral and dorsal longitudinal sulci, and similar to KNM-WS 65325 is thicker medially and more flattened laterally. There is a remnant of enamel at its very tip (Figure 4.9G), and signs of heavy wear on the dorsal and ventral sides of the distal end, mostly transversely oriented.

The cheek tooth dentition is typical of *Archaeobelodon*. Intermediate molars are trilophed. Upper and lower third molars have four loph(id)s. The m3s of KNM-WS 12660 exhibit four lophs and a strong postcingulid formed of two stout conelets. Length and width of the left and right m3s is 169.9 mm × 68.5 mm and 169.5 mm × 72.0 mm, respectively. The lophids of these molars are well spaced, with a lamellar frequency of 3.0. The length of the left and right m2 alveoli are 92.5 mm and 87.3 mm. In m3, the pretrite posterior accessory conules are much larger than the pretrite anterior accessory conules, present in association with all four lophids; in lophids 1 and 2, the posterior accessory conules reach obliquely across the transverse valleys. Mesoconelets are smaller than outer, main conelets and in lophids 3 and 4 the pretrite main conelet is positioned more posteriorly than the posttrite main conelet. There are no posttrite accessory conules.

M3 of KNM-WS 12660 is similar to the m3s. The fourth loph is transversely narrower than the first three lophs and is formed of four stout conelets, with three conelets forming a postcingulum behind it (Figure 4.9H). There are no posttrite accessory conules and only a small amount of cementum in the floor of the transverse valleys. In lophs 1 and 2, the large pretrite posterior accessory conules are angled obliquely across the transverse valley and contact the rounded anterior accessory conules of the next loph. Left and right M3s measure L = 150.2 mm × W = 70.3 mm and L = 143.7 mm × W = 72.9 mm, respectively; the worn right M2 is 100.1 mm in length and 57.7 mm in width. A newly discovered m3 in a dentary fragment, B064, from Buluk, is very similar to the type m3s but in addition exhibits diminutive posttrite accessory conules.

Distinctive of this new archaeobelodont species, and more exaggerated in *Protanancus* spp., is the pseudoanancoidy of the half-loph(id)s, in which the mesoconelets of the pre- and posttrite sides meet transversely but the main conelets of each side (particularly the pretrite half-lophids in m3) are reflected more posteriorly. The effect of this arrangement creates an occlusal configuration that is chevroned, paralleling the condition in choerolophodonts. As well, *Archaeobelodon* sp. nov. is characterized by very large pretrite accessory conules that reach obliquely across the crown toward the posttrite side, dominantly posterior in lower molars (Figure 4.9I) and anterior in upper molars (Figure 4.9H).

Intermediate molars may have weaker expression of amebelodont features than those of third molars. For instance, KNM-WS 65411 from Buluk, a left m1 in a dentary fragment, looks similar to m1s of *Gomphotherium*, with formation of pretrite trefoil occlusal enamel wear patterns, but its first posterior accessory conule reaches obliquely across the transverse valley in the amebelodont manner. These molars of *Archaeobelodon* sp. nov. are distinguishable from even the small molars of *Protanancus* sp. nov. from Buluk by the less dynamic development of accessory conules in size and distribution, particularly on the posttrite side of the crown, in the former species.

A small sample of tooth and tusk fragments accompanied by a handful of manus elements from a single individual was recovered from the terminal early Miocene site of Nachola, Kenya (Table 4.1) and attributed to "*Archaeobelodon* cf. *filholi*" (Pickford et al., 1987). The lower tusks are of flattened lozenge shape, which suggests that they are similar to the Buluk tusks and aids in the identification of the species. The postcranial elements are indicative of an animal that is more cursorial than succeeding graviportal proboscideans (Pickford et al., 1987). Fossils at the site derive primarily from shales and silts laid down in littoral, lacustrine settings, and aside from the amebelodont include the remains of deinotheres, rodents, hyraxes, rhinos, anthracotheres, multiple suiform species, giraffoids, tragulids, cervoids, and a predominance of hominoids (*Kenyapithecus* and *Nyanzapithecus*) alongside primitive cercopithecoids. In addition, the site has produced bivalves, fish, crocodiles, turtles, snakes, and birds. This fauna is distinct from that of Buluk, indicating that *Archaeobelodon* occupied diverse habitats, though contrasts in depositional environments likely also contributed to the faunal differences.

Tassy (1986) reported evidence of the retention of at least upper and lower third and fourth premolars in "*Archaeobelodon* aff. *filholi*." P3 is small (L = 32 mm, W = 25.1 mm), ovoid in occlusal view, and has a low cingular band around the entire crown. The paracone is superficially doubled and the protocone is much smaller and lower. Both the metacone and hypocone are diminutive and very low; the hypocone is "captured" within the cingular band. P4 is larger (L = 40.5 mm, W = 35.5 mm), bilophodont with four cusps, and has pronounced pretrite central conules. The p3 is less well preserved but resembles *Archaebelodon* p3 from Rusinga, with two strong anterior cusps and a posttrite central conule. The more recent proboscidean sample from Buluk has produced additional amebelodont premolars, including another small P3 (L = 31.7 mm, W = 28.4 mm), but with much greater development of pre- and posttrite accessory conules and larger protocone and hypcone. A small P4 (L = 33.8, W = 33.0) also has pronounced development of pre- and posttrite accessory conules. Additionally, a new trilophed dp3 specimen attributed to *Archaeobelodon* sp. nov. is incomplete but it is possible to ascertain overall size as L = 54.5 mm, W = 22.6 mm, and to see that the development of pretrite anterior and posterior accessory conules is as in permanent molars of the species, large and reaching obliquely across the crown (in lophid 2).

Some gnathodental specimens from the lower part of the Dam Formation at Ad Dabtiyah, Saudi Arabia that were originally placed in *Gomphotherium cooperi* (Gentry, 1987a) and later included in "*G. angustidens libycum*" (Sanders and Miller, 2002; Sanders et al., 2010a) may instead belong in this amebelodont species (see above).

PROTANANCUS Arambourg, 1945
PROTANANCUS MACINNESI Arambourg, 1945

(Tables 4.1 and 4.2; Figures 4.1, 4.10, and 4.11)

Partial Synonymy—*Trilophodon angustidens kisumuensis* (in part) MacInnes, 1942; *T. angustidens kisumuensis* (in part) Arambourg, 1945; *T. angustidens* cf. *kisumuensis* (in part) Hooijer, 1963; *Gomphotherium angustidens* (in part) Maglio, 1973:fig. 22; *Platybelodon kisumuensis* (in part) Tobien, 1973a; *Pl. kisumuensis* (in part) Coppens et al., 1978; *Protanancus macinnesi* Tassy, 1979a; *Platybelodon* Shipman et al., 1981; *T. angustidens* Boaz, 1994

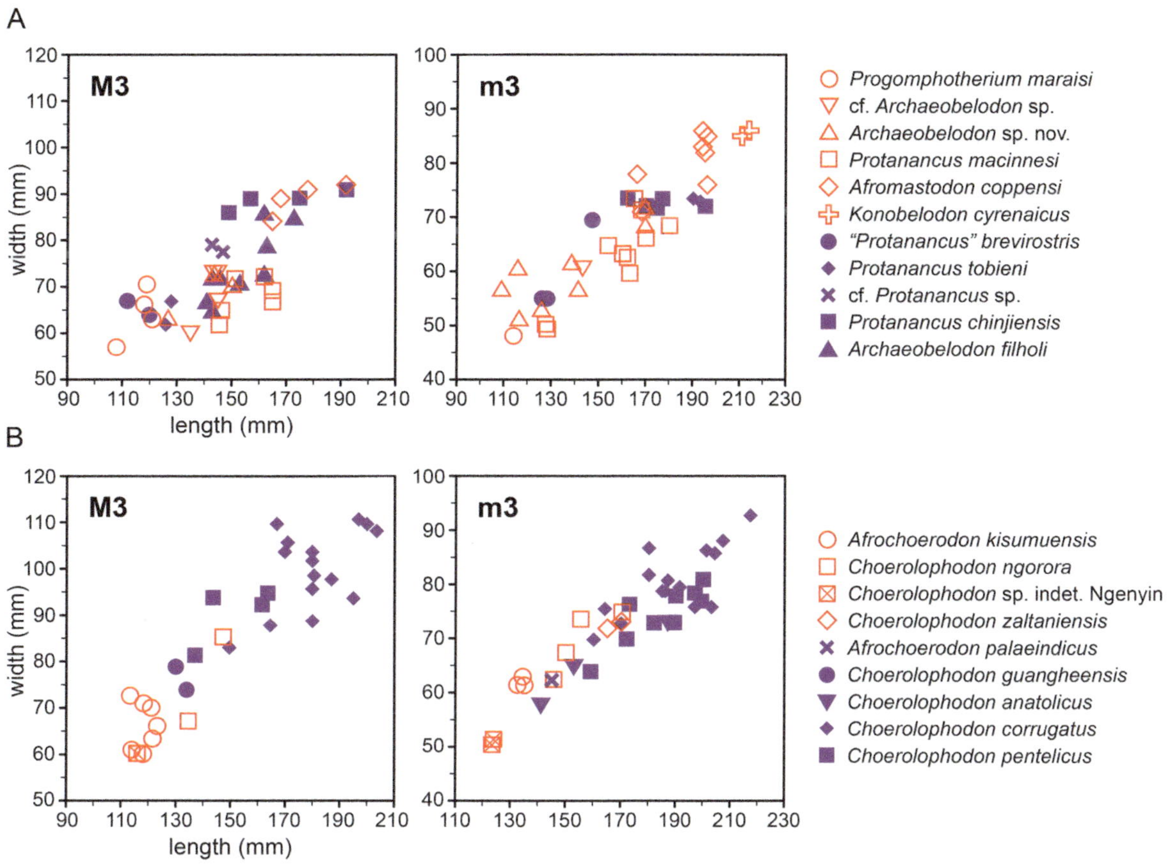

FIGURE 4.10 Bivariate plots of cheek tooth crown length versus width in amebelodonts and choerolophodonts. Comparative dimensions supplementing original measurements are from Forster Cooper (1922), Gaziry (1976, 1987a, b), Tiercelin et al. (1979), Tassy (1983a, b, 1985, 1986), Suwa et al. (1991), Pickford (2001, 2003a), Sanders and Miller (2002), Sanders (2003), Wang and Deng (2011), and Wang et al. (2015). African taxa represented by open symbols. (A) M3 and m3, amebelodonts. (B) M3 and m3, choerolophodonts.

Protanancus macinnesi is a terminal early to middle Miocene African species (Table 4.1), with a good dental representation alongside *Afrochoerodon kisumuensis* and *Prodeinotherium hobleyi* at its type site of Maboko, Kenya (Tassy, 1986). It is also documented at a handful of other similar-aged sites in Kenya and the Western Rift, and in a slightly more advanced form at Ft. Ternan, Kenya (Table 4.1), where it co-occurs with a primitive morph of *Choerolophodon ngorora* and *P. hobleyi*. Its generic etymology refers to Arambourg's (1945) belief that the lateral offset of its molar half-loph(id)s ("pseudoanancoidy") signified an ancestral position to the late Miocene-Pliocene gomphotheriid *Anancus*, and the species nomen is in honor of the numerous contributions to East African paleontology by Donald MacInnes.

This species is differentiated from *Progomphotherium* and *Archaeobelodon* by molars of larger size (except at Buluk; Figure 4.10), more dorsoventrally compressed, relatively broader lower incisors, m3s with five lophids, stronger expression of pretrite accessory conules, and a stronger tendency for transverse dislocation or pseudoanancoidy of pre- and posttrite half-loph(id)s. It also differs from *Archaeobelodon* in having a relatively shorter face, from *Platybelodon* by lacking dentinal tubules in lower tusks that are not as dorsoventrally compressed and bilaterally expanded, from *Protanancus chinjiensis* of the Siwaliks Series in South Asia by having a shorter face, smaller molars, and less pseudoanancoid dislocation of third molar half-loph(id)s, and from *Afromastodon coppensi* of Namibia by its smaller, narrower M3s and wider, flatter lower incisors. Posttrite accessory conules are small and may be irregularly distributed on molar crowns. Intermediate molars are, as in other early amebelodont species, trilophed.

Analysis of paleosols from Maboko provide evidence of distinctly different paleoenvironments at the site, including bushland, wooded grassland, and early-successional riparian woodland, across which a cohort of early cercopithecoid monkeys and at least five species of apes were distributed, with some evidence for semiterrestrial adaptations, along with a galagid strepsirrhine (Retallack et al., 2002). Cercopithecoids are more abundant than hominoids at Maboko (Pickford and Kunimatsu, 2005). These authors report a rich mammalian fauna accompanying the primates and proboscideans, including canids and viverrids, aardvarks, hyraxes, rhinos, chalicotheres, hippos, sanitheriids, suids, tragulids, bovids, and climacoceriids, along with numerous species of birds, crocs, snakes, turtles, amphibians, millipedes, and gastropods. $^{40}Ar/^{39}Ar$ dating reveals the age of the capping phonolite of the Maboko Fm. to be 13.80±0.04 Ma and provides a date of 14.71±0.16 Ma for a tuff within the formation (Feibel and Brown, 1991). Older fossiliferous beds at the site are >14.71 Ma but cannot be more precisely dated. Maboko was unique among other significant middle Miocene eastern African sites in having a semiarid to arid paleoclimate in which some taxa were adapting to terrestriality in more open conditions while others persisted in more wooded refugia (Retallack et al., 2002). It is tempting to think that choerolophodonts and amebelodonts exploited the former, while deinotheres utilized the latter.

Kipsaramon, in the Tugen Hills, Kenya is similar in age and faunal composition to Maboko, including the association of *Prodeinotherium hobleyi*, *Protanancus macinnesi*, and *Afrochoerodon kisumuensis* (Behrensmeyer et al., 2002). Fossils are derived from sediments comprising the Muruyur Beds that were deposited near the western margin of a lake, some of which were fluvially transported. Based on the faunal composition and evidence from stable isotopes, the presence of both forest and more open environments are documented at the site, including some of the earliest evidence for C_4 grasses in the local ecosystem (Behrensmeyer et al., 2002). $^{40}Ar/^{39}Ar$ single crystal radiometric dating of volcanic rocks bracketed the main bone bed between 15.8 and 15.6 Ma and younger deposits between 15.6 and 15.4 Ma (Behrensmeyer et al., 2002 and references therein); alternatively, Pickford and Kunimatsu (2005) date the Muruyur Fm. at ca. 14.5 Ma. The Kipsaramon fauna is diverse and includes ostracods, molluscs, crocs, turtles, rhinos, hippos, sanitheres, suids, giraffids, tragulids, rodents, creodonts, carnivores, hyraxes, cercopithecoids, and a small cohort of apes (Winkler, 1992; S. Ward et al., 1999; Behrensmeyer et al., 2002; Pickford and Kunimatsu, 2005; Tsujikawa and Pickford, 2006; Morales and Pickford, 2008). *Protanancus* and *Afrochoerodon* are particularly well represented dentally. The pairing of these genera may be ecologically meaningful because it is documented both in eastern Africa and in the Siwalik Series in South Asia.

The Ft. Ternan beds are younger than the fossil horizons at Maboko (Table 4.1). K/Ar and $^{40}Ar/^{39}Ar$ dating of feldspars and biotites in the Ft. Ternan sequence calibrate the age of the fossil-bearing sediments as 13.7±0.3 Ma (Pickford et al., 2006). Earlier dating efforts at the site yielded slightly older geological age estimates (e.g., Shipman et al., 1981). Vertebrate taxa at the sites are diverse but many are only known from single or few occurrences (Shipman et al., 1981, Pickford et al., 2006). These taxa include several hominoid species (but not cercopithecoids), lorises, crocs, birds, tragulids, antelopes, a variety of carnivores, rodents, lagomorphs, sciurids, anomalurids, hyraxes, enigmatic pecorans with possible cervoid affinities, aardvarks, listriodont suids, rhinos, primitive hippos, elephant shrews, and choerolophodonts and deinotheres alongside *Protanancus*. The composition and characteristics of the Ft. Ternan fauna were compared with faunas from diverse modern African habitats to test hypotheses about their paleoecology, and it was concluded that the paleoenvironment of the site was predominantly open country with forest close by but poorly represented in the mammalian site assemblage (Shipman, 1986). However, ecomorphological study of bovid postcrania from Ft. Ternan suggests that the primary habitat was woodland (Kappelman, 1991). Dental isotope analyses of bovids, giraffids, rhinos, suids, and proboscideans (deinotheres and choerolophodonts, but not the amebelodonts) from Ft. Ternan indicate that they had pure C_3 diets, with no evidence for the presence of Serengeti-like C_4 wooded grasslands (Cerling et al., 1997a). Nonetheless, the study did not rule out the possibility that C_3 grasses were present at the site.

In the sample from Maboko, m3s of *Protanancus* have five lophids and a postcingulid of one or two conelets (Figure 4.11A). In several specimens, the fifth lophid is either constituted by an inflation of the postcingulid or is inferior in size to the first four lophids and has a diminutive postcingulid behind it. Crowns of these molars are usually modestly curved longitudinally. There are strong anterior and particularly posterior pretrite accessory conules; the posterior conules are angled across the transverse valleys toward the posttrite side. There are also distinct anterior posttrite accessory conules present, though not always at every half-lophid. Pretrite half-lophids may be slightly angled posteriorly at their buccal edges, giving the impression of pseudoanancoidy of half-lophids in worn molars, but the diminutive mesoconelets of each half-lophid side are transversely aligned, unlike the condition in choerolophodont molars. Main conelets are dominant and bulbous, enamel is thick, and cementum is sparsely distributed in the floors of transverse valleys.

M3s have four lophs and usually simpler occlusal morphology than m3s (Figure 4.11B). The first three pretrite half-lophs have anterior and posterior accessory conules, with the anterior conules larger in size; the fourth pretrite half-loph usually has an anterior accessory conule. Posttrite accessory conules are very irregularly present or may be completely absent. Mesoconelets are much smaller than main, outer conelets and are aligned across the midline, but in the posterior lophs the main conelets may be laterally offset, giving the impression of transverse dislocation of half-lophs. Loph formula is usually x4x, but M3 specimen M 32769 lacks a fourth loph and has a loph formula of x3x.

The lower incisors of *P. macinnesi* are broader and more flattened than those of *Progomphotherium*, *Archaeobelodon*, and *Afromastodon*. The i2s from Maboko are dorsoventrally compressed, with a substantially wide dorsomedial longitudinal sulcus (Figure 4.11C) and a more modest ventrolateral sulcus; the tusks were apparently socketed in the lower jaw on an angle to one another. Cross-sectional indices for the i2s (H/W×100) reflect their flattening, 36, 44, and 50, with widths of 83.5mm, 80.0mm, and 77.0mm, respectively (Tassy, 1986). Lower incisors from Nyakach and Alengerr, Kenya attributed to the species are similarly compressed and wide (Tassy, 1986). There are polished wear facets on the dorsal and ventral sides of the distal end of these tusks, indicating that they were used primarily for slicing through tough vegetation

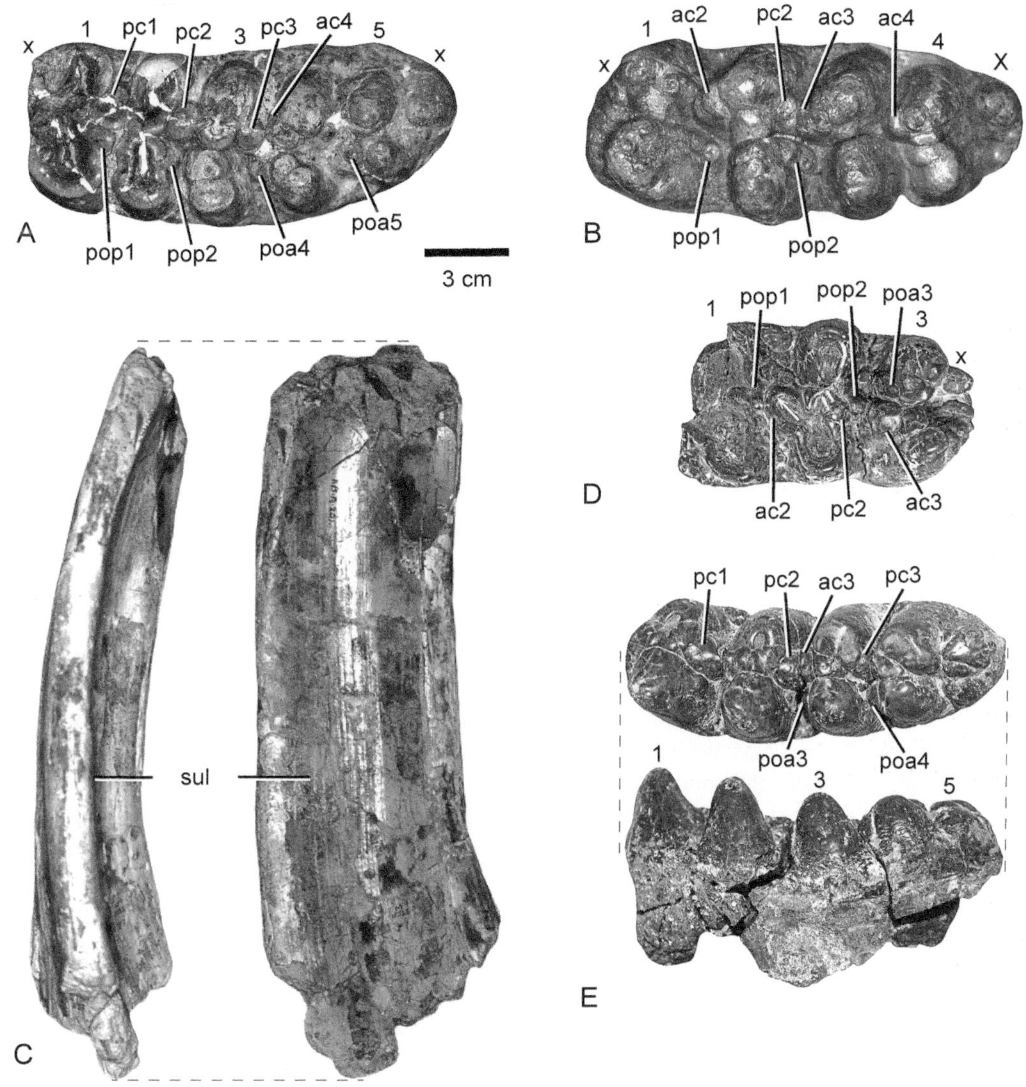

FIGURE 4.11 Dentition of early to middle Miocene African *Protanancus* spp. Abbreviations: ac, anterior pretrite accessory conule; pc, posterior pretrite accessory conule; poa, anterior posttrite accessory conule; pop, posterior posttrite accessory conule; sul, dorsal longitudinal sulcus; x, pre- or postcingulum(id); X, large postcingulum or incipient loph; 1, 2, 3, …, loph(id) number counted from the anterior of the crown. All specimens to the same scale. (A, B, D, E) anterior to the left. (C) anterior to the top. (A) Right m3, M 15538 from Maboko, Kenya, occlusal view, *Protanancus macinnesi*. (B) Right M3, M 15525 (type) from Maboko, Kenya, occlusal view, *Protanancus macinnesi*. (C) Right i2, M 15532 from Maboko, Kenya, medial and dorsal views, *Protanancus macinnesi*. (D) Left M2, KNM-WS 65227 from Buluk, Kenya, occlusal view, *Protanancus* sp. nov. (small). Note oblique angle of long axis of anterior pretrite accessory conule 2 (white arrow). (E) Right m3, KNM-WS 65257 from Buluk, Kenya, occlusal and lateral views, *Protanancus* sp. nov. (small).

(Lambert, 1992). Upper incisors are massive (KBA 110 from Maboko measures L = 1,055 mm with basal diameters of 105 mm × 95 mm [Sanders et al., 2010]), with a lateral enamel band and torque about the longitudinal axis (MacInnes, 1942).

Several recently collected molars from Buluk closely resemble molars of *P. macinnesi* except for smaller size (Figure 4.10) and are more complex in occlusal morphology than is typical for *A. tassyi*. These specimens may represent a new species antecedent to *P. macinnesi*. KNM-WS 65227, a left M2 with a loph formula of x3x, has an immense anterior pretrite accessory conule on loph 2 that reaches obliquely across the transverse valley to communicate with a large posterior accessory conule on loph 1 (Figure 4.11D). These block the transverse valley completely. Similarly, loph 3 has an immense pretrite accessory conule anterior to the half-loph that nearly touches the posterior accessory conule of pretrite half-loph 2. Posttrite half-loph 3 has multiple anterior accessory conules that contact the posterior accessory conule of posttrite half-loph 2. A complex of small accessory conules fills the transverse valley between posttrite half-lophs 1 and 2. Pre- and posttrite half-lophs are composed of large bulbous main conelets and smaller but distinct mesoconelets. The postcingulum is formed of four small conelets. Crown dimensions are L=96.3 mm, W=54.6 mm.

KNM-WS 65257 from Buluk is a complete right m3 with a lophid formula of x5x, thick enamel (4.7 mm), and a relatively very narrow crown (L=128.2 mm, W=49.3 mm). It is very brachyodont (H=39.0), but its hypsodonty index does not seem especially low (HI=79) because the crown is so narrow. Lophids 1 and 2 have large posterior pretrite accessory conules that communicate across the transverse valleys with smaller anterior pretrite accessory conules of lophids 2 and 3, respectively. Lophid 3 has a small posterior pretrite conule and lophid 4 has small anterior and posterior pretrite conules. Lophid 3 also has a diminutive posttrite anterior accessory conule. The postcingulid is formed of a single large conelet. Main conelets are dominant and bulbous, compared to much smaller mesoconelets, in each of the more anterior half-lophids. The crown is longitudinally curved, concave to the buccal side.

KNM-WS 65798, another right m3 from Buluk, is similar to KNM-WS 65257 in occlusal organization and size (L=128.0, W=50.4 mm), with even greater expression of posttrite accessory conules (posterior to half-lophids 1–3 and anterior to half-lophid 4). In KNM-WS 65798, the fifth lophid is formed of three large conelets and is transversely narrower than lophids 1–4, without an appreciable postcingulid posterior to it (Figure 4.11E). The crown is brachyodont (H=40.4 mm), with a hypsodonty index of 74. The extreme narrowness of these molars, the distinction of the mesoconelets, immense size of pretrite accessory conules, complexity of pre- and posttrite accessory conule distribution in the M2 (resembling molar M 32769 of *P. macinnesi* from Maboko), and formation of five lophids in m3 confirm the presence of a second amebelodont species at Buluk, signaling the beginning of the *Protanancus* lineage. Consistent with their position as early examples of the species, Buluk *Protanancus* molars are smaller than those from elsewhere (Figure 4.10).

It seems likely that *Protanancus* evolved from an archaeobelodont ancestor, exaggerating the molar occlusal features present in *Archaeobelodon* (large, obliquely-leaning pretrite accessory conules, distribution of posttrite accessory conules, pseudoanancoidy of half-loph(id)s), further flattening and broadening lower incisors, and adding a lophid to m3 in conjunction with molar size increase. The co-occurrence of putative ancestor-descendants at Buluk (*Archaeobelodon* sp. nov., *Protanancus* sp. nov.) supports a possible case of peripatric or parapatric speciation. The evidence is reasonable for the initial appearance of the genus in Africa and subsequent spread in the middle Miocene to South Asia (Tassy, 1983a; Abbas et al., 2016), Asia (Saegusa et al., 2005; Wang et al., 2015), and Europe (Tassy, 1986; Markov and Vergiev, 2010). Fossil material attributed to *Protanancus brevirostris* from early Miocene sediments of the Linxia Basin, China (Wang et al., 2015) is on grounds of molar occlusal morphology more likely evolved from an earlier form of amebelodont; contrary to the authors, its cranium is not autapomorphic but in some details reminiscent of palaeomastodont crania (e.g., dorsal convergence of temporal lines, relatively short rostrum, small nasal aperture, straight tusk alveoli). Along with the presence of *Archaeobelodon filholi* in Europe (Tassy, 1984), these observations suggest multiple migrations of amebelodonts out of Africa to Eurasia, during the early and middle Miocene.

AFROMASTODON Pickford, 2003a
AFROMASTODON COPPENSI Pickford, 2003a

(Tables 4.1 and 4.2; Figures 4.1, 4.8, and 4.10)

Partial Synonymy—*Gomphotherium* cf. *angustidens* Corvinus and Hendey, 1978; Gomphotheriidae, gen. et sp. indet. Hendey, 1978a; *Protanancus macinnesi* Tassy, 1985, 1986

This species was recovered from the middle Miocene site of Arrisdrift, Namibia (Table 4.1; Pickford, 2003a), and was named after the famous French anthropologist Yves Coppens, whose vast scientific accomplishments include helping to contemporize the taxonomy and phylogeny of African fossil proboscideans (Coppens et al., 1978). The site is important as the most abundant vertebrate assemblage of the middle Miocene in the southern region of the continent (Corvinus and Hendey, 1978; Pickford, 1994). It is situated on terrace deposits of the Orange River constituted of gravel deposits with silt lenses from a river channel cut into bedrock; the fossils were recovered from an exploratory mining pit and subsequent excavation (Hendey, 1978a; Corvinus and Hendey, 1978; Pickford, 1994, 2003a). The area today is very dry with sparse plant cover, but was more humid with more substantial vegetation during the middle Miocene deposition of sediments at Arrisdrift (Corvinus and Hendey, 1978; Pickford, 2000). Among the vertebrate taxa recovered are a wealth of hyraxes, deinotheres, macroscelids, suids, pecorans, bovids, tragulids, palaeomerycids, stem giraffoids, lagomorphs, several families of rodents, rhinos, hyaenodontid creodonts, felids, stenoplesictids, mustelids, viverrids, and amphicyonids, alongside birds, fish, amphibians, crocodiles, turtles, and indeterminate squamates (Hendey, 1978a; Corvinus and Hendey, 1978; Meylan and Auffenberg, 1986; Pickford, 1994; Morales et al., 1999, 2001a, b (Morales et al., 2001a, b), 2003a, b; Pickford, 2003a, b; Sánchez et al., 2018). This assemblage is characteristic of middle Miocene African faunas (Corvinus and Hendey, 1978) following repeated episodes of Eurasian immigration into the continent, and stands in stark contrast to the largely endemic African faunas of the preceding epochs. However, based on the pig species at the site, Pickford (1995) suggested an age of 17.8–17.2 Ma for Arrisdrift. This is too old, as one of the species, *Namachoerus moruoroti*, while known from Moruorot, Kenya, constrained radiometrically between 17.5 and 16.8 Ma (Adrian et al., 2018), is also reported from the younger sites of Buluk, Maboko, and Nyakach (Table 4.1), and the balance of biochronological correlation of the entire fauna points to a middle Miocene age (Corvinus and Hendey, 1978).

The sequence of *Progomphotherium* and *Afromastodon* from Auchas and Arrisdrift parallels the East African succession of *Archaeobelodon* and *Protanancus*, but there are reasons to believe that the morphological advances evidenced in each pair may be convergent. Molars of *A. coppensi* differ from those of *Progomphotherium* by their larger size (Figure 4.10), greater number and independence of mesoconelets, greater transverse offset of pre- and posttrite half-loph(id)s, and greater occlusal complexity of accessory conules. This is similar to the more derived condition of *P. macinnesi* molars compared with those of early archaeobelodonts.

Unlike *P. macinnesi*, however, in *A. coppensi* the lower incisors are oval rather than flattened in cross-section and lack longitudinal sulci and grooving. Pickford (2003a) depicted the widest dimension of these lower tusks as horizontally transverse; cross-sectional breadth varies from 35.5 to 62.5 mm and height from 29.8 to 52.4 mm, yielding tusk indices (H/W ×100) that range from 68 to 84. The upper tusk of *A. coppensi* is *Gomphotherium*-like, exhibiting a D-shaped cross-section with a flat enamel band running the length of its outer surface.

Afromastodon coppensi is further distinguished from *P. macinnesi* by greater size and relative width of its molars (Figure 4.10), and by its m3s having four rather than five lophids. It is more similar in size and molar proportions to *Protanancus chinjiensis* of South Asia, though has less pronounced pseudananacoidy of molar half-loph(id)s (Sanders et al., 2010a). Intermediate molars are trilophed, and there may be a trace of cementum in the floor of molar transverse valleys. An m2, AD 206'95, has a low, anteroposteriorly narrow precingulid and more prominent postcingulid of two conelets. Length is 118.2 mm and the greatest width at the third lophid is 70.4 mm (Pickford, 2003a). Each half-lophid has a large, bulbous main conelet and well-differentiated smaller mesoconelet. The mesoconelets are aligned transversely straight with one another. Pretrite half-lophid 1 has a moderate-sized anterior accessory conule and a tripled array of posterior accessory conules (the middle one in the series is immense) that extend slightly obliquely into the transverse valley. These contact the anterior accessory conule of pretrite half-lophid 2. Pretrite half-lophid 2 also has a tripled posterior accessory conule array, and posttrite half-lophid 2 has a modest, free posterior accessory conule. Pretrite half-lophid 3 has equal-sized anterior and posterior accessory conules and posttrite half-lophid 3 has a small posterior accessory conule.

An M2, PQAD 1065, is as complex in occlusal morphology (Figure 4.8F). It has more of a shelf-like "ribbon" of a precingulum, and multiple conelets comprise its postcingulum. Length is 118.2 mm and the greatest width is 67.4 mm (Pickford, 2003a). It is unique in that its second and third lophs are oblique relative to the long axis of the crown; in each of these lophs, the main posttrite conelet is offset posteriorly from the main pretrite conelet, and the mesoconelets are obliquely aligned, providing the molar with a pseudoanancoid occlusal appearance. Pretrite half-lophs 1 and 2 have anterior and posterior accessory conules; posttrite accessory conules are found posterior to lophs 1–3.

M3 and m3 have four loph(id)s each, with good expression of pre- and posttrite accessory conules, though unlike in *P. macinnesi*, there is slightly less of an exaggerated oblique extension of pretrite accessory conules toward the posttrite side of the crown. M3s vary in length from 165 to 192 mm and width from 84 to 92 mm; m3s vary in length between 166 and 196 mm and in width between 71 and 86 mm (Pickford, 2003a). In m3 PQAD 257, main conelets are well differentiated from their smaller mesoconelets and may be shifted a little posterior to the mesoconelets (pseudoanancoidy) (Figure 4.8G). All four pretrite half-lophids have anterior and posterior accessory conules (doubled posterior conules in half-lophids 2 and 3), and posttrite half-lophids 2–4 sport accessory conules, as well. Lophid formula in m3 ranges from x4x to x4. In M3 specimen AD 600'00, the fourth loph is formed of a transverse series of smaller conelets without pretrite accessory adornment; pretrite half-lophs 1–2 exhibit anterior and posterior accessory conules that no doubt in wear would have contributed to trefoil enamel figures; and posttrite half-loph 1 has a posterior accessory conule. Loph formula of M3 is x4x.

The crowns of these molars are as complex, in a slightly different manner, as those of *P. macinnesi*, though they lack a fifth lophid in m3 and are accompanied by *Gomphotherium*-like upper and lower incisors. This could indicate a regionally-distinct pattern of parallel evolutionary trajectory of amebelodont molar occlusal morphology between eastern and southern Africa.

Complicating this scenario is the occurrence of *A. coppensi*-like gomphotheres at Burji, Ethiopia, of latest early Miocene or early middle Miocene age (Suwa et al., 1991; WoldeGabriel et al., 1991). The fossils derive from a thin red sandstone horizon in close approximation to dated volcanics, and include a number of proboscidean postcranial elements as well as a partial M3 and dentary fragment with a nearly complete m3, as well as a large portion of a lower tusk. The m3 has four lophids and is preceded in the jaw by an alveolus for m2. Each half-lophid is composed of a large, rounded main conelet accompanied by a much smaller mesoconelet. The half-lophids are transversely straight or nearly so, with no sign of chevroning or pseudoanancoidy. Pretrite half-lophid 1 has a small anterior accessory conule and larger posterior accessory conule; pretrite half-lophids 2 and 3 appear to have small anterior accessory conules, but it is not clear if they have posterior conules because a thin layer of cementum covers much of the crown. There is a distinct, small accessory conule posterior to posttrite half-lophid 2 and posttrite half-lophid 3 may also have a posterior accessory conule, but it is obscured by cementum. A cementum covering is typical of choerolophodont molars, which led Suwa et al. (1991) to conclude that the specimen belongs to a primitive species of *Choerolophodon* (=*Afrochoerodon*).

The specimen, however, shows no chevroning of half-lophids, no advancement of the pretrite mesoconelet anterior to the posttrite half-lophid, and no choerodonty. In addition, the association of a lower tusk with this specimen is atypical of choerolophodonts, which generally lack these teeth. The relative narrowness of the m3 (L=155.5 mm, W=59.5 mm) is a feature of amebelodont molars, and the presence of small posttrite accessory conules is consistent with that identification. Although Suwa et al. (1991) correctly mentioned that amebelodonts of similar geological age do not usually possess rounded lower tusks, middle Miocene *Afromastodon coppensi* has rounded lower tusks. This led Pickford (2003a) to align the Burji proboscidean with the Namibian amebelodont, "cf. *Afromastodon coppensi*." The Burji specimen is +630 mm in length, with midlength cross-sectional dimensions of W=76 mm and H=64.5 mm.

PLATYBELODON Borissiak, 1928
PLATYBELODON SP. Maglio, 1969a
(Table 4.1; Figure 4.1)

Partial Synonymy—*Platybelodon* sp. cf. *P. grangeri* Van Couvering and Van Couvering, 1976; ?*Platybelodon* Coppens et al., 1978; *Platybelodon* sp. Tassy 1986

The early Miocene site of Loperot in the Lokichar Basin (Table 4.1), to the southwest side of Lake Turkana, Kenya produced an intriguing lower incisor fragment suggesting the presence of the shovel-tusked amebelodont *Platybelodon* (Maglio, 1969a). This incisor is severely compressed dorsoventrally and its height/width ratio (14.6 mm/78.5 mm), or compression index, is 19, below the range of incisor indices of all other African amebelodonts but comparable to values for Asian and North American *Platybelodon* and *Torynobelodon* lower incisors (Osborn and Granger, 1931, 1932; Osborn, 1936; Guan, 1996). The cross-sectional profile of the incisor is shallowly concave dorsally and nearly horizontal across its ventral surface. Most importantly, the transverse section of the incisor reveals the presence of a complex system of fine dentinal tubules and rods ("dentinal systems"), invested within and surrounded by laminar dentine (Maglio, 1969a; Coppens et al., 1978), diagnostic of *Platybelodon*. The outer surface of the incisor is sheathed in strongly laminated dentine that superficially resembles enamel. Other remains of platybelodont lower incisors have been recovered from Loperot more recently (A. Grossman, pers. comm.), confirming the original report of this taxon by Maglio (1969a).

Lower incisors of platybelodonts were hypothesized to have functioned for digging, scooping, shoveling, and dredging soft vegetation in swampy or lowland aquatic settings (Osborn, 1936; Coppens et al., 1978). This interpretation was based on observations of beveling along the distal ends of platybelodont lower incisors, with the complex of dentinal systems encased in laminated dentine providing resistance to bending during these activities (Osborn and Granger, 1931, 1932; Tobien, 1986; see Lambert, 1990). However, analysis of wear surfaces of *Platybelodon* lower tusks reveals that they were more likely to have been employed as scythes to cut through tough vegetation, rather than shoveling soft plant matter (Lambert, 1992).

The Lokone Fm. at Loperot has also produced a diverse vertebrate fauna, including crocs, tortoises, fish, cercopithecoid and hominoid primates, rodents, carnivores, creodonts, hyraxes, primitive rhinos, anthracotheres, tragulids, giraffids, archaic suids, and a ziphiid whale that must have navigated its way upriver from the Indian Ocean; other proboscidean taxa reported (without description) are deinotheres, indeterminate gomphotheriids, and "cf. *Archaeobelodon*" (Hooijer, 1971; Mead 1975; Grossman et al., 2014). The archaeobelodont from Loperot is claimed to be "uniquely" shared with Songhor, Kenya (Liutkus-Pierce et al., 2019), which is highly unlikely (see Table 4.1). Although the Loperot cercopithecoids are reported to resemble the species from Buluk, Grossman and colleagues (2014) believed that biochronological comparative methods indicate an age similar to the Hiwegi Fm. of Rusinga, Kenya, ~18 Ma. Brown and McDougall (2011) also reported that Loperot dates to the early Miocene. A minimum age of about 17 Ma is provided by whole rock $^{40}Ar/^{39}Ar$ analysis of basalt samples from above the exposed strata at Loperot (Liutkus-Pierce et al., 2017, 2019). Grossman et al. (2014) reconstructed the habitats of the fossil beds at Loperot to represent a large river near forest and dense woodland, with some evidence for grass and more open areas. Analysis of $\delta^{13}C$ values measured on rhizoliths in the Loperot section indicates oscillation over time between a purely C_3 riparian woodland to a water-stressed C_3-dominated environment and back to a riparian woodland (Liutkus-Pierce et al., 2019).

The Loperot platybelodonts comprise the only African record of this amebelodont clade and indicate an unexpectedly early and temporally condensed diversification of the Amebelodontinae, probably rooted in Africa (Sanders et al., 2010a). However, the genus is reported from even older sediments (20–19 Ma) in the Tiejianggou Fm., Gansu Province, China (Wang and Deng, 2011). The taxon dispersed quickly through Asia (Guan, 1996), and reached North America by the late Miocene, with a cohort of other amebelodonts (Frick, 1933; Osborn, 1936; Fisher, 1996; Lambert, 1996; Lambert and Shoshani, 1998).

CHOEROLOPHODONTINAE Gaziry, 1976

Choerolophodonts are a distinct branch of elephantimorphs that made their first appearance in the early Miocene of Africa, as part of the rapid diversification of the Gomphotheriidae. They are generally distinguished by a mandible with a long, gutter-like symphysis lacking lower tusks and a narrow cranium with streamlined zygomatic arches, strong facial development, and distally upcurved tusks without extensive enamel bands (Konidaris et al., 2014a, b). Congruent with the elongation of the splanchnocranium, orbits are positioned near the posterior end of the last molars; intermediate molars are trilophed, premolars (P2, P3/p3, P4/p4) may be lost, and accessory conules and pretrite mesoconelets of molars are large and tall (Tassy, 1985, 1986). In addition, their molars are characterized by ptychodonty, choerodonty, chevroning, and in more derived forms, considerable cementodonty. Chevroning of half-loph(id) pairs involves a transverse offset of half-loph(id)s with the pretrite mesoconelet and accompanying anterior accessory conule of each pair projecting anterior to the posttrite mesoconelet and to its own abaxial main conelet (Tassy, 1985, 1986). Mesoconelets may be fused with central accessory conules; both are relatively high and large.

After their early Miocene origin, choerolophodonts quickly made it out of Africa into South Asia, Asia, eastern Europe, and the eastern Mediterranean region; the group particularly flourished in peri-Mediterranean settings, where it persisted until the end of the Miocene (Tassy, 1977b, 1983b, 1985; Pickford, 2001; Sanders, 2003; Wang and Deng, 2011; Konidaris et al., 2014a, b; Tibuleac, 2014; Abbas et al., 2018). Dental carbon isotope analyses consistently indicate that choerolophodonts were browsers, including results on specimens from Buluk, Maboko, and Ft. Ternan, Kenya (e.g., Cerling et al., 1997a, 1999; Sanders et al., 2020). Isotope data for middle-late Miocene *Choerolophodon corrugatus* from the Siwalik sequence in the Potwar Plateau, Pakistan yield similar results (Semprebon et al., 2015). Accordingly, associated paleoecological data suggest that African choerolophodonts inhabited mostly closed, wet forest and woodland habitats (Savage and Hamilton, 1973; Pickford, 1985; Jacobs and Kabuye, 1987; Sanders and Miller, 2002; Tsujikawa, 2005a), although at Ft. Ternan, Kenya they may have existed in more open conditions (Evans et al., 1981; Shipman et al., 1981; Shipman, 1986).

AFROCHOERODON Pickford, 2001
AFROCHOERODON KISUMUENSIS (MacInnes, 1942)

(Tables 4.1 and 4.2; Figures 4.1, 4.10, and 4.12)

Partial Synonymy—*Trilophodon angustidens kisumuensis* (in part) MacInnes, 1942:51; plate 3, figures 13, 17, 18; plate 4, figures 5–8; plate 6, figure 7; *Protanancus* (in part) Arambourg, 1945:491; *Trilophodon angustidens kisumuensis* (in part) Leakey, 1967:14; *Platybelodon kisumuensis* (in part) Tobien 1973a:261; *Choerolophodon kisumuensis* Tassy, 1977b:2488; *Afrochoerodon kisumuensis* Pickford, 2001; *Afrochoerodon* aff. *kisumuensis* Pickford, 2005a

The oldest choerolophodont species was originally named as a subspecies of "*Trilophodon* (=*Gomphotherium*) *angustidens*" after the Kenyan county (Kisumu) where its middle Miocene type locality of Maboko is located (MacInnes, 1942). Unfortunately, MacInnes included fossils belonging to *Protanancus* in "*T. a.*" *kisumuensis*. Tobien (1973a) raised *kisumuensis* to species level, but Tassy (1977b) was the first to properly sort the majority of the fossils correctly (see Pickford, 2001; Sanders et al., 2010a), including the type cranium (M 15524), and to recognize their true choerolophodont identity. It is distinguished by the primitive aspect of its craniodental morphology, including molars below the size range of other subfamilials, simple occlusal structure with little or no choerodonty, ptychodonty, or cementum, occasional modest chevroning of half-loph(id)s, M3s with as few as three lophs, large anterior pretrite accessory conules and smaller, irregularly expressed posterior pretrite accessory conules.

The type cranium is fragmentary. Supplemented by a more complete specimen from the middle Miocene site of Cheparawa, Kenya (BAR 219'99; Pickford, 2001), however, it is possible to gain a comprehensive appreciation of the species' cranial morphology. The orbits are positioned low above the maxillary alveolus, the basicranium is short and weakly raised, the face is anteroposteriorly relatively short (upper tusks emerge immediately anterior to the M2s) but steeply inclined, and the zygomatic arches are short, non-flaring, and massive. Temporal fossae are extensive coincident with the close convergence of the temporal lines behind the orbits. Tusk alveoli are widely separated distally. The occipital planum is approximately as wide as it is high, and its width represents the greatest breadth of the cranium.

An undescribed cranium from Buluk, KNM-WS 12874, probably belongs in this species, as well. It has a greatest length from the posterior of the condyles to the anterior of the tusk alveoli of 820 mm. Preservation is good on the ventral side, but it is heavily weathered and eroded on its dorsal aspect (Figure 4.12A). The cranium has M2–3 on both sides and no alveolar sockets anterior to those molars, evidence of horizontal tooth displacement. The M2s are heavily worn; on the right side this molar measures L=78.4 mm, W=59.6 mm. These intermediate molars are trilophed. The M3s have three good lophs each and a heel suggesting a nascent fourth loph (Figure 4.12B). They closely resemble the type M3s of the species from Maboko, Kenya (Figure 4.12C). The third molars are relatively small, L=126.3, W=64.9 mm and L=123.6 mm, W=67.0 mm, right and left sides, respectively. The pretrite anterior accessory conules are very large compared to the posterior accessory conules, and there is a small posttrite posterior accessory conule associated with loph 1. Advancement of the pretrite half-loph ahead of the posttrite half-loph is pronounced in the third loph, constituting chevroning. Molar enamel is thick and there is no cementum.

Pickford (2001) erected *Afrochoerodon* for the early to middle Miocene *kisumuensis* species based on salient differences between crania from the primitive form at Cheparawa and more advanced *Choerolophodon* species from Eurasia. *Afrochoerodon* is also appropriately applied to early Miocene choerolophodonts from the Bugti Beds, Pakistan ("*C.*" *palaeindicus*; Forster Cooper, 1922; Raza

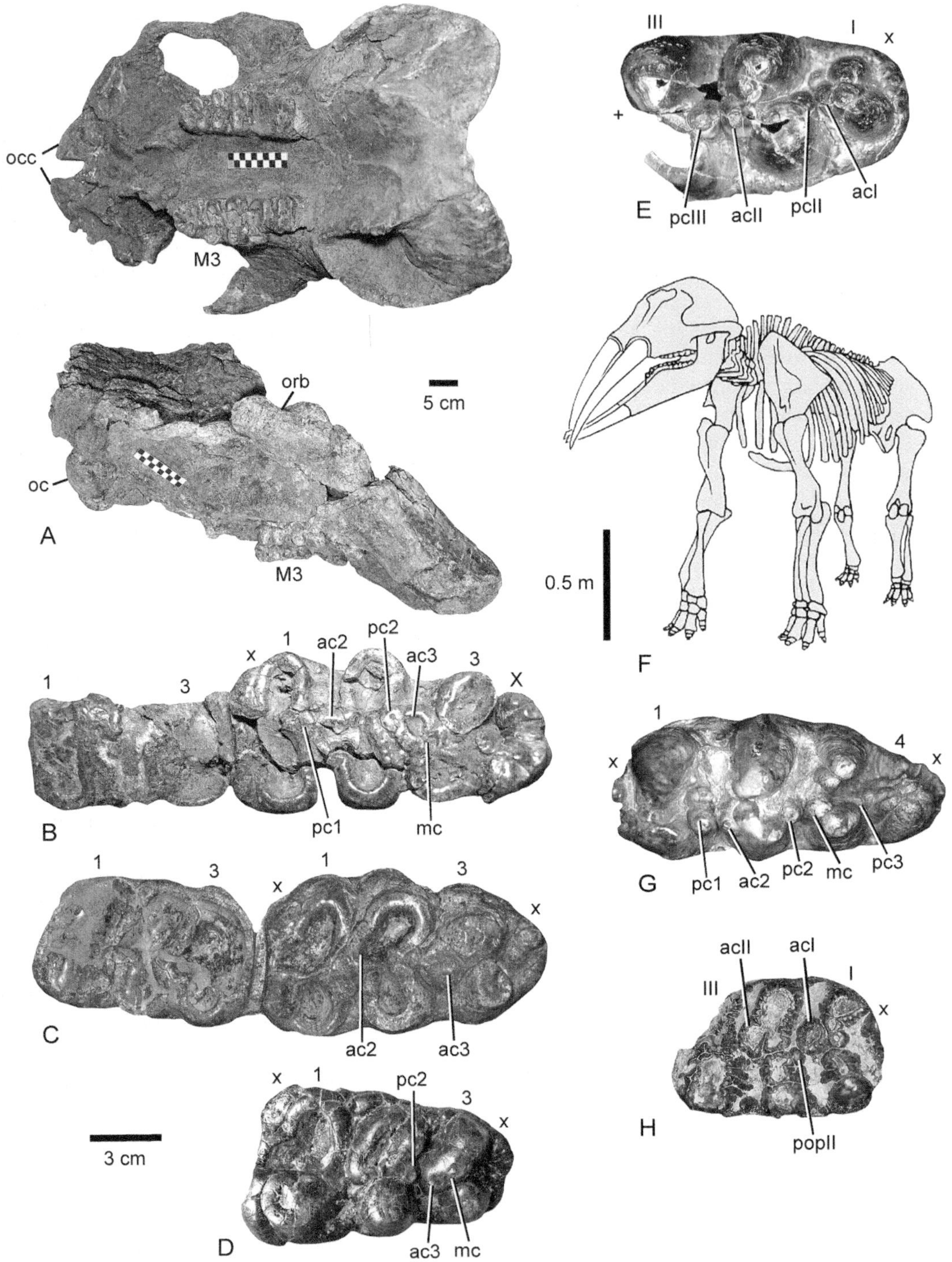

FIGURE 4.12 Cranial and dental specimens of early to middle Miocene African choerolophodonts. Abbreviations: ac, anterior pretrite accessory conule; M, upper molar; mc, mesoconelet; oc, occipital condyle; occ, occipital condyles; orb, orbit; pc, posterior pretrite accessory conule; pop, posterior posttrite accessory conule; x, pre- or postcingulum(id); X, large postcingulum or incipient loph; 1, 2, 3, …, +, indicates missing morphology. loph(id) number counted from the anterior of the crown; I, II, III, …, loph(id) number counted from the posterior of the crown. (A) Anterior to the right; (B–H) anterior to the left. (B–E, G, H) to the same scale. (A) Partial cranium missing much of the upper face and vertex, KNM-WS 12874 from Buluk, Kenya, ventral and lateral views, *Afrochoerodon kisumuensis*. (B) Left M2–3, KNM-WS 12874 from Buluk, Kenya, occlusal view, *Afrochoerodon kisumuensis*. (C) Right M2–3, M 15524 (type) from Maboko, Kenya, occlusal view, *Afrochoerodon kisumuensis*. (D) Right M3, DPC 14584 from Wadi Moghara, Egypt, occlusal view, *Afrochoerodon kisumuensis*. (E) Partial ?left M3, no # from Gebel Zelten, Libya, occlusal view, *Choerolophodon zaltaniensis*. (F) Skeletal reconstruction, oblique lateral view, *Choerolophodon* sp. indet. (dwarfed species) from Ngenyin, Kenya. (Image courtesy of M. Pickford [modified by B. Miljour from Pickford, 2004:fig. 3]). (G) Left m3, KNM-FT 3835 from Ft. Ternan, Kenya, occlusal view, primitive morph *Choerolophodon ngorora*. (H) Nearly complete right M1, KNM-NA from Nakali, Kenya, occlusal view, advanced morph *Choerolophodon ngorora*. Note the folding of the enamel and strong offset of pre- and posttrite half-lophids.

and Meyer, 1984; Tassy, 1985, 1986; Sanders and Miller, 2002) that appear dentally indistinguishable from *A. kisumuensis* (Sanders and Miller, 2002), and early middle Miocene choerolophodonts from Chios Island, Greece, originally placed in *C. chioticus* (Tobien, 1980a; Pickford, 2001), which are the oldest representatives of the subfamily in Europe. The latter species resembles *A. kisumuensis* in the low position of the orbit, convergence of temporal lines, steeply inclined face and downward emergence of upper tusks, and weak expression of cementum, ptychodonty, and choerodonty (Konidaris et al., 2014b).

The most primitive dental morphology of *A. kisumuensis* is found in isolated specimens from the early Miocene site of Wadi Moghara, Egypt (Sanders and Miller, 2002). Included in the modest site sample is a small-sized M3 (L=113.4mm; W=72.6mm) (DPC 14584) showing three lophs, posterior tapering of the occlusal outline, weak cementodonty and ptychodonty, very low, bulbous cusps, thick enamel (5.0–6.0mm), and a low cingular rim along the lingual edge of the crown (Figure 4.12D). Posttrite half-lophs are formed of single conelets; pretrite half-lophs 2 and 3 are angled obliquely and forward of their posttrite half-loph pairs, providing nascent chevroning. Accessory conules posterior to pretrite half-lophs 2 and 3 are modest in size. Trilophed m2, M2, dP3, and dp3 choerolophodont specimens from Wadi Moghara are similar in these aspects to the M3; the m2 lacks cementum and has an even more prominent buccal cingular shelf, as does the dp3.

CHOEROLOPHODON Schlesinger, 1917
CHOEROLOPHODON ZALTANIENSIS Gaziry, 1987a

(Table 4.1; Figures 4.1, 4.10, and 4.12)

Partial Synonymy—*Afrochoerodon zaltaniensis* Pickford, 2001; *Choerolophodon zaltaniensis* Pickford, 2004

The species is named after its occurrence at Gebel Zelten, located in Central Libya. This site has a history of paleontological exploration dating back to the 1930s, though serious fossil collecting was not undertaken until the 1960s (Arambourg, 1961, 1963; Arambourg and Magnier, 1961; Savage and White, 1965; Savage and Hamilton, 1973), initiated by Magnier's 1960 expedition (Magnier, 1962). Initial stratigraphy and mapping of Gebel Zelten was done by Magnier (1962), followed by more intensive stratigraphic, geomorphological, and sedimentological work by Selley (1967, 1969). Fossiliferous depositional environments at the site comprise continental lagoons, estuaries, tidal flats, and fluviatile coastal plains, from which derive an impressive vertebrate fauna, including a few elasmobranch and actinopterygian fish, turtles and a more abundant variety of crocodylians, ratites, herons and storks, sea cows, multiple rodent lineages, lagomorphs, bats, hyraxes, several forms of rhino, multiple suid species, anthracotheres, giraffids and sivatheres, tragulids, climacoceratids, early bovids, creodonts, and carnivorans (canids and felids) (Arambourg, 1961, 1963; Savage and White, 1965; Savage and Hamilton, 1973; Gaziry, 1987a; Mlíkovsky, 2003; Wessels et al., 2003; Agrasar, 2004b; Morales et al., 2010; Sánchez et al., 2010). In addition, a diverse proboscidean assemblage featuring abundant deinotheres, rare mammutids, multiple gomphotheriine species, and choerolophodonts has been recovered from Gebel Zelten (Gaziry, 1987a; Sanders, 2008a). Amebelodonts are notably absent. Biostratigraphic correlation of the fauna suggests an age of ca. 16.5 Ma (Pickford, 1991a), just slightly older than Buluk in East Africa (Table 4.1).

Choerolophodonts are represented at Gebel Zelten by only a small number of molars, first recognized and named by Gaziry (1987a). The initial species sample is all from a single locality, but a few specimens from other localities at Gebel Zelten have since been identified (Sanders, 2008a). Molars of *C. zaltaniensis* are readily distinguishable from those of *Afrochoerodon* by their more complex distribution of conules (choerodonty) and stronger expression of ptychodonty and development of cementum (see Gaziry, 1987a). Chevroning of half-loph(id)s, anterior advancement of pretrite mesoconelets, and oblique alignment of pretrite half-loph(id)s relative to the long axis of the molar crowns are also more advanced in *C. zaltaniensis* over the condition of *Afrochoerodon* molars (Figure 4.12E; Sanders, 2008a). Although in 2001 Pickford relegated *C. zaltaniensis* to *Afrochoerodon* (and also felt that it could be synonymous with eastern African "*A.*" *ngorora*), by 2004 he reversed himself and returned the species to *Choerolophodon*, doubtless because of the progressive qualities of its molar occlusal morphology.

The cranium is unknown for this species, and mandibular remains are too fragmentary to estimate symphyseal angulation or gauge absence of lower tusks, though the symphyseal gutter is broad. Choerolophodont species with more complete mandibles all lack i2s (Tassy, 1985, 1986). Intermediate molars are trilophodont with chevroning limited to the last loph(id), and M3s have four lophs with a low postcingulum formed of four conelets tightly appressed to the last loph; chevroning is well expressed in lophs 2–4. Molar size is modest but greater than in *Afrochoerodon*, with M3s measuring L=138.0mm and W=74.0–76.0mm (Gaziry, 1987a). These molars are very low-crowned (HI=53).

CHOEROLOPHODON SP. INDET. [DWARFED SPECIES]

(Table 4.1; Figures 4.1, 4.10, and 4.12)

Partial Synonymy—*Choerolophodon pygmaeus* Pickford, 2004; *Gomphotherium pygmaeus* (in part) Sanders et al., 2010a

Choerolophodonts from Ngenyin, Kenya pose a nomenclatural nightmare, as they were unfortunately referred to the same species as an enigmatic lower molar from Kabylie, Algeria (FSL 213759; Pickford, 2004) that has its own tortuous taxonomy, starting with "*Mastodon angustidens* mut.

asc. *pygmaeus*" (Depéret, 1897). The use of "*Mastodon*" at that time was partially equivalent to what is now called "*Gomphotherium*." Since then, as summarized by Pickford (2004) and Tassy et al. (2013), FSL 213759 has been referred to as a gomphothere, pygmy gomphothere, palaeomastodont, supernumerary tooth, a nomen dubium, and a choerolophodont (e.g., Osborn, 1936; Tobien, 1973a; Welcomme, 1993; Roger et al., 1994; Sanders et al., 2010a; Tassy et al., 2013). The molar preserves three lophids but likely originally had four lophids. It would be small even if complete, has very diminutive mesoconelets, anteroposteriorly broad transverse valleys, and an abundant covering of cementum. Tobien's (1973a) decision to place the Kabylie molar in *Choerolophodon* was influenced by its development of thick cementum, a taxonomic assignment that was followed by Pickford (2004). To the contrary, Sanders et al. (2010a) failed to see any sign of chevroning on the crown and placed it in *Gomphotherium*. Welcomme (1993) and Tassy et al. (2013) hypothesize that the overall morphology of FSL 213759 is abnormal and therefore that it could represent a supernumerary molar.

CT-scan imaging of the tooth reveals that it does not have choerolophodont-like occlusal morphology (Tassy et al., 2013). Indeed, the transverse alignment of its mesoconelets, small size of the mesoconelets, and development of pretrite anterior and posterior accessory conules more closely resemble the condition in molars of *Gomphotherium*, though the degree of cementum covering does not. If it is not an abnormal molar, it might represent a derived new gomphotheriine (in which case "*Gomphotherium pygmaeus*" would be appropriate), but it certainly is not a choerolophodont. Pickford (2004) and Sanders et al. (2010a) were in error to lump this specimen with the Ngenyin choerolophodont sample—together, they are morphologically heterogeneous (as was pointed out by Tassy et al., 2013), and consequently consideration of the Ngenyin choerolophodonts should be restricted solely to the material from that site. Moreover, molars of the small gomphothere species from Gebel Zelten, Libya that Pickford (2004) placed in *C. pygmaeus* belong instead in *Gomphotherium* and are very different from those of the Ngenyin choerolophodont.

Choerolophodont remains from Ngenyin represent the partial skeleton of a single individual (Figure 4.12F) preserved in lacustrine sediments that had been exposed at the surface prior to being covered by clay and silt. There are few other vertebrate remains at the site, other than fish teeth, small crocodile remains, and pieces of rhinos and bovids (Pickford, 2004). The teeth of this individual are small compared with those of *C. ngorora* (Tassy, 1986), which has also been recovered from Mb. A of the Ngorora Fm. in the Tugen Hills (Table 4.1). An M3, Bar 809'02, measures L = 116.0 mm, W = 60.2 mm, and the m3s (Bar 805'02) have dimensions of L = 123.4–124.0 mm and W = 50.4–51.4 mm (Pickford, 2004). Intermediate molars of this individual are trilophodont and third molars have four loph(id)s. In the third molars, chevroning is modest and limited to the second and third loph(id)s. Choerodonty, ptychodonty, and development of cementum are all moderate. Based on tooth size, Pickford (2004) estimates that this individual was about 1.5 m tall at the shoulder. Postcranial remains include a partial manus and pes, and hindlimb bones, all of which are diminutive in comparison with a typical gomphothere. Small size can be accounted for in part by the juvenile status of this individual. Enamel wear figures are present to the middle of the crown in M2 and third molars were just showing some apical wear at death. The hypothesis that the Ngenyin specimens represent a pygmy choerolophodont species is well supported by the very small size of its molars and postcrania (Pickford, 2004).

CHOEROLOPHODON NGORORA (Maglio, 1974)
(Table 4.1; Figures 4.1, 4.10, and 4.12)

Partial Synonymy—*Gomphotherium ngorora* Maglio, 1974; *G. ngorora* Bishop and Pickford, 1975; *Gomphotherium* Andrews and Walker, 1976:300; *Choerolophodon ngorora* Tassy, 1977b; "*Gomphotherium*" *ngorora* Coppens et al., 1978; *C. ngorora* Tassy, 1986; *C. ngorora* Nakaya, 1993; *Afrochoerodon ngorora* Pickford, 2001; *Afrochoerodon* sp. nov. [specimens from Mbs. A-D, Ngorora Fm.] Pickford, 2004; *C. ngorora* [specimens from Mb. E, Ngorora Fm.] Pickford, 2004; *Choerolophodon* sp., Tsujikawa, 2005b

This choerolophodont species occurs in middle to early late Miocene sites in eastern Africa and is divisible into primitive and advanced morphs, primarily based on successive occurrences within the Ngorora Fm. of the Tugen Hills, Kenya (Tables 4.1 and 5.1). The type material is from the advanced morph, which is easier to distinguish from *Afrochoerodon kisumuensis* and *Choerolophodon zaltaniensis*. Pickford (2001) felt that *C. zaltaniensis* and *C. ngorora* are synonymous, but if so this would only apply to the primitive subsample of *C. ngorora* from the late middle Miocene. In addition, Pickford (2004) hypothesizes speciation between the early and later subsamples of the *C. ngorora* lineage, in which case the advanced morph should retain the species nomen "*ngorora*." The relationships of North African choerolophodonts of similar age to *C. ngorora* (Table 4.1) remain to be established, as they have not been formally described. North Africa may have been the last refuge of choerolophodonts in Africa; the taxon did not survive into the second half of the late Miocene on the continent (Tassy, 1986). Because they developed complex molars with a robust covering of cementum, which should have provided adaptability to more diverse diets, there is no obvious reason for the demise of choerolophodonts in Africa. The timing of their disappearance after 9 Ma may coincide with an abrupt shift in vegetation and increase in C_4 plants soon thereafter (see Cerling et al., 1993; Kingston et al., 1994), and with the emergence of a cohort of new proboscidean rivals, including tetralophodont interlopers from Eurasia, stegodonts from Asia, and elephants.

Choerolophodon ngorora co-occurs with *Protanancus macinnesi* and *Prodeinotherium hobleyi* at middle Miocene Ft. Ternan, Kenya, and choerolophodonts of similar age were recovered with *Gomphotherium*, *Tetralophodon*, *Zygolophodon*, and *Prodeinotherium* at Gebel Cherichera, Tunisia, and with *Tetralophodon* at early late Miocene Djebel Krechem el Artsouma, Tunisia (Tables 4.1 and 5.1), but not with the succeeding late Miocene proboscidean assemblages of elephants, stegodonts, *Deinotherium*, and anancine gomphotheres. The species is distinguished from *A. kisumuensis* by having larger molars with stronger chevroning and more complex occlusal features (Figure 4.12G), and from *C. zaltaniensis* by exhibiting rugose, folded enamel and usually greater development of choerodonty. M3s have four lophs and m3s variably develop a fifth lophid. Intermediate molars are trilophodont. Upper molars tend to conserve a pretrite accessory conule posterior to loph 2, and the anterior pretrite accessory conules tend to remain free of their associated mesoconelets (Tassy, 1986). Cementum may be thick, and enamel tubercles may be present in transverse valleys (see Tassy, 1986:fig. 27). Other molar features of the species include: mesoconelets small but higher than main conelets; anterior accessory conules are diminished or lost, but posterior accessory conules may be massive in size. Deciduous premolars are very similar to those of Eurasian choerolophodonts (Tassy, 1986); for example, dp3 has cusps offset from one another and displays large central accessory conules. Primitive and advanced morphs (Figure 4.12G and H) of the species are primarily differentiated by degree of enamel folding, size, and degree of hypsodonty.

There are no crania known for the species, but upper tusks are curved upward, lack an enamel band, and in cross-section are ovoid to slightly flattened dorsoventrally. The mandibular symphysis is elongate and strongly downturned; however, there is no evidence for lower tusks being deployed in *C. ngorora*. A small sample of postcrania from Ft. Ternan, Kenya has been assigned to the species, with morphological similarities to skeletal remains of Eurasian choerolophodonts (Tassy, 1986). These include carpals, tarsals, and long bones. The femur has a prominent third trochanter and medial epicondyle, and the calcaneum is modest in size but heavily constructed, with a robust tuber calcanei and enlarged fibular facet. The humerus has a reduced trochlear separation between the condyles, and is distinctive for it high, massive epicondyles.

SUMMARY

1. Great climatic, ecological, geological, and faunal turbulence characterized the early to middle Miocene of Africa. Major global climate shifts included a significant early Miocene cooling event, a subsequent climatic optimum of high temperatures between about 17 and 15 Ma, and another major cooling event between 14.8 and 14.1 Ma, with numerous smaller climatic perturbations within these intervals.
2. Opening of a cold circumpolar current around Antarctica led to steeper longitudinal temperature gradients, which likely acted as an ecological catalyst to drive Eurasian mammals southward into Afro-Arabia once geographic conditions permitted these movements.
3. Geologically, substantial uplift of the East African and Ethiopian plateaus initiated, and movement of the Afro-Arabian plate into the Eurasian Plate led to docking of Afro-Arabia with Anatolia. This intercontinental collision permitted repeated episodes of significant immigration of Eurasian mammals into Afro-Arabia, which swamped and in some instances replaced older endemic taxa inhabiting the supercontinent.
4. Afro-Arabian proboscideans appear to have been evolutionarily unfazed by these upheavals and faunal movements, undergoing radiation of elephantimorphs (particularly Elephantida) into a diversity of mammutids, gomphotheriines, choerolophodontines, and amebelodontines. Although proboscidean taxonomic diversity was high during the early and middle Miocene, morphological disparity was reduced, with skeletal bauplans of elephantimorphs very similar.
5. Elephantimorph success was associated with the adaptive advantages of horizontal tooth displacement, which slowed the appearance of molars into the occlusal platform, thereby extending lifespan and perhaps creating opportunities for more breeding cycles. Conversely, the last of archaic proboscideans such as barytheres and primitive elephantiforms appear to have had last, brief appearances at the beginning of the Neogene. Other salient morphological traits of elephantimorphs in the early Neogene include trends for increasing body size, larger teeth, greater tusk hypertrophy, a greater number of loph(id)s in third molars, and greater complexity of half-loph(id) configuration and development of more conelets and accessory conules, increasing grinding surface area and masticatory efficiency of molars.
6. A number of elephantimorph lineages underwent parallel progressive changes that accentuated features of their molar occlusal configurations, with mammutids exhibiting greater development of zygodont crests and crescentoids, and taxa in Elephantida evolving more pronounced pseudoanancoidy, choerolophodonty, chevroning, thicker cementum, and greater distribution of accessory conules throughout their molar crowns. As a result, these lineages are highly useful for biochronological comparison of sites; for example, mammutids evolved through the series *Losodokodon-Eozygodon-Zygolophodon* and gomphotheriid amebelodonts evolved from *Progomphotherium* to archaic *Archaeobelodon*,

and to *Protanancus* via a more progressive form of *Archaeobelodon*. The ages of these lineage segments have become increasingly well-calibrated.

7. Deinotheres, which had appeared in the early Oligocene, thrived in forested conditions of the early and middle Miocene in Africa, evidenced by superb fossil records at a number of localities. Although they lacked the advantages of horizontal tooth displacement, instead employing all of their cheek teeth simultaneously, nonetheless they had become firmly entrenched in a specialized hyper-browsing niche, thus avoiding competition with elephantimorph proboscideans.
7. The early Miocene also witnessed the first extensive emigrations of proboscideans out of Africa to Eurasia, with mammutids and gomphotheres departing prior to the migration of deinotheres off the continent. These taxa quickly made their way throughout the Old World, "seeding" numerous adaptive radiations of proboscideans.
8. Elephants were not yet present in the early and middle Miocene, but the success of transplanted *Gomphotherium* in Eurasia was important for the eventual origins of elephants in Africa. Gomphotheriines comprised the ancestral group for Eurasian tetralophodontines that in turn migrated back into Africa at the close of the middle Miocene, providing the phylogenetic roots for the Elephantidae.

5 Late Miocene
The Rise of Elephants

Nature's great masterpiece, an elephant; the only harmless great thing.

—John Donne, "The Progress of the Soul" (1612)

INTRODUCTION

The late Miocene African proboscidean fossil record is notable for a sharp turnover in taxa, constituting replacement of mammutids as well as the gomphotheriid radiation of early and middle Miocene choerolophodontines, amebelodontines, and gomphotheriines by immigrant tetralophodontines, anancine gomphotheres, stegodonts, and most notably by a novel endemic array of newly evolved elephants. The apparent precipitousness of this replacement is exaggerated by the dearth of late middle and early late Miocene proboscidean sites on the continent (Table 5.1). Proboscideans of the late Miocene are particularly well represented in eastern Africa but reported from most regions of the continent and Arabia (Figure 5.1).

The first phase of this episodic proboscidean replacement was initiated by the immigration into Africa of one or more species of *Tetralophodon*, possibly as early as the latest middle Miocene and certainly by the start of the late Miocene (Table 5.1; Tsujikawa, 2005b; Lihoureau et al., 2015; Geraads et al., 2019), followed in the second half of the late Miocene by the inception of elephants evolving in Afro-Arabia from tetralophodontines, and the contemporaneous arrival from Eurasia of invasive stegodonts, anancine gomphotheres (Petrocchi, 1954; Beden, 1975; Coppens et al., 1978; Tassy, 1986, 2003; Kalb and Mebrate, 1993; Sanders, 1999, 2008b, 2022; Vignaud et al., 2002; Ambrose et al., 2003; Saegusa and Hlusko, 2007; Sanders et al., 2010a), and a possible "sweepstake" dispersal of an advanced amebelodont from Eurasia into North Africa (Gaziry, 1982, 1987b; Sanders, 2008b). The latter immigration episode was facilitated by the beginning of the Messinian Crisis and closing off of the Mediterranean Sea, which provided land bridges between Africa and Eurasia via the Gibraltar Strait and the Gulf of Aden (Rögl, 1999). Conversely, mammutids, gomphotheriines, endemic amebelodontines, and choerolophodontines disappeared from the continent, while *Prodeinotherium hobleyi* evolved into the larger *Deinotherium bozasi* (Harris, 1976, 1978a; Sanders et al., 2010a).

The rise of elephants and the coincident decline of other proboscideans in Africa during the second half of the late Miocene is associated with dramatic climatic, geological, and ecological change. These changes involved a global shift from warm greenhouse to colder ice-house regimes including the establishment and expansion of permanent ice sheets in Antarctica (Zachos et al., 2001), and a steepening of climatic gradients between the poles and equator (Gussone et al., 2004). In southwestern Africa, there is evidence for late Miocene-Pliocene decline in precipitation volume, onset of major aridification, and intensification of seasonality, which has been linked in part with a transfer of the main moisture source for Africa from the Atlantic to the Indian Ocean and initiation of strong Atlantic Meridional Overturning Circulation following shoaling of the Central American Seaway (Dupont et al., 2013). Changes in faunal composition and adaptive strategies of herbivores during the late Miocene were related in part to the transformation of vegetation, which, in turn, was associated with a number of factors, including (a) the latitudinal position of the intertropical convergence zone and its effect on convectional rainfall, (b) the degree of isolation of East Africa from the Atlantic Ocean and shift of moisture content in the region from the Indo-Pacific due to displacement eastward of the Congo Air Boundary, (c) changes in topography (such as the rise of rift shoulders) that cause rain-shadows and orographic rain, and (d) increase in lake-effect rainfall as rifting broadened the size of basins (Linder, 2017).

Especially in eastern Africa, the late Miocene rise of Western and Kenya Rift shoulders and mountains altered rainfall patterns by creating orographic barriers, which would have led to rain-shadows and xerification to the east and high-precipitation areas west of the uplift, adding to local aridification effects of global climate change (Chorowicz, 2005; Spiegel et al., 2007). This would have led to fragmentation of forests and produced more heterogenous landscapes to the east of these barriers, including fostering the spread of C_4 grasslands (Pickford, 1990; Linder, 2017). Increasing rainfall seasonality (featuring warm-season precipitation) favored the spread of C_4 photosynthesizing grasses and grasslands during the second half of the late Miocene, leading to the establishment of partially open grassy woodlands and savanna environments (Cerling et al., 1997b; Strömberg, 2011). Increased seasonality during the latter part of the late Miocene represented a major and fairly abrupt alteration in global terrestrial ecology, particularly in the tropics (Cerling et al., 2005). C_4 plants, notably grasses, have higher water-use efficiency and are more competitive than plants using a C_3 photosynthesizing pathway in dry forests, savannas, grasslands, and deserts, and perform better in ecosystems subject to strong warm-season precipitation (Tipple and Pagini, 2010). They perform better than C_3 plants in conditions

DOI: 10.1201/b20016-5

TABLE 5.1
Major Occurrences of Late Miocene Afro-Arabian Proboscidea

Taxon	Occurrence (Site, Locality)	Stratigraphic Unit	Age	Key References
Proboscidea Illiger, 1811				
Suborder incertae sedis				
Deinotherioidea Bonaparte, 1845				
Deinotheriidae Bonaparte, 1845				
Deinotheriinae Bonaparte, 1845				
Deinotherium bozasi	Tugen Hills, Kenya	Ngorora Fm.	Within 13.0–8.5 Ma interval	Harris (1983), Hill (2002), and Pickford et al. (2009)
	Nakali, Kenya	Nakali Fm.	9.9–9.8 Ma	Harris (1978) and Nakatsukasa et al. (2007)
	Samburu Hills, Kenya	Lower, Upper Mbs., Namurungule Fm.	9.5 Ma	Nakaya et al. (1984), Sawada et al. (1998), and Tsujikawa et al. (2005a, b)
	Shuwaihat 1, Abu Dhabi, United Arab Emirates (type)	Baynunah Fm.	8.0–6.0 interval, probably >6.5 Ma	Whybrow and Hill (1999), Bibi et al. (2013), Tassy (1999), and Sanders (2022)
	Lothagam, Kenya	Lower and Upper Mbs., Nawata Fm.	at least 7.4–5.0 Ma	McDougall and Feibel (2003) and Harris (2003)
(*Deinotherium* sp. nov. A)	Toros Menalla, Chad		7.0–6.0 Ma (alt. 7.0 Ma)	Mackaye (2001) and Le Fur et al. (2014)
	Middle Awash Valley, Ethiopia	Adu-Asa Fm., Kuseralee Mb., Sagantole Fm.	~6.3–5.6 Ma; ~5.6–5.2 Ma	Kalb et al. (1982a, b), Renne et al. (1999), WoldeGabriel et al. (2001), Haile-Selassie et al. (2004)
	Tugen Hills, Kenya	Lukeino and Chemeron (=Mabaget) Fms.	6.2–5.6 Ma; 5.6–1.6 Ma	Bishop et al. (1971), Harris (1977b), Hill et al. (1985, 1986), Deino et al. (2002), and Hill (2002)
	Kanam East, West, Central, Kendu Bay, Kenya	Kanam Fm.	Late Miocene to early Pliocene	Kent (1942), MacInnes (1942), Pickford (1986a), and Ditchfield et al. (1999)
	Bala, Homa Peninsula, Kenya	Kanam Fm.	Late Miocene to early Pliocene	Pickford (1986a)
	Wadi Natrun, Egypt		Late Miocene or early Pliocene	Bernor and Pavlakis (1987) and Pickford et al. (1995)
Elephantiformes Tassy, 1988				
Elephantimorpha Tassy and Shoshani, 1997 in Shoshani et al., 1998				
Mammutida Tassy and Shoshani, 1997 in Shoshani et al., 1998				
Mammutidae Hay, 1922				
Zygolophodon sp. indet.	?Menacer (ex-Marceau), Algeria		?late Miocene	Arambourg (1959), Thomas and Petter (1986), and Pickford (2007b)
Elephantida Tassy and Shoshani, 1997 in Shoshani et al., 1998				
Gomphotheriidae Hay, 1922				
Amebelodontinae Barbour, 1927				
Gen. et sp. indet.	Samburu Hills, Kenya	Lower Mb., Namurungule Fm.	9.5 Ma	Tsujikawa (2005)
Konobelodon cyrenaicus	Sahabi, Libya (type)	Unit U-1, Lower Mb. U, Sahabi Fm.	Late Miocene equivalent to Toros Menalla (alt. early-middle Pliocene)	Boaz (1982), Gaziry (1982, 1987b), Heinzelin and El-Arnauti (1982), Warny et al. (2003), Boaz et al. (2008), Sanders (2008b), and El Shawaihdi et al. (2016)

(*Continued*)

TABLE 5.1 (*Continued*)
Major Occurrences of Late Miocene Afro-Arabian Proboscidea

Taxon	Occurrence (Site, Locality)	Stratigraphic Unit	Age	Key References
Choerolophodontinae Gaziry, 1976				
Choerolophodon ngorora, advanced morph	Tugen Hills, Kenya (type)	Mb. E, Ngorora Fm.	10.5 Ma	Maglio (1974) and Tassy (1986)
	Mbagathi (Kirimun), Kenya	Kirimun Fm.	?early late or	Pickford (1981) and Pickford et al. (2009)
			?middle Miocene (alt.?early Miocene)	Ishida and Ishida (1982) and Tassy (1986)
	Nakali, Kenya	Nakali Fm.	9.9–9.8 Ma	Aguirre and Alberdi (1974), Aguirre and Leakey (1974), Tassy (1986), Pickford (2003a), and Nakatuskasa et al. (2007)
Choerolophodon sp. indet.	Gebel Krechem el Artsouma, Tunisia	Segui Fm.	Late Miocene, ~11 Ma	Geraads (1989)
Tetralophodontinae Van der Maarel, 1932				
Tetralophodon sp.	Bir el Ater 3, Algeria	Nementcha Fm.	Late middle Miocene or early late Miocene	Lihoureau et al. (2015)
Tetralophodon sp. indet.	Gebel Cherichera, Tunisia	Beglia Fm.	~13–11 Ma	Bergounioux and Crouzel (1956) and Robinson (1974)
	?Bled Douarah, Tunisia	Beglia Fm.	~13–11 Ma	Robinson (1974), Robinson and Black (1974), and Geraads (1989)
(Gen. et sp. indet. tetralophodont form)	Tugen Hills, Kenya	Mb. D, Ngorora Fm.; Ngerngerwa Fm.	~11 Ma; 10.5–10.0 Ma	Tassy (1979a, 1986) and Pickford et al. (2009)
(*Stegotetrabelodon* n. sp.?)	Middle Awash Valley, Ethiopia	Chorora Fm.	11.0–10.0 Ma	Tiercelin et al. (1979) and Geraads et al. (2002)
(*Tetralophodon* cf. *longirostris*)	Gebel Krechem el Artsouma, Tunisia	Segui Fm.	Late Miocene, ~11 Ma	Geraads (1989)
	?Gebel Sémène, Algeria		?late Miocene	Bergounioux and Crouzel (1956)
	?Zidania, Morocco		?late Miocene	Coppens et al. (1978)
	?Smendou, Algeria		?late Miocene	Coppens et al. (1978)
	?Sinda River, Democratic Republic of Congo	Sinda Beds	?late Miocene	Hooijer (1963) and Madden (1977, 1982)
Tetralophodon sp.	Skoura, Ouarzazate Basin, Morocco	Upper Mb., Aït Kandoula Fm.	Late Miocene	Zouhri et al., (2012) and Geraads et al. (2019)
Tetralophodon sp. nov.	Samburu Hills, Kenya	Lower, Upper Mbs., Namurungule Fm.	9.5 Ma	Nakaya et al. (1984, 1987), Sawada et al. (1998), and Tsujikawa (2005a, b)
("elephantid")	?Nakali, Kenya	=Kabarsero Fm.	~9.5 Ma	Saegusa et al. (2014)
Tetralophodon sp. indet. ("Elephantidae, primitive form, gen. et sp. incertae sedis")	Kisegi (Kisegi-Nyabusosi area), Uganda	Kakara Fm.	Late Miocene,?~9.0 Ma	Pickford et al, (1993) and Tassy (1995)
Tetralophodon sp. indet.	Hamra 5, Abu Dhabi, United Arab Emirates	Baynunah Fm.	Late Miocene, in the interval ~8.0–6.0 Ma, probably >6.5 Ma	Sanders (2022)
Anancinae Hay, 1922				
Anancus kenyensis	Lothagam, Kenya	Lower and Upper Mbs., Nawata Fm.	7.4–5.0 Ma	McDougall and Feibel (2003) and Tassy (2003)
	Toros Menalla, Chad		7.0–6.0 Ma (alt. 7.0 Ma) (alt. 7.24–6.26 Ma)	Mackaye (2001), Lebatard et al. (2008), and Hautier et al. (2009)

(*Continued*)

TABLE 5.1 (*Continued*)
Major Occurrences of Late Miocene Afro-Arabian Proboscidea

Taxon	Occurrence (Site, Locality)	Stratigraphic Unit	Age	Key References
	Tugen Hills, Kenya	Mpesida Beds	~7.0–6.0 Ma	Tassy (1986) and Kingston et al. (2002)
		Lukeino Fm.	6.2–5.6 Ma	Hill et al. (1985, 1986), Tassy (1986), and Hill (2002)
	Middle Awash Valley, Ethiopia	Adu-Asa Fm. Kuseralee Mb., Sagantole Fm.	~6.3–5.6 Ma ~5.6–5.2 Ma	Kalb and Mebrate (1993), Renne et al. (1999), Haile-Selassie (2001), WoldeGabriel et al. (2001), and Haile-Selassie et al. (2004)
	Lemudong'o, Kenya	Lemudong'o Fm.	6.1–6.0 Ma	Ambrose et al. (2003) and Saegusa and Hlusko (2007)
	Manonga Valley, Tanzania	Ibole Mb., Wembere-Manonga Fm.	~5.5–5.0 Ma	Harrison and Baker (1997) and Sanders (1997)
	Kanam East (type) and West, Kenya	Kanam Fm.	Late Miocene or earliest Pliocene	MacInnes (1942), Tassy (1986), and Ditchfield et al. (1999)
	Kossom Bougoudi, Chad		5.26±0.29 Ma	Mackaye (2001), Lebatard et al. (2008), and Hautier et al. (2009)
Anancus petrocchii	Sahabi, Libya (type)	Upper Mb. U or Mb. V, Sahabi Fm.	Late Miocene equivalent to Toros Menalla (alt. early-middle Pliocene)	Petrocchi (1943, 1954), Boaz (1982), Heinzelin and El-Arnauti (1982), Warny et al. (2003), Boaz et al. (2008), Sanders (2008b), and El Shawaihdi et al. (2016)
	?Toros Menalla, Chad		7.0–6.0 Ma (alt. 7.0 Ma) (alt. 7.24–6.26 Ma)	Mackaye (2001) and Lebatard et al. (2008)
Anancus osiris	Wadi Natrun, Egypt	Qaret el-Muluk Fm.	Late Miocene (alt. early Pliocene) (alt. late Pliocene)	Coppens et al. (1978), Geraads (1982), Thomas et al. (1982), and El-Shahat et al. (1997)
	?Kossom Bougoudi, Chad		5.2 Ma	Mackaye (2001), Lebtard et al. (2008), and Schuster et al. (2009)
Anancus capensis	Langebaanweg, South Africa (type)	Quarry E, Quartzose Sand Mb. and Pelletal Phosphorite Mb., Varswater Fm.	Latest Miocene-earliest Pliocene, ca. 5.0 Ma	Hendey (1981) and Sanders (2006, 2007)
Stegodontidae Osborn, 1918				
Stegodon sp. nov.	Toros Menalla, Chad		7.0–6.0 Ma (alt. 7.0 Ma)	Mackaye (2001), Vignaud et al. (2002), and Le Fur et al. (2009, 2014)
Stegodon kaisensis "Nkondo stage"	Sahabi, Libya	Unit U-1, Sahabi Fm.	Late Miocene equivalent to Toros Menalla (alt. early-middle Pliocene)	Petrocchi (1943, 1954), Boaz (1982), Gaziry (1982, 1987b), Heinzelin and El-Arnauti (1982), Warny et al. (2003), Boaz et al. (2008), Sanders (2008b), and El Shawaihdi et al. (2016)

(*Continued*)

TABLE 5.1 (*Continued*)
Major Occurrences of Late Miocene Afro-Arabian Proboscidea

Taxon	Occurrence (Site, Locality)	Stratigraphic Unit	Age	Key References
(*Stegodon* cf. *S. kaisensis*)	Tugen Hills, Kenya	Mpesida Beds (Upper Mpesida Beds)	6.4 Ma (alt. ~7.0 Ma)	Sanders (1999), Kingston et al. (2002), and Doman (2017)
	Shoshomagai 2, Inolelo 3, Manonga Valley, Tanzania	Ibole Mb., Wembere-Manonga Fm.	ca. 5.5–5.0 Ma	Harrison and Baker (1997)
	Kossom Bougoudi, Chad		~5.3 Ma (alt. 6.0–5.0 Ma) (alt. 5.2 Ma)	Brunet et al. (2000), Zazzo et al. (2000), Brunet (2001), Mackaye (2001), Fara et al. (2005), and Schuster et al. (2009)
	"Kaiso Village," south of the Howa River, probably Nkondo area, Uganda (type)	?Nkondo Fm.	~5.0 Ma	Hopwood (1939), Cooke and Coryndon (1970), Coppens et al. (1978), Sanders (1990), and Tassy (1995)
	Kaiso Central; Kisegi Wasa/N. Nyabrogo; Nyawiega site I, Uganda	Lower Kaiso Fm. (=Nkondo Fm.)	~5.0 Ma (alt. 6.0 Ma)	Cooke and Coryndon (1970), Sanders (1990, 1999), and Pickford et al. (1993)
	Kazinga-Kisenyi area, Kazinga Channel, Uganda	Kazinga Beds	~5.0 Ma	MacInnes (1942), Cooke and Coryndon (1970), Sanders (1990), Pickford et al. (1993), and Tassy (1995)
	Nkondo-Kaiso area, Uganda	Nkondo Fm.	ca. 5.0 Ma	Pickford et al. (1993) and Tassy (1995)
Elephantidae Gray, 1821				
Stegotetrabelodontinae Aguirre, 1969				
Stegotetrabelodon emiratus	Al Gharbia Area, Abu Dhabi, United Arab Emirates (type)	Baynunah Fm.	8.0–6.0 Ma interval, probably >6.5 Ma	Whybrow and Hill (1999), Bibi et al. (2013), Tassy (1999), and Sanders (2022)
Stegotetrabelodon sp. indet.	Toros Menalla, Chad		7.0–6.0 Ma (alt. 7.0 Ma)	Mackaye (2001) and Le Fur et al. (2014)
Stegotetrabelodon syrticus	Cessaniti, southern Italy		Late Tortonian, >7.0 Ma	Ferretti et al. (2003a)
	Sahabi, Libya (type)	Unit U-1, Mb. U, Sahabi Fm.	Late Miocene equivalent to Toros Menalla (alt. early-middle Pliocene)	Petrocchi (1943, 1954), Boaz (1982), Gaziry (1982, 1987), Heinzelin and El-Arnauti (1982), Warny et al. (2003), Boaz et al. (2008), Sanders (2008b), and El Shawaihdi et al. (2016)
Stegotetrabelodon orbus	Lothagam, Kenya (type)	Lower, Upper Mbs., Nawata Fm.; Apak Mb., Nachukui Fm.	At least 7.4–6.5 Ma; 6.5–5.0 Ma; 5.0–4.2 Ma	Maglio (1970a, 1973), Maglio and Ricca (1977), McDougall and Feibel (2003), and Tassy (2003)
	Tugen Hills, Kenya	Mpesida Beds (Upper Mpesida Beds)	~7.0 Ma (alt. 6.4 Ma)	Hill et al. (1985, 1986), Kingston et al. (2002), Pickford et al. (2009), and Doman (2017)
		Lukeino Fm. (Upper Lukeino Fm.)	6.2–5.6 Ma (5.9–5.7 Ma)	Hill et al. (1985, 1986), Tassy (1986), Hill (2002), and Doman (2017)
(*Stegotetrabelodon* sp. indet.)	?Lemudong'o, Kenya	Lemudong'o Fm.	6.1–6.0 Ma	Ambrose et al. (2003) and Saegusa and Hlusko (2007)
(*Stegotetrabelodon* sp. indet.)	Manonga Valley, Tanzania	Ibole Mb., Wembere-Manonga Fm.	~5.5–5.0 Ma	Harrison and Baker (1997) and Sanders (1997)

(*Continued*)

TABLE 5.1 (*Continued*)
Major Occurrences of Late Miocene Afro-Arabian Proboscidea

Taxon	Occurrence (Site, Locality)	Stratigraphic Unit	Age	Key References
(*Stegotetrabelodon* sp. indet.)	Kanam East and Central, Kenya	Kanam Fm.	Late Miocene to early Pliocene	MacInnes (1942) and Ditchfield et al. (1999)
	Nyabusosi, Uganda	Kakara Fm., Lower and Upper Oluka Fm.	Late Miocene	Pickford et al. (1993) and Tassy (1995)
Elephantinae Gray, 1821				
Stegodibelodon schneideri	Toros Menalla, Chad (type)		7.0–6.0 Ma	Coppens (1972), Mackaye (2001), Mackaye et al. (2005), Lebtard et al. (2008), and Schuster et al. (2009)
	Kolinga 1, Chad		Late Miocene or early Pliocene	Coppens (1972) and Mackaye et al. (2005)
Primelephas korotorensis	Lothagam, Kenya (type)	Lower and Upper Mbs., Nawata Fm.	At least 7.4–5.0 Ma	Maglio (1970b, 1973), Maglio and Ricca (1977), Leakey et al. (1996), McDougall and Feibel (2003), and Tassy (2003)
	Toros Menalla, Chad		7.0–6.0 Ma (alt. 7.0 Ma)	Brunet et al. (2000), Mackaye (2001), Lebtard et al. (2008), and Mackaye et al. (2008)
("cf. *Stegodibelodon schneideri*") ("*Stegodon* cf. *St. kaisensis*")	Adu Dora North, Adu Dora South, Saitune Dora, Middle Awash Valley, Ethiopia	Adu Mb., Adu-Asa Fm.	6.2 to ≥5.77 Ma	Kalb and Mebrate (1993), WoldeGabriel et al. (2001), Saegusa and Haile-Selassie (2010), and Sanders et al. (2010a)
	"Galili Area," Mulu Basin, Ethiopia	Asa Koma Mb., Adu-Asa Fm.	5.77–5.44 Ma	Haile-Selassie (2000) and WoldeGabriel et al. (2001)
(*P. korotorensis saitune*)	Saitune Dora, Middle Awash Valley, Ethiopia	Asa Koma Mb., Adu-Asa Fm.	5.77–5.44 Ma	Haile-Selassie and WoldeGabriel (2009) and Saegusa and Haile-Selassie, (2009)
(*P. korotorensis gomphotheroides*)	Asa Koma, Alayla, Amba East, Adu Dora, Kuseralee Dora, Middle Awash Valley, Ethiopia	Asa Koma Mb., Adu-Asa Fm. and Kuseralee Mb., Sagantole Fm.	5.77–5.44 Ma; 5.2 Ma	Haile-Selassie and WoldeGabriel (2009), Hart et al. (2009), and Saegusa and Haile-Selassie (2009)
	Manonga Valley, Tanzania	Ibole Mb., Wembere-Manonga Fm.	~5.5–5.0 Ma	Harrison and Baker (1997) and Sanders (1997)
	Kossom Bougoudi, Chad		5.2 Ma	Mackaye (2001), Mackaye et al. (2008), and Schuster et al. (2009)
	Nyabusosi, Uganda	Lower Oluka Fm.	Late Miocene	Pickford et al. (1993) and Tassy (1995)
	North and South Nyabrogo, Uganda	Lower horizons, Kaiso Fm.	Late Miocene-early Pliocene	Cooke and Coryndon (1970)
Loxodonta cookei (*Loxodonta* nov. sp. A de Lukeino)	?Toros Menalla, Chad		7.0–6.0 Ma (alt. 7.0 Ma) (alt. 7.24–6.26 Ma)	Mackaye (2001) and Lebatard et al. (2008)
	Tugen Hills, Kenya	Lukeino Fm.	6.2–5.6 Ma	Hill et al. (1985, 1986), Tassy (1986), Hill (2002), and Sanders (2007)
		Lower Chemeron Fm. (=Mabaget Fm.)	5.3–4.0 Ma	Hill et al. (1985, 1986) and Deino et al. (2002)
	Nkondo-Kaiso area, Uganda	Nkondo Fm.	~5.0 Ma (alt. ~6.0 Ma)	Pickford et al. (1993) and Tassy (1995)
	Langebaanweg, South Africa (type)	Quarry E, Quartzose Sand Mb., Varswater Fm.	Latest Miocene or earliest Pliocene, ~5.0 Ma	Hendey (1981) and Sanders (2006, 2007)

(*Continued*)

TABLE 5.1 (*Continued*)
Major Occurrences of Late Miocene Afro-Arabian Proboscidea

Taxon	Occurrence (Site, Locality)	Stratigraphic Unit	Age	Key References
cf. *Loxodonta* sp.	Amba and Kuseralee, Middle Awash Valley, Ethiopia	Kuseralee Mb., Sagantole Fm.	5.2 Ma	Saegusa and Haile-Selassie (2009)
Mammuthus subplanifrons	Tugen Hills, Kenya	Lukeino Fm.	6.2–5.6 Ma (alt. 6.06–5.6 Ma)	Hill et al. (1985, 1986), Tassy (1986), Hill (2002), Sanders (2007), and Doman (2017)
		?Kuseralee and Aramis Mbs., Sagantole Fm.	5.77–5.18 Ma (alt. ca. 4.4 Ma) (alt. 5.2 Ma)	Mebrate (1983), Kalb and Mebrate (1993), Renne et al. (1999), and Haile-Selassie (2001)
	Langebaanweg, South Africa	Quarry E, Quartzose Sand Mb., Varswater Fm.	Latest Miocene-early Pliocene, ca. 5.0 Ma	Maglio and Hendey (1970), Hendey (1981), and Sanders (2006, 2007)
	?Nyawiega, Uganda	Lower Kaiso Beds (=Nkondo Fm.)	ca. 5.0 Ma (alt. 6.0 Ma)	Cooke and Coryndon (1970) and Pickford et al. (1993)
	?Nkondo-Kaiso area, Uganda	Lower Kaiso Beds (=Nkondo Fm.)	ca. 5.0 Ma (alt. 6.0 Ma)	Pickford et al. (1993) and Tassy (1995)
?*Mammuthus* sp. indet. probably nov. ("*Loxodonta* nov. sp. C")	Toros Menalla, Chad		7.0–6.0 Ma (alt. 7.0 Ma) (alt. 7.24–6.26 Ma)	Mackaye (2001) and Lebtard et al. (2008)

?=attribution, age, or occurrence uncertain; alt., alternative; alternate taxonomic names are in parentheses.

of high temperatures, greater irradiance, low atmospheric CO_2, high salinity, aridity, and waterlogging, and therefore are better adapted for tropical-subtropical lowlands (Sage, 2004; Strömberg, 2011). In contrast, C_3-dominated grasslands dominate at higher latitudes and altitudes, and in ecosystems with cool-season precipitation (Strömberg, 2011).

In eastern Africa in particular, a significant spread of plants using a C_4 photosynthesizing pathway is noted from about 10 Ma and expanded substantially in importance in the Plio-Pleistocene with increasing aridification (Cerling et al., 1997b; Jacobs et al., 1999; Uno et al., 2011; Ségalen et al., 2007; Levin, 2015). Prior to 10 Ma, African flora is interpreted as mainly composed of non-rainforest trees, shrubs, and C_3 grasses, with no C_4 grasses at a regional scale (Linder, 2017), although some East African early and middle Miocene sites locally may have fostered substantial amounts of C_4 grasses (e.g., Lukens et al., 2017). The success of C_4 plants in the late Miocene of Africa involved not only the physiological advantage of C_4 grasses under low atmospheric CO_2 to replace C_3 grasslands, greater seasonality and the flammability of C_4 grasses promoted fires that facilitated the invasion of forests and the spread of fire-resistant savanna trees (which helped expand the places into which savannas can penetrate) (Schreiter et al., 2012).

C_4 grasses date to the Paleogene and evidently arose multiple times (Strömberg, 2011). Although evidence from isotopic investigations of paleosol carbonates shows that C_4 plants were part of local ecosystems in eastern Africa by or prior to the first half of the late Miocene, and that C_4 plant biomass was not insignificant during the African late Miocene, grasslands and savannas did not become truly prominent or regionally primary ecosystems until the early Pleistocene (Cerling, 1992). The earliest evidence for grasslands in East Africa is documented at early Miocene Bukwa, Uganda (Linder, 2017). It is important to note that these changes in plant cover were not the same everywhere in Africa or even in the greater Turkana Basin region during the late Miocene. For example, the lower Omo Valley of Ethiopia remained more forested than other locations in the Turkana Basin during the late Miocene and early Pliocene (Bobe, 2006). And, not all proboscideans reacted to the availability of new C_4 resources in the same manner everywhere, as can be seen by the preference of elephants for browse in the late Miocene Mpesida Beds and Lukeino and lower Chemeron (or Mabaget) fms. of the Tugen Hills, Kenya (Roche et al., 2013; Doman, 2017), even as conspecifics elsewhere and sympatric anancine gomphotheres were grazing (e.g., Levin et al., 2008).

Late Miocene fossil mammal sites are documented in eastern, northern, southern, and Central Africa, as well as in Arabia, but the most common, securely dated, and intensely studied mammal-bearing occurrences of the time interval are East and Central African (Figure 5.1; Table 5.1). Mammalian communities were comprised of Families characteristic of modern African faunas, though many extant genera had not yet appeared by the late Miocene. Nonetheless, there was a major mammalian faunal upheaval in Africa from the early to late parts of the late Miocene.

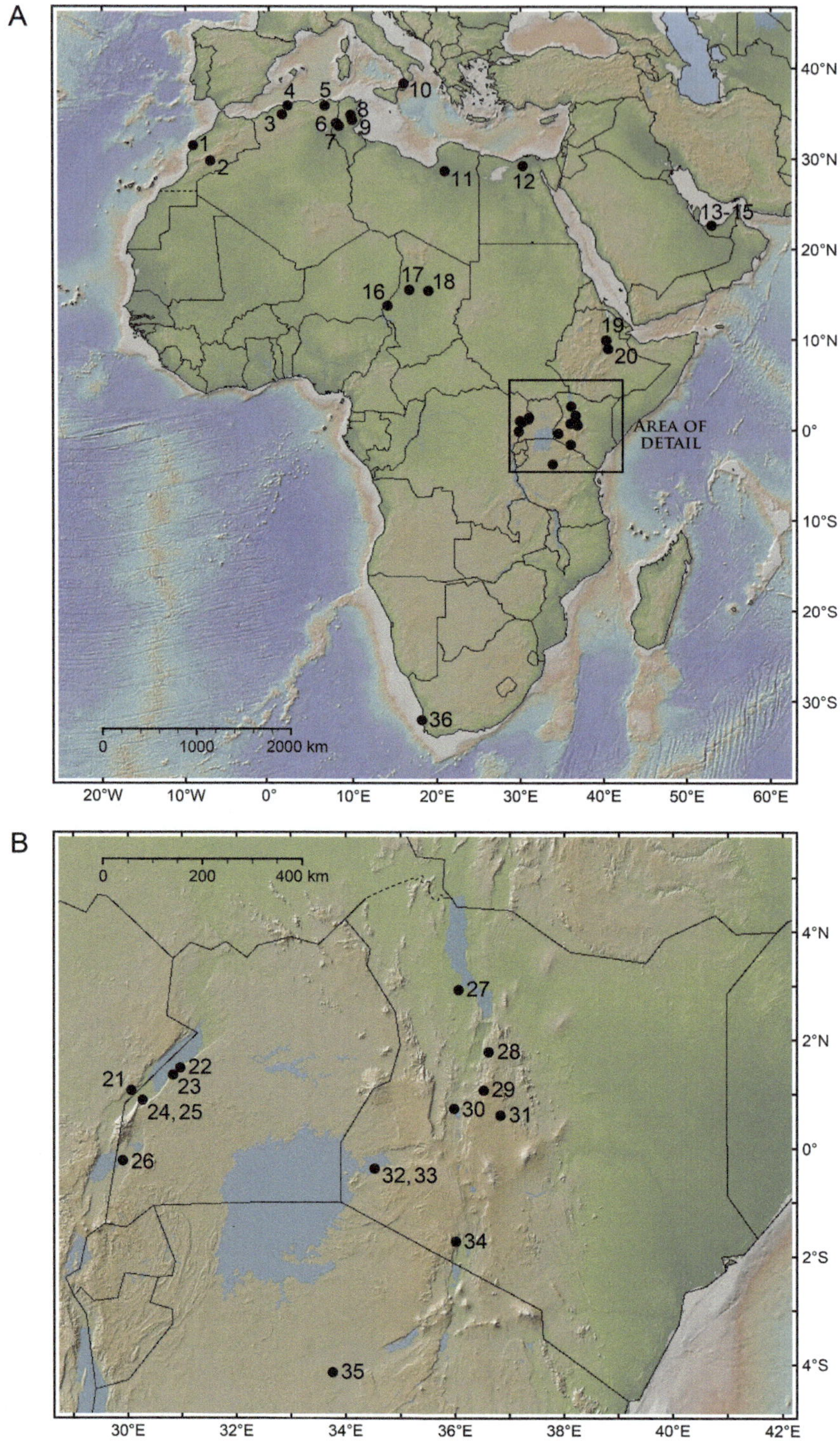

FIGURE 5.1 Maps showing the location of late Miocene sites, (A, B). Area of detail, (B). 1, Zidania, Morocco; 2, Skoura, Ouarzazate Basin, Morocco; 3, Gebel Sémène, Algeria; 4, Menacer (ex-Marceau), Algeria; 5, Smendou, Algeria; 6, Bir el Ater 3, Algeria; 7, Bled Douarah, Tunisia; 8, Gebel Cherichera, Tunisia; 9, Gebel Krechem el Artsouma, Tunisia; 10, Cessaniti, southern Italy; 11, Sahabi, Libya; 12, Wadi Natrun, Egypt; 13, Shuwaihat 1, Abu Dhabi, United Arab Emirates; 14, Hamra 5, Abu Dhabi, United Arab Emirates; 15, Al Gharbia Area, Abu Dhabi, United Arab Emirates; 16, Kolinga 1, Chad; 17, Toros Menalla, Chad; 18, Kossom Bougoudi, Chad; 19a, Middle Awash Valley, Ethiopia; 19b, Amba and Kuseralee, Middle Awash Valley, Ethiopia; 19c, Saitune Dora, Middle Awash Valley, Ethiopia; 19d, Asa Koma, Alayla, Amba East, Adu Dora, Kuseralee Dora, Middle Awash Valley, Ethiopia; 20, "Galili Area," Mulu Basin, Ethiopia; 21, Sinda River, Democratic Republic of Congo; 22a, "Kaiso Village," south of the Howa River, probably Nkondo area, Uganda, b, Nkondo-Kaiso area, Uganda; 22c, Kaiso Central, Kisegi Wasa/N. Nyabrogo; Nyawiega site I, Uganda; 23, Nyawiega, Uganda; 24a, Kisegi (Kisegi-Nyabusosi area), Uganda, b, Nyabusosi, Uganda; 25, North and South Nyabrogo, Uganda; 26,

(*Continued*)

FIGURE 5.1 *(Continued)* Kazinga-Kisenyi area, Kazinga Channel, Uganda; 27, Lothagam, Kenya; 28, Samburu Hills, Kenya; 29, Nakali, Kenya; 30, Tugen Hills, Kenya; 31, Mbagathi (Kirimun), Kenya; 32, Bala, Homa Peninsula, Kenya; 33a, Kanam East, West, Central, Kendu Bay, Kenya, b, Kanam East and West, Kenya, c, Kanam East and Central, Kenya; 34, Lemudong'o, Kenya; 35a, Manonga Valley, Tanzania, b, Shoshomagai 2, Inolelo 3, Manonga Valley, Tanzania; 36, Langebaanweg, South Africa.

The Nawata Formation at the Kenyan site of Lothagam provides a representative sampling of mammalian faunal composition and diversity in the latter half of the late Miocene, including bovids (such as the ancient impala *Aepyceros* and boselaphin *Tragoportax*; in total there are at least nine bovid tribes and 17 species in the Lothagam sequence [Uno et al., 2011]), suids such as *Nyzanzachoerus*, hexaprotodont hippos (*Archaeopotamus*), cercopithecoid monkeys (predominantly *Parapapio*), equids (*Eurygnathohippus*), rhinos (*Brachypotherium* was more common than the more derived genera *Ceratotherium* and *Diceros*), carnivores (including archaic amphicyonids), deinotheres, anancine gomphotheres, and some of the most primitive and oldest elephants (*Stegotetrabelodon orbus* and *Primelephas korotorensis*) (Cerling et al., 2005; Bobe, 2011). It is during this interval in the second half of the late Miocene that hominins also first appeared (Pickford and Senut, 2001; Senut et al., 2001; Brunet et al., 2002; Haile-Selassie and WoldeGabriel, 2009), perhaps adaptively advantaged by environmental changes that were accelerated or promoted by the activities of elephants (Sanders, 2020).

Isotopic sampling of dental enamel of mammals from Lothagam, as well as from older late Miocene sites such as the Samburu Hills, Kenya, and taxa from early and middle Miocene African localities, reveals a major late Miocene faunal turnover and shift from animals with predominantly browse-based to largely C_4-grass-based diets, in concert with the opening of habitats and greater availability of this food resource (Cerling, 1992; Cerling et al., 1997b, 1999, 2005; Bobe, 2011; Leakey et al., 2011; Uno et al., 2011; Saarinen, 2019). Equids (by 9.9 Ma) and rhinos (*Ceratotherium*; by 9.6 Ma) were the earliest taxa to significantly exploit C_4 graze, followed by hippos in the *Archaeopotamus* lineage and some bovids by 7.4 Ma; primitive elephants, at least in some areas, became dedicated grazers by 6.5 Ma (and were eating a substantial amount of C_4 graze before then), and suids in the *Nyanzachoerus-Notochoerus* lineage adopted C_4-dominated diets between 6.5 and 4.2 Ma (Kingston, 1999; Cerling et al., 2005; Bobe, 2011; Uno et al., 2011). The early commitment to grazing by mega-herbivore keystone taxa, particularly elephants, coincided with and may have contributed to opening up heterogenous habitats that favored bipedality and terrestrial foraging strategies of the earliest hominins (Sanders, 2020).

Evolution of two morphological complexes or features were critical to the successful transition from a browsing ancestry to grazing in elephants. The first evolutionary innovation was the reorganization of cusps, conelets, and conules that are arranged within half-loph(id)s of gomphotheriid molars into transverse lamellae or "plates" by fusion of these structures, consolidated together by cementum (Lister, 2013; Saegusa, 2020). Roth (1989) detailed the derived process by which elephant molars develop: (a) dental epithelium of cheek teeth anticipate lamellae (or "plates") separated by transverse valleys by forming a series of deep folds or pockets; (b) lamellae begin as individual unconnected hollow units of enamel surrounding dentine and a pulp cavity, growing from the apex downward; (c) lamellae thicken, fill with dentine, and unite at their bases; and (d) lamellae are consolidated into a functional cheek tooth by an external covering of cementum. Masticatory action of these molars was transformed from Phase I compression of the food bolus and Phase II transverse grinding to a mechanism that de-emphasized Phase I and accentuated horizontal shearing in fore-aft proal jaw motion, evenly across flat occlusal surfaces (Saegusa, 2020).

The second set of innovations defining elephants comprised evolutionary modifications of the cranium and mandible associated with driving this masticatory mechanism, and include morphological features that enhance uniform occlusal force across the molars and provide a powerful anterior power stroke for the lower jaw (Maglio, 1972a). Changes to the elephant skull from ancient gomphotherine morphology generally involved higher and more anteroposteriorly shortened crania, loss of lower tusks, and foreshortening of mandibular symphyses (Maglio, 1973). The consequence of these changes moved the center of mass of the cranium posteriorly, vertically aligning the temporalis muscle with the center of mass, where it can more effectively maintain constant even pressure on molars in occlusion and does not interfere with the action of the superficial masseter and pterygoid muscles driving the forward power stroke of the lower jaw (Maglio, 1972a, 1973).

Modifications of the elephant cranium appear to have preceded those of the cheek teeth (Sanders et al., 2021). For this reason, the feeding behavior and dental adaptations of late Miocene elephants were "out of phase" (Lister, 2013). Although grit from feeding low to the ground and opal phytoliths characteristic of C_4 grasses adaptively favor the evolution of hypsodont molars with greater numbers of lamellae and abundant cementum, earliest elephant dentitions were not substantially different from those of *Tetralophodon* (Sanders, 2022); there are not many more lamellae than loph(id)s, cementum covers but does not completely infill molar transverse valleys, molar crowns are brachyodont, lamellae are widely spaced anteroposteriorly and remained pyramidal in shape in sagittal cross-section, symphyses are very to moderately elongate, and stegotetrabelodontine elephants retained long lower incisors (Maglio, 1973; Sanders et al., 2010a). Molars of this kind are usually associated with browsers, not grazers. More advanced elephants that appear in the Pliocene, with higher-crowned molars formed of a greater number of lamellae, thicker cementum, and higher lamellar frequencies (and consequently more cutting surfaces available for mastication) were morphologically more coordinated with their feeding behavior (Lister, 2013). Also, the cranial adaptations predicted by Maglio (1972a) to have evolved in elephants for grazing are difficult to confirm in the earliest species, as no complete elephant crania have been recovered from late Miocene sites (Sanders et al., 2021).

Nonetheless, the adaptations observed in Miocene elephants reveal an aspect of life history that would have competitively favored elephants over most other proboscideans—a consequence of raising the cranium for reorientation of muscle fibers would be to provide greater vertical space for development of longer molars that could be deployed more slowly into occlusion, thus extending lifespans and increasing the number of breeding cycles. Cranial architecture in some species of *Tetralophodon* suggests that this arrangement predated the origin of elephants.

In addition, evolution of a "block-like molar crown of parallel lamellae held together by cement[um]" (Lister, 2013:332) was not only a defining adaptation of elephants, but also evolved independently in Stegodontidae along with proal jaw motion, around the same time as in elephants, though not in association with feeding on C_4 grasses and perhaps not with grazing (Saegusa, 2020). Despite possessing this adaptation, stegodonts apparently lacked the genetic and developmental capacity to evolve molar hypsodonty, and although their morphology was sufficient to coexist with primitive elephants in the African late Miocene, this inability was evidently a key factor in their demise on the continent during the Pliocene as they began to compete with more advanced elephants and anancine gomphotheres for graze (Sanders et al., 2010a). Similarly, Saegusa (2020) provided a plausible hypothesis for anancine extinction in Africa during the early Pleistocene, suggesting that the organization of their molar half-loph(id)s was less efficient than the molar plates of elephants as a grazing mechanism, at a time when numerous mammals were competing for grasses (Sanders et al., 2010a; Sanders, 2020).

SYSTEMATIC PALEONTOLOGY

DEINOTHERIIDAE Bonaparte, 1845
DEINOTHERIINAE Bonaparte, 1845
DEINOTHERIUM Kaup, 1829
DEINOTHERIUM BOZASI Arambourg, 1934a

(Tables 5.1, 5.2, 6.1, and 7.1; Figures 3.9, 5.1, and 5.2)

Partial Synonymy—*Deinotherium giganteum* Joleaud, 1928; *D. bozasi* Arambourg, 1934a, b; *D. hopwoodi* Osborn, 1936; *D. giganteum* var. *bozasi* Dietrich, 1942; *D. bozasi* MacInnes, 1942; *Prodeinotherium* sp. Nakaya et al., 1984; *Deinotherium* sp. nov. A Mackaye, 2001

Deinotherium bozasi was named in honor of Robert Du Bourg de Bozas, who led an expedition to the Omo, Ethiopia in the early 1900s to study the geography, ethnography, and zoology of the region. The first fossil of the species reported was an isolated upper molar from the Omo, incorrectly identified by Haug (1911) as a mandibular molar. Initially, it was attributed to the European species *D. giganteum* by Joleaud (1928), and was not recognized as a part of a separate African species until later (Arambourg, 1934a, b). Although Arambourg (1947:249) mentioned this molar as the type specimen, it is now missing. A mandible with tusks and molars from the Omo, Ethiopia has been designated as a lectotype replacement for this specimen and is housed in the Museum National d'Histoire Naturelle in Paris (see Harris, 1976).

The temporal extent of *D. bozasi* ranges from the late Miocene to the end of the early Pleistocene (Tables 5.1, 6.1, and 7.1), succeeding that of *Prodeinotherium hobleyi*, from which it is presumed to have evolved (Harris, 1976, 1978a). The first appearance of *D. bozasi* is in the early part of the late Miocene, either in the Ngorora Fm. in the Tugen Hills, Kenya or at Nakali, Kenya (Figure 5.2A; Harris, 1978a; 1983b; Sanders et al., 2010a). *Deinotherium bozasi* is not documented outside of Afro-Arabia and appears to be more closely connected phylogenetically with *P. hobleyi* than with other species of *Deinotherium*, posing an unresolved systematic and classificatory nightmare (Harris, 1983b). Unlike its predecessor *P. hobleyi*, *D. bozasi* is known from much more limited site-samples, mostly comprised of isolated teeth and readily recognizable tooth fragments. Primarily recovered from East African localities, the species has been documented as far south as Malawi (Bromage et al., 1995) and Mozambique (Harris, 1977a), in Central Africa (Mackaye, 2001), and apparently made it into North Africa, as well (Remy, 1976).

The morphology of *D. bozasi* has been well described by Harris (1976, 1978a, 1983b). It is distinguished from *P. hobleyi* by its larger size, including dentition (Figure 3.9), a longer, relatively higher cranium with a shorter, narrower vault, more laterally projecting orbits, smaller rostral protuberances, and greater projection of paroccipital processes. *Deinotherium giganteum*, of similar skeletal magnitude, is claimed to have reached body masses up to 20,000 kg, according to Christiansen (2004). Cranial length of *D. bozasi* exceeds 1,100 mm. The mandibular symphysis is relatively shorter than in *P. hobleyi* but more abruptly downturned, and its lower tusks are curved slightly more below the symphysis. In addition, the nasal aperture of *D. bozasi* is more retracted than in *P. hobleyi*, its superiormost margin reaching above and behind the position of the orbits, and the rugosity of the rostral walls of its maxillae extend onto the cranial vault, features indicative of the possession of a more powerful trunk. According to the compelling reconstruction of deinothere facial morphology by Markov and colleagues (2001), however, it is unlikely that the trunks of these proboscideans were long and elephant-like. In their reconstruction, the lower lip was located below the tusks (see Figure 5.2B); the tusks were used to pull foliage close to the face, where the proboscis could place it in the mouth; a long muscular tongue then manipulated the food in position for mastication.

Despite these differences, the overall skull morphology of *D. bozasi* is closer to *P. hobleyi* than to that of *Deinotherium* species outside of Afro-Arabia. For example, its cranium rostrum is more steeply downturned than in *D. giganteum*, and its nasal aperture and rostral trough are narrower (Sanders et al., 2010a). To Harris (1976), this indicated an ancestral-descendant relationship of *P. hobleyi-D. bozasi*.

Convergent on elephantimorph proboscideans, deinotheres, including *D. bozasi*, possess skeletal adaptations for graviportal support of their massive bodies, including pillar-like limb bones, relatively short manus and pes

TABLE 5.2
Adult Dental Formulae and Incisor Dominance in Late Miocene African Proboscidea

Taxon	Upper Permanent Dentition	Largest Upper Incisor	Lower Permanent Dentition	Largest Lower Incisor
Deinotherium bozasi	P3–4, M1–3	—	i2, p3–4, m1–3	i2
Konobelodon cyrenaicus	I2, ?P3–4, M1–3	I2	i2, ?p3–4, m1–3	i2
Choerolophodon ngorora (advanced morph)	I2, ?P, M1–3	I2	?i, ?p, m1–3	?
Anancus kenyensis	I2, ?P, M1–3	I2	?p, m1–3	—
Tetralophodon spp.	I2, P3–4, M1–3	I2	i2, p3–4, m1–3	i2
Stegodon kaisensis	I2, M1–3	I2	m1–3	—
Stegotetrabelodon spp.	I2, P3–4, M1–3	I2	i2, p3–4, m1–3	i2
Stegodibelodon schneideri	I2,?P, M1–3	I2	?p, m1–3	—
Primelephas korotorensis	I2, P3–4, M1–3	I2	p3–4, m1–3	—
Loxodonta cookei	I2, P3–4, M1–3	I2	p3–4, m1–3	—
Mammuthus subplanifrons	I2, ?P, M1–3	I2	?p, m1–3	—

? = presence uncertain

elements, compact lumbar regions, and broad innominates (Figure 5.2B). However, there are several differences in the postcranium unique to deinotheres, such as reversed curvature of the anterior cervical vertebral zygapophyses to facilitate greater maneuverability and downward flexion of the skull (perhaps associated with the orientation and function of the tusks), and an astragalus with a prominent posteromedial projection, squared in outline and narrow, with a convex tibial facet that has a long medial edge and a concave navicular facet (Harris, 1978a, 1983b). Proportions of the metapodials and their inferred orientation in *D. bozasi* suggest that it was a more cursorial animal than *P. hobleyi* (Harris, 1983b).

The dentition of *D. bozasi* is little different from that of earlier deinotheres, demonstrating the morphological and evolutionary conservatism of the group. All adult teeth (Table 5.2) were in occlusion at the same time without being ejected from the cheek tooth row (Figure 5.2C), and the strong ectoloph(id)s of premolars, which acted as guides to restrict lateral motion of the dentition in occlusion, contrast with the distinctive lophodonty of the molars. Reflecting these distinctions, the deinothere dentition has been referred to as "bilophodont-bifunctional," with dP2/dp2-dP3/dp3 acting as the anterior dental battery and dP4/dp4 as the posterior battery in juvenile individuals, and P3/p3-P4/p4 as the anterior battery (Figure 5.2D and E) and M1/m1-M3/m3 as the posterior battery in adults (Harris, 1976, 1978a). Postcingulae(ids) of M2–3 are generally anteroposteriorly more restricted than in *P. hobleyi*. The ectolophid in p3 of *D. bozasi* is better developed than in *P. hobleyi* (Figure 5.2D; Harris, 1976). Molars are very low-crowned and lophodont, with "tapiroid" cutting facets that were initially subvertical in orientation and extended across the crown (Figure 5.2F and G); deciduous fourth premolars and first molars are trilophodont. As established by von Koenigswald (2014, 2016a), the beveled cutting edges of molar loph(id)s were used primarily in Phase I vertical shearing; these edges constantly re-honed themselves by their action through the life of an individual (Harris, 1975). Stable isotope analyses of dental enamel confirm that with few exceptions (see Lohrke, 2017), *D. bozasi* maintained a C_3 browsing diet from its origin in the late Miocene until its extinction in the early Pleistocene (Cerling et al., 1999; 2015; Levin et al., 2008; Uno et al., 2011), contrasting with the increasing commitment to C_4 grazing in African elephantimorphs over the same interval (Cerling et al., 2005). The persistence of deinotheres in Africa until the close of the early Pleistocene (Table 7.1) was much younger than in South Asia or Europe, suggesting the retention of favored wooded or forested areas (perhaps gallery forests along rivers) despite overall trends for greater aridity and more open habitats in eastern Africa.

Mackaye (2001) erected a new, unnamed species for the deinothere teeth recovered from Toros Menalla, Chad. The main traits enumerated in support of this classification are: small size of the dentition compared with *D. bozasi*, lack of development of an entoloph between the protocone and hypocone in P3 and P4, and absence of a metaloph in P3 and P4. The evidence for differentiating the Toros Menalla specimens from *D. bozasi*, however, is weak. As pointed out by Harris (1983b), deinothere dentition increased in size over time, and while the Toros Menalla cheek teeth are small compared with early Pleistocene deinothere premolars and molars, they are in the size range of earlier samples of *D. bozasi* from sites such as Lothagam and Kanapoi (Harris, 2003; Sanders, 2020). The normal condition for P3–4 in deinotheres is for the protocone and hypocone to be separated. P3 often lacks a fully realized metaloph in deinotheres, and the transverse ridge extending from the hypocone toward the metocone in the P4 of *Deinotherium* may not comprise a complete loph. For now, it is preferable to envision *D. bozasi* as having evolved anagenetically and subsume Mackaye's (2001) species into it.

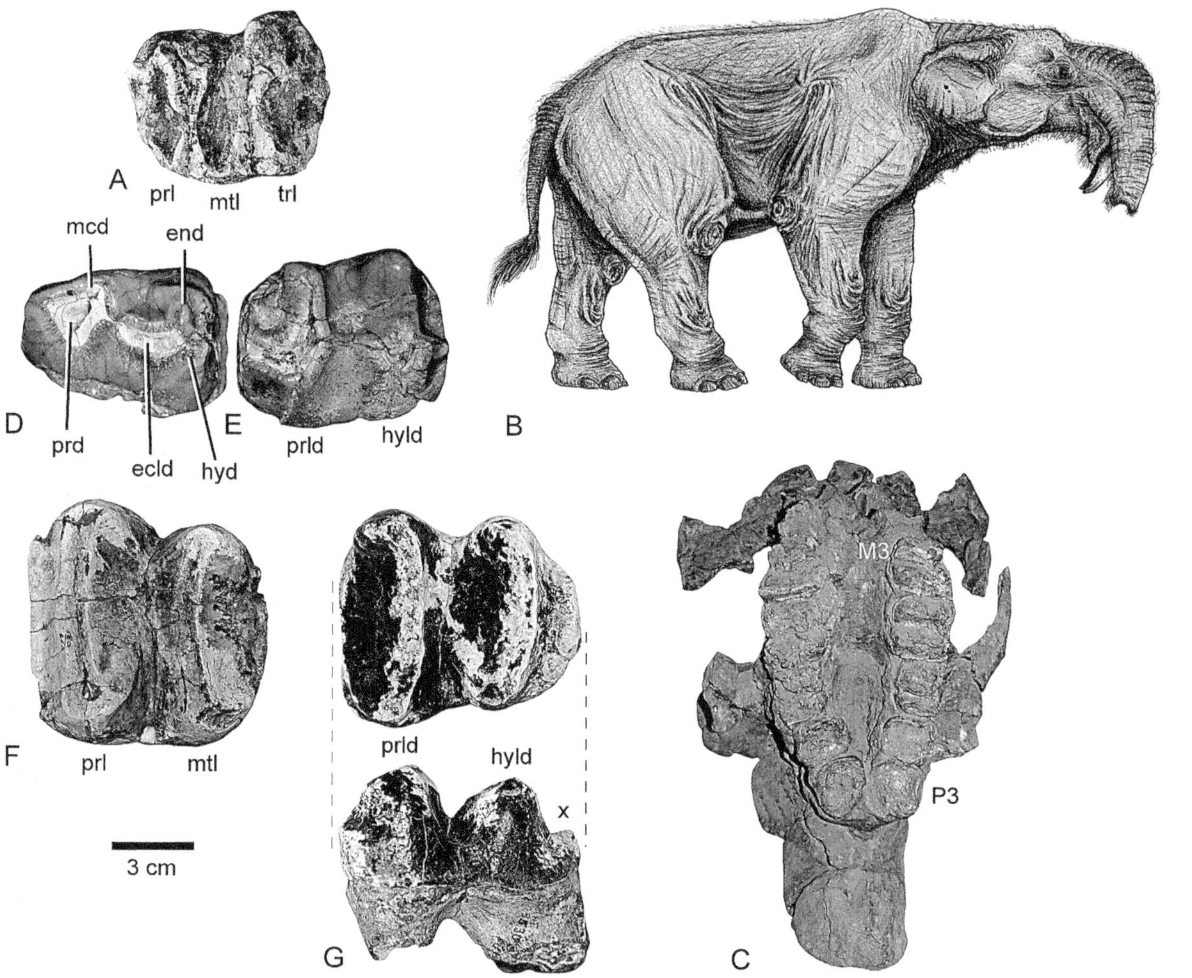

FIGURE 5.2 Fossils and life reconstruction of late Miocene-early Pleistocene *Deinotherium bozasi*. Abbreviations, ecld, ectolophid; end, entoconid; hyd, hypoconid; hyld, hypolophid; M, upper molar; mcd, metaconid; mtl, metaloph; P, upper premolar; prd, protoconid; prl, protoloph; prld, protolophid; trl, tritoloph; x, postcingulid. (A, D–G) to the same scale; anterior to the left. (A) Left dP4, KNM-NA 263 from Nakali, Kenya, possibly the oldest known example of the species, occlusal view. (B) Life reconstruction of *D. bozasi*, lateral view. Cranial morphology is based on the reconstruction of Markov et al. (2001). Artwork by A. Babut (University of Michigan Museum of Paleontology). (C) Partial cranium, KNM-ER 1087 from Koobi Fora, Kenya, ventral view. Length of the cranium is 1,150 mm. Anterior at the bottom. (D) Left p3, LAET 75-2032 from Laetoli, Tanzania, occlusal view. (E) Left p4 (same individual as the p3), LAET 75-2032 from Laetoli, Tanzania, occlusal view. (F) Left M2, KNM-KP 401 from Kanapoi, Kenya, occlusal view. (G) Left m3, KNM-KP 30152 from Kanapoi, Kenya, occlusal and lateral views.

GOMPHOTHERIIDAE Hay, 1922
AMEBELODONTINAE Barbour, 1927
GEN. ET SP. INDET. [Samburu Hills]

(Table 5.1; Figures 5.1 and 5.3)

Synonymy—*Choerolophodon* sp. Tsujikawa, 2005a

A single elephantimorph specimen, right and left m1–2 from the Lower Mb. of the Namurungule Fm. in the Samburu Hills, Kenya (KNM-SH 38455) was allocated to *Choerolophodon* sp. by Tsujikawa (2005a). The geological unit is dated to the early part of the late Miocene (Table 5.1; Tsujikawa, 2005b). It is easy to see the reason for the assignment of these teeth to the Choerolophodontinae, as the pretrite half-lophids appear to be obliquely angled anterior to their posttrite half-lophids, resulting in a form of chevroning. However, closer examination of this specimen reveals that its occlusal pattern is very similar to molar morphology in derived forms of the amebelodont *Protanancus*.

The first molar is best preserved on each side and exhibits a lophid formula of x3x, clearly eliminating it from consideration as *Tetralophodon*, which is more common at the site. The posttrite half-lophids are formed of two large conelets that are aligned orthogonally to the long axis of the crown. Posttrite half-lophid 2 has a distinct accessory conule posterior to it. The pretrite half-lophids are more elaborate, with coarsely folded, thick enamel, and are also composed of at least two conelets, a small mesoconelet and larger outer main

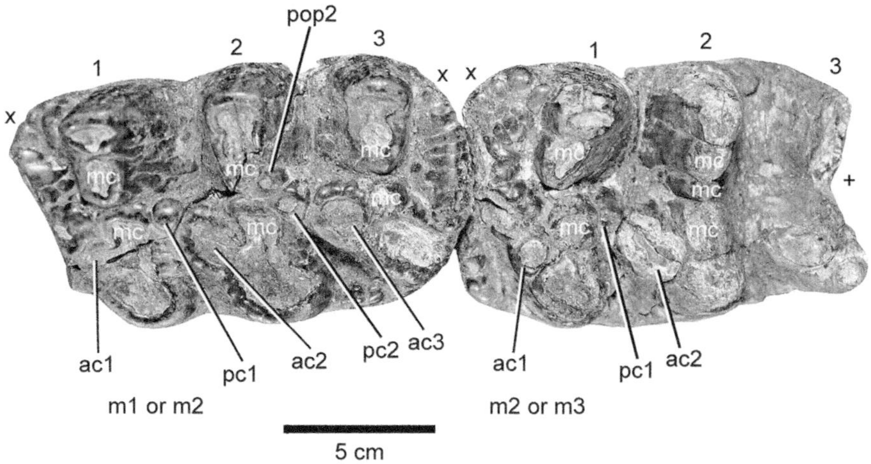

FIGURE 5.3 First and second or second and third lower molars of Amebelodontinae gen. et sp. indet., KNM-SH 38455, from late Miocene Samburu Hills, Kenya. Anterior to the left. Abbreviations: ac, pretrite anterior accessory conule; mc, mesoconelet; m1, m2, m3, first, second, third molar; pc, pretrite posterior accessory conule; pop, posttrite posterior accessory conule; x, pre- or postcingulid; 1, 2, 3 ..., lophid number; +, indicates missing morphology. Note that the mesoconelets are aligned straight transversely, and the large size and medial angulation of the pretrite anterior accessory conules.

conelet. Typical of amebelodonts but not choerolophodonts, the pre- and posttrite mesoconelets are transversely aligned with one another—the pretrite mesoconelet is not advanced anterior to the posttrite mesoconelet in any of the lophids. The impression of chevroning is imparted by the pseudo-anancoidy of the pretrite main conelets, which are positioned posterior to their accompanying mesoconelets.

Each pretrite half-lophid has an immense anterior accessory conule, which is angled toward the median line of the crown. Pseudoanancoidy and the size and angulation of the pretrite anterior accessory conules are traits of some amebelodonts, particularly *Protanancus*. The anterior accessory conules are fully incorporated into the enamel wear figures whereas the posterior pre- and posttrite accessory conules remain independent entities even in moderately heavy occlusal use. Although Tsujikawa (2005a) described this specimen as having thick cementum (and interpreted this to mean it was a grazer), a choerolophodont trait, there is no more than just a trace of cementum on these teeth. Dimensions of the teeth are: m1s, L=124.2–128.3 mm, W=78.6=79.5 mm, ET=5.0–5.3 mm; m2s, L=130.6+–134.9+, W=88.1–91.5 mm, H=69.7 mm (HI=76). Alternatively, these molars could be moderate-sized m2–3, as labeled in the NMK collections, with the third molars missing lophids posteriorly. The Samburu Hills amebelodont is possibly the last known occurrence of the *Protanancus* lineage in Africa, and adds a transitional aspect to the fauna from the site.

KONOBELODON Lambert, 1990
KONOBELODON CYRENAICUS (Gaziry, 1987)
(Tables 5.1 and 5.2; Figure 5.1)

Partial Synonymy—*Amebelodon* sp. Gaziry, 1982; "*Amebelodon* sp." Tassy, 1986; *Amebelodon cyrenaicus* Gaziry, 1987; *Amebelodon* (*Konobelodon*) *cyrenaicus*, Lambert and Shoshani, 1998; "*Mastodon*" *grandincisivus* Tassy, 1999; *Amebelodon cyrenaicus* Sanders et al., 2010a; *Konobelodon* sp. Konidaris et al., 2014c; ?*Konobelodon* Konidaris et al., 2014c; *Konobelodon cyrenaicus* Lambert, 2016

Konobelodon cyrenaicus from Sahabi, Libya is the geologically youngest known amebelodontine gomphothere from Africa (Table 5.1). The fossiliferous vertebrate units at the site in the Sahabi Fm. are dated to the late Miocene, possibly equivalent to Toros Menalla, Chad in age (Warny et al., 2003; Boaz et al., 2008; El Shawaihdi et al., 2016; but see Heinzelin and El-Arnauti, 1982, who believed these units to be early-middle Pliocene in age), based in part on marine biostratigraphic correlation (Muftah et al., 2008). Depositional environments at the site range from shallow marine to broad deltaic channels and fluvial settings (Boaz, 1982; Heizelin and El-Arnauti, 1982). Along with this shovel-tusked gomphothere, the mammalian component of Sahabi's diverse vertebrate fauna also incorporates cetaceans, sirenians, primates, carnivores (notably including ursids), hippos, anthracotheres, suids, equids, rhinos, bovids (multiple tribes), giraffes, shrews, squirrels, murids, gerbils, a primitive species of elephant (*Stegotetrabelodon syrticus*), a stegodont (cf. *Stegodon kaisensis*), and an anancine gomphothere (*Anancus petrocchii*) (Boaz et al., 1982; Sanders, 2008b). Thus, even with the winnowing of gomphotheriid diversity regionally in Africa during the late Miocene, individual sites still featured taxonomically differentiated assemblages of proboscideans, in the example of Sahabi representing four distinct taxa.

Although amebelodont remains were not recovered in the course of Petrocchi's large-scale paleontological survey of Sahabi in the 1930s, a small sample of isolated molars and tusk fragments of these proboscideans, collected between 1978 and 1981 by the International Sahabi Research Project, was described by Gaziry (1982, 1987), leaving no doubt that shovel-tuskers were present at the site. In the absence of postcrania, it is difficult to estimate body mass for *K. cyrenaicus*. Body size of closely related North American *Amebelodon* is indicated to range between approximately 3,000–4,500 kg and to be about 2.3–2.7 m in height at the shoulder, roughly equivalent to extant Asian elephants in magnitude (Lambert 1990; Christiansen, 2004).

Molars assigned to the species include a trilophodont m1, tetralophodont m2, and m3 with six lophids. These teeth do not display a tendency for pseudo-anancoid offset of their pre- and posttrite half-lophids. In addition to its three lophids, the m1, specimen 1P63A, has a large postcingulid and abundant cementum infilling its transverse valleys. Dimensions of the tooth are L=157 mm, and H=57 mm, indicating a brachyodont crown. The m2, 75P11A, is plastically deformed and incomplete, but clearly has four lophids and traces of cementum in its transverse valleys. This molar is 170 mm in length.

More details can be garnered from the m3, specimen 200P15A. It has a lophid formula of x6x, is relatively narrow and brachyodont, and is subdivided into half-lophids by a narrow median sulcus. It preserves anterior and posterior accessory conules and trefoil patterns on the pretrite (buccal) half-lophids, lacks posttrite accessory conules, and has cementum covering the sides of the lophids (but not infilling the anteroposteriorly narrow transverse valleys). The posttrite half-lophids are composed of two subequal-sized conelets, whereas the pretrite half-lophids have only a single, large conelet. Dimensions of the m3 are: L=e.211 mm, W=85 mm at the third lophid (it may have been wider at the laterally broken second lophid), and H=64 mm, yielding a hypsodonty index of +75 (slightly worn).

An alleged second amebelodont m3, 155P11A, was said by Gaziry (1987) to be missing its sixth lophid and postcingulid, but in the accompanying photographic figure (Gaziry, 1987:fig. 5), this unerupted molar crown appears to be complete with very low pre- and postcingulids closely appressed to the first and last lophids, and therefore has a lophid formula of x5x. Its dimensions are L=214 mm; W=86 mm; and H=86 mm, yielding a greater hypsodonty index of 100. Posttrite half-lophids have two conelets that may be subdivided into three apical digitations, and the pretrite half-lophids are formed of a large, outer main conelet and a diminutive mesoconelet, with anterior and posterior accessory conules (the posterior conules may be doubled). While Gaziry maintained that the widest point of the crown is at lophid 3, the figure provided of the specimen shows it to be widest anteriorly, at lophid 1, unlike the condition in m3 specimen 200P15A. It is difficult to reconcile 200P15A and 155P11A as m3s of the same species, despite their similarity in overall crown dimensions. Either they belong to different taxa, or perhaps 155P11A is an M3 of the same taxon as 200P15A, though the strong anterior angulation of its lophids in lateral crown view mediates against the latter interpretation.

There is no ambiguity about the affinity of the lower tusk fragments that Gaziry (1982, 1987) assigned to this species, however. They are dorsoventrally flattened and transversely broad, with ventral and dorsal sulci, characteristic of the lower incisors of "shovel-tusked" amebelodonts (Sanders, 2008b). The most complete specimen, 481P34A, a right i2, has a length of +420+ mm, a width of 127 mm, and height of 44 mm, yielding a compression index (H×100/W) of 35, trending toward the highly derived, hyper-flattened lower tusks of Asian and North American *Platybelodon* and *Torynobelodon* (Osborn and Granger, 1931, 1932; Osborn, 1936; Guan, 1996; Sanders et al., 2010a). In contrast, they are slightly to markedly more dorsoventrally compressed than lower tusks in more primitive amebelodonts such as *Archaeobelodon*, *Afromastodon*, and *Protanancus* (Sanders, 2008b). Abrasive marks are limited to the ventral side of the Sahabi tusks, indicating that they were used for scooping or shoveling rather than for stripping vegetation or scraping bark (Barbour, 1930; Lambert, 1992).

In cross-section, the Sahabi lower tusks notably exhibit large dentinal tubules and rods, surrounded by dentinal laminae. Lambert (1990) subdivided *Amebelodon* into two subgenera, with *A.* (*Konobelodon*) distinguished from *A.* (*Amebelodon*) by tetralophodont second molars, third molars with six loph(id)s, and most especially by lower incisors with dentinal tubules and rods. Thus, the Sahabi amebelodont belongs in the *A.* (*Konobelodon*) group, which has been raised to generic status (e.g., Gheerbrant and Tassy, 2009; Konidaris et al., 2014c; Lambert, 2016). This seems reasonable, based on lower incisor morphological criteria.

Further useful taxonomic service was provided by Konidaris and colleagues (2014c), who resolved the identity of most late Miocene tetralophodont proboscideans from western Asia and eastern Europe that had been classified as "*Mastodon*" *grandincisivus* as *Konobelodon*. Although "*Mastodon*" *grandincisivus* has priority over *Konobelodon*, it is a wastebasket taxon because not all proboscideans referred to it are amebelodonts. For that reason, it is preferable to place the Sahabi amebelodonts in *Konobelodon cyrenaicus*. Moreover, though the presence of dentinal tubules and rods in its lower incisors suggests closer affinity of *Konobelodon* with *Platybelodon*, *Torynobelodon*, and the newly described Chinese amebelodont *Aphanobelodon zhaoi* (Wang et al., 2016), it has been hypothesized that this could be a case of convergence (Sanders et al., 2010a). This interpretation is confirmed by the recent work of Lambert (2016), who showed that the shape and distribution of rods and development of typical gomphothere-like dentinal laminae in the lower tusks differ from the condition in the *Platybelodon* clade and concluded that *Konobelodon* and *Amebelodon* are closer sister taxa (and, consequently, that

Konobelodon is not derived from *Platybelodon*, contra Konidaris et al., 2014c). Nonetheless, given the distribution of the genus, it is likely that *Konobelodon* was a late Miocene emigrant from Eurasia into Africa (see Konidaris et al., 2014c).

TETRALOPHODONTINAE Van der Maarel, 1932
TETRALOPHODON Falconer, 1857
(Tables 5.1, 5.2, and 5.3; Figures 5.1 and 5.4)

Tetralophodon is ideally positioned morphologically, temporally, and geographically (Table 5.1) to be ancestral to elephants (Sanders et al., 2010a; Wang et al., 2017b). Markov (2008) recognized this and succinctly noted that the "distinction between 'derived *Tetralophodon*' and 'primitive *Stegotetrabelodon*' is becoming increasingly technical." Occurrences are all Old World, throughout Eurasia (e.g., Tassy, 1985; Fukuchi et al., 2007; Markov, 2008; Thasod et al., 2012; Wang et al., 2017b) and in Africa (Sanders et al., 2010a); attribution of the genus to North American sites more likely records parallel development of a different lineage, which has been placed in *Pediolophodon* (Lambert, 2007). It is probable that ancestors of African members of the genus evolved in and emigrated from Eurasia, as the genus accompanies an anthracothere of Asian derivation at the locality of its potentially earliest occurrence in Africa (Lihoureau et al., 2015). European *Tetralophodon longirostris* may be the oldest species of the genus, originating during the Astaracian (MN zonation 7–8), between about 13.0 and 12.5 Ma and possibly surviving until MN 12 times, ca. 7.0 Ma (Bergounioux and Crouzel, 1956; Mazo and Montoya, 2003; Tibuleac et al., 2015). Because of dating uncertainties with some North African occurrences of the genus, it is difficult to know if *T. longirostris* is the source for one of the rare incidences of immigration of proboscideans into Africa, but clearly, it is a late middle Miocene to late Miocene taxon (Göhlich, 1999; Göhlich and Huttunen, 2009). *Tetralophodon atticus*, from Turolian sites in western Asia and Europe, is also a possible source for the ancestry of elephants (Markov, 2008). The genus likely evolved from a species of trilophodont *Gomphotherium* and is a close sister taxon to *Anancus*, as well (Coppens et al., 1978; Van der Made and Mazo, 2003; Tibuleac et al., 2015).

The bunodont, brachyodont-to-subhypsodont molars of *T. longirostris* have been interpreted as useful for browsing foliage, fruits, and lignified parts of woody plants, in woodland settings (Tibuleac et al., 2015). Analysis of dental isotopes in *Tetralophodon* from late Miocene Nakali and the Samburu Hills, Kenya is concordant with this paleoecological interpretation, indicating predominantly browsing diets (Cerling et al., 1999, 2003; see also Uno et al., 2011).

Body form in *Tetralophodon* is well documented from a partial skeleton of *T. longirostris* from the late Miocene of Spain (Alberdi, 1971). The resemblance between the limb elements and those of the primitive elephant *Stegotetrabelodon syrticus* from Sahabi, Libya (Petrocchi, 1954) is striking, though the Spanish specimens are even more robust and from a shorter individual than the Sahabi elephant. The femur has a shorter neck than stegotetrabelodont femora, but with the same orientation of the head, low greater trochanter, close spacing and downward orientation of the condyles, diminutive lesser trochanter, and pronounced third trochanter, although it is shorter (950 mm) and the diaphysis is broader relative to length. The length of the tibia is 600 mm, yielding a crural index of 63, close to that of the Sahabi stegotetrabelodont (Petrocchi, 1954) and slightly greater than that of the Baynunah Fm. stegotetrabelodont from Abu Dhabi (Sanders, 2022); its anteroposterior diaphyseal diameter, at 235 mm, indicates it is more robust than tibiae of stegotetrabelodonts. As in stegotetrabelodonts, the humerus is heavily-built, with a strong, projecting greater tuberosity, broadly flaring deltoid and supinator crests, and a large, upwardly-facing head. The humerus and ulna are shorter than in the Sahabi elephant, 770 mm and 820 mm, respectively, but similar in length proportions to one another. The radius was obviously in a fixed pronated position on the ulna. A proboscidean with these long bone dimensions would probably have stood approximately 250 mm tall at the shoulder, and an estimate of 3,629 kg would be reasonable for body mass.

TETRALOPHODON SP. Lihoreau et al., 2015
[Bir el Ater 3]
(Tables 5.1 and 5.3; Figure 5.1)

A modest sample of cheek teeth referable to *Tetralophodon* were recovered from the locality of Bir el Ater 3, Algeria, possibly comprising the earliest evidence of the genus in Africa (latest middle Miocene or early late Miocene; Lihoreau et al., 2015). The simple crown morphology of these teeth supports this biochronological interpretation. The sample includes dp4, p4, and m2. Specimen UONM-15, the dp4, has a lophid formula of x4x, with low, pyramidal lophids divided by a distinct, narrow median sulcus. The crown length is >76.0 mm, and the greatest width, at the base of the lophids, is 43.5 mm. Each half-lophid is constructed of an outer main conelet and a similar-sized mesoconelet. Half-lophids are arranged transversely straight across the crown. There are no posttrite accessory conules but each pretrite half-lophid has a small anterior accessory conule, and the first three pretrite half-lophids have tiny posterior accessory conules that in wear contribute to poorly-formed trefoil enamel figures. Enamel is unfolded, rugose, and moderately thick (ET = 3.0 mm). There is no cementum. In dimensions and morphology, the specimen resembles a tetralophodont dP4 from Mb. D of the Ngorora Fm., Kenya (Lihoreau et al., 2015).

The p4, UONM-16, is bilophodont and very low crowned, with a prominent precingulid and smaller postcingulid closely appressed to lophid 2, formed of two conelets. Pretrite half-lophids are formed of an outer main conelet and a smaller mesoconlet; posttrite half-lophids are formed of a single conelet. Each pretrite half-lophid has an associated posterior accessory conule. There is no cementum. Dimensions of the

TABLE 5.3
Cheek Tooth Dimensions of African *Tetralophodon* Specimens

Taxon/Specimen/Tooth	Loph(id) Formula	LF	L	W	H	HI	ET
Tetralophodon sp. nov. (Samburu Hills)							
KNM-SH 12307 l. M2	x4x	3.0	150.2	82.6	63.8	77	—
KNM-SH 12307 r. M2	x4x	3.0	148.4	82.4	—	—	—
KNM-SH 12308 l. M1[d]	x4x	3.8	106.3	61.8	47.5	77	4.2
KNM-SH 12309 r. M1	+3x	—	+81.4	62.8	38.1	—	3.6[d]
KNM-SH 12310 l. P4	x3x	3.6[c]	56.7	49.1	+29.2	—	3.3[d]
KNM-SH 12311 r. P4	x3x	3.6[c]	55.4	48.7	+28.1	—	2.8[d]
KNM-SH 12312 r. P3	x2x	5.3[c]	36.0	35.5	Worn	—	2.1–2.5
KNM-SH 12313 r. p3	x2x	6.1[c]	32.7	24.0+	17.3	—	2.5[d]
KNM-SH 12373 l. m2[d]	x4x	2.3	176.0	91.4	34.5	37	10.9
KNM-SH 12380 partial r. m2	+3x	3.0	+118.0	92.7	Worn	—	9.6[d]
KNM-SH 15858 l. M2	x4X		166.0	100.0	—	—	—
r. M2	x4X		190.0	101.0	—	—	—
l. M3	x6x		244.0	114.0	—	—	—
r. M3	x6x		237.0	121.0	—	—	—
KNM-SH 15781 dp3	x2x	—	45.3	32.8	—	—	—
KNM-SH 15782 dp2	x2	—	32.0	21.1	Worn	—	thin
Tetralophodon sp. (Bir el Ater 3)[e]							
UONM-15, l. dp4	x4x	—	+76.0	43.5	Worn	—	3.0
UONM-16, r. p4	x2x	—	53.4	35.7	—	—	3.0
UONM-17, r. m2	x4x	—	147.0	68.0	Worn	—	6.0–10.0
Tetralophodon sp. indet. (Chorora, Ethiopia)[c]							
AL 1000-1, r. m3	6x	—	182.0	78.0	—	—	—
Tetralophodon cf. *longirostris* (Djebel Krechem el Artsouma, Tunisia)[a]							
OM I-43, M2	x4x	—	156.0	88.0	45.0	51	—
OM I-21, p4	x2x	—	56.0	49.0	—	—	—
OM I-II, P3	x2x	—	44.0	41.0+	—	—	—
Tetralophodon sp. (Kakara Fm., Kisegi, Uganda)[b] ("Elephantidae, forme primitive, gen. et sp. incertae sedis)							
l. M3	x5x	2.6	225.6	111.6	61.0	61–68	Thick
Tetralophodon sp. (Skoura)[f]							
FSC-Sk-180, P4	x2x	—	55.0	57.5	—	—	3.0–4.0
M1	4x	—	108.5	67.5	—	—	5.0–6.0
M2	x4x	—	149.0	84.5	63.0	75	5.0+
cf. *Tetralophodon* sp. indet. (Baynunah Fm., Hamra, Abu Dhabi)[g]							
l. M2	x4X	2.9	e. 172.0	e. 105.0	Worn	—	5.0–7.0
r. M2	x4X	2.75	e. 170.0	e. 95.0	Worn	—	5.0–7.0

(Continued)

TABLE 5.3 (*Continued*)
Cheek Tooth Dimensions of African *Tetralophodon* Specimens

Taxon/Specimen/Tooth	Loph(id) Formula	LF	L	W	H	HI	ET
l. M3	x5X	2.6	e. 230.0	e. 110.0	Worn	—	5.0–7.0
r. M3	x5x	2.5	e. 210.0	e. 105.0	Worn	—	6.0–7.0

All Dimensions in mm. Abbreviations: e., estimated; ET, enamel thickness; H, height; HI, hypsodonty index, H/W×100; l., left; L, length; LF, lamellar frequency, number of loph(id)s per 100 mm; M/m, upper or lower molar; N, number of specimens; P/p, upper or lower premolar; r., right; s.d., standard deviation; W, width; x, indicates an anterior or posterior cingulum(id); X, indicates a very large anterior or posterior cingulum(id); +, indicates missing morphology.
[a]Geraads, 1989; [b]Tassy, 1995; [c]Geraads et al., 2002; [d]Nakaya et al., 2005a; [e]Lihoreau et al., 2014; [f]Geraads et al., 2019; [g]Sanders, 2022.

specimen are L=53.4 mm and W=35.7 mm. Lihoreau et al. (2015) stated that the specimen is differentiated from p4s of archaic elephants in the genus *Stegotetrabelodon* by the presence of an accessory conule posterior to pretrite half-lophid 1, but p4s of *Stegotetrabelodon* from the Baynunah Fm., Abu Dhabi, United Arab Emirates have posterior swellings in this position.

Specimen UONM-17, a very worn right m2, is tetralophodont with a lophid formula of x4x. Each half-lophid has two similar-sized conelets. On the pretrite side, posterior accessory conules are associated with lophids 1–3 and anterior accessory conules are associated with lophids 1–4. They contribute to imperfectly formed trefoil enamel wear figures. The postcingulid is formed of three conelets and tightly appressed on the last lophid. The crown is brachyodont, and in lateral view lophids are pyramidal in shape, which is reflected by the low lamellar frequency between 2.5 and 3.0. Enamel is unfolded and very thick, 6–10 mm. Dimensions of the molar are L=147.0 mm and W=68.0 mm. Cementum is not in evidence. Half-lophids are arranged transversely straight across the crown, except for lophid 4 where they are angled posterolaterally at their outer aspects.

Fossil-bearing sediments of the Nementcha Fm. are fluvial to fluvial-deltaic sandstones alternating with red silts and gritstones (Lihoreau et al., 2015). The anthracothere *Libycosaurus algeriensis* has also been recovered at Bir el Ater 3. Establishing a more precise geological age for the locality is dependent on which hypothesis is accepted regarding the phylogenetic systematics of *Libycosaurus* (Lihoreau et al., 2015).

TETRALOPHODON SP. NOV. Tsujikawa, 2005a [Samburu Hills]

(Tables 5.1–5.3; Figures 5.1 and 5.4)

Partial Synonymy—*Tetralophodon longirostris* Bergounioux and Crouzel, 1956; *Trilophodon angustidens* cf. *kisumuensis* Hooijer 1963:32, plate I, figure 2; *Tetralophodon* sp. indet. Madden, 1977; ?*Tetralophodon* cf. *longirostris* Coppens et al., 1978:346; Proboscidea indet. Tiercelin et al. 1979:257; *Stegotetrabelodon grandincisivum* Madden, 1982; *Tetralophodon* sp. Nakaya et al., 1984; *Tetralophodon* sp. Nakaya et al., 1987; *Tetralophodon* cf. *longirostris* Geraads, 1989; *Stegotetrabelodon* n. sp.? Geraads et al., 2002:114

The best assemblage of Afro-Arabian *Tetralophodon* fossils is from the late Miocene site of Samburu Hills, Kenya. Proboscidean fossils derive from the Upper and Lower Mbs. of the Namurungule Fm., biochronologically dated to the early part of the late Miocene (Nakaya et al., 1984) and radiometrically by K-Ar analysis to 9.5 Ma (Tsujikawa, 2005a, b). The Namurungule Fm. is composed of mudstones, sandstones, and gravels interdigitated with beds of mudflows and tuffaceous sediments (Tsujikawa, 2005b); underlying and overlying pumice beds were dated to 9.57±0.22 Ma and 9.47±0.22 Ma, respectively (Sawada et al., 1998), providing the most concise age for a *Tetralophodon* occurrence in Africa. The mammalian assemblage for the site gives a good window into African early late Miocene faunal composition, comprised of the hominoid primate *Samburupithecus kiptalami*, two species of thryonomyid rodents, felid, hyaenid, and amphicyonid or ursid carnivores, calicotheres, hipparionine equids, possibly four species of rhino, two species of nyanzachoere suids, primitive hippos, two species of giraffids, tragulids, gazelles and other bovids (boselaphines and reduncines), hyraxes, along with deinotheres, an amebelodont, and tetralophodonts (Nakaya et al., 1984, 1987; Tsujikawa, 2005a, b). Composition of the fauna shows a strong departure from early-middle Miocene African faunas while resembling assemblages from other early late Miocene East African sites (Tsujikawa, 2005b), but also shows a strong similarity to late Miocene faunas from the South Asian Siwalik Series and Eurasian sites such as Samos and Pikermi, Greece, and Maragheh, Iran (Nakaya et al., 1987). Paleoecological interpretation of the Samburu Hills fauna suggests that the Upper Mb. of the Namurungule Fm. was likely an open environment such as grassland or savanna, whereas the Lower Mb. had a more wooded environment (Tsujikawa, 2005a).

The composition of the *Tetralophodon* sample from Samburu Hills, which Tsuijikawa (2005a) considered to belong to a new species, includes a partial cranium and isolated gnathodental specimens that furnish an abundance of morphological details. Nakaya et al. (1984, 1987) provided cursory descriptions of these specimens, but made numerous

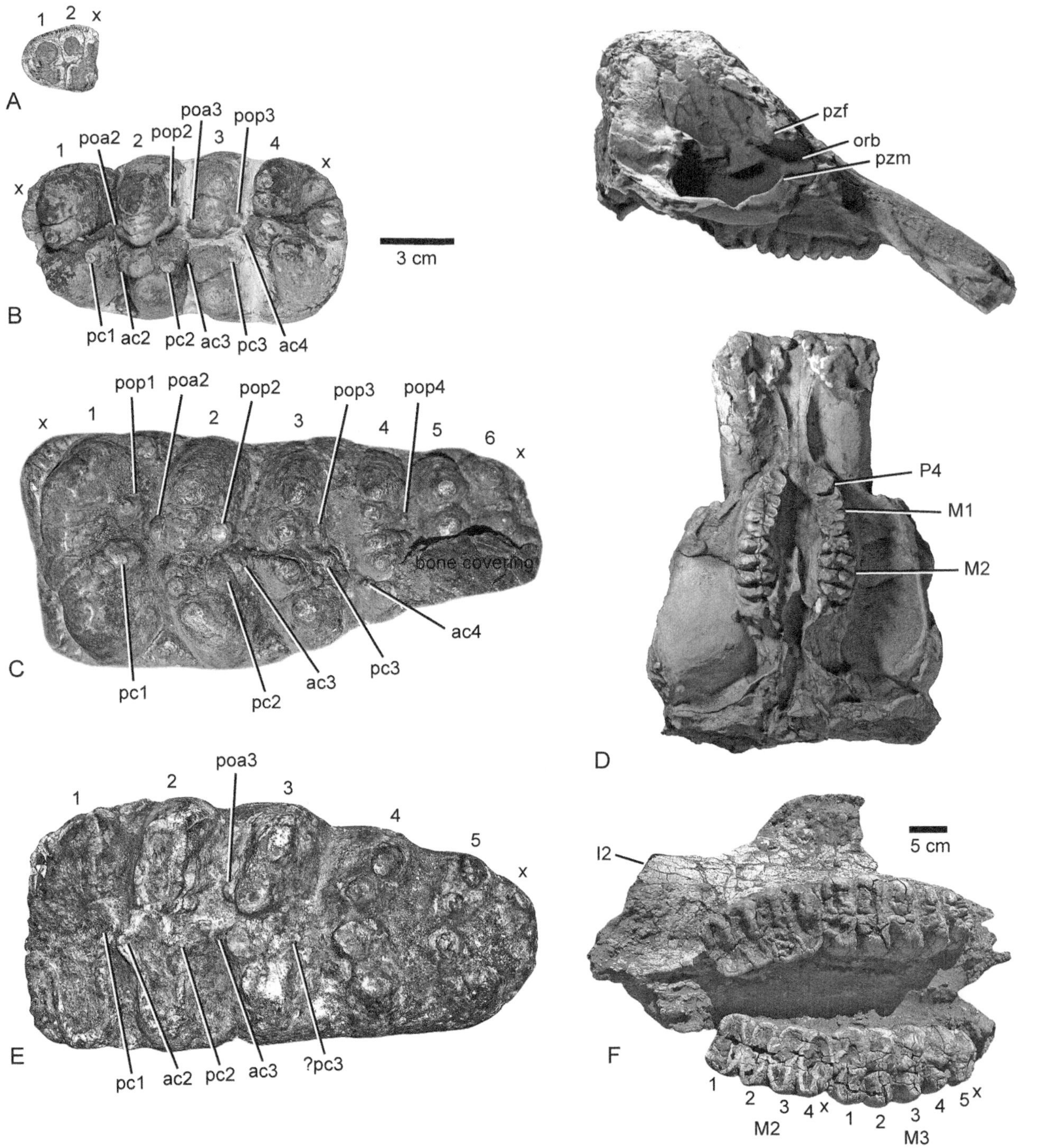

FIGURE 5.4 Craniodental specimens of African *Tetralophodon* spp. Abbreviations: ac, pretrite anterior accessory conule; I, upper incisor (tusk); M, upper molar; orb, orbit; P, upper premolar; pc, pretrite posterior accessory conule; poa, posttrite anterior accessory conule; pop, posttrite posterior accessory conule; pzf, zygomatic process, frontal; pzm, zygomatic process, maxilla; x, pre- or post-cingulum; 1, 2, 3 ..., loph counted from the anterior end of the crown. (A–C, E) to the same scale. (A–C, E, F) anterior to the left. (D) lateral view, anterior to the right; ventral view, anterior to the top. (A) Left P3, KNM-SH 12312 from the Samburu Hills, Kenya, occlusal view. (B) Right M2, KNM-SH 12307 from the Samburu Hills, Kenya, occlusal view. (C) Emergent left M3 in crushed cranium, KNM-SH 15858 from the Samburu Hills, Kenya, occlusal view. Note that the maxillary bone occludes part of the posterior end of the molar. (D) Juvenile cranium with left and right P4-M2, FSC-SK-180 from Skoura, Morocco, lateral and ventral views (Images of FSC-SK-180 courtesy of D. Geraads). (E) Left M3, KI 64'92 from the Kisegi-Nyabusosi area, Uganda, occlusal view. (F) Palate with left and right M2–3, AUH 899 from Hamra, Abu Dhabi, United Arab Emirates, occlusal (ventral) view. Not to the same scale as (A–C, D, or E). (Photograph courtesy of F. Bibi.)

errors in tooth position identifications. Correct identifications of specimens to tooth position are KNM-SH 12307–12313 (same individual), left and right M2s (not M1s), left and right M1s (not P4s), left and right P4s (not P3s), right P3 (not P2) and right p3 (not p2!), respectively; KNM-SH 12373 and 12380 (same individual), left dentary fragment with m2 and partial right m2; KNM-SH 15858, partial cranium with right and left M2–3; KNM-SH 15779, partial M1 or M2; KNM-SH 15781, dp3; KNM-SH 15782, dp2. Dimensions of these specimens are provided in Table 5.3.

The lower dp2, KNM-SH 15782, is a simple tooth composed of two lophids, longer than wide and is well worn, particularly posteriorly. It is broadest posteriorly. The anterior loph is formed of two conelets and shows a trace of a narrow median sulcus. KNM-SH 12312 is a left P3 that is semi-triangular in occlusal view (Figure 5.4A), with a loph formula of x2x. Each loph has two stout conelets, well worn. There is no sign of cementum or accessory conules. KNM-SH 12310 and 12311 are P4 antimeres that are more rounded in occlusal shape and are considerably more complex than the P3 in the configuration of crown features. They are also substantially larger premolars, but very brachyodont. Their loph formula is x3x, with anterior and posterior cingulae tightly appressed to the lophs, and lophs 2 and 3 not well separated from one another. Each side of loph 3 has two-three conelets, and there is some transverse offset of half-lophs. Only the first loph has a strong anterior accessory conule. There is no cementum and enamel appears thick.

Molars of this taxon reveal a wealth of morphological detail. Right and left M2s, KNM-SH 12307, are tetralophodont with loph formulae of x4x and low lamellar spacing formulae. The crowns of these molars are of a broad "bread loaf" shape occlusally, with their posterior moieties slightly wider than anteriorly (Figure 5.4B). Each half-loph is formed of two conelets, with mesoconelets close in size to the outer main conelets. These molars are very low crowned and are broadest at their bases. Accessory conules are present on both pre- and posttrite sides of the crown. For example, in the right M2, lophs 1–3 have pretrite anterior and posterior accessory conules, posttrite anterior and posterior accessory conules associated with loph 2, a posttrite posterior accessory conule on loph 3, and a small posttrite anterior conule on loph 4. There is no cementum. These molars most closely resemble those of *Tetralophodon* from the Baynunah Fm., Abu Dhabi. Lower m2s (KNM-SH 12373, 12380) are also tetralophodont, with massively thick enamel (Table 5.3), and though worn exhibit trefoil enamel patterns on their pretrite sides, and a distinct but very narrow median sulcus. Half-lophids are transversely in a straight line with each other. These molars exhibit no cementum.

The cranial specimen (KNM-SH 15858) is heavily weathered and incomplete, limiting its use for comparative study. However, the dimensions of M2-M3s on each side (Table 5.3) suggest that it belonged to a massive animal. The cranium has an estimated height of 900 mm (Nakaya et al., 1987) and its basicranial angle is steep, calculated at 60° (Nakaya, 1994). These features are derived compared with the lower cranial height and basicranial angulation in *T. longirostris* (Nakaya, 1994), and are more reminiscent of the high angulation of the basicranium in *Paratetralophodon* from the Siwaliks Series, Pakistan (Tassy, 1983b), in contrast with the lower basicranial angles in tetralophodontine cranial specimens from Skoura, Morocco (Geraads et al., 2019) and Hamra, Abu Dhabi (2022). The raised braincase and steep basicranial angle of KNM-SH 15858 anticipates the morphological reorganization comprising anteroposterior compression and dorsoventral elevation of the cranium in elephants (Maglio, 1972a). The breadth of KNM-SH 15858 across the zygomatic arches is estimated at 700 mm, but this seems artificially low due to the bilateral compressive deformation of the specimen.

The M2s of KNM-SH 15858 are the anteriormost molars preserved in the cranium, and are tetralophodont with a loph formula of x4X; the distal cingulum in each M2 is prominent, particularly lingually. Enamel is thick and unfolded, and enamel wear figures show that the pre- and posttrite moieties of each loph maintain their independence despite heavy occlusal attrition. Occlusal morphology is very similar to the Hamra tetralophodont M2s, with complex development of anterior and posterior accessory conules on the pre- and posttrite sides of lophs, throughout the crown. The M3s are basically unworn as they are still in vertical position, and had not yet rotated into occlusion at the time of death. They have a loph formula of x6x. Lophs are very brachyodont and are broadest at their bases. In the lateral view, the lophs are pyramidal in shape. A narrow median sulcus subdivides the lophs into pre- and posttrite sides. Accessory conules are present posterior to posttrite half-lophs 1–4 and anterior and posterior to pretrite half-lophs 1–4 (Figure 5.4C). There is only a trace of cementum. Based on comparison with extant African elephants, the wear and emergence of M2–3 in this individual is equivalent to stage XVIII in Laws' system (1966), consistent with an age of 30±2 years.

TETRALOPHODON SP. Geraads et al., 2019 [Skoura]

(Tables 5.1 and 5.3; Figures 5.1 and 5.4)

Synonymy—cf. *Tetralophodon* sp. Zouhri et al., 2012

The most complete cranium of African *Tetralophodon* is from the Upper Mb. of the Aït Kandoula Fm. near Skoura village and a road pass at Tizi N'Tadderht in the Ourzazate Basin of Morocco. The geological setting of the specimen and additional proboscidean material from the site is composed of alluvial deposits including sandstones and conglomerates (Zouhri et al., 2012; Geraads et al., 2019). Along with the occurrence of *Tetralophodon* sp. at the site, other faunal constituents include an ostrich, terrestrial chelonians, crocodilians indistinguishable from the modern Nile crocodile, felids, several species of hipparionine horses, elasmothere and ceratothere rhinos, indeterminate suids, giraffids, and a new caprine bovid (Geraads et al., 2012, 2019; Cirilli and Zouhri, 2018). Biochronological interpretation of the fauna suggests that the sediments date to the early part of the late Miocene.

The cranium, FSC-SK-180, was recovered by a French-Moroccan team in 2013. It is from a subadult individual, with right and left P4-M1 in wear and M2s nearly emerged from their crypts (Figure 5.4E), equivalent to Tassy's (2013) stage XIII or XIV for *Gomphotherium angustidens*. It is one of the best-preserved proboscidean crania of the late Miocene in Africa, and was comprehensively described by Geraads and colleagues (2019). The specimen is missing its left tusk, part of the right tusk, the occipital condyles, and was vertically compacted in situ, which has dorsoventrally foreshortened the vault and reduced the angulation and height of the basicranium above the level of the palate. Nonetheless, it is preserved well enough to provide a reasonably accurate understanding of cranial morphology in a young, possibly female, individual of this species. The greatest length of the cranium is 960 mm, of which the rostral length comprises 625 mm. The greatest width is 640 mm, and the width across the orbits is 560 mm.

In dorsal view, the cranium has a relatively broad, anteroposteriorly very truncated vault with an elongated rostrum. It is wider than in *Gomphotherium* crania between the temporal lines. The braincase is transversely and anteroposteriorly flattened to convex. The nasals overhang the broad, laterally downturned nasal aperture. The external incisive fossa is relatively broad distally, shallow, and deepens considerably just below the nasal aperture. The premaxillae are largely straight and only diverge slightly at their distal ends. Small dimensions of the tusk alveoli and the remnant of the right tusk correlate with the cranium belonging to a young, possibly female, individual. The broadest aspect of the cranium in this view is at the posterior extent of the zygomatic arch, whereas the orbits are narrower than the occipital region or the zygomatic arch. A broad premaxillary planum of bone is present on each side anterior to the nasal aperture, medial to the frontal struts that run to the orbit, and lateral to the upper extent of the external incisive fossa.

Laterally, the cranium appears only modestly raised (Figure 5.4D), similar to the condition in *G. angustidens* (see Tassy, 2013). The supraorbital process is weak but the ventral postorbital process, formed on the zygomatic arch, is more salient. Orbital size is small relative to the overall area of the temporal fossa. The orbit is located very posteriorly, above the M2 on each side, differing from its more anterior position in modern elephants. The face and rostrum run in almost a straight line from the braincase, angled downward at about 45°. The zygomatic arch is straight and not heavy. Emergence of the M2s appears to be at a low angle, in association with the modest height of the cranium.

Posteriorly, the occipital planum is rounded superiorly and laterally, widest above the level of the squamosal portion of the zygomatic arch, vertical in orientation, and flattened with no sign of bossing. The fossa for the nuchal ligaments is located just above half-height of the planum, deep, and is divided by a strong midline ridge.

In ventral view, the palate is shallow, elongated, and parabolic in shape, widest at the anterior end of the M2s (Figure 5.4D). The basicranium is short, the glenoids are biconcave and broad, and the postglenoid processes form the posterior wall of the auditory canal. The internal choanae comprise a narrow, diamond-shaped opening.

P4 is squared in shape, with dimensions of L = 55.0 mm and W = 57.5 mm. Enamel thickness is 3.0–4.0 mm. Its loph formula is 2x. M1 has a loph formula of 4x. Half-lophs are formed of a main outer conelet and mesoconelet of nearly the same size. Small anterior and posterior pretrite accessory conules contribute to imperfectly formed trefoil enamel wear figures. Dimensions provided for one of the M1s are L = 108.5 mm and W = 67.5, with enamel thickness of 5.0–6.0 mm. There are traces of cementum in the transverse valleys. M2 is also tetralophodont, with a formula of x4x. The postcingulum is low, formed of four small conelets, and is tightly appressed to the last loph. Only the first loph is in wear. Dimensions of one of the M2s are L = 149.0 mm and W = 84.5 mm; enamel thickness is greater than 5.0 mm. Pretrite half-lophs 1–3 have anterior and posterior accessory conules and pretrite half-loph has an anterior accessory conule. There are low, tiny anterior accessory conules on posttrite half-lophs 1–3, tightly appressed to the mesoconelets. On each side of the lophs, the main conelets are larger and higher than the mesoconelets.

Compared with *Tetralophodon* sp. nov. from the Samburu Hills, Kenya the Skoura cranium exhibits a more primitive morphology. For example, the basicranial angle was likely much lower (estimated at 30°), compared with 60° in the Samburu Hills cranium. Greatest similarity of the Skoura cranium in this regard is with *T. longirostris* from Europe; however, the teeth of the Skoura specimen are more derived in having greater anteroposterior compression of lophs, a less pronounced transverse wear gradient across molar crowns, more enlarged mesoconelets, and reduced size of accessory conules (Geraads et al., 2019). The Skoura molars differ from those of the Baynunah Fm. *Tetralophodon* sp. indet. palate in having weaker development of accessory conules, Morphological contrasts between the Skoura, Samburu, and Baynunah material reflect differences in geological age and/or presence of multiple lineages of *Tetralophodon* on the continent. Although the Skoura cranium seems to be too primitive to represent a direct ancestor of elephants, it provides a baseline for understanding the coordinated changes that occurred in the reorganization of the cranium during the evolution of gomphotheres into elephants. In comparison with the Skoura cranium, the cranial anatomy of early elephants appears to have involved raising of the basicranium well above the level of the palate, which would have angled the occipital planum forward at its superior margin; lifting the cranial vault would have foreshortened the neurocranium, and, in turn, this may have led to a greater downward angulation of the viscerocranium (and consequent straighter line between the nuchal crest and distal end of the tusk sheaths).

A partial mandible with a badly preserved, incomplete right m3 was also recovered from Skoura, not in association with the cranium (Zhouri et al., 2012). The symphysis is broken but housed large tusks in moderately downturned alveoli. Little can be gleaned from the broken m3 except that its lophids were transversely straight with no sign of chevroning, and that its dimensions were ~200 mm long and 80 mm wide. The length of the partial mandible is approximately 600 mm (Geraads et al., 2019). In addition, the Skoura site produced an isolated broken upper tusk, about 750 mm in length, ovoid, and with cross-sectional dimensions of 110×83 mm, composed of concentric dentine and with no mention of an enamel band (Geraads et al., 2019).

TETRALOPHODON SP. INDET. [Kisegi]

(Tables 5.1 and 5.3; Figures 5.1 and 5.4)

Synonymy—*Tetralophodon* Pickford et al., 1993; Elephantidae, primitive form, gen. et sp. incertae sedis Tassy, 1995

An isolated M3 (KI 64'92) from the late Miocene Kakara Fm. at Kisegi, Uganda, in the Western Rift (Table 5.1), has been considered as morphologically intermediate between molars of gomphotheres and elephants (Tassy, 1995). The specimen has a loph formula of x5x, and, similar to the molars of tetralophodonts and stegotetrabelodonts, is very broad and low-crowned (Table 5.3). The last three lophs in particular are separated into pre- and posttrite sides by a very narrow median sulcus (Figure 5.4E). Each loph is formed of four conelets, two on each side, of equal size and higher toward the midline. In occlusal view, the last three lophs are anteriorly convex. The first loph is heavily worn and exhibits a trefoil pattern, incorporating anterior and posterior accessory conules into the enamel loop, on the pretrite side. The second loph also incorporates anterior and posterior accessory conules into the pretrite side of the enamel loop. Tiny pretrite accessory conules are associated with the third loph but are fused to the loph and not free apically. A small posttrite accessory conule is present on the anterior face of loph 3. In lateral view, the lophs are pyramidal in shape, low, and squat in appearance. Although Tassy (1995) felt that the lophs are anteroposteriorly compressesd, the very low lamellar frequency (Table 5.3) shows that they are not. Cementum is present in the transverse valleys but does not cover the lophs.

Features believed to support affinity with elephants include equal wear on the pre- and posttrite sides of the crown, greatest crown height toward the midline, attenuation of the median sulcus, and anteroposterior compression of lophs into plate form (Tassy, 1995). The primitive features shared with tri- and tetralophodont gomphotheres include low crown height, formation of trefoil enamel wear figures, presence of a posttrite accessory conule, and equal-sized main, outer coneletes and mesoconelets, as well as a low number of lophs for an M3. Except for the distinctive uniform wear across the crown, none of the supposed elephant-like features of KI 64'92 are compelling evidence for placement of the specimen in the Elephantidae. The pattern of wear across the crown suggests an elephant-like fore-aft chewing motion, but the construction of the lophs and their low number are very gomphothere-like. Moreover, the mesiodistal wear gradient is not strong, indicating that the angle of emergence probably was not very steep, hinting that the cranium from which this molar derived was not particularly raised and therefore may not have been anteroposteriorly compressed in the manner of elephant crania. In this scenario, it is tempting to posit that changes in masticatory behavior preceded the evolution of craniodental morphological adaptations in derived tetralophodonts. Alternatively, Saegusa (1996a) listed propalinal occlusal motion as a trait defining elephants, which would support Tassy's (1995) interpretation of KI 64'92 as belonging to Elephantidae. It is worth noting, however, that stegodonts convergently evolved this chewing motion (Saegusa, 1996a), so some care should be exercised in framing the adaptation as an "elephant" trait.

TETRALOPHODON SP. INDET. (Geraads et al., 2002) [Chorora]

(Tables 5.1 and 5.3; Figure 5.1)

Synonymy—Proboscidea indet., *Mastodon* sp. longirostrine group, "close to Elephantidae and particularly the Stegotetrabelodontinae" Tiercelin et al., 1979; *Stegotetrabelodon* n. sp. Geraads et al., 2002

Another example of African *Tetralophodon* that bridges the phylogenetic and morphological gap between gomphotheriids and elephantids occurs at the late Miocene site of Chorora, Ethiopia (Table 5.1). This occurrence consists of a right m3 (AL 1000–1), with a lophid formula of 6x. The specimen is distinguished from Afro-Arabian stegotetrabelodont m3s by the greater girth and height of its main conelets relative to its mesoconelets, fewer lophids, and a less plate-like construction of each lophid (Geraads et al., 2002). It preserves a covering of cementum around its lophids and exhibits large pretrite accessory conules posterior to the first five lophids, a strong posttrite accessory conule posterior to lophid 1, and tiny posterior posttrite accessory conules in association with the other lophids. The greatest width of the crown is at its base, at lophid 3. There is longitudinal curvature of the crown, convex to the lingual side. Bilaterally, the lophids slope inward toward their apices. Each lophid is formed of 3–5 conelets; three low, stout conelets that are closely appressed to the lophid 6 form a postcingulid rather than a nascent seventh lophid. The estimated crown length is 182 mm, and the width is approximately 78 mm.

The Chorora Fm. sediments that produced the specimen and other faunal elements include diatomites, pumice, tuffs, clays, sand, and gravel; dating of rhyolithes bracketing the fossiliferous layer yields an age of about 10.5 Ma (Tiercelin et al., 1979; Geraads et al., 2002). The fauna is distinctive of the early part of the late Miocene, along with the proboscidean specimen critically including the equid *Hipparion* cf. *primigenium*, a probable sivatheriine giraffe, the primitive hippopotamid *Kenyapotamus*, a bovid species recalling tragelaphines, a small neotragine bovid, chalicotheres, dicerotine rhinos, the viverrid carnivore *Herpestides*, and the rare felid *Machairodus* (more typical of the latter part of the late Miocene in Africa) (Geraads et al., 2002). The *Hipparion* from Chorora is one of the earliest well-dated occurrences of this genus in Africa (see Rook et al., 2019) marking the near isochrony of the immigration event of this equid taxon into the Old World (Geraads et al., 2002).

TETRALOPHODON SP. INDET. Geraads, 1989 [Gebel Krechem el Artsouma]

(Tables 5.1 and 5.3; Figure 5.1)

Synonymy—Tetralophodon cf. longirostris Geraads, 1989

A few isolated cheek teeth attributable to *Tetralophodon* were recovered from late Miocene Gebel Krechem el Artsouma, in accompaniment with an isolated fragment of a choerolophodont molar and a modest fauna of other vertebrates typical of the early part of the late Miocene (Geraads, 1989). The most informative of these specimens is a left M2 with a lophid formula of x4x. This molar has a relatively low, simple crown showing no cementum and thick enamel. Overall size is modest among molars of generally massive-sized *Tetralophodon* individuals (Table 5.3). Lophids are composed of four conelets, and the postcingulid has a small number of low conelets tightly appressed on the last lophid. Pretrite half-lophids 1 and 2 have formed enamel trefoils with wear, evidently incorporating anterior and posterior accessory conules; the posttrite half-lophids 1 and 2 are less heavily worn (transverse wear gradient indicative of Phase II lateral motion of teeth in mastication) but also exhibit a hint of a trefoil pattern, and both pre- and posttrite half-lophids 3–4 have distinct posterior accessory conules attached to them. There is a distinct but very transversely compressed median sulcus spanning the distance of the crown. This specimen resembles *Tetralophodon* M2s from the Samburu Hills, Kenya. Geraads (1989) felt that the molar has greater morphological similarity to *T. longirostris* than to *T. atticus*.

A large (Table 5.3), worn p4, OM 1–21, has a lophid formula of 2x, with a much wider posterior lophid. The postcingulum is well expressed on the lingual side in particular and is as tall as the last lophid, to which it is closely appressed. Primitively, it exhibits a large accessory conule in the transverse valley between lophids. The smaller size of the postcingulum and presence of the impressive accessory conule in OM 1–21 distinguishes it from the stegotetrabelodont p4 from Mpesida, Kenya, KNM-MP 47.

A possible P3, OM I/II, is very reminiscent in size (Table 5.3) and morphology of a P3 from the Samburu Hills, Kenya, KNM-SH 12312, with a lophid formula of x2x. The posttrite half-lophids are each formed of large, single conelet, and pretrite half-lophid 1 exhibits a trefoil wear pattern, suggesting incorporation of small anterior and posterior accessory conules. The specimen is widest posteriorly and has a broad interproximal facet for contact with dP4. These specimens could almost pass as antimeres in their similarity. They are distinguished from stegotetrabelodont upper premolars by the presence of accessory conules and fewer conelets contributing to each half-loph.

TETRALOPHODON SP. INDET. Sanders, 2022 [Hamra]

(Tables 5.1 and 5.3; Figures 5.1 and 5.4)

Because the Baynunah Fm. has been dated biochronologically to the latter half of the late Miocene, in the interval of 8.0–6.0 Ma (and likely >6.5 Ma in age) (Bibi et al., 2013), the recovery of a *Tetralophodon* palate with right and left M2–3 at the locality of Hamra 5 was unexpected (Sanders, 2022). In the last half of the late Miocene, primitive elephants had elsewhere replaced *Tetralophodon* throughout Afro-Arabia (Sanders et al., 2010a), and indeed, stegotetrabelodonts are the main proboscidean inhabitants of sites in the Baynunah Fm. (Sanders, 2022). Other proboscideans from the Baynunah Fm. include rare deinotheres and a single specimen of another, indeterminate gomphotheriid (Sanders, 2022). The Hamra specimen, AUH 899, is the fragmentary remnant of a cranium, preserving mostly a palate with dentition (Figure 5.4F). The palate is narrow, broadest between the anterior end of the M3s and most constricted transversely across the anterior aspect of the M2s, and the left zygomatic is rooted low on the maxilla at the anterior end of M3. A slight lateral flare of the left maxilla at its anteriormost preserved end suggests an outward angulation of the tusk sheaths at their bases.

The M2s are heavily worn but preserve a loph formula of x4x, with a substantial postcingulum. A strong interproximal facet is present on the anterior face of each molar, indicating the former presence of M1s, but the alveolar bone that housed these molars has been resorbed completely. M2–3 wear in this individual is consistent with age-grade stage XIX out of 23 stages, in Tassy's (2013) scheme for *Gomphotherium angustidens*, and age-grade stage XIX of Laws' (1966) scheme for modern African elephants, equivalent to fully adult individuals approximately 32 years old. M2 lophs are composed of four to five conelets; enamel is very thick (Table 5.3) and unfolded, and only traces of cementum remain in the floor of the transverse valleys in these molars and the M3s. Lamellar frequency indicates anteroposteriorly broad loph spacing (Table 5.3). Enamel wear figures are like those of gomphotheriines, with the formation of pretrite trefoils, achieved by incorporation of strong anterior accessory conules into lophs 1–4 and posterior accessory conules into lophs 1–2.

Like the M2s, the M3s are large and broad for their tooth position with exceptionally thick enamel and very low lamellar frequencies (Table 5.3). Each M3 has five full lophs and on the left side, the postcingulum comprises a nascent sixth loph. Wear is heaviest on the lingual (pretrite) side of the crown in the M2–3s. The pre- to posttrite wear gradient indicates a chewing mechanism similar to that of gomphotheres, with Phase II lateral motion of the molars (von Koenigswald, 2016). In addition, the greater wear on the anterior lophs of AUH 899 than in M3 of *Tetralophodon* sp. nov. from the Samburu Hills, Kenya (KNM-SH 15858), even though both have the same degree of wear across the M2s, suggests a steeper angle of eruption in the Samburu Hills specimen.

The Hamra specimen is more comparable to *T. longirostris* from Europe in this regard (see Nakaya, 1994). Anterior and posterior accessory conules are present on lophs 1–4 on the pretrite side, forming trefoil enamel outlines in more worn half-lophs 1–2; the posttrite side of the crown exhibits a strong posterior accessory conule behind loph 1 and a small anterior accessory conule is fused to loph 3.

Considered as a group, Afro-Arabian *Tetralophodon* appears to have immigrated into the continent from Eurasia as early as the close of the middle Miocene, and certainly by the beginning of the late Miocene. The closest resemblance of the most primitive morphologies in this assemblage is with *T. longirostris*. There is some evidence for minor differences among the constituents of the Afro-Arabian assemblage and with *T. longirostris*, suggesting regional differentiation of the genus once it arrived on the continent. The proximity of the genus geographically and temporally to the first elephants suggests that tetralophodonts evolved into elephants in Afro-Arabia, soon after 8.0 Ma. This idea has been supported by "transitional" tetralophodont forms at sites such as Kisegi, Chorora, and Nakali (Table 5.1; Tassy, 1995; Geraads et al., 2002; Saegusa et al., 2014).

ANANCINAE Hay, 1922
ANANCUS Aymard in Dorlhac, 1855

Anancine gomphotheres originated in Eurasia as early as MN 11–12 in the Turolian, during the second half of the late Miocene (Tassy, 1986; Markov, 2008; Konidaris and Roussiakis, 2019). South Asian *Anancus perimensis* is documented to be older than 8.5 Ma in MN 11, and European *A. lehmanni* to be about as old as 7.5 Ma, predating the oldest evidence for *Anancus* in Africa. Anancine gomphotheres quickly made their way to Africa after their origin in Eurasia (Coppens et al., 1978; Tassy, 1985; Tobien et al., 1988), and their first appearance on the continent is contemporaneous with the immigration of stegodonts and the origin of elephants (Tassy, 1986). Anancine gomphothere occurrences are ubiquitously sympatric with early elephants in Africa (Sanders et al., 2010a). At least one phylogenetic analysis of the genus suggests that African anancines are paraphyletic and immigrated into the continent over multiple episodes (Tassy, 1986). Anancinae is monogeneric (*Anancus*) and was widely distributed throughout the Old World during the late Miocene to the end of the late Pliocene, probably derived from *Tetralophodon* (Tobien, 1973a; Coppens et al., 1978; Mebrate and Kalb, 1985; Tassy, 1985, 1986, 1996a; Tobien et al., 1988; Metz-Muller, 1995; Kalb et al., 1996a, b; Shoshani, 1996; Göhlich, 1999; Markov, 2008; Sanders et al., 2010a). In Africa, anancine gomphotheres apparently persisted into the early Pleistocene (Arambourg, 1970; Geraads and Amani, 1998; Geraads, 2002; Sahnouni et al., 2002), although in East Africa their last record is mid-Pliocene in age (Tassy, 1986; Kalb et al., 1996b; Sanders, 2007, 2011). Occurrence of the genus in Africa is cosmopolitan, stretching the length and (in North Africa) the breadth of the continent, encompassing the species *A. kenyensis*, *A. ultimus*, *A. petrocchii*, *A. capensis*, and *A. osiris*.

Anancus is defined by short, wide crania with domed vaults, very raised basicrania, enlarged tympanic bullae, straight upper tusks lacking enamel, brevirostrine mandibles without lower tusks, tetralophodont or pentalophodont intermediate molars, loss of premolars in most species, and anancoidy of half-loph(id)s, such that pretrite half-lophs are offset anterior to posttrite half-lophs and posttrite half-lophids are transversely offset anterior to pretrite half-lophids (Petrocchi, 1954; Coppens, 1967; Coppens et al., 1978; Tassy, 1985, 1986; Kalb and Mebrate, 1993; Sanders, 2007, 2011; Sanders et al., 2010a). In addition, there is a tendency for pretrite posterior accessory conules of upper molars to be reduced and for reduction and fusion of the pretrite mesoconelet with the pretrite anterior accessory conule in lower molars (Tassy, 1986).

Although anancine gomphotheres may have shared with elephants descent from *Tetralophodon* ancestors, the highly apomorphic anancoidy of their molars makes it unlikely that they have any direct phylogenetic connection with the origin or subsequent evolution of elephants. Based on the bunolophodont, brachyodont-to-subhypsodont condition of their molars, anancine gomphotheres were morphologically interpreted as having been browsing forest forms (e.g., Smart, 1976). From the time of their arrival in Africa until their extinction on the continent, however, anancine gomphotheres appear to have been C_4 grazers or mixed-feeders with a substantial intake of C_4 grasses (Cerling et al., 1999, 2003; Zazzo et al., 2000; Harris et al., 2003; Semaw et al., 2005; Kingston and Harrison, 2007; Levin et al., 2008; Uno et al., 2011; Roche et al., 2013; Drapeau et al., 2014; Manthi et al., 2017) except in southern Africa where C_3 grasses are inferred to have been prevalent (Franz-Odendaal et al., 2002; Groenewald et al., 2020), and therefore in many ecosystems they competed with early elephants (and toward the latter half of their existence on the continent, with stegodonts) for forage. In contrast, there is evidence that early Pleistocene European *Anancus arvernensis* partitioned resources when sympatric with mammoths by shifting from grass-dominated mixed feeding to browsing diets that included fruits, seeds, bark, and twigs (Rivals et al., 2014).

Elephants, with their molar plate ridges oriented to make efficient use of cutting surface in proal horizontal shearing, had a surface area advantage over the oblique shearing motion of anancine gomphotheres, which only covered two-thirds of the same cutting surface, primarily working against the obliquely-oriented lower buccal and upper lingual half-loph(id) enamel loops (Saegusa, 2020). Elephants increased their masticatory advantage as they steadily evolved greater numbers of plates in their molars over time. Further benefits were gained by elephants evolving greater hypsodonty and thicker cementum, extending the durability of individual molars, presumably contributing to longer lifespans and concomitant greater number of breeding cycles. In the context of these advantages, the failure to evolve significantly more loph(id)s, thick cementum, and hypsodont molar crowns (for unknown genetic and developmental reasons) comparable to the condition of elephant molars may well have contributed to the extinction of anancine gomphotheres in Africa.

ANANCUS KENYENSIS (MacInnes, 1942)
(Tables 5.1, 5.2, 5.4, and 6.1; Figures 5.1, 5.5, and 5.8)

Partial Synonymy—*Pentalophodon sivalensis kenyensis* MacInnes, 1942; *Anancus arvernensis* subsp. Dietrich, 1943; *A. arvernensis* subsp. Arambourg, 1945; *Pentalophodon sivalensis kenyensis* Arambourg, 1945; *A. kenyensis* Arambourg, 1947; *A. kenyensis* Leakey, 1967; *A. osiris* (in part), Coppens et al., 1978; *A. kenyensis* "*kenyensis* morph" Tassy, 1986; *A. kenyensis* "*kenyensis* morph" Brunet et al., 2000; *Anancus* cf. *A. kenyensis* Haile-Selassie et al., 2004; *Anancus* sp. indet., Haile-Selassie et al., 2004; *Anancus kenyensis* Sanders, 2011

Anancus kenyensis is conceived of and constituted differently in a number of distinct taxonomic schema. For example, Tassy (1986) included North African *A. petrocchii* in the species as an advanced morph, contrasted with a more primitive "*kenyensis* morph." Alternatively, based on progressive changes in occlusal complexity, loph(id) number, enamel folding, and expression of cementum in molar samples from the late Miocene-early Pliocene Adu-Asa and Sagantole Fms. of the Middle Awash, Ethiopia, Kalb et al. (1982a, b) subdivided the species into four time-successive stages, "A–D" (Kalb et al., 1982b). Later, these stages were revised to include stages A–C from the Adu-Asa Fm. and Haradaso Mb. and Beearyada beds of the Sagantole Fm. in *A. kenyensis* and stage D from the Kalaloo beds of the Sagantole Fm. in a very derived new (but unnamed) species of *Anancus* (Mebrate and Kalb, 1985). Subsequently, stage D was better defined and named as "*Anancus* sp. Sagantole-type," but unfortunately, anancine gomphotheres from Langebaanweg, South Africa were included in stage C (Kalb and Mebrate, 1993). In this study, *Anancus* sp. (Sagantole-type) included specimens from the Beearyada beds and the superjacent Kalaloo beds at the top of the Sagantole Fm. As a result of Kalb and Mebrate's meticulous study, a comprehensive and detailed description of a time-successive series of African anancine gomphotheres emerged, suggesting the anagenetic evolution of these proboscideans in eastern (and Central) Africa.

There is considerable dental morphological variation in the lineage (see Saegusa and Haile-Selassie, 2009), no matter how it is partitioned, which is not surprising given its geochronological extent (Tables 5.1 and 6.1). Sanders (2011) taxonomically formalized the contrasts in morphology by recognizing *A. kenyensis* for Tassy's (1986) primitive *A. kenyensis* "*kenyensis* morph," or Kalb and Mebrate's (1993) stages A-B and part of stage C, characterized by tetralophodont intermediate molars (Figure 5.5A and B), modest expression of anancoidy, little or no deposition of crown cementum, and limited occlusal complexity. *Anancus ultimus* was erected to encompass the advanced *A. kenyensis* "*petrocchii* morph" (but not including the species *A. petrocchii*), or part of stage C and *A. kenyensis* "Sagantole-type," defined by stronger anancoidy, greater expression of cementum, pentalophodonty of intermediate molars, third molars with six-seven loph(id)s, and greater occlusal complexity including occasionally folded enamel and accessory conules sometimes doubled and often continued to the posterior half of molar crowns (Sanders, 2011). *Anancus kenyensis* is further diagnosed by a short, high cranium with a strongly raised basicranium, large tympanic bullae, and short mandibular symphysis lacking lower tusks. In addition, adult upper tusks are straight and lack enamel bands. Although Hautier et al. (2009) list loss of premolars as a trait of *A. kenyensis*, Tassy (1986, 2003) described P4s of *A. kenyensis* from Kaperyon and Lothagam, Kenya. These have dimensions of L=54.0 mm, W=48.0 mm, and L=45.2 mm, W=44.3 mm, respectively.

Fossils of *A. kenyensis* sensu Sanders (2011) comprise one of the better species samples of African proboscideans. Intermediate molars from the lower fossil beds at the nominotypical site of Kanam, Kenya, including the type M2 (M15400), have four loph(id)s, very thick enamel, only traces of cementum, weak ananoidy, and simple crown morphology (Figure 5.5B). M2 has pretrite accessory conules posterior to lophs 1–3 and posttrite accessory conules expressed as enamel folds posterior to lophs 1–3.

Molar specimens from the late Miocene Lukeino Fm. in the Tugen Hills, Kenya are also primitive in morphology. Lower third molars vary in lophid formulae from x5x to 6x, exhibit weak anancoidy, very little or no cementum, have thick, unfolded enamel (usually, ET>5 mm), pretrite accessory conules posterior to the first three or four lophids, and smaller posttrite accessory conules posterior to the first

TABLE 5.4
Comparative Aspects of African Anancine Gomphothere Molars

Species	Intermediate Molar Loph(id) Formula	Third Molar Loph(id) Formulae	Degree of Anancoidy	Crown Complexity	Enamel Folding
Anancus kenyensis	Tetralophodont	5–6	Weak	Simple	Absent-coarse
A. ultimus	Pentalophodont	6–7	Pronounced	Complex	Absent-coarse-fine
A. petrocchii	Pentalophodont	6–7	Weak	Moderate-complex	Absent
A. osiris	Tetralophodont	5–6	Pronounced	Simple	Absent-coarse
A. capensis	Tetralophodont	6–7	Pronounced	Complex	Coarse-fine

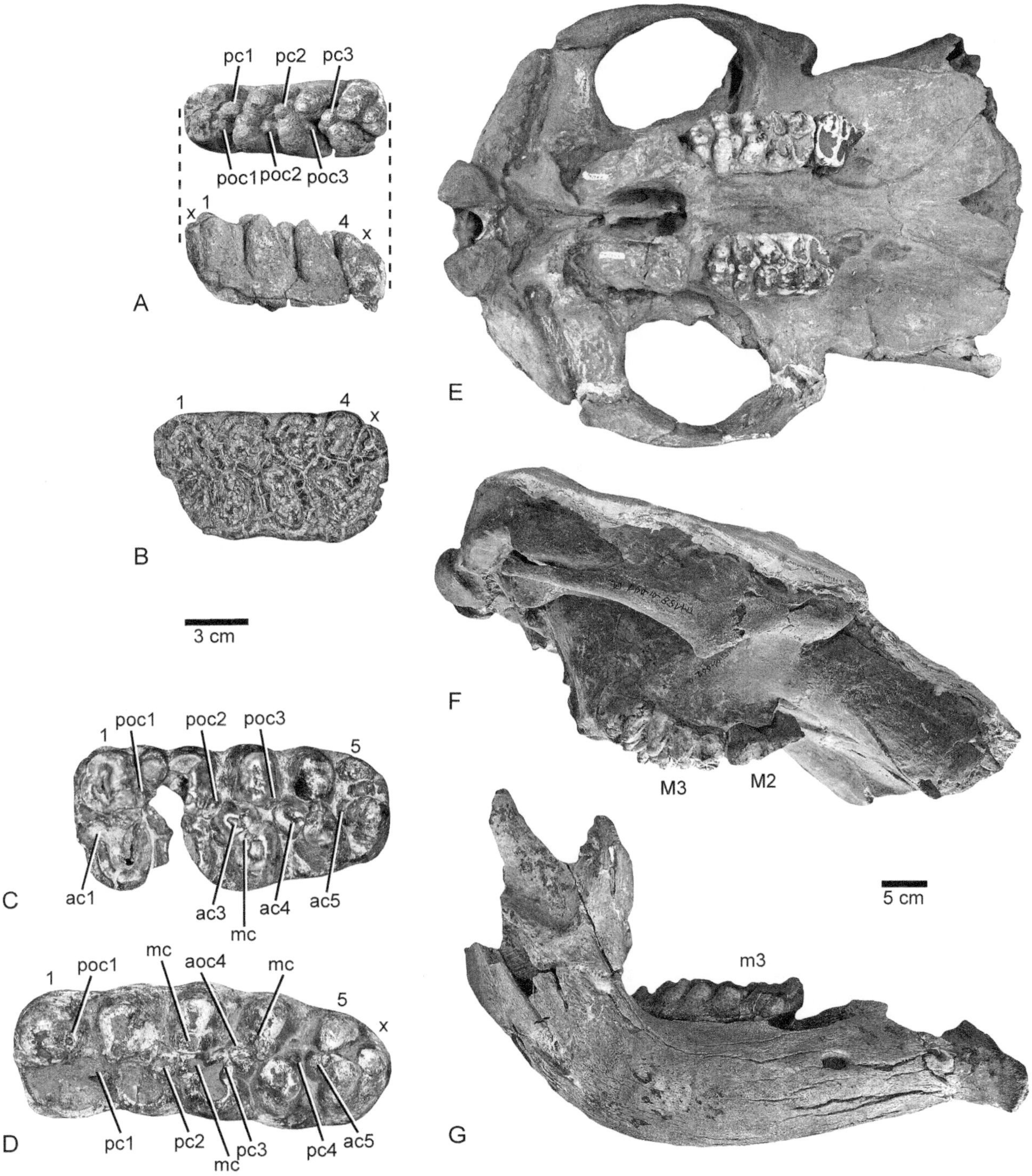

FIGURE 5.5 Molars and skull elements of *Anancus kenyensis*. Abbreviations: ac, pretrite anterior accessory conule; aoc, posttrite anterior accessory conule; M/m, upper/lower molar; mc, mesoconelet; pc, pretrite posterior accessory conule; poc, posttrite posterior accessory conule; x, pre- or postcingulum (id); 1, 2, 3, …, loph(id) number counted from the anterior end of the crown. (A–D) anterior to the left; (E–G) anterior to the right. (A) Right m1, WM 1800/92 from the Manonga Valley, Tanzania, occlusal and lateral views. (B) Left M2, M15400 from Kanam, Kenya (type), occlusal view. (C) Left M3, KNM-LU 57 from the Lukeino Fm., Tugen Hills, Kenya, occlusal view. (D) Left m3, KNM-LU 57 from the Lukeino Fm., Tugen Hills, Kenya, occlusal view. (A–D) to the same scale. (E) Cranium, TM 158-01-01 from Toros Menalla, Chad, Ventral view. (F) Cranium, TM 158-01-01 from Toros Menalla, Chad, right lateral view. Note that the dorsal surface of the cranium is extensively damaged and missing bone. (G) Mandible, TM 158-01-01 from Toros Menalla, Chad, right lateral view. (E–G) approximately to the same scale. (Images in [E–G] courtesy of L. Hautier [Hautier et al., 2009:figs.2a, c, 3b].)

two-four lophids (Figure 5.5C and D). M3s in this assemblage have loph formulae of x5x, weak anancoidy, low lamellar frequencies (LF=3.5–3.75), only traces of cementum, thick, unfolded enamel, anterior accessories conules associated with pretrite half-lophs 1–4 or 1–5, and more limited expression of posttrite posterior accessory conules either absent or limited to the mesial half of the crown. Third molars from higher in the Tugen Hills sequence, out of the lower Chemeron Fm., tend to have six loph(id)s but also possess thick, unfolded enamel (ET>5.0 mm), brachyodont crowns, weak anancoid arrangement of half-loph(id) pairs, and posttrite half-lophid accessory conules limited to the mesial half of the crown. Both pre- and posttrite half-loph(id)s tend to have mesoconelets to accompany their main, outer conelets.

Only very limited dental samples of *A. kenyensis* have been recovered from Tanzania, from the late Miocene Manonga Valley and Lower Laetolil Beds at Esere 1 (=Endolelo) (Sanders, 1997, 2011). In the Manonga Valley sample, a mandible (WM 1800/92) has a brevirostrine symphysis lacking tusks, and its dp4s and m1s are tetralophodont (Figure 5.5A). These Manonga anancines co-existed with stegotetrabelodonts, stegodonts, deinotheres, and *Primelephas korotorensis*. The Lower Laetolil Beds specimens are valuable because they indicate a time interval for the replacement of *A. kenyensis* by *A. ultimus*. From Esere 1, specimens collected by Louis Leakey in 1935 include an M3 (BM(NH) 32958) that has only five lophs, traces of cementum in its transverse valleys, and accessory conules only expressed to the middle of the crown. An intermediate molar of the same individual (Endo:LS BKE 35), is very worn but definitely tetralophodont. An incomplete astragalus from Esere 1 (EP 1671/98) has a more saddle-shaped tibial articular surface than is typical for elephants and probably belongs with these teeth. However, intermediate molars from Kakesio, Tanzania, also from the Lower Laetolil Beds, are more progressive in being pentalophodont, and third molars from the site have seven lophs, marking their identity as *A. ultimus* (Sanders, 2011). At Esere 1, *A. kenyensis* co-existed with *Loxodonta cookei*, whereas at Kakesio *A. ultimus* was sympatric with the more advanced *L. exoptata*. The Lower Laetolil Beds cover the interval 4.4–3.85 Ma (Drake and Curtis, 1987; Deino, 2011).

The best gnathodental sample of *A. kenyensis* is documented from the Asa Koma and Kuseralee (=Rawa) Mbs. of the Adu-Asa Fm. in the Middle Awash, Ethiopia (Saegusa and Haile-Selassie, 2009), dated to 5.77–5.44 Ma and 5.2 Ma, respectively (Haile-Selassie and WoldeGabriel, 2009). The faunas from these members are robust and include the early hominin *Ardipithecus kadabba*, along with colobine and papionine cercopithecoids, soricids, leporids, sciurid, murid, hystricid, and thryonomyid rodents, canids, ursids, mustelids, viverrids, herpestids, hyaenas, felids, tubulidentates, hyraxes, hipparionine equids, rhinos, suids, hippos, palaeotragine, giraffine, and sivatheriine giraffes, eight bovid tribes, and a rare but welcome diverse assemblage of birds (Haile-Selassie and WoldeGabriel, 2009). The molar sample exhibits a great amount of morphometric variation, though not outside the range known for other elephantimorph species. Anancoidy ranges from weak to strong. Intermediate molars are tetralophodont. Third molars vary from having five to six loph(id)s. In the expression of accessory conules, Saegusa and Haile-Selassie (2009) felt that the Asa Koma and Rawa Mbs. molars, particularly m3 with very distinct posterior posttrite accessories 1–3 and anterior pretrite accessory 2, are more derived than those of Lukeino, Mpesida, Lemudong'o, and the Nawata Fm. of Lothagam, Kenya. Middle Awash mandibles have robust corpora with a gentle curve along their ventral surface, proportionally deeper alveolar segments than in elephants or stegodonts, deep masseteric fossae, a short symphyseal spout lacking tusks, tall rami, laterally-flared coronoid processes, and anteriorly sloping condyles. Mental foramina are small. In these members, anancine gomphotheres co-existed with deinotheres, as well as the elephant genera *Primelephas*, *Loxodonta*, and *Mammuthus* (Haile-Selassie and WoldeGabriel, 2009).

A smaller sample of *A. kenyensis* molars is reported from late Miocene Toros Menalla, Kossom Bougoudi, and early Pliocene Kollé, Chad (Mackaye, 2001). The Kollé specimens are anomalously geologically young (~4.0 Ma), perhaps reflecting asynchrony of evolution of the lineage in different basins. A Kollé palate (KL4.96.008) carries tetralophodont right and left M2s; the postcingulum in each constitutes a nascent fifth loph. Anancoidy is weak but crown morphology is complex with doubled posttrite accessory conules. Dimensions of the M2s are L=150.0 mm in each, and W=72.5–75.7 mm. A fragment of an M3 from Kollé (KL5.98.81) also has doubled posterior posttrite accessory conules and well-developed mesoconelets. An m2 from older, late Miocene Kossom Bougoudi (KB4.00.001) is tetralophodont and is small (L=131.0 mm, W=76.0 mm), but has marked anancoidy and coarsely folded, very thick enamel (ET=6.0–7.0 mm). Overall, the Chadian *A. kenyensis* sample, particularly from late Miocene Toros Menalla, has broadly spaced loph(id)s, very thick enamel, low crown height, weak anancoidy, regular presence of posterior posttrite accessory conules in upper molars, absence of posttrite accessory conules in lower molars, and is characterized by tetralophodont intermediate molars. It would be extraordinary if *A. kenyensis* was sympatric with *A. petrocchii* and *A. osiris* at Toros Menalla and with *A. osiris* at Kossom Bougoudi and Kollé, as suggested by Mackaye (2001). These anancine assemblages would benefit from restudy in the context of the variation documented in the abundant *A. kenyensis* sample from the Adu-Asa Fm. (Saegusa and Haile-Selassie, 2009).

Skull morphology is particularly well documented by specimens recovered from the Lukeino Fm. in the Tugen Hills, Kenya (Tassy, 1986) and Toros Menalla, Chad (Hautier et al., 2009) (Table 5.1). KNM-LU 975 (not -LU 795 as in Tassy, 1986) from the Lukeino Fm. features a largely complete mandible with m3s, a short, deep corpus, brevirostrine symphysis with a blunt spout, high ramus with vertical anterior and posterior margins, and a substantially elevated, transversely broad condyle contrasted with a considerably lower coronoid process that is slightly laterally everted. The corpus is broadest posteriorly (144.5 mm) and the anterior margin of the ramus is forward of the distal end of m3. A large mental foramen is located just anterior to m3, about mid-height on the corpus. The greatest length of the mandible is 630 mm, and the height of the condyles above the level of the m3s is 140–143 mm (Tassy, 1986). The internal width of the symphyseal gutter is 48 mm. The m3s vary in length from 175 to 180 mm and both are relatively narrow at 78.7 mm. Enamel is unfolded and quite thick at ET=6.0–6.4 mm. Anancoidy is mild. The crowns are very worn, but it is possible to ascertain the presence of posttrite posterior accessory conules in lophids 2–4, which appear to articulate with or nearly touch pretrite anterior accessory conules of lophids 4–5. The lophid formula is either 5Xx or 6x, with the nascent sixth lophid composed of two conelets and the postcingulid of one conelet.

The associated cranium of KNM-LU 975 is distinguished by its markedly raised and steeply angled basicranium, resulting in the external auditory meatus and condyles being elevated well above the level of the palate. Its occipital condyles project modestly posteriorly, are relatively large, and face a little more ventrally than posteriorly; it also has a wide, short rostrum that is downturned at a greater angle than the neurocranium, a broad external incisive fossa, and laterally-flared zygomatic arches with narrow jugal segments. The tusk alveoli are sub-parallel and not widely flaring, and in cross-section they are flattened and ovoid. There is no anterior convergence of the M3s and the palate is relatively wide. The back of the palate is posterior to the M3s. The glenoids are shallow, laterally broad, anteroposteriorly narrow, and their anterior margins are mesial to the posterior margins of the choanae. There are no post-glenoid fossae. Auditory bullae are large. Much of the dorsal surface of the cranium is weathered away, making it difficult to assess overall cranial height, but the raised basicranium suggests that the skull was higher than in more primitive gomphotheres. The M3s are pentalophodont with a strong postcingulum composed of six conelets, crown morphology is simple, and anancoidy is very weak. The crown length for each is 175.0 mm and widths are 76.3–76.6 mm, enamel thickness varies from 5.6 to 6.6 mm, and lamellar frequency is low, LF=3.5. The crown is worn but each loph has a salient pretrite anterior accessory conule and only loph 1 has a posterior accessory conule. Posterior posttrite accessory conules are present in lophs 1–3, which differentiates the M3s of the cranium from the even simpler morphology of the type M3 of *A. osiris* (which also exhibits stronger anancoidy).

Specimen TM158-01-01 from Toros Menalla confirms and adds important information about the morphological condition of *A. kenyensis* skulls (Figure 5.5E and F; Hautier et al., 2009). It also has suffered damage to the dorsum of its brain case, but like KNM-LU 975 the basicranium is raised well above the level of the palate, and the occipital condyles are large and project slightly behind the occipital planum. The greatest length of the cranium is 860 mm (Hautier et al., 2009). The rostrum is short and very broad, but the tusk alveoli indicate that the tusks were sub-parallel in placement. The zygomatic arches are short and flare laterally; the width across the zygomatics is 620 mm. The cranium retains the remnant of the right M2 and both M3s. Most importantly, the occipital planum is very well preserved. It is low and very broad, with the exoccipitals and supraoccipitals stretched out laterally to form a semi-rectangular occipital shape. The posterior of the choanae is distal to the anterior of the articular glenoid. The M3s have either four lophs and a prominent postcingulum, or five lophs. Anancoidy is not pronounced. Pretrite half-lophs 1–4 have prominent anterior accessory conules and pretrite half-lophs 1–2 have swellings representing posterior accessory conules. On the posttrite side, half-lophs 1–3 exhibit posterior accessory conules. The greatest length of these M3s is 155 mm; width varies from 82 to 86 mm, and crown height from 56 to 58 mm (HI=65–70); enamel is very thick, ET=6.0 mm (Hautier et al., 2009).

The associated mandible is one of several of *A. kenyensis* from Toros Menalla (Figure 5.5G; Hautier et al., 2009). Its corpora and symphysis are well preserved, but the rami are posteriorly damaged. The symphyseal spout is robust and more prominent than in the mandible of KNM-LU 975, and the symphysis is primitively elongated (L=155 mm), with a broad gutter. There are no lower tusk alveoli. The width across the anterior of the rami is 407.5 mm. The corpora, including the spout, are long (390.5 mm), transversely narrow, and relatively low with a uniform height throughout (127–128 mm). The overall length of the mandible is 542.5 mm. The ramus is relatively quite high but not expanded anteroposteriorly or at the angle. A prominent mental foramen is located below the m2 on each side, a little above the mid-height of the corpus. The mandibular coronoid process is anteroposteriorly narrow superiorly, everted laterally, and not much lower in height than the articular process. Of the dentition, only the right m3 is well preserved, with five lophids and dimensions of L=186 mm, W=74.8 mm, H=60.7 mm (yielding a hypsodonty index of 81), and ET=5.5 mm. Anancoidy is mild and occlusal morphology is uncomplicated.

Within the subfamily, the phylogenetic relationships of *A. kenyensis* remain contentious. For Tassy (1986), *A. kenyensis* is a sister species to Asian *A. sivalensis*, united by the regular presence of posttrite accessory conules in upper molars, tendency for m2 to have posttrite accessory conules, and m2 pentalophodonty—but this latter character would only apply to an expanded definition of *A. kenyensis* that includes *A. petrocchii*. Saegusa and Haile-Selassie (2009) felt that *A. osiris* could have derived from *A. kenyensis* by simplification of its occlusal morphology, and that *A. sivalensis* is more similar to *A. arvernensis* than to *A. kenyensis*. Sanders (2008b, 2011) and Sanders et al. (2010a) retain *A. petrocchii* as a species separate from *A. kenyensis*, and place the derived stages of the *A. kenyensis* lineage into *A. ultimus*. Alternatively, Hautier et al. (2009) suggest synonymizing *A. osiris* with *A. kenyensis* but appear to accept *A. petrocchii* as a separate species. This hypothesis is not robustly supported by a detailed comparison of M3s

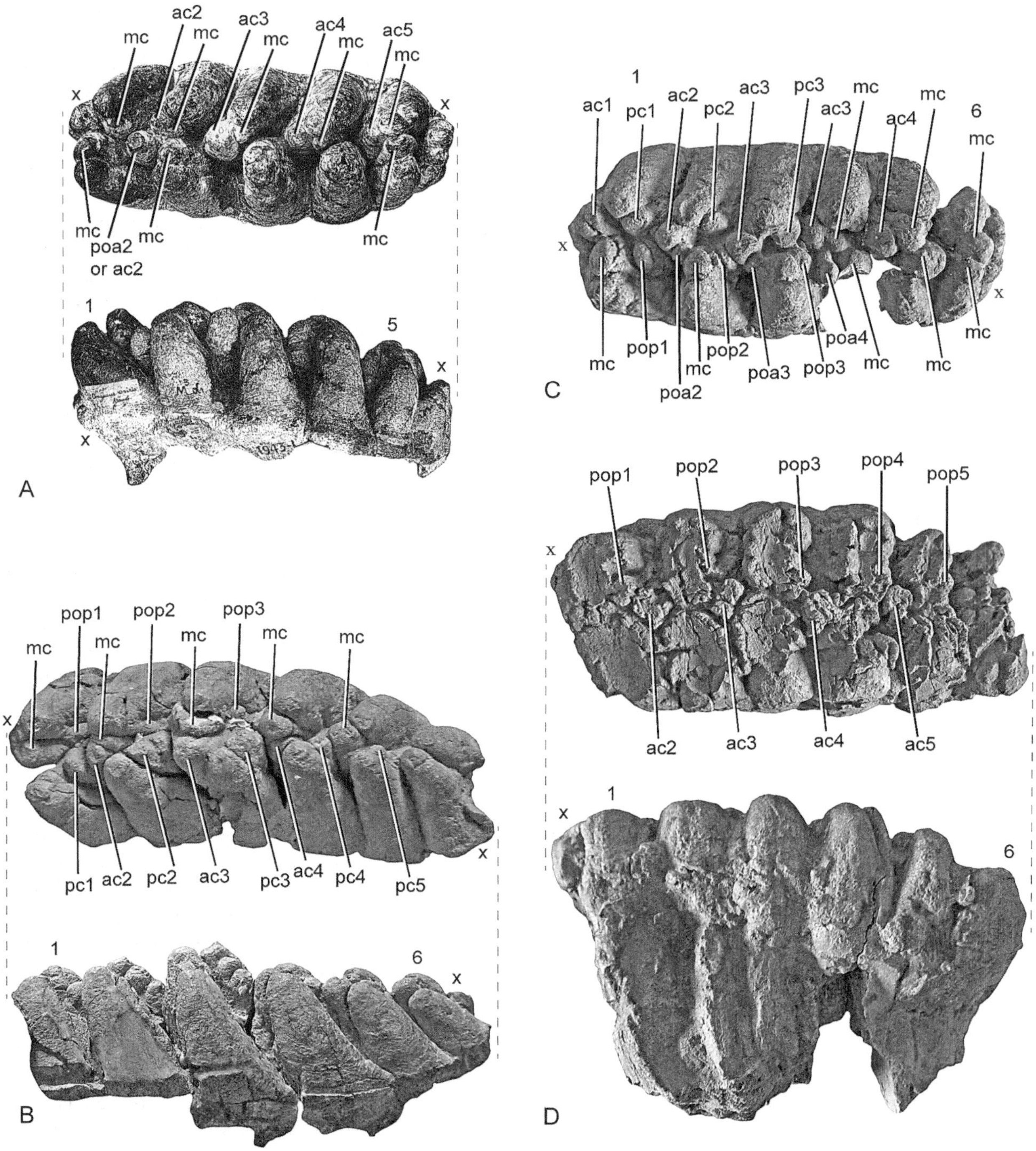

FIGURE 5.6 Third molars of *Anancus osiris* and *Anancus petrocchii*, not to exact scale. Abbreviations: ac, pretrite anterior accessory conule; mc, mesoconelet; pc, pretrite posterior accessory conule; poa, posttrite anterior accessory conule; pop, posttrite posterior accessory conule; x, pre- or postcingulum (id); 1, 2, 3, …, loph(id) number counted from the anterior end of the crown. Anterior to the left. (A) M3, specimen 1943-1 (type) from Wadi Natrun, Egypt, occlusal and lateral views, *Anancus osiris*. The length of the molar is 197 mm. (From Arambourg [1970:fig. 20] copyright Publications Scientifiques du Museum national d'Histoire naturelle, Paris.) (B) M3, proboscidean #1 from Sahabi, Libya, occlusal and lateral views, *Anancus petrocchii*. (C) M3, proboscidean #2 from Sahabi Libya, occlusal view, *Anancus petrocchii*. Pretrite anterior accessory conules are equivalent to Arambourg's (1970) "tubercles intermediares de neoformation." (D) M3, proboscidean #6 from Sahabi, Libya, occlusal and lateral views, *Anancus petrocchii*. Pretrite anterior accessory conules are equivalent to Arambourg's (1970) "tubercles intermediares de neoformation." (Images in [B–D] courtesy of N. T. Boaz.)

of these species, and most site-samples of *A. osiris* appear to be too young to belong to *A. kenyensis* (Tables 5.1, 6.1, and 7.1).

ANANCUS OSIRIS Arambourg, 1945

(Tables 5.1, 5.4, 6.1, and 7.1; Figures 5.1, 5.6, and 5.8)

Partial Synonymy—*Mastodon* Andrews, 1907; *Mastodon arvernensis* Deperet et al., 1925; *Anancus arvernensis* Dietrich, 1943; *Anancus (Mastodon) arvernensis* Ennouchi, 1949; *Anancus kenyensis* Hautier et al., 2009

Anancus osiris dates primarily to the late Pliocene-early Pleistocene and constitutes the last occurrence of the genus in Africa (Table 7.1). However, its record at Wadi Natrun, Egypt may be late Miocene in age, and there is some thought its presence at Lac Ichkeul, Tunisia dates to the late Miocene, as well (Coppens et al., 1978; Geraads, 1983; Thomas et al., 1982; Pickford et al., 1995). It is fitting that *Anancus osiris* was named after the Egyptian god of the afterlife, considering that the type derives from near the pharaonic mortuary site of Gizeh (Arambourg, 1945).

The type specimen (1943-1) is a right M3 composed of five lophs with a low postcingulum of five conelets tightly appressed to the last loph (Figure 5.6A). Dimensions are L=197.0 mm and W=84.0 mm (Arambourg, 1945). Anancoidy is pronounced and its tall, pyramidal lophs are subhypsodont, but the crown lacks cementum and is simple in occlusal morphology. Posttrite half-loph 1 is formed of two stout conelets, and the outer main conelet and mesoconelet of pretrite half-loph 1 are similarly stout and equal-sized, but the pretrite mesoconelet is advanced anterior to the posttrite half-loph (anancoidy). There is a large accessory conule posteromedial to pretrite main conelet 1. Posttrite half-loph 2 is formed of two conelets transversely oriented, and is paired with pretrite half-loph 2 formed of a large outer main conelet and a mesoconelet set mesial to the posttrite side. A prominent anterior pretrite accessory conule 2 is set oblique to the mesoconelet. Loph 3 is similar in configuration to loph 2, except that the posttrite mesoconelet is fused to its outer main conelet, and the main conelet along with the mesoconelet and anterior accessory conule of the pretrite side are mesial to the posttrite half-loph. Arambourg (1945) referred to these accessory conules as "tubercles intermédiaires de néoformation." Loph 4 is similar to loph 3 in morphology. In loph 5, both half-lophs are formed of a main, outer conelet, and mesoconelet; only anterior pretrite accessory conule 5 is substantially mesial to posttrite half-loph 5. The pretrite accessory conules contact the posttrite mesoconelets anterior to them. The type site is not well dated. Arambourg (1970) suggested its age as Lower Villafranchien; if so, it is late Pliocene.

Molars of geologically younger individuals attributed to the species may exhibit cementum (Pickford, 2003a) and five and a half or six loph(id)s in third molars paired with basic occlusal morphology, as in specimens from Aïn Boucherit, Algeria (Arambourg, 1970). Second molars, in contrast, may have pretrite accessory conules throughout the crown and also display posttrite accessory conules (Arambourg, 1970). Intermediate molars are tetralophodont. Thus, the species is characterized by intermediate molar tetralophodonty, simple third molar occlusal morphology, conspicuous anancoidy, and tall loph(id)s (Table 5.4).

A fragmentary palate from Grombalia, Tunisia has an anteriorly broken M2 with worn lophs but clearly strong anterior pretrite accessory conules; M3 has five lophs and its morphology is nearly identical with the type M3 (Arambourg, 1970:pl. II, fig. 4). Dimensions of the M3 are L=193.0 mm, W=83.0 mm, HI=66.

A broken m2 from Fouarat, Morocco (no. 1949-16) has simple occlusal morphology, with half-lophids formed of single, large conelets. Pretrite half-lophids have anterior and posterior accessory conules, and at least some posttrite half-lophids have anterior accessory conules. Anancoidy is marked, but there is no cementum.

A large m3 (L=225.0 mm, W=85.0 mm) from Aïn Boucherit, Algeria (no. 1956-4) exhibits weaker anancoidy and little to no development of accessory conules on its posttrite side, but it has six lophids and a relatively high crown for an anancine gomphothere (HI=88). Posterior accessory conules are present in association with pretrite half-lophids 1–4 (doubled behind half-lophid 1) (Arambourg, 1970:pl. I, fig. 3). This degree of hypsodonty is similar to that of an incomplete m3 (no. 1949-11) from Lac Ichkeul Tunisia (HI=87). This m3 has unremarkable anancoidy, particularly anteriorly, and is absent of accessory conules on its posttrite half-lophids. Pretrite half-lophids 2–5 have very prominent anterior accessory conules ("tubercles intermédiaires de néoformation") and a posterior pretrite accessory conules on lophid 1–2. Posttrite half-lophids are exceptionally simple, formed of only single conelets. There are only five lophids composing the crown of this specimen (Arambourg, 1970:fig. 22). Arambourg (1970) also listed a number of other molars from Aïn Boucherit, which may be the youngest anancine gomphothere site in Africa (Geraads and Amani, 1998; Geraads, 2002; Sahnouni et al., 2002). It is worth mentioning that an M2 (no. 1956-4) is tetralophodont, that a putative p4 (no. 1954-13) is in fact a dp4, and that a heavily worn m3 (no. 1949-5) has six lophids, with the sixth lophid comprised of a single large conelet, and exhibits a very simple crown composition arranged in a distinctive checkerboard pattern (Arambourg, 1970:pl. I:fig. 2).

Tassy (1986) considered *A. osiris* to be a sister species of *A. arvernensis*, to the exclusion of *A. kenyensis*+*A. sivalensis*, united by marked anancoidy and enhancement of crown height. These characters are liable to homoplasy and therefore not reliable synapomorphies. In addition, the occlusal morphology of *A. arvernensis* evolved to become considerably more complex than that of *A. osiris* (see Garrido and Arribas, 2014). On the contrary, Hautier et al. (2009) followed the suggestion of Cooke and Corydon (1970) to conclude that *A. osiris* is a junior synonym of *A. kenyensis*. In support of their hypothesis, Hautier and colleagues provide the example of Toros Menalla, Chad specimen TM158-01-01, which they feel has lower molars resembling those of *A. osiris* but upper molars with *A. kenyensis* traits (but see above). This observation notwithstanding, the distinctions of M3 in *A. osiris* from those of *A. kenyensis*, *A. ultimus*, *A. capensis*, and *A. petrocchii* are sufficient to maintain it as a valid species. *Anancus petrocchii* was hypothesized as the terminal derivation of *A. osiris* (Coppens, 1965; Coppens et al., 1978), but the younger geological age range of the latter and the less pronounced anancoidy and crown height of the former indicate that this is untenable.

A small handful of molar specimens from late Miocene Kossom Bougoudi and early Pliocene Kollé, Chad (Lebtard et al., 2008) were attributed by Mackaye (2001) to *A. osiris* and described as being sympatric with *A. kenyensis* at those sites. These specimens are strongly anancoid with simple crown morphology and include tetralophodont m1 specimen KB26.98.016, and m2 specimens KL9.98.054 and KL9.98.025. The m1 has dimensions of L=96.0mm, W=46.0mm, H=36.5mm, and ET=3.3mm; the m2 dimensions vary from L=129.0–154.0mm, W=72.0–75.0mm, H=58.0–60.0mm, and ET=3.0–5.0mm. If *A. osiris* is present in this sample, it would represent the extension of the species from North into Central Africa. Along with tetralophodonty and pronounced anancoidy, features supporting identification of these specimens with *A. osiris* include absence of lower molar pretrite mesoconelets, relatively high crowns, and contact between pretrite posterior accessory conules and posttrite mesoconelets (Mackaye, 2001).

In contrast to the *A. kenyensis*–*A. ultimus* lineage in eastern Africa, which underwent progressive evolutionary changes to molar occlusal features over time, including the addition of loph(id)s, development of cementum, folding of enamel, and greater complexity of mesoconelets and accessory conules, the northern and Central African *A. osiris* lineage appears to have remained relatively unchanged over a similar or greater span of geological time.

ANANCUS PETROCCHII Coppens, 1965

(Tables 5.1, 5.2, and 5.4; Figures 5.1, 5.6, and 5.8)

Synonymy—*Pentalophodon sivalensis* Petrocchi 1943, 1954; *Anancus* (*Pentalophodon*) *petrocchii* Coppens, 1965; *A. osiris* (in part) Arambourg, 1970; *A. petrocchii* Coppens et al., 1978; *Anancus kenyensis petrocchii*-morph Tassy, 1986; *A. petrocchii* Sanders et al., 2010a

Documented initially at the North African late Miocene site of Sahabi, Libya, the species was named for its discoverer, Carlo Petrocchi (Coppens, 1965). The original sample includes the type mandible with m3s (no. 1), a mandible with ruined molars (no. 2), a left dentary with m3 (no. 3), a left dentary with m3 (no. 4), a right dentary with m2–3 (no. 5), a right dentary with m2–3 (no. 6), a right dentary with m2 and m3 in crypt (no. 7), and a complete, isolated m3 (no. 8) (Petrocchi, 1954). The Sahabi anancine gomphotheres, including previously undescribed specimens collected in the 1930s, are all from localities in the western area of exposure of the Sahabi Fm., from either Upper Mb. U or Member V, and may be a million years or more younger than other proboscideans from the site collected primarily from unit U-1 of Mb. U (Boaz et al., 2008).

This species is distinguished from *A. kenyensis*, *A. capensis*, and *A. osiris* by its pentalophodont intermediate molars, and from *A. ultimus* by its simpler m3 occlusal morphology and massiveness of its pyramidal loph(id)s (Table 5.4). Its m3s are relatively narrow and range to a substantially larger size than m3s of other African congeners. *Anancus petrocchii* was included in Tassy's (1986) advanced phase of *A. kenyensis*, *A. kenyensis petrocchii*-morph (=*A. ultimus* of Sanders [2011] in part), but elsewhere morphological differences between a wide sampling of *A. ultimus* and *A. petrocchii* were maintained as sufficient to distinguish them as separate species (Sanders et al., 2010a; Sanders, 2011).

An emended diagnosis for *A. petrocchii* by Metz-Muller (2000) included upper molars with loss of contact between pretrite half-loph(id)s across the transverse valleys, primary contact between pre- and posttrite elements across the transverse valleys, and strong anancoidy; m3 with six lophids, reduced mesoconelets, variable development of posttrite accessory conules, simple crown structure, and pretrite accessory conules in multiple in anterior lophids; and m2 low-crowned and either tetra- or pentalophodont. Some of these features, notably intermediate molar tetralophodonty, are only valid if the small sample from Toros Menalla, Chad is included in the species (see Mackaye, 2001). Not all of these features appear to be valid for the species; for example, crown structure is not as simple as in *A. osiris* or many specimens of *A. kenyensis*, anancoidy is not pronounced, and if specimens from Toros Menalla, Chad are included in *A. petrocchii*, m3 may have as many as seven lophids. An m2 (TM21.97.001) and m1 (TM159.01.005) from the site attributed to the species (Mackaye, 2001) are both tetralophodont, though each is missing some morphology. The m2 is anomalously small for the species (L=125.0mm, W=70.0mm); the m1 is said to be a perfect fit for an m2–3 (TM29.97.003) from the site that is described as resembling m2–3 from Sahabi (Mackaye, 2001). Unfortunately, neither of the tetralophodont intermediate molars from Toros Menalla were figured by Mackaye (2001).

The m2 of TM29.97.003, however, is pentalophodont and therefore would be incongruous with tetralophodont m1 specimen TM159.01.005. The m3 has a lophid formula of x7x and is large, with dimensions of L=267.0mm and W=91.2mm. Anancoidy is not pronounced. It has thick enamel (5.5–6.0mm) and is mesodont in height (H=75.0; HI=82). The m2 has a lophid formula of x5x and is much larger than TM21.97.001, closer to the dimensions of m2 in the Sahabi sample: L=156.5mm, W=74.0mm, ET=4.8mm. In the m3, the postcingulid is formed of four stout conelets and is closely appressed to lophid 7. Lophid 7 has a large pretrite main conelet with a small apparent anterior accessory conule associated with it; it is mesial to the large conelet comprising the posttrite half-lophid. Lophid 6 has a large conelet comprising the pretrite half-lophid and a very large anterior accessory conule located medially ("tubercle intermédiaire de néoformation"); posttrite half-lophid 6 is composed of one large main conelet, located transversely across from the pretrite half-lophid. Lophid 5 has a pretrite half-lophid formed of a single large conelet; the posttrite half-lophid is angled forward and just slightly mesial to the pretrite half-lophid, associated with a centrally-positioned anterior accessory conule or what could be the posttrite mesoconelet. Pretrite half-lophid 4 has a large main conelet that is fused to a smaller mesoconelet and is distally offset from the large conelet and accessory conule or mesoconelet of posttrite half-lophid 4. Pretrite half-lophid 3 has two conelets and is strongly obliquely oriented to the main axis of the crown; the posttrite half-lophid has a large main conelet and a centrally-located accessory conule or mesoconelet, mesial to the pretrite side. Lophid 2 is distinguished from the more posterior lophids by the small posterior accessory conule fused to the conelet of its pretrite half-lophid, the presence of a posterior posttrite accessory conule, and a central anterior accessory conule located in a straight line with the main and mesoconelets of the posttrite side, which is aligned mesial to the pretrite half-lophid. The same morphology is observable in lophid 1, which is fronted by a few small conelets comprising the precingulid. There is no cementum. The accompanying pentalophodont m2 has a "checkerboard" pattern of alternating half-lophids but its crown is even simpler, with no evident accessory conules.

Mackaye (2001) listed specimen TM134.01.015, a mandible with right and left m3s, as belonging to *A. petrocchii*. Its third molars are anomalously narrow and at least the m3 on the right side has seven lophids. The presence of seven lophids is more derived than the typical composition of m3s in *A. kenyensis*, which Mackaye (2001) also identified at Toros Menalla; this is probably his primary reason for assigning m3 specimens TM134.01.015 and TM29.97.003 to *A. petrocchii*. Molars of *A. petrocchii* are usually relatively narrow, and it is possible that the small size of the m3s of TM134.01.015 are attributable to sexual dimorphism. The mandible lacks rami but has a brevirostrine symphysis without lower tusks, and is widest at the posterior end of its molars (Mackaye, 2001:pl.5).

Previously undescribed molars from Sahabi include an m3 (Sahabi proboscidean #1) that differs from TM29.97.003 by greater crown complexity, less pronounced anancoidy, and composition of only six lophids (Figure 5.6B). Crown height appears mesodont but no measurements are available for this specimen. Except for the last lophid, which is simply formed of a stout conelet in its pre- and posttrite sides, the lophids have pretrite and accessory conules (pcl–5, ac2-?4; pocl–3) and large posttrite mesoconelets centrally located in lophids 2–5, crowding occlusal space among and between the half-lophids. Relative to the long axis of the molar, the main conelets of posttrite half-lophids 2–4 are just slightly anterior in position to the main conelets of their accompanying pretrite half-lophids. There is no trace of cementum, but the molar was still forming in crypt at time of death. Only the posterior portion of m2 is preserved, unworn, or lightly worn.

Lower third molars of *A. petrocchii* reported by Petrocchi (1943, 1954) range in length from 221.0 to 290.0 mm, probably evidencing sexual dimorphism. Crown height in m3 no. 8 yields a mesodont hypsodonty index of 75. All of the m3s in Petrocchi's (1954) sample have six lophids. As the current accession in Libya of these specimens is uncertain, comparative study has been hampered by possible oversimplification of m3 occlusal morphology in Petrocchi's (1943, 1954) drawings (for example, Petrocchi, 1954:fig.20B, which depicts each lophid as composed of a single outer main conelet on the pre- and posttrite side accompanied by a large central conule). More detailed drawings of *A. petrocchii* m3s from Petrocchi (1954) suggest that anancoidy is greatest in lophids 2–4, and that pretrite half-lophid 1 has doubled posterior conules. In addition, there are large central accessory conules ("tubercles intermédiaire de néoformation") anterior to lophids 4–5. Posttrite half-lophids 2–3 possess prominent mesoconelets. The possible liberty taken by Petrocchi in depicting these teeth makes the recovery of the undescribed *A. petrocchii* molars particularly valuable for more accurately assessing molar morphology of the site sample. Petrocchi's (1954) photographs of specimens are not of high quality, but from them, it is clear that the lophids are of massive construction, and m2s are pentalophodont.

Upper third molars of *A. petrocchii* were not described by Petrocchi (1943, 1954), but there are a few in the recovered collection. Right M3 specimen #2 exhibits complex crown morphology (Figure 5.6C). It has a loph formula of x6x and pretrite anterior and posterior accessory conules through lophs 1–4. Loph 5 also has a pretrite anterior accessory conule. In addition, it has posttrite anterior and posterior accessory conules associated with lophs 1–3. Anancoidy is weak in this specimen and there is no cementum. Although anancoidy is more accentuated in the type M3 of *A. osiris*, it is less derived than the Sahabi M3 in having five lophs and no accessory conules posterior to its pretrite half-lophs or on the posttrite side.

A more heavily worn left M3, specimen #6, also has six lophs, weak anancoidy, and no cementum (Figure 5.6D). Dentine is deeply excavated within the enamel loops. Anterior accessory conules are present centrally, associated with pretrite half-lophs 2–5 (they are doubled in lophs 2–4). An enamel fold posterior to posttrite half-loph 1 may constitute an accessory conule. Occlusal wear has forced the enamel loops of the first three half-lophs into close proximity and may have obliterated or obscured some conules. Mesoconelets of posttrite half-lophids 1–4 are transversely stretched. Although occlusal morphology of M3 specimen #4 is poorly preserved, lateral view of the molar shows it has a loph formula of 6x.

Mandibles and dentaries of *A. petrocchii* from Sahabi are distinctive morphologically (see Petrocchi, 1954). They share with other anancine gomphotheres the traits of the absence of lower tusks and a short symphysis. However, the Sahabi specimens have long corpora that are externally of nearly equal height through their lengths, exhibit two large mental foramina (one lateral to the anterior end of the symphysis and the other anterolateral to the mesial

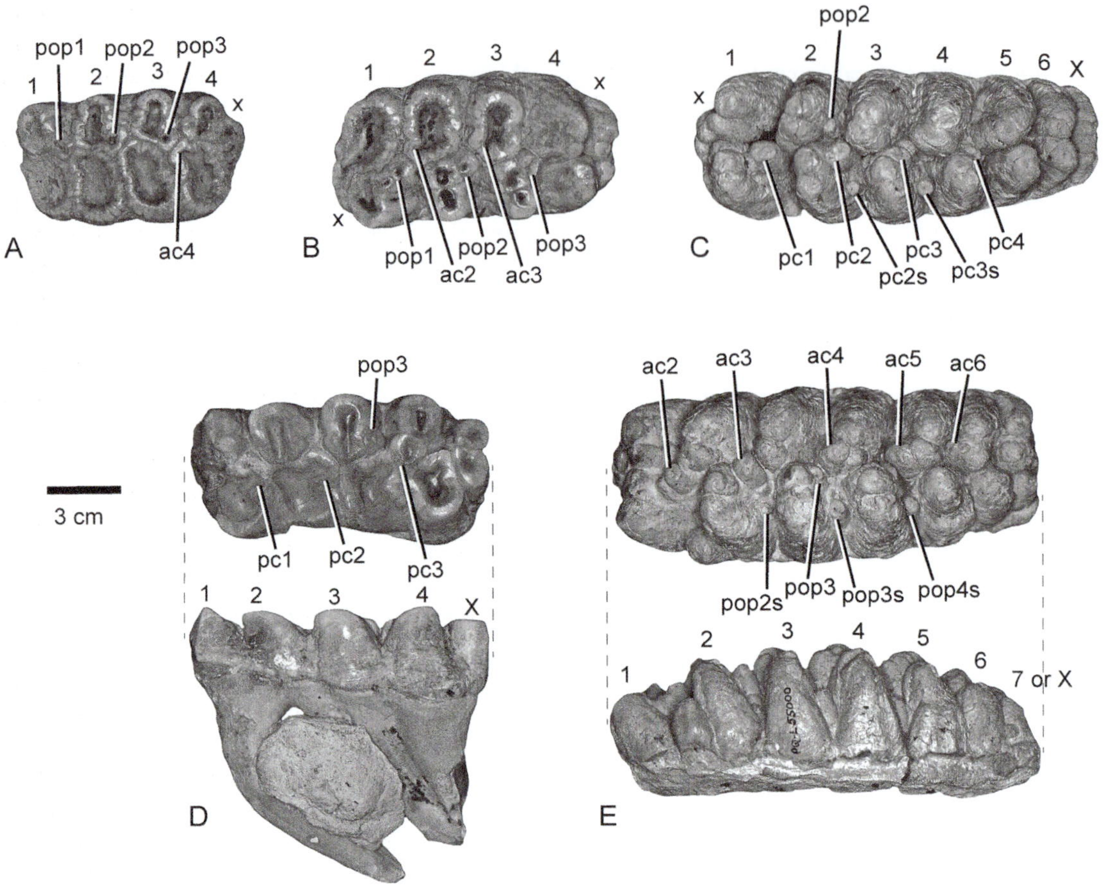

FIGURE 5.7 Molars of *Anancus capensis* from the latest Miocene-earliest Pliocene Langebaanweg, South Africa. Abbreviations: ac, pretrite anterior accessory conule; pc, pretrite posterior accessory conule; pcs, supplemental pretrite posterior accessory conule; pop, posttrite posterior accessory conule; pops, supplemental posttrite posterior accessory conule; x, pre- or postcingulum (id); X, large postcingulum (id) or incipient loph(id); 1, 2, 3, ..., loph(id) number counted from the anterior end of the crown. (A–E) to the same scale. Anterior to the left. (A) Left M1, PQ-L 41692, occlusal view. (B) Right M2, PQ-L 41692, occlusal view. (C) Left m3, PQ-L 41692, occlusal view. (D) Left m2, PQ-L 41018, occlusal and lateral views. (E) Right M3, PQ-L 55000, occlusal and lateral views.

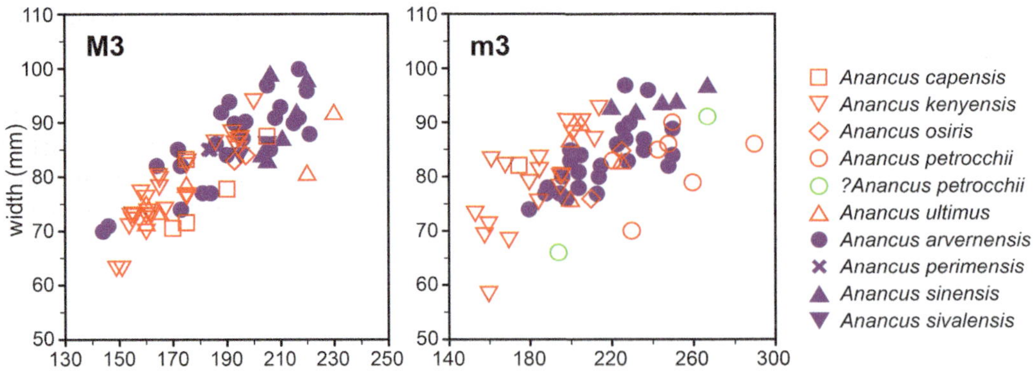

FIGURE 5.8 Bivariate plots of length and width in third molars of *Anancus* spp. Dimensions (in part) from Arambourg, (1945, 1970), Petrocchi (1954), Tassy (1986, 1995), Tobien et al. (1988), Boeuf (1992), Metz-Muller (1995), Mackaye (2001), Sanders (2007, 2011, 2018), and Hautier et al. (2009).

end of the molar row, both about halfway up the corpus in position), have an excavated deep masseter muscle insertion area superiorly on the ramus and a long, oblique superficial masseter muscle insertion that usually has a buttress or torus posterior to it, an anteroposteriorly long, blunt and slightly everted coronoid process separated from the condyle by an elongate, shallow coronoid notch, and there is no prominent gonial angle. The elongation of the corpus may be associated with the great length of the m3s in this species. Mandible no. 1 has the greatest width across the rami of 565 mm, anteroposterior length of the ramus is 270 mm, and the height of the corpus posteriorly and anteriorly is 155 mm and 140 mm, respectively. Mandible no. 2 has the greatest width of 550 mm and the anterior height of the corpus is 200 mm. Left dentary no. 3 has a long corpus (720 mm) and a corporal height of 230 mm. Left dentary no. 4 has a corpus length of 700 mm and a height of 220 mm. Preserved length of right dentary no. 5 is 650 mm and height is 250 mm. Incomplete right dentary no. 6 has a corpus height of 230 mm. Right dentary no. 7 has a corpus height of 240 mm. The size of the jaws suggests that similar to the stegotetrabelodonts from the site, the Sahabi anancine gomphotheres were very large proboscideans.

ANANCUS CAPENSIS Sanders, 2007

(Tables 5.1, 5.2, and 5.4; Figures 5.1, 5.7, and 5.8)

Synonymy—Gomphotheriidae incertae sedis Hendey, 1970:p. 125; Gomphotheriidae gen et sp. indet. Hendey, 1976:p. 236; *Anancus* sp. Coppens et al., 1978:pp. 346, 348; *Anancus* sp. Hendey, 1981:p. 51; cf. *Anancus* sp. Mol, 2006:p. 190; *Anancus* sp. nov. Sanders, 2006:-p. 196; *Anancus capensis* Sanders, 2007:p. 2

This latest Miocene-earliest Pliocene anancine gomphothere species is only known from the ‘E’ Quarry at Langebaanweg, South Africa, deriving from the Quartzose and Pelletal Phosphate Members of the Varswater Formation (Hendey, 1976, 1981). It was named for its geographic occurrence (Sanders, 2007). *Anancus capensis* had been recognized for some time as a novel species (e.g., Coppens et al., 1978; Hendey, 1981), but was only formally named more recently (Sanders, 2007). An interesting mix of primitive and advanced dental traits characterize *A. capensis*. For example, intermediate molars are tetralophodont (Figure 5.7A, B, and D), molars are low-crowned (M3 HI = 64–81), have low lamellar frequencies (LF = 3.5–4.0), and develop very thick enamel (M3 ET = 4.5–6.0 mm), but are otherwise distinguished from species such as *A. kenyensis* and *A. osiris* by much greater occlusal complexity featuring coarsely undulated to folded enamel in wear figures, occasional multiplication of accessory conules, and distribution of pre- and posttrite accessory conules to the third or fourth loph. In addition, M3s exhibit an incipient seventh loph, with a loph formula of x6X, and anancoidy is conspicuous (Figure 5.7C and E). Molar size is modest in the species: M1 L = 90.6–93.8 mm, W = 57.5–62.0 mm; M2 L = 117.8 mm, W = 65.0 mm; M3 L = 170.0–205.0 mm, W = 70.5–89.0 mm.

Only one anancine gomphothere molar is known from Baard’s Quarry at Langebaanweg, and it differs from molars of *A. capensis* by its laterally-elongated, anteroposteriorly compressed posttrite half-lophids (Hendey, 1978, 1981), but it is too fragmentary to assign more precisely than to *Anancus* sp. (Sanders, 2007). Other southern African anancine gomphotheres, from Karonga and Uraha, Malawi, were allotted to *A. kenyensis* and *Anancus* sp. (Mawby, 1970; Bromage et al., 1995), but given the distinctions of *A. capensis* they should be studied anew to reassess their affinities.

The unique mix of primitive and derived features in *A. capensis* suggests an endemic southern African evolution of the genus different from that in eastern Africa, where increasing crown complexity over time was accompanied by pentalophodonty of intermediate molars. Unlike many of its East and Central African congeners, *A. capensis* did not incorporate C_4 grasses into its diet (Groenewald et al., 2020). These authors also suggested that *A. capensis* may have avoided competition for resources with the sympatric elephants *Loxodonta cookei* and *Mammuthus subplanifrons* by occupying more wooded parts of the ecosystem.

STEGODONTIDAE Osborn, 1918

Stegodontidae (stegodontids) is comprised of the archaic genus *Stegolophodon* (“*S.*”; stegolophodonts) and the more advanced genus *Stegodon* (“*St.*”; stegodonts). The meaning of “stegodon” is “roof-toothed,” referring to the peaked transverse plate ridges characteristic of the occlusal surfaces of molars. Typically, as exemplified by more advanced stegodont species, molars are formed of brachyodont plates with numerous apical digitations, enamel may be rugose, there are no accessory conules or they are limited to the anteriormost plate, intermediate molars have at least five plates, and enamel occlusal surfaces show a step-like wear pattern (“Stüfenbildung”) produced by the differential attrition of their harder inner and softer outer surfaces (Saegusa et al., 2005). In addition, apical digitations may be small and buccolingually compressed. There are no lower tusks in stegodonts, and their mandibular symphyses are short. *Stegodon* crania show wide morphological variation between species, but generally are raised, anteroposteriorly compressed, have relatively narrow, downturned rostra with narrow external incisive fossae, and often support immense, elongate tusks that are set in parallel in their alveoli close to one another at least proximally, and that lack enamel bands and are mostly straight, curving lightly upward distally (Saegusa, 1987). Little to nothing is documented about stegodont postcranial skeletons in Africa, but typically these proboscideans had stout bodies with robust limbs and reached impressive sizes; an individual of *St. zdanski* was reconstructed by volumetric methods as having a body mass of 12.7 tons and a shoulder height of 387 cm (Larramendi, 2016).

Stegodontidae is documented ranging in time from early or early middle Miocene *S. nasaiensis* of Thailand (Tassy et al., 1992: Saegusa, 2020 and references therein)

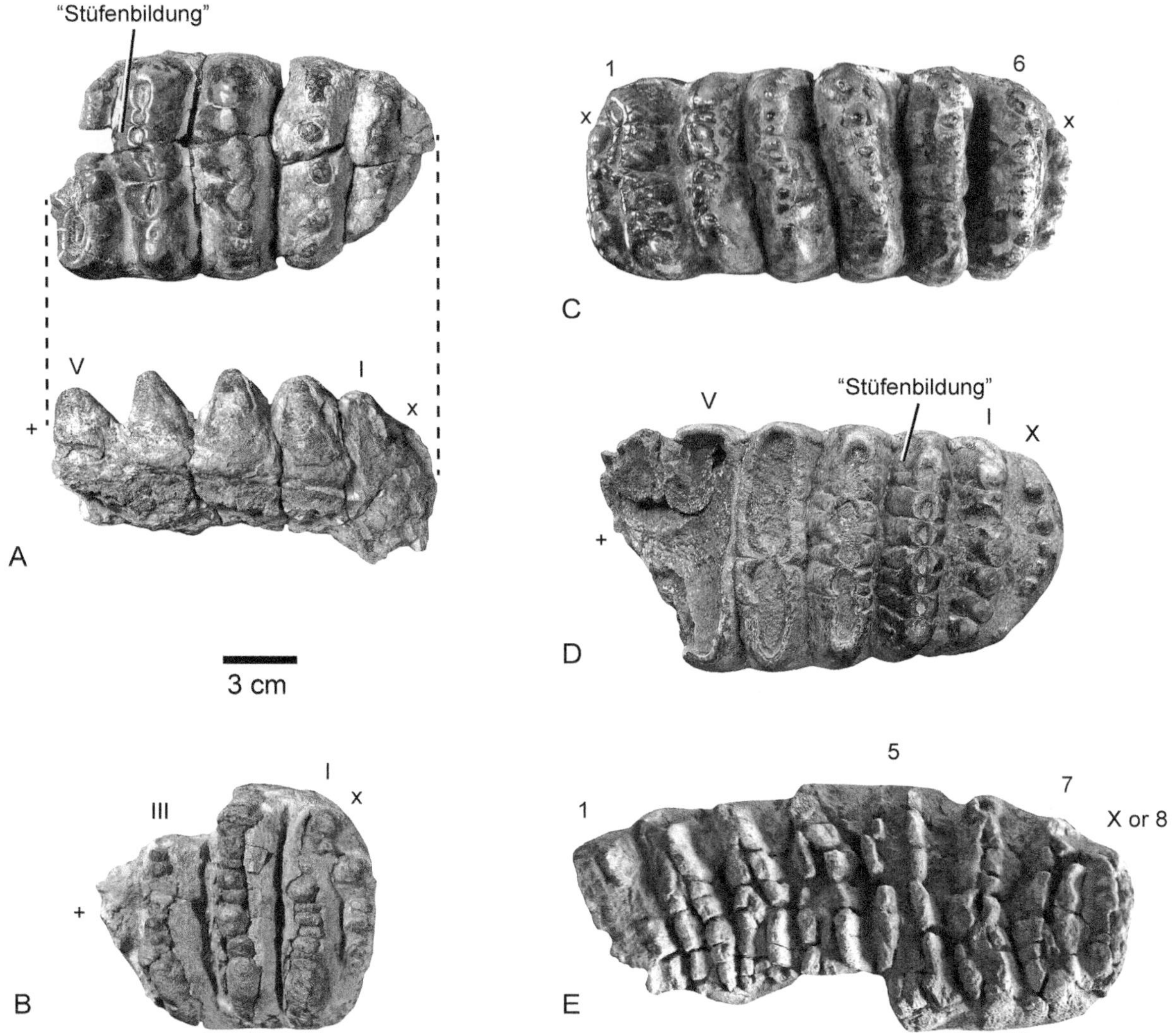

FIGURE 5.9 Selected molars of late Miocene *Stegodon kaisensis*. Abbrevations: x, pre- or postcingulum; X, large postcingulum or incipient plate; 1, 2, 3, ..., plate number counted from the anterior end of the crown; I, II, III, ..., plate number counted from the posterior end of the crown; +, missing morphology. "Stüfenbildung" refers to distinctive stepped molar enamel of stegodonts. Anterior to the left. All specimens approximately to the same scale. (A) Partial right M3, WM 1232/92 from the Manonga Valley, Tanzania, occlusal and lateral views. (B) Partial left M3, KNM-MP 46 from the Mpesida Beds, Tugen Hills, Kenya, occlusal view. (C) Left M2, M 15408 from the Kazinga Channel, Uganda, occlusal view. (D) Partial right M3, KN 92'88 from the Kazinga Beds, Kazinga-Kisenyi area, Uganda, occlusal view. (E) Damaged left m3, proboscidean #5 from Sahabi, Libya, occlusal view. (Image courtesy of N. T. Boaz.)

to Holocene *St. orientalis* of China (Ma and Tang, 1992; Saegusa, 1996b). *Stegolophodon* from the Honya Fm. of Japan is dated to 17.6–17.2 Ma and may be even older than the Thailand stegolophodonts (Yanagisawa, 2011; Saegusa, 2020). In impact on ecosystems and faunas, and geographic and temporal distribution, Stegodontidae is an Asian group with incursions into South Asia and Africa (Coppens et al., 1978; Saegusa et al., 2005). African occurrences of *Stegodon*, dating from the late Miocene to the late Pliocene, are comparatively rare (Tables 5.1 and 6.1).

Although the earliest occurrence of *Stegodon* was claimed to be from ca. 7.0 Ma in the Mpesida Beds of the Tugen Hills, Kenya (Figure 5.9B; Sanders, 1999), its derivation from the Upper Mpesida Beds is now dated to 6.4 Ma (Doman, 2017). The oldest occurrence of stegodonts in Africa may be from Toros Menalla, Chad in the interval 7.0–6.0 Ma (Mackaye, 2001; Le Fur et al., 2014). However, older localities in Yunnan, China with forms transitional between *Stegolophodon* and *Stegodon* have since been correlated to 9.0 Ma, and advanced stegolophodonts in late Thailand and Myanmar, stratigraphically succeeded by primitive, morphologically similar stegodonts in late Miocene beds, support the hypothesis of an Asian origin and immigration into Africa of *Stegodon* (Saegusa et al., 2005; Takai et al., 2006; Sanders et al., 2010a; Saegusa, 2018). African proboscideans reported as stegolophodonts (Petrocchi, 1954; Singer and Hooijer, 1958; Hooijer, 1963) were subsequently re-identified as early elephants (e.g., Maglio and Hendey, 1970; Maglio, 1973; Coppens et al., 1978).

Stegodontids have also been considered close relatives of mammutids (Maglio, 1970a; 1973; Tobien, 1975; Coppens et al., 1978), or assigned to the Elephantidae (Osborn, 1918; Simpson, 1945; Kalb and Mebrate, 1993; Kalb et al., 1996). Nonetheless, monophyly of Stegodontidae has been weakly defended by parsimony analyses based on dental traits and more powerfully supported by chronostratigraphic succession in the fossil record (Saegusa, 1987, 1996b; Tassy, 1990b, 1996a; Shoshani, 1996; Shoshani et al., 1998; Saegusa et al., 2005). Moreover, the consensus view based on evidence from chronostratigraphic succession and dental characteristics is that stegodontids belong in the Elephantida and not Mammutida (Tassy and Darlu, 1986; Tassy, 1990b, 1996a; Kalb and Mebrate, 1993; Shoshani, 1996; Shoshani et al., 1998).

Morphological features for proal jaw movement during mastication in elephants are convergent on those present among stegolophodonts by the early Miocene, and elephants and stegodonts also independently evolved molars comprised of plates with cementum invested between them, although in stegodonts, because of some physiological, functional, or developmental constraint, the evolution of molar hypsodonty was suppressed (Saegusa, 2020). Because of the extreme brachyodonty of their molars, stegodonts have long been thought of as forest-dwelling browsers (e.g., Osborn, 1921, 1942). Isotopic analysis of dental enamel from late Miocene South Asian stegodonts indicates that they were eating C_3 vegetation (Cerling et al., 1999). Their close-set, lightly-curved tusks and stocky bodies were thought to be forest-dwelling adaptations (Osborn, 1921). However, dental isotopic values of stegodonts from Central African sites reveal that over time they became eclectic feeders with shifting dietary preferences, ranging from browsers to grazers during the late Miocene, to mixed feeding and grazing during the early Pliocene, to preferential grazing during the mid-Pliocene (Zazzo et al., 2000), as local ecosystems became richer in C_4 plants between 5.3 and 3.0 Ma (Fara et al., 2005). The tendency to increase commitment to C_4 resources, however, unfortunately occurred during an interval when increasing numbers of mammalian taxa were also beginning to exploit and compete for the same plants, including elephants and anancine gomphotheres. In this context, the failure to evolve hypsodont molars proved catastrophic and by the end of the Pliocene stegodonts had vanished from the African continent.

STEGODON Falconer and Cautley, 1847
STEGODON SP. NOV. Mackaye, 2001
(Tables 5.1 and 5.2; Figure 5.1)

Mackaye (2001) placed several fragmentary molar specimens from Toros Menalla, Chad in a new stegodont species distinct from *Stegodon kaisensis*. As at other fossiliferous areas of the Chad Basin, Toros Menalla has produced a rich vertebrate assemblage, including the oldest known hominin, *Sahelanthropus tchadensis* (Vignaud et al., 2002; Le Fur et al., 2009, 2014). The Toros Menalla area at ~7 Ma was characterized by extensive habitat diversity, featuring well-watered, dominantly open landscapes with short-grass plains, as well as woodlands-bushlands and floodplains. These paleoenvironments supported a host of proboscideans aside from stegodonts, including anancine gomphotheres, deinotheres, and multiple species of primitive elephants (Table 5.1; Mackaye, 2001). Among the other mammals from the site are cercopithecoid monkeys, anthracotheres, bovids from at least six tribes, multiple genera of giraffids, hexaprotodont hippos, hipparionine horses, suids, rodents, lagomorphs, tubulidentates, and a wide variety of carnivores (amphicyonids, hyaenas, felids, herpestids, mustelids, viverrids, and canids). There is some indication that the faunal community of Toros Menalla had its closest biogeographic connections with North Africa (e.g., Lihoreau et al., 2006).

The Toros Menalla specimens represent possibly the oldest occurrence of stegodonts in Africa (Table 5.1). Mackaye considered the molars of the new species to be archaic in construction, with very thick enamel (6.7–7.5 mm), plates formed of few large-sized conelets, low lamellar frequency (3.3), low crown height (HI=52–57), retention of accessory conules posterior to the first few plates, and abundant cementum. These molars are large, with the greatest widths approaching 100 mm. It is not certain that all of these molar specimens are those of stegodonts, though several have a transversely straight arrangement of up to seven conelets or apical digitations composing their plates that is typical for the taxon.

In addition, Mackaye (2001) assigned a left M3 fragment from Mpesida, Kenya, KNM-MP 46, that had been referred to *Stegodon* cf. *St. kaisensis* (Sanders, 1999), to this new species. Like the Toros Menalla molars, the Mpesida molar is low-crowned (H=59 mm), broad (W=101.7 mm), has thick enamel (ET=6.5 mm), and its plates are well-spaced (LF=4.0). Cementum abundantly covers the plate faces and invests the apices of the conelets, but does not fill its transverse valleys. However, KNM-MP 46 differs from the Toros Menalla molars in having up to 11 apical digitations in its plates, and most of these are small and buccolingually compressed. Therefore, it should not be included in Mackaye's (2001) proposed new stegodont species and is a better fit with *St. kaisensis*.

STEGODON KAISENSIS Hopwood, 1939
(Tables 5.1, 5.2, and 6.1; Figures 5.1 and 5.9)

Partial Synonymy—*Stegodon kaisensis* Hopwood, 1939; *St. fuchsi* MacInnes, 1942; *St. syrticus* Petrocchi 1943, 1954; *St. kaisensis* Cooke and Coryndon, 1970; *Stegotetrabelodon orbus* (in part) Maglio, 1973:18; *Primelephas gomphotheroides* (in part) Maglio, 1973; *Stegodon* sp. Beden, 1975; *St. kaisensis* "Nkondo stage" and "Warwire stage" Tassy, 1995; *Primelephas gomphotheroides* (in part) Sanders, 1997; *Stegodon* cf. *St. kaisensis* Sanders, 1999; Elephantidae sp. indet. Sanders, 2008b; *Stegotetrabelodon syrticus* (in part) Sanders et al., 2010a

Most African stegodont fossils have been assigned to *Stegodon kaisensis* (Sanders et al., 2010a). *Stegodon kaisensis* was named for Kaiso Village, near its type site in Uganda (Hopwood, 1939). The species first appeared in Africa as an Asian immigrant during the second half of the late Miocene and persisted until the end of the Pliocene. It is better represented in Western Rift and Central African localities than its rarer, more easterly occurrences (Tables 5.1 and 6.1). *Stegodon kaisensis* was diagnosed as possessing intermediate molars with six-seven plates, low lamellar frequencies of LF=3–4, moderately thick and crenulated enamel, very low crowns, no median sulci except at the very anterior end of molars, plates formed of between seven and 10 conelets, deep grooves separating apical digitations that reach the bottom of plates, and salient step-like wear configuration of enamel wear figures ("Stüfenbildung") combined with weak or no enamel folding (Cooke and Coryndon, 1970; Mackaye, 2001; Sanders et al., 2010a). Some of these traits require emendation as the fossil record of the species has expanded. As well, some molars of *St. kaisensis* exhibit bilateral compression of individual apical digitations, a classic feature of stegodonts.

Tassy (1995) subdivided the species into time-successive primitive "Nkondo" and advanced "Warwire" stages. Included in the "Nkondo stage" sample are the initial collections of Western Rift specimens (Cooke and Coryndon, 1970; Sanders, 1990) and a more recent sample from the region (Tassy, 1995) (Figure 5.9C and D), as well as specimens from the late Miocene Wembere-Manonga Fm., Tanzania (Harrison and Baker, 1997; Sanders, 1997) and from late Miocene-early Pliocene Kossom Bougoudi and Kollé in Chad (Brunet et al., 2000; Brunet, 2001; Mackaye, 2001). Because there are no skulls known for African stegodonts, taxonomic decisions about these proboscideans are based on teeth, mostly isolated molars. Third molars of the "Nkondo stage" are wide (as broad as 117 mm), with thick enamel (as great as 6.0 mm), have low lamellar frequencies (LF=3.0–4.0), thin coatings of cementum, and are very brachyodont (third molar HI=43–55).

In contrast, molars of the "Warwire stage" are said to be more progressive with more apical digitation per plate (10–13), smaller size of apical digitations, greater hypsodonty (third molar hypsodonty indices ranging into the 60s and absolutely greater crown heights), and thicker cementum (Tassy, 1995). Third molars of this stage have up to nine plates and are also wide (up to 116 mm). Specimens assigned to this stage derive from latest Miocene-Pliocene sites in Chad, Tanzania, Uganda, Democratic Republic of Congo, and Ethiopia (Tables 5.1 and 6.1; Heinzelin, 1957; Beden, 1975, 1976; Peters, 1990; Sanders, 1990, 2005, 2011; Pickford et al., 1993; Tassy, 1995; Brunet, 2001; Mackaye, 2001; Fara et al., 2005). However, if Mackaye's (2001) assignment of specimens from Kossom Bougoudi and Kollé to these stages is correct, because they exhibit significant morphometric overlap between them the phylogenetic scheme of successive, evolving stages would be difficult to apply taxonomically or for biochronological correlation. Molars from the Middle Awash, Ethiopia (Kalb and Mebrate, 1993; Tassy, 1995) and Koula and Kolinga, Chad (Coppens, 1965, 1967) that were referred to species of *Stegodon* have been re-assigned to *Primelephas "gomphotheroides"* (now *korotorensis*; Haile-Selassie, 2001).

A single molar, m2 specimen KB 17.96.001 from Kossom Bougoudi, is attributed to *St. kaisensis* "Warwire stage." It is large, with dimensions of L=212 mm and W=87.5 mm, and has broad plate spacing (LF=3.5) and thick enamel (ET=4.0 mm). Its plate formula is x7x. However, this molar is better attributed to the "Nkondo stage" of the species; its absolute molar height is only 47 mm (HI=54) and it has six-eight conelets per plate. In addition, its conelets are more robust than is typical for molars of *St. kaisensis* "Warwire stage." The stegodont sample from Kollé presents a greater challenge: third molars from the site attributed to the "Nkondo stage" are primitive in absolute crown height (H=44.0–55.0 mm) and hypsodonty indices (HI=42–50), and have only a moderate number of conelets per plate (six-eight), but are composed of up to nine plates, which is advanced for this stage. The subset of the sample from the site attributed to the "Warwire stage" has third molars with only eight plates, but greater absolute crown heights (62.3–70.0 mm) and hypsodonty indices (55–62), and up to seven-nine conelets per plate. It is incongruous to posit concurrent "stages" of the same species (sensu Mackaye, 2001); it is more likely that the Kollé *St. kaisensis* sample represents a coherent, variable, evolving population.

The paleoenvironment of Kossom Bougoudi represents fluctuations between ephemeral streams and lakes bordered by gallery forests, with repeated desiccation episodes (Brunet et al., 2000). The aquatic habitats are evidenced by diverse crocodylians, snakes and lizards that prefer watery settings, and wading birds, as well as multiple fish families. Similar to Toros Menalla, the habitats of Kossom Bougoudi supported a great variety of proboscideans (stegodonts, primitive elephants, and anancine gomphotheres; Table 5.1; Mackaye, 2001) and numerous other mammalian species typical of latest Miocene African beds, such as hipparionine equids, the suid *Nyanzachoerus*, and hexaprotodont hippos (Brunet et al., 2000). The fauna also includes camels, giraffes, bovids from possibly five tribes, rhinos, tubulidentates, lagomorphs, rodents, and carnivores. The Kollé fauna also indicates mosaic paleoenvironments in the early Pliocene Chad Basin, with bodies of fresh water evidenced by trionychid turtles, multiple species of fish, and crocs, alongside terrestrial mammals that preferred grasslands and woodlands (Brunet et al., 1998). Aside from stegodonts, anancine gomphotheres and multiple species of elephants are documented from Kollé (Mackaye, 2001). Other mammals include rhinos, hipparionine equids, hexaprotodont hippos, two species of giraffe, several tribes of bovids, a few carnivores, and the suid *Nyanzachoerus*. The quality of the fossil assemblages from these sites, documenting a good sampling of faunal diversity, is tremendously valuable for establishing the heterogenous nature of paleoenvironments in Central Africa during the latest Miocene-early Pliocene, and reinforces the observation that it was not unusual for ecosystems to carry multiple species of proboscideans during this time.

There is also considerable morphological variation in specimens attributed to *St. kaisensis* "Nkondo stage" from other sites. For example, the type lower molar M15170 from Kaiso Village, Uganda has four heavy conelets per plate that apically subdivide into seven or eight digitations, a low lamellar frequency of LF=4.0, and thick unfolded or undulated (but not folded) enamel. Its transverse valley is V- rather than Y-shaped in lateral profile. The postcingulid is formed of three large conelets. Another lower molar fragment from the same site, BMNH 15171, has the same morphology, with only remnants of cementum remaining and V- but not Y-shaped transverse valleys. Its enamel is occlusally undulated but not folded but laterally exhibits fine crenulations. Kaiso Village localities evidently cover a wide range of ages, as an elephant molar attributable to *Elephas recki shungurensis* (BMNH 12639) was recovered from the site, which correlates with a late Pliocene age. The stegodont specimens from Kaiso Village contrast with Kazinga Channel (Uganda) molars, of similar age (Table 5.1) but more advanced in appearance, particularly a large right M2, BMNH 15408 (Fuchs, 1934), originally the type of *St. fuchsi* (MacInnes, 1942). This specimen, only worn on its first plate, has a plate formula of x6x, with dimensions of L=191.0 mm and W=91.3 mm. It has broad anteroposterior plate spacing (LF=4.0) and is very low crowned (HI=56). Its transverse valleys are very pinched off at their bases (Y-shaped) and plates are formed of numerous fine conelets (between nine and 11). In occlusal view it is rectangular. Enamel is ornamented laterally with fine, hippo-like rugosities. The postcingulum is formed of five stout conelets.

A worn distal fragment of an M1 from the same site, BMNH 15407, has coarsely folded enamel and exhibits a "Stüfenbildung" wear pattern on its plate faces. Similarly, a partial M1 from Central Kaiso (Nyawiega), Uganda (BMNH 25162) also exhibits up to 11, bilaterally compressed apical digitations per plate. An m1 (NK 460'86) with a plate formula of x6x, from the Nkondo Fm. of Uganda, has 10–11 small conelets per plate. Its dimensions are L=123.1 mm, W=61.2 mm, and a very low crown height of 34.7 mm. An anteriorly heavily worn M3 (NK 92'88) from the same formation has seven-eight equal-sized conelets per plate, a plate formula of 7x, and is extremely low-crowned. It exhibits classic "Stüfenbildung" stepped wear of its plate faces, cementum thinly covering the plates and transverse valleys, and has thick enamel (4.9–5.2 mm) with horizontal striations. Worn plates exhibit coarse enamel folding. Its dimensions are LF=3.75, L=217.8 mm, and W=100.4 mm.

Additional specimens attributed to *St. kaisensis* "Nkondo stage" include an unpublished stegodont m2 of large size (L=>200 mm; W=>80 mm), WM 244/06 from Shoshomagai 2 in the late Miocene Wembere-Manonga Fm., Tanzania. This specimen has extremely low-crowned plates, a plate formula of x6x, no accessory conules or median sulcus, four-five main conelets subdivided apically into as many as seven digitations, and a thin coating of cementum covering the plate faces but not filling the transverse valleys. An incomplete M3, WM 1232/92, from Inolelo 3 in the same geological unit, was originally published as "*Primelephas gomphotheroides*" (Sanders, 1997), but Saegusa (pers. comm.) correctly pointed out that it exhibits the classic "Stüfenbildung" stepped wear of plates (Figure 5.9A) and belongs in *St. kaisensis*. Plates are very low-crowned (HI=57) and widest at their bases (W=99.3 mm), cementum covers plate faces but does not infill transverse valleys, enamel is thick (ET=4.9–5.6), plates are well-spaced (LF=3.4), and each plate has five main conelets that are apically subdivided into seven digitations. Conversely, molars of mandibular specimen KL89-5 (thought to be m1s) and an isolated molar (KL113-1, thought to be an m1) from the late Miocene Asa Mb. of the Adu-Asa Fm., Middle Awash, Ethiopia were assigned to "*Stegodon* cf. *Stegodon kaisensis*" (Kalb and Mebrate, 1993), but have since been re-assigned to *Primelephas korotorensis* (="*P. gomphotheroides saitunensis*") and re-identified as dp4s (Saegusa and Haile-Selassie, 2009),

An intriguing specimen that provisionally belongs to the "Nkondo stage" of *St. kaisensis* is a moderately worn left M3 from Sahabi, Libya placed by Petrocchi (1943, 1954) in "*Stegodon syrticus*," later attributed to "Elephantidae sp. indet." (Sanders, 2008b), and then subsumed into *Stegotetrabelodon syrticus* (Sanders et al., 2010a). The greatest dimensions of the specimen are L=302 mm and W=124 mm. However, this molar is morphologically contrasted with stegotetrabelodont M3s from the site by its greater number of plates (plate formula x8x), absence of accessory conules, lack of a median sulcus, construction of plates from numerous conelets, and Y-shaped configuration of transverse valleys in lateral view. It should be properly returned to *Stegodon*. An unpublished left m3 from the site, "Sahabi Proboscidean #5," has not been accurately measured but preserves at least seven plates with a nascent eighth plate (Figure 5.9E). It is missing parts of plates 1–4 on the buccal side and has suffered some swelling that produced a series of cracks through its enamel wear figures. As with Petrocchi's M3, its plates have no accessory conules, are transversely straight, lack median sulci, and do not show signs of an outsized central pillar. It is further evidence of *Stegodon* at Sahabi.

Stegodont molars have been found as evident manuports or objets d'art in late Pleistocene human habitations at Ishango, Democratic Republic of Congo (Peters, 1990; Sanders, 1990), likely transported from late Pliocene Lusso Beds localities, perhaps from the nearby site of Kanyatsi (Heinzelin, 1957). Only a very few stegodont molars have been recovered outside of Central Africa and the Western Rift, including a single fragment of a dP4 (EP 1197/98) from the Laetolil Beds, Tanzania with up to 13 bilaterally compressed, small apical digitations per plate and abundant cementum in its transverse valleys (Sanders, 2011). This specimen, a few molars from the Omo Shungura sequence, and the isolated molars of the Lusso and Sinda Beds of the Western Rift (Table 6.1) are all that is known of *St. kaisensis* "Warwire stage" beyond the Chad collections, evincing a steep decline of African stegodonts at the close of the Pliocene.

ELEPHANTIDAE Gray, 1821

The extant species of elephants (*Loxodonta africana*, *L. cyclotis*, and *Elephas maximus*) and their fossil antecedents comprise the Elephantidae. The Family has been documented from sites across Africa, Eurasia (including the Indonesian archipelago), and North America (Osborn, 1942; Maglio, 1973; Shoshani, 1998; Zhang, 2020). For most of their history, elephants were considerably more diverse taxonomically and geographically widespread (Todd and Roth, 1996; Shoshani and Tassy, 1996). Elephantidae is commonly subdivided into Stegotetrabelodontinae, here recognized as accommodating primitive elephants with paired sets of upper and lower tusks, and Elephantinae, those elephants usually with a single set of upper tusks (although there have been atavistic exceptions; Hooijer, 1954; Maglio, 1973), and in more advanced species loss of premolars and shortening of the mandibular symphysis (Sanders, 2017).

There are a number of alternative classifications of elephants, including that of Kalb and Mebrate (1993), who included *Stegodon* and *Stegolophodon* in Elephantidae and placed them, with *Stegotetrabelodon* and *Stegodibelodon,* in an informal group, "Stegomorphs," outside of Elephantinae (compare with Table 1.1). Macakaye (2001) also included *Stegodibelodon* with *Stegotetrabelodon* implicitly in a Subfamily separated from Elephantinae, but the subfamilial placement of *Selenitherium* is unclear in his dissertation and subsequent publication of the taxon (Mackaye et al., 2005). *Stegodibelodon* was also included in Stegotetrabelodontinae by Shoshani and Tassy (2005), who further subdivided Elephantinae into *Primelephas*+Loxodontini (=*Loxodonta*) +Elephantini (=*Palaeoloxodon, Elephas*, and *Mammuthus*), based on craniodental similarities of mammoths, Asian elephants, and palaeoloxodonts and their contrasts with African elephants.

Elephants share features with other large-bodied proboscideans—mammutids and tetralophodontines, for instance—such as graviportal postcranial adaptations, trunks, tusks, horizontal tooth displacement, pneumatized crania, and probably infrasonic hearing and vocalizations, but are specially characterized by formation of molars composed of parallel "plates" or lamellae that are consolidated by cementum, relatively high, anteroposteriorly compressed crania, and fore-aft horizontal shearing in mastication (Maglio, 1972a, 1973; Roth, 1989; Saegusa, 2020). These features are also shared by stegodonts, but paleontological evidence indicates that they evolved convergently during the late Miocene, with stegodonts deriving from stegolophodonts in Asia and elephants from tetralophodontines in Africa (Saegusa, 1987, 1996b, 2020; Saegusa et al., 2005; Wang et al., 2017b; Sanders et al., 2010a; Sanders, 2022; but see Kalb et al., 1996a). Craniodental features diagnostic of elephants have been hypothesized as adaptations for grazing (Maglio, 1972a, 1973; Saegusa, 2020; Sanders et al., 2021); from the morphology of the earliest elephants, these adaptations progressed over geological time by loss of lower tusks, foreshortening of the mandible, greater molar hypsodonty, increased numbers of molar plates, enamel folding, and thicker cementum, some of which occurred independently among different elephant lineages (Sikes, 1967; Aguirre, 1969a; Maglio, 1973; Froehlich and Kalb, 1995, Todd and Roth, 1996; Sanders et al., 2010a; Lister, 2013).

STEGOTETRABELODONTINAE Petrocchi, 1954
STEGOTETRABELODON Petrocchi, 1941

(Tables 5.1, 5.2, and 6.1; Figures 5.1 and 6.1)

Although Aguirre (1969a, b) is acknowledged as erecting the subfamily Stegotetrabelodontinae, Petrocchi named the Stegotetrabelodontidae in 1954, and therefore has priority for the subfamilial nomen. As with some of their tetralophodont predecessors, stegotetrabelodonts were impressively massive in size, both in inferred body weight and reconstructed height. They are the most archaic elephants, retaining tetrabelodonty, elongated mandibular symphyses, and premolars. Their molars are composed of only a few plates (tetra- or pentalophodont intermediate molars), and are also primitive in having thick enamel, brachyodont loph(id)s organized into nascent, pyramidal plates that are widely spaced anteroposteriorly from one another and formed of a small number of conelets, large, independent accessory conules (often throughout the length of the crown), V-shaped transverse valleys, and strong median longitudinal sulci (Maglio, 1973; Coppens et al., 1978; Sanders et al., 2010a). Morphologically, stegotetrabelodont molars are barely more advanced than those of tetralophodonts, which are composed of loph(id)s with distinct pre- and posttrite moieties. Nonetheless, stegotetrabelodont molars exhibit elephant traits such as formation of plates, loss of trefoil enamel wear figures, obliteration of median sulci with moderate wear, and consequent formation of enamel loops spanning the width of the crown. Transformation of loph(id)s into plates involved mesoconelets becoming larger and closer to equivalence in size with outer main conelets, decrease of disparity in height between outer- and mesoconelets, less separation between pre- and posttrite sides of the crown, straighter transverse arrangement of conelets, and reduction of transverse wear gradients as the result of adoption of fore-aft, rather than lateral, Phase II movement of cheek teeth in mastication (von Koenigswald, 2016a).

There are no complete stegotetrabelodont crania; however, there are indications that their skulls were raised and anteroposteriorly compressed in the manner of more advanced elephants (Maglio, 1973; Tassy, 1999:fig. 18.1; Sanders, 2022; contra Maglio, 1972a). This morphological configuration of the skull would accord with the efficient function of molar plates in fore-aft horizontal shearing (Maglio, 1972a). Although lower tusks have periodically been documented in fossil elephants other than stegotetrabelodonts (Hooijer, 1954; Maglio, 1973; Sanders, 2017), those cases appear atavistic as secondary occurrences, rather than primitive retentions. The Stegotetrabelodontinae remains definable by the presence of upper and lower tusks, while the loss of lower tusks is a synapomorphy of the Elephantinae (Sanders et al., 2010a; Sanders, 2017).

Dental isotopic investigations reveal that at least some populations of stegotetrabelodonts were primarily grazers, a behavior shared with other early elephants (Cerling et al., 1999, 2003; Kingston, 1999; Uno et al., 2011; Uno and Bibi, 2022). Isotopic studies of early elephants in the Mpesida Beds and Lukeino and Lower Chemeron (=Mabaget) fms., however, show that in the Tugen Hills, Kenya during the interval of ~7.0–5.6 Ma they were browsers or mixed-feeders (Roche et al., 2013; Doman, 2017). Even with regional variation in feeding preferences, elephants were among the first African mammalian groups to behaviorally adapt to grazing, associated with a worldwide late Miocene pattern of increased aridity and seasonality (Pagani et al., 1999) and widespread expansion of C_4 ecosystems (Cerling et al., 1993; Jacobs et al., 1999). As pointed out by Lister (2013), however, grazing diets of late Miocene elephants such as stegotetrabelodonts were out-of-phase with morphological apparatuses for masticatory function, particularly molar brachyodonty.

Phylogenetically, the most obvious close sister taxon and potential ancestor of stegotetrabelodonts is Tetralophodontinae. Stegotetrabelodonts have often been suggested as ancestral to elephantine elephants (e.g., Maglio, 1973; Coppens et al., 1978; Beden, 1983), but alternatively it has been hypothesized that they were evolutionary "side branches" in elephant phylogeny (Maglio, 1970b; Tassy and Debruyne, 2001; Sanders, 2004). The similarity of *S. orbus* molars with those of the contemporaneous elephantine *Primelephas korotorensis* suggests that they were closely related and possibly shared a common ancestor. Alternatively, stegotetrabelodonts were placed in a phylogenetic sequence with late Miocene-early Pliocene *Stegodibelodon schneideri*, based primarily on symphyseal reduction and loss of lower tusks in the latter taxon (Coppens, 1972). Primitive aspects of *Stegotetrabelodon* species have previously led to their classification as gomphotheres (e.g., Petrocchi, 1934, 1941; Aguirre, 1969a, b), but the plate-like construction of their molars was sufficient for Maglio (1970b, 1973), Tassy (1999), and Ferretti et al. (2003a) to recognize them as elephants.

The group is strictly Afro-Arabian and late Miocene to early Pliocene in age (Sanders et al., 2010a). Remains of stegotetrabelodonts have been recovered from sites in Libya (Petrocchi, 1943, 1954; Gaziry, 1982, 1987b; Sanders, 2008b), Uganda (Tassy, 1995), Tanzania (Sanders, 1997), Kenya (MacInnes, 1942; Maglio, 1970b, 1973; Maglio and Ricca, 1977; Hill et al., 1985, 1986; Tassy, 1986, 2003; Hill, 2002; Saegusa and Hlusko, 2007), Abu Dhabi (Tassy, 1999; Sanders, 2022), Chad (Mackaye, 2001; Le Fur et al., 2014), and from sediments geologically of African continental derivation in southern Italy (Ferretti et al., 2003a). Tassy (1999) robustly parried hypotheses allocating Eurasian and South Asian fossils to various species of *Stegotetrabelodon* and found none of them credible (e.g., "*Stegotetrabelodon maluvalensis*" [Sarwar, 1977]; "*Stegotetrabelodon gigantorostris*" [Tobien, 1980b]; "*Stegotetrabelodon* sp." and "*Stegotetrabelodon primitium*" [Zhou and Zhang, 1983]; "*Stegotetrabelodon exoletus*" (Tobien et al., 1988]). Thus, the origin of elephants can be considered an Afro-Arabian event.

STEGOTETRABELODON EMIRATUS Sanders, 2022
(Tables 5.1 and 5.2; Figures 5.1 and 5.10)

Synonymy—*Stegotetrabelodon syrticus Tassy*, 1999; *Stegotetrabelodon syrticus* Mackaye, 2001; *Stegotetrabelodon syrticus emiratus* Khalaf, 2010 [nomen nudum]; *Stegotetrabelodon sp.* indet. Sanders et al., 2010a; *Stegotetrabelodon emiratus* Sanders, 2022

Stegotetrabelodon emiratus, from sites in the Baynunah Fm., Emirate of Abu Dhabi, United Arab Emirates, exhibits some of the most primitive dental traits among stegotetrabelodonts and may be the oldest species of the genus. The species nomen, which derives from the Arabic "imārah" referring to a territory of an Emir, is repurposed from Khalef (2010), who considered the Baynunah Fm. material to belong to the North African species *S. syrticus* (as did Tassy, 1999) and used it as a subspecific name. He did not provide a diagnosis or type specimen for this subspecies, however, making "*emiratus*" a nomen nudum in that usage. The species has the same adult dental formula as other congenerics (Table 5.2), but is distinguished from them by having cheek tooth plate formulae different from *S. orbus*, *S. syrticus*, or both (dp4 = 5; m1 = 5; m3 = 8), and more brachyodont molars with greater expression of anterior as well as posterior accessory conules (Sanders, 2004, 2008b, 2022).

The Baynunah Fm. can be dated biochronologically to between 8.0 and 6.0 Ma and is likely >6.5 Ma in age (Bibi et al., 2013). Recent paleomagnetostratigraphic correlation is consistent with an age of between 7.7 and 7.0 (Bibi et al., 2022), which would support claims for *S. emiratus* to be the oldest known elephant. It is one of only a small handful of late Miocene sedimentary sequences in the Arabian Peninsula and the most important of those in terms of its mammalian faunal remains. The mammalian assemblage is taxonomically diverse, comprising soricids, rodents, carnivores, perissodactyls, artiodactyls, and cercopithecoid primates (de Bruijn and Whybrow, 1994; Barry, 1999; Bishop and Hill, 1999; de Bruijn, 1999; Eisenmann and Whybrow, 1999; Gentry, 1999a, b; Hill and Grundling, 1999; Boisserie, 2005; Kraatz et al., 2009; Gilbert et al., 2014; Boisserie et al., 2017) alongside the proboscideans, which besides stegotetrabelodonts include deinotheres (Whybrow et al., 1990), tetralophodonts, and an indeterminate gomphothere originally but inaccurately designated as "*Stegotetrabelodon grandincisivus*" (Madden et al., 1982). Of these proboscideans, only *S. emiratus* is abundant, with the sites of Ruwais, Shuwaihat, Ras Dubay'ah, Jebel Mimiyah, Umm al Ishtan, and Hamra yielding a good sample that together represents most skeletal elements of the species (Stewart, 2005). Together, the evidence accumulated for *S. emiratus* constitutes the most comprehensive

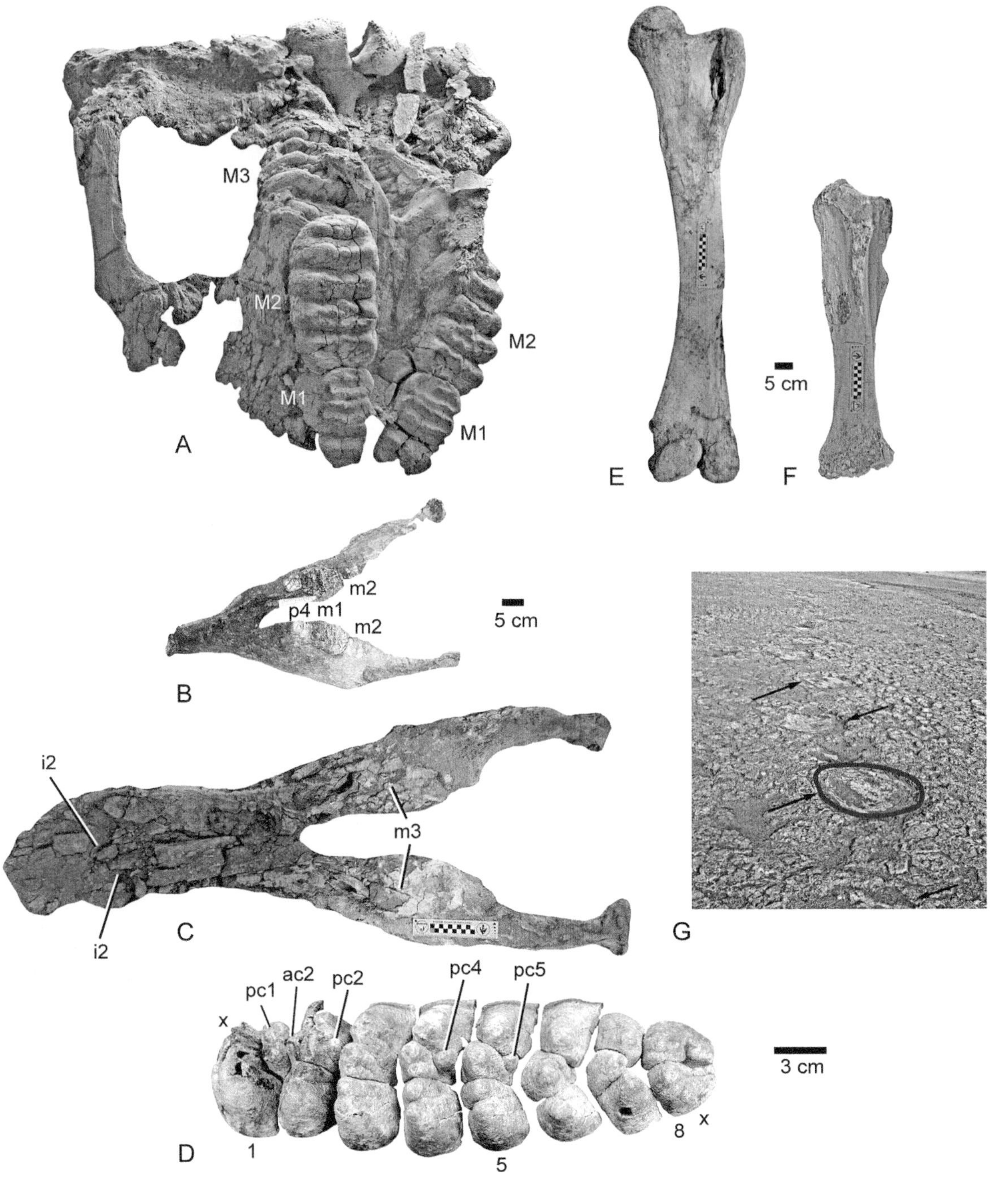

FIGURE 5.10 Fossil remains of late Miocene *Stegotetrabelodon emiratus* from the Baynunah Fm., Abu Dhabi, United Arab Emirates. Abbreviations: ac, anterior accessory conule; i, lower incisor; M/m, upper/lower molar; p, lower premolar; pc, posterior accessory conule; x, pre- or postcingulum; 1, 2, 3, …, plates counted from the anterior end of the crown. (A) Right and left M2s of the cranium are 101 mm in length, for scale. (B, C) to the same scale. (E, F) to the same scale. (A) Partial cranium, AUH 502 from Shuwaihat, ventral view. M3 is emergent from the crypt. Anterior at the bottom. (B) Juvenile mandible, AUH 1171 from Hamra, occlusal view. Note that m1 is heavily worn as p4 was emergent and m2 was beginning to emerge from its crypt. (C) Adult mandible, AUH 1737 from Ruwais Central, occlusal view. (B and C) anterior to the left. (D) Right m3 (type), AUH 456 from Ras Dubay'ah, occlusal view. Anterior to the left. (E) Right femur, AUH 1740 from Um al Ishtan, posterior view. (F) Left tibia, AUH 1740 from Um al Ishtan, same individual as the femur, anterior view. (G) Ichnofossils (footprint series attributed to *S. emiratus*) from Mleisa 1, part of 14 trackways at the site from a minimum of 13 individuals, evidence of herd composition and behavior. The first four footprints are indicated by alternating arrows; the second footprint is outlined to highlight its shape.

record of the paleobiology of late Miocene elephants, including size, social structure and behavior, diet, and dental development.

Among the most informative specimens recovered from the Baynunah Fm. is a partial cranium (AUH 502; Figure 5.10A) with an associated damaged mandible (AUH 503), found at locality Shuwaihat 6. The cranium preserves right and left M1–2 and an emerging M3, as well as left and right I2s. The M1s are heavily worn, M2s are worn into enamel demi-loops anteriorly, and M3 has well-formed plates but had not descended from its crypt at the time of death. The tooth emergence and wear pattern matches Laws' (1966) age category of XVII, about 28 years of age.

Unfortunately, the cranium is heavily weathered on its dorsal aspect, making it impossible to ascertain cranial height. Nonetheless, it appears to have been anteroposteriorly compressed and raised. Where morphology is preserved, the cranium resembles the cranium of *S. syrticus* from Sahabi, Libya. The basicranium is raised well above the level of the palate at an angle of about 50–60°, the condyles are small and do not seem to project markedly posterior to the occipital planum (though this may be the result of crushing in this area of the cranium), and the cranium is very broad, particularly as can be inferred across the zygomatic arches (e. 820 mm) and low across the occipital planum (see Tassy, 1999). The palate is shallow and broader posteriorly between the back of the m2s (W = 75.0 mm). The glenoid articulation is transversely broad and saddle-shaped; conversely, anteroposteriorly it is convex and narrower. The zygomatic appears to be set at a right angle to its squamosal and maxillary roots, enclosing a square-shaped temporal foramen. Unfortunately, as the specimen is currently on display in the Environment Agency in Abu Dhabi, it could not be more closely studied or directly measured.

The upper tusks are extraordinary in length for a subadult, and very slender for their length: the left I2 is broken at both ends yet preserves a length of 760 mm, with a width of only 84.5 mm and a dorsoventral height of 76.4 mm. The right I2 is nearly complete distally but damaged proximally and must have been longer, as the pulp cavity is missing; it is 1030 mm in length and has cross-sectional dimensions of W = 82.4 mm and H = 70.0 mm. These tusks are rounded in cross-section, lack enamel bands and sulci, and are nearly straight. In comparison, an isolated upper tusk from the locality of Ruwais Central, AUH 1738, on display in the Environment Agency in Abu Dhabi, is 2.54 m in length and relatively very slender for its size. The slender build of *S. emiratus* tusks makes it difficult to envision their effectiveness in interindividual agonistic interactions, but their lengths suggest that they might have been impressive in threat and mating displays.

M1 is tetralophed with plates worn into complete enamel loops and has a small, low postcingulum. M2 is pentalophed with a small, low postcingulum, and has both pre- and posttrite accessory conules. It is not possible to ascertain the complete number of plates present in M3. Molar plates are formed of four conelets each and mesoconelets are nearly as large, and are as high, as the outer main conelets. Molar dimensions provided by Tassy (1999) show these teeth to be large and broad. M2s range in length from 174.4 to 176.9 mm and width around 101 mm, and the left M3 has a width of 103 mm. Lamellar frequency of the molars is low, 2.9–3.0, and their hypsodonty indices are very low (46–60).

A very crushed mandible (AUH 503) missing its rami is associated with the cranium, with a well-preserved right i2 still in its alveolus. It is impossible to determine the degree of downward angulation of the symphysis because the mandible has been considerably flattened. Symphyseal length is estimated at 370 mm (Tassy, 1999). The lower tusk is oval in cross-section, lacks enamel, and has a slight outward longitudinal torque and sulcus along its dorsal surface. This latter feature is sometimes present in tetralophodontine i2s (Osborn, 1936:fig. 320). The tusk projects well anterior to the symphysis and appears to be quite slender. Anterior cheek teeth are missing or have been broken away, but m3s are apparent on both sides in their crypts. Each has at least seven low, massive pyramidal-shaped plates visible in lateral view, but other details could not be ascertained as the mandible is on display with its cranium.

Other mandibular remains contribute further information about the lower jaw morphology of the species. AUH 1171, a mandible from the site of Hamra missing parts of its rami, has a steeply downturned symphysis with a deep trough. The downward angulation of the symphysis is impressive for such a young individual. The specimen has p4-m1 and the first two or three plates of m2 emergent on each side (Figure 5.10B); the presence of worn m1s with emergent m2s correlates with Laws' (1966) age group XI, 14–16 years old in modern African elephants. Dentaries of an even younger individual, AUH 1779 from the site of Jebel Mimiyah, are very damaged but preserve p3s emergent from their crypts, worn dp4s, and right m1 just emergent from its crypt. Specimen AUH 475, from Jebel Barakah, comprises a left dentary in multiple pieces, with alveoli for or traces of i2, p4, and m1, with a partial m2 in its crypt, suggesting an age of 14–16 years at death, compared with modern African elephants (Laws, 1966). Laterally, there are three mental foramina; the anteriormost is the largest and opens anteriorly, alongside the symphysis. It is obvious from this specimen that the mandibular canal does not communicate directly with the tusks alveolus, except by small foramina that are infilled with sediment. A heavily eroded but nearly complete mandible from Ruwais Central, AUH 1737, is missing its dentition but the size of the posterior alveolus (L = 220 mml; W = 200 mm) suggests that it housed an m3, and there are more anterior alveoli on both sides for smaller molars, presumably m2s (Figure 5.10C). In modern African elephants, this arrangement correlates with age group XXIII individuals of about 43 years old (Laws, 1966). The mandible is immense, with a length of 1,500 mm (symphyseal and incisor alveoli L = 700 mm). The maximum width across the right i2 alveolus is 110 mm, consistent with the socketing of a slender tusk that is characteristic of these stegotetrabelodonts. The symphyseal/incisor alveolus portion of the mandible is 300 mm at its greatest width. It lacks the bilateral

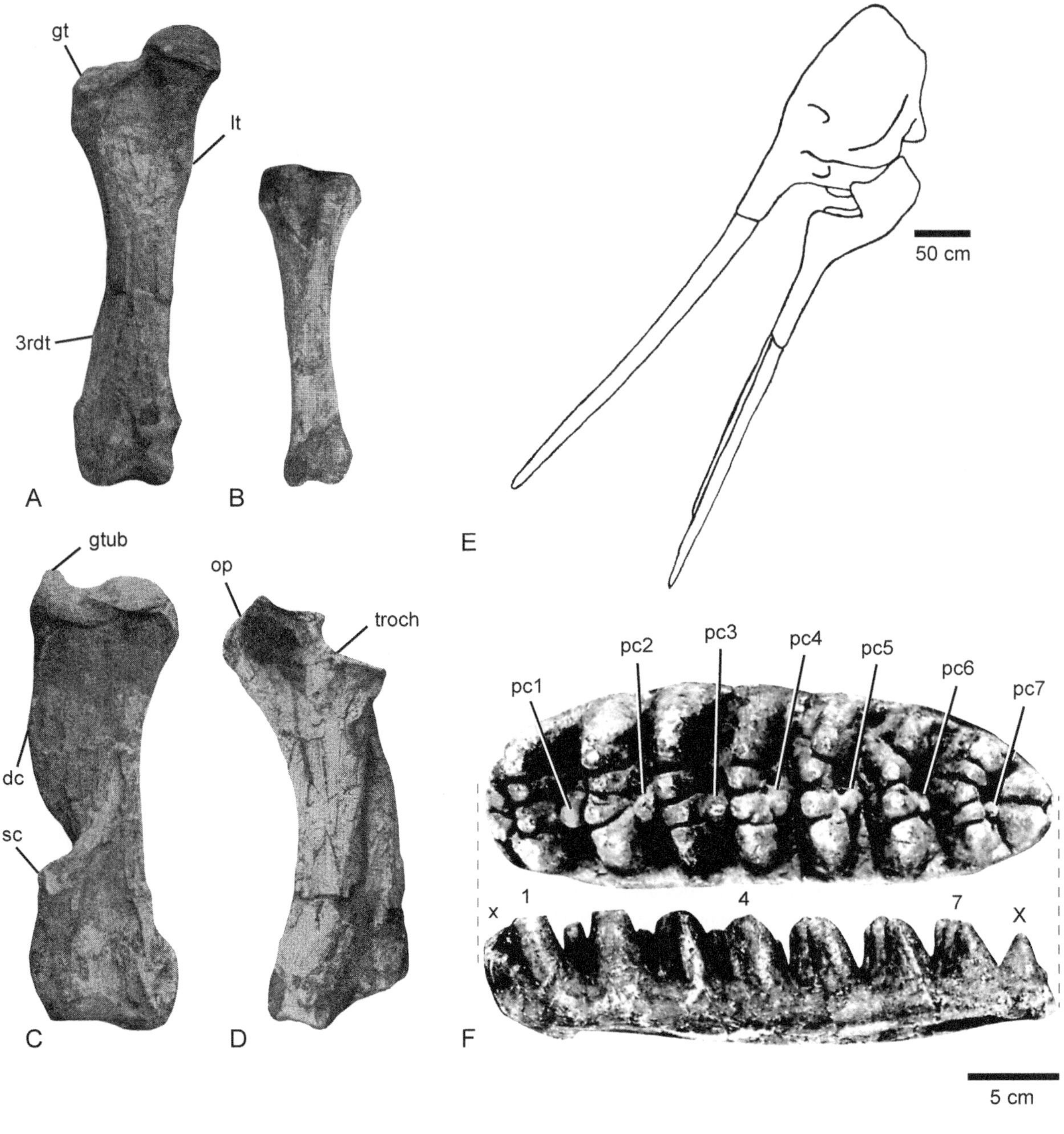

FIGURE 5.11 Skull, dental, and postcranial specimens of late Miocene *Stegotetrabelodon syrticus* from Sahabi, Libya. Abbreviations: dc, deltoid crest; gt, greater trochanter; gtub, greater tuberosity; lt, lesser trochanter; op, olecranon process; pc, posterior accessory conule; sc, supinator crest; troch, ulnar trochlear surface; x, precingulid; X, large postcingulid or nascent additional plate; 1, 2, 3, …, plate number counted from the anterior end of the crown; 3rdt, third trochanter. (A–D) to approximately the same scale. (A) Femur "e," anterior view. Length is 1,130 mm. (B) Tibia "d," anterior view. Length is 700 mm. (C) Humerus "c," posterior view. Length is 1,100 mm. (D) Ulnar and radius "c," medial view. Ulnar length is 1,040 mm. (E) Reconstruction of type skull, lateral view. Note that the dorsal surface of the cranium is speculative as it was badly damaged by weathering in the original specimen. (Reproduced from the cover illustration of Garyounis Bulletin Special Issue No. 4, 1982, with permission of the editor, N. T. Boaz.) (F) m3, unnumbered, occlusal and lateral views, "*Stegotetrabelodon lybicus*." Anterior is to the left. (All fossil specimen images are from Petrocchi, 1954:tav. XVII, XVIII, XV, XVI, with permission of the L'Accademia Nazionale delle Scienze detta dei XL).

flare of the symphysis of shovel-tusked amebelodonts. The spacing between the posteriorly-facing condyles is 360 mm; on the left side, the condyle is 140 mm wide.

The cheek tooth sample for *S. emiratus* provides the best overview of onotogenetic dental development available for stegotetrabelodonts. By examining the different dental series of the species as a composite, it is possible to establish that tooth emergence and displacement was dp2-dp3-dp4-p3-m1-p4-m2-m3, retaining the ancient pattern of gomphotheres (Tassy, 1996c). In addition, analyses of dental carbon isotope composition show that there was a strong graze component to the diet of this species (Kingston, 1999;

Uno, in press). Lower dp4 has a plate formula of x5X and is large (L=102.0mm; W=49.9mm). It is robustly constructed with coarsely-folded enamel and massive low plates that are pyramidal in cross-section. Correspondingly, transverse valleys are V-shaped. There are as many as six conelets comprising each plate. Median sulci subdivide the crown into pre- and posttrite sides. In wear, anterior and posterior pretrite accessory conules, present in association with each plate, contribute to the formation of trefoil enamel figures, but there is no development of loxodont "<>" wear figures.

p3 has a lophid formula of x3x, and is very diminutive (L=33.8–35.2mm; W=23.0–23.6mm). It is morphologically similar to p3 in *S. orbus*. The protolophid has a single conelet on each side of the crown; the hypolophid has two conelets on each side; and the tritolophid has a single pretrite conelet and two posttrite conelets. The conelets of the pre- and posttrite sides are separated by a median sulcus.

p4 is longer than wide, with a lophid formula of x2X. The crown is subdivided by a median sulcus. Dimensions vary from L=38.6–39.4mm and W=32.7–39.0mm. As pointed out by Tassy (1999), the lophids do not form true plates. The protolophid is formed of two conelets; the hypolophid has two large conelets, each subdivided into two apical digitations. Each half-lophid has a posteriorly projecting enamel lobe, or posterior accessory conule. The greatest resemblance is with a stegotetrabelodont p4 from the Mpesida Beds, Tugen Hills, Kenya (Sanders, 2022). The p4 of *S. emiratus* is distinguished from those of gomphotheres by stronger postcingulid development and from *Primelephas korotorensis* p4s by only nascent construction of a tritolophid.

In occlusal view, P4 has a more square-shaped crown, with low, anteroposterior pre- and postcingulids closely appressed to the lophs. Loph formula is x2x. The anterior loph is highest. Cusps are bulbous and in lateral view appear pyramidal. Minute traces of cementum remain. Dimensions are L=49.7mm and W=48.5mm; crown height is low at 32.5mm. The Baynunah Fm. specimen (AUH 560) is close to P4s of *S. syrticus* and *S. orbus* in size and morphology (Sanders, 2017).

Lower first molars are known for this species for the first time, in association with the recently-collected juvenile mandible AUH 1171. Although heavily worn, with the first two plates nearly obliterated occlusally, it can be observed that posterior accessory conules are incorporated into the enamel wear figures of plates 3–5, from which it may be inferred that they were present throughout the length of the crown. The anterior interproximal facet is broad and embraces the newly emergent p4. The specimen is broadest posteriorly, typical of proboscidean m1s. Plate formula is x5x. Dimensions are L=117.1–119.4mm and W=72.0–72.3mm; enamel is thick at 4.2–4.5mm. Compared with m1s of *S. syrticus*, in *S. emiratus* m1 is shorter, relatively broader, and has an additional plate.

The type m3, AUH 456 from Ras Dubay'ah is comprised of a series of plates that are assembled into a complete molar (Figure 5.10D). It is large and wide, as is typical for stegotetrabelodont third molars (L=277.0mm; W=109.1mm). Enamel is very thick (ET=6.6mm) and crown height is extremely brachyodont (HI=56). Lamellar frequency (3.0) reflects the wide spacing of the massive, pyramidal plates by V-shaped transverse valleys. Accessory conules are present posterior to plates 1–6 and anterior to plates 2–3. Each plate is formed of between only three to five conelets. In occlusal view, the more posterior plates are curved into a mesial convexity, with their central conelets each forming a stout pillar. A very narrow longitudinal sulcus is present through plates 2–5. The crown is less wide posteriorly, so that plate 8 is transversely narrower than the other plates, and the postcingulid is a single low conelet closely appressed to the last plate. Except for the presence of eight plates, AUH 456 has a more primitive look than the molars of other stegotetrabelodonts.

Postcranial remains of *S. emiratus* demonstrate that these elephants were exceptionally tall and heavy, as large as or exceeding the size of the biggest extant African elephants (Sanders, 2022). Their height is consistent with carrying very long (up to 2.5m), straight, downturned upper and lower tusks. Specimen AUH 1740, from Um al Ishtan, includes a well-preserved femur and tibia (Figure 5.10E, F). The femur is quite long and slender (L=1,420mm from head to distal condyle), as in modern African elephants (Smuts and Bezuidenhout, 1994), correlating with a shoulder height of 370cm and body mass of about 10,000–11,000kg, much heavier and taller than the average African bull elephant. The femoral head is 320mm measured on the curve; its neck is 380mm wide and short; the diaphysis tapers down to 157mm in width at midshaft and is only 95mm thick anteroposteriorly at the same place in the shaft, constituting substantial flattening of the element; and the distal medial condyle is broader than the lateral condyle, W=156mm versus 128mm, respectively. Regression analyses of these dimensions using Christiansen's (2004) database yield disparate but heavy body mass estimates, from 6,875 to 15,612kg. The AUH 1740 femur differs from femora of extant elephants in having its head oriented more directly proximal than proximomedial, less pronounced size disparity between medial and lateral condyles, and a shorter and less defined neck.

Length of the AUH 1740 tibia is 870mm, 60% of femoral length, which is similar to proportions in modern African elephants (Larramendi, 2016). As with the femur, the tibia is relatively slender compared with its length, with a midshaft transverse width of 127mm. It is proximally 260mm wide and distally 210mm wide. An unusual feature of this element is a projecting flange on the proximolateral side of the shaft, which may have increased the surface area of the extensor digitorum muscle or enhanced stability of the tibiofibular articulation. Regression analysis of tibial length in proboscideans (Christiansen, 2004) yields a body mass of 8,413kg, based on the dimensions of AUH 1740. Other postcranial elements of *S. emiratus*, such as the scapula, pelvis, and podials, are morphologically similar to those of modern elephants but generally more robust.

The most remarkable evidence of *S. emiratus*, attributed to this species because it is by far the most common proboscidean in the Baynunah Fm., is comprised of ichnofossils—trackways—from the site of Mleisa 1 that document striking details about social behavior, herd composition, and body size (Bibi et al., 2012). These footprints are extensive, including 14 trackways made by a minimum of 13 individuals, covering a distance of 190 m in parallel within a narrow corridor of 20–30 m, indicating a herd traveling together (Bibi et al., 2012). The herd is interpreted as having a social structure like that of modern elephants, with adult females, juveniles, and an infant. The track of a larger, presumably solitary male elephant, crosses these footprints and covers a distance of 260 m. Preservation of the trackways is in a silty-to-sandy carbonate level that was deposited in shallow fresh to brackish water, based on analysis of associated ostracod fossils (Mazzini et al., 2013; Mazzini and Kovacova, 2022). The trackways are the oldest evidence of modern-type sex-segregation of adult males and females and complex herd structure in the proboscidean fossil record (Sanders, 2022).

Estimates of body mass from inferred stride lengths of individual trackways at Mleisa 1 range from >500 kg to nearly 5,500 kg, with the largest elephants in the herd exceeding the body size of extant female elephants, and places the body size of the lone male within the range of modern adult male elephant size (Bibi et al., 2012). Elephant footprint dimensions also may be used to calculate the body mass and shoulder height of fossil proboscideans (Western et al., 1983; Lee and Moss, 1995; Roberts et al., 2008; Schanz et al., 2013). Some of the footprints in the Mleisa 1 trackways exceed those of modern elephants in size, but if the scaling relationship is the same as in the modern sample, *S. emiratus* individuals must have been taller than their modern confamilials of the same sex and age grade, reaching ≥350 cm in shoulder height. Close transverse spacing of footprints within trackways and the proximal orientation of the femoral head relative to the shaft correlates with a narrow-gauge stance in these elephants from the Baynunah Fm. (Sanders, 2022).

Paleoecological reconstruction of the localities in the Baynunah Fm. indicates the presence of a highly seasonal, semi-arid environment, with trees but also C_4 grasslands in open habitats (Bibi et al., 2022). The most updated dental isotopic analysis of proboscideans from the site shows *S. emiratus* to have been a mixed feeder predominantly eating C_4 plants (Uno and Bibi, 2022).

STEGOTETRABELODON SYRTICUS Petrocchi, 1941
(Tables 5.1 and 5.2; Figures 5.1 and 5.11)

Partial Synonymy—*Stegotetrabelodon syrticus* Petrocchi, 1941, 1943, 1954; *Tetrabelodon* Petrocchi, 1943; *Stegotetrabelodon lybicus* Petrocchi 1943, 1954; *Stegolophodon sahabianus* Petrocchi 1943, 1954; *Stegotetrabelodon lybicus* Maglio, 1973; *Stegolophodon sahabianus* Gaziry, 1982; *Stegotetrabelodon lybicus* Gaziry, 1982, 1987b; *Stegotetrabelodon syrticus* Tassy, 1995; *Stegotetrabelodon syrticus* Sanders, 2008b

Stegotetrabelodon syrticus is known in North Africa from the site of Sahabi, Libya, estimated to derive from late Miocene sediments (Table 5.1; Warny et al, 2003; Boaz et al., 2008; El Shawaihdi et al., 2016; but see Heinzelin and El-Arnauti, 1982). In addition, a mandible with m3s from Cessaniti and dated to the late Tortonian (>7.0 Ma) is also attributed to *S. syrticus*. Though the site is in modern-day Italy, its region of derivation was geologically part of northern Africa during the late Miocene (Ferretti et al., 2003a). Sahabi was first recognized to have paleontological potential in the early 1930s by Ardito Desio, but the first vertebrate collections and accounts of any note were made by Carlo Petrocchi, who undertook field survey and excavation over a period of six years between 1934 and 1939 and deserves credit for understanding the importance of the site for the study of African mammalian evolution (Rook, 2008).

Among the remains of *S. syrticus* recovered from Sahabi is a good sample of postcrania, including isolated limb bones and a partial skeleton (that should be designated the lectotype specimen) collected during the first field campaign of Petrocchi in 1934 that comprises limb and axial elements. These have been described and their dimensions provided in Petrocchi (1954). The Sahabi postcranials are generally more robust than the same elements in *S. emiratus*, but this may be due to age-grade differences, with *S. syrticus* remains representing older, fully adult individuals. Femora from Sahabi range in length from 1,080 to 1,470 mm and transverse midshaft width of 89–116 mm. Their heads are primarily oriented upward and only slightly medially; the greater trochanter is rugose but low, and expression of the lesser trochanter (below the short neck) and third trochanter (very distally placed along the medial margin of the shaft) is variable (Figure 5.11A). The condyles extend about the same distance at the distal end and face primarily downward, but their articular surfaces curve posteriorly and would have permitted flexion of the lower leg on the femur.

Tibial lengths range from 650 to 950 mm and transverse midshaft widths vary from 106 to 147 mm. Relative lengths of the tibiae seem slightly longer than in *S. emiratus*, about 65% of the femoral length (Figure 5.11B). A lone fibula from Sahabi belongs with the type skeleton (and largest femur and tibia); it is 885 mm in length.

The Sahabi humeri are particularly robust and nearly as long as the femora, with lengths ranging from 1,000 to 1,100 mm. The head is large and upward facing, surmounted by a strong greater tubercle that extends more proximally than the head (Figure 5.11C). Distally, the trochlea and capitulum run together and are very broad, facing primarily downward. The upper forelimb must have been extremely powerful and heavily muscled; the deltoid crest flares broadly laterally, and the more distally located supinator crest is equally impressive in the degree of its flaring and the extent of its muscle attachment area. The disposition of these crests relative to one another provides a prominent torsion of the humeral shaft.

The radius appears to have been fixed in pronation on the ulna. The ulnae of the Sahabi stegotetrabelodont are in some instances longer than the humeri, with ulnar length ranging from 1,040 mm to 1,500 mm (Petrocchi, 1954). The ulnar trochlea faces upward, and the olecranon process is anteroposteriorly broad but reflected posteriorly and not very long (Figure 5.11D). Together with scapular length of about a meter, the additive length of forelimb elements is impressive and supports the reconstruction of the shoulder height of the largest individuals to be well over 300 cm.

Body mass estimates, based on allometric analysis of limb bone lengths (see Christiansen, 2004) range from 4,128 to 11,259 kg, exceeding the normal upper limits for extant elephants. Femoral length correlates with shoulder heights of up to ~450 cm in the largest individuals; humeral length yields shorter correlative shoulder heights of up to ~360 cm (see Larramendi, 2016).

The Sahabi *S. syrticus* skull (unnumbered), associated with a partial skeleton, is of greatest value in defining the species, and constitutes the type specimen (reconstruction in Figure 5.11E). The cranium is weathered away dorsally, such that the braincase, nasals, orbital region, face, and upper part of the occipital planum are missing. This is unfortunate because it prevents an assessment of the degree of anteroposterior compression and the height of the skull. Nonetheless, Petrocchi (1943, 1954) estimated the length of the skull, including the upper tusks, as nearly 3.6 meters in length. The length of the portion of the tusks that protrudes from their alveoli is 2,200 mm, leaving an estimated length of the cranium as 1,400 mm. The tusks are rounded in cross-section, lack enamel bands, and have a greatest transverse diameter of 140 mm and dorsoventral dimension of 135 mm, relatively thin for their length. They are only slightly curved upwardly toward their distal ends and are not widely flared from one another.

Although much of the occipital planum is absent, it is possible to observe that it is very broad (970 mm), is not inflated, has diminutive occipital condyles (left condyle L = 125 mm × W = 130 mm), and has a shallow fossa for the nuchal ligament that extends low on the planum. The estimated distance across the zygomatic arches is even broader, at 1,030 mm. If the cranium is rotated so that the upper molars are horizontal, the occipital planum appears to be angled forward toward its dorsal extent, and the condyles would be projecting posteriorly in this orientation.

The mandible associated with the cranium is virtually complete. Including the lower tusks, it is about 3 meters in length (Figure 5.11E; Petrocchi, 1943, 1954). Its most salient features are the low, anteroposteriorly broad rami, a relatively short corpus that is much higher anteriorly than posteriorly, and an impressively elongate symphysis that is broadest distally and housed lower tusks set close to one another in their alveoli. In ventral view, the corpora are angled abruptly on the symphysis—it is obvious that tusks develop and are socketed in the symphyseal alveoli and not in the mandibular canals and their terminal antechambers in the corpora. On each side, the exit for blood vessels and nerves from their plexuses in the antechamber is marked by the position of the mental foramen at mid-height at the anteriormost aspect of the corpus. In the lateral view, the symphysis is angled downward strongly from the jaw. Each coronoid process is low, small, slightly everted to the lateral side, and separated from the condyle (at the same height) by a long, shallow concavity. Insertion areas on the rami appear larger for the masseter than temporalis muscles.

The lower tusks are perfectly preserved, and are narrow, straight, lack enamel bands, and must have protruded impressively downward in their correct anatomical position. They are oval in cross-section, higher than wide, and project 1,685 and 1,615 mm, left and right sides, respectively. Dimensions at their alveolar insertions are W = 70 mm and H = 100 mm. Tusks of such narrow dimensions relative to their impressive length would appear highly vulnerable to shattering if used in agonistic encounters and may have had more of a role in sexual selection for display.

M3s of the cranium have a plate formula of 6X (large postcingulum), and the greatest dimensions of L = 230–250 mm and W = 120–121 mm. Enamel is thick and unfolded, and has formed enamel loops anteriorly with wear. Along the length of the crown, stout accessory conules are associated with plates 1–5 at their posterior aspects. The m3s are also worn, with very thick enamel forming complete loops anteriorly. The greatest dimensions of these molars are L = 260–280 mm and W = 121–116 mm, left and right, respectively. Plates 1–5 exhibit stout posterior accessory conules that must have been apically free. The plate formula of m3 is 7X, with the postcingulid heel comprising a nascent eighth plate. Cementum is distributed around the plates in the transverse valleys. In lateral view, plates are pyramidal in shape, anteroposteriorly, and transversely widest at their bases. Plates appear to be composed of three-four conelets, of approximately equal size and height. Longitudinally, the m3s are convex lingually.

An m3 placed in "*Stegotetrabelodon lybicus*" (Petrocchi, 1943, 1954) is unworn, with a plate formula of x7X; as in the m3s from the skull specimen, the postcingulid is large with two large conelets, and constitutes a nascent eighth plate (Figure 5.11F). The length of the specimen is 315 mm and the greatest width is 121 mm. Crown height is relatively low, 73 mm, yielding a hypsodonty index of 60. Each plate is formed of four-five conelets of approximately equal size and is accompanied by distinct, lower posterior accessory conules that are free apically. In the lateral view, the plates are angled forward and pyramidal in shape, and the transverse valleys are correspondingly V-shaped. The specimen may have been forming in crypt at the time of death, as there is no cementum preserved on the crown. In occlusal view, the median sulcus persists only as a very narrow trace through each plate, and the distinction between pre- and posttrite sides has been essentially obliterated.

A nearly complete right (not left; contra Petrocchi 1943, 1954) M3 attributed to "*Stegolophodon sahabianus*" (Petrocchi, 1943, 1954) is similar to the M3s in the Sahabi skull and belongs in *S. syrticus*. This specimen is also unworn and may have been forming in crypt at the time of death, as it lacks cementum. It preserves the precingulum and six plates, lacking only the postcingulum (x6+). Plates are composed of five conelets of equal size, and in lateral view are pyramidal with correspondingly V-shaped transverse valleys. Very low accessory conules are present at the outer lingual edge of transverse valleys 1, 2, 4, and 5. In addition, plates 1–2 have strong posterior accessory conules, and plate 3 has a strong anterior accessory conule. Preserved crown length is 243+ mm and the greatest width is 125 mm. Crown height is low, only 78 mm, which produces a hypsodonty index of 62.

Stegotetrabelodon syrticus primitively retained P3–4 (and probably p3–4, as well) (Gaziry, 1987b). These permanent premolars are very robust, simple teeth that exhibit two plates each and between two to four conelets per loph (Sanders, 2017). P3 specimen 87P33A is damaged but has a greatest width of 28.0 mm and displays anterior and posterior accessory conules in association with loph 2. There are three P4s from the site, 52P16A, 18P21A, 421P34A, which vary in size from L=45.0–54.0 mm and W=41.0–44.0 mm. They are all brachyodont teeth. Given that permanent premolars vertically replace deciduous premolars and would have posed potential interference with horizontally displaced adult molars (Sanders, 2017), they must have had short existences in the jaws and therefore been physiologically expensive to develop, relative to their masticatory value. Nonetheless, the size of *S. syrticus* permanent premolars suggests that while they were in alveoli they were a valuable part of the occlusal platform.

Stegotetrabelodon syrticus was also recognized from the late Miocene Central African site of Toros Menalla, Chad (Mackaye, 2001; Mackaye et al., 2005; Le Fur et al., 2014). Stegotetrabelodonts were subsequently included in the rich Toros Menalla mammalian faunal list as part of a proboscidean assemblage additionally comprised of deinotheres, anancine gomphotheres, loxodont elephants, and stegodonts (Le Fur et al., 2014). Features of the lower molars include low, pyramidal plates formed of four-six conelets, thick, unfolded enamel (ET=5.0–6.0 mm), distinct median sulci, V-shaped transverse valleys, and broad plate spacing (LF=2.9–3.4). Central conelets, either single or doubled, form prominent pillars. These pillars are accentuated by the incorporation of single or double posterior accessory conules into enamel wear figures along most of the plates. The postcingulid is formed of two stout conelets. The plate number of the m3s varies from x7x to 8x. Abundant cementum is present in the transverse valleys but does not cover the plates. These molars are very brachyodont and large, as is typical for stegotetrabelodonts: (n=4) L=236.0–276.0 mm, W=95.0–112.0 mm. Stegotetrabelodont m1s from the Toros Menalla sample are similar in morphology: Low-crowned (HI=50–51), pyramidal plates separated by V-shaped transverse valleys, with broad plate spacing and thick enamel (ET=3.4 mm), and a plate formula of x5x or x6 when complete; the postcingulid has nearly the same crown height as the plates and forms a nascent sixth plate. The most complete m1, TM112.00.033, has dimensions of L=117.0, W=62.0, and H=31.7 mm. The subdivision of conelets into many apical digitations and the bilateral compression of these conelets are convergent on stegodont plate construction (see Mackaye, 2001:pl. 11). M3s of the sample (n=3) are much like the m3s with at least seven and up to eight plates, and have equally thick enamel (ET=5.0–7.0 mm), broadly-spaced plates (LF=3.1–3.4), and are large (L=225.0+ to 252.0+ mm; W=95.0–123.0 mm). Crown height is brachyodont and ranges up to 77.0 mm. The conelet composition of plates and the presence of accessory conules are similar to the condition in the m3s.

Because of slightly greater hypsodonty and differences in plate numbers between the Chad specimens and Sahabi stegotetrabelodont molars (plate number, m1=4; m3=7; M3=6 [Sanders, 2008b]), the Toros Menalla stegotetrabelodont should be maintained as a separate species. This is critical for establishing accurate biogeographic connections rather than assuming a faunal relationship with adjacent North Africa. The Sahabi molars are more primitive than those from Toros Menalla. In plate numbers for m1 and m3, the Toros Menalla stegotetrabelodonts resemble *S. emiratus* more than *S. syrticus* (Sanders, 2022).

STEGOTETRABELODON ORBUS Maglio, 1970b
(Tables 5.1, 5.2, and 6.1; Figures 5.1 and 5.12)

Partial Synonymy—*Primelephas gomphotheroides* (in part) Coppens et al., 1978; cf. *Primelephas gomphotheroides* Tassy, 1986; cf. *Stegotetrabelodon* Sanders, 1997; Elephantidae gen. et sp. indet. (in part) Tassy, 2003; Elephantidae gen. et sp. incertae sedis A (in part) Tassy, 2003; *Elephas nawataensis* (in part) Tassy, 2003:fig. 8.8, pl. 4; *Elephas* cf. *E. ekorensis* (in part) Tassy, 2003:fig. 8.8, pl. 5; Elephantidae gen. et sp. indet. Tassy, 2003:fig. 8.5, pl. 4.5; Elephantidae gen. et sp. indet. Saegusa and Hlusko, 2007

This species is the eastern African representative of Stegotetrabelodontinae, and is best documented from localities at the Kenyan site of Lothagam and in the Tugen Hills (Table 5.1). In the lower part of the Nawata Fm. at Lothagam, its geological age is calculated at least as old as 7.4 Ma (McDougall and Feibel, 2003), and in the Mpesida Beds of the Tugen Hills it is dated to ca. 7.0 or 6.4 Ma (Hill et al., 1985, 1986; Kingston et al., 2002; Doman, 2017). The age of these occurrences suggests that these eastern African stegotetrabelodonts have as reasonable a claim as North African and Arabian congenerics to the title of the oldest elephant species. The name "*orbus*," meaning "childless," was given to the species in Maglio's (1970b) remarkable paper naming four new species of early elephant, to denote his hypothesis that it was an evolutionary dead end without an ancestral role for subsequent elephants. In East Africa, the species co-occurs at several sites with the primitive elephantine species *Primelephas korotorensis* and the anancine gomphothere *Anancus kenyensis* (Table 5.1; Maglio, 1970b, 1973; Hill et al., 1985, 1986; Tassy, 1986, 2003; Sanders, 1997; Hill, 2002; Sanders et al., 2010a); it is probably ecologically significant that *S. orbus* is not present in Ethiopian late Miocene localities that otherwise have produced these elephantines and anancines (Table 5.1; see Mebrate, 1983; Kalb and Mebrate, 1993; Haile-Selassie, 2001; Saegusa and Haile-Selassie, 2009), but the implication of stegotetrabelodont absence from these localities remains obscure.

Stegotetrabelodon orbus is documented by abundant isolated dental remains, a few poorly preserved crania and mandibular specimens, and a very small sample of postcrania. Its dental sample provides the largest collection of cheek teeth among stegotetrabelodont species. The adult dental formula reflects the retention of lower tusks and adult third and fourth premolars (Table 5.2; Sanders, 2017, 2022); adult cheek teeth were preceded in upper and lower jaws by second-fourth deciduous premolars. Plate formulae for the species are dP2/dp2 = 3x/--; dP3/dp3 = x3x/x3x; dP4/dp4 = --/--; P3/p3 = --/x3x; P4/p4 = x2x/x3x; M1/m1 = --/--; M2/m2 = x5x/5x-x6; M3/m3 = x6x-(7)/7x-x7xx-(8) (Sanders et al., 2010a).

Two incomplete, damaged crania were collected from the Nawata Fm. at Lothagam. The first, KNM-LT 26318, is badly weathered and fragmented from crushing. It lacks its braincase, occipital region, face, basicranium, and most of the rostrum. The rostrum is strongly downturned, with closely set, non-divergent tusk alveoli. Its orbit has a robust zygomatic process of the frontal. It also has anteriorly worn M3s with six full plates and pre- and postcingulae. The molar emergence and wear patterns in this individual cannot be matched in modern African elephants, but by the time M2 has been ejected from the alveolus and all plates are emerged, extant elephants are fully adult and mature (Laws, 1966). Plates are composed of four-five conelets, and it is possible that the first two enamel wear figures incorporate posterior accessory conules. Enamel is thick, 4–5 mm, and plates are widely spaced anteroposteriorly. Typical of stegotetrabelodont third molars, width is >100 mm and the overall size is big (L = 250–260 mm).

The second cranium, KNM-LT 26319, is best preserved on its ventral side, and is from a younger individual, with right and left M2 in wear and M3s emergent. The widest part of the cranium is estimated to have been at the squamosal end of the broken zygomatic arch. The rostrum is relatively narrow in ventral view, downturned, and its tusk alveoli are parallel to one another. The basicranium is angled high above the level of the palate. Right and left glenoids are anteroposteriorly convex and transversely saddle-shaped, with a deep postglenoid sulcus behind them. Each M2 has a plate formula of 5x. The right M2 has thick enamel (ET = 5–6 mm), lamellar frequency of 4.0, and is large (L = 156.4 mm, W = 92.7 mm). Accessory conules are incorporated into the posterior sides of each plate. Lightly or unworn plates of the emerging M3s show that each plate is formed of five equal-sized conelets, and at least plates 1–2 have posterior accessory conules. The features of the molars are consistent with the assignment of the specimen to *S. orbus*. An additional half-plate is wedged between plates 2 and 3 in the M3s. The bilateral symmetry of this "fabricational noise" makes it more likely that it is the result of a genetic variant rather than the influence of anterior pressure on the molars due to resistance of the M2s to the horizontally advancing M3s as they rotated into the alveolus (see Roth, 1989). Tooth emergence suggests the identification of this individual to age-grade XIX in Laws' (1966) system, about 32 years of age at death, or in the interval of 28–34 years in the revision of the technique by Lee et al. (2011).

Mandibular remains are equally rare for this species. A right dentary with partial symphysis and right and left m2s from the Tugen Hills, KNM-TH 32836, has a preserved total length of 590 mm, of which 180 mm comprises the symphyseal spout. The downturned symphysis is crushed anteriorly, making it impossible to mark the presence or absence of tusk alveoli, but nonetheless, it is more extensive than in extant elephants. The height of the corpus below the anterior of m2 is 151.6 mm. There are three mental foramina, one at mid-corpus height anterior to m2, and two lateral to the symphysis, of which the anteriormost opens forward.

The type specimen, KNM-LT 354, includes a left dentary with partial symphysis and lower tusks (Maglio, 1970b). It was described by Maglio and Ricca (1977). It is missing its articular condyle, tip of the coronoid process and the distal end of the incisive alveoli of the symphysis. The dentary has a long corpus, prominent attachment area for the medial pterygoid muscle, and is high and strongly downsloped anterior to the molar alveolar row. Its ramus has a robust anterior edge for insertion of the temporalis muscle tendon and a broad concavity for a deep masseter muscle, but the medial temporalis fossa is shallower than that in extant African elephants. The preserved portion of the symphysis is 413 mm in length but may have been 100 mm longer originally. The symphysis and jaw appear relatively delicate to support elongated lower tusks. Each tusk is socketed to within 40 mm of the posterior of the symphysis and is narrowly separated from one another by a thin median lamina of bone. Anteriorly, the mandibular canal is described as opening laterally by a short channel to the mental foramen, and presumably medially by a passage to the tusk alveolus. The thick coating of plaster over the tusks in their alveoli makes it impossible to confirm the morphology of the antechamber of the mandibular canal. The right tusk measures 995 mm in length (604 mm exposed outside of the incomplete alveolus) and has maximum transverse dimensions of 61 × 72 mm. In cross-section, the tusks are laterally compressed until near their distal ends, where they are nearly circular. There is a distinct longitudinal concavity along the dorsomedial aspect of the left tusk, and proximally the tusks

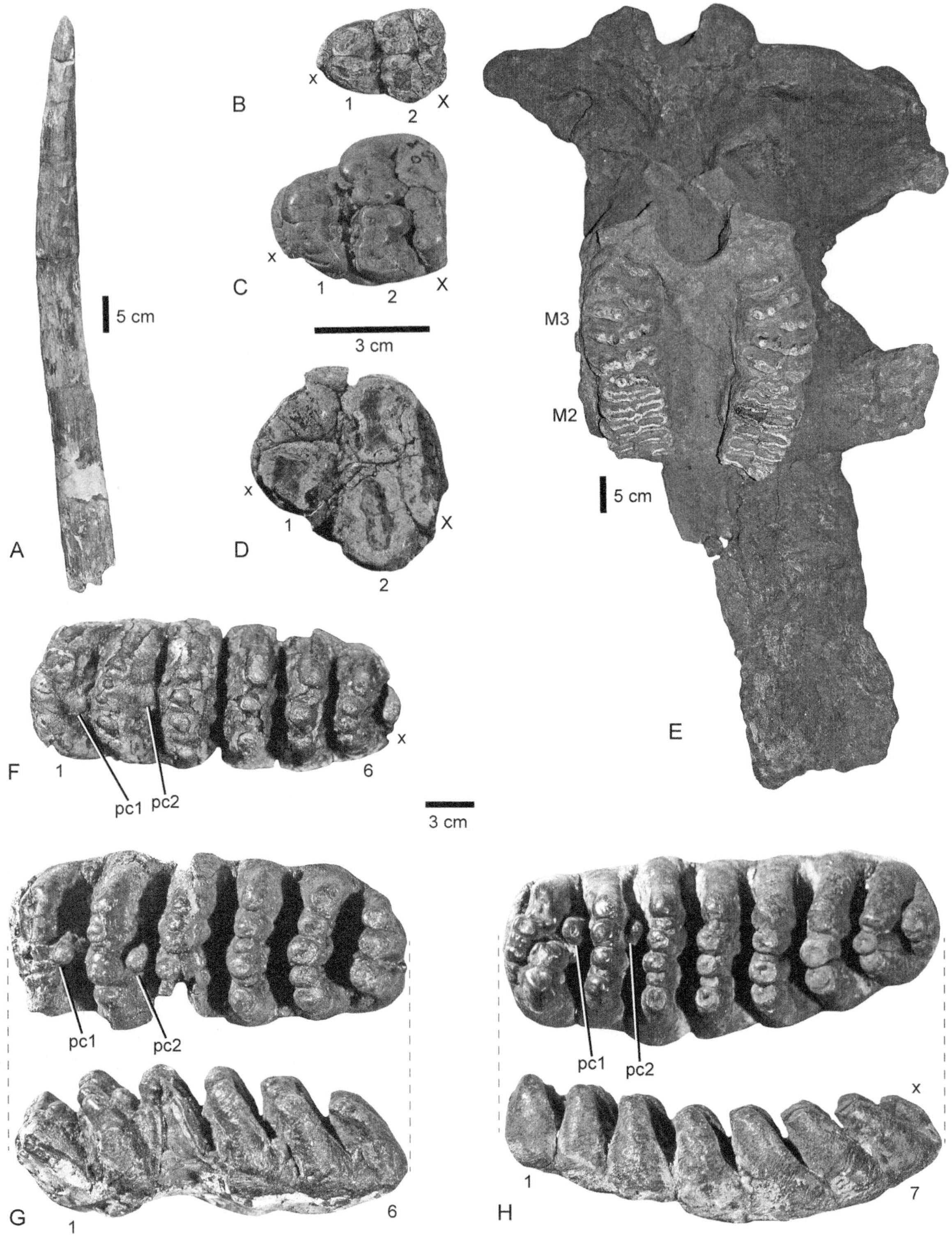

FIGURE 5.12 Cranial and dental specimens of late Miocene *Stegotetrabelodon orbus*. Abbreviations: M, upper molar; pc, posterior accessory conule; x, pre- or postcingulum(id); X, large pre- or postcingulum, incipient loph(id); 1, 2, 3 . . ., plate number counted from anterior of crown. (A, E) to the same scale; (B–D) to the same scale; (F–H) to the same scale. (A) Lower tusk missing part of the proximal end, KNM-LT 354 (type) from Lothagam, Kenya, lateral view. (B) p3, KNM-LT 26339 from Lothagam, Kenya, occlusal view. Anterior to the left. (C) p4, KMM-MP 47 from the Mpesida Beds, Tugen Hills, Kenya, occlusal viw. Anterior to the left. (D) P4, KNM-LT 343 from Lothagam, Kenya, occlusal view. Anterior to the left. (E) Partial cranium, KNM-LT 26319 with right and left M2–3 from Lothagam, Kenya, ventral view. Anterior at bottom of page. (F) Right M3, KNM-LT 354 (type) from Lothagam, Kenya, occlusal view. Anterior to the left. (G) Left M3, KNM-LT 359 from Lothagam, Kenya, occlusal and lateral views. Anterior to the left. (H) Right m3, KNM-LT 359 from Lothagam, Kenya, occlusal and lateral views. Anterior to the left.

display numerous longitudinal grooves. A lower tusk fragment of a juvenile (KNM-LU 121) from Lukeino has cross-sectional dimensions of 35.6×46.9 mm and sports a similar dorsomedial sulcus and longitudinal grooves. Distally, the type tusks are lightly curved upward. A distal lower tusk fragment from Lothagam, KNM-LT 24213, exhibits a rounded cross-sectional profile and is slightly upturned. Its transverse dimensions are also modest, 62.7×78.0 mm.

M3s in *S. orbus* (n=8), including in the type specimen (KNM-LT 354) and the Lothagam crania, usually have six plates, low lamellar frequencies (2.75–3.25), thick enamel (4.7–7.0 mm), are broad relative to length (W=104.3–110.5 mm; L=207.5–254.4 mm), and have relatively low crown heights (HI=66–71). In lateral view, plates are pyramidal in shape with corresponding V-shaped transverse valleys separating them, and are covered in cementum, although transverse valleys are not completely infilled. Posterior accessory conules are present posterior to the first two plates and in several M3s extend as posterior as plate 4. Lower third molars (n=13), including in the type specimen, exhibit seven plates and are constructed similar to the M3s, with three–six conelets per plate, accessory conules posterior to the first two plates that in some specimens occur as posteriorly as plate 5, have low lamellar frequencies (LF=2.9–3.5), thick enamel (4.4–7.4 mm), are brachyodont (HI=66–68), and are generally broad relative to length (W=89.0–111.0 mm; L=209.5–276.7 mm).

Several M3s from Kanam, Kenya have seven plates, but otherwise exhibit features consistent with identification as *Stegotetrabelodon*, including low lamellar frequencies, broad plates, a small number of robust conelets comprising each plate, and persistence of median clefts, or sulci. An m3 from the site has posterior accessory conules throughout the crown (Sanders et al., 2010a). Along with a stegotetrabelodont m3 with eight plates from the Oluka Fm. of the Western Rift, Uganda (Tassy, 1995), these specimens provide potential evidence that stegotetrabelodonts continued to evolve, if modestly, prior to their demise. This could also explain the morphological oddity of the m2s in KNM-LT 350 (see below), with six plates, as museum accession notes suggest that this specimen may derive from the Pliocene Apak Mb. of the Nachukui Fm. at Lothagam, younger than most teeth in the species sample.

The first and second molars are very similar in morphology to M3/m3, but there are no complete first molars in which to assess plate numbers. M2s from the Nawata Fm., including in the type specimen, all (n=4) are pentalophodont, are low crowned, have anteroposteriorly broadly spaced plates (LF 3.25–4.0), very thick enamel (ET=5.0–7.0), and usually have posterior accessory conules incorporated into enamel loops of plates 1–4 with wear.

Lower second molars associated with mandibular remains from Tugen Hills, KNM-TH 32836, are very broad (92.7–99.6 mm), with thick enamel (ET=3.5–5.8 mm), and have five full plates. Accessory conules are associated posteriorly with plates 1–3. There is an accessory half-plate on the buccal side of the right m2 posterior to plate 1. The plates are widest basally and are separated by V-shaped transverse valleys that are coated with cementum. Lower second molars from the Nawata Fm. (n=5) are similar in morphology to these m2s, and have metrics that range from L=154.5–173.0 mm, W=90.0–102.6 mm, ET=3.5–6.0, LF=3.4–3.5, and HI=57–66. M2/m2s usually have five plates, but KNM-LT 350 has right and left m2s with six plates. KNM-LT 350 was originally placed in *S. orbus* (Maglio, 1973 [mislabeled as "KNM-LT 342"]; Maglio and Ricca, 1977), later was transferred to *Primelephas "gomphotheroides"* (=*korotorensis*) (Tassy, 1986, Kalb and Mebrate, 1993; Froehlich and Kalb, 1995), and subsequently was placed in Elephantidae gen. and sp. indet. (Tassy, 2003). Contra Tassy (2003), the dimensions and occlusal features of KNM-LT 350 (and -LT 342) fit well with other *S. orbus* m2s, including that of the type specimen, in lamellar frequency, expression of accessory conules, plate shape, and broadening of the crown posteriorly. Features used to argue against inclusion of KNM-LT 350 in *S. orbus* include the development of the postcingulid into a sixth plate, anterior concavity of plates, thinner enamel (this seems untenable), and number of apical digitations (up to eight) in its plates.

An isolated m2 from Lemudong'o, Kenya (KNM-NK 42396) was attributed to a new, indeterminate species of elephant (Saegusa and Hlusko, 2007), but the evidence for this is not compelling. It shares with *S. orbus* features such as a plate formula of 5x, pyramidal plate shape, V-shaped transverse valleys, and posterior accessory conules throughout the crown, and therefore is retained in the species.

There is scant evidence of permanent premolars in *S. orbus*, but the assignment of p3s, p4, and a P4 to the species suggests that, as in other stegotetrabelodont species, upper and lower permanent third and fourth premolars were likely retained. The p4, from the Mpesida Beds in the Tugen Hills, KNM-MP 47, is a robust specimen with two lophids, measuring L=47.4 mm×W=42.1 mm. Each loph is formed of between two and five conelets, and there are pre- and posttrite accessory conules posterior to lophid 2. It is brachyodont. The p3s, KNM-LT 26339 and -LT 26329, derive from the lower Nawata Fm. at Lothagam and each has a plate formula of x3x. Size varies from L=33.5mm and 32.8mm×W=24.6 mm and 25.7 mm. Lophids have between two to four conelets and are brachyodont (Sanders, 2017). P4 specimen KNM-LT 343, also from the lower Nawata Fm., is relatively large as in *S. emiratus* and *S. syrticus*, with dimensions of L=50.4 mm×W=52.0 mm. It has thick enamel (ET=3.6 mm), a plate formula of x2x, and each loph is formed of two-four conelets. The size of these premolars suggests that they contributed meaningfully to the masticatory function of the dental battery before being ejected by horizontal displacement of emerging molars (Sanders, 2017).

Postcranial remains of the species are sparse, with only an atlas and humerus identified as *S. orbus* by Maglio and Ricca (1977). The humerus, KNM-LT 369, is from a moderately tall elephant, with a length of 960 mm. Proboscideans with similar humeral lengths can be reconstructed as having shoulder heights of close to 300 cm (Larramendi, 2016).

As in the Sahabi humeri, the diaphysis is longitudinally torqued, the greater tubercle projects prominently proximally, and the head and distal articulation are oriented vertically for graviportal support. However, it is not as massive as the Sahabi humeral specimens, with less flaring deltoid and supinator crests. The olecranon fossa is shallow, indicating that the ulna had a short, posteriorly reflected olecranon process and that flexion and extension of the forelimb on the humerus were not powerful or extensive.

Although it has been hypothesized that stegotetrabelodonts may not have been ancestral to later elephants (e.g., Maglio, 1970b; Tassy and Debruyne, 2001; Sanders, 2004), the features that define the group are overwhelmingly primitive and therefore do not strongly support this view (Coppens et al., 1978). The argument that stegotetrabelodonts could not have been ancestral to the *Loxodonta* clade, for example, because species of *Loxodonta* preserve the development of anterior and posterior accessory conules from the ancient gomphotherian pretrite trefoil patterns, whereas *S. orbus* and *S. syrticus* lost their anterior accessory conules (Sanders et al., 2010a), is overcome by the presence of these accessories in molars of *S. emiratus* (Sanders, 2022). Moreover, it is not known how readily or by what mechanism accessory conules may be developmentally suppressed or re-initiated in elephants. In addition, the cheek teeth of *S. orbus* are little differentiated from those of the coeval and often sympatric elephantine *Primelephas korotorensis* and can be difficult to sort between these species when incomplete, so it is possible that at least some elephants derived from or share close common ancestry with stegotetrabelodonts. Gaziry (1987b) proposed incorporating *S. orbus* into *S. syrticus*, an idea that Sanders et al. (2010a) felt merited consideration, but in the absence of more information about cranial morphology and morphometric aspects of intermediate molars, and because the molars are not quite as primitive as in the Sahabi stegotetrabelodont, it is reasonable to maintain *S. orbus* as a separate species.

ELEPHANTINAE Gray, 1821

In the current classification, Elephantinae includes elephant species lacking lower tusks (except for atavistic occurrences). Other features attributed to the Subfamily include a minimum number of third molar plates = seven (Kalb and Mebrate, 1993; Kalb et al., 1996a), mandibular symphysis reduced in length, and molar median sulci diminished or absent (Sanders et al., 2010a). However, although lacking lower tusks, *Selenetherium* and *Stegodibelodon* possess prominent symphyseal spouts, and *Selenetherium* has only six third molar plates and exceptionally brachyodont molar crowns (Mackaye, 2001; Mackaye et al., 2005). Thus, distinctions are blurred between Stegotetrabelodontinae and Elephantinae. Nonetheless, if lower tusks were lost just once, in an ancestral elephantine, then despite a weak morphological contrast between stegotetrabelodonts and basal elephantines, African Elephantinae includes *Primelephas, Stegodibelodon, Selenetherium, Mammuthus, Elephas, Palaeoloxodon*, and *Loxodonta* (and perhaps also *Phanagoroloxodon* [Zhang, 2020]).

The late Miocene fossil record of elephantine elephants shows their near-simultaneous origin alongside stegotetrabelodonts (Maglio, 1970b, 1973; Sanders et al., 2010a). The rapid subsequent evolution and wide geographic dispersals of elephantine elephants make them especially useful for biochronological correlation (Maglio, 1973).

STEGODIBELODON Coppens, 1972
STEGODIBELODON SCHNEIDERI Coppens, 1972

(Tables 5.1 and 5.2; Figure 5.1)

Synonymy—*Stegodibelodon schneideri* Beden, 1985:-p. 26; *Loxodonta schneideri* Beden, 1985:fig. 2

This species is represented at late Miocene-early Pliocene sites in the Chad Basin (Coppens, 1972; Mackaye, 2001; Mackaye et al., 2005) and was claimed to occur in late Miocene beds of the Middle Awash, Ethiopia (Kalb and Mebrate, 1993; Kalb and Froehlich, 1995). The species is named for the hydrogeologist who first discovered the fossil beds of the type site, and the genus nomen refers in part to the presence of a single set of (upper) tusks.

The type specimen of *S. schneideri* is from Menalla, Chad (Table 5.1) and includes a cranial fragment, postcranial elements (forelimb bones, tarsals, carpals, vertebrae, and ribs), and most importantly, a mandible with right and left m3. Coppens (1972) erected and placed *Stegodibelodon* in the Stegotetrabelodontinae in an evolutionary series that he envisioned progressed from *Stegotetrabelodon syrticus-Stegotetrabelodon orbus-Stegodibelodon schneideri*. *Stegodibelodon* is also included in the Stegotetrabelodontinae by Kalb and Mebrate (1993). They reasoned that the convex-convex shapes of the anterior and posterior margins of enamel loops in *S. schneideri* resembled those in stegotetrabelodont molars and are not found in "true" (elephantine) elephant molars. Mackaye et al. (2005) added to this phylogenetic scenario by positing that early Pliocene *Selenetherium kolleensis* (Table 6.1) was intermediate between this succession of taxa and a clade composed of *Primelephas* + *Stegodon*, primarily based on the reduction of the elongate, tusk-bearing symphyses of stegotetrabelodonts to the shorter, tuskless anterior symphyseal spouts of *Primelephas* and *Stegodon*. Although the type mandible of *Stegodibelodon* lacks lower tusks, it has a very long, projecting spout (L = 460 mm) that is 105 mm wide at its distal termination. This is extraordinarily long for tuskless symphyseal projection and constitutes a significant fraction of a mandible that is about a meter long. However, because it lacks lower tusks, *Stegodibelodon* belongs to the Elephantinae (see Tobien, 1978b), as the absence of lower tusks should be considered a synapomorphy of the subfamily. Although Beden (1979, 1983, 1985) considered *S. schneideri* to be a primitive loxodont elephant based on putative molar similarities between *S. schneideri* and *L. adaurora* (in his view, *Loxodonta* derived from the stegotetrabelodont lineage and *Elephas*+*Mammuthus* evolved from *Primelephas*), no synapomorphies have been identified to support this hypothesis (Mackaye, 2001), and molars of *L. adaurora* are very different morphometrically from those of *S. schneideri*.

The type mandible of *S. schneideri* has a long symphysis that projects via a prominent spout horizontally forward, nearly in line with the corpus. A mandible (KL176-1) from the Adu Mb. of the Adu-Asa Fm. from the Middle Awash, Ethiopia, that has lost its molars postmortem, was assigned to cf. *S. schneideri* because it also has an elongate, tuskless symphyseal spout (Kalb and Mebrate, 1993). However, this spout is deflected downward and is shorter (~300 mm) than in the type mandible, the overall morphology of the mandible closely matches that of *Primelephas korotorensis* jaws, and although no other specimens of *S. schneideri* have been recovered from these Middle Awash beds, *P. korotorensis* is abundant in the Adu Mb. For these reasons, KL176-1 is referrable to *P. korotorensis* rather than *S. schneideri* (Saegusa and Haile-Selassie, 2010; Sanders et al., 2010a). The left m3 of the type specimen has dimensions of L = 277 mm and W = 118 mm, and exhibits a plate formula of x7x, thick enamel (ET = 5–6 mm), and broadly spaced plates (LF = just >3.0) (Coppens, 1972). In the occlusal view, enamel wear figures are strongly convex-convex, but it is difficult to ascertain the presence of accessory conules because the occlusal platform is worn. An upper tusk from the same site is lightly curved and is large, with an overall length of 1.8 m and a greatest diameter of 155 mm at its base. Upper third molars from Kolinga 1, Chad also have plate formulae of x7x, thick, unfolded enamel (ET = 5 mm), broad plate spacing (LF = 2.5–3.0), are low crowned (HI = 71–79), and are massive in size (L = 300 mm, W = 110 mm) (Coppens, 1972).

Fossils from late Miocene Toros Menalla and early Pliocene Kollé, Chad also have been attributed to *S. schneideri* (Table 5.1; Mackaye, 2001; Mackaye et al., 2005). Included in the expanded assemblage from Toros Menalla is an incomplete mandible, TM127.01.001, with low, broad corpora (H = 170 mm; W = 185 mm), a long, anteriorly broken symphysis (L = +180 mm), no sign of lower tusks, and m3s. The m3s each have an estimated plate formula of x7x or 8x, are long, wide, and very low crowned. Enamel is thick and unfolded, plates are pyramidal and are broadly separated by V-shaped transverse valleys, and each plate is comprised of a low number of conelets (four-five). A well-marked median sulcus is preserved throughout the crowns. An M3 associated with the mandible also has seven plates, very thick enamel (ET = 7.0), and well-spaced plates (LF = 2.9). Palatal remains from Toros Menalla (TM160.01.005) carry M1-M2s, with plate formulae of x5x and x6x, very low crowns, thick, undulated enamel, low lamellar frequencies, and are large for their tooth positions (Mackaye, 2001). These molars also have transversely rectilinear plates. It is not possible to unequivocally ascertain the presence of accessory conules in these teeth. It is also difficult to differentiate this subsample from contemporaneous *Primelephas korotorensis*.

The dental sample from Kollé assigned to the species is more impressive in a number of specimens (Mackaye, 2001). However, compared with molars from Menalla, Kolinga 1, and Toros Menalla, these teeth are more derived. Although it is possible that they are evidence of a more advanced segment of the *S. schneideri* lineage, their attribution to the species is questionable. For example, M3 specimens KL6.99.001 and.002 have plate formulae of x9x, M3 KL5.98.001 has at least eight plates, M2 specimens KL19.96.005 and KL5.96.002a have plate formulae of x7x, and M1 KL5.98.082 has six plates. Nonetheless, they are otherwise primitive in having low lamellar frequencies, thick enamel, and very low crowns (HI = <60). These molars have few conelets per plate (usually five in the M3s), and plates are primarily transversely rectilinear. Transverse cross-sectional plate profiles are broadest basally, have sloping sides, and generally resemble those of *P. korotorensis*, from which it is insignificantly differentiated in molar morphology, except by plate number.

PRIMELEPHAS Maglio, 1970b
PRIMELEPHAS KOROTORENSIS (Maglio, 1970b)

(Tables 5.1, 5.2, and 6.1; Figures 5.1, 5.13, and 5.14)

Partial Synonymy—*Stegodon korotorensis* Coppens, 1965:343; *Stegodon kaisensis* Coppens, 1967:I, fig. 2; *Mammuthus (Archidiskidon) subplanifrons* (in part) Cooke and Coryndon, 1970:123; *Primelephas gomphotheroides* Maglio, 1970b; P. *gomphotheroides* Maglio, 1973; *P. korotorensis* Maglio, 1973; *P. gomphotheroides* Coppens et al., 1978; *P. korotorensis* Coppens et al., 1978; *Stegotetrabelodon orbus* Kalb and Mebrate, 1993:40–44; *Mammuthus subplanifrons* (in part) Kalb and Mebrate 1993: 59–63; *Stegodibelodon schneideri* Kalb and Mebrate, 1993:44–46; cf. *P. gomphotheroides* Tassy, 1995; *P. korotorensis* Mackaye, 2001; *P. gomphotheroides* Mundinger and Sanders, 2001; *Elephas nawataensis* Tassy, 2003; *Stegodibelodon schneideri* Haile-Selassie et. al., 2004; *P. gomphotheroides* Sanders, 2007; *P. korotorensis* Mackaye et al., 2008; *E. nawataensis* Saegusa and Haile-Selassie, 2009; *P. gomphotheroides gomphotheroides* Saegusa and Haile-Selassie, 2009; *P. gomphotheroides saitunensis* Saegusa and Haile-Selassie, 2009; *P. korotorensis* Sanders et al., 2010a

The taxonomic homogeneity and legitimacy of late Miocene-early Pliocene *Primelephas korotorensis* (Tables 5.1 and 6.1) have been questioned (e.g., Coppens et al., 1978; Beden, 1979, 1983, 1985), and the species has been considered by some to be a "wastebasket" taxon (e.g., Mackaye, 2001). Study of the original hypodigm, however, demonstrates that it is a well-delineated stem elephantine species (Sanders et al., 2010a). Thus, it was aptly named "first elephant" by Maglio (1970b) in his type paper. Although Maglio (1970b) originally named the species *Primelephas* "*gomphotheroides*," he also named a second species

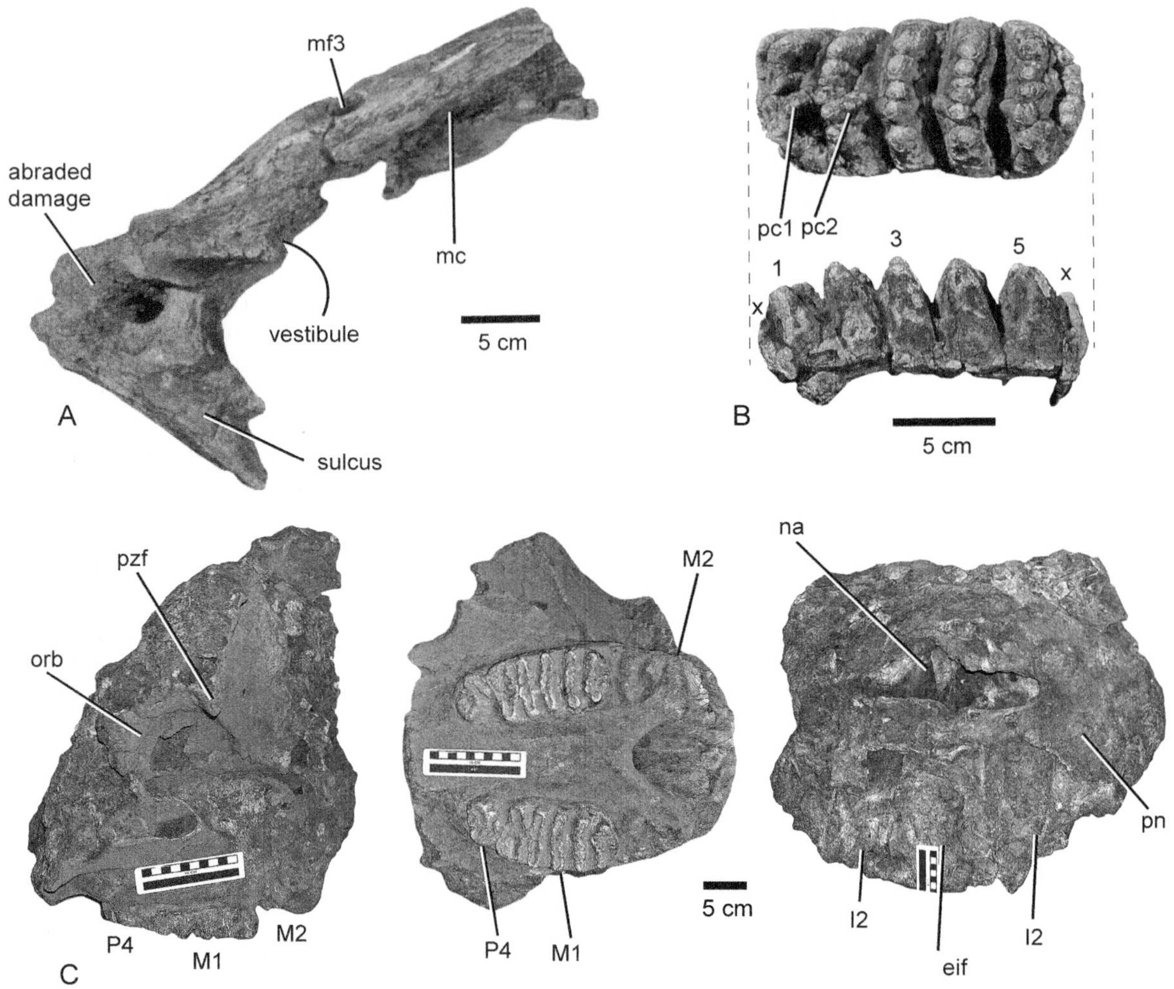

FIGURE 5.13 Mandibular, cranial, and dental specimens of late Miocene *Primelephas korotorensis*. Abbreviations: eif, external incisive fossa; I2, alveolus for upper tusk; mc, mandibular canal; mf3, posteriormost mental foramen (of three); M1, M2, upper first, second molars; na, nasal aperture; orb, orbit; pc1, pc2, posterior accessory conule 1, 2; pn, nasal process of the premaxilla; pzf, zygomatic process of the frontal; P4, upper permanent fourth premolar; x, pre- or postcingulum; 1, 2, 3, …, plate number counted from the anterior end of the crown. (A) Symphyseal specimen KNM-LT 358 articulated with right partial dentary specimen KNM-LT 378 of the same individual from Lothagam, Kenya. The anterior spout is broken away and the symphysis has a perforation. Sulci are the anteriormost ends of the lower part of the mandibular canal. Note that the dentary fragment anteriorly encloses the mandibular canal, forming a vestibule rather than an extension of a tusk alveolus. Anterior to the left. (B) Right M2, KNM-LT 358 from Lothagam, Kenya, occlusal and lateral views. Anterior to the left. (C) Partial cranium with right and left P4-M1 and M2s in crypt, NY 46'92 from Kisegi-Nyabusosi, Uganda, left lateral, ventral, and anterior views.

P. korotorensis (Maglio, 1970b, 1973), based on fragmentary molars originally assigned to "*Stegodon korotorensis*" (Coppens, 1965). While these specimens are elephant and not stegodont molars, they are not morphologically discernable enough from *P. "gomphotheroides"* molars to warrant unique specific status (Sanders, 1997, 2004; Mundinger and Sanders, 2001), and therefore have priority (Mackaye, 2001; Mackaye et al., 2008; Sanders et al., 2010a).

The original diagnosis distinguished the species by its very low crown height, loss or diminution of median clefts (sulci), pyramidal-shaped plates widely separated by V-shaped transverse valleys, moderately to thick, unfolded enamel, and a relatively modest mandibular symphysis hypothesized to support small but prominent incisors in some individuals (possibly males) (Maglio, 1970b). This diagnosis was emended to differentiate the species from contemporaneous (and in some localities sympatric) *Stegotetrabelodon orbus* by generally narrower molars, *absence* of lower tusks, restriction of accessory conules to the anterior end of m3s, less robust molar crown features, and (in occlusal, view) more transversely rectilinear plates (Sanders et al., 2010a). Since the species was first named, an abundance of additional fossils attributed to *P. korotorensis* have been recovered, including a fragmentary cranium and mandibles (Mebrate, 1983; Tassy, 1986, 1995, 2003; Kalb and Mebrate, 1993; Sanders, 1996, 1997; Haile-Selassie, 2000, 2001; Mackeye et al., 2008). These fossils have expanded the morphological accounting and variability of the species, but suggest that it is worth inquiring whether all of the expanded sample comprising *P. korotorensis* is sufficiently homogeneous to be considered the same species.

The hypothesized presence of small lower tusks in *Primelephas* was considered important to delineate the taxon as a phylogenetic intermediate between stegotetrabelodonts and more advanced elephantine elephants lacking i2s (Maglio, 1970b, 1973; Coppens et al., 1978). The supposed insertion of lower tusks into the apparently foreshortened symphysis of *Primelephas* created a problem that Maglio (1970b) resolved by arguing that the root of each lower incisor was socketed in the mandibular canal. As a result, the position of the lower incisors was alleged to be "no longer appressed in the midline of the symphysis …, but [to] lie on either side of the 'spout' formed by the downturned symphysis" (Maglio, 1970b:11). In this scenario, the lower incisors were considered to be developed and socketed in the anterior chamber of the mandibular canal. Although the anterior chamber exists in elephants and other proboscideans without lower tusks and forms the terminus of the mandibular canal, it is apparent that this chamber was incorrectly assumed to be functionally related to tusk development. Maglio further assumed that these cavities or chambers represented vestigial incisive chambers in early species of *Loxodonta* and *Elephas*, such as *Loxodonta adaurora* (Maglio, 1970b:19).

This evolutionary scenario is controversial (see Tassy, 1986, 1999; Kalb et al., 1996; Sanders, 1997; Mackaye et al., 2008). The supposition of a brevirostrine mandible with lower incisors in *Primelephas* relies heavily on the morphology of a weathered symphyseal fragment, KNM-LT 358 (Figure 5.13A). Although Kalb et al. (1996) believed that this fragment belonged to *S. orbus*, its associated molars are all from a single *Primelephas* individual (e.g., Figure 5.13B; see Tassy, 2003), and their interpretation is thus unreasonable. On the dorsum of each side of the symphyseal fragment (KNM-LT 358; Figure 5.13A) is a shallow sulcus, imagined by Maglio (1970b) and Maglio and Ricca (1977) to be the floor of an incisive alveolus extended into the mandibular canal. However, re-examination of the specimen showed that it articulates perfectly with a fragment of the right mandibular corpus (KNM-LT 378) from the same locality (Figure 5.13A) Mundinger and Sanders, 2001; Sanders, 2004). The corresponding pieces enclose the mandibular canal anteriorly and comprise a chamber, or vestibule, which communicates laterally to the outside of the jaw via two mental foramina, but which exhibits no passage forward for an incisor (Figure 5.13A). This vestibule is the anterior termination of the mandibular canal (see also Tassy, 1999).

The presence of an anteriorly closed mandibular vestibule is the normal condition in proboscidean jaws, including elephants, regardless of whether they have lower tusks or not. For example, in lower-tusked mastodons (*Mammut americanum* and *M. borsoni* [Tobien et al., 1988:fig. 61]) and "*Stegotetrabelodon*" *maluvalensis* (Sarwar, 1977:fig. 54), lower incisors are socketed in incisive alveoli that continue nearly to the posterior end of the symphysis, *medial* to the anterior chamber of the mandibular canal. The chamber and alveolus are separated from one another by a wall of bone. Nerves and vascular supply presumably reach the lower tusks via a network of barely discernable, minute foramina. As in extant elephants, the mandibular vestibule of KNM-LT 358/378 must have housed vascular and nerve plexuses that supplied the lower face and lips.

Despite such strong evidence to the contrary in other proboscideans, Maglio and Ricca (1977:18) reported the type specimen of *S. orbus* (KNM-LT 354) to have communication between the mandibular canal and incisive alveolus by a large opening. Damage to the symphysis, anterior corpus, and left incisor makes it difficult to assess this claim, particularly as they are oddly offset in the current reconstruction and liberally repaired with plaster. A small lower tusk from Lukeino, Kenya, accessioned as *Primelephas* (KNM-LU 121), is probably from a juvenile individual of *S. orbus* (Tassy, 1986). It preserves a length of 158.6 mm and a width and height of only 35.6×46.9 mm, with an ovoid to piriform cross-section and dorsomedial longitudinal sulcus. The morphological features of this tusk closely match those of other stegotetrabelodont i2s.

A number of mandibles of *P. korotorensis* are now known, and all show evidence of moderately projecting, blunt symphyseal spouts without accompanying tusks or tusk alveoli. These specimens include TM177-01-043 from Toros Menalla, Chad (Mackaye et al., 2008), and KL176-1, KL89-5, and STD-VP-2/86 from the Middle Awash, Ethiopia (Saegusa and Haile-Selassie, 2009). The adult mandible with right and left m3s, TM177-01-043, is the most informative of these: its symphysis, including the anteriorly-projecting spout, is 170 mm in length and about 70 mm in width; the mandible is otherwise large with a width across the posterior corpora of 400 mm, the ramus is 320 mm anteroposteriorly, posterior width of each corpus is 150 mm, and anterior height of the corpus is 145 mm. In this specimen, the spout is slightly downturned, there are two mental foramina in each dentary, and the condyle is rounded and robust. KL176-1, probably originally misidentified as "*Stegodibelodon schneideri*" (see above) is similar morphologically with the same angulation of the symphysis, though its spout is longer than in the Toros Menalla specimen. STD-VP-2/86 has the same type of robust condyle as in TM177-01-043, even though it is from an ontogenetically younger individual. In addition, these and other mandibular remains from the Middle Awash exhibit a restricted mandibular angle that does not extend much posteriorly or ventrally, relatively anteroposteriorly long rami, and restricted muscular insertion area on the ramus for the deep masseter muscle. Where preserved, condyles sit high above the level of the alveolus. Older individuals have more downturned symphyseal spouts. Aside from the differences in symphyseal morphology and absence of tusks, mandibles of *Primelephas* are readily distinguished

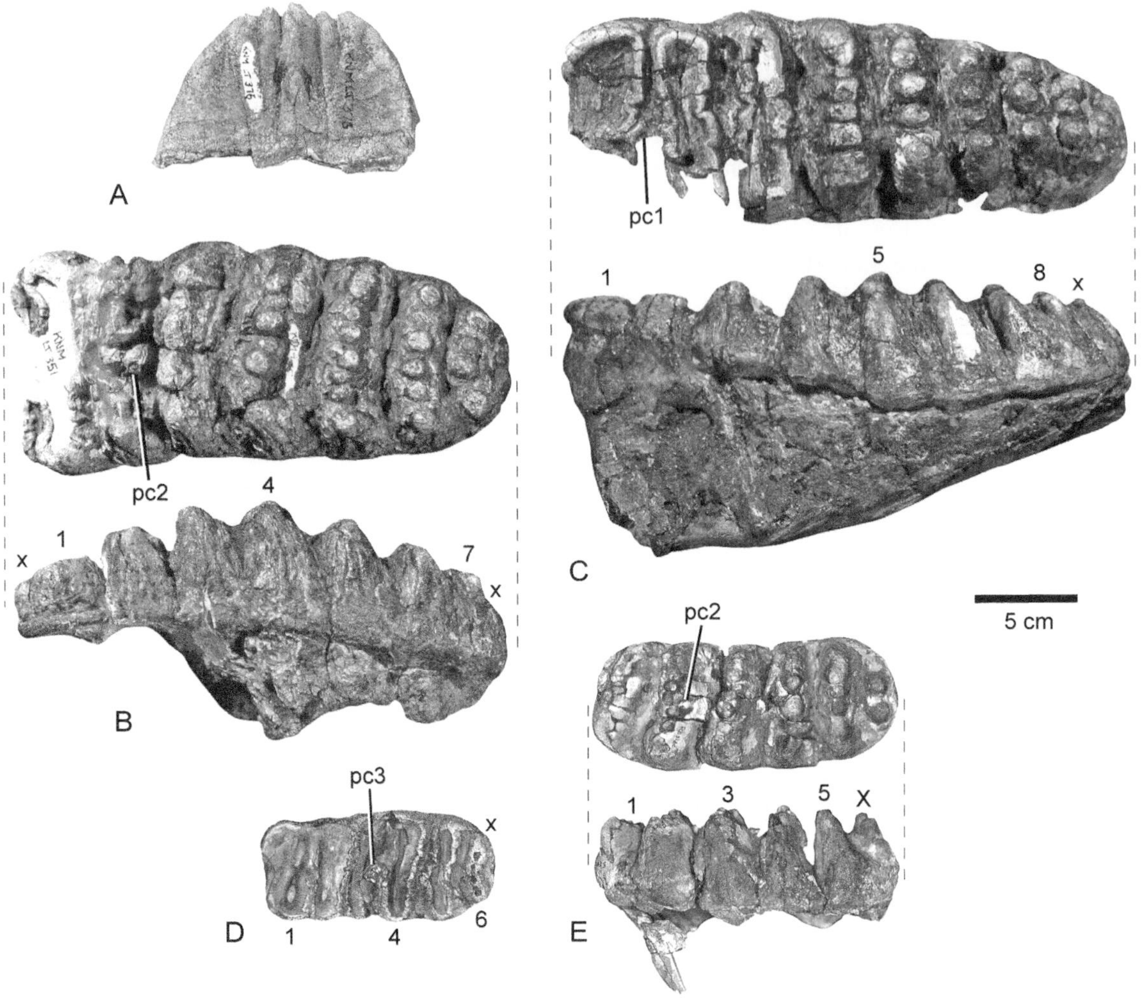

FIGURE 5.14 Molar specimens of late Miocene *Primelephas korotorensis*. Abbreviations: pc, posterior accessory conule; x, pre- or postcingulum(id); X, large postcingulid; 1, 2, 3 …, plate number counted from the anterior end of the crown. Specimens to the same scale. Anterior to the left in (B–H). (A) Third molar plate KNM-LT 376 from Lothagam, Kenya, in transverse cross-sectional view. Note the composition of five apical digitations or conelets. (B) Left M3, KNM-LT 351 (type of *Primelephas* "*gomphotheroides*") from Lothagam, Kenya (Maglio, 1970b), occlusal and lateral views. (C) Left m3, KNM-LT 351 (type of *Primelephas* "*gomphotheroides*") from Lothagam, Kenya (Maglio, 1970b), occlusal and lateral (reversed) views. (D) Right dp4, KNM-LT 23783 (from the type series of "*Elephas nawataensis*") from the Upper Mb., Nawata Fm., Lothagam, Kenya (Tassy, 2003), occlusal view. (E) Right m1, KNM-LT 23783, same individual as the right dp4, occlusal and lateral views. as the right dp4.

from stegotetrabelodont mandibles by higher condyles, more restricted gonial angles, and relatively longer (post-symphyseal) corpora.

On dental criteria, a heavily weathered juvenile partial cranium, NY 46'92 (Figure 5.13C), from the late Miocene Lower Oluka Fm. in the Kisegi-Nyabusosi area, Uganda, is referrable to *P. korotorensis* (Tassy, 1995). Most of its rostrum is missing, as is the braincase posterior to the parietofrontal suture, and the zygomatic arches (Figure 5.13C). It primitively retains P4s, along with M1s in wear and M2s undescended from their crypts (Figure 5.13F). Tooth development and emergence are consistent with an age of between 12.5 and 14.8 years at the time of death (Laws, 1966; Jachmann, 1988; Lee et al., 2011). Notable features of the cranium include broad, inflated nasal processes of the premaxillae, lateral to the nasal aperture, a small nasal aperture that is not downturned laterally, a supposed anteroposteriorly concave frontal surface, weakly divergent tusk alveoli, a massive zygomatic process of the frontal, loss of the dorsal opening of the infraorbital foramen, shallow maxillary alveolus, steep angle of emergence of the M2s permitting inference of a high basicranium, and anteroposteriorly short face (Figure 5.13C; Tassy, 1995). If correct, the concavity of the frontal could indicate a sister-group relationship with *Elephas* and *Mammuthus*, which share this feature, but the apparent autapomorphic status of other character states in this cranium, such as the inflated nasal processes, indicate that this feature could also be homoplasic with these crown elephant genera. *Primelephas* has alternatively been posited as phylogenetically central to other elephantine elephants (Todd and Roth, 1996), as the sister taxon to a more restricted clade of *Elephas*+*Mammuthus* (Beden, 1983; Tassy, 1995), or as an early side branch of elephantines separate from the origination of crown elephant genera (Tassy and Debruyne, 2001; Sanders, 2004).

More recently, fossils from the Middle Awash, Ethiopia were subdivided into two subspecies, *P. gomphotheroides gomphotheroides* and *P. gomphotheroides saitunensis* (Saegusa and Haile-Selassie, 2009). On the basis of priority, these subspecies should be placed in *P. korotorensis*. In this taxonomic scheme, *P. korotorensis gomphotheroides* is diagnosed by intermediate lower molars with accessory central conules posterior to the anterior two or three lophids, m3 with a distinct accessory central conule posterior only to lophid 1, M3 with no or only a weak accessory central conule posterior to loph 1, second molars with six loph(id)s, and third molar plate formulae varying from x7x-x8x. *Primelephas korotorensis saitunensis* is diagnosed by m3s with accessory central conules posterior to lophids 1 and 2, pentalophodont intermediate molars, and M3/m3 plate formulae of x7x/x8x, respectively. Most of the specimens assigned to *P. korotorensis saitunensis* derive from a single small sedimentary lens at the Middle Awash locality of Saitune Dora, including a wealth of postcranial skeletal remains that await description (Saegusa and Haile-Selassie, 2009).

The *Primelephas korotorensis* sample from the Middle Awash derives from the Asa Koma Mb. of the Adu-Asa Fm. and Kuseralee Mb. of the Sagantole Fm., radiometrically dated to 5.77–5.44 Ma and 5.2 Ma, respectively (Haile-Selassie and WoldeGabriel, 2009), contributing the largest collection known for the species, as part of a substantial mammalian assemblage that also includes soricids, cercopithecoids, early hominins in the genus *Ardipithecus*, lagomorphs, several families of rodents, canids, ursids, mustelids, viverrids, herpestids, hyaenas, felids, tubulidentates, hyraxes, hipparionines, rhinos, suids, hippos, several tribes of giraffids, and a large diversity of bovids, along with deinotheres, anancine gomphotheres, and early species of *Loxodonta* and *Mammuthus*. In addition, these beds preserve one of the best late Miocene collections of fossil birds in Africa (Louchart et al., 2008).

Between the Middle Awash samples, the fossil material from Lothagam, Kenya and collections from the Chad Basin, Manonga Valley, Tanzania, and the Tugen Hills, Kenya, the dentition of *Primelephas* is abundantly well represented in the fossil record. The adult dental formula preserves upper and lower third and fourth premolars but no lower tusks (Table 5.2). Plate formulae are: dP2/dp2, x2-x3x/x3x; dP3/dp3, x4x/x4x; dP4/dp4, x5x/x5 to 6x; P3/p3, x3/x3-x3x; P4/p4, x3x/x-x3; M1/m1, x5x/x5x-x6; M2/m2, x5x-6x/x6x; M3/m3, x7x-8x/7x-x8x (Mackaye et al., 2008; Saegusa and Haile-Selassie, 2009; Sanders et al., 2010a). Plates are formed of thick, unfolded enamel, and are low and widest at the base (Figure 5.14A). Third molars usually do not exceed 100 mm in width and generally are narrower than those in stegotetrabelodont species, in samples from East Africa, but *Primelephas* molars from the Chad Basin appear larger. *Primelephas* molars are usually comprised of plates of four-five conelets that may be superficially subdivided into up to nine apical digitations. In lateral view, plates are pyramidal and anteroposteriorly well-separated by V-shaped transverse valleys. Cementum may coat plates up to their apices but does not fill the transverse valleys (Maglio and Ricca, 1977).

The retention of permanent premolars in *Primelephas* is now well documented (Sanders, 2017). The third premolars of the species are small and little advanced over their condition in stegotetrabelodonts, but upper and lower fourth premolars of *Primelephas* are more derived in their development of three full loph(id)s, larger size, and a greater amount of cementum. P3/p3 range in size from L=36.1–40.8 mm, W=28.7–31.2 mm/L=26.4–33.4 mm, W=21.8–25.2 mm, and P4/p4 measure L=+43.0–54.4 mm, W=45.9–54.4 mm/L=44.2–46.5/W=40.0–40.3 mm (Saegusa and Haile-Selassie, 2009; Sanders, 2017). They tend to develop three loph(id)s, which is a little more advanced than permanent premolars in stegotetrabelodonts but more primitive than those of *Loxodonta cookei* (Sanders, 2017). These teeth emerge vertically from their crypts, and although there is some sign that they caused dental crowding and malocclusion because of the action of molars progressing horizontally forward along the jaws, their development indicates that they had masticatory value (Sanders, 2017).

In molars of the type series from Lothagam, intermediate molars of *Primelephas korotorensis* may have posterior accessory conules throughout the crown, and in third molars, posterior accessory conules are restricted to the anterior end of the crown (Figure 5.14B, C). Alternatively, in the *Primelephas* sample from the Manonga Valley, Tanzania most of the third molars lack accessory conules (Sanders, 1997) and are strongly convergent on the crown morphology of stegodont molars, which have been recovered from the same beds (Table 5.1). Based on these differences, Saegusa and Haile-Selassie (2009) suggested that the Manonga Valley *Primelephas* sample might belong to a third subspecies of *P. gomphotheroides* (=*korotorensis*). Two undescribed third molars from the Chemeron Fm. in the Tugen Hills, Kenya also might belong to this third subspecies. KNM-BC 377 is a right M3 with a plate formula of x7x, dimensions of L=250 mm, W=99 mm, and H=57 mm. It is primitive in having a lamellar frequency of only 3.75, very thick enamel (ET=5.5–6.4 mm), and a hypsodonty index of 57. Although the anterior two plates are in wear, there is no apparent cementum. Plates are composed of seven conelets each, without any accompanying accessory conules. The overall crown shape is loaf-like, with a broad postcingulum. This specimen is highly convergent on primitive *Stegodon*. A left m3, KNM-BC 378, which could belong to the same individual, is worn on the first four plates, has a plate formula of x7x, and is equally primitive in having a low lamellar frequency of 3.75 and thick enamel (ET=5.1–5.7 m). Its dimensions are quite similar to those of KNM-BC 377, L=250 mm, W=97 mm, and its plates are formed of between four and seven conelets. Cementum is invested in transverse valleys but does not cover the plates. It is concave occlusally and curved longitudinally. Crown height is very brachyodont and it is also convergent on the morphology of primitive *Stegodon* molars.

Elephas nawataensis was named by Tassy (2003:figs. 8.7 and 8.8a–d) based on several specimens from the late Miocene Upper Mb. of the Nawata Fm. at Lothagam, Kenya, including the holotype mandible with right and left dp4-m1 (Figure 5.14D, E), and an isolated distal fragment of an m2 (KNM-LT 23783). It was diagnosed by its moderately short, horizontally straight symphysis that has a "true elephantine beak" (Tassy, 2003:344), the presence of accessory conules only behind plate 1 in the m1s, and dp4s with six plates.

However, the species was subsumed into *Primelephas korotorensis* by Sanders et al. (2010a). The elephantine beak of "*E. nawataensis*" is similar to that of *Primelephas korotorensis* (Mackaye et al., 2008; Saegusa and Haile-Selassie, 2009) and shows the morphology expected of a juvenile of the species. In proboscideans, symphyseal projection often is more downturned with age, which is what occurred in ontogenetically older *Primelephas* individuals. Other aspects of mandibular morphology in KNM-LT 23783 also resemble *Primelephas* jaws from the Middle Awash (Saegusa and Haile-Selassie, 2009).

In addition, dp4 is poorly known in different site-samples of *P. korotorensis*, and therefore variability is not clearly understood for this tooth. A single dp4 in crypt with its anterior crown hidden under the distal end of the dp3 (KNM-LU 694) is known from Lukeino, Kenya; there are no complete dp4s reported from Lothagam, the Manonga Valley, or the Toros Menalla area. KNM-LU 694 appears to have five plates and exhibits accessory conules posterior to the first three plates. Lower deciduous fourth premolars from the Middle Awash are pentalophodont and exhibit accessory conules posterior to either the first two or three plates (Saegusa and Haile-Selassie, 2009). *Primelephas* intermediate molars usually have posterior accessory conules behind multiple plates (plates 1–3 in Middle Awash m1 specimens), but this can occasionally be limited to just the first plate (Saegusa and Haile-Selassie, 2009). In KNM-LT 23783, dp4 has accessory conules at least posterior to plate 3 and m1 at least posterior to plate 2 (Figure 5.14D, E). Plate 6 in the dp4 is smaller than the rest of the plates in this deciduous premolar. The five plates composing the m1 of KNM-LT 23783 and the rest of the morphology of this molar are typical for *Primelephas*. The features used to diagnose *E. nawataensis* are inadequate to support erecting this species and placing it in *Elephas*, particularly in the absence of cranial apomorphies which would be desirable to demonstrate the affinity. For now, *E. nawataensis* is conservatively recognized as a variant of *P. korotorensis*.

There are few described postcranial specimens of *P. korotorensis*, except for an associated humerus and femur of an adult individual from Lothagam (KNM-LT 370) (Maglio and Ricca, 1977). In the humerus, only the distal half is preserved and shows that, compared with modern African elephant humeri, the deltoid ridge is more prominent and extends more distally, and the lateral supracondylar ridge extended more proximally on the diaphysis. This morphology is more similar to that in humeri of *Stegotetrabelodon*. The nearly complete femur is similar to that of *Loxodonta adaurora* from Kanapoi, but has a larger head and longer neck, which has a greater angle from the shaft. Also, the medial condyle is longer and posteriorly more extensive. Dimensions of the femur are L=1,211 mm, with a transverse midshaft diameter of 164 mm and anteroposterior midshaft diameter of 120 mm. The anteroposterior diameter of the head is 170 mm. Although this evidence is scant, it suggests that the body size of *P. korotorensis* was smaller than the larger individuals of *Stegotetrabelodon*, and approximated the height and weight of *L. africana* adult males.

Limited carbon isotope sampling of dental enamel in *P. korotorensis* from the Apak Mb. of the Nachukui Fm. and Upper Nawata Fm. at Lothagam indicates that it was primarily eating C_4 grasses (Cerling et al., 1999; Uno et al., 2011). In contrast, elephants of the Lukeino Fm. and lower Chemeron Fm. (=Mabaget Fm.) that were sampled for dental enamel isotopes (Roche et al., 2013) produced results consistent with either mixed feeding or browsing. However, it is worth noting that only one elephant specimen was sampled from the Lukeino Fm. and only four specimens were sampled from the Chemeron Fm. in this study, and the authors did not report which species were sampled. A single specimen of *P. korotorensis* from the lower part of the Lukeino Fm. sampled for dental carbon isotopes in a different study (Doman, 2017) revealed evidence of mixed feeding. Overall, *P. korotorensis* is an elephant species whose dental adaptations (brachyodonty, modest cementum, low number of plates) were out of phase with significant incorporation of C_4 grasses in its diet (Lister, 2013), as were other of its proboscidean contemporaries (e.g., anancine gomphotheres, stegotetrabelodonts). Although its relationship to other confamilials is unclear, nonetheless *P. korotorensis* provides a grade of morphological organization slightly more advanced than that of stegotetrabelodonts and primitive enough (gnathodentally) to represent the ancestral condition for elephantine elephants.

LOXODONTA F. Cuvier, in F. Cuvier and Geoffry Saint-Hilaire, 1825 (anonymous emendation, 1827)

The genus *Loxodonta* dates from the late Miocene to extant forest and savanna African elephants (Coppens et al., 1978; Sanders et al., 2010a). The initial fossil record of the genus temporally overlaps with occurrences of the earliest stem elephantids and elephantines (Table 5.1), suggesting a very rapid diversification of early elephants. Compared with *Elephas, Palaeoloxodon,* and *Mammuthus*, the geographical and morphological trajectories of *Loxodonta* were more conservative, with modest morphological evolution limited solely to African taxa (Coppens et al., 1978).

Distinctive features of *Loxodonta* include molar enamel wear figures of lozenge or loxodont shape ("<>") (F. Cuvier, 1825), lateral body profile swaybacked with the highest point at the shoulder (Laursen and Bekoff, 1978), skull with distally flared premaxillae, frontoparietal convex in all directions, borders of the temporal fossae rounded, no occipitoparietal inflation or "bossing," no superior midline depression in the nuchal crest, occipital planum generally vertical or angled forward superiorly, occipital condyles projecting posteriorly, tusks not torqued and gently curved upward at their distal ends, mandibles with long, straight corpora showing prominent symphyses lacking lower tusks, rounded, upward-directed condyles, and very extensive masseteric fossae, and molars with modest numbers of plates and crown height in the most derived forms not exceeding twice the width (Maglio, 1973; Beden, 1983).

Despite these distinctions, taxonomic recognition of fossil representatives of the genus was delayed until the 1960s (Coppens, 1965). A small number of species are now widely accepted, alongside extant forest (*L. cyclotis*) and savanna (*L. africana*) elephants: late Miocene-early Pliocene *L. cookei* (Sanders, 2007), Plio-?Pleistocene *L. adaurora* (Maglio, 1970b; Beden, 1983), Pliocene-?Pleistocene *L. exoptata* (Dietrich, 1941; Coppens, 1965; Beden, 1987b; Sanders, 2011), and Pleistocene *L. atlantica* (Pomel, 1879; Maglio, 1973). Although *L. adaurora* was the dominant elephant of the Turkana Basin and environs during the early and mid-Pliocene (Sanders, 2018), for much of the late Pliocene and Pleistocene elephants encompassed in the "*Elephas recki* complex" (see chapter 7) ecologically supplanted loxodont elephants in eastern Africa, and consequently, loxodont elephants may have retreated to peripheral refugia in northern and southern Africa, as well as to West-Central African forests (Sanders et al., 2010a). Documented as grazers or mixed feeders with a strong C_4 component to their diets during the Miocene and Pliocene, loxodont elephants that replaced species of the "*Elephas recki* complex" in the latter part of the middle Pleistocene re-emerged in East Africa as preferential browsers (Koch et al., 1995; Cerling et al., 1999, 2003; Zazzo et al., 2000; Harris et al., 2003; Schoeninger et al., 2003; Kingston and Harrison, 2007).

Morphology-based cladistic analyses have identified *Loxodonta* as the sister taxon to an *Elephas+Mammuthus* (*+Palaeoloxodon*) clade (e.g., Kalb and Mebrate, 1993; Shoshani, 1996; Tassy, 1996a; Shoshani et al., 1998; Ferretti, 2016; Zhang, 2020). The results of genomic studies have been more complex or contradictory, either with *Loxodonta* maintained as the sister taxon to other elephants (Yang et al., 1996; Ozawa et al., 1997; Krause et al., 2006; Rogaev et al., 2006; Rohland et al., 2007), or alternatively presenting evidence for a *Loxodonta+Mammuthus* clade (Hagelberg et al., 1994; Noro et al., 1998; Barriel et al., 1999; Thomas et al., 2000; Thomas and Lister, 2001; Debruyne et al., 2003a). Although Thomas et al. (2000) felt that morphological features linking *Elephas* with *Mammuthus* are homoplasic, nonetheless there are no traits discernable from the fossil record that would support a more recent common ancestry of *Loxodonta* and *Mammuthus* to the exclusion of *Elephas* (Mundinger and Sanders, 2001; Sanders, 2004; Sanders et al., 2010a) or *Palaeoloxodon*. While the late Miocene fossil record of *Mammuthus* is poor and essentially lacking for *Elephas*, assuming cladogenesis of crown elephant lineages would indicate, from the presence of loxodont elephants during this time, that each of these lineages arose sometime in the 7–6 Ma interval. Rather unexpectedly, study of ancient DNA in *Palaeoloxodon antiquus* placed it as a closer sister taxon to *L. cyclotis* than *L. africana*, with all three species comprising a group that is the sister clade to *Elephas+Mammuthus* (Meyer et al., 2017), despite a lack of obvious morphological characters that would support this hypothesis. The hypothesis was elaborated on by Palkopoulou et al. (2018), with genetic data suggesting a deep common ancestry of *Palaeoloxodon* with *Loxodonta* in Africa, and subsequent substantial genetic contribution of forest elephants to the *Palaeoloxodon* lineage. Unfortunately, although *Palaeoloxodon* has been identified from early Pleistocene African sediments (Larramendi et al., 2020), there is no fossil record for *L. cyclotis* (Sanders et al., 2010a).

LOXODONTA COOKEI Sanders, 2007

(Tables 5.1, 5.2, and 6.1; Figures 5.1 and 5.15)

Synonymy—*Mammuthus subplanifrons* (in part) Hendey, 1976:236; *Mammuthus subplanifrons* (in part) Hendey, 1981:51, 56; Elephantinae indet., cf. *Loxodonta* Tassy, 1986:115–117, pl. 14, fig. 6; *Loxodonta* sp. 'Lukeino stage' Tassy, 1995:237–239, fig. 5, pl. 4, figs. 1–4; *Loxodonta* sp. nov. A (in part) Mackaye, 2001; *Loxodonta* sp. nov. Sanders, 2006:196

The oldest established species of the African elephant genus *Loxodonta* is late Miocene-earliest Pliocene *L. cookei* (Table 5.1), named in honor of the late Basil Cooke, for his important and voluminous contributions to African mammalian paleontology, notably including studies of suids and elephants. *Loxodonta cookei* is an eastern and southern African species known primarily from isolated dentition that is best documented from the Pelletal Phosphate Mb. of the Varswater Fm. at Langebaanweg, South Africa (Sanders, 2006, 2007). There, it co-occurs with the anancine gomphothere species *Anancus capensis* in the Phosphate Mb., but the mammoth *Mammuthus subplanifrons* from the site is only reliably known from the Quartzose Sand Mb. of the Varswater Fm. (Hendey, 1976, 1981; Sanders, 2006, 2007). In the geologically older Lukeino Fm. of the Tugen Hills, Kenya (Table 5.1), an m3 described by Tassy (1986), KNM-LU 67 (Figure 5.15A), is attributable to *L. cookei* and co-occurs with the archaic elephants *Stegotetrabelodon orbus* and *Primelephas korotorensis* (Sanders et al., 2010a). This

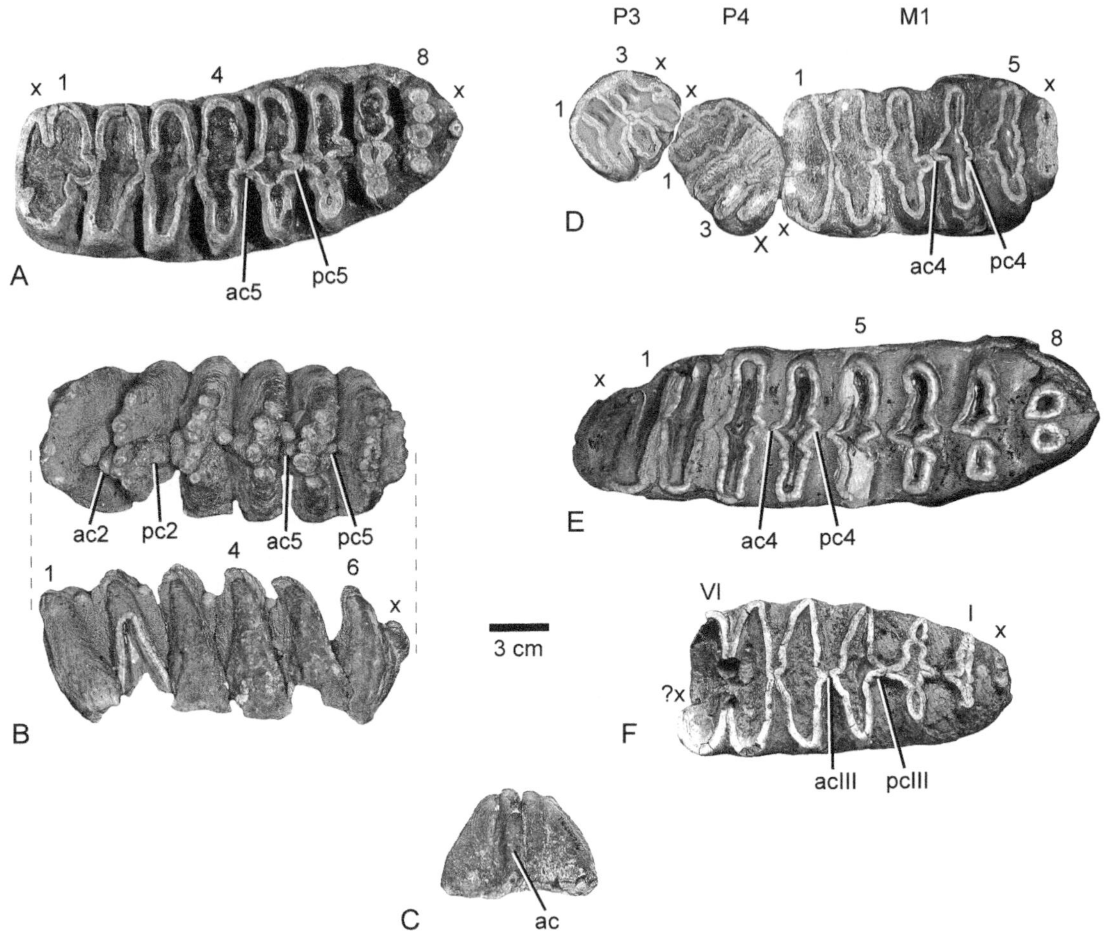

FIGURE 5.15 Cheek tooth specimens of late Miocene *Loxodonta cookei*. Abbreviations: ac, anterior accessory conule; M, upper molar; P, permanent premolar; pc, posterior accessory conule; x, anterior or posterior cingulum (id); X, large posterior cingulum or nascent plate; 1, 2, 3, …, plate number counted from the anterior of the crown; I, II, III, …, plate number counted from the posterior of the crown. (A, B, D, E) anterior to the left. All specimens approximately to the same scale. (A) Right m3, KNM-LU from the Lukeino Fm., Tugen Hills, Kenya, occlusal view. (B) Left M2, SAM-PQ-L 45618 from Langebaanweg, South Africa, occlusal and lateral views. (C) Molar plate, SAM-PQ-L 45784 from Langebaanweg, South Africa, anterior view. Note prominent accessory conule closely appressed to the anterior face of the plate, at a lower height than the plate conelets. (D) Right P3-M1, SAM-PQ-L 45616 from Langebaanweg, South Africa, occlusal view. (E) Right m3, SAM-PQ-L 45788 from Langebaanweg, South Africa, occlusal view. (F) Anteriorly broken left M3, NK 62'89 from the Nkondo Fm., Nkondo-Kaiso area, Uganda, occlusal view. Note the development of loxodont sinuses ("<>") in enamel wear-figures.

is an important occurrence because, along with undescribed loxodont elephant remains from Toros Menalla, Chad (Brunet et al., 2000), it establishes that the phylogenetic split between *Loxodonta* and *Elephas+Mammuthus* must have occurred at least by the interval 7.0–5.6 Ma.

Morphologically and metrically, the cheek teeth of *L. cookei* anticipate the dentition of both *L. adaurora* (sharing a strong development of central pillars in molar plates and robust accessory conules (Figure 5.15B) and transverse plate profile (Figure 5.15C), as well as retention of permanent premolars) and elephants in the clade that include *L. exoptata*, *L. africanavus*, and the extant *L. africana* and *L. cyclotis* (by nascent development of median loxodont sinuses, or "<>" occlusal shape of molar plates, a feature more fully realized in these species). *Loxodonta cookei* is distinguished from other congeners by fewer molar plates, particularly in posterior molars, lower hypsodonty, and retention of permanent premolars (except for *L. adaurora*, which also retains premolars), and from non-loxodont elephants by the tendency for enamel wear figures to form incipient median loxodont sinuses. Plate formulae for the species are: dP2/dp2=--/3x; dP3/dp3=x4/x4x; dP4/dp4=x5x or x6/x6x; P3/p3=x3x–x4/x3x–x4x; P4/p4=x4/x4x; M1/m1=5x–x6/x5x; M2/m2=x5x–x6x/x7x; M3/m3=7–8/x7x–x8x.

The permanent premolars are well formed in this species and are the largest known for elephants. P3s are rounded and p3s are more ovoid; permanent fourth premolars are larger and relatively more elongate, with better expression of accessory conules. In specimen SAM-PQ-L 45616, P3 has dimensions of L=57.3 mm, W=53.6 mm, and ET=2.3–2.9 mm; P4 dimensions are L=69.3 mm, W=52.0 mm, and ET=2.6–3.0 mm (Sanders, 2007). Third and fourth premolars both exhibit cementum. Specimen SAM-PQ-L 45616 from Langebaanweg shows that P3-P4 were in occlusion simultaneously with M1; these teeth all have similar degrees of wear. The specimen also reveals the potential conflict between vertically emerging premolars and the horizontal displacement mechanism of the emerging molars. In this specimen, the stresses created by the anterior shift of M1 have evidently caused the premolars to misalign anteroposteriorly (Figure 5.15D) and may have resulted in their malocclusion. It reveals a possible selective pressure for loss of premolars in elephants and other proboscideans that employed horizontal tooth displacement of their molars (Sanders, 2017).

Second molars are advanced over the condition in *Stegotetrabelodon* and *Primelephas*. M2 has six plates, m2 has seven plates, and these molars are subhypsodont to hypsodont. The type M2, SAM-PQ-L 45627, has a plate formula of x6x, lamellar frequency of 4.0, and dimensions of L=202.0 mm, W=92.0 mm, and H=83.0 mm (yielding a HI of 90). Lower second molars SAM-PQ-L 46561+50863+46220 and SAM-PQ-L 50862 have plate formulae of x7x, lamellar frequencies of 4.0, and dimensions of L=210.0 mm, W=81.2 mm, and H=83.0 mm (yielding HI of 102); these molars are almost certainly antimeres (Sanders, 2007).

The third molars of the species are mesodont (HI≥75≤99), with thick enamel (3.3–6.2 mm) and abundant cementum between the plates. Lamellar frequency is low, 3.5–4.2. Lower third molars from Langebaanweg have eight plates (Figure 5.15E) and dimensions ranging from L=260.0–290.0 mm, W=91.4–93.5 mm, and H=76.0 (resultant HI=81). Third molars from the Lukeino and Chemeron fms. in the Tugen Hills, Kenya, including KNM-LU 67, have seven-eight plates, LF=4.0, and are smaller than the Langebaanweg specimens, with widths of 85.0–92.2 mm, L=217.8–235.0 mm, and ET=3.8–4.0 (Sanders, 2007). An M3 from the Nkondo Fm. in the Nkondo-Kaiso area, Uganda, was claimed to have seven or eight plates (Tassy, 1995, Sanders, 2007), but the tight fold of the missing anterior loop on the first preserved enamel loop suggests it was a precingulum, and therefore plate formula appears to be x6x (Figure 5.15F). Its LF=3.5, and dimensions are also relatively modest, with L=+204.0 mm and W=86.6 mm.

In the lateral view, molar plates of *L. cookei* form tall, narrow pyramids separated by correspondingly V-shaped transverse valleys (Figure 5.15B). This aspect of plate shape is distinctly different from the parallel-sided plates and U-shaped transverse valleys of *L. adaurora* (Sanders, 2020). In the anterior view, the transverse cross-section of plates is broadest at their bases, and curves evenly inward toward their apices (Figure 5.15C). Median conelets are slightly higher than lateral conelets. Accessory conules are present anterior and posterior to plates throughout the crown. In occlusal view, accessory conules are incorporated into enamel wear figures to form loxodont sinuses. Because conules are lower in height than plate conelets, their contribution to median expansions of enamel loops is only prominent in worn molar crowns. Third and fourth deciduous premolars are miniature versions of molars; second deciduous premolars are diminutive, triangular teeth with three closely appressed plates.

Archaic elephants and other representatives of the genus *Loxodonta* from Mio-Pliocene East and Central Africa that have been sampled for dental isotopes appear to have been grazers or mixed feeders with a high proportion of C_4 plants in their diets (Cerling et al., 1999, 2003; Zazzo et al., 2000, Harris et al., 2003; Schoeninger et al., 2003; Kingston and Harrison, 2007; Uno et al., 2011). However, the ecological regime of latest Miocene-earliest Pliocene Langebaanweg, featuring strong seasonal winter rainfall, favored plants with C_3 photosynthesizing pathways, including grasses, so that even grazers would exhibit a C_3 signal in isotope studies of their dentition. Accordingly, Langebaanweg herbivores universally show diets dominated by C_3 vegetation (Franz-Odendall et al., 2002; Lehmann et al., 2016). Investigation of carbon isotopes in the dentition of proboscideans from Langebaanweg, however, reveals that *L. cookei* and *M. subplanifrons* incorporated a small amount of C_4 graze in their diets, which differentiated their feeding preferences from those of the pure C_3 diets of anancine gomphotheres from the site (Groenewald et al., 2020). This study further concluded that the anancine gomphotheres probably browsed in shady, wooded parts of the ecosystem, whereas the Langebaanweg elephants spent more time in open habitats and had more flexible diets.

Mackaye (2001) placed a number of isolated molars from late Miocene Toros Menalla, Chad into "*Loxodonta* nov. sp. A de Lukeino," and included some specimens later allocated by Sanders (2007) to *Loxodonta cookei* in this taxon (Sanders, 2007). The most important of the Chad specimens are left and right m3s from individual TM74.00.001 and an incomplete M3, TM31.97.001. Metrically, they could belong in *L. cookei*: the m3s have eight plates, are mesodont (HI=72–74), enamel is thick (ET=5.0 mm), lamellar frequency is low (LF=3.7), and overall dimensions are similar (L=233–246 mm; W=88–90 mm; H=67–68 mm); enamel thickness of the M3 is as great as 5.6 mm and lamellar frequency is low (LF=3.2).

The Chadian m3s have plates that are widest basally and taper strongly toward the apex. In lateral view, the plates are tall but pyramidal, and transverse valleys vary from V- to U-shaped. Cementum nearly fills the transverse valleys but does not cover the plates. In the anteriormost three plates, anterior and posterior accessory conules are incorporated into enamel wear figures, but do not form loxodont sinuses (not "<>"). In the left m3, posterior accessory conules continue to plate 5 and in the right m3 they occur as distally as plate 6, but anterior accessory conules are only present as distally as plate 3. Plate height is significantly lower than in molars from Toros Menalla attributed to "*Loxodonta* sp.

nov. C," and the m3s differ from molars of *Primelephas korotorensis* by the greater expression through their crowns of posterior accessory conules and the presence of anterior accessory conules (Mackaye, 2001). M3 of "*Loxodonta* sp. nov. A de Lukeino" has coarsely folded enamel wear figures that appear to incorporate small posterior accessory conules as distally as plate 4 (offset from the midline) and have strong central conelets or pillars that may be doubled; it is less clear if there are any anterior accessory conules incorporated in these wear figures. This differs from the condition of upper molars of *L. cookei* from Langebaanweg, which have anterior and posterior accessory conules centrally located throughout their crowns. Similarly, third molars of *L. cookei* from Langebaanweg and the Tugen Hills have well expressed anterior and posterior accessory conules throughout their crowns. At Langebaanweg, molars of *L. cookei* primarily resemble the Chadian specimens dimensionally and by weak or absent expression of loxodont sinuses. Nonetheless, support for recognition of the Chad specimens as belonging to *Loxodonta* or having affinity with *L. cookei* is tenuous at best.

A very few molar specimens from the Adu-Asa Fm. (and one from the Kuseralee Mb. of the Sagantole Fm.) of the Middle Awash, Ethiopia, have also been assigned to an undetermined species of *Loxodonta* (Saegusa and Haile-Selassie, 2009), penecontemporaneous with other specimens in *L. cookei* (Table 5.1). Incomplete M3 AME-VP-1/93 preserves part of five distal plates and a small postcingulid. Its plates appear to be composed of five conelets, with the central conelet dominant and anteroposterior extensive such that the central conelets of the plates are nearly in contact with one another in the midline. The structure of the enamel wear figures can be described as "propeller-shaped," with the lateral arms curved back at their edges. The Kuseralee Mb. specimen (KUS-VP-1/112), an m3, is more tenuously assigned to the genus. It has at least seven plates, composed of three–five conelets, and has cementum distributed in the transverse valleys. Although it sports a posterior accessory conule distal to the penultimate plate, there are no anterior accessory conules and no sign of development of loxodont sinuses in the enamel wear figures. Contra Saegusa and Haile-Selassie (2009), it bears comparison with stegotetrabelodont m3s, or with m3 of *Primelephas*.

MAMMUTHUS Brookes, 1828

Perhaps the fossil proboscideans to have caught the popular imagination to the greatest degree are mammoths, particularly the woolly mammoth, *Mammuthus primigenius* (Lister and Sher, 2001; Lister and Bahn, 2007). Genetic studies generally support the closer relationship of mammoths to Asian elephants (*Elephas*) than to the African species (e.g., Ozawa et al., 1997; Noro et al., 1998; Capelli et al., 2006; Krause et al., 2006; Rogaev et al., 2006; Brandt et al., 2012; Palkopoulou et al., 2018s; but see Barriel et al., 1999; Thomas et al., 2000; Debruyne et al., 2003a), and estimate that *Mammuthus*+*Elephas* diverged from *Loxodonta* in the late Miocene around 7.8 Ma and from one another circa 6.9 Ma (Rohland et al., 2007; Murata et al., 2009). The paleontological record accords with these estimates and shows the earliest evidence of *Mammuthus* in late Miocene Africa (Table 5.1; Sanders, 2006, 2007). In addition, *Mammuthus* shares a greater number of craniodental features with *Elephas* than with *Loxodonta* (e.g., Maglio, 1973; Shoshani et al., 2007; Ferretti, 2016).

Despite their early appearance in Africa, mammoths were not ultimately successful on the continent and were relegated primarily to northern Africa by the mid-Pliocene. Mammoths disappeared from Africa during the early Pleistocene (Tables 6.1 and 7.1). Following migration via a Levantine corridor to Eurasia during the mid-late Pliocene (Lister and van Essen, 2003; Lister, 2004; Markov, 2004; Lister et al., 2005), the first immigrants dentally closely resembled African mammoths of the same geological age. Mammoths reached North America by the early Pleistocene, thriving there until the end of the epoch (Dudley, 1996; Fisher, 1996; Lister, 1996; Lister and van Essen, 2003). The phylogenetic history of mammoths in Eurasia and North America was complex and involved multiple episodes of speciation and cross-species introgression, resulting in profound morphological alteration of craniodental anatomy (Lister, 2001, 2017; Lister et al., 2005; Lister and Bahn, 2007; Lister and Sher, 2015). Mammoths persisted in the Northern Hemisphere into the Holocene, but suffered climate-based loss of habitat, restriction of geographical ranges, and population collapse that accelerated through the late Pleistocene, leaving them vulnerable to human hunting pressures and contributing to their extinction (Nogués-Bravo et al., 2008 and references therein).

Mammoths underwent significant morphological changes to their skulls and dentition during their evolution, in parallel to other elephant genera, including heightening and anteroposterior compression of the cranium, shortening of the mandible, thinning of dental enamel, and increase in molar hypsodonty and plate numbers (Lister et al., 2005), associated with a greater commitment to grazing in at least part of the lineage (Lister, 2004). Their crania are characterized by dorsally expanded parietals, vertically concave, transversely convex frontoparietal surfaces, strong temporal constriction, widely separated orbits, and premaxillary tusk sheaths that are proximally closely spaced and distally outwardly curved (Maglio, 1973; Sanders et al. 2010a). Unfortunately, while these features comprise a salient complex for identification of mammoths, there is a paucity of described mammoth crania from the African fossil record.

MAMMUTHUS SUBPLANIFRONS (Osborn, 1928)

(Tables 5.1, 5.2, and 6.1; Figures 5.1 and 5.16)

Partial Synonymy—*Archidiskodon subplanifrons* Osborn, 1928; *A. andrewsi* Dart, 1929; *A. proplanifrons* Osborn, 1934; *Mammuthus* (*Archidiskodon*) *scotti* Meiring, 1955; *Stegolophodon* sp. Singer and

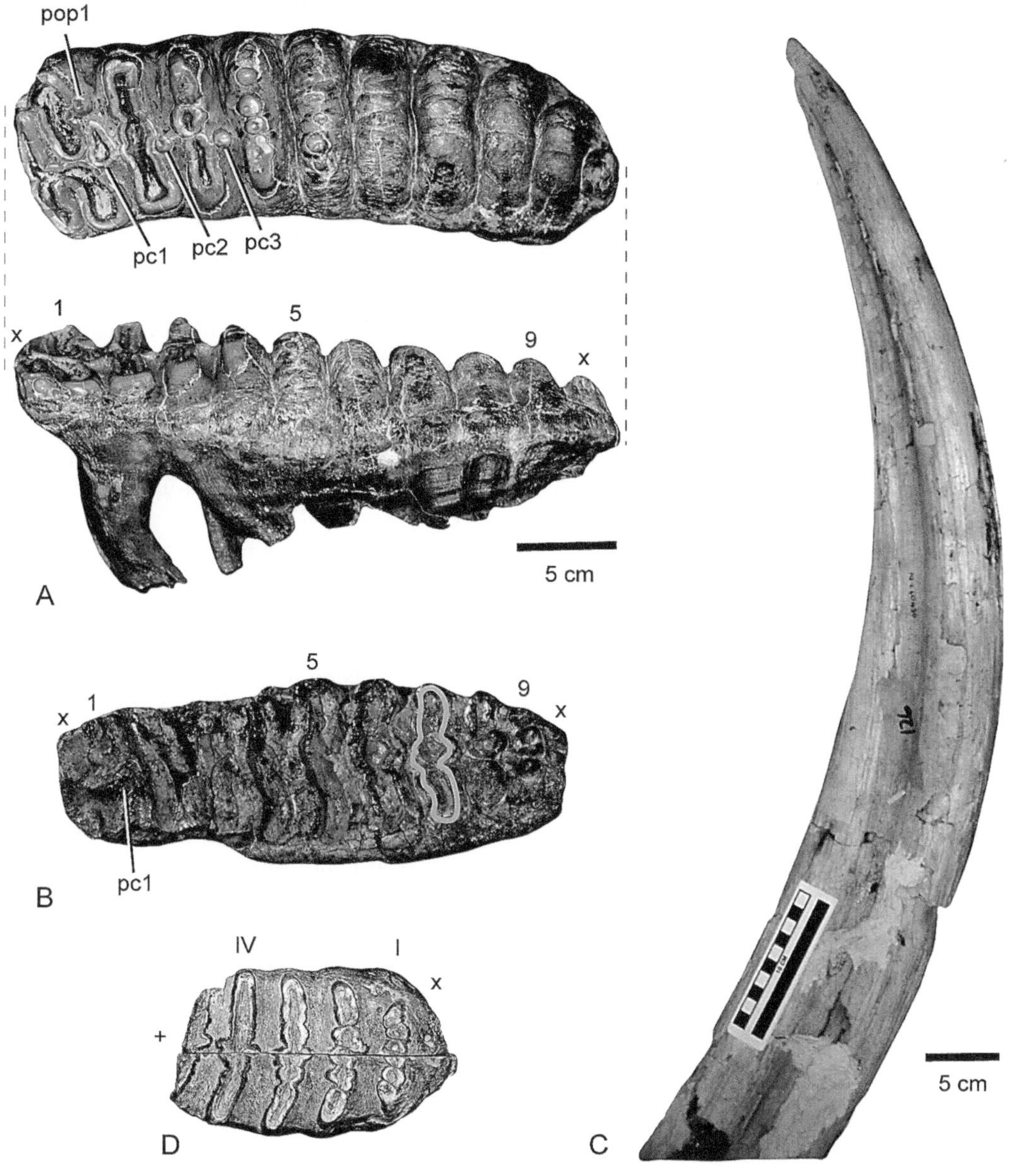

FIGURE 5.16 Dental specimens of late Miocene *Mammuthus subplanifrons*. Abbreviations: pc, pretrite posterior accessory conule; pop, posttrite posterior accessory conule; x, pre- or postcingulum(id); 1, 2, 3, …, plate number counted from the anterior of the crown; I, II, III, …, plate number counted from the posterior of the crown; +, indicates missing morphology. (A, B, D) to the same scale and anterior to the left. (A) Left m3, SAM-PQ-L 12723 from Langebaanweg, South Africa, occlusal and lateral views. (B) Left m3, KNM-LU 7597 from the Lukeino Fm., Tugen Hills, Kenya, occlusal view. Plate 7 is outlined to highlight the dominant central pillar and propeller shape of the enamel wear-figure; compare with plate 3 in SAM-PQ 12723. (C) Upper tusk, SAM-PQ-L 40430 from Langebaanweg, South Africa, lateral view. Note the spiral to the longitudinal sulcus. (D) Partial right m3, MMK 3920 (type) from the Vaal River, South Africa, occlusal view. The midline separation of the crown is where the specimen was sagittally sectioned.

Hooijer, 1958; *Mammuthus* (*Archidiskodon*) *subplanifrons* Cooke and Coryndon, 1970; *Mammuthus subplanifrons* Maglio and Hendey, 1970; *M. subplanifrons* Maglio, 1973; *Primelephas gomphotheroides* (in part) Tassy, 1986:pl. 12, fig. 3; *Loxodonta* nov. sp. C (in part) Mackaye, 2001; *Primelephas gomphotheroides saitunenesis* (in part:M3 specimen ASK-VP-3/16) Saegusa and Haile-Selassie, 2009; cf. *Mammuthus* sp. Saegusa and Haile-Selassie, 2009

The oldest known mammoth is *Mammuthus subplanifrons*, from late Miocene-early Pliocene eastern and southern Africa (Table 5.1). It is represented primarily by isolated gnathodental specimens, and unfortunately no crania have been recovered of the species. The type specimen of Pliocene *Loxodonta adaurora*, KNM-KP 385, which includes a skull and partial postcranial skeleton, was erroneously assigned by Larramendi (2016) to *M. subplanifrons*, but this specimen exhibits no credible synapomorphies to link it with mammoths. Based on shared characters that are certainly homoplasic (number of plates, brachyodonty) in her phylogenetic systematic analyses, Todd (2010a) recovered *L. adaurora* as the sister taxon and ancestor of *Mammuthus*. While they *do* homoplasically share a similar number of molar plates and both are brachyodont (HI<100), *L. adaurora* has higher molar crowns reaching subhypsodonty, greater expression of posterior and anterior accessory conules through the length of the crown, and (in lateral view) parallel-shaped rather than pyramidal plates, while exhibiting none of the synapomorphies typical of the cranium among mammoth species, but some shared with *Loxodonta*. Thus, there is no credible evidence to support these hypotheses.

Together with the lack of diagnostic crania, the dental sample of *M. subplanifrons* is very small, and therefore its identification with the genus *Mammuthus* is very tenuous.

A partial mandible with molars (SAM-PQ-L 12723) was recovered from the Quartzose Mb. of the Varswater Fm. at Langebaanweg, South Africa and placed in *M. subplanifrons* by Maglio and Hendey (1970). It was originally assigned to *Stegolophodon* sp. by Singer and Hooijer (1958), which is partly explicable by the dearth of primitive elephant specimens for comparison at the time. The mandibular corpus is long and slender, with a high ramus that has a smaller masseteric fossil than is usually present in loxodont elephants. The anteriorly broken symphysis was evidently long. It is possible to see that the anterior chambers of the mandibular canal open externally through mental foramina. Lower incisor buds never develop in these chambers in proboscideans, contradicting the opinion of Maglio and Hendey (1970) that they must have housed vestigial lower incisors. Thus, there is no evidence for lower tusks in this species. The accompanying third molars (Figure 5.16A) are large (L=325.0 mm; W=101.1–110.5 mm) and distinctive, with nine brachyodont plates (HI=67–69) that are composed of thick enamel (ET=4.8–5.4 mm). Lamellar frequency is low (LF=3.2–3.8). The plates are, in lateral view, bulbous and pyramidal in shape. In occlusal view, the plates are transversely straight with up to seven apical digitations, and may sport large central conelets but wear into simple enamel loops and not loxodont sinuses. Cementum covers the plates but does not infill the transverse valleys. Accessory conules are present posterior to the first three plates.

Several third molar specimens from the Kuseralee Mb. of the Sagantole Fm., Middle Awash, Ethiopia are morphometrically compatible with the Langebaanweg specimens (see Mebrate, 1983; Kalb and Mebrate, 1993; Haile-Selassie, 2001; Sanders, 2007). These teeth are firmly dated to the late Miocene (Table 5.1; Renne et al., 1999). They are exceeded in age only by an isolated m3 from the Lukeino Fm. in the Tugen Hills, Kenya, KNM-LU 7597 (Table 5.1; Hill, 2002), that was originally placed by Tassy (1986) in *Primelephas*, though it has nine full plates and more closely resembles the Langebaanweg m3 by its size and presence of dominant central pillars and propeller-shaped enamel wear-figures (Figure 5.16B; Sanders, 2007).

There is little to support inclusion of *Mammuthus* among the species assemblage from Virginia, South Africa except for several upper tusks, including a massive I2 (L=2,489 mm; midpoint transverse dimensions 155.0×175.0 mm) that displays the spiral twisting typical of the genus (Meiring, 1955; Maglio, 1973). This large tusk was discovered associated with an m3 that is comparable with the Middle Awash and Langebaanweg molars of *M. subplanifrons* (Sanders, 2007). In addition, a smaller, isolated upper tusk from Langebaanweg (SAM-PQ-L 40430) (L=+830 mm; midpoint dimensions 58.0×103.0 mm) also displays a distinct longitudinal torque and curves upward distally (Figure 5.16C; Sanders, 2007).

As presently constituted, *M. subplanifrons* is certainly a wastebasket taxon (Sanders et al., 2010a). In addition, the species has had a slightly tortuous taxonomic history. It was initially named "*Archidiskodon*" *subplanifrons* based on an isolated molar from the Vaal River terraces of South Africa (Osborn, 1928). *Archidiskodon* as understood at the time primarily consisted of species that are now placed in *Mammuthus*, and was defined by features that are characteristic of that genus; however, it also included species that belong to several other lineages of elephant (see Osborn, 1942). It is unfortunate that the type specimen of *M. subplanifrons* (MMK 3920) from the Vaal River, South Africa ("*Archidiskodon subplanifrons*" [Osborn, 1928]), the distal five plates of an m3, differs significantly from other molars of the species and instead resembles a primitive loxodont elephant molar in having midline anterior and posterior accessory conules incorporated into enamel wear figures (though not deployed into well-developed loxodont sinuses) (Figure 5.16D; Sanders, 2006, 2007). This is true of other species from the Vaal River that have been synonymized into *M. subplanifrons*, including "*Archidiskodon proplanifrons*" (Osborn, 1934) and "*A. andrewsi*" (Dart, 1929). Specimens of these taxa should be removed from the species hypodigm and a new type will need to be selected, most rationally the associated m3 and I2 from Virginia, South Africa (A2882).

An associated upper tusk fragment with substantial longitudinal torque and an m3 with nine plates (ALA-VP-2/309a and b), from the late Miocene Asa Koma Mb. of the Adu-Asa Fm., Middle Awash, Ethiopia have been referred to cf. *Mammuthus* sp. by Saegusa and Haile-Selassie (2009) because they interpret occlusal morphology in the m3 as including both anterior and posterior accessory conules, which would differentiate it from *M. subplanifrons*. However, while there are clearly independent accessory conules posterior to plates 1–3, anteroposterior expansions in enamel wear figures appear instead to be the expression of the large central conelet(s) in each plate. Therefore, these specimens can be accommodated in *M. subplanifrons*.

?*MAMMUTHUS* SP. INDET. PROBABLY NOV.

(Table 5.1; Figure 5.1)

Synonymy—*Loxodonta* nov. sp. C Mackaye, 2001

Molars of a single individual (TM190.01.01) from late Miocene Toros Menalla, Chad were allocated to "*Loxodonta* nov. sp. C" by Mackaye (2001). These include M2s with plate formulae of x6x, exhibiting thick enamel (ET = 5.0–5.5 mm), low lamellar frequencies (LF = 3.2), and dimensions of L = 186–190 mm, W = 112 mm. The M3s have plate formulae of x8x and are immense (L = 319–324 mm, W = 119–124 mm). Their enamel is massively thick (ET = 6–7 mm) and lamellar frequency is quite low (LF = 2.7). Conversely, crown height is subhypsodont (H = 108–113, HI = 91). Posterior accessory conules are limited to the first one or two plates, and plate 1 of each M3 has an anterior accessory conule. Plates are formed of five to seven conelets, and in lateral view are parallel-faced with U-shaped transverse valleys. Cementum covers the plates but does not infill the transverse valleys. Lower third molars are longitudinally curved concave to the buccal side, and exhibit posterior accessory conules in association with plates 1–2. The m3s have plate formulae of x9x. They are also immensely large, with dimensions of L = 346–352 mm, W = 122–125 mm, H = 110 mm. These dimensions yield hysodonty indices of 88–90. Enamel is thick (ET = 5–6 mm) and lamellar frequency is low (LF = 2.8). Despite the thick enamel and low lamellar frequencies, these molars are distinguished from those of more primitive elephants such as stegotetrabelodonts and *Primelephas korotorensis* by their greater numbers of plates and higher crowns.

Mammuthus subplanifrons was synonymized into "*Loxodonta* nov. sp. C" by Mackaye (2001), but this is untenable. First, there are no obvious loxodont apomorphies in the Chad sample attributed to this taxon, and therefore no reason to place the specimens in *Loxodonta*. Second, as has been pointed out by Sanders et al. (2021), the molars of "*Loxodonta* nov. sp. C" are larger, higher-crowned, and have lower lamellar frequencies than molars of *M. subplanifrons*. In addition, the lower third molars of *M. subplanifrons* from Langebaanweg have posterior accessory conules to the third plate, Conversely, some similarities bear examination: in third molars, plate formulae are the same, plates are coated but transverse valleys are not infilled by cementum, and apically plates are transversely orthogonal to the long axis of the crown and formed of up to seven-eight conelets. Worn second molars from Toros Menalla attributed to "*Loxodonta* nov. sp. C" (Mackaye, 2001:pl. 24) are strikingly similar to worn molars of *M. subplanifrons* from Langebaanweg, South Africa (Sanders, 2007). It is possible that Toros Menalla "*Loxodonta* nov. sp. C" belongs to a more hypsodont species of *Mammuthus*.

SUMMARY

1. The late Miocene interval in Africa was a time of substantial turnover or replacement of proboscidean cohorts, marked by the precipitous decline of the great gomphothere radiation and mammutids that populated the continent during the early and middle Miocene. The timing and conditions of this turnover are poorly understood as there is a dearth of late middle Miocene sites in Africa.
2. Gomphotheriines, amebelodontines, choerolophodontines, and mammutids were supplanted by immigrant proboscidean taxa arriving from Eurasia. Tetralophodonts arrived during the earliest phase of the late Miocene, and anancine gomphotheres and stegodonts made their way onto the continent later. These migrations may have been facilitated by the Messinian Crisis, in which the Mediterranean was closed up, creating corridors for the movement of terrestrial animals.
3. The most remarkable proboscidean event of the interval was the evolution of elephants in Afro-Arabia during the second half of the late Miocene. Their ancestry can be traced to the Tetralophodontinae. Tetralophodonts are documented in the same Afro-Arabian regions as the earliest elephants, coincident with or immediately prior to the first appearance of the Elephantidae. Postcranial skeletal morphology and dental features of tetralophodonts and the earliest elephants are very similar, as well.
4. Deinotheres, survivors from the African Oligocene, underwent a less remarkable transformation than elephantimorph proboscideans, with the presumed endemic evolution of *Prodeinotherium hobleyi* into *Deinotherium bozasi*. Differences between these genera are largely size-related, and the lineage maintained the same skeletal bauplan, with only minor differences in the anatomy of the skull, postcranium, and dentition. This evolutionary conservatism may be attributed to having been well adapted to a specialized browsing niche.
5. The late Miocene underwent substantial climatic, geological, and ecological change, featuring colder ice-house regimes globally, expansion of permanent ice sheets in Antarctica, and steeper climate gradients between the poles and equator. In Africa, the effects of these changes manifested as greater seasonality, with a shift of primary moisture sources from the Atlantic to the Indian Ocean. As well, the rise of rift shoulders created rain shadow effects, particularly in eastern Africa, that led to the fragmentation of forests, the establishment of more heterogenous landscapes, and

the spread of C_4 grasses and grassy woodlands. C_4 grasses were superbly adapted to the novel conditions of the late Miocene in Afro-Arabia.

6. Mammalian faunas of this time were more progressive than earlier in the Miocene, with a greater number of modern families, but were still composed of many archaic genera. Hominins originated during this interval, perhaps in association with or driven by ecological changes that were fostered by the earliest elephants.
7. Late Miocene African elephants included archaic forms (e.g., *Stegotetrabelodon, Primelephas*) penecontemporaneous with stem members of crown elephant lineages (*Loxodonta, Mammuthus*). This indicates that elephants underwent rapid evolutionary differentiation during this interval.
8. In response to ecological changes across African landscapes, a number of mammalian groups shifted their feeding behavior to include substantial amounts of C_4 graze, with equids, rhinos, and elephants as the first adoptees of these new resources. As competition for grasses and selective pressures for efficient processing of graze ramped up, adaptive experiments in cranial, mandibular, and dental forms likely played a role in the radiation of new elephant genera.
9. Morphological adaptations characterizing the rise of elephants include reorganization of the gomphotherian bunolophodont molar pattern and lateral rotary mastication to molars composed of a series of transverse plates consolidated with cementum, along with anteroposterior compression and greater height of the cranium, reconfiguring the action of chewing muscles to drive molar plates across one another as in scissors with multiple blades, in horizontal fore-aft shearing mastication. These changes have been directly invoked as adaptations to grazing. The craniodental morphology of late Miocene elephants was poorly synchronized with their feeding behavior, however, as their molar plates were few, low-crowned, and not closely packed; in some species, the center of mass of the lower jaw remained too anterior to be highly efficient at proal mastication, because of the retention of long symphyses and (in stegotetrabelodonts) elongate lower tusks.
10. Many late Miocene Afro-Arabian sites hosted sympatry of multiple species of elephants, alongside stegodonts, anancine gomphotheres, deinotheres, and in a few exceptional sites, amebelodonts or tetralophodonts. The common co-occurrence of so many megaherbivores in Afro-Arabian ecosystems is testimony to the observation that there are few or no modern ecological analogs to late Miocene sites on the greater continent.

6 Early Pliocene
Proboscidean Relay Interval

The appearance of a new genus, often resulting from invasion (an allochthonous form), causes competition with existing, ecologically similar genera and leads to the demise of one (usually the autochthonous form). The result of this is . . . an evolutionary relay.

—**David M. Raup and Larry G. Marshall (1980)**

INTRODUCTION

The early Pliocene may be thought of as a relay interval in proboscidean evolution, during which survivors of the modest archaic elephant radiation of the African late Miocene (e.g., *Stegotetrabelodon orbus*, *Stegodibelodon schneideri*, *Primelephas korotorensis*, *Mammuthus subplanifrons*, *Loxodonta cookei*) were replaced or succeeded by slightly more advanced elephants (e.g., *Loxodonta adaurora*, *L. exoptata*, "*Elephas*" *ekorensis*, early forms of the "*Elephas recki* complex," more advanced mammoths), some of which may have provided more direct phylogenetic sources for crown elephant taxa. Additionally, the interval witnessed the evolution of a more progressive species in the *Anancus kenyensis* lineage (Gomphotheriidae), the last of the gomphotheres in eastern Africa, and a more advanced stage of *Stegodon kaisensis* (Stegodontidae), which vanished soon thereafter (Tassy, 1995; Sanders et al., 2010a; Sanders, 2011). Among proboscideans, only deinotheres passed through the interval unchanged, undoubtedly because of their specialized adaptations to a narrow browsing niche (see Harris, 1975, 1978) that remained immune to the climatic and ecological upheaval that affected much of the continent during this time.

Global deep-sea oxygen isotope records as proxies for temperature and ice volume reflect a sharp and generally continuous fall in temperatures and increase in ice-sheet composition in Antarctica and the Arctic commencing at the start of the Pliocene (Zachos et al., 2001). The impact on regional and local ecosystems in Afro-Arabia was undoubtedly closely related with increase in C_4 resources and decrease of canopy cover, as evidenced by a strong shift in pedogenic carbonate values through the early Pliocene in the Turkana and Awash Basins (Levin et al., 2011), and by the coincident skewed preference for C_4 grasses in the diets of herbivores in eastern Africa, as revealed by dental isotopic analysis (Uno et al., 2011). During the early Pliocene, there was a steady decline in browsing and concomitant increase in mixed feeding and grazing among large herbivores in eastern Africa (Cerling et al., 2015). Open C_4 grasslands and savannas were not dominant features of the early Pliocene in Africa, but woody cover was generally <40% at this time (Cerling et al., 2011a, b). Although elephants and anancine gomphotheres exploited C_4 resources from the beginning of their presence in Afro-Arabia during the late Miocene (e.g., Cerling et al., 1999, 2003; Kingston, 1999; Zazzo et al., 2000; Uno et al., 2011), other mammal lineages became grazers later, many in the early Pliocene (see Cerling et al., 2003, for example). At Lothagam, Kenya, there was drastic faunal turnover between the late Miocene and early Pliocene, including an influx of new bovid taxa adapted to more open habitats (Leakey and Harris, 2003), indicating a change in ecological conditions and greater competition for grazing resources that impacted selection on elephantimorph proboscideans for evolution of more efficient and effective adaptations for feeding close to the ground and eating silica-laden C_4 grasses. The types of habitats evidenced by isotopic analyses, in eastern Africa in the early Pliocene, comprised grassy woodlands, shrublands, or bushlands and featured contributions from woody plants, forbs and herbs, and C_4 grasses (Cerling et al., 2013). Definitively bipedal hominins such as *Australopithecus anamensis* and *A. afarensis* first evolved during this interval in these habitats (e.g., Johanson et al., 1978; Ward et al., 1999, 2001; Ward, 2002; Haile-Selassie et al., 2010, 2019). There is isotopic evidence that these hominins shifted their diets over time to consume a wider range of food items than sympatric large-bodied mammals, including C_4 plants (Cerling et al., 2013; Wynn et al., 2013, 2016). Proboscideans of the early Pliocene may have competed with these other mammals for resources, but their activities also assisted in the maintenance and spread of mosaic woodland habitats that provided selective advantages for these early bipedal hominins (Schoeninger et al., 2003; Sanders, 2020).

While archaic grazing elephants in the late Miocene, such as stegotetrabelodonts, *Primelephas*, and *Stegodibelodon* appeared morphologically out-of-phase with their feeding behavior (Lister, 2013), low-crowned molars composed of few plates and sparse cementum, and long mandibles with protuberant symphyses (some with long lower tusks), "new" elephants of the early Pliocene evolved more progressive molars with a greater number of plates, thicker coating of cementum, and greater hypsodonty (Sanders et al., 2010a; Sanders, 2020). The cranium of at least one of these elephant species was precociously advanced for efficient proal shearing mastication of grasses (Sanders et al., 2021). Anancine gomphotheres in several regions exhibited parallel trends to increase the effectiveness of their molars for grazing, by increasing crown height, modest addition of loph(id)s, addition of thicker cementum, exaggeration of anancoidy, and

DOI: 10.1201/b20016-6

greater distribution of accessory conules in transverse valleys (Sanders, 2007, 2008, 2011; Saegusa, 2020).

Although elephants in the "*Elephas recki* complex" are first documented in this interval, they did not become numerous or ecologically dominant until the late Pliocene, and the taxonomic and phylogenetic issues surrounding that group warrant separate and cohesive consideration. Thus, they are discussed together with their late Pliocene-late Pleistocene cohort in Chapter 7.

SYSTEMATIC PALEONTOLOGY

GOMPHOTHERIIDAE Hay, 1922
ANANCINAE Hay, 1922
***ANANCUS ULTIMUS* Sanders, 2011**
(Tables 5.4, 6.1, and 6.2; Figures 5.8, 6.1, and 6.2)

Partial Synonymy—*Trilophodon angustidens* cf. *kisumuensis* (in part) Hooijer, 1963; *Anancus osiris* Coppens, 1965; *A. osiris* Servant-Vildary, 1973; *A. kenyensis* (in part) Coppens et al., 1978; *A. kenyensis* (in part; *A. kenyensis* "*petrocchii*-morph") Tassy, 1986; *Anancus* sp. (Sagantole-type) Kalb and Mebrate, 1993; *Anancus* cf. *Anancus* sp. (Sagantole-type) Sanders, 1997; *A. kenyensis* Harris and Leakey, 2003; *A. kenyensis* (in part) Tassy, 2003; *A. kenyensis* Mackaye et al., 2005; *A. osiris* Mackaye et al., 2005; *A. kenyensis* Kingston and Harrison, 2007

As the species name implies ("most extreme in time or sequence"), *Anancus ultimus* represents the epitome of anancine gomphothere evolution in Africa and the last appearance of the genus in eastern Africa (Tables 5.4 and 6.1; Sanders, 2011). It is an early to mid-Pliocene taxon, present at its type site of Laetoli and environs in w Lower and Upper Laetolil Beds, dated to 4.3–3.6 Ma (Deino, 2011; Sanders, 2005, 2011). The species encompasses Tassy's (1986) *A. kenyensis* "*petrocchii* morph" (not including the species *A. petrocchii*) and part of stage C and *A. kenyensis* "Sagantole-type" of Kalb and Mebrate (1993). It is defined by strong anancoidy, thick development of cementum, pentalophodonty of intermediate molars (Figure 6.2A), third molars with six-seven loph(id)s, and complex occlusal features including occasionally folded enamel and accessory conules on both sides of the crown, often to the posterior half of molars and sometimes doubled (Sanders, 2011). As part of a lineage starting with late Miocene-early Pliocene *A. kenyensis*, and occupying an interval from about 5.0 to 3.6 Ma (Table 6.1), *A. ultimus* exhibits considerable variability of cheek tooth morphology, with geologically younger representatives of the species exhibiting more advanced features than older specimens. Although it is tempting to view the evolution of the lineage as an example of anagenesis, potential overlap of 500 Kyr between *A. kenyensis* and *A. ultimus* (Tables 5.1 and 6.1) suggests that other modes of evolution, such as parapatric or "budding" speciation, should be considered.

The variation in molar features of *A. ultimus* can be summarized from different site samples: Specimens from the lower Chemeron Fm. in the Tugen Hills, Kenya exhibit thick, unfolded or coarsely folded enamel, weak anancoidy, unimpressive development of posttrite accessory conules, and a maximum of six loph(id)s in third molars (Figure 6.2B); in contrast, third molars from the Lower Laetolil Beds at Kakesio, Tanzania have seven lophs, coarsely folded enamel, doubled posterior accessory conules, and greater expression of accessory conules throughout their crowns (Figure 6.2C), whereas the type m3 (EP 197/05) from the Upper Laetolil Beds at Laetoli has a much more derived anancine morphology, with seven lophids, exaggerated anancoidy, strongly folded enamel, and a complex occlusal pattern of anterior and posterior accessory conules on both pre- and posttrite sides and throughout the length of the crown (Figure 6.2D; Sanders, 2011). The morphology of this specimen is reminiscent of some of the molars from the Aramis and Beidareem Mbs. of the Sagantole Fm. in the Middle Awash, Ethiopia (Kalb and Mebrate, 1993). The complex anancine crown morphology from Aterir, Kenya (Figure 6.2E) is like that from the Upper Laetolil Beds and suggests that the site is closer in age to <4.0 Ma than approximately 4.5 Ma (see Bishop et al., 1971; Hill, 1994). While M3s from Kakesio, Tanzania have loph formula of 7x–x7x and dimensions of L = 205.0–220.0 mm, W = 80.8–90.0 mm, and ET = 4.8–5.5 mm, the geologically younger (Table 6.1) type m3 from Laetoli has a lophid formula of x7x and is small by comparison, L = 160.0 mm, W = 79.6 mm (Sanders, 2011).

Although a partial upper molar referrable to *A. ultimus* from the Upper Laetolil Beds, LAET 81-75, was collected by Mary Leakey's team, it remained unreported at the time, and it was not until the activities of the Eyasi Plateau Expedition that these gomphotheres were identified from Laetoli and nearby sites (Sanders, 2011). As with anancine gomphotheres from other East and Central African sites (Cerling et al., 1999, 2003; Zazzo et al., 2000; Harris et al., 2003; Semaw et al., 2005), molars of *A. ultimus* from Kakesio, Tanzania indicate that individuals of the species had predominantly C_4-based diets or were mixed feeders (Kingston and Harrison, 2007; Kingston, 2011). At these sites, *A. ultimus* is frequently found with other proboscideans, including stegodonts, deinotheres, and elephants, and may have competed for resources with some of them (consequently undergoing selection that favored evolution of features associated with more effective grazing, such as thicker cementum, more loph(id)s, greater crown complexity, and higher crowns).

ELEPHANTIDAE Gray, 1821
ELEPHANTINAE Gray, 1821
***SELENETHERIUM* Mackaye et al., 2005**
***SELENETHERIUM KOLLEENSIS* Mackaye et al., 2005**
(Table 6.1; Figure 6.1)

Synonymy—*Selenetherium kolleensis* nov. gen. nov. sp. Mackaye, 2001

TABLE 6.1
Major Occurences of Early Pliocene Afro-Arabian Proboscidea

Taxon	Occurrence (Site, Locality)	Stratigraphic Unit	Age	Key References
Proboscidea Illiger, 1811				
Suborder incertae sedis				
Deinotherioidea Bonaparte, 1845				
Deinotheriidae Bonaparte, 1845				
Deinotheriinae Bonaparte, 1845				
Deinotherium bozasi	Tugen Hills, Kenya	Lower Chemeron Fm. (=Mabaget Fm.)	5.3–4.5 Ma (alt. ~5.0–4.0 Ma) (alt. 4.6–3.8 Ma)	Hill et al. (1985, 1986), Deino et al. (2002), and Roche et al. (2013)
	Manonga Valley, Tanzania	Tinde Mb., Wembere-Manonga Fm.	~5.0–4.5 Ma	Harrison and Baker (1997), and Sanders (1997)
	As Duma, Gona Western Margin, Ethiopia	Sagantole Fm.	<5.2–>4.6 Ma (alt. 4.5–4.3 Ma)	Semaw et al. (2005), Levin et al. (2008), and Quade et al. (2008)
	Galili, southern Afar Depression, Ethiopia	Dhidinley, Godiray, and Shabeley Laag Mbs., Mount Galili Fm.	<4.43–>3.94 Ma; <3.94 Ma; 3.97±0.03–<3.87±0.02 Ma) (alt. ~4.5–3.5 Ma)	Kullmer et al. (2008), and Hujer et al. (2015)
	Aramis, Middle Awash, Ethiopia	Haradaso Mb., Aramis Mb., Beearyada beds, Kalaloo beds, Sagantole Fm.	>4.4 Ma, early Pliocene	Kalb et al. (1982), and WoldeGabriel et al. (2001)
	Tobène, Senegal		4.4–4.0 Ma	Lihoreau et al. (2021)
	Koobi Fora, Kenya	Lonyumun Mb., Koobi Fora Fm.	4.38–4.28 Ma (alt. >3.97 Ma)	F. Brown (pers. comm.), and Brown and McDougall (2011)
		Allia Bay, below or within the Moiti Tuff;	4.2–3.9 Ma	Coffing et al. (1994), Leakey et al. (1995), and Bobe (2011)
		Koobi Fora, between Suregei Complex and Tulu Bor Tuff	~4.0–3.44 Ma	Harris (1976, 1983a, b)
	Lothagam, Kenya	Apak Mb., Nachukui Fm.	~4.2 Ma (alt. 5.0–4.0 Ma)	Harris (2003)
	Omo, Ethiopia	Mursi Fm.	~4.2 Ma (alt. >4.15 Ma; alt. >3.99 Ma)	Arambourg et al. (1969), Beden (1975, 1976), Brown and McDougall (2011), and Drapeau et al. (2014)
	Kanapoi, Kenya	Kanapoi Fm.	4.2–4.1 Ma	Feibel (2003), Harris et al. (2003), McDougall and Brown (2008), Brown and McDougall (2011), and Sanders (2020)
	South Turkwel, Kenya		Early Pliocene	J. Harris (pers. comm.)
	Ekora, Kenya	Kanapoi Fm.	Slightly <4.0 Ma	Patterson et al. (1970), and Behrensmeyer (1976)
	West Turkana, Kenya	Kataboi Mb., Nachukui Fm.	3.97–3.44 Ma	Harris et al. (1988b)
	Laetoli, Tanzania	Upper Unit, Laetolil Beds	3.85–3.63 Ma	Drake and Curtis (1987), Harris (1987a), Deino (2011), and Sanders (2011)
	Chiwondo Beds, Malawi	Unit 3A	3.75–2.0 Ma	Schrenk et al. (1993)

(Continued)

TABLE 6.1 (*Continued*)
Major Occurences of Early Pliocene Afro-Arabian Proboscidea

Taxon	Occurrence (Site, Locality)	Stratigraphic Unit	Age	Key References
Elephantiformes Tassy, 1988				
Elephantimorpha Tassy and Shoshani, 1997 in Shoshani et al., 1998				
Elephantida Tassy and Shoshani, 1997 in Shoshani et al., 1998				
Gomphotheriidae Hay, 1922				
Anancinae Hay, 1922				
Anancus kenyensis	Tugen Hills, Kenya	Lower Chemeron Fm.	5.3–4.5 Ma; late Miocene-early Pliocene	Roche et al. (2013)
	Nkondo, Uganda	Nkondo Fm.	~5.0–4.0 Ma	Tassy (1995)
	Middle Awash, Ethiopia	Haradaso Mb., Sagantole Fm.	~5.0–4.4 Ma (probably about 4.8 Ma)	Kalb and Mebrate (1993), and Renne et al. (1999)
	As Duma, Gona Western Margin, Ethiopia		4.5–4.3 Ma	Semaw et al. (2005)
	Esere 1=Endolele, Tanzania	Lower Unit, Laetolil Beds	ca. 4.4–3.85 Ma	Drake and Curtis (1987), and Sanders (2011)
	Galili, southern Afar Depression, Ethiopia	?Lasdanan Mb., Galili Fm.	>5.37±0.04–4.43 Ma; late Miocene-early Pliocene	Kullmer et al. (2008), and Hujer et al. (2015)
	Karonga, Uraha, Malawi	Chiwondo Beds, Unit 2	early Pliocene, >4.0 Ma (alt. ≥4.0–~3.75 Ma)	Mawby (1970), Schrenk et al. (1993), Bromage et al. (1995), and Sandrock et al. (2007)
Anancus ultimus (*Anancus* sp.)	As Duma, Gona Western Margin, Ethiopia	Sagantole Fm.	early Pliocene; <5.2 Ma–>4.6 Ma	Semaw et al. (2005), Levin et al. (2008), and Quade et al. (2008)
	Tugen Hills, Kenya	Lower Chemeron Fm.	~5.0–4.0 Ma (alt. 4.6–3.8 Ma)	Hill et al. (1985, 1986), and Deino et al. (2002)
	Kollé, Chad		5.0–4.0 Ma	Brunet (2001)
	Nkondo-Kaiso Area, Nyawiega, Uganda	Nyawiega Mb., Nkondo Fm.	5.0–4.0 Ma	Cooke and Coryndon (1970), and Tassy (1995)
	Kiloleli; Ngofila; Beredi South, Manonga Valley, Tanzania	Kiloleli Mb., Wembere-Manonga Fm.	~4.5–4.0 Ma	Harrison and Baker (1997), and Sanders (1997)
	Aterir, Kenya	Aterir Beds	~4.5 Ma (alt. slightly <4.0 Ma)	Bishop et al. (1971), and Hill (1994)
	Galili, southern Afar Depression, Ethiopia	Dhidinley and Shabeley Laag Mbs., Mount Galili Fm.	<4.43–>3.94 Ma; 3.97 ±0.03–<3.87 ±0.02 Ma	Kullmer et al. (2008), and Hujer et al. (2015)
	Middle Awash, Ethiopia	Aramis, Beidareem, and Adgantole Mbs., Sagantole Fm.	4.4–4.3 Ma	Kalb and Mebrate (1993), Renne et al. (1999), and White et al. (2006)
	Kakesio, Tanzania	Lower Unit, Laetolil Beds	4.3 Ma	Drake and Curtis (1987), Harris (1987b), Hay (1987), and Sanders (2005, 2011)
	Lothagam, Kenya	Apak Mb., Nachukui Fm.	~4.2 Ma (alt. 5.0–4.0 Ma)	McDougall and Feibel, (2003), and Tassy (2003)
	Kanapoi, Kenya	Kanapoi Fm.	4.2–4.1 Ma	Feibel (2003), Harris et al. (2003), McDougall and Brown (2008), Brown and McDougall (2011), and Sanders (2020)

(*Continued*)

TABLE 6.1 (*Continued*)
Major Occurences of Early Pliocene Afro-Arabian Proboscidea

Taxon	Occurrence (Site, Locality)	Stratigraphic Unit	Age	Key References
	Allia Bay, Kenya	Below or within the Moiti Tuff	4.2–3.9 Ma (alt. 3.97 Ma)	Leakey et al. (1995), Coffing et al. (1994), and Bobe (2011)
	Sinda River, Lower Semliki, Democratic Republic of Cong	Sinda Beds	~4.1 Ma	Hooijer (1963), Yasui et al. (1992), and Boaz (1994)
	Ekora, Kenya	Kanapoi Fm.	Slightly <4.0 Ma	Behrensmeyer (1976), and Kalb and Mebrate (1993)
	Laetoli, Tanzania (type)	Upper Unit, Laetolil Beds	3.85–3.63 Ma	Drake and Curtis (1987), Harris (1987b), Deino (2011), and Sanders (2011)
Anancus osiris	?Kollé, Chad		~4.0 Ma	Mackaye (2001), Lebtard et al. (2008), and Schuster et al. (2009)
Anancus sp. indet.	Aramis, Middle Awash, Ethiopia	Haradaso Mb., Sagantole Fm.	~4.4 Ma	WoldeGabriel et al. (1994)
	Omo, Ethiopia	Mursi Fm.	~4.2 Ma (alt. >4.15 Ma)	Beden (1976), Feibel et al. (1989), and Drapeau et al. (2014)
	Karonga, Uraha, Malawi	Chiwondo Beds, stratigraphic Unit 2	≥4.00–3.75 Ma	Bromage et al. (1995), and Sandrock et al. (2007)
Stegodontidae Osborn, 1918				
Stegodon kaisensis ?"Nkondo stage"	Ouadi Derdemy, Chad		?early Pliocene	Coppens (1965)
Stegodon kaisensis "Warwire stage" (*Stegodon* sp. indet.)	Locality 14, Sinda River, Lower Semliki, Democratic Republic of Congo	Upper Mb., Sinda Beds	~4.1 Ma	Yasui et al. (1992)
(*Stegodon kaisensis* "Nkondo stage" and "Warwire stage")	Kollé, Chad		5.0–4.0 Ma	Brunet et al. (1998), Mackaye (2001), Fara et al. (2005), and Schuster et al. (2009)
	Laetoli, Tanzania	Upper Unit, Laetolil Beds	3.8–3.5 Ma	Drake and Curtis (1987), and Sanders (2005, 2011)
	Nkondo-Kaiso Region, Uganda	Warwire Fm.	~3.5–3.0 Ma	Pickford et al. (1993), and Tassy (1995)
	Koro Toro, Chad		3.5 Ma (alt. 3.5–3.0 Ma)	Brunet (2001), Mackaye (2001), Fara et al. (2005), Lebtard et al. (2008), and Schuster et al. (2009)
	Omo, Ethiopia	Upper Mb. B, Shungura Fm.	2.95–2.85 Ma	Beden (1975, 1976), and Alemseged (2003)
	Ishango 11 and Senga near Sn1, Upper Semliki, Democratic Republic of Congo	Lusso Beds	Late Pliocene	de Heinzelin (1957), Peters (1990), and Sanders (1990)
Elephantidae Gray, 1821				
Stegotetrabelodontinae Aguirre, 1969				
Stegotetrabelodon orbus	Lothagam, Kenya	Apak Mb., Nachukui Fm.	5.0–4.2 Ma (alt. 5.0–4.0 Ma)	McDougall and Feibel (2003), and Tassy (2003)

(*Continued*)

TABLE 6.1 (*Continued*)
Major Occurences of Early Pliocene Afro-Arabian Proboscidea

Taxon	Occurrence (Site, Locality)	Stratigraphic Unit	Age	Key References
Elephantinae Gray, 1821				
?*Stegodibelodon schneideri*	Kollé, Chad		5.0–4.0 Ma	Mackaye (2001), Mackaye et al. (2005), and Schuster et al. (2009)
Selenetherium kolleensis	Kollé, Chad		5.0–4.0 Ma	Mackaye (2001), Mackaye et al. (2005), and Schuster et al. (2009)
Primelephas korotorensis ("*Stegodon*" *korotorensis*)	Koulà, Chad (type)		?early Pliocene	Coppens (1965), Maglio (1970b, 1973), Coppens et al. (1978), and Mackaye et al. (2008)
	Kolinga, Chad		?early Pliocene	Maglio (1973), Mackaye (2001), and Mackaye et al. (2008)
	Lothagam, Kenya	Apak Mb., Nachukui Fm.	5.0–4.2 Ma (alt. 5.0–4.0 Ma)	McDougall and Feibel (2003), Tassy 2003), and Sanders et al. (2010a)
	Kollé, Chad		4.0 Ma (alt. 5.0–4.0 Ma)	Mackaye (2001), Mackaye et al. (2008), and Schuster et al. (2009)
Loxodonta sp. cf. *Loxodonta cookei*	Lothagam, Kenya	Apak Mb., Nachukui Fm.	5.0–4.2 Ma (alt. 5.0–4.0 Ma)	McDougall and Feibel (2003), Tassy (2003), and Sanders et al. (2010a)
	Endolele=Esere 1, Tanzania	Lower Laetolil Beds	~4.4–3.85 Ma	Sanders(2005, 2011), and Deino (2011)
Loxodonta adaurora adaurora	?Kanam East, Kenya	Kanam Fm.	Latest Miocene-earliest Pliocene	MacInnes (1942), and Ditchfield et al. (1999)
	?Bugoma, Kaiso Central, Nyawiega, Uganda	Lower Kaiso Beds (=Nkondo Fm.)	ca. 5.0 Ma (alt. 6.0 Ma)	Cooke and Coryndon (1970), Sanders (1990), Pickford et al. (1993), and Boaz (1994)
	Tugen Hills, Kenya	Lower Chemeron Fm.	Early Pliocene (within 5.3–4.0 interval)	Maglio (1973), Hill et al. (1985, 1986), and Deino et al. (2002)
	Kubi Algi, Kenya	Kubi Algi Fm.	4.55 Ma (tuffs dated to 4.6 Ma and 3.9 Ma)	Maglio (1973), Fitch and Miller (1976), Abell et al. (1982), and Tassy and Debruyne (2001)
	Lothagam, Kenya	Apak Mb., Nachukui Fm.	5.0–4.2 Ma (alt. 5.0–4.0 Ma)	McDougall and Feibel (2003), Tassy (2003), and Sanders et al. (2010a)
("*Loxodonta* cf. *adaurora*")	?Galili, southern Afar Depression, Ethiopia	Dhidinley and Shabeley Laag Mbs., Mount Galili Fm.	<4.43–>3.94 Ma; 3.97 ±0.03–<3.87 ±0.02 Ma (alt. in the interval ~4.5–3.5 Ma)	Kullmer et al. (2008), and Hujer et al. (2015)
	Middle Awash Valley, Ethiopia	Aramis and Beidareem Mbs., Sagantole Fm.	4.4–4.3 Ma	Mebrate (1983), Kalb and Mebrate (1993), and Renne et al. (1999)
	Ileret, Kenya	Lonyumun Mb., Koobi Fora Fm.	4.38–4.28 Ma (alt. >3.97 Ma) (alt. >4.3 Ma, possibly >4.5 Ma)	F. Brown (pers. comm.), Brown and McDougall (2011), and Sanders et al. (2021)

(Continued)

TABLE 6.1 (*Continued*)
Major Occurences of Early Pliocene Afro-Arabian Proboscidea

Taxon	Occurrence (Site, Locality)	Stratigraphic Unit	Age	Key References
	Allia Bay, Kenya	Below or within the Moiti Tuff (below Allia Tuff)	4.2–3.9 Ma	Maglio (1971, 1972), Beden (1983), and Leakey et al. (1995)
		below Moiti Tuff	(alt. 3.97 Ma)	Coffing et al. (1994), and Bobe (2011)
		(alt. Moiti and Lokochot Mbs., Koobi Fora Fm.)	(alt. 3.9–3.4 Ma)	Beden (1983), Harris (1983), and Brown (1994)
	Kanapoi, Kenya (type)	Kanapoi Fm.	4.2–4.1 Ma	Maglio (1970b, 1973), Maglio and Ricca (1977), Feibel (2003), Harris et al. (2003), McDougall and Brown (2008), Brown and McDougall (2011,) and Sanders (2020)
	Omo, Ethiopia	Mursi Fm.	~4.2 Ma (alt. >3.99 Ma) (alt. >4.15 Ma)	de Heinzelin (1983), Beden (1987a), Feibel et al. (1989), and Drapeau et al. (2014)
	West Turkana, Kenya	Kataboi Mb. Nachukui Fm.	4.10–3.36 Ma	Harris et al. (1988)
	Chiwondo Beds, Malawi	Unit 2	≥4.0–~3.75 Ma in the interval	Sandrock et al. (2007)
		Unit 3A	of 3.75–2.0 Ma	Schrenk et al. (1993)
	Ekora Kenya	Ekora Fm.	Slightly <4.0 Ma	Maglio (1970b), Behrensmeyer (1976), Kalb and Mebrate (1993), and Sanders et al. (2021)
("cf. *Loxodonta adaurora*")	?Aralee Issie and Mesgid Dora, Woranso-Mille, Ethiopia	Woranso-Mille Beds, Afar Series	3.76–3.72 Ma	Deino et al. (2010), and Sanders and Haile-Selassie (2012)
	South Turkwel, Kenya		Early Pliocene	J. Harris (unpublished data)
	Eshoa, Kenya		Early Pliocene	J. Harris (unpublished data)
Loxodonta exoptata	?Manonga Valley, Tanzania	Kiloleli Mb., Wembere-Manonga Fm.	~4.5–4.0 Ma	Harrison and Baker (1997), and Sanders (1997)
("*Loxodonta* cf. *exoptata*")	?Galili, southern Afar Depression, Ethiopia	Dhidinley and Shabeley Laag Mbs., Mount Galili Fm.	<4.43–>3.94 Ma; 3.97 ±0.03–<3.87 ±0.02 Ma (alt. in the interval ~4.5–3.5 Ma)	Kullmer et al. (2008), and Hujer et al. (2015)
	Kakesio (and Emboremony and Noiti), Tanzania	Lower Laetolil Beds	4.36–3.85 Ma	Drake and Curtis (1987), Harris (1987b), Hay (1987), Sanders (2005, 2011), and Deino (2011)
	Karonga, Uraha, Mwimbi, Malawi	Chiwondo Beds, stratigraphic Unit 2	≥4.0–3.0 Ma (alt. ≥4.0–3.75 Ma)	Bromage et al. (1995), and Sandrock et al. (2007)
	Laetoli, Tanzania (type)	Upper Laetolil Beds	3.85–3.63 Ma	Arambourg (1942), Beden (1987b), Drake and Curtis (1987), Harris (1987b) Harrison (2002), Sanders (2005, 2011), and Deino (2011)

(*Continued*)

TABLE 6.1 (*Continued*)
Major Occurences of Early Pliocene Afro-Arabian Proboscidea

Taxon	Occurrence (Site, Locality)	Stratigraphic Unit	Age	Key References
	Kanapoi, Kenya	Kanapoi Fm.	4.2–4.1 Ma	Maglio (1970a, b, 1973), Feibel (2003), Harris et al. (2003), McDougall and Brown (2008), Brown and McDougall (2011), and Sanders (2020)
	Omo, Ethiopia	Mb. A, Shungura Fm.	3.60–3.44 Ma	Beden (1987a), and Levin et al. (2011)
"*Elephas*" *ekorensis* ("*Elephas*" cf. *ekorensis*)	Galili, southern Afar Depression, Ethiopia	Shabeley Laag Mb., Mount Galili Fm.	3.97±0.03–<3.87±0.02 Ma (alt. in the interval ~4.0–3.5 Ma)	Kullmer et al. (2008), and Hujer et al. (2015)
	Ekora, Kenya (type)	Kanapoi Fm.	Slightly <4.0 Ma	Maglio (1970a, 1973), Maglio and Ricca (1977), Behrensmeyer (1976), and Sanders et al. (2021)
	Allia Bay, Kenya	Below Moiti Tuff	3.97 Ma	Coffing et al. (1994), and Bobe (2011)
	?Woranso-Mille, Ethiopia	Woranso-Mille Beds, Afar Series	3.82–3.57 Ma	Deino et al. (2010), and Sanders and Haile-Selassie (2012)
	Chiwondo Beds, Malawi	Unit 3A	in the interval of 3.75–2.0 Ma	Schrenk et al. (1993)
Elephas recki brumpti (alt. "*Phanagoroloxodon planifrons*") (alt. *Elephas brumpti*)	Middle Awash, Ethiopia	Aramis Mb., Sagantole Fm.	3.9–3.4 Ma (alt. 4.4 Ma) (alt. 4.1–3.8 Ma)	Kalb and Mebrate (1993), Kalb (1995), Renne et al. (1999), and Zhang (2020)
	Galili, southern Afar Depression, Ethiopia	Shabeley Laag Mb., Mount Galili Fm.	3.97±0.03–<3.87±0.02 Ma (alt. in the interval ~4.0–3.5 Ma)	Kullmer et al. (2008), and Hujer et al. (2015)
	South Turkwel, Kenya		Early Pliocene	J. Harris (pers. comm.)
(alt. "*Elephas* sp." and "*E. ekorensis*")	Aralee Issie, Mesgid Dora, and Ramidele Dora, Woranso-Mille, Ethiopia	Woranso-Mille Beds	3.82–3.72 Ma	Deino et al. (2010), Sanders and Haile-Selassie (2012), and Zhang (2020)
Mammuthus subplanifrons	Uraha, Malawi	Chiwondo Beds, stratigraphic Unit 2	≥4.0–3.75 Ma (alt. ≥4.0 Ma)	Sandrock et al. (2007), Schrenk et al. (1993), Mawby (1970), and Bromage et al. (1995)
	?Vaal River, South Africa (type)	Middle Terrace	?early Pliocene	Osborn (1928, 1934)
	Virginia, Orange Free State, South Africa		?early Pliocene	Meiring (1955), and Maglio (1973)
	?Ishasha River, Virunga National Park, Democratic Republic of Congo	Kaiso Group sediments	Early Pliocene	Vanoverstraeten et al. (1990)
?*Mammuthus africanavus* (*Loxodonta africanavus*)	Kollé, Chad		5.0–4.0 Ma	Mackaye (2001), Fara et al. (2005), and Schuster et al. (2009)
cf. *Mammuthus* sp. "Hadar-type"	?Megsid Dora, Woranso-Mille, Ethiopia	Woranso-Mille Beds, Afar Series	3.76–3.72 Ma	Deino et al. (2010), and Sanders and Haile Selassie (2012)

?=attribution, age, or occurrence uncertain; alt., alternative; alternate taxonomic names are in parentheses.

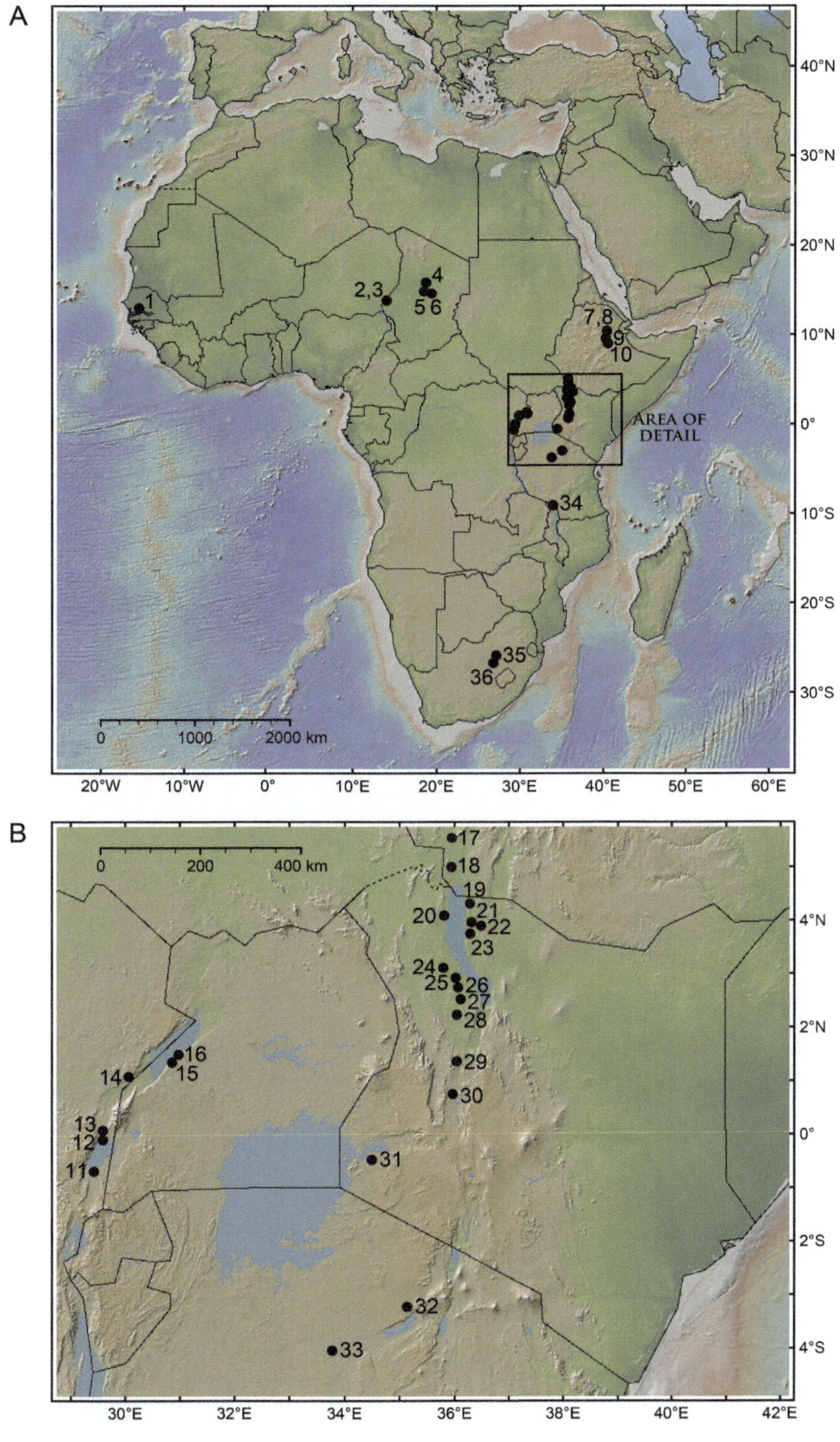

FIGURE 6.1 Maps showing location of early to mid-Pliocene sites, (A, B). Area of detail, (B). 1, Tobène, Senegal; 2, Kolinga, Chad; 3, Koulà, Chad; 4, Kollé, Chad; 5, Koro Toro, Chad; 6, Ouadi-Derdemy, Chad; 7, Aralee Issie and Mesgid Dora, Woranso-Mille, Ethiopia; Woranso-Mille, Ethiopia; 8, As Duma, Gona Western Margin, Ethiopia; 9, Aramis, Middle Awash, Ethiopia; Middle Awash, Ethiopia; Middle Awash Valley, Ethiopia; 10, Galili, southern Afar Depression, Ethiopia; 11, Ishasha River, Virunga National Park, Democratic Republic of Congo; 12, Ishango, Democratic Republic of Congo; 13, Upper Semliki, Democratic Republic of Congo; 14, Sinda River, Lower Semliki, Democratic Republic of Congo; 15a, Nkondo-Kaiso Area, Nyawiega, Uganda; 15b, ?Bugoma, Kaiso Central, Nyawiega, Uganda; 16, Kaiso-Nkondo, Uganda; 17, Omo (Mursi), Ethiopia; 18, Omo (Shungura), Ethiopia; 19, Ileret, Kenya; 20, West Turkana, Kenya; 21, Koobi Fora, Kenya; 22, Kubi Algi, Kenya; 23, Allia Bay, Kenya; 24, South Turkwel, Kenya; 25, Lothagam, Kenya; 26, Eshoa, Kenya; 27, Ekora, Kenya; 28, Kanapoi, Kenya; 29, Aterir, Kenya; 30, Tugen Hills, Kenya; 31, Kanam, Kenya; 32a, Laetoli, Tanzania; 32b, Esere 1=Endolele, Tanzania; 32c, Kakesio, Tanzania; Kakesio (and Emboremony and Noiti), Tanzania; 33a, Manonga Valley, Tanzania; 33b, Kiloleli; Ngofila; Beredi South, Manonga Valley, Tanzania; 34a, Karonga, Uraha, Malawi; 34b, Karonga, Uraha, Mwimbi, Malawi; 34c, Uraha, Malawi; 34d, Chiwondo Beds, Malawi; 35, Vaal River, South Africa; 36, Virginia, Orange Free State, South Africa.

Included among the rich and diverse proboscidean assemblages of the Chad Basin is a small sample of what was named as a new genus and species from early Pliocene Kollé (Table 6.1; Mackaye et al., 2005). The genus was named after the diagnostic quarter-moon-shaped molar enamel wear-figures that are anteriorly and posteriorly bowed convexly, and the species nomen honors the site of discovery. Importantly, the holotype is a mandible with m3s preserving the symphyseal region, which is narrow and forms a projecting, nearly horizontal spout ("bec") that is longer than in *Primelephas korotorensis*. There are no lower tusks and no sign of alveoli for them.

The mandibular corpus is much higher at the anterior end of the molar row (270 mm) than posteriorly (145 mm), and flares broadly across the distal end of the m3s (>500 mm). The symphysis, including its projecting spout, is moderately elongated (287 mm), and terminates distally as a rounded, procumbent prow. Overall, the mandible is relatively short, broad, and stout.

The absence of lower tusks, form of the spout, and curious shape of the enamel wear figures are crucial for diagnosing this species from stegotetrabelodonts, as the molars are remarkably primitive in many features. For example, third molar lamellar frequency is very low (outside of the usual range for *Primelephas* and at the low end for *Stegotetrabelodon*), LF = 2.1–3.0; enamel is unfolded and very thick, ranging from 5.5 to 8.0 mm; and crown height is very low (Mackaye et al., 2005: table 2). In lateral view, plates are bulbous and pyramidal in shape, and transverse valleys are correspondingly V-shaped. Additionally, third

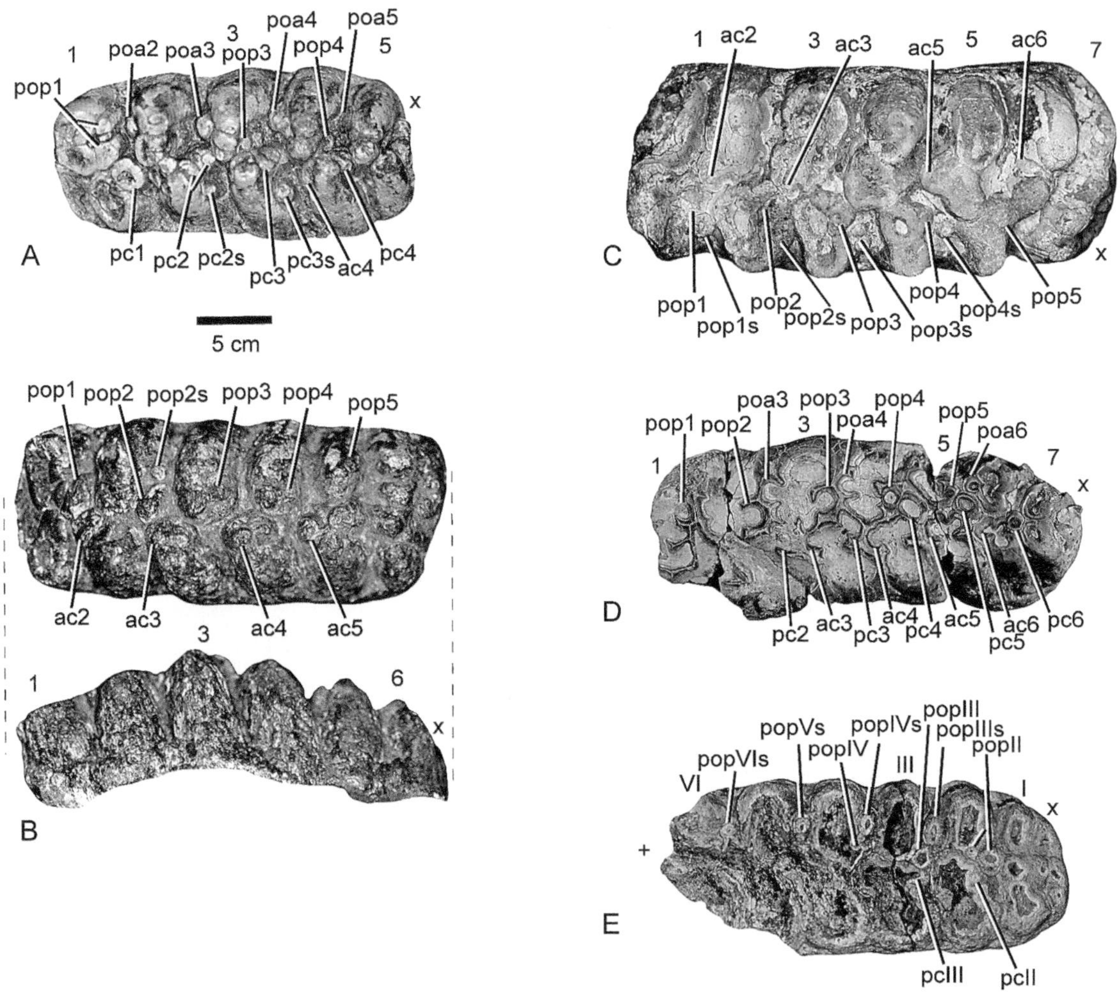

FIGURE 6.2 Molars of early to mid-Pliocene *Anancus ultimus*. Abbreviations: ac, anterior pretrite accessory conule; pc, posterior pretrite accessory conule; pcs, supplemental posterior pretrite accessory conule (e.g., "pc3s"); poa, anterior posttrite accessory conule; pop, posterior posttrite accessory conule; pops, supplemental posterior posttrite accessory conule (e.g., "pop2s" or "pop IVs"); x, pre- or postcingulum(id); 1, 2, 3, …, loph(id) counted from the anterior end of the crown; I, II, III, …, loph(id) counted from the posterior end of the crown; +, indicates missing morphology. Anterior to the left. (A–E) to the same approximate scale. (A) Left m2, KNM-KP 384 from Kanapoi, Kenya, occlusal view. (B) Left M3, KNM-BC 33 from the Lower Chemeron Fm., Tugen Hills, Kenya, occlusal and lateral views. (C) Right M3, KK/82-292 from Kakesio, Tanzania, occlusal and lateral views. (D) Left m3, EP 197/05 (type) from the Upper Laetolil Beds, Laetoli, Tanzania, occlusal view. (E) Anteriorly broken left M3, KNM-AT 20 from Aterir, Kenya, occlusal view.

molars have only six plates. Cementum is sparse in the transverse valleys. Plates appear to be formed of five stout conelets. It is difficult to ascertain evidence for accessory conules in association with the plates; plates have massive central conelets that in wear may obscure any contribution from small accessory conules. Despite the very primitive aspect of its molars, its lack of lower tusks suggests that *Selenetherium* belongs in the Elephantinae. Cladistic analyses consistently place it as a sister taxon to *Primelephas* (but also to *Stegodon*; Mackaye et al., 2005). It co-occurs at Kollé with stegodonts, anancine gomphotheres, and multiple species of primitive elephants, in a unique proboscidean cohort (Mackaye, 2001; Mackaye et al., 2005) that underscores the vast differences between ancient and modern African ecosystems.

LOXODONTA F. Cuvier, in F. Cuvier and Geoffroy Saint-Hilaire, 1825 (anonymous emendation, 1827) *LOXODONTA ADAURORA ADAURORA* Maglio, 1970b

(Tables 6.1 and 6.2; Figures 6.1 and 6.3)

Partial synonymy—*Elephas* cf. *meridionalis*, MacInnes, 1942; *Elephas* (*Archidiskodon*) cf. *planifrons*, Arambourg, 1947; primitive member of the *Elephas recki* series Patterson, 1966; *Loxodonta adaurora* Maglio, 1970b; *L. adaurora adaurora* Beden, 1983; *Protoloxodonta adaurora* Garutt, 1986; *Mammuthus* lineage Todd, 2010a; *Mammuthus subplanifrons* Larramendi, 2016; *L. adaurora adaurora* Sanders, 2020

The best documented proboscidean of the African early Pliocene is *Loxodonta adaurora*, its species nomen connoting "dawn elephant" because at the time of its first description it was the oldest species attributed to the genus *Loxodonta* (Maglio, 1970b). It is an eastern-southeastern African species, with a range extending from Ethiopia to Malawi (Table 6.1; Sanders et al., 2010a). The species was subdivided by Beden (1983) into time-successive subspecies, Pliocene *L. adaurora adaurora* (including the original species hypodigm named by Maglio [1970b]) and early Pleistocene *L. adaurora kararae* (Tables 6.1 and 7.1). *Loxodonta adaurora* is particularly well represented at sites in and near the greater Turkana Basin, especially its type site of Kanapoi, Kenya and appears to have been the dominant proboscidean of the basin during the early Pliocene (Maglio, 1970b, 1971, 1972, 1973; Maglio and Ricca, 1977; Coppens et al., 1978; Beden, 1987a; Harris et al., 1988a, 2003; Sanders, 2020).

Kanapoi is situated to the southwest of Lake Turkana, in the lower Kerio River valley (Figure 6.1). The site is confirmed as early Pliocene in age (Table 6.1) by radiometric dating of basalts and tuffs interpolated in its sedimentary sequence (Feibel, 2003; Harris et al., 2003; McDougall and Brown, 2008; Brown and McDougall, 2011). Successive teams working at the site since paleontological fieldwork was initiated there in the 1960s have recovered a rich vertebrate fossil assemblage, including fish, chelonians, crocodiles, birds, rhinos, equids, hippos, proboscideans, suids, giraffes, bovids, and strepsirhine and anthropoid primates (Patterson et al., 1970; Behrensmeyer, 1976; Harris et al., 2003; Stewart, 2003). The anthropoid primates notably include the type of the oldest species of the hominin genus *Australopithecus*, *A. anamensis* (Leakey et al., 1995; Ward et al., 1999, 2001; Manthi et al., 2012). The proboscideans accompanying *L. adaurora* at Kanapoi are taxonomically diverse and include two other species of elephant, "*Elephas*" *ekorensis* and *Loxodonta exoptata*, a gomphothere, *Anancus ultimus*, and a deinothere, *Deinotherium bozasi* (Harris et al., 2003; Sanders, 2020). *Elephas recki* also may occur high in the Kanapoi beds (Sanders, 2020). At Allia Bay, Kenya *L. adaurora* also co-occurs with a very diverse vertebrate fauna including fish, turtles, crocodiles, and a wealth of mammal species, including *A. anamensis*, "*Elephas*" sp. (probably "*E.*" *ekorensis*), and *Deinotherium bozasi* (Coffing et al., 1994; Leakey et al., 1995). As at Kanapoi, cercopithecines are far more common than colobine monkeys at Allia Bay, indicating similar availability of terrestrial paleoenvironments at these sites (Bobe, 2011).

Sediments at Kanapoi are primarily fluvial and deltaic (Harris and Leakey, 2003). Study of paleosols and stable carbon isotopes indicate a mosaic of paleoenvironments at the site during the early Pliocene, including relatively open low tree-shrub savanna with sparse annual grasses and forbs (C_4 plants contributed between 15 and 45% to soil CO_2) and gallery woodland along river channels in an arid to semi-arid climate regime (Wynn, 2000). Proboscidean fossils from Kanapoi have been recovered from fluvial and lacustrine depositional environments that include floodplain and channel habitats, mostly from a narrow time interval no greater than 100 ka (Wynn, 2000; Feibel, 2003). *Loxodonta adaurora*, *L. exoptata*, and "*E.*" *ekorensis* from the site are documented as mixed C_3–C_4 feeders with a strong component of grazing whose isotopic signals do not overlap with those of sympatric deinotheres (hyperbrowsers) and gomphotheres (either with more grazing or more browsing, depending on the individual specimen; Manthi et al., 2017), indicating that they exploited different paleoenvironments. Other studies of dental isotopes confirm that *L. adaurora* was a grazer or mixed feeder that ate a substantial amount of C_4 plants (Cerling et al., 1999, 2011a; Schoeninger et al., 2003). That *L. adaurora* was a mixed feeder is not a surprise because there is no evidence for widespread open grasslands during the early Pliocene at sites where it occurred, such as Koobi Fora or West Turkana, though there is a trend in the basin through the interval 4.0–3.5 Ma and younger for decreasing woody canopy cover and increasing importance of C_4 vegetation (Levin et al., 2008).

At Allia Bay, the depositional environment is floodplain of an ancient, meandering Omo River associated with a belt of gallery forest, grassland, woodland, and bushland, which explains the presence of a wide assortment of closed-habitat and dry-savanna inhabitants alongside aquatic animals (Coffing et al., 1994; Schoeninger et al., 2003). The co-occurrence of elephants and hominins on the same landscape is probably not coincidental, but instead may reflect the disruptive effect of elephants on closed, wooded conditions, resulting in a proliferation of heterogenous habitats (Laws, 1970; Western, 1989; Kerley et al., 2008; but see Dublin et al., 1990) that favored the unique positional abilities of early bipeds (Schoeninger et al., 2003; Sanders, 2020).

Loxodonta adaurora is also documented in the Mursi Fm. and from Omo Shungura Mb. B, Ethiopia from localities dated to >4.15 to 2.85 Ma (Heinzelin, 1983; Beden, 1987a; Feibel et al., 1989; Alemseged, 2003; WoldeGabriel et al., 2005). A more advanced subspecies of *L. adaurora* occurs higher in the Omo sequence (Beden, 1987a). Age control of the Mursi Fm. is established by an overlying basalt, and the age of the deposits make them contemporaneous with Kanapoi. Lithology of the sediments indicates deltaic, prodeltaic, alluvial, and low energy littoral environments responsible for their deposition (Drapeau et al., 2014). Although the Mursi Fm. has been generally underappreciated, its age during the interval when australopithecines were first evolving makes it valuable for recording the range of habitat conditions and faunas at that time. The fauna is diverse but modest in numbers of individuals, including fish, several species of crocodiles, turtles, a very large number of suids (primarily *Nyzanzachoerus*), hippos, giraffes, bovids (no open-country alcelaphine and antilopine individuals), rhinos, hipparionine equids, carnivores, and a small assemblage of proboscideans (*L. adaurora*, anancine gomphotheres, deinotheres, and indeterminate elephantids) (Bobe and Behrensmeyer, 2004; Bobe, 2011; Drapeau et al., 2014). Importantly, the Mursi Fm. has yielded one of the few, fragmentary crania known for *L. adaurora*, associated with cheek teeth (Beden, 1987a). Strangely, although the Mursi vertebrate sample is composed of approximately 300 individuals, no primates have been recovered from the unit (Bobe, 2011).

In his original description of *L. adaurora*, Maglio (1970b) fortuitously had the better part of an entire skeleton of an immense male individual to designate as the type specimen (KNM-KP 385). The upper and lower third molars are in place by themselves and in wear to the last plate; the expression of wear through the occlusal platform is equivalent to Laws' (1966) age group XXV of modern African elephants, with a mean of 47±2 years old. This individual was a mature bull at the time of death. Maglio (1970b, 1973) and Maglio and Ricca (1977) relied heavily on a geologically much younger (Feibel, 2003; McDougall and Feibel, 2003), incomplete cranium of the species from Lothagam, Kenya (KNM-LT 353) from the Kaiyumung Mb. of the Nachukui Fm. to supplement the damaged type specimen (missing much of its vault and occiput) for cranial description and reconstruction. Unfortunately, KNM-LT 353 is now missing from the National Museums of Kenya.

Maglio (1970b, 1973) diagnosed *L. adaurora* as having low-crowned molars (HI ≤100), with unfolded (though sometimes coarsely undulated), relatively thick enamel (3–5 mm) and very large anterior and posterior accessory conules (free at their apices) (Figure 6.3A) that contribute to prominent sinuses in the midline of enamel loops with wear. Despite the strong development of accessory conules, enamel wear figures on the crown do not form the loxodont sinuses ("<>") that typify extant African elephants (Figure 6.3B; Sanders et al., 2010a; Sanders, 2020). Lamellar frequency in the molars of the species is low (2.6–4.4 in third molars), due to ample spacing of thick molar plates. Upper third molars were counted to have 8–10 plates and lower m3s as having 10–11 plates. In addition, plates are broadest at their bases, with convex, rounded lateral borders that converge apically and anteroposterior faces that also converge at their apices (Beden, 1983, 1987a). Lateral profiles of plates, however, are not pyramidal and transverse valleys are correspondingly U-shaped in lateral view (Figure 6.3A). Dental formulae include evidence for primitive retention of permanent premolars, notably in juvenile mandible KNM-KP 30616 from Kanapoi, Kenya (Sanders, 2017, 2020) and comprehensive review establishes plate formulae of dP2/dp2, 3/?4; dP3/dp3, --/6; dP4/dp4, 8/7; P3/p3, 2/3; M1/m1, --/6; M2/m2, 6–7/7; M3/m3, 8–10/10 (Sanders, 2020). Adult dental formula for the species is likely 1-0-2–3/0-0-2–3 (Table 6.2). Molars of *L. adaurora* are distinguished from those of older, more primitive elephants such as stegotetrabelodonts, *Primelephas korotorensis*, and

TABLE 6.2
Adult Dental Formulae and Incisor Dominance in Early Pliocene African Proboscidea

Taxon	Upper Permanent Dentition	Largest Upper Incisor	Lower Permanent Dentition	Largest Lower Incisor
Deinotherium bozasi	P3–4, M1–3	—	i2, p3–4, m1–3	i2
Anancus kenyensis	I2, ?P3–4, M1–3	I2	?p3–4, m1–3	—
Anancus ultimus	I2, M1–3	I2	m1–3	—
Loxodonta adaurora adaurora	I2, P3–4, M1–3	I2	p3–4, m1–3	—
Loxodonta exoptata	I2, M1–3	I2	m1–3	—
"*Elephas*" *ekorensis*	I2, M1–3	I2	m1–3	—
Elephas brumpti	I2, M1–3	I2	m1–3	—
Mammuthus sp. "Hadar type"	I2, M1–3	I2	m1–3	—

C/c = canine; I/I = incisor; M/m = molar; P/p = premolar; ? = presence unknown.

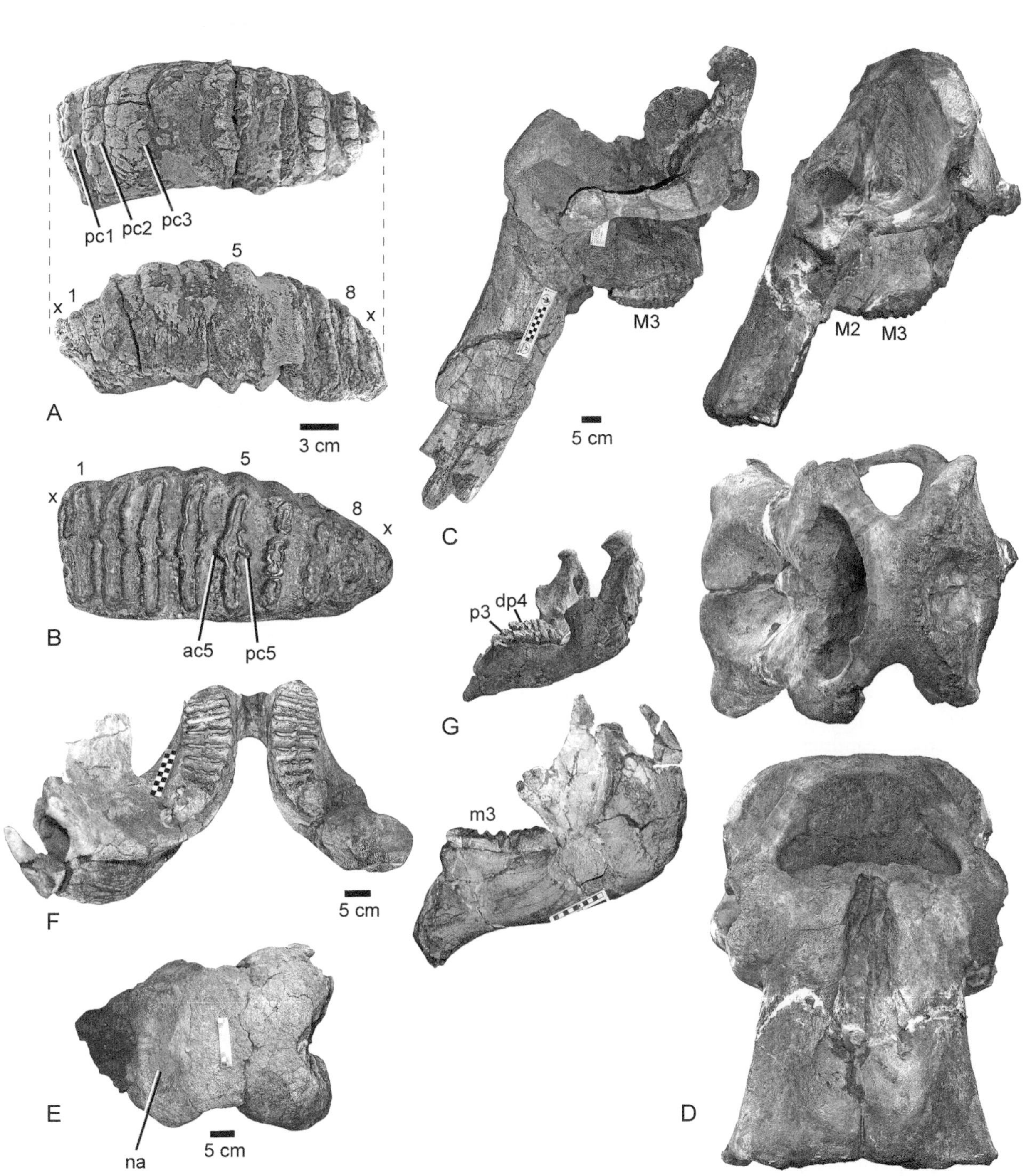

FIGURE 6.3 Skull and molar specimens of early Pliocene *Loxodonta adaurora* from Kanapoi (KP) and Ileret (ER), Kenya. Abbreviations: ac, anterior accessory conule; dp, lower deciduous premolar; M/m, upper/lower molar; na, nasal aperture; p, lower permanent premolar; pc, posterior accessory conule, x, pre- or postcingulum; 1, 2, 3, …, plate number counted from the anterior of the crown. (A, B) to the same scale; (C, D) to the same approximate scale; (F and G) to the same approximate scale. (A) Left M3, KNM-KP 383 from Kanapoi, Kenya, occlusal and lateral (reversed) views. Anterior to the left. (B) Left M3, KNM-KP 385 (type) from Kanapoi, Kenya, occlusal view. Anterior to the left. (C) Partial cranium with left and right M3, KNM-KP 385 (type) from Kanapoi, Kenya, lateral view. Anterior to the left. (D) Cranium with left and right M2-3, KNM-ER 63642 from Ileret, Kenya, lateral, dorsal, and anterior views. Anterior to the left in the lateral and dorsal views. (Scanned images courtesy of T. G. Ibui). (E) Infant or young juvenile partial cranium, KNM-KP 66240 from Kanapoi, Kenya, dorsal view. (F) Mandible with left and right m3, KNM-KP 385 (type) from Kanapoi, Kenya, posterodorsal and lateral views. Anterior to the top in the dorsal view and to the left in the lateral view. (G) Juvenile mandible with left and right dp4 and emerging p3 and m1, KNM-KP 30616 from Kanapoi, Kenya, oblique lateral view. Anterior to the left.

L. cookei by having a slightly greater number of plates, higher crowns, thinner enamel, a greater covering of cementum, and more U-shaped transverse valleys, with correspondingly more parallel-sided plates in lateral view. Molars of *Mammuthus subplanifrons* from Langebaanweg, South Africa, Middle Awash, Ethiopia, and Tugen Hills, Kenya, which best characterize the species, have slightly fewer, thicker, and more pyramidal-shaped plates (eight-nine plates in m3 as opposed to 10), are lower-crowned than molars of *L. adaurora*, and also have accessory conules limited to the posterior sides of plates at the anterior end of their crowns (Sanders, 2006, 2007).

Although Beden (1987a) felt that Kanapoi specimens KNM-KP 383 (Figure 6.3A), 390, and 401 were more appropriately placed in "*E.*" *ekorensis*, 383 and 390 are metrically and morphologically typical of *L. adaurora*—for example, the former comprises in part M3s with only eight plates, low LF of 3.25, and hypsodonty indices of 95–100, and the latter is an M3 with nine plates, a low LF of 4.0, and thick, unfolded enamel (ET=4.3–5.3). KNM-KP 401 is a deinothere molar!

The crania of the Kanapoi *L. adaurora*-type specimen (KNM-KP 385; Figure 6.3C), Lothagam specimen KNM-LT 353 and extant African elephants share features in common such as a broad, biconvex cranial roof, blunter demarcation between the frontoparietal area and temporal fossae than in *Elephas*, distal flaring of tusk alveolae and their separation by a deep, broad, tear-drop-shaped external incisive fossa, flat occipital planum angled forward, projecting occipital condyles, high position of the nuchal ligament fossa on the occiput, absence of parietal bosses and lack of inflation of the occipital planum (Maglio, 1970b, 1973; Grubb et al., 2000).

In addition, a partial cranium of *L. adaurora* from Koobi Fora, KNM-ER 346, importantly preserves the morphology of the posterior palatal region, showing that the arrangement of its constituent bones is shared with that in crania of extant African elephants but not of other elephants, e.g., the posterolateral extension of the pterygoid process of the palatine covers the posterior edge of M3 and is not posteriorly overlapped by the alisphenoid (Beden, 1983). This same arrangement is present in a partial cranium of *L. adaurora* from the Mursi Fm., Omo, Ethiopia, Y.S. 69 3213 (Beden, 1987a). These features readily distinguish *L. adaurora* from *Elephas*, *Palaeoloxodon*, and *Mammuthus* but are shared by the modern species of *Loxodonta*. Accordingly, cladistic analyses of elephants based on cranial characteristics recover *L. adaurora* and *L. africana* as a basal clade among elephantine species (Ferretti, 2016; Zhang, 2020).

Cranial differences of *L. adaurora* from skulls of living African elephants may in part be attributable to the greater massivity of body size of the fossil species. In *L. adaurora*, these include tusk sheaths that are more prominently vertically-oriented, more salient frontoparietal struts lateral to the nasal opening, deeper maxillae and maxillary fossae, more anteriorly-placed orbits, higher position of the nasal opening relative to the orbits, lower level of the greatest width of the occiput, a more elongate nuchal fossa, straighter, narrower basioccipital, flatter, narrower auditory bullae, and longer, more posteriorly and outwardly swept temporal fossae (Maglio 1970b, 1973; Maglio and Ricca, 1977; Beden, 1983). The occipital planum is broadest just above the level of the external auditory meati, the occipital condyles are relatively small, and the nasal opening is large, wide, and downturned laterally.

These morphological details are confirmed by the recent find of an extraordinarily complete cranium of a large male *L. adaurora* from the Lonyumun Mb. of the Koobi Fora Fm. at Ileret, Kenya, KNM-ER 63642 (Figure 6.3D), older geologically than the Kanapoi sample with the greatest confirmed antiquity of any fossils assigned to the species (Sanders et al., 2021; Table 6.1). This immense specimen has a greatest length from the nuchal crest to anteriormost point on the premaxillae of more than 1,360 mm and has an approximate bizygomatic width of 915 mm—remarkably, it is slightly *smaller* than the massive type cranium, though this may be accounted for by ontogenetic age differences, as the type individual was a mature bull. Heavily worn M2s and well-preserved emergent M3s are present in the cranium, consistent with Laws' (1966) age group XX for extant African elephants, fitting a mean of 34 ± 2 years old. Absence of nasal bones in this specimen shows that the species has a short frontal vault segment, as in extant African elephants, with the nasals evidently contributing strongly to the length of the forehead in the type cranium. The overall impression provided by this superb specimen is of an enormous, very broad, anteroposteriorly compressed (acrocephalic), raised cranium with a disproportionately long lower face enclosed by divergent tusk alveoli that are widely spaced around a deeply excavated external incisive fossa (Sanders et al., 2021).

The cranium of a young juvenile elephant from Kanapoi, KNM-KP 66240 (Figure 6.3E, is accessioned as *L. adaurora* in the National Museums of Kenya. It preserves the cranial vault on its dorsal aspect, but the dentition is broken away and most of the rostrum is missing. Nonetheless, enough remains to see that the external incisive fossa was broad and deep and the tusk alveoli were small, widely spaced, and not divergent proximally. The cranial roof is broad and convex both anteroposteriorly and transversely. The parietals are more swollen laterally than in adult crania. It is also relatively long from the nasal aperture to the nuchal margin. The nasal aperture is ovoid, modest in size, located slightly above the level of the orbits, and transversely straight without downturn of its lateral margins. As in the adult cranial form of *L. adaurora*, it has robust frontoparietal struts angled inward as they pass posteriorly, such that greatest cranial constriction occurs near the nuchal margin of the cranial roof. The nuchal planum is concave superiorly in the midline (accentuated by the nuchal ligament fossa), giving the nuchal crest a strongly curved appearance medially. If this specimen belongs to *L. adaurora*, it suggests that substantial reorganization of the cranium took place ontogenetically, involving lateral and downward migration of the orbits, pulling the lateral margins of the nasal aperture with them, and greater bilateral expansion than lengthening of the cranial roof.

The type mandible (KNM-KP 385; Figure 6.3F) has a long corpus that is highest anteriorly, with a robust, downturned symphyseal spout, a laterally everted ramus that is anteroposteriorly narrower at its base than at the level of the coronoid process and exhibits a strong angle of ascent from the coronoid process to the condyle, which is set high above the level of the coronoid process. Most of these features are shared by the exquisitely preserved mandible of juvenile mandible KNM-KP 30616 (Figure 6.3G), though at its young age its symphyseal spout is more lightly built. With dp4s in wear and m1s (and p3s) emergent, -KP 30616 is closest to age group VI individuals of young modern African elephants that are about four years old (Laws, 1966).

A nicely-preserved mandible with m1–2s in place and m3s in crypt from Kanapoi, KNM-KP 32542, exhibits some features consistent with an attribution to *Loxodonta*, including a long corpus, everted ramus, long mandibular symphysis, corpus widest posteriorly, a shallow masseteric fossa limited in area, and ramus much shorter at its base than at the level of the coronoid process. The specimen has been placed in *L. adaurora* (Harris et al., 2003; Sanders, 2020), but it differs from the type mandible in having a lower, less angled superior margin of the ramus between the coronoid process and condyle, a larger, blunter anterior margin of the coronoid process, a straighter inferior margin of the anterior corpus, and it seems unlikely to have had a symphysis of the same spout-like projection. Thus, the attribution of this mandible to the species is questionable. An Allia Bay mandible of *L. adaurora*, KNM-ER 346, is more similar to the type mandible in having a rounded coronoid process with a steeper ascent to the condyle (Beden, 1983); conversely, a demi-mandible with m1–2 from the Omo Mursi Fm. that is placed in *L. adaurora* (Y. 7.5) has a short, strong symphysis that projects horizontally, more like that of KNM-KP 32542 (Beden, 1987a). It is difficult to know how much significant age difference between the KNM-KP 385 and -KP 32542 contribute to their morphological distinctions; the plate emergence and wear pattern in the latter indicates it belongs in Laws' (1970) age group XIV, approximately 22 ± 2 years at death, more than two decades junior to the type individual. The m3s of KNM-KP 32542 remain largely hidden in their crypts and their definitive identification requires future CT-scanning of the mandible.

Body mass of the type individual, KNM-KP 385, was estimated as nine tons by Larramendi (2016), although he unaccountably assigned it to *Mammuthus subplanifrons*, without a single craniodental trait to support such an attribution. Correspondingly, the identification of *L. adaurora* as a mammoth in Todd's (2010a) elephant phylogeny is untenable, as there are no mammoth synapomorphies in the cranium, incisors, or molars of the species. Compared with *M. subplanifrons*, *L. adaurora* molars are higher-crowned, more parallel-sided, and less pyramidal in lateral view and develop accessory conules anterior and posterior to plates throughout the crown, rather than being limited to the posterior side of the first few plates (see Sanders, 2006, 2007, 2020).

Many *L. adaurora* postcranial features support placement in *Loxodonta*. Most anatomical features of the manus and pes elements resemble those of *L. africana*, although signs of carrying greater body weight are evident in the larger size, robustness, and broader articular contacts of these elements than in *L. africana* (Maglio and Ricca, 1977). In addition, while they are slightly heavier in construction, the femur and tibia of *L. adaurora* closely resemble those of the African savanna elephant in details such as low height of the femoral head relative to the greater trochanter, position of the femoral lesser and third trochanters, and the course of the anterior muscle ridge of the tibia for m. tibialis cranialis. The fibula of *L. adaurora* is much more robustly constructed distally than in living African elephants, although otherwise similar in morphology. The contrast of the tall, stout spinous processes of the anterior thoracic vertebrae (T2–5) that terminate in massive tuberosities, with the much shorter spinous processes of more posterior thoracic vertebrae (Maglio and Ricca, 1977), is consistent with the classic swaybacked profile characteristic of living African elephants. In the type individual, lengths of the humerus, ulna, and radius are reported to be 116.2 cm, 90.0 cm, and 98.5 cm, respectively (Maglio and Ricca, 1977), but this seems inaccurate as the ulna should not be shorter than the radius. If these ulnar and radial measurements are transposed, the ratio of ulnar to humeral and radius to humeral lengths in *L. adaurora* are 85 and 77, respectively, very similar to mean values for relative ulnar and radial lengths of 89 and 78 for *L. africana* (calculated from the database in Christiansen, 2007).

It has been claimed that limb proportions of *L. adaurora* support placement of the species in *Mammuthus subplanifrons* (Larramendi, pers. comm.). However, there are no postcranial remains known for *M. subplanifrons*, and the evidence from other mammoths is equivocal. For example, femur and tibial lengths in adult KNM-KP 397, attributed to *L. adaurora*, yield a relative tibial length ratio of 57 (Maglio and Ricca, 1977), which is slightly lower than a mean of 61 for *L. africana* (calculated from Christiansen, 2007) and identical to the ratio for *M. imperator*—but the ratio for *M. imperator* is affected strongly by a single outlier, and otherwise this mammoth species is identical to extant African elephants, as well as to mean ratios for *M. "meridionalis"* and *M. primigenius* of 59 and 60, respectively. Thus, there is no compelling evidence from limb proportions to support re-assignment of *L. adaurora* into *Mammuthus*. Moreover, skeletal proportions appear to be less reliable and more labile than cranial characteristics for taxonomic assessment in fossil elephants. *Loxodonta adaurora* appears to be a good constituent of the genus but the lack of loxodont sinuses in its molars suggests it is not on the main line of evolution to the living African elephants.

LOXODONTA EXOPTATA (Dietrich, 1941)

(Tables 6.1 and 6.2; Figures 6.1, 6.4, and 6.5)

Partial synonymy—*Palaeoloxodon antiquus recki* Hopwood, 1936; *Archidiskodon exoptatus* Dietrich, 1941, 1942; *Elephas antiquus recki* Hopwood in Kent, 1941; *Elephas* aff. *planifrons* Hopwood in Kent,

1941; *Archidiskodon planifrons nyanzae* (in part) MacInnes, 1942; *Elephas* aff. *planifrons* Arambourg, 1943; *E. recki* Arambourg, 1947; *E.* cf. *planifrons* Arambourg, 1947; *Archidiskodon subplanifrons* (in part) Cooke, 1960; *Elephas* (*Archidiskodon*) *exoptatus* (in part) Cooke, 1960; *E. recki* (in part) Leakey, 1965; *Loxodonta africanavus exoptata* Coppens, 1965; *E.* (*Archidiskodon*) *exoptatus* Arambourg, 1969b; *Loxodonta* sp. (in part) Maglio, 1969b; *E. recki* (in part) Maglio, 1969b; *Loxodonta* sp. "C" (in part) Maglio, 1970a; *E. recki* (in part) Maglio, 1970a; *L. adaurora* (in part) Maglio, 1970b; *Mammuthus* (*Archidiskodon*) *recki* (in part) Cooke and Coryndon, 1970; *L. adaurora* (in part) Maglio, 1973; *E. recki* (in part) Maglio, 1973; *L. adaurora* (in part) Coppens et al., 1978; *L. exoptata* Beden, 1987b; *L. exoptata* Sanders, 2011

Loxodonta exoptata, whose species nomen means "longed for," is best known from Laetoli and adjacent sites of Kakesio, Noiti, and Emboremony, Tanzania (Table 6.1; Sanders, 2011). Collections of primarily isolated molar, deciduous premolar, and tusk remains were initiated by Louis Leakey and colleagues in the East African Archaeological Expedition of 1934–1935, with more substantial additions by Kohl-Larsen's explorations of 1938–1939, Mary Leakey's team in the 1970s and early 1980s, and by the Eyasi Plateau Expedition led by Terry Harrison between 1998 and 2005. Most of the fossils derive from the Upper Laetolil Beds, radiometrically dated to between 3.85 and 3.62 Ma, but a few specimens are provenienced to the Lower Laetolil Beds (such as those from Kakesio) and Upper Ndolanya Beds, dated to 4.36–3.85 Ma and ~2.66 to 2.30 Ma, respectively (Deino, 2011).

Along with the occurrence of *L. exoptata*, *L. cookei*, *Deinotherium bozasi*, *Anancus kenyensis*, *A. ultimus*, and a rare find of *Stegodon kaisensis*, the Upper and Lower Laetolil Beds and Upper Ndolanya Beds at Laetoli and environs have produced evidence of a diversity of vertebrates, including amphibians, snakes, crocs and tortoises, at least 11 families of birds, elephant shrews, aardvarks, eight rodent families, hares, shrews, six families of carnivores, equids, chalicotheres, multiple species of rhinos, suids, giraffes, a wealth of bovid taxa, camels, galagos, several species of cercopithecid monkeys, and notably the hominin species *Australopithecus afarensis* and *Paranthropus aethiopicus* (Harrison, 2011; Su and Harrison, 2015). Laetoli is the type site of *A. afarensis* (Johanson et al., 1978) and is renowned for the preservation of hominin trackways showing bipedal progression, as well as tracks of other mammals (Leakey, 1987; Robbins, 1987; White and Suwa, 1987).

These beds are unusual in that they primarily entomb fossils in aeolian and tuffaceous deposits. Paleoecological reconstructions of the Upper Laetolil Beds indicate that the early Pliocene environment at the site was a mosaic of woodlands, shrub-, bush-, and grasslands, as well as forest, crossed by seasonal watercourses and proximate to springs (Su and Harrison, 2015). The absence of large rivers or lakes may have seasonally compromised the water supply at Pliocene-aged Laetoli. Isotopic study of dental enamel and eggshell suggests that, particularly during the interval of the Upper Laetolil Beds, Laetoli had low rainfall regimes (Kingston, 2011). Because *L. exoptata* is common at Laetoli but rare elsewhere, it is tempting to hypothesize that it may have been a specialist capable of dealing with extreme conditions. However, carbon and oxygen isotope analyses of *L. expotata* teeth from Laetoli show that these elephants were acting as dietary generalists, opportunistically feeding on C_3 and C_4 vegetation (albeit with a strong component of C_4 grasses) from a mosaic of habitats; in addition, they were obligate drinkers (Kingston, 2011). From this perspective, it is doubtful that *L. exoptata* possessed special adaptations to survive water-stressed ecosystems, but perhaps was forced into surviving as an eclectic feeder by local ecological circumstances or was relegated by competition with other large mammals to sub-optimal habitats.

The handful of elephant fossils collected from the "Vogel River" (Laetolil) beds by the initial Leakey expedition were described by Hopwood (1936), which he initially attributed to a single species, *Palaeoloxodon antiquus recki*. Soon thereafter, he added recognition of a second species in the sample, *Elephas* aff. *planifrons* (Hopwood in Kent, 1941). The first researcher to use the current species name was Dietrich, in 1941, who placed all of the elephant fossils into a single variable species, *Archidiskodon exoptatus*. Despite the subsequent apparent recognition of three species in the Kohl-Larsen sample (*Archidiskodon exoptatus*, *Palaeoloxodon antiquus*, and *Loxodonta africana*), in a more comprehensive description of the elephant fossils from the Kohl-Larsen expedition, Dietrich (1942) continued to assign the sample to a single species, *Archidiskodon exoptatus*. It would not be until much later that the species was recognized as *Loxodonta* (Coppens, 1965). The contrasting interpretations of Hopwood and Dietrich laid the foundation for a complex taxonomic history for the species and a more contemporary debate about the composition of the larger site sample.

Kohl-Larsen's collection was composed of 108 cheek tooth specimens, many of which are very worn or fragmentary; of these, Dietrich (1941) identified 12 M1/9 m1; 5 M2/12 m2; and 4 M3/24 m3, supplemented by deciduous premolar specimens. In a prelude to his taxonomic reorganization of the Elephantidae in 1973, Maglio (1969b) re-examined the Kohl-Larsen elephant molar collection, along with the specimens from the first Leakey expedition, and distinguished two species based on morphological differences and preservation appearances, which he placed into a group "A" and group "B." Group A, provenienced to higher beds at the site, was characterized by molars with thin, folded enamel, hypsodonty indices ranging from 120 to 200, third molars with 11–13 plates, and weak or absent development of anterior accessory conules ("anterior sinuses") in enamel wear figures. These specimens were placed in his "*Elephas recki* stage II" (Maglio, 1969b, 1973). In addition, because Dietrich had not designated a type specimen for "*A. exoptatus*," Maglio (1969b) selected an m3 (IPUB no. Z. 94–96) as the lectotype and subsumed it into *E. recki* as a junior synonym. Group B, provenienced to lower beds at the site, was denoted by molars with thicker, smooth enamel, hypsodonty indices ranging from 85 to 110, third molars with 9–11 plates, and development of prominent anterior sinuses in enamel wear figures. These molars Maglio (1969b) initially placed in *Loxodonta* sp. and later incorporated into *Loxodonta adaurora* (Maglio, 1973).

Although several images provided by Maglio (1969b:pl. III, fig. 7 and pl. IV, fig. 16) reveal strong contrasts in transverse cross-sectional profiles, as do several molar cross-sections illustrated by Dietrich (1942: pl. XI, figs. 73, 74), none of the molars measured in Maglio's study are complete, and therefore the plate numbers assigned to molars in each group are conjectural. Moreover, enamel wear figures of some specimens assigned to *E. recki* exhibit salient anterior and posterior midline projections resembling the morphology of molars assigned to Group B (for example, Maglio, 1969b: pl. II, fig. 6).

Addition of significant numbers of fossils by the Mary Leakey expeditions of the 1970s and 1980s led to a broader study of the entire Laetoli elephant collection, including the new material, by Beden (1987b). His meticulous analysis led to the conclusion that the entire sample from the Laetolil Beds comprised a morphologically homogeneous assemblage of molars from a single species, *Loxodonta exoptata*. Most of the relevant specimens derive from the Upper Laetolil Beds. It was Beden's hypothesis that *L. exoptata* was not part of the *Loxodonta adaurora* lineage but instead represented an ancestor of the extant African elephants.

Study of the elephant material collected by the Eyasi Plateau Expedition along with the fossils of the Leakey expeditions (but not directly the specimens from the Kohl-Larsen expedition as they were inaccessible due to construction on the Humboldt Museum in Berlin) was undertaken by Sanders (2011). He made a distinction between more primitive elephant material from the Lower Laetolil Beds at Endolele (Esere 1), referred to *Loxodonta cookei*, and specimens from other localities in the Lower Laetolil Beds and the Upper Laetolil and Ndolanya Beds. These latter specimens were placed in a single species, *Loxodonta exoptata*, in agreement with the view of Beden (1987b). Thus, no single author has studied firsthand the entirety of the Laetoli elephant fossil assemblage, and there are starkly alternative hypotheses about the taxonomic composition of these fossils.

Diagnostic features provided by Sanders (2011) include slightly hypsodont molars with moderately thick enamel, plates spaced more closely than in archaic elephants, and contribution of anterior and posterior accessory conules to the formation of loxodont median sinuses throughout the length of molar crowns. This latter feature is critical for consideration of the relationship of the species; it is important to note that the main criterion used for creating the species *exoptatus* was the similarity of worn occlusal enamel figures in the anterior molars to those in *Loxodonta africana* (Dietrich, 1941). It is critical to observe that because accessory conules in molars of *L. exoptata* are lower than conelets in their associated plates, and have their greatest circumference at mid-height, loxodont sinuses may not be apparent or may form "propeller" shapes with the center of enamel loops rounded and the lateral segments of the loops anteroposteriorly compressed in light occlusal wear (Figure 6.4A–D; Beden, 1987a; Kalb and Mebrate, 1993; Sanders, 1997, 2011). In medium occlusal wear, loxodont sinuses form (Figure 6.4E), whereas in heavy wear, as accessory conules taper at their bases, enamel wear figures are transversely rectilinear and simple, lacking median sinuses (see Figure 6.4F; Sanders, 2011:fig. 9.9).

Loxodonta exoptata is distinguished from *L. cookei* by molars with more plates, greater hypsodonty, tighter anteroposterior spacing of plates, and with the widest transverse points higher on the crown, and is differentiated from *L. adaurora adaurora* by molars with more and narrower plates, more pronounced development of median sinuses, and less robust plate construction. Additionally, *L. exoptata* differs from these species by apparent loss of permanent premolars.

Plate formulae for the species are as follows: dP2/dp2, x3x-x4/x3x-x4x; dP3/dp3, x5x-6x/x6x; dP4/dp4,?x6-x7x/?7; M1/m1, --/x7x-x8 (?9); M2/m2, 8x-9x/8–9x; M3-m3, 11–12/11–12 (Beden, 1983, 1987b; Harris et al., 2003; Sanders et al., 2010a; Sanders, 2011). There is no record of recovery of permanent premolars for this species, despite the collection of numerous, similar-sized deciduous premolars from the fossil beds at Laetoli. In fact, the Laetoli elephant sample is dominated by deciduous premolars, suggesting a disproportionate number of calves and juveniles in the mortality profile of *L. exoptata* unless the proportions are taphonomic artifacts of differential preservation in open conditions prior to being covered by ash (Sanders, 2011). Deciduous premolars comprise 65% (138/213) of identifiable cheek tooth specimens (of the entire sample, dP2/dp2 comprise 14%; dP3/dp3 29%; dP4/dp4 22%; M1/m1 11%, M2/m2 13%, and M3/m3 11%) (Sanders, 2011). The small percentage of second and third molars show mature and senescent adults to be underrepresented in the Laetoli elephant assemblage. The greater number of deciduous third and fourth premolars in the sample evokes a scenario in which young, weaned elephants in drought conditions could not reach far enough with their trunks into water holes dug by older elephants to obtain water. Among African elephants, individuals are weaned by the age of 4–5 years (Moss, 1992), by which time they have shed their deciduous second premolars and their deciduous fourth premolars are completely erupted (Sikes, 1967). Nursing calves hydrate from milk and presumably have greater protection from arid conditions than older juveniles (Dudley et al., 2001; Foley et al., 2008; Woolley et al., 2008). Age-grade mortality of Laetoli elephants is indicative of attritional mortality due to repeated drought conditions (Sanders, 2011). This could have been exacerbated by carnivoran predation, which is documented in extant African elephants to increase during times of drought (Loveridge et al., 2006).

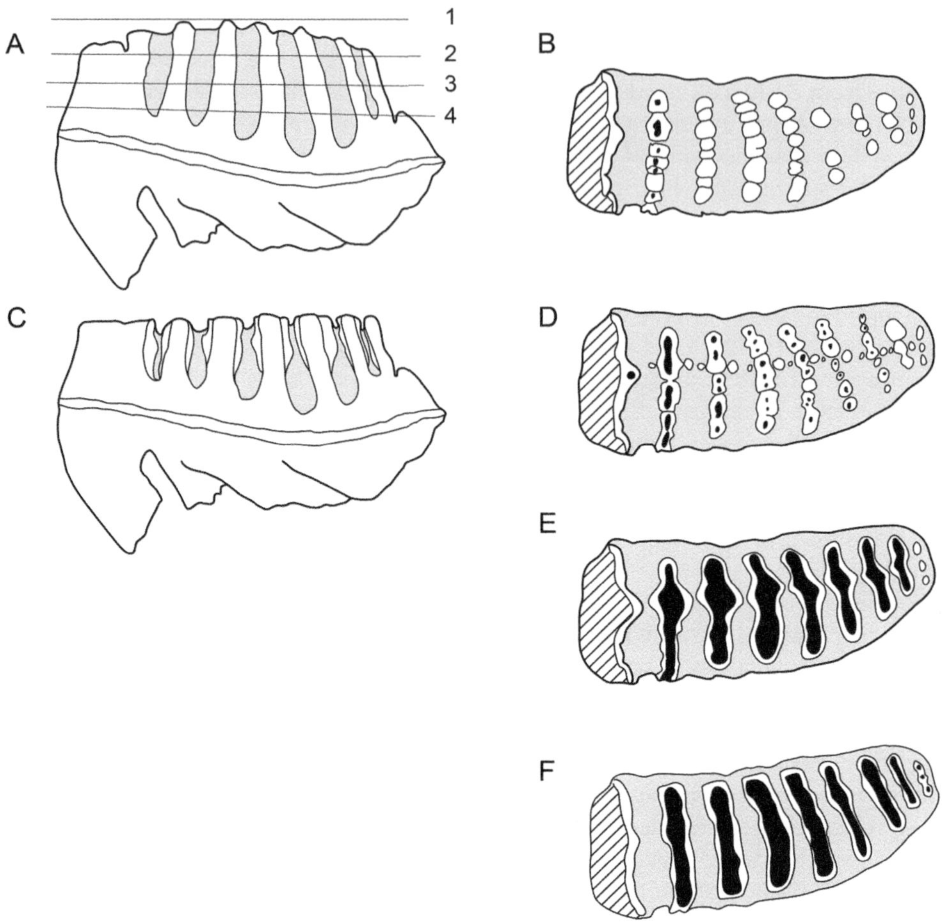

FIGURE 6.4 Diagram contrasting variation in early to mid-Pliocene *Loxodonta exoptata* molar occlusal patterns at different stages of wear. Based on right M2 specimen EP 519/98 from the Upper Laetolil Beds at Laetoli, Tanzania. Anterior to the left. (A) Lateral view, lightly worn molar with lines indicating horizontal planes through the tooth at different levels; 1 = light wear, 2 = moderate wear, 3 = mid-height of molar crown, 4 = heavy wear. (B) Occlusal view, lightly worn molar equivalent to plane 1. Note that individual conelets are visible but accessory conules are not. (C) Lateral view, moderately worn molar equivalent to plane 2. (D) Occlusal view, moderately worn molar equivalent to plane 2. Note the appearance of anterior and posterior accessory conule apices associated with each plate. (E) Occlusal view, molar crown worn to mid-height, equivalent to plane 3. Note the incorporation of accessory conules into enamel wear figures and the resulting development of midline loxodont sinuses. (F) Occlusal view, heavily worn molar equivalent to plane 4. Note that this plane is below the level of the base of accessory conules and the consequent rectilinear shape of enamel wear figures.

No skulls are known for *L. exoptata*, but a partial mandible (OMO 20.1.67.919) from Mb. A of the Ethiopian Omo Shungura Fm. preserves a right dentary, anteriorly broken symphysis, and partial m3. Member A is radiometrically constrained between 3.60 and 3.44 Ma (Levin et al., 2011). Two other partial third molars of *L. exoptata* are reported from Mb. A of the Shungura Fm. (Beden, 1987a). The m3 associated with the jaw exhibits nascent loxodont sinuses with the incorporation of anterior and posterior accessory conules into enamel wear figures, and slightly undulated enamel. The most striking feature of the mandible is the prominent anterior elongation of the symphysis, even with its broken tip. This spout is slightly downturned but reveals no sign of lower tusk alveoli. There is a small mental foramen at mid-height at the base of the symphysis, anterior to a larger mental foramen situated at mid-height on the corpus just anterior to the m3.

Several third molars from the Upper Laetolil Beds are complete enough to warrant description (Sanders, 2011). M3 specimen EP 1618/00 collected by the Eyasi Plateau Expedition preserves a plate formula of +12x and has dimensions of L = 290.0 mm, W = 97.5 mm, H = 140.0 mm (yielding HI = 144), and ET = 3.5 mm (Figure 6.5A). M3 specimen LAET 80–6403 collected by Mary Leakey's team has a plate formula of 11 and dimensions of L = 260.0 mm, W = 82.0 mm, and ET = 3.0–3.5 mm (the crown is too worn to calculate height). An m3, LAET 76–4558 collected by Mary Leakey's team has a plate formula of 11x (13 in Beden, 1987b) and dimensions of L = 250.0 mm, W = 93.2 mm, H = 107 mm (yielding HI = 107), and ET = 3.0–4.0. Cross-sectional plate shape can be observed in broken molars and isolated plates (e.g., specimen 78–4927) and is approximately parallel-sided with the greatest width about one-third

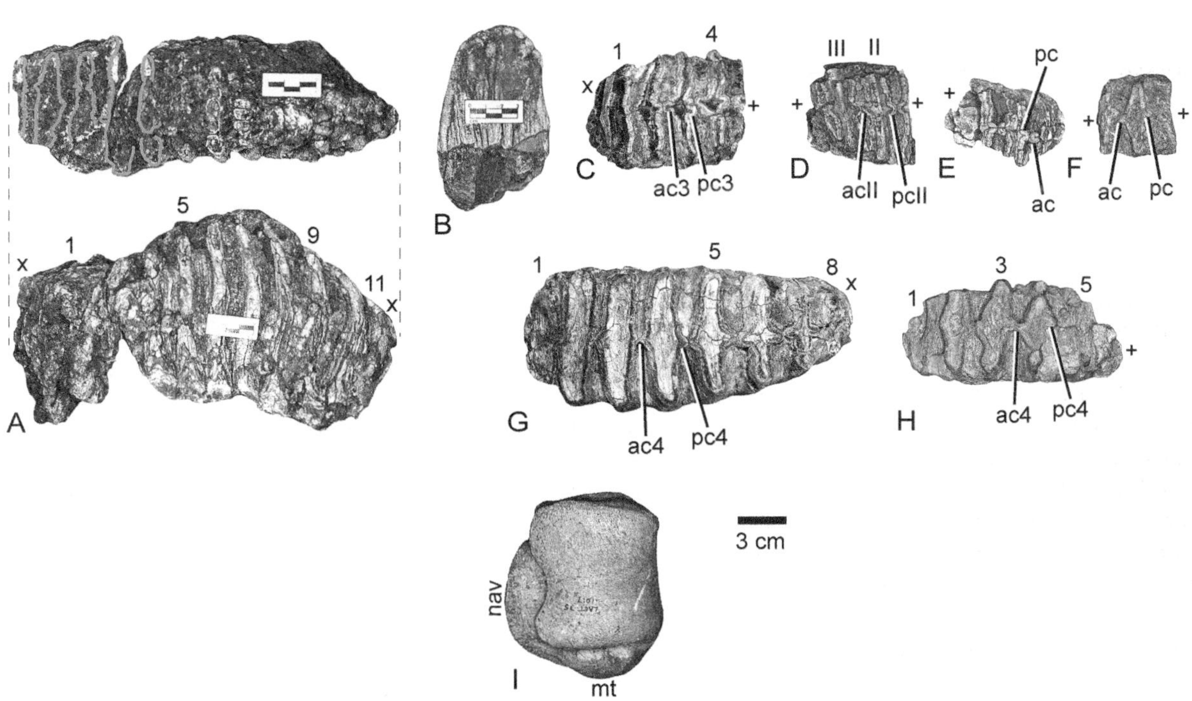

FIGURE 6.5 Molar and postcranial fossil specimens of early to mid-Pliocene *Loxodonta exoptata*. Abbreviations: ac, anterior accessory conule; mt, median tubercle; nav, navicular articular surface; pc, posterior accessory conule; x, pre- or postcingulum(id); 1, 2, 3, …, plates counted from the anterior of the crown; I, II, III, …, plates counted from the posterior of the crown; +, indicates missing morphology; (A, C–I) anterior to the left. All specimens to the same scale. (A) Right M3, EP 1618/00 from the Upper Laetolil Beds, Laetoli, Tanzania, occlusal and lateral views. In lateral view, note the parallel-sided plates and thick infilling of cementum in U-shaped transverse valleys. (B) Molar plate, 78-4927 from the Upper Laetolil Beds, Laetoli, Tanzania, Anterior view. (C) Partial M2, 75-521 from the Upper Laetolil Beds, Laetoli, Tanzania, occlusal view. (D) Partial m2, LOL AS 12 VI 35 from Laetoli, Tanzania, occlusal view. (E) Partial m1, EP 560 from the Upper Laetolil Beds, Laetoli, Tanzania, occlusal view. (F) Partial m1, 75-488 from the Upper Laetolil Beds, Laetoli, Tanzania, occlusal view. (G) Right M2, KNM-WT 16458 from West Turkana, Kenya, occlusal view. (H) Partial M2, 75-3818 from the Upper Ndolanya Beds, Laetoli, Tanzania, occlusal view. (I) Right astragalus, 75-1017 from the Upper Ndolanya Beds, Laetoli, Tanzania, dorsal view.

the height from the cervix (Figure 6.5B). Although hypsodonty indices and enamel thicknesses overlap with those of *L. adaurora kararae*, the greater number of plates and tendency to develop loxodont sinuses with wear in molars of *L. exoptata* make it clear that these are distinct taxa. The expression of anteroposteriorly projecting midline sinuses in the wear figures of *L. exoptata* (Figure 6.5C–H) resembles the morphology of *L. cookei* molars and anticipates the shape of enamel wear figures in extant African elephants, possibly evidencing a segment of a lineage connecting archaic and modern loxodont elephants.

Postcrania is scant for *L. exoptata* and includes isolated elements, primarily dense, compact podials that survive weathering well. These include a right astragalus (LAET-75-1017; Figure 6.5I), right navicular (LAET-76-18-263), right ectocuneiform (LAET-75-1033), left unciform (LAET-76-18-308), distal radius (LAET-76-18-177), manual sesamoid (EP 4023/00), thoracic vertebral body (EP 3031/00), fragmentary magnum (EP 1227/00), sacrum (EP 1317/98), right scaphoid (EP 1318/98), left trapezoid (EP 1316/98), proximal phalanx (LAET 75-1250), right patella (EP 1422/04), left patella (EP402/04), left metatarsal IV (LAET 78-4909), and left metacarpal II (EP 210/03). Comparative study of the fossil postcranial sample indicates greater similarity to postcrania of *L. africana* than those of other extant or recent elephants (Beden, 1987b).

MAMMUTHUS Brookes, 1828
MAMMUTHUS SP. "HADAR TYPE"

(Table 6.1; Figures 6.1 and 6.6)

Synonymy—*Mammuthus* sp. "Hadar-type" White et al., 1984; cf. *Mammuthus* sp. "Hadar-type" Sanders and Haile-Selassie, 2012

The presence of a mammoth morphometrically intermediate between *M. subplanifrons* and *M. africanavus* was established by field identification of skulls, tusks, and partial skeletons in Sidi Hakoma and Denen Dora member sediments of the Hadar Fm. in Ethiopia (Beden, 1979, 1985; White et al., 1984; Kalb and Froehlich, 1995). They are similar morphometrically to contemporaneous *M. rumanus* of Europe, the first mammoth out of Africa (Lister and van Essen, 2003; Markov and Spassov, 2003;

Radulescu et al., 2003; Lister et al., 2005; Markov, 2012), and probably provided the phylogenetic source for that migration event in the mid-Pliocene. It is possible that the earliest European mammoths and these African intermediates are conspecific.

The Hadar specimens attributed to *M.* sp. "Hadar-type" have never been described, but several molars from Woranso-Mille, Ethiopia from the early Pliocene interval 3.8–3.6 Ma (Table 6.1) were tentatively assigned to this taxon and discussed in detail in Sanders and Haile-Selassie (2012). These are associated right and left M3s (MSD-VP-3/039). The left M3 is complete (Figure 6.6A), with a plate formula of x9x and dimensions of L=237.8mm, W=89.5mm, H=82.0mm (yielding HI=91), and ET=4.0mm (Sanders and Haile-Selassie, 2012). Lamellar frequency is 4.0. In occlusal view, these M3s are only very slightly internally concave longitudinally; in lateral view, they are very convex. Cementum partially covers the plates and infills the transverse valleys. In lateral view, plates are parallel to one another and accordingly transverse valleys are U-shaped. Distally, the crown tapers markedly. In cross-section, the plates are straight on one side and angled upward and inward on the other side, and broadest near the base (Figure 6.6B). There is a large central accessory conule posterior to plate 1, but cementum obscures the possible presence of other accessory conules. Although the Woranso-Mille molars are similar in hypsodonty and plate number to M3s of *L. adaurora*, the distal taper of the crown, cross-sectional plate shape, narrower breadth, and less robust build of these molars are distinctive. They are distinguished from *M. subplanifrons* molars by higher lamellar frequency, greater hypsodonty, and thinner enamel, and from molars of *M. africanavus* by lower crown height, thicker enamel, more widely spaced plates, and plates less parallel-sided in cross-sectional view (Arambourg, 1970; Maglio, 1973; Sanders et al., 2010a).

At Woranso-Mille, *M.* sp. "Hadar-type" co-occurs with a variety of other proboscideans, including deinotheres, anancine gomphotheres, "*Elephas*" *ekorensis*, *E. brumpti*, and possibly *L. adaurora*, alongside an extremely diverse vertebrate fauna comprised of fish, crocs, turtles, birds, rodents, lagomorphs, insectivores, aardvarks, equids, rhinos, several lineages of giraffes, multiple subfamilies of suids, at least seven bovid tribes, felids, canids, herpestids, viverrids, hyaenas, galagos, cercopithecoid monkeys, and australopithecine hominins (Haile-Selassie et al., 2010). The sediments are fluviatile and paleoecological reconstruction is of a mosaic of habitats representing a riverine gallery forest extended laterally to closed and open woodland and grassland (Haile-Selassie et al., 2010), which could support both deinotheres and elephants. Dietary habits of these mammoths have not yet been investigated, and too little fossil material has been recovered to say much about the nature of their phylogenetic relationship to geologically older or younger mammoth species, except that their molars are morphometrically intermediate between those of *M. subplanifrons* and *M. africanavus*.

ELEPHAS Linnaeus, 1758

The first formally recognized elephant taxa were *Elephas* and its type species *E. maximus* (Linnaeus, 1758), the extant Asian elephant. However, Linnaeus did not distinguish

FIGURE 6.6 Fossil molar specimens of early Pliocene *Mammuthus* sp. "Hadar-type" from Woranso-Mille, Ethiopia. Abbreviations: pc, posterior accessory conule; x, pre- or postcingulum; 1, 2, 3, …, plate number counted from the anterior of the crown. Molars to the same approximate scale. (A) Left M3, MSD-VP-3/039, occlusal and lateral views. Anterior to the left. (B) Anterior of partial right M3, MSD-VP-3/039, transverse cross-sectional view. Antimere of the left M3.

between the Asian and African elephants, and one of the syntypes of his new species, a foetal specimen, was shown by DNA analysis to be an African elephant (a skeleton housed in Florence, Italy has been designated as the lectotype of *E. maximus*) (Cappellini et al., 2014). With the radiation of species in *Palaeoloxodon* raised from subgeneric to full generic status, relegation of African *E. nawataensis*, *E. celebensis*, and *E. planifrons* to other genera, and the dissolution of the "*E. recki* complex" (see Chapter 7; Markov and Saegusa, 2008; Sanders et al., 2010a; Larramendi et al., 2017, 2020; Meyer et al., 2017; Palkopoulou et al., 2018; Zhang, 2020), the proposed fossil composition of *Elephas* is now considerably reduced from Maglio's (1973) conception of the genus in his revision of the Elephantidae. Nonetheless, the paleontological evidence points to Africa for the geographic origin of *Elephas* (Maglio, 1973; Sanders et al., 2010a), and results of genomic studies indicate the timing of the origin of the genus to be late Miocene, around 6.7 Ma (e.g., Rohland et al., 2007; Murata et al., 2009). Aside from the early record of the genus represented in Africa, possibly by "*Elephas*" *ekorensis* and by the older part of the "*E. recki* complex" (see Chapter 7), notable other members of the genus include Pleistocene *E. hysudricus* and late Pleistocene *E. hysudrindicus*, which are closely related and possibly ancestral to *E. maximus* (Maglio, 1973; Zhang, 2020). The genus dispersed from Africa to the Middle East and Asia (including South Asia) in the Pliocene or early Pleistocene (Lindsay et al., 1980; Van der Made, 2010), but vanished from the continent during the Pleistocene after a lengthy episode of ecological dominance (see Manthi et al., 2019). *Elephas hysudricus* apparently persisted in or reentered the Middle East to co-occur with *Palaeoloxodon recki* during the late middle Pleistocene (at Shishan Marsh, Jordan [Pokines et al., 2018]).

Elephas is now limited to a single surviving species (*E. maximus*) that is widely distributed across Asia and South Asia (Shoshani and Eisenberg, 1982). However, its range is very fragmented and populations of the species are declining precipitously (Shoshani and Eisenberg, 1982; Sukumar and Santiapillai, 1996; Fleischer et al., 2001; Blake and Hedges, 2004). The species has been on the IUCN Red List as endangered since 1986, because of loss of habitat, range fragmentation, and poaching, and in 2019 the overall census of wild *E. maximus* was estimated between 48,000 and 51,000 individuals (Williams et al., 2020). *Elephas maximus* is divisible into three subspecies (Shoshani and Eisenberg, 1982; Sukumar and Santiapillai, 1996), with evidence of genetic variation suggesting organization into two major clades that experienced extensive gene flow between them in the past (Fernando et al., 2000; Fleischer et al., 2001; Vidya et al., 2005).

Skull morphology of extant *E. maximus* provides information useful for identification of fossil congenerics. The cranium is high, anteroposteriorly compressed, with a flat or concave forehead, sharp delineation of the vault from the temporal fossae, and distinct occipitoparietal "bosses" separated by a median depression in the nuchal crest; premaxillaries are parallel, closely spaced, downward pointing, and separated by an elongated, narrow triangular external incisive fossa; the widest point of the cranium is across the occipitoparietal region; the junction of the premaxillaries/maxillaries is U-shaped and not notched; the orbital fossae are small and located well below the nasal aperture; the nuchal fossa is located lower on the occipital planum than in African elephants; in posterior view, the occipital planum is laterally angled rather than rounded (Maglio, 1973; Coppens et al., 1978; Todd, 2010b). Mandibular corpora are relatively short, broad, and very convex laterally; mandibular rami are anteroposteriorly very broad, marked by extensive masseteric fossae, and have coronoid processes that are longitudinally curved and angle inwardly; the condyles face forward and are angled inwardly; the symphysis is very short and angles downward (Todd, 2010b). Dentally, derived species of *Elephas* have very hypsodont molars with numerous plates covered by abundant cementum, thin, plicated enamel, and accessory conules that either form larger folds in occlusal enamel wear figures or are absent (though they may be prominent in some fossil forms, such as "*E. recki shungurensis*").

"ELEPHAS" EKORENSIS Maglio, 1970b

(Tables 6.1 and 6.2; Figures 6.1 and 6.7)

Partial Synonymy—*Elephas africanavus* Arambourg et al., 1969; primitive *Elephas recki* Howell et al., 1969; Ekora species "D" Maglio, 1970a:331; *Loxodonta adaurora* Coppens and Howell, 1974:2275; *Elephas recki* Coppens and Howell, 1974:2275; *Elephas recki* stage I Beden, 1975:704.

"*Elephas*" *ekorensis* was named for its type site by Maglio in his momentous 1970b publication that recognized two novel genera and four new species of elephant. The type specimen, right and left M3s of the same individual from Ekora, Kenya (KNM-EK 424), and additional fossils from Ekora and the nearby site of Kanapoi, comprise the original species hypodigm. Although the fossiliferous sediments at Ekora occur above a basalt dated at Kanapoi to ca. 4.1 Ma (Brown and McDougall, 2011) and the bulk of fossils recovered from Kanapoi are dated between 4.2 and 4.1 Ma from horizons below this tuff (see above), the faunal remains from these sites have been considered essentially contemporaneous (Patterson et al., 1970; Behrensmeyer, 1976). However, "*E.*" *ekorensis* specimens from Kanapoi described in the type paper by Maglio (1970b) derive from the top of the Kanapoi beds and may be somewhat younger than most of the type series of *L. adaurora*. The sediments at Ekora consist of coarse clastics along with fluvial and lacustrine deposits; these depositional environments are attested to by the site occurrence of fish, turtles, and crocodilians (Behrensmeyer, 1976). Fossils from Ekora mentioned in the original description of the species (Maglio, 1970b) were collected nearby a junction of the Kerio and Kalabata Rivers by a Museum of Comparative Zoology (Harvard University) expedition to the area in 1967 led by Bryan Patterson. In addition to the aquatic taxa, the Ekora fauna also includes rhinos, suids, a gomphothere, and *Loxodonta adaurora* (Patterson et al., 1970). The diversity of vertebrates and mix of closed- and open-habitat faunal elements with aquatic animals similarly characterizes habitat heterogeny at Allia Bay, Kenya around 3.9 Ma, where "*E.*" *ekorensis* also occurs (Coffing et al., 1994; Schoeninger et al., 2003).

Dentally, "*E.*" *ekorensis* was originally distinguished from contemporaneous *L. adaurora* by a greater number of plates in M3 (11 versus between 8 and 10), crown height of molars modestly greater than width yielding hypsodonty indices of 100–115 (Maglio, 1973) or 100–125 (Maglio, 1970b), weak folding of enamel loops near the bases of plates, marked narrowing of M3s posteriorly, and slightly thinner, more closely spaced molar plates (Figure 6.7A; Maglio, 1970b, 1973; Maglio and Ricca, 1977; Beden, 1987a). Mackaye (2001) reported crown heights as great as 106 mm and hypsodonty indices reaching HI = 147 in a small sample of molars from Koro-Toro, Chad that he assigned to the species. Beden (1987a) added that in the molars of "*E.*" *ekorensis* plates have strong anterior and posterior median accessory conules that are cylindrical throughout their height and may be apically free, cross-sectional plate profiles are not as bulbous as those of *L. adaurora* because the lateral conelets are more powerfully expressed and greatest width is located at one-quarter to one-third height of the crown (Figure 6.7B), plates are composed of few conelets, greatest width is at the base of the crown, width of third molar crowns reduces steadily posterior to plate 4, cementum is abundant and covers unworn molar crowns, U-shaped transverse valleys are anteroposteriorly broader than the plates themselves (Figure 6.7A), and enamel is thick and folded only grossly or irregularly in very worn plates. Based on teeth assigned to the species, primarily from the Usno Fm. of the Omo, Ethiopia, which is older than 3.1 Ma and extends back at least to 4.18 Ma (Kidane et al., 2004), early Pliocene Kanapoi and Ekora, Kenya (Maglio, 1973; Harris et al., 2003; Sanders, 2020), and the Kaiso Fm., Uganda (Hopwood, 1926, Maglio, 1973), plate formulae for "*E.*" *ekorensis* appear to be dP2/dp2, x3/--; dP3/dp3, 6–6x/x5x; dP4/dp4, 8x/8x or 9; M1/m1, x7x/--; M2/m2, 9-x9/x9; M3/m3, x11x/x11x–x12x. There are no known permanent premolars or lower tusks of this species (Table 6.2). Plate counts in second and third molars are consistently slightly higher than those in *L. adaurora* (Sanders, 2020). In third molars, lamellar frequency is low, between about 3.2 and 4.8, and enamel is moderately thick, ranging from 3.2 to 4.2. Anterior and posterior faces of plates are parallel (Beden, 1987a). "*Elephas*" *ekorensis* was not a small elephant; the largest of the type M3s measures over 300 mm in length, and is nearly 100 mm in width, with a crown height of 118 mm (Sanders, 2020: table 2).

It is worth noting that information about plate number and hypsodonty in m3 of "*E.*" *ekorensis* was originally based on specimens from Kwenichum in the Tugen Hills, Kenya and the Kaiso Fm. in Uganda (Hopwood, 1926; Maglio, 1973; Maglio and Ricca, 1977), and not from sites that produced the original hypodigm of the species. While there is little reason to think that a species with 11 plates in M3 would not have at least 12 plates in its m3s, the mixed-site assemblage of specimens comprising "*E.*" *ekorensis* has some potential problems. For example, the m3s from the Kaiso Fm. (M 12639), with 12 plates on the left side and strong posterior accessory conules exposed by sectioning on the right side (Figure 6.7C, D), could instead belong to *E. brumpti* (formerly the earliest stage of the "*E. recki* complex" [Sanders, 2020]). A left dentary and symphysis with molars from Kanapoi, Kenya (KNM-KP 30471) has an m3 with plate formula of x11x and dimensions of L = 284.7 mm, W = 88.2 mm, and H = 95.8 mm (yielding a HI = 108), with LF = 4.35 (Figure 6.7E; Sanders, 2020). The symphysis is notable for its long, pointed spout (lacking lower tusks) (Figure 6.7F).

There are no derived features in the dentition of "*E.*" *ekorensis* to indicate that it belongs in *Elephas*. Molars of the species, if incomplete and worn, are difficult to differentiate from those of *L. adaurora* (Sanders, 2020). The weight of the species assignment to *Elephas* relies on features of an incomplete cranium from Ekora, KNM-EK 422, missing the lower extent of the rostrum, tusks, and cheek teeth. In the type paper, the cranium was described as having parietals that are expanded laterally and dorsally (small parieto-occipital bosses) with a mid-sagittal depression between them, strong frontoparietal flanges that form a sharp angle between the roof of the cranium and temporal fossae, and tusk alveolae widely separated but parallel to one another (Maglio, 1970b). These features, except for the width of the external incisive fossa, are characteristic of extant *Elephas maximus* (Todd, 2010b), and according to Beden (1987a) are also shared by "*Elephas recki*" and *E. planifrons*, which led him to suggest that "*E.*" *ekorensis* is ancestral to those taxa. "*Elephas*" *ekorensis* is also differentiated from *L. adaurora* by this roster of features.

Other aspects of cranium KNM-EK 422 include a cranial roof wider than long, very large nasal aperture not downturned laterally, dorsal border of the orbit projecting causing the orbital rim to face downward, circum-orbital torus robust and with a ventral process lateral to the infraorbital canal, deep, wide external incisive fossa, flaring zygomatic arches, greatest width of the nuchal planum above the ear region, dorsal parieto-frontal region flattened and not concave (because the parieto-occipital bosses were not inflated anteriorly), posteriorly inflated occiput, and braincase raised, anteroposteriorly flattened, and in line with the rostrum (according to Maglio, 1970b, 1973; Maglio and Ricca, 1977; Beden, 1987a).

It is unfortunate that this specimen was heavily damaged in shipping subsequent to study outside of Kenya, because published photographs and illustrations of it (e.g., Maglio, 1970b: fig. 4A and pl. VII, fig. 21; Maglio, 1973: fig. 11A and pl. VIII, fig. 3; Maglio and Ricca, 1977: pl.2, fig. 4) do not convincingly show incipient parieto-occipital bossing or mid-sagittal depression of the nuchal crest, and it is difficult to determine if the tusk alveoli were truly parallel

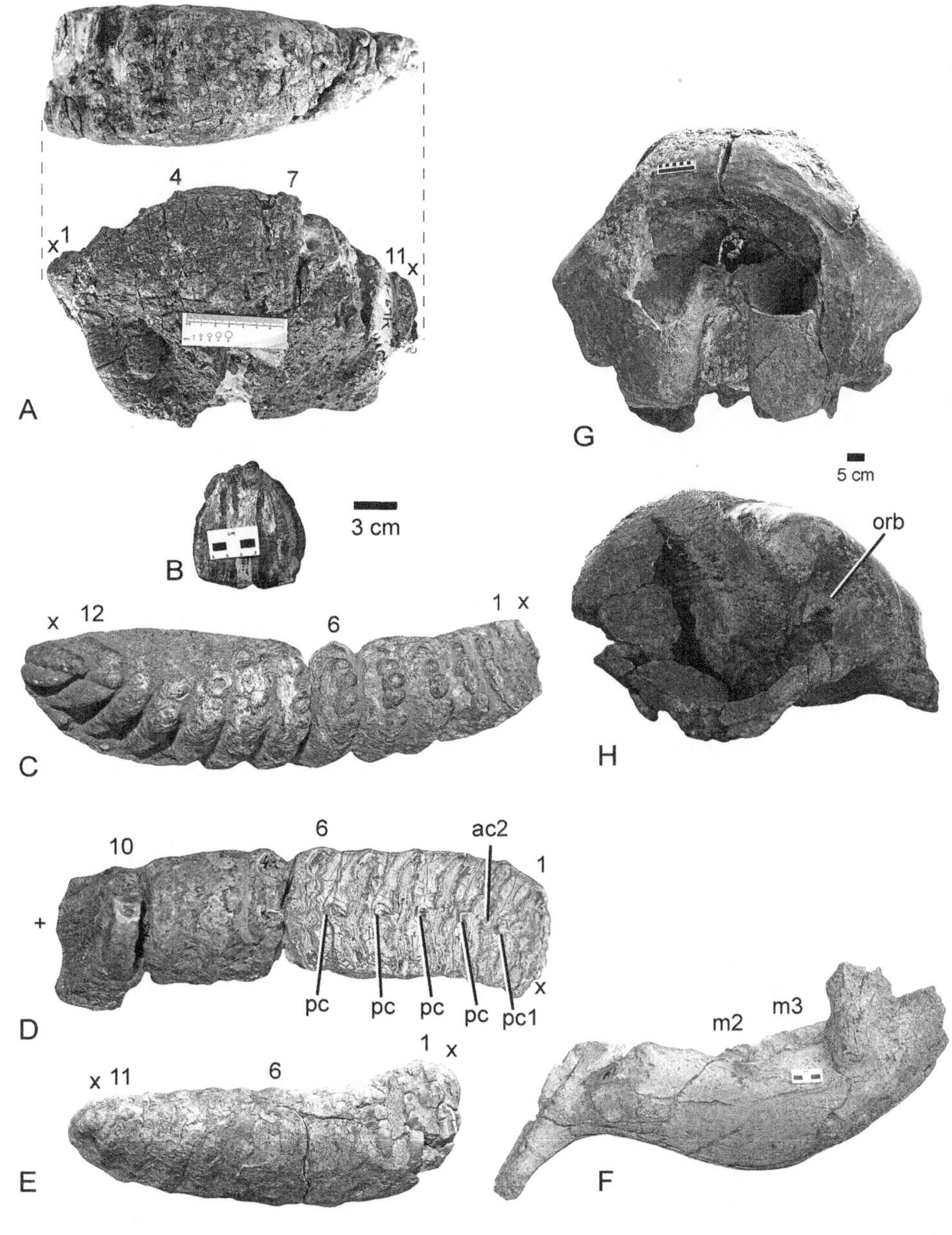

FIGURE 6.7 Molar, dentary, and cranial fossils attributed to early to ?mid-Pliocene "*Elephas*" *ekorensis*. Abbreviations: ac, anterior accessory conule; m, lower molar; orb, orbit; pc, posterior accessory conule; x, pre- or postcingulum(id); 1, 2, 3, …, plate number counted from the anterior of the crown. +, indicates missing morphology; (A–E) to the same approximate scale. (G, H) to the same scale. (A) Right M3, KNM-EK 424 (type) from Ekora, Kenya, occlusal and lateral views. (B) Molar plate, KNM-KP 30236 from Kanapoi, Kenya, anterior view. (C) Left m3, M 12639 from the Kaiso Fm., Uganda, occlusal view. Anterior to the right. (D) Partial right m3, M 12639 from the Kaiso Fm., Uganda (antimere of the left m3). Anterior to the right. Note the presence of large posterior accessory conules in the sectioned segment of the molar. (E) Left m3, KNM-KP 30471 from Kanapoi, Kenya, occlusal view. Anterior to the right. (F) Left dentary with symphysis and m2-3, KNM-KP 30471 from Kanapoi, Kenya, lateral view. Anterior to the left. (G) Partial cranium, KNM-EK 422 from Ekora, Kenya, anterior view. (H) Partial cranium, KNM-EK 422 from Ekora, Kenya, right lateral view. Anterior to the left.

(in fact, Maglio and Ricca [1977] emended the description of the specimen to say that the tusks were slightly divergent). Similarly, there are no published posterior views of the cranium to assess the degree of posterior inflation of the occiput or to gauge the level of the greatest width of the braincase. This specimen was reportedly missing in the NMK collections in Nairobi. The only match for it in the collections lacks a specimen tag and has "KNM KP" written on it. However, there are no published or cataloged cranial specimens from Kanapoi that match this mystery specimen, which is heavily plastered dorsally and approximates the specimen described as the Ekora cranium of "*E.*"

ekorensis by Maglio (1970b). This partial cranium exhibits no bossing, no mid-sagittal depression, and the posterolateral aspect of its temporal fossa resembles that of *L. adaurora* (Figure 6.7G and H).

Although a second cranium of "*E.*" *ekorensis* was reported from near Kubi Algi, Kenya and dated to about 4.5 Ma, purportedly representing a more primitive stage of the species (the M3s of the specimen have only 9–10 plates and a low hypsodonty index of 103; Maglio, 1971, 1973), it has since been re-assigned to *L. adaurora kararae* (Beden, 1983). A cranium, (erroneously) listed as "*E. ekorensis*" by Maglio (1973) under the field number KNM-SB (1969 FS/160), is in fact a palate of *L. adaurora adaurora* with the museum accession number KNM-ER *346*, from the Kubi Algi Fm.; the cranium Maglio actually studied (while still in preparation) is KNM-ER *347*, derived from a horizon just below the KBS Tuff in Area 130 of Koobi Fora, Kenya (Beden, 1983) and dated to about 1.9 Ma (Brown and Feibel, 1986), which would make it appreciably younger than Maglio's (1971, 1972) early Pliocene age estimate. Beden (1983) felt that the morphology of KNM-ER 347 differs significantly from that of the "*E.*" *ekorensis* cranium from Ekora (KNM-EK 420) and is closer to but dentally more progressive than the cranium of *L. adaurora adaurora*, and therefore assigned it to a new subspecies, *L. adaurora kararae*.

A juvenile cranium of very small size with right and left dp2–3 and dp4 still in crypt from Kanapoi, Kenya (KNM-KP 30197) was allocated to *E. ekorensis* by Harris et al. (2003). DP2 is very small, with three plates and dimensions of L=23.6mm and W=21.3mm; the dP3s have six plates and dimensions of L=68.7–70.5mm, W=43.4mm, and ET=1.7mm (Sanders, 2020). It is unfortunate not to have a confirmed adult cranium of "*E.*" *ekorensis* for comparison to facilitate ontogenetic study of cranial growth and proportions.

There are no published records or descriptions of postcrania of "*E.*" *ekorensis*. What is known about the behavior of this species relates to its feeding preferences. Dental isotopic analyses indicate grazing or mixed feeding incorporating a substantial amount of graze (Cerling et al., 1999, 2011a; Schoeninger et al., 2003). At Kanapoi, carbon isotope values indicative of mixed feeding for "*E.*" *ekorensis* are indistinguishable from those for *L. adaurora* (Manthi et al., 2019). Although there are no derived features to link "*E.*" *ekorensis* with *Elephas*, and little evidence exists to starkly differentiate it from contemporaneous *L. adaurora* except for plate number(s) and degree of hypsodonty, the dental similarity of its molars with those of basal members of the "*Elephas recki* complex" indicate the emergence of the genus in Africa. The relatively late development of *Elephas* synapomorphies suggests that although genetic research specified a late Miocene origin for the genus (e.g., Rohland et al., 2007; Murata et al., 2009), reliable identification of fossils older than the latest Pliocene-earliest Pleistocene to the genus may be difficult.

SUMMARY

1. Climatic conditions of the early Pliocene are marked by a fall in global temperatures and concomitant increase in the formation of polar ice sheets, contributing in Africa to changing ecological circumstances, such as decreased canopy cover and spread of C_4 resources, including grasses. However, large-scale open grasslands were not generally a feature of this interval.
2. Bipedal australopithecine hominins first appeared in Africa during this interval, with postcranial and positional behavioral adaptations advantageous for exploiting unevenly distributed resources across heterogenous, partly open landscapes. These sorts of habitats may have been promoted or maintained by the activities of elephants.
3. African herbivores adaptively responded to ecological changes by steadily decreasing overall browsing and showing greater preferences for C_4 grazing.
4. The early Pliocene was also a "relay" interval for proboscideans, with archaic elephant species such as stegotetrabelodonts persisting alongside more progressive species that exhibited greater synchronization between morphological features more finely adapted for fore-aft horizontal mastication and feeding behavior featuring substantial amounts of C_4 graze. These features include higher-crowned molars with more plates and thicker coatings of cementum. Archaic loxodont elephants (*Loxodonta cookei*) were replaced by more advanced congeners (*L. adaurora*, *L. exoptata*), and primitive mammoths (*M. subplanifrons*) were succeeded by the more progressive *M.* sp. "Hadar type," which appears to have emigrated out of Africa into Europe as ancestral stock for *M. rumanus*. Although *Elephas* is not incontestably documented in late Miocene African sites, candidates for the presence of the genus are first known in Africa during the early Pliocene, including "*E.*" *ekorensis* and a primitive form of the "*Elephas recki* complex."
5. Simultaneously, anancine gomphotheres and stegodonts in eastern Africa evolved more advanced dental morphologies, consistent with greater effectiveness at eating C_4 grasses. In anancine gomphotheres, these morphological changes involved developing slightly greater numbers of loph(id)s, higher molar crowns, more complex occlusal configurations with more accessory conules and stronger anancoidy, and a greater amount of cementum covering molars; stegodonts evolved a slightly greater number of plates and marginally higher molar crowns. However, anancines and stegodonts did not develop hypsodonty, and anancine gomphothere shearing was less comprehensive across

occlusal surfaces than in elephants; as a result, they were evidently outcompeted by elephants. Anancines went extinct in eastern Africa by the end of the interval, and stegodonts died out in Africa within the late Pliocene.

6. The dominant elephant of the interval, documented by abundant fossil remains in the greater Turkana Basin, including in the Omo, was *Loxodonta adaurora*. This elephant grew to sizes much larger than modern elephants, had sub-hypsodont molars of a modest number of plates, and a surprisingly modern cranium that was very tall and anteroposteriorly compressed. Conversely, it is primitive in its retention of permanent premolars. Features of the skull and molars, such as the splayed tusk alveoli of the rostrum, lack of occipitoparietal swelling or bossing, and development of strong anterior and posterior accessory conules, are shared with extant African elephants, but the cheek teeth of this species did not form loxodont sinuses ("<>") in occlusal enamel wear figures. These sinuses are found in another loxodont elephant species of the interval, *L. exoptata*, indicating its probably closer relationship with extant African elephants. *Loxodonta exoptata* is more poorly known than *L. adaurora*, except in the Laetolil Beds, where the predominance of infant and juvenile teeth suggest a mortality profile consistent with repeated episodes of drought. The species is more hypsodont and has more plates in its molars than *L. adaurora*.
7. "*Elephas*" *ekorensis* appeared in this interval and was initially identified to genus by an incomplete cranium claimed to belong to the species and exhibit typical *Elephas* features (such as nascent occipitoparietal swelling and bossing, as well as closely-spaced, parallel tusk alveoli), but these traits cannot be confirmed. However, molars attributed to the species are slightly to moderately hypsodont with a greater number of plates than in *L. adaurora,* and closely resemble those of early representatives of the "*E. recki* complex." Thus, the species may provide the oldest fossil evidence for the genus *Elephas*. The "*E. recki* complex," which likely includes several different genera and species, also first appeared in this interval. This group may have given rise to the *Elephas* and *Palaeoloxodon* lineages of Eurasia, and comprised the ecologically and numerically dominant elephants of the mid-Pliocene to late middle Pleistocene of Africa, although the paleontological and morphological evidence for its role in the ancestry of *Palaeoloxodon* conflicts with the findings of paleogenomic investigations, as is discussed in Chapter 7.

7 Late Pliocene-Holocene *The Rise and Fall of the "Elephas Recki Complex"*

A revision of the presently available material referred to the species *Elephas recki* Dietrich, 1915 ... leads up to distinguish a succession of five subspecies within the species: *E. r. brumpti* n. subsp.; *E. r. shungurensis* n. subsp., *E. r. atavus* Arambourg, 1947; *E. r. ileretensis* n. subsp. and *E. r. recki* Dietrich, 1915. They together represent a monophyletic lineage ...

—Michel Beden (1980)

[T]he validity of Beden's (1980) five subspecies has been questioned recently by several authors. These studies suggest that several species, and possibly even genera, may be included in Beden's *Elephas recki* samples.

—Haruo Saegusa and W. Henry Gilbert (2008)

INTRODUCTION

The final episode of African proboscidean evolution documented in the fossil record dates from the mid-Pliocene to Holocene and was accompanied and driven by vast changes in global climate, regional ecology, and geomorphology (continued rifting and uplift of East African rift shoulders [Maslin et al., 2014 and references therein; Macgregor, 2015]). Accumulation of sediments and formation of lakes associated with rifting accounts for the greatest density of sites in eastern Africa during his interval, although proboscideans were present across the continent, and at least some species became pan-African in distribution (e.g., possibly *Palaeoloxodon recki*, *Palaeoloxodon jolensis*, *Loxodonta africana*; see Figure 7.1; Table 7.1).

Evidence for periodic warming and cooling cycles punctuated by extreme climate transients was documented from the deep-sea stable isotope record in landmark research by Zachos et al. (2001). Following a global warming trend that peaked in the middle Miocene, major ice-sheet expansion occurred in Antarctica, correlated with gradual cooling. Following subtle mid-Pliocene warming, cooling initiated again in the late Pliocene, accompanied by the onset of northern hemisphere glaciation. Zachos et al.'s (2001) study further established accelerated periodicity and greater amplitude of global temperature change, within generally greater accumulation of ice and decreasing temperatures, during the Pleistocene. These conditions contributed to the establishment of aridification across Africa since the late Miocene, which is thought to have played a role in the expansion of C_4 plants (grasses and sedges) (Edwards et al., 2010). Although Blumenthal et al. (2017) failed to confirm *trending* for long-term aridification in the Omo-Turkana Basin of eastern Africa, they did document mid-to-late Pliocene arid and mesic conditions associated with woodland/bushland/shrubland in Turkana, and a transition to relatively more open conditions, including wooded grasslands, regionally in the early Pleistocene. Conditions in the Turkana Basin at the end of the Pleistocene into the Holocene continued to be mesic to arid. For the late Pliocene-Pleistocene, Cerling et al. (2005) and Blumenthal et al. (2017) documented long-term increase of C_4 grazers among artiodactyls, perissodactyls, and proboscideans and concomitant decrease in the proportion of C_3 browsers. The latter authors speculated that increasing contrasts in rainfall seasonality rather than intensified aridity were more important factors driving changes in plant cover and feeding preferences. Cerling et al. (2015) noted that from ~4.1 to 2.35 Ma, the Turkana Basin was dominated by a greater percentage of mixed C_3-C_4 feeders than is found in modern African ecosystems, whereas from ~2.35 to 1.0 Ma, the landscape was dominated by a greater proportion of bovid and non-ruminant grazers than occur in modern environments.

A model of instability across the late Pliocene-Holocene proposed a hypothesis in which extreme regional ecological variability, including aridity and rainfall, during transitions from presence of lakes to absence of lakes and driven by climate change, would have caused speciation and extinction events in eastern Africa (Maslin et al., 2014). Nonetheless, no matter how the complex factors regulating regional ecologies and plant cover are interpreted, pedogenic and biogenic carbonate records indicate that C_4-dominated ecosystems became more important in Africa during the late Pliocene-Pleistocene and that open landscapes dominated by high C_4 biomass emerged in the early Pleistocene (Ségalen et al., 2007). Habitats shifted from grassy woodlands/shrubland/bushland between 4.0 and 3.0 Ma to wooded grasslands between 2.0 and 1.4 Ma (Cerling et al., 2013). Pure C_4 grasslands, however, appear to have been a middle Pleistocene phenomenon (Cerling, 1992). Certainly, in eastern Africa, topographic changes associated with rifting formed orographic barriers producing rain-shadow effects to their east that contributed to conditions conducive to the spread of C_4 plants (Spiegel et al., 2007). As pointed out by Bobe (2006) and Levin

DOI: 10.1201/b20016-7

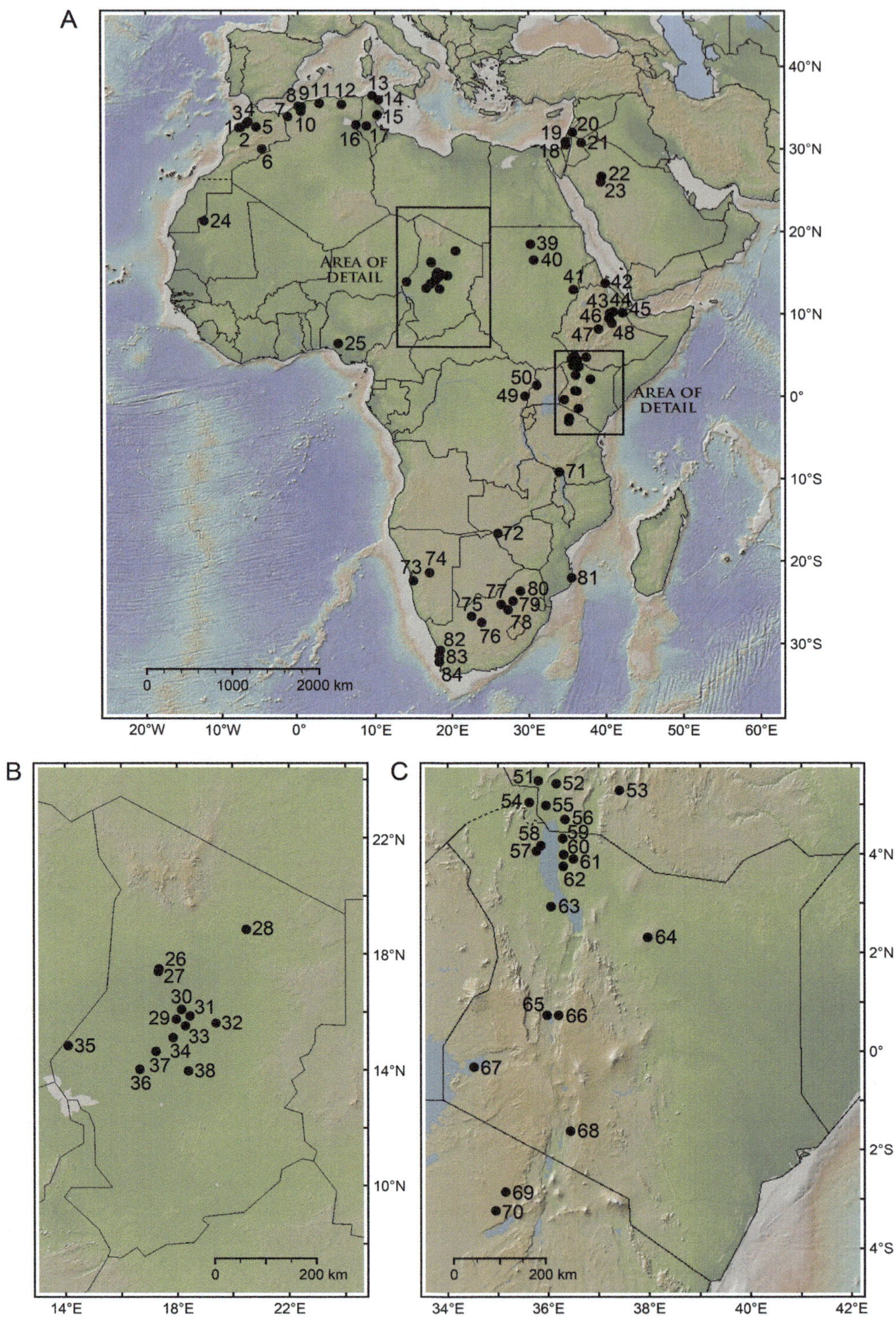

FIGURE 7.1 Maps showing the location of mid-Pliocene to Holocene sites, (A-C). Areas of detail, (B, C). 1a, Sidi Abderrahmane, Morocco, b, Sidi Abder Rahmane (Casablanca), Morocco; 2a, Ahl al Oughlam, Morocco, b, Ahl al Oughlam, Casablanca, Morocco; 3a, Salé, Morocco, b, Salé Village, Morocco; 4a, Oued Akrech, Morocco; 4b, Fouarat, Morocco; 5, Aïn Maarouf, Morocco; 6, Anchrif Quarry, Taghrout, Morocco; 7, Mansourah (Constantine), Algeria; 8, Port de Mastaganem, Algeria; 9, Djebel Bel Hacel, Algeria; 10a, Ternifine (Tighenif), Algeria; 10b, Sablière de Palekao, Algeria;

(Continued)

FIGURE 7.1 ***(Continued)*** 11a, Gué de Constantine, Algeria, b, Gué de (Oued) Constantine, Algeria; 12a, Beauséjour Farm, Algeria; 12b, Aïn Boucherit, Algeria; 12c, Aïn Hanech, Algeria; 13a, Lac Ichkeul, Tunisia; 13b, Gebel Melah, Tunisia; 13c, Ferryville (Menzel Bourguiba), Tunisia; 14a, Hamada Damous, Tunisia; 14b, Grombalia, Tunisia; 15, El Douira, Tunisia; 18, Revadim Quarry, Israel; 19, Holon, Israel; 20, Gesher Benot Ya'aqov, Israel; 21, Shishan Marsh 1 (SM1), Azraq Basin, Jordan; 22, Locality #3 (Thomas locality 3), Nafud Desert, Saudi Arabia; 23, Ti's al Ghadah, Nafud Desert, Saudi Arabia; 24, Zouerate, Mauritania; 25, Idanre, Nigeria; 26, Yayo, Angamma Region, Chad; 27, Kichi-Kichi, Chad; 28a, Ounianga, Kebir, Chad, b, Ounianga, Kebir I, Chad; 29, Bochianga, Chad; 30, Toungour, Chad; 31, Koro-Toro, Chad; 32, Ouadi-Derdemy, Chad; 33, Kouba, Chad; 34a, Tchitcheli Brahm Padelanga, Chad; 34b, Beurkia, Chad; 35, Koulà, Chad; 36, Fantrassou, Chad; 37, Salal, Chad; 38, Goz-Kerki, Chad; 39, Kaddanarti, northern Sudan; 40, Wadi Howar, Sudan; 41a, Butana, Bridge, Sudan; 41b, Khashm el Girba, Sudan; 42a, Dioli and Aalad areas, Buia, Eritrea; 42b, Mulhuli-Amo, Buia area, northern Danakil Depression, Eritrea; 42c, Dandero, northern Danakil Depression, Eritrea; 43a, Middle Awash, Ethiopia, b, Middle Awash Valley, Ethiopia, c, Matabaietu Dora, Middle Awash Valley, Ethiopia, d, Matabaietu D'ar, Wilti Dora, Middle Awash, Ethiopia, e, Dakanihyalo, Matabaietu D'ar, Wilti Dora, Meadura Dora, Middle Awash, Ethiopia; 43f, Hadar, Ethiopia, g, Maka'amitalu, Hadar, lower Awash Valley, Ethiopia; 43h, Busidima-Telalak region, Afar, Ethiopia; 44, Megenta area, Lower Awash basin, Afar Rift, Ethiopia; 45a, Gafalo, Republic of Djibouti; 45b, Baragoli, Republic of Djibouti; 45c, Haïdalo, Republic of Djibouti; 46a, Herto, Ethiopia; 46b, Dikika (DIK-1), Ethiopia; 47, Melka Kunture (Garba IV), Ethiopia; 48, Bouri, Ethiopia; 49, Upper Semliki River, Democratic Republic of Congo; 50a, Kikigati, Kaiso Village, Uganda, b, Kaiso Village and Behanga I, Western Rift, Uganda; 51, Omo (Kibish Fm.), Ethiopia; 52, Omo (Usno Fm.), Ethiopia; 53, Konso, Ethiopia; 54, Natodomeri, South Sudan, administered by Kenya; 55, Omo (Shungura Fm.), Ethiopia; 56, Fejej FJ-1 Site, Ethiopia; 57, West Turkana, Kenya; 58a, Nadung'a 4, Kenya, b, Nadung'a 4, West bank of Lake Turkana, Kenya; 59a, Ileret, Kenya, b, Areas 1A, 6, 6A, 8, 8A, 11, Ileret, Kenya; 60a, Koobi Fora, Kenya, b, Koobi Fora, Kenya (e.g., Areas 105, 117), Upper beds, Kubi Algi, Kokoi Horst, Koobi Fora, Kenya, c, Areas 103, 104, Koobi Fora, Kenya, d, Area 123, Koobi Fora, Kenya; 61, Kubi Algi, Kenya; 62, Allia Bay, Kenya; 63, Lothagam, Kenya; 64, Marsabit Road, Kenya; 65a, Locality 90/91, Tugen Hills, Kenya, b, Tugen Hills, Kenya, locality 91; 66, Chesowanja, Kenya; 67, Kanjera, Kenya; 68, Site 15, Olorgesailie, Kenya; 69, Olduvai Gorge, Tanzania; 70, Laetoli, Tanzania; 71a, Karonga, Malawi; 71b, Mwanganda's Village, Malawi; 71c, Uraha, Malawi; 71d, Uraha, Malema, Malawi, e, Malema, Malawi; 71f, Uraha, Mwimbi, Malawi; 72, Left bank, Zambezi River near Victoria Falls, Zambia; 73, Namib IV, Namibia; 74a, Windhoek Zoo Park Garden, Namibia, b, Zoo Park, Namibia; 75, Kathu Pan, South Africa; 76, Power's Site, South Africa; 77, Bolt's Farm, South Africa; 78, Vaal River, South Africa; 79a, Swartkrans, South Africa; 79b, Sterkfontein, South Africa; 80, Makapansgat, South Africa; 81, Praia de Morrungusu (sic), Mozambique (alt. Morrungulo); 82a, Duinefontein 2 (=Melkbosstrand), South Africa, b, Duinefontein 2 (=Melkbosstrand), southwestern Cape Province, South Africa; 83, Elands Bay Cave, South Africa; 84a, Elandsfontein, South Africa, b, Bone Circle assemblage, Elandsfontein, South Africa.

et al. (2011), however, not all parts of Africa, even eastern Africa, responded the same way to climate fluctuations, with, for example, the lower Omo Valley in Ethiopia and coastal and montane forests of East Africa remaining significantly more wooded than other parts of the region during the late Pliocene and Pleistocene. The study of modern elephants (see below) shows that they contribute substantially to the maintenance of open conditions; the success of grazing elephants in the "*Elephas recki*" group from the mid-Pliocene through the Pleistocene may have contributed to the replacement of species of *Australopithecus* by hominins that favored more open-country habitats and utilized a greater amount of C_4 resources, such as *Kenyanthropus* (3.6–3.0 Ma), *Paranthropus* (2.5–1.4 Ma), and *Homo* (2.3 Ma to the present), at least in the greater Turkana Basin (Cerling et al., 2013).

Elephantids responded to increased availability of C_4 resources and competition with growing numbers of grazers of other taxa (e.g., horses, suids, bovids, hippos, rhinos) by evolving adaptations for more effective mastication of these plants and to counter the challenges to tooth wear and mortality posed by uptake of grit and opal phytoliths from feeding on C_4 grasses low to the ground. In elephants, the first phase of this adaptive response had been achieved at least by the early Pliocene with the modification of skull to change the center of mass of the lower jaw, shortening the dentary and dispensing with lower tusks, and making a more efficient proal masticatory mechanism for fore-aft horizontal shearing by raising and anteroposteriorly shortening the cranium to re-orient chewing muscles (Maglio, 1972a; Sanders et al., 2021). Additional adaptations included increased numbers of molar plates, greater relative crown height (hypsodonty), and thicker cementum covering molar crowns (cementodonty) (Saarinen, 2019). These steps were documented in Chapter 6 for more advanced "relay" species of elephants, such as *Loxodonta adaurora* and "*Elephas*" *ekorensis* (Maglio, 1970b, 1973; Beden, 1983, 1987a; Sanders, 2020), that overlapped in time with archaic, brachyodont elephants (e, g, *Primelephas korotorensis*, *Stegotetrabelodon orbus*), and for more advanced hypsodont species in the "*Elephas recki* complex." During the late Pliocene-Pleistocene, these features evolved to extremes in species of the "*E. recki* complex" and its descendants, with trends in cheek tooth morphology documented for significant increases in hypsodonty, plate number, lamellar frequency, frequency and amplitude of enamel folding, and cementum (see Beden, 1980; Sanders and

TABLE 7.1
Major Occurrences of Late Pliocene-Pleistocene Afro-Arabian Proboscidea

Taxon	Occurrence (Site, Locality)	Stratigraphic Unit	Age	Key References
Proboscidea Illiger, 1811				
Suborder incertae sedis				
Deinotherioidea Bonaparte, 1845				
Deinotheriidae Bonaparte, 1845				
Deinotheriinae Bonaparte, 1845				
Deinotherium bozasi	Omo, Ethiopia	Usno and Shungura Fms.	3.6–2.7 Ma; 3.6–1.6 Ma	Arambourg (1934a, b, c), Arambourg et al. (1969a), Beden (1975, 1976), Brown et al. (1985), Feibel et al. (1989), and Alemseged (2003)
	West Turkana, Kenya	Lomekwi Mb., Nachukui Fm.	3.44–2.53 Ma	Harris et al. (1988b) and McDougall et al. (2012)
	Hadar, Ethiopia	Sidi Hakoma, Denen Dora, Kada Hadar Mbs., Hadar Fm.	3.42 to <2.94 Ma	Taieb et al. (1976), Kalb et al. (1982), White et al. (1984), Bonnefille et al. (2004), and Wynn et al. (2016)
	Koobi Fora, Kenya	Tulu Bor, Burgi, KBS, and Okote Mbs., Koobi Fora Fm. ("Zone C and *Notochoerus scotti* and *Metridiochoerus andrewsi* zones" between Tulu Bor and Koobi Fora Tuffs)	3.44–1.49 Ma	Harris (1976, 1983a, b)
	Middle Awash Valley, Ethiopia	Matabaietu Fm.	Late Pliocene	Kalb et al. (1982a, b)
	Praia de Morrungusu (sic), Mozambique (alt. Morrungulo)		Late Pliocene or early Pleistocene	Harris (1977a) and Pickford (2012)
("*Deinotherium* sp.")	Uraha, Malema, Malawi	Chiwondo Beds, stratigraphic Unit 3	Early Pleistocene	Sandrock et al. (2007)
	Olduvai Gorge, Tanzania	Beds I and II Olduvai Fm.	2.038–1.353 Ma (alt. 1.87–1.6 Ma)	Osborn (1936), Leakey (1965), Hay (1976), Tamrat et al. (1995), Diez-Martin et al. (2014), and McHenry and Stanistreet (2018)
	Konso, Ethiopia	Interval 1, Konso Fm.	~1.91 Ma	Suwa et al. (2003)
	Chesowanja, Kenya		>1.42 Ma	Bishop et al. (1971), Hooker and Miller (1979), and Beden (1985)
	Marsabit Road, Kenya		Early Pleistocene	Gentry and Gentry (1969) and Harris (1978a)
	Kanjera, Kenya	Kanjera Fm.	1.1–1.0 Ma	Behrensmeyer et al. (1995)
Elephantiformes Tassy, 1988				
Elephantimorpha Tassy and Shoshani, 1997 in Shoshani et al., 1998				
Elephantida Tassy and Shoshani, 1997 in Shoshani et al., 1998				
Gomphotheriidae Hay, 1922				
Anancinae Hay, 1922				
Anancus osiris	Grombalia, Tunisia		Late Pliocene	Arambourg (1970)
	Lac Ichkeul, Tunisia		Late Pliocene (alt. ~6.0–5.0 Ma)	Arambourg (1952, 1970) and Pickford et al. (1995)
	Ferryville (Menzel Bourguiba), Tunisia		Late Pliocene	Depéret et al. (1925) and Coppens et al. (1978)
	Hamada Damous, Tunisia		Late Pliocene	Fournet (1971)
	Gebel Melah, Tunisia		Late Pliocene	Fournet (1971)

(*Continued*)

TABLE 7.1 (*Continued*)
Major Occurrences of Late Pliocene-Pleistocene Afro-Arabian Proboscidea

Taxon	Occurrence (Site, Locality)	Stratigraphic Unit	Age	Key References
	Fouarat, Morocco		Late Pliocene	Arambourg (1970)
	Ahl al Oughlam, Casablanca, Morocco		Early Pleistocene, ~2.5 Ma	Geraads and Metz-Muller (1999) and Geraads (2002)
	Aïn Boucherit, Algeria		Early Pleistocene 2.32 Ma (alt. ca. 2.0 Ma) (alt. late Pliocene)	Arambourg (1970), Geraads and Amani (1998), Geraads (2002), and Sahnouni et al. (2002)
("*Anancus*")	Aïn Hanech, Algeria	Aïn Hanech Fm.	Between 1.95 and 1.78 Ma (alt. ~1.2 Ma)	Arambourg (1970), Maglio (1973), Sahnouni and Heinzelin (1998), Geraads et al. (2004), and Sahnouni et al. (2002, 2004, 2011)
Stegodontidae Osborn, 1918				
Stegodon kaisensis	Omo, Ethiopia	Mb. B, Shungura Fm.	3.44–2.97 Ma	Beden (1976) and Levin et al. (2011)
Elephantidae Gray, 1821				
Elephantinae Gray, 1821				
Loxodonta adaurora adaurora	Lothagam, Kenya	Kaiyamung Mb., Nachukui Fm. (=Unit 3)	~3.5–3.0 Ma	Coppens et al. (1978), Feibel (2003), and McDougall and Feibel (2003)
	Omo, Ethiopia	Mb. B, Shungura Fm.	3.44–2.97 Ma	Sanders et al. (2021) and Levin et al. (2011)
	Hadar, Ethiopia	Sidi Hakoma and Denen Dora Mbs., Hadar Fm.	3.42–3.20 Ma	White et al. (1984), Bonnefille et al. (2004), and Wynn et al. (2016)
	West Turkana, Kenya	Lower Lomekwi Mb., Nachukui Fm.	3.44 to ~3.3 Ma	Harris et al. (1988a, b) and McDougall et al. (2012)
Loxodonta adaurora kararae	Omo, Ethiopia	Mb. E, Shungura Fm.	2.4–2.36 Ma	Heinzelin (1983), Beden (1987a), and Alemseged (2003)
	Karonga, Malawi	Chiwondo Beds, stratigraphic Unit 3A	2.4–2.3 Ma	Bromage et al. (1995)
	?Middle Awash, Ethiopia	Matabaietu Fm.	~2.3–2.0 Ma	Kalb and Mebrate (1993)
("*Phanagoroloxodon planifrons*")	Koobi Fora, Kenya (type)	Upper Burgi Mb., Koobi Fora Fm.	2.0–1.89 Ma	Beden (1983), Feibel et al. (1989), and Zhang (2020)
Loxodonta exoptata	West Turkana, Kenya	Lower Lomekwi Mb., Nachukui Fm.	3.44 to ~3.3 Ma	Harris et al. (1988b) and McDougall et al. (2012)
	Hadar, Ethiopia	Denen Dora Mb., Hadar Fm.	3.24–3.20 Ma	White et al. (1984) and Wynn et al. (2016)
	Matabaietu Dora, Middle Awash Valley, Ethiopia	Matabaietu Fm.	Late Pliocene or early Pleistocene	Kalb and Mebrate (1993)
	Laetoli, Tanzania (type)	Upper Ndolanya Beds	2.66 Ma	Beden (1987b), Sanders (2005, 2011), and Deino (2011)
Loxodonta atlantica atlantica	Ternifine (Tighenif), Algeria (type)		1.07–0.99 Ma (alt. 700 Ka; middle Pleistocene)	Pomel (1879), Osborn (1942), Maglio (1973), Coppens et al. (1978), Geraads et al. (1986), Kowalski and Rzebik-Kowalska (1991), Hublin (2001), and Geraads (2016)
	Gué de (Oued) Constantine, Algeria	Maison Carrée Fm.	?middle Pleistocene	Arambourg (1952), Maglio (1973), Coppens et al. (1978), and Manthi et al. (2019)
	Sablière de Palekao, Algeria		Middle to late Pleistocene	Maglio (1973)

(*Continued*)

TABLE 7.1 (*Continued*)
Major Occurrences of Late Pliocene-Pleistocene Afro-Arabian Proboscidea

Taxon	Occurrence (Site, Locality)	Stratigraphic Unit	Age	Key References
	Sidi Abderrahmane, Morocco		Middle Pleistocene	Maglio (1973), Coppens et al. (1978), Kowalski and Rzebik-Kowalska (1991), Raynal et al. (2004a), and Youcef (2020)
	Aïn Maarouf, Morocco		?0.7 Ma (alt. middle Pleistocene)	Geraads et al. (1992), Hublin (1992), Geraads (2002), and Youcef (2020)
Loxodonta africana ("*Loxodonta atlantica angammensis*")	Omo, Ethiopia	Mbs. D3, F, Shungura Fm.	2.44±0.05 Ma and 2.32–2.27 Ma	Maglio (1973), Beden (1979, 1987a), Sanders et al. (2010a), McDougall et al. (2011), and Zhang (2020)
	Middle Awash Valley, Ethiopia	?Wehaietu Fm.	In the interval 0.8–0.2 Ma	Kalb and Mebrate (1993)
	Ounianga, Kebir I, Chad		?middle Pleistocene	Joleaud and Lombard (1933a, b) and Coppens (1967)
	Kanjera, Kenya	Apoko Fm.	~0.5 Ma	Plummer and Potts (1989) and Behrensmeyer et al. (1995)
	Duinefontein 2 (=Melkbosstrand), South Africa		270 Ka	Klein et al. (2007)
	Omo, Ethiopia	Mb. III, Mb. IV, Kibish Fm.	130–74 Ka <10 Ka	Butzer et al. (1972), Assefa et al. (2008), and Brown et al. (2012)
	Elands Bay Cave, South Africa		13,600–7,900 y BP	Klein et al. (2007)
	Bone Circle assemblage, Elandsfontein, South Africa		12,000–10,000 y BP	Klein and Cruz-Uribe (1991)
("*Loxodonta atlantica angammensis*")	Yayo, Angamma Region, Chad (subspecies type)		?Holocene <10,160–9,260 y BP	Coppens (1965), Maglio (1973), Tillet (1985), Sanders et al. (2010a), and Zhang (2020)
("*Loxodonta africana zulu*")	Windhoek Zoo Park Garden, Namibia	Spring deposits	7241–6991 y BP	Brook et al. (2014)
	Wadi Howar, Sudan	Gravel bar deposits	5430±180 y BP	Rodrigues et al. (2000)
	Ounianga Kebir, Chad		Age unknown; Quaternary	Coppens (1965)
	Northeast (not northwest) of Koro-Toro, Chad		Age unknown; Quaternary	Coppens (1965)
	Unnamed site, 15 km south-southwest of Koro-Toro, Chad		Age unknown; Quaternary	Coppens (1965)
	Fantrassou, Chad	Alluvial grits of Bahr el Ghazal	Age unknown; Quaternary	Coppens (1965)
	Salal, Chad	Alluvial grits of Bahr el Ghazal	Age unknown; Quaternary	Coppens (1965)
	Tchitcheli Brahm Padelanga, Chad	Alluvial grits of Bahr el Ghazal	Age unknown; Quaternary	Coppens (1965)
	Beurkia, Chad		Age unknown; Quaternary	Coppens (1965)
	Kouba, Chad		Age unknown; Quaternary	Coppens (1965)
	Bochianga, Chad	Low terrace, North bank of Bahr el Ghazal	Age unknown; Quaternary	Coppens (1965)

(*Continued*)

TABLE 7.1 (*Continued*)
Major Occurrences of Late Pliocene-Pleistocene Afro-Arabian Proboscidea

Taxon	Occurrence (Site, Locality)	Stratigraphic Unit	Age	Key References
	Toungour, Chad	Fluviolacustrine formation	Age unknown; Quaternary	Coppens (1965)
	Kichi-Kichi, Chad		Age unknown; Quaternary	Coppens (1965)
Loxodonta cyclotis	Idanre, Nigeria		Geological age unknown, probably Holocene	Ogunfolakan and Olayemi (2008)
Loxodonta sp.	Konso, Ethiopia	Intervals 1, 4, Konso Fm.	~1.91 Ma, ~1.43–1.41 Ma	Suwa et al. (2003)
(probably *Loxodonta africana*)	Zoo Park, Namibia	Later Stone Age lithic context; possible elephant butchery site	5,200 y BP	MacCalman (1967) and Clark and Haynes (1970)
Mammuthus sp. "Hadar-type"	Hadar, Ethiopia	Sidi Hakoma and Denen Dora Mbs., Hadar Fm.	3.42–3.20 Ma	White et al. (1984), Beden (1979, 1985), Bonnefille et al. (2004), and Wynn et al. (2016)
Mammuthus africanavus ("*Elephas africanavus*")	Lac Ichkeul, Tunisia (type)		Mid-late Pliocene	Arambourg (1970) and Coppens et al. (1978)
	Aïn Brimba, Tunisia		Mid-late Pliocene	Arambourg (1970)
	Kebili, Tunisia		Mid-late Pliocene	Maglio (1973) and Coppens et al., (1978)
	Gebel Melah, Tunisia		Late Pliocene	Fournet (1971)
	Garet et Tir, Algeria		Mid-late Pliocene	Arambourg (1970) and Coppens et al. (1978)
	Oued Akrech, Morocco		Mid-late Pliocene	Arambourg (1970) and Coppens et al. (1978)
	Fouarat, Morocco		Mid-late Pliocene	Arambourg (1970)
("*Loxodonta africanavus*")	Goz-Kerki, Chad (70 km SE of Koro-Toro)		~3.5–3.0 Ma	Coppens (1965), Brunet (2001), and Fara et al. (2005)
("*Loxodonta africanavus*")	Koro-Toro, Chad		3.5 Ma (alt. 3.5–3.0 Ma)	Mackaye (2001), Fara et al. (2005), Lebtard et al. (2008), and Schuster et al. (2009)
("*Loxodonta africanavus*")	Koulà, Chad		~3.5–3.0 Ma	Coppens (1965), Maglio (1973), Coppens et al. (1978), Brunet (2001), and Fara et al. (2005)
("*Loxodonta africanavus*")	Ouadi-Derdemy, Chad		~3.5–3.0 Ma	Coppens (1965), Maglio (1973), Coppens et al. (1978), Brunet (2001), and Fara et al. (2005)
("*Loxodonta africanavus*")	Toungour, Chad (30–50 km W of Koro-Toro)		~3.5–3.0 Ma	Coppens et al. (1965), Maglio (1973), Coppens et al. (1978), Brunet (2001), and Fara et al. (2005)
	Aïn Boucherit, Algeria	Oued Boucherit Fm.	Early Pleistocene, 2.32 Ma (alt. ~2.0 Ma)	Arambourg (1970), Coppens et al. (1978), Geraads and Amani (1998), Sahnouni and Heinzelin (1998), Geraads (2002), and Sahnouni et al. (2002)
?*Mammuthus meridionalis* ("*Elephas moghrebiensis*") ("*Elephas recki ileretensis*")	Aïn Hanech, Algeria	Aïn Hanech Fm.	Between 1.95 and 1.78 Ma (alt. ~1.2 Ma)	Arambourg (1970), Maglio (1973), Sahnouni and Heinzelin (1998), Geraads and Metz-Muller (1999), Geraads et al. (2004), and Sahnouni et al. (2002, 2004, 2011, 2013)

(*Continued*)

TABLE 7.1 (*Continued*)
Major Occurrences of Late Pliocene-Pleistocene Afro-Arabian Proboscidea

Taxon	Occurrence (Site, Locality)	Stratigraphic Unit	Age	Key References
("*Elephas moghrebiensis*")	Djebel Bel Hacel, Algeria		Early Pleistocene, equivalent in age to Aïn Hanech	Arambourg (1970)
"*Elephas*" *ekorensis*	Koro-Toro, Chad		3.5 Ma (alt. 3.5–3.0 Ma)	Mackaye (2001), Fara et al. (2005), Lebtard et al. (2008), and Schuster et al. (2009)
	Hadar, Ethiopia	Sidi Hakoma and Denen Dora Mbs., Hadar Fm.	3.42–3.20 Ma	White et al. (1984), Kalb and Mebrate (1993), and Wynn et al. (2016)
	Omo, Ethiopia	?Usno Fm.	Late Pliocene, 3.31–2.97 Ma	Beden (1987a), but see Kalb and Mebrate (1993)
Elephas recki brumpti (alt. "*Phanagoroloxodon planifrons*") (alt. "*Elephas brumpti*")	Omo, Ethiopia (subspecies type)	Upper Mb. A, Mb. B, Shungura Fm.	~3.5–3.07 Ma	Beden (1980, 1985, 1987a), Feibel et al. (1989), Alemseged (2003), Brown and McDougall (2011), McDougall et al. (2012), and Zhang (2020)
	?Goz-Kerki, Chad (70 km SE of Koro-Toro)		~3.5–3.0Ma	Coppens (1965), Maglio (1973), Brunet (2001), and Fara et al. (2005)
	?Koulà, Chad		~3.5–3.0Ma	Coppens (1965), Maglio (1973), Brunet (2001), and Fara et al. (2005)
	Koro-Toro, Chad		3.5 Ma (alt. 3.5–3.0 Ma)	Mackaye (2001), Fara et al. (2005), Lebtard et al. (2008), and Schuster et al. (2009)
	?Ouadi-Derdemy, Chad		~3.5–3.0Ma	Coppens (1965), Maglio (1973), Coppens et al. (1978), Brunet (2001), and Fara et al. (2005)
	Koobi Fora, Kenya	Below Tulu Bor Tuff, Lokochot Mb., Koobi Fora Fm.	>3.44 Ma (alt. ~3.50–3.36 Ma)	Beden (1980, 1983, 1985), Harris (1983a), Brown (1994), and McDougall et al. (2012)
	Dikika (DIK-1), Ethiopia	Sidi Hakoma Mb., Hadar Fm.	3.42–3.24 Ma	Wynn et al. (2006, 2016)
	Bolt's Farm, South Africa		~3.4–2.9 Ma (alt. ~5.0–4.0 Ma)	Maglio (1973), Cooke (1993), and Sénégas and Avery (1998)
	West Turkana, Kenya	Lower, middle, and upper Lomekwi Mb., Nachukui Fm.	3.36–2.52 Ma (to ?~3.0 Ma)	Harris et al. (1988a, b) and MacDougall et al. (2012)
	Allia Bay, Kenya	Below Hasuma Tuff; in Allia Tuff; between Allia and Hasuma Tuffs, Tulu Bor Mb., Koobi Fora Fm.	~3.25–3.07 Ma	Beden (1983) and McDougall et al. (2012)
	Hadar, Ethiopia	Sidi Hakoma and Denen Dora Mbs., Hadar Fm.	3.42–3.24 Ma; 3.24–3.20 Ma	White et al. (1984), Beden (1985), Kalb and Mebrate (2003), Bonnefille et al. (2004), and Wynn et al. (2016)
	Omo, Ethiopia	Usno Fm.	Late Pliocene, 3.31–2.97 Ma	Beden (1987a) and Kalb and Mebrate (1993)
	Kubi Algi, Kenya	Below Hasuma Tuff, Tulu Bor Mb., Koobi Fora Fm. (="Zone B, Kubi Algi Fm.")	>3.07 Ma (alt. ~3.50–3.36 Ma)	Beden (1980, 1983, 1985, 1987a), Abell et al. (1982), Harris (1983a), Brown (1994), Tassy and Debruyne (2001), and McDougall et al. (2012)

(*Continued*)

TABLE 7.1 (*Continued*)
Major Occurrences of Late Pliocene-Pleistocene Afro-Arabian Proboscidea

Taxon	Occurrence (Site, Locality)	Stratigraphic Unit	Age	Key References
Elephas recki shungurensis (alt. "*Elephas shungurensis*")	Hadar, Ethiopia	Kada Hadar Mb., Hadar Fm.	3.2 to <2.94 Ma	White et al. (1984), Beden (1985), and Wynn et al. (2016)
	Koobi Fora, Kenya (e.g., Areas 105, 117), Upper beds, Kubi Algi, Kokoi Horst, Koobi Fora, Kenya	Zone C, *Notochoerus scotti* zone; between Tulu Bor and KBS tuffs	Probably upper Tulu Bor and Burgi Mbs., Koobi Fora Fm., ~3.07 to >1.87 Ma (Biozonation zone C is upper Tulu Bor-lower Burgi Mbs., and *N. scotti* zone is upper Burgi Mb.)	Beden (1980, 1983, 1985), Harris (1983a), Brown and Feibel (1986), and McDougall et al. (2012)
	Omo, Ethiopia (subspecies type)	Mb. C-Lower Mb. F, Shungura Fm.	3.07 to ~2.3 Ma	Beden (1980, 1985, 1987a) and McDougall et al. (2012)
	Kikigati, Kaiso Village, Uganda	Upper Kaiso Fm., cf. Mbs. C-Lower F, Shungura Fm.	~3.07 to ~2.3 Ma	Cooke and Coryndon (1970) and Sanders (1990)
	Locality 90/91, Tugen Hills, Kenya	Chemeron Fm.	~3.0–2.4 Ma (alt. ~2.85–2.40 Ma)	Beden (1985) and Hill (1999)
	West Turkana, Kenya	Upper Lomekwi, Lokalalei, and Kalochoro Mbs., Nachukui Fm.	In the interval ~3.0–1.88 Ma	Harris et al. (1988b) and McDougall et al. (2012)
	Bouri, Ethiopia	Hata Mb., Bouri Fm.	2.5 Ma	Heinzelin et al. (1999)
(?*E. recki shungurensis*)	Ahl al Oughlam, Morocco		~2.5 Ma	Geraads and Metz-Muller (1999) and Geraads (2006)
	Matabaietu D'ar, Wilti Dora, Middle Awash, Ethiopia	Matabaeitu Fm.	In the interval ~2.5–2.0 Ma; 2.4–2.0 Ma (based on biochronology)	Kalb and Mebrate (1993) and Kalb (1995)
	Malema, Malawi	Chiwondo Beds, stratigraphic Unit 3A	~2.4–2.3 Ma	Bromage et al., (1995) and Sandrock et al. (2007)
(?*E. recki shungurensis*)	Salé, Morocco		?early Pleistocene	Arambourg (1970)
Elephas recki atavus (alt. *Elephas atavus*) (alt. *Elephas* [*Palaeoloxodon*] *recki atavus*)	?West Turkana, Kenya	Kalochoro Mb., Nachukui Fm.	2.35–1.90 Ma	Brugal et al. (2003)
	Maka'amitalu, Hadar, lower Awash Valley, Ethiopia	Busidima Fm.	~2.35 Ma	Rowan et al. (2022)
("*Elephas* sp.")	Omo, Ethiopia (subspecies type)	Upper Mb. F, Mb. G, Shungura Fm.	~2.3 to ~1.9 Ma	Arambourg (1943), Beden (1980, 1985, 1987a), McDougall et al. (2012), and Zhang (2020)
	Upper Semliki River, Democratic Republic of Congo	Lusso Beds	~2.3 to ~1.9 (equivalent to Upper Mb. F., Mb. G, Shungura Fm.)	Sanders (1990)
("*E. shungurensis/atavus*")	Uraha, Malawi	Chiwondo Beds, stratigraphic Unit 3A	In the interval 2.4–1.8 Ma	Bromage et al. (1995) and Sandrock et al. (2007)

(*Continued*)

TABLE 7.1 (*Continued*)
Major Occurrences of Late Pliocene-Pleistocene Afro-Arabian Proboscidea

Taxon	Occurrence (Site, Locality)	Stratigraphic Unit	Age	Key References
	Olduvai Gorge, Tanzania	Bed I, Lower Bed II	2.01–1.8 Ma, 1.8–?1.5 Ma	Beden (1980, 1985, 1987a) and Deino et al. (2021)
("*Elephas recki* cf. *atavus*")	Fejej FJ-1 Site, Ethiopia	Unit III	1.90 Ma	Moullé et al. (2001)
	Konso, Ethiopia	Intervals 1–3, Konso Fm.	1.9–1.7 Ma (alt. 1.91–1.50 Ma)	Suwa et al. (2003)
	Koobi Fora and Ileret, Kenya	Below and above the KBS Tuff; Upper Burgi Mb., Koobi Fora Fm.	± 1.87 Ma; ~1.9 Ma	Beden (1980, 1983, 1985) and McDougall et al. (2012)
	Area 123, Koobi Fora, Kenya	Between the KBS and Lower Ileret Tuffs	1.87–1.53 Ma	Gathogo and Brown (2006)
Elephas recki ileretensis = *Palaeoloxodon recki ileretensis*=*Elephas* (*Palaeoloxodon*) *recki ileretensis*	?West Turkana, Kenya	Kaitio Mb., Nachukui Fm.	1.88–1.6 Ma	Harris et al. (1988a, b)
	Omo, Ethiopia	?Mb. J, Mb. K, Lower Mb. L, Shungura Fm. (alt. ?Upper Mb. G, ?Mb. H)	?1.76 Ma; 1.53–1.185 Ma (alt. <1.945 Ma to 1.76 Ma)	Beden (1980, 1985, 1987a) and McDougall et al. (2012)
("*Elephas recki ileretensis*") ("*Elephas* (*Palaeoloxodon*) *recki ileretensis*")	Baragoli, Republic of Djibouti	Hardened compact clay strata, Gobaad region	In the interval 1.6–1.3 Ma	Berthelet (2001) and Berthelet and Chavaillon (2001)
("*Palaeoloxodon recki ileretensis*")	Gafalo, Republic of Djibouti	Gobaad region	In the interval 1.6–1.3 Ma	Faure and Guérin (1997)
	Kaddanarti, northern Sudan	Coarse pebble deposit	In the interval 1.6–0.5 Ma	Chaix et al. (2020)
	Olduvai Gorge, Tanzania	Upper Bed II	~1.6–1.14 Ma (alt. ?1.75 Ma)	Hay (1976), Beden (1980, 1985), Brown (1994), Tamrat et al. (1995), and Deino et al. (2021)
	Areas 1A, 6, 6A, 8, 8A, 11, Ileret, Kenya	Just below Lower Ileret Tuff up to Chari Tuff, Okote Mb. of the Koobi Fora Fm.	>1.53–1.38 Ma	Beden (1983), Brown and Feibel (1986), and McDougall et al. (2012)
	Areas 103, 104, Koobi Fora, Kenya	Immediately below Koobi Fora tuff (alt. below Okote tuff); slightly below Chari Tuff; "*Metridiochoerus compactus* zone" (equivalent to Mb. L, Shungura Fm.) (alt. above Okote Tuff), Okote Mb., Koobi Fora Fm.	>1.49 Ma (alt. >1.64 Ma) <1.38 Ma (alt. <1.64 Ma)	Beden (1980, 1983, 1985), Harris (1983a), Brown and Feibel (1986), and McDougall et al. (2012)
	Chesowanja, Kenya		>1.42 Ma	Hooker and Miller (1979) and Beden (1985)
Elephas recki recki = *Palaeoloxodon recki recki*=*Elephas* (*Palaeoloxodon*) *recki recki* (alt. "*Elephas recki ileretensis/recki*")	Konso, Ethiopia	Intervals 4–6, Konso Fm.	~1.43–1.39 Ma (alt. 1.5–1.3 Ma)	Suwa et al. (2003)
	Ileret, Kenya	Chari Tuff (alt. above Okote Tuff), Chari Mb., Koobi Fora Fm.	1.38 Ma (alt. <1.64 Ma)	Beden (1980, 1983, 1985), Brown and Feibel (1986), McDougall (2012)

(*Continued*)

TABLE 7.1 (*Continued*)
Major Occurrences of Late Pliocene-Pleistocene Afro-Arabian Proboscidea

Taxon	Occurrence (Site, Locality)	Stratigraphic Unit	Age	Key References
	West Turkana, Kenya	Nariokotome Mb., Nachukui Fm.	~1.3–1.0 Ma	Harris et al. (1988a, b)
("*Elephas* (*Palaeoloxodon*) *recki recki*")	Haïdalo, Republic of Djibouti	Gobaad Fm.	In the interval 1.2–0.5 Ma	Chavaillon et al. (1990)
	Dakanihyalo, Matabaietu D'ar, Wilti Dora, Meadura Dora, Middle Awash, Ethiopia	Lower Dakanihyalo Mb.; Equus beds; Meadura Mb., Lower Wehaietu Fm.	~1.2–1.0 Ma; ~500–400 Kyr	Kalb and Mebrate (1993)
	Omo, Ethiopia	Upper Mb. L, Shungura Fm. (alt. Mbs. K, L)	<1.185 Ma (alt. 1.53 to <1.185 Ma)	Beden (1980; 1985, 1987a), Feibel et al. (1989), Alemseged (2003), and McDougall et al. (2012)
("*Palaeoloxodon recki*")	Olduvai Gorge, Tanzania (subspecies type) (type)	Beds III, IV	1.14–0.82 Ma (disappears before Masek beds, a little before 0.6 Ma)	Beden (1980, 1985, 1987a), Deino et al. (2021), Larramendi et al. (2020), and Zhang (2020)
("*Elephas* (*Palaeoloxodon*) *recki recki*")	Bouri, Ethiopia	Daka Mb., Bouri Fm.	~1.0 Ma (<1.043 Ma; >0.78 Ma)	Saegusa and Gilbert (2008) and WoldeGabriel et al. (2008)
(alt. "intermediate between *E. recki ileretensis* and *E. r. recki*")	Dioli and Aalad areas, Buia, Eritrea	Dioli and Aalad beds, Upper Danakil Fm.	~1.0 Ma, late early Pleistocene	Ferretti et al. (2003b)
("*E. recki recki*")	Butana Bridge, Sudan	Butana Bridge Synthem	Late early Pleistocene or early middle Pleistocene	Abbate et al. (2010)
("*Palaeoloxodon* cf. *P. recki*")	Mulhuli-Amo, Buia area, northern Danakil Depression, Eritrea	Maebele Synthem	~1.0 Ma, late early Pleistocene	Delfino et al. (2018)
	Site 15, Olorgesailie, Kenya	Upper Mb. 1, Olorgesailie Fm.	0.99 Ma (alt. in the interval ~1.0–0.5 Ma) (alt. 0.93–0.70 Ma)	Beden (1980, 1985), Bye et al. (1987), Potts (1989), Deino and Potts (1990), and Potts et al. (1999)
	Melka Kunture (Garba IV), Ethiopia	Melka Kunture Fm.	>0.78 Ma, ~1.0 Ma (alt. <1.79 Ma)	Beden (1980, 1985), Bocherens et al. (1996), Raynal et al. (2004b) (but see Morgan et al., 2012)
	Busidima-Telalak region, Afar, Ethiopia	?Wehaietu Fm.	In the interval 0.8–0.2 Ma	Alemseged and Geraads (2000)
("*Palaeoloxodon recki recki*") (alt. "*Palaeoloxodon antiquus*")	Gesher Benot Ya'aqov, Israel		0.78 Ma	Goren-Inbar et al. (1994, 2000), Saegusa and Gilbert (2008), and Rabinovich et al. (2012)
	Middle Awash, Ethiopia	Bodo Mb., Wehaietu Fm.	In the interval 0.74–0.55 Ma	Kalb and Mebrate (1993), Clark et al. (1994), Kalb (1995), and Rightmire (1996)
(?*Elephas recki recki*) (alt. "*Palaeoloxodon antiquus*")	Revadim Quarry, Israel	Unit 2, Area B	<0.78 Ma; between ≥0.5 and 0.3 Ma (alt. 780–300 Ka)	Rabinovich et al. (2012) and Haynes (2020)
	Nadung'a 4, Kenya	Purported elephant butchery site	~700 Ka	Delagnes et al. (2006)

(*Continued*)

TABLE 7.1 (*Continued*)
Major Occurrences of Late Pliocene-Pleistocene Afro-Arabian Proboscidea

Taxon	Occurrence (Site, Locality)	Stratigraphic Unit	Age	Key References
	Kathu Pan, South Africa	Acheulian deposit	Middle Pleistocene, in interval 0.7–0.4 Ma	Klein (1984, 1988)
	Power's Site, South Africa	Lower terrace, Vaal River sequence, Rietputs Fm.	Middle Pleistocene, similar in age to Kathu Pan	Klein (1988)
	Kanjera, Kenya	Apoko Fm.	~0.5 Ma	Thomas et al. (1998), Plummer and Potts (1989) and Behrensmeyer et al. (1995)
("*Palaeoloxodon* cf. *recki*")	Ti's al Ghadah, Nafud Desert, Saudi Arabia	Unit 5	0.5 Ma (alt. 0.5–0.3 Ma) (alt. 318–280 Ka)	Thomas et al. (1998), Stimpson et al. (2016), Mufarreh et al. (2017), and Stewart et al. (2019)
("cf. *Palaeoloxodon* (*recki*)")	Locality #3 (Thomas locality 3), Nafud Desert, Saudi Arabia		419±39 Ka to 286±30 Ka	Stimpson et al. (2016)
	Namib IV, Namibia	Acheulean context	~350 Ka (alt. in interval 0.7–0.4 Ma) (alt. late Lower or early Middle Pleistocene)	Shackley (1980), Klein (1984, 1988), and Shackley et al. (1985)
(?*Elephas recki recki*) (alt. "*Palaeoloxodon antiquus*")	Holon, Israel		201–198 Kyr (base of sequence 240±29 Kyr)	Chazan et al. (2001) and Davies and Lister (2007)
("*Palaeoloxodon* cf. *recki*")	Shishan Marsh 1 (SM1), Azraq Basin, Jordan	Levels 7b and 8	266±40 Kyr (level 8); 125±12 Kyr	Pokines et al. (2018)
Elephas recki subsp. indet. ("*E. r. brumpti* or *E. r. shungurensis*")	Makapansgat, South Africa	Mb. 4	<3.11 Ma (alt. 2.7–2.5 Ma)	Cooke (1993), Partridge (2000), and Partridge et al. (2000)
	Sterkfontein, South Africa	Mb. 4	2.8–2.4 Ma (alt. 2.5–1.5 Ma)	Cooke (1993) and Berger et al. (2002)
("*Elephas* sp. indet.")	Uraha, Mwimbi, Malawi	Chiwondo Beds, Unit 3	Early Pleistocene	Sandrock et al. (2007)
	Marsabit Road, Kenya		Early Pleistocene, "well before Upper Bed II of Olduvai"	Gentry and Gentry (1969)
("*Elephas* sp.")	Mansourah (Constantine), Algeria		Early Pleistocene	Chaid-Saoudi et al. (2006)
("*Elephas* cf. *recki*")	Salé Village, Morocco		Oldowan context, corresponding with upper levels of Aïn Hanech	Arambourg (1970)
("*Elephas* sp.")	Swartkrans, South Africa	Hanging Remnant Mb. 3	~1.8–1.5 Ma (alt. ~1.5–1.0 Ma)	de Ruiter (2003)
	Dandero, northern Danakil Depression, Eritrea	Upper Danakil Fm.	~1.0 Ma	Shoshani et al. (2001c)
("*Elephas recki*")	Nadung'a 4, West bank of Lake Turkana, Kenya	Nachukui Fm.	1.0–0.7 Ma	Delagnes et al. (2006)

(*Continued*)

TABLE 7.1 (*Continued*)
Major Occurrences of Late Pliocene-Pleistocene Afro-Arabian Proboscidea

Taxon	Occurrence (Site, Locality)	Stratigraphic Unit	Age	Key References
("*Loxodonta atlantica zulu*")	Elandsfontein, South Africa		~1.0–0.6 Ma	Maglio (1973) and Smith and Stynder (2015)
("*E.* cf. *recki*")	Megenta area, Lower Awash basin, Afar Rift, Ethiopia		Middle Pleistocene; Acheulean archeological context	Sahle et al. (2019)
("*Elephas recki*")	Anchrif Quarry, Taghrout, Morocco		Pleistocene; Acheulean archeological context	Marinheiro et al. (2014) and Marinherio (2015)
("cf. *Elephas*/*Loxodonta*")	Mwanganda's Village, Malawi		282 Ka (alt. 1.2–0.6 Ma) (associated stone tools possibly younger)	Clark and Haynes (1970), Kaufulu (1990), and Wright et al. (2014)
Elephas jolensis =*Palaeoloxodon jolensis* ("*E. pomeli*")	Gué de Constantine, Algeria		?middle Pleistocene	Arambourg (1952, 1960), Maglio (1973), and Coppens and Gaudant (1976)
	Aïn Maarouf, Morocco		Early middle Pleistocene	Geraads and Amani (1997)
("*Archidiskodon broomi*") ("*Archidiskodon hanekoni*") ("*Archidiskodon transvaalensis*") ("*Archidiskodon sheppardi*")	Vaal River, South Africa	Lower (younger) terraces and gravels	Late middle Pleistocene	Dart (1927, 1929), Osborn (1928), Haughton (1932), Maglio (1973), and Coppens et al. (1978)
	Left bank, Zambezi River near Victoria Falls, Zambia	Older Falls gravels	Late middle or late Pleistocene	Cooke and Clark (1939)
	Natodomeri, South Sudan, administered by Kenya	Mb. I, Kibish Fm. Mb. II, Kibish Fm.	205–95 Ka 130 Ka at top of Mb. II and KHS tuff in Mb. II dated to 154±7 Ka	McDougall et al. (2005), Brown and Fuller (2008), Brown et al. (2012), and Manthi et al. (2019)
	Herto, Ethiopia	Upper Herto Mb., Bouri Fm.	~160 Ka	Clark et al. (2003)
	Sidi Abder Rahmane (Casablanca), Morocco		?close to middle Pleistocene/late Pleistocene boundary	Arambourg (1960), Maglio (1973), Coppens and Gaudant (1976), Coppens et al. (1978), and Raynal et al. (2004a)
	Duinefontein 2 (=Melkbosstrand), southwestern Cape Province, South Africa		Late Pleistocene, Middle Stone Age archeological contex	Hendey (1967) and Klein (1976)
	Beausejour Farm, Algeria (type)		?late Pleistocene	Pomel (1895), Maglio (1973), and Coppens and Gaudant (1976)
	Port de Mastaganem, Algeria		Late Pleistocene	Arambourg (1960), Maglio (1973), Coppens and Gaudant (1976), and Coppens et al. (1978)
	Zouerate, Mauritania		Late Pleistocene	Maglio (1973) and Coppens et al. (1978)

(*Continued*)

TABLE 7.1 (*Continued*)
Major Occurrences of Late Pliocene-Pleistocene Afro-Arabian Proboscidea

Taxon	Occurrence (Site, Locality)	Stratigraphic Unit	Age	Key References
("*Elephas recki* vel. *E. iolensis*")	Khashm el Girba, Sudan	KGS 1 and 2 Subsynthems	Late middle Pleistocene and late Pleistocene	Abbate et al. (2010)
	El Douira, Tunisia		90–75,000 y BP	Coppens and Gaudant (1976) and Marinheiro et al. (2014)
	Kaiso Village and Behanga I, Western Rift, Uganda	?Rwebishengo Beds	Latest Pleistocene	Cooke and Coryndon (1970), Sanders (1990) and Pickford et al. (1993)

?=attribution, age, or occurrence uncertain; alt., alternative; alternate taxonomic names are in parentheses.

Haile-Selassie, 2012). As the number of shearing surfaces was increased by unit area, enamel consequently decreased substantially in thickness.

From the inception of elephants in the late Miocene to the disappearance of the last of the "*Elephas recki* complex" and its African descendants, third molar plate numbers increased from 6 to 21. An equally impressive aspect of the reorganization of elephant molars as long-lasting horizontal shearing instruments was the increase of their hypsodonty, transformed over a nearly seven-Myr span from low, broad transverse profiles of squat, pyramidal plates separated by broad V-shaped transverse valleys to relatively very tall, parallel-sided transverse plate cross-sections constructed of straight, anteroposteriorly narrow plates spaced by anteroposteriorly tight, cementum-filled U-shaped transverse valleys. Hypsodonty indices over the same seven-Myr span increased from less than 70 to well over 200. The benefit of hypsodont molars, to keep them "in play" for more time against a tough wear regimen, provided their bearers with a competitive advantage to live longer and pass on a greater number of offspring.

Mammalian teeth achieve hypsodonty in a variety of developmental pathways (Koenigswald, 2011). In elephants, molars become high-crowned via multi-cusped hypsodonty, in which the teeth are formed of isolated cusps comprising columns united at their bases. As teeth become taller, their stability is maintained by the addition of greater amounts of cementum. Augmentation of crown height is achieved through extension of Phase I development, wherein the occlusal surface is constructed and vertically elongated, and these molars are characterized by constant position of the occlusal surface. It is not known why elephants were able to engage this developmental pathway, but not stegodonts. Other mammalian examples of this form of molar hypsodonty are desmostylians and some suids. Elephants also employ a Phase III–dominated form of hypsodonty in the hypertrophy of their tusks, which involves continuous extension of the dentine development phase of ontogeny.

"Winners" in the final episode of proboscidean fossil occurrences were those species that exhibited the capacity for rapid phyletic increase in plate number, lamellar frequency, hypsodonty, and enamel folding; "losers" that were slower to develop these adaptations were geographically and ecologically marginalized, and most either migrated off the continent, or went extinct. For example, with only very rare exceptions (Lohrke, 2017), deinotheres remained browsing specialists in forested, closed-canopy conditions (Harris, 1975, 1978a), until they disappeared from Africa near the end of the early Pleistocene because their habitats became too fragmented or unreliable due to seasonality and arid conditions. African stegodonts and anancine gomphotheres became mixed feeders or grazers (Cerling et al., 1999, 2003; Zazzo et al., 2000; Harris et al., 2003; Semaw et al., 2005; Kingston and Harrison, 2007; Levin et al., 2008; Uno et al., 2011; Roche et al., 2013; Drapeau et al., 2014; Manthi et al., 2017), but failed to evolve hypsodonty and plate numbers to the same extent as elephants, contributing to their competitive demise in Africa during the late Pliocene and early Pleistocene, respectively; mammoths were largely relegated to the northern edge of the continent by the end of the late Pliocene and vanished from Africa during the early Pleistocene (Maglio, 1973). By emigrating to Eurasia during the mid-Pliocene and onward to North America during the early Pleistocene, however, mammoths were able to thrive in new conditions and undergo multiple speciations (Dudley, 1996; Fisher, 1996; Lister, 1996, 2001; Lister and van Essen, 2003; Lister et al., 2005; Lister and Bahn, 2007; Markov, 2012).

More vexing is the Pleistocene biogeography of loxodont elephants, which contemporaneously survive as African savanna (*L. africana*) and forest (*L. cyclotis*) species. Possible successors to Pliocene *L. adaurora* persisted in eastern Africa into the early Pleistocene (Beden, 1987a), dentally paralleling early members of the "*E. recki* complex," but it is unlikely that they contributed to the evolution of modern African elephant species. Pliocene *L. exoptata* is morphologically a better candidate to have evolved into *L. africana* during the start of the early Pleistocene (Beden, 1987b; Sanders, 2011), but the fossil record of the lineage following speciation is extremely poor, and the territories of the ancestors of modern African elephants for most of the Pleistocene are unknown. Perhaps they retreated to montane habitats and the forests of Central and West Africa. A close phylogenetic cousin, *L. atlantica*, had a brief existence in North Africa during the end of the early to middle Pleistocene (Zhang and Lister, 2017), but was also marginalized away from the center of the continent. Its extinction may be linked to pan-African expansion of the highly advanced, terminal continental descendants of the "*E. recki* complex," *Palaeoloxodon jolensis*, or the same ecological factors that contributed to the extinction of specialist grazers in African *Palaeoloxodon*, but this is also speculative.

For most of the late Pliocene-Pleistocene, then, the "winners" and most widely distributed proboscideans on the continent were constituents of the "*E. recki* complex," classically subdivided into successive stages (Maglio, 1973) or subspecies ("*brumpti*," "*shungurensis*," "*atavus*," "*ileretensis*," and "*recki*") (Beden, 1980). Progressive changes in hypsodonty, plate number, lamellar frequency, enamel thickness and folding in molars of these stages intimated anagenetic evolution of the lineage (Maglio, 1973; Beden, 1980; Sanders and Haile-Selassie, 2012). Initial challenges to these schema suggested that they were based on inappropriate application of the subspecies concept and failed to account for temporal overlap of stages and excessive metric variability within stages (e.g., Todd, 2001, 2005; Ferretti et al., 2003b). These complaints can largely be dismissed by revised biostratigraphy and realization that (a) the stages or "subspecies" are arbitrary lineage subdivisions delineated by boundaries of geological units, and (b) metric variance of the subspecies is strongly affected by sample size—subspecies with larger dental samples do not exceed variability observed in modern elephant populations. More profound reasonings for recognizing potential cladogenesis within the "*E. recki* complex" and for taxonomic reorganization of the group have been generated by studies that examined the comparative cranial morphology of the subspecies in a systematic framework (see discussion below; Saegusa and Gilbert, 2008; Zhang, 2020). These studies suggest that, at the very least, the geologically youngest stages ("*E. recki ileretensis*" and "*E. recki recki*") appear (along with "*E. jolensis*") to belong in *Palaeoloxodon* as precursors of the late Pleistocene Eurasian radiation of that clade, and should be differentiated at the species and/or genus level from geologically older "*E. recki brumpti*" and "*E. recki atavus*." Morphologically, "*Elephas recki brumpti*" does not belong in *Palaeoloxodon*, whereas it remains unresolved whether (a) "*E. recki atavus*" belongs in *Elephas* (Zhang, 2020) or *Palaeoloxodon* (Saegusa and Gilbert, 2008), and if so, (b) whether *Palaeoloxodon* is a subgenus of *Elephas*. Paleogenomic analyses further complicate these considerations, indicating that *Palaeoloxodon* is part of the *Loxodonta* rather than *Elephas* clade (Meyer et al., 2017; Palkopoulou et al., 2018).

Despite these unresolved conundrums, it is likely that at different intervals, successive members of the "*E. recki* complex" provided ancestors for both extant Asian elephants and the impressive diversification of Eurasian *Palaeoloxodon*. They adaptively tracked Plio-Pleistocene ecological changes both behaviorally and morphologically (Maglio, 1970b, 1973; Beden, 1980; Cerling et al., 1999; Sanders and Haile-Selassie et al., 2012) to become committed grazers (see Manthi et al., 2019) and the dominant elephants in Africa through numerous intervals of climatic instability. Thus, it is surprising that they suddenly vanished from the continent during the late Pleistocene, replaced by loxodont elephants from peripheral areas that today are mixed feeders with strong browsing preferences.

The demise of *Palaeoloxodon* in Africa, and with it the "*E. recki* complex," was not an isolated phenomenon but part of an extinction event that occurred across the length of Africa during the late Pleistocene to Holocene that extirpated many other large-bodied ungulate grazers, including species in the suid lineages *Kolpochoerus* and *Metridiochoerus*, and the giant buffalo *Syncerus antiquus*, an example of "taxonomic winnowing of grassland specialists" (Faith, 2014). The reasons for this involved the increased amplitude of climate variability over the last million years, and associated substantial alterations in the structure and productivity of grassland habitats (Faith, 2014 and references therein). In the case of replacement of *Palaeoloxodon* by *Loxodonta*, a harsh dry, cold climate interval marked by intense episodes of drought around the start of the late Pleistocene is implicated (Cohen et al., 2007; Scholz et al., 2007; Tierney et al., 2017).

SYSTEMATIC PALEONTOLOGY

ELEPHANTIDAE Gray, 1821
ELEPHANTINAE Gray, 1821
***LOXODONTA* F. Cuvier, in F. Cuvier and Geoffroy Saint-Hilaire, 1825 (anonymous emendation, 1827)**
***LOXODONTA ADAURORA KARARAE* Maglio, 1970b**
(Tables 7.1 and 7.2; Figures 7.1 and 7.2)

Partial Synonymy—*Elephas ekorensis* (in part) Maglio, 1973; *Loxodonta adaurora* Coppens and Howell, 1974; *Loxodonta* sp. Harris, 1978; *Loxodonta adaurora kararae* Beden, 1979, 1983; *Phanagoroloxodon planifrons* Zhang, 2020

In his doctoral dissertation, Beden (1979) subdivided *Loxodonta adaurora* into an early to late Pliocene subspecies "*L. a. adaurora*" and a late Pliocene to primarily early Pleistocene subspecies "*L. a. kararae*," and later codified the taxonomy more formally in publication (Beden, 1983). The subspecies nomen is derived from the Karari escarpment at Koobi Fora, Kenya where the subspecies type specimen was collected. The subspecific type is a partial cranium with M3s, KNM-ER 347, from just below the KBS Tuff at Koobi Fora, datable to a little older than 1.87 Ma (McDougall et al., 2012). It is missing its frontoparietal region, apex of the vault, and upper part of the occipital planum, as well as much of the rostrum (Figure 7.2A). This cranium was originally allocated by Maglio (1971) to *Elephas ekorensis*. At the time, the specimen had not been completely prepared, and Maglio's taxonomic assignment was biased by an erroneous assumption of greater antiquity of the specimen, based on a mistaken field number identification (pertaining instead to *L. a. adaurora* specimen KNM-ER 346) and consequent belief that the cranium derived from the Kubi Algi Fm. (Beden, 1983).

Salient features of KNM-ER 347 include the low and anterior position of the orbits very mesial to the M3s, high

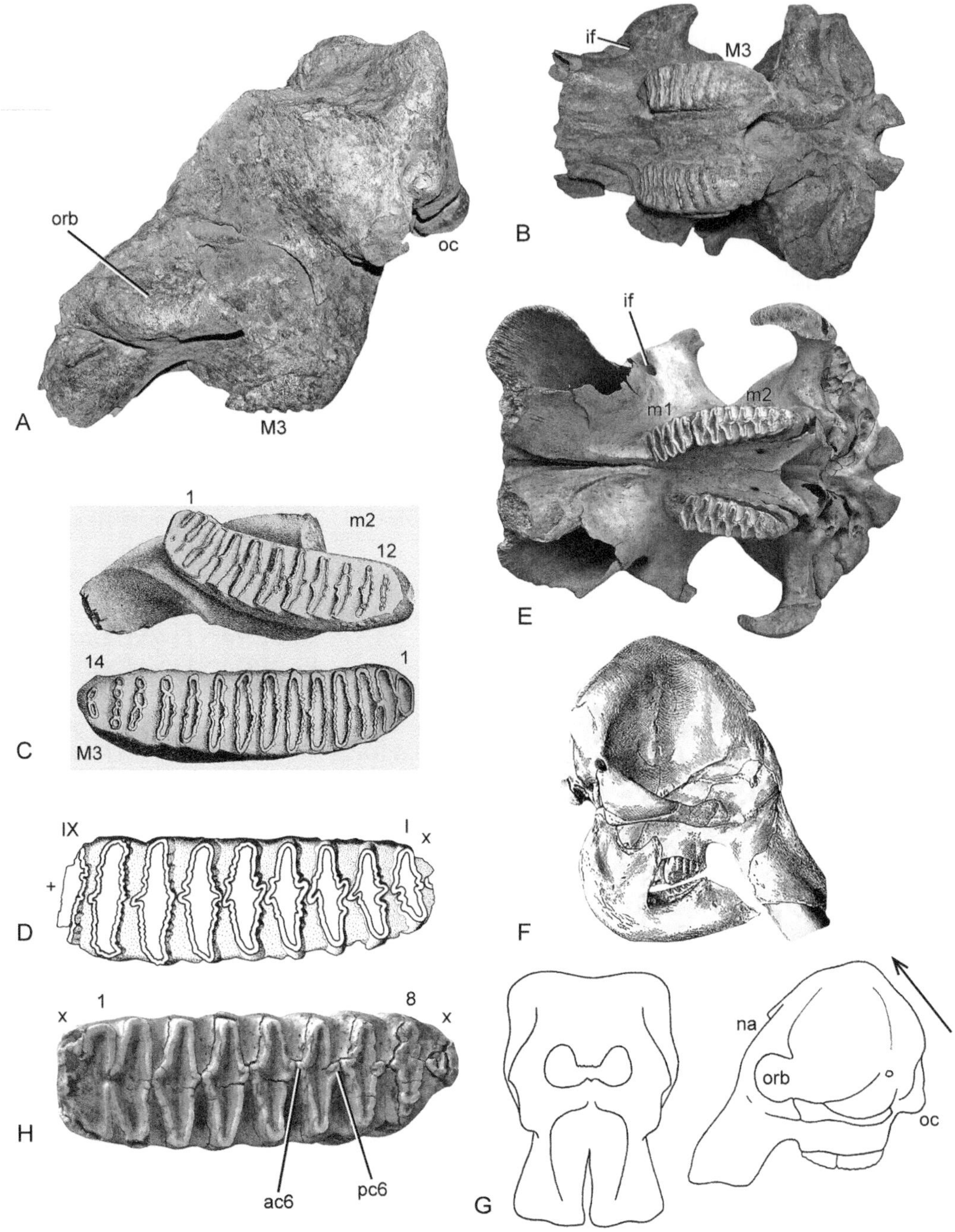

FIGURE 7.2 Skull and molar specimens of fossil and extant *Loxodonta*. Abbreviations: ac, anterior pretrite accessory conule; if, infraorbital foramen; M/m, upper/lower molar; na, nasal aperture; oc, occipital condyle; orb, orbit; pc, posterior pretrite accessory conule; x, pre- or postcingulum(id); +, indicates missing morphology; 1, 2, 3, …, plate number counted from the anterior of the crown; I, II, III, …, plate number counted from the posterior of the crown. +, indicates missing morphology. Specimens not to relative scale. (A) Partial cranium, KNM-ER 347 (type) from Koobi Fora, Kenya (just under the KBS Tuff), lateral view, *Loxodonta adaurora kararae*. anterior to the left. (Image courtesy of J. Saarinen and P. Mbatha.) (B) Cranium with right and left M3, KNM-ER 347 (type) from Koobi Fora, Kenya, *L. adaurora kararae*, ventral view. Anterior to the left. (C) m2 (co-type) and M3, from Ternifine, Algeria, occlusal view, *Loxodonta atlantica*. Anterior to the left for the m2 (top) and to the right for the M3 (bottom). (After Pomel [1895:pl. VIII, figs. 1 and 3]). (D) Partial M3, L 161-19a from Mb. D, Shungura Fm., Omo, Ethiopia, occlusal view, *Loxodonta africana*. Anterior to the left. (From Beden [1987a:fig. 8], with permission of the Museum National d'Histoire naturelle, Paris.) (E) Cranium, ventral view, extant specimen of *Loxodonta africana*. Anterior to the left. (F) Drawing of skull, AMNH 3283, lateral view, extant specimen of *Loxodonta africana*. Anterior to the right. (From Osborn [1942:fig. 1060], with permission of the American Museum of Natural History.) (G) Schematic of cranium, anterior and lateral views, extant specimen of *Loxodonta africana*. Anterior to the left in the lateral view. Note the anterosuperior angulation of the occipital planum. (From Maglio [1970b:fig. 2B], with permission of the Museum of Comparative Zoology, Harvard University.) (H) m2, occlusal view, extant specimen of *Loxodonta africana*. Note the formation of loxodont sinuses by incorporation of anterior and posterior accessory conules into the midline of the enamel wear-figures.

elevation of the occipital condyles above the palatal plane, small size and modest posterior projection of the occipital condyles, flat, anterosuperiorly oriented occipital surface, absence of parietal swelling and no superolateral bossing of the parietals, deep temporal fossae nearly orthogonal to the palate, and straight tusk alveoli that are strongly divergent and separated by a triangular external incisive fossa (Beden, 1983). In addition, Beden (1983) noted that the position and form of the damaged nasal orifice suggests that the frontal was wide and anteroposteriorly short with a flat or convex surface. The fossa for the nuchal ligament is deep and high. External auditory meati are at the same vertical level as the occipital condyles.

The M3s of the type cranium were still emerging posteriorly at time of death, with the anteriormost six plates in occlusal wear (Figure 7.2B); they are estimated to have a plate formula of x10. Approximate age equivalence with living African elephants is with Laws' (1966) stage XXIII, or 43±2 years old at death. In occlusal view, the worn plates are transversely rectilinear, composed of three to six conelets, display rounded central pillars without significant median anteroposterior expansion, and show no development of loxodont sinuses. Dimensions of the molars are L=200–205 mm, W=81–82 mm, H=90 mm (yielding hypsodonty indices of 109–111), and enamel thicknesses of 2.8–3.5 mm (Beden, 1983). Plates are more closely spaced than in *L. a. adaurora*, with lamellar frequencies of 4.5–5.0. Hypsodonty and enamel thickness are also more progressive than in *L. a. adaurora*, and the absence of obvious anterior and posterior accessory conules in *L. a. kararae* also differentiate it from its putative ancestor. Modest size of the M3s indicates that the individual may have been a female. Other molars from the same horizon as the type at Koobi Fora, and from Mbs. B2 and E of the Omo Shungura Fm., dated to about 3.4 and 2.4–2.3 Ma (Brown and McDougall, 2011), respectively, exhibit similar morphology, with m3 having as many as 12 plates (Beden, 1983, 1987a). The M3 from Mb. B2, although considerably older than the type M3s, is metrically similar, with a plate formula of x9 and dimensions of L=225 mm, W=98 mm, ET=3.0–3.5 mm, and LF=4.2 (Beden, 1987a). It appears to have five to seven conelets per plate and lacks obvious expression of accessory conules.

Features such as the anterosuperior angulation of the nuchal planum, the high position of the fossa for the nuchal ligament, the lack of inflation of the occipital region and absence of parietal swelling (no "bossing"), indications that the frontal was convex or flat, the posterior protrusion of diminutive occipital condyles, straight, strongly divergent tusk alveoli, and the level of the external auditory meati at the same height as the occipital condyles suggest affinity to *Loxodonta* and differ from the morphology of crania the "*Elephas recki* complex." These features were sufficient to have led Beden (1983, 1987a) to his decision to erect *L. a. kararae* to accommodate the cranium and a few other specimens from Koobi Fora and the Omo Shungura Fm. sequence in Ethiopia as intermediate between *L. a. adaurora* and extant *L. africana* (but closer to the former morphologically). However, the lack of loxodont sinuses and poor or absent development of median accessory conules differ from the molar morphology of extant African elephants and other fossil members of the clade.

Alternatively, Zhang (2020) placed KNM-ER 347 (and *L. adaurora kararae*) in a new binomial combination, *Phanagoroloxodon planifrons*, along with fossil crania from the late Pliocene-early Pleistocene Upper Siwalik Series in South Asia and with African *Elephas recki brumpti*, including a cranium from unit Mb. B11 of the Shungura Fm. in the Omo, Ethiopia, specimen L.1.70. He supported this hypothesis by interpreting the tusk configuration of KNM-ER 347 as non-divergent, implying that they are artificially splayed due to preservational asymmetry of the cranium (although proximal configuration of the tusk alveoli, especially on the left side, indicates real divergence), by the very low, anterior position of the orbits, particularly relative to the height of the occipital condyles, and by the more robust development of the post-temporal aspects of the cranium, compared with typical *L. a. adaurora*.

Nevertheless, although there is only weak support for its inclusion in *Loxodonta*, it is unlikely that KNM-ER 347 belongs to the same species as Omo specimen L.1.70 or in *Phanagoroloxodon planifrons*. The Omo cranium, which is a million years older than KNM-ER 347, differs from it in having nascent bossing of the occipitoparietals, less difference in elevation of the orbits and occipital condyles, truly parallel tusk alveoli, and distinctly more laterally protuberant orbits. KNM-ER 347 resembles a cranium of the type species of *Phanagoroloxodon* (Pliocene-aged *Ph. mammontoides*) from Russia by shared features of low, anteriorly placed orbits, an uninflated occipital planum, high position of the fossa for the nuchal ligament, and deep temporal fossa nearly orthogonal to the alveolar plane (see Garutt, 1995). However, *Ph. mammontoides* has a proven sagittal depression in the margin of the nuchal crest and differs from KNM-ER 347 by possessing a slightly concave frontal surface longitudinally and transversely and tusk sheaths set closer together that diverge just at their distal ends. Curiously, the tusks of *Ph. mammontoides* were reported to be spirally curved, in the manner of mammoth incisors (Garutt, 1995), which complicates the relationship of this genus to crown elephant genera and the appropriateness of referring *Elephas planifrons* to *Phanagoroloxodon* (Zhang, 2020).

Although the frons of *E. planifrons*, or *Ph. planifrons*, is broad and flattened, and has a very raised basicranium and location of the occipital condyles well above the level of the palate, as well as an anteriorly located orbit, it differs from the morphology of KNM-ER 347 by a vertically straighter basicranium, more downturned, parallel tusk sheaths, less posteriorly protruded occipital condyles, and orbits relatively taller and anteroposteriorly shorter, more in line with the chord from the nuchal crest to the tusk alveoli, and more laterally flared. In addition, the occipitoparietals in South Asian crania of *Ph. planifrons* form distinct lateral bosses separated by a midline depression; cranial "bossing" seems unlikely in KNM-ER 347 and damage to the vault makes it

TABLE 7.2
Adult Dental Formulae and Incisor Dominance in Late Pliocene-Pleistocene African Proboscidea

Taxon	Upper Permanent Dentition	Largest Upper Incisor	Lower Permanent Dentition	Largest Lower Incisor
Deinotherium bozasi	P3–4, M1–3	—	i2, p3–4, m1–3	i2
Loxodonta adaurora kararae	I2, M1–3	I2	m1–3	—
Loxodonta atlantica	I2, M1–3	I2	m1–3	—
Loxodonta exoptata	I2, M1–3	I2	m1–3	—
Loxodonta africana/cyclotis	I2, M1–3	I2	m1–3	—
"*Elephas recki* complex"[a] including *Elephas brumpti*, *E. shungurensis*, *E. atavus*, *Palaeoloxodon recki ileretensis*, and *P. recki recki*	I2, M1–3	I2	m1–3	—
Palaeoloxodon jolensis	I2, M1–3	I2	m1–3	—
Mammuthus africanavus	I2, M1–3	I2	m1–3	—
Mammuthus meridionalis	I2, M1–3	I2	m1–3	—

C/c = canine; I/I = incisor; M/m=molar; P/p = premolar.

[a] There is evidence that *Elephas planifrons* retained permanent premolars in its dentition (Sanders, 2017). Zhang (2020) placed *Elephas planifrons* together with *E. brumpti* in *Phanagoroloxodon planifrons*, which if correct would suggest that permanent premolars should eventually be recovered for *E. recki brumpti*.

impossible to document the presence or absence of a midline nuchal depression. While the molars of "*L. adaurora kararae*" and *Elephas* (or *Phanagoroloxodon*) *planifrons* are equally conservative in forming simple enamel wear figures (often, a "dash-dot-dot-dash" wear pattern) without significant contribution of accessory conules, and overlap in plate number (see Maglio, 1973), the cranial evidence for inclusion of KNM-ER 347 in *Phanagoroloxodon planifrons* is not compelling. The proposed conspecificity of "*L. adaurora kararae*" and South Asian *Ph. planifrons* also poses a biogeographic problem, as they have no intermediate occurrences to link them—a conundrum alluded to by Zhang (2020:pp. 107–108).

As with all other elephant taxa of the mid-Pliocene through Pleistocene interval, there are no lower tusks or permanent premolars known for *L. adaurora kararae* (Table 7.2). However, permanent premolars have been recovered for South Asian *Ph. planifrons* (Sanders, 2017), suggesting that if Zhang (2020) is correct about the specific affinities of *L. adaurora kararae* and *E. recki brumpti*, they should also have permanent premolars.

LOXODONTA ATLANTICA (Pomel, 1879)
(Tables 7.1 and 7.2; Figures 7.1 and 7.2)

Partial Synonymy—*Elephas atlanticus* Pomel, 1879; *Elephas antiquus recki* MacInnes, 1942:94; *Palaeoloxodon atlanticus* Osborn, 1942; *Elephas* (*Archidiskodon*) *recki* Arambourg, 1947:253–269; *Elephas atlanticus* Arambourg, 1947; *Elephas pomeli* (in part) Arambourg, 1952; *Elephas atlanticus* Arambourg, 1960; *Elephas recki* stage 4 Cooke and Maglio, 1972:164; *Loxodonta atlantica* (in part) Maglio, 1973; *Paraloxodonta atlantica* Garutt, 1986

Loxodonta atlantica atlantica is a middle-late Pleistocene North African taxon (Table 7.1), distinguished by its great body size and, for the genus, derived condition of its molars. The species was subdivided regionally by Maglio (1973) into *L. atlantica atlantica* for the type series from North Africa, and *L. atlantica zulu* for material from southern Africa. Coppens (1965) and Beden (1987a) recognized a third subspecies from Central and East African sites, *L. atlantica angammensis*, which suggested a pan-African distribution of the species. Of these subspecies, only *L. atlantica atlantica* is accepted uncritically as belonging to the species, and therefore use of the subspecific nomenclature may be unnecessary.

Molar plate formulae for *L. atlantica* third molars exceed counts for molars of extant African elephants and other members of the genus. Proportions of molars are relatively narrow and very hypsodont (Maglio, 1973). Occlusal wear figures form loxodont sinuses ("<>") with irregularly folded enamel; notably, median segments of enamel loops may be bifurcated into Y-shaped projections. A cranium from Ternifine, Algeria was reported to be rounded and resemble that of extant African elephants (Maglio, 1973; Coppens et al., 1978), but it is heavily damaged. Occipital condyles of the specimen are relatively very large while the tusk alveoli were described as small (Maglio, 1973), although they do not appear disproportionately diminutive. Maglio (1973) and Coppens et al. (1978) depicted the cranium as having less divergent tusk alveoli with narrower premaxillaries than in *L. africana*. Subsequent study by Zhang and Lister (2017) showed that this cranium is more anteroposteriorly foreshortened than the crania of extant African elephants, with greater constriction at the proximal end of the premaxillaries and narrower across the exoccipitals, and that the tusk alveoli are downward facing and impressively laterally flaring as in other species of the genus. In addition, Zhang (2020) suggested that the external incisive fossa of this cranium (MNHN F.TER 006) is oval-shaped, although it is incomplete and poorly preserved. The mandible of this species has deep, extensive masseteric fossae and a relatively salient but short, mostly forwardly-directed symphyseal spout (Zhang and Lister, 2017).

Molar specimens studied by Maglio (1973) and attributable to the species include an M3 from Ternifine (MNHN no No.) with 14 plates and dimensions of L=327.8 mm, W=89.3 mm, H=154.1 mm (yielding HI=172), ET=3.4 mm, and LF=4.4; an M2 from Ternifine (MNHN 25-2) with 10 plates and dimensions of L=220.0 mm, W=82.8 mm, H=125.1 mm (HI=+154), ET=2.7, and LF=4.5; the cotype m3 from Ternifine (MNHN no accession number) with 15 plates and dimensions of L=340.0 mm, W=79.2 mm, H=140.1 mm (HI=175), ET=3.6 mm, and LF=3.9; and an m2 from Constantine, Algeria (MNHN no accession number) with 12 plates and dimensions of L=+258.6 mm, W=70.3 mm, H=122.6 mm (HI=184), ET=3.3 mm, and LF=5.1. Pomel (1895) figured an m2 (cotype) with 11 or 12 plates and dimensions of L=240 mm, W=60 mm, and an M3 with 13 or 14 plates and dimensions of L=324 mm and W=84 mm (Figure 7.2C). Molar plate formulae for the North African assemblage of the species are M1/m1, 8–9/7–8; M2/m2, 9–11/11–12; M3/m3, 14/10–16 (Maglio, 1973; Zhang and Lister, 2017).

At the middle to late Pleistocene interval during which these elephants were present in North Africa, representatives of the "*Elephas recki* complex" were the dominant elephants in eastern Africa, and the more peripheral continental distribution of *L. atlantica atlantica* may constitute an example of competitive ecological exclusion (Coppens et al., 1978). Both groups were convergent in their hypsodonty, enamel folding, and increased numbers of plates. *Loxodonta atlantica* appears morphologically to have been a grazing specialist whose disappearance temporally parallels that of elephants in the "*E. recki* complex," probably for the same global environmental and ecological reasons (Sanders et al., 2010a; Manthi et al., 2019), or because of the pan-African distribution of *Palaeoloxodon jolensis*. It is possible that *L. atlantica* was foraging primarily on C_3 grasses, however, as there is evidence at middle Pleistocene Ternifine Quarry for C_3 grasslands (Bocherens et al., 1996). The paleobiogeography of *L. atlantica* suggests that it evolved via allopatric speciation by long-distance dispersal from the lineage that produced the extant African elephants. The archeological contexts within which the species samples were recovered are all Acheulian (Youcef, 2020).

A putative second subspecies of *L. atlantica*, "*L. atlantica angammensis*," was named after its subspecific type locality of Angamma Yayo in Chad (Coppens, 1965). Specimens comprising the site sample include an incomplete cranium of a young individual with dP3–4 in occlusion and M1 forming in the crypt, associated with a partial skeleton, a partial cranium with M2, and a maxillary fragment with M2. The M2 in the partial cranium is immense for its tooth position, with dimensions of L=250 mm, W=85 mm. Its plate formula is x8x. Plates are broadly spaced (LF=3.0–3.5) and enamel is very little folded and not very thick (ET=3.0 mm). Salient loxodont sinuses are present in enamel wear figures; the anterior median expansions of these in several plates overlap next to the posterior expansions of adjacent enamel loops (Coppens, 1965). The M2 of the maxillary fragment has nine plates and measures L=205 mm and W=71 mm. Enamel thickness is 2.5–3.0 mm and the plates are broadly spaced (LF=3.5–4.0). The molar is very hypsodont, with height of the seventh plate reaching 155 mm. The anterior plates are in wear and exhibit very strong anterior and posterior accessory conules incorporated into loxodont sinuses.

The cranium of the young individual has not been adequately described and its measurements have not been published. It is sheared on its left side, exposing the brain cavity and pneumatization of the frontoparietal vault. The bases of the tusk alveoli are strongly downturned below the nasal aperture, which is at the level of the top half of the orbit (observable on the right side). The frontoparietal vault is convex transversely and anteroposteriorly. Inclination of the occipital planum is anterior at its superior margin. Morphologically, the cranium appears indistinguishable from those of young extant African elephants. The dP3 of the cranium has a plate formula of 5x and dimensions of L=62 mm, W=32 mm, and ET=1.0. Lamellar frequency of the dP3 is between 8.5 and 9.0. A dp4 from dentaries associated with the cranium exhibits eight plates and has dimensions of L=103 mm, W=41 mm, and thin enamel measuring 1.0 mm. Its lamellar frequency is 7.25. Loxodont sinuses are evident in the enamel wear figures (Coppens, 1965).

The age of these fossils and their faunal associations are uncertain. Angamma Yayo produced a heavily weathered hominin craniofacial fragment that Coppens (1961) named "*Tchadanthropus uxoris*" and considered to be an australopithecine. However, Cooke (1978) stated that this attribution was "problematic," and Wood (2002) wrote that the cranial fragment is that of a modern human that superficially resembles an australopithecine face because of its wind-scoured preservation. A small form of the extant hippo species *Hippopotamus amphibius* was also recovered from the site (Cooke, 1978; Tillet, 1985). The base of the Angamma Formation has been dated radiometrically at its base to between 10,160 and 9,260 y bp, which would make these faunal elements, including the *L. atlantica angammensis* specimens, early Holocene in age, but one caution to any definitive dating of the fauna is that the fossils were discovered in what is likely a secondary context (Tillet, 1985).

Additional, geologically older fossil specimens from the Omo Shungura Fm. were subsequently added to the subspecies (Beden, 1979, 1987a). The Omo material includes deciduous premolars (Omo 175.2.73.1576 and 175.2.73.939) and upper third molars from Mb. D (Figure 7.2D; L 161.19a and b), and an incomplete cranium with an M3 from Mb. F (Omo 1E.67.1036), dated to ~2.44 Ma and between 2.32 and

2.27 Ma, respectively (McDougall et al., 2011). The M3s from Mb. D are anteriorly incomplete, but Beden (1987a) estimated that originally they were composed of 12 plates. They are very narrow (W=70 mm), have modestly thick enamel (ET=2.4–3.1 mm), and plates are more closely spaced than in earlier forms of *Loxodonta* (LF=4.2–4.8). Cementum is well distributed in the transverse valleys but does not cover the apices of the plates as they are all worn. Enamel is finely to coarsely folded, particularly in median parts of wear figures; enamel folding is more strongly and regularly expressed toward the base of the plates. Notably, the enamel wear figures form true loxodont sinuses ("<>"), with salient anterior and posterior projections that incorporate accessory conules. In plates IV-VI, counted from the posterior, the anterior median fold is bifid. The anterior projections of the loxodont sinuses are angled to the lateral side of the posterior projections of the preceding wear figures.

The M3 of the cranium from Mb. F has a plate formula of x12x, with the anteriormost 10 plates in wear. The crown is hypsodont (HI=154) and narrow, given dimensions of L=267 mm, W=80 mm, and H=123 mm. Enamel is modestly thick, ET=2.3–2.8 mm, and plates are even more closely spaced than in Mb. D, with the lamellar frequency reaching 5.0. Cementum abundantly fills the transverse valleys. With wear, the enamel remains relatively unfolded and the enamel loops form more anteroposteriorly compressed loxodont sinuses, with salient anterior and posterior projections that may contact the adjacent plates across the transverse valleys. The loxodont sinuses are asymmetrical, with anterior and posterior projections both angled to the lingual side of the crown. The crown narrows distally. If the Angamma Yayo and Omo samples comprise a single taxon, plate formulae are: dP3/dp3, 5x/5x; dP4/dp4, --/x8; M2/m2, x8x–9/--; M3/m3, x12x/--, overlapping with the molar plate formulae for extant African elephants (Roth, 1992).

Omo cranium IE.67.1036 preserves only part of the palate, upper rostrum, and orbital region on the right side (Beden, 1987a). The anterior border of the nasal aperture is located at the level of the zygomatic process of the frontal, a little higher than is typical for modern African elephant crania. Otherwise, the preserved morphology is similar to that of extant African elephant crania; for example, the external incisive fossa is distally broad, proximally deep, and triangular in overall shape, widening distally. The minor differences between the Omo specimens and extant *L. africana* in nasal aperture height and folding of enamel in some molars were thought sufficient by Beden (1987a) to differentiate them at the species level. In contrast, Maglio (1973) and Coppens et al. (1978) subsumed the subspecies into *L. africana*. More recently, Sanders et al. (2010a) remarked that with the exception of some enamel folding, the subspecies shares no special similarity to *L. atlantica atlantica* and were therefore dubious about its inclusion in *L. atlantica*. Moreover, Zhang (2020) also argued that such differences could be accommodated in the range of variation observed in extant African elephants, and consequently he included "*L. atlantica angammensis*" in *L. africana*. This is a reasonable interpretation of the Omo and Angamma Yayo fossils and provides evidence that the fossil record of *L. africana* extends to at least the beginning of the early Pleistocene.

A supposed third subspecies of *L. atlantica*, "*L. atlantica zulu*," was erected by Maglio (1973), based on fossils from southern Africa. The first iteration of this taxon was made by Scott (1907), who followed a classificatory scheme that placed African elephants into a subgenus of *Elephas* ("*Elephas* (*Loxodon*) *africanus*") and promoted a phylogeny in which modern African elephants were hypothesized to have evolved from a southern African fossil form, "*Elephas* (*Loxodon*) *zulu*." The fossils on which these ideas were based were right and left m3s, PU 11548, from Windhoek Spring Deposit, Namibia. Dimensions of one of these molars are L=265 mm, W=85 mm, and H=81 mm in wear (Scott, 1907). Plate formula appears to be x11x or slightly greater. At least the first nine plates incorporate small but prominent posterior accessory conules into their occlusal wear figures and several plates also exhibit tiny anterior accessory conules in the midline. With heavy wear, there may be modest midline anteroposterior expansion of the occlusal wear figures, but there is no formation of true loxodont sinuses. In plates worn to full enamel loops, enamel is folded along the length of the loops.

Later, Osborn (1942) changed the species name to "*Loxodonta zulu*," and Maglio (1973) added isolated fossil molars from Elandsfontein, South Africa to the taxon's hypodigm and absorbed it into *L. atlantica* as a subspecies. The identity of this taxon has also been confounded by its referral to "*Loxodonta* (*Palaeoloxodon*) *antiquus recki*" (Singer and Crawford, 1958). More recently, following reexamination of the material from Elandsfontein, of middle Pleistocene age (1.0–0.6 Ma; Braun et al., 2013), Smith and Stynder (2015) reaffirmed *L. atlantica zulu* as a valid subspecies. Maglio (1973) confused PU 11548 as an M3, but of his m3s accounted for from Elandsfontein, specimen SAM 14220 has a plate formula of 12x and dimensions of L=216 mm, W=80 mm, H=106 mm (yielding HI of 140), ET=2.3 mm, and LF=6.5. Overall, plate formulae provided for the subspecies are m1, 10; m3, x11x-13x; M2, 12; and M3, 12 (Scott, 1907; Osborn, 1942; Maglio, 1973). Despite the confidence expressed by these authors to recognize these molar specimens as a subspecies of *L. atlantica*, the occlusal morphology does not support those assignments, with no true loxodont sinuses formed in occlusal wear. Thus, only the northern African sample of *L. atlantica* is recognized with certainty as belonging to the species.

LOXODONTA AFRICANA (Blumenbach, 1797)
(Tables 7.1 and 7.2; Figures 7.1 and 7.2)

Partial Synonymy—*Elephas africana* Blumenbach, 1797; *E. capensis* Cuvier, 1798; *Elephas priscus* Goldfuss, 1821; *Loxodonta africana* Gray, 1843; *E. (Loxodonta) oxyotis* Matschie, 1900; *E. africanus capensis* Lydekker, 1907; *E. africanus knochenhaueri*

Lydekker, 1907; *L. africana africana* Heller and Roosevelt, 1914; *L. prima* Dart, 1929; *L. africana* var. *obliqua* Dart, 1929; *L. africana pharoahensis* Deraniyagala, 1948; *L. atlantica angammensis* Coppens, 1965; *L. atlantica angammensis* Beden, 1979, 1987a

Loxodonta africana encompasses the extant African savanna elephants and their immediate lineal ancestors whose fossil record can be traced to the beginning of the early Pleistocene (Table 7.1; see above). Results of genomic studies have provided abundant evidence supporting separation of African savanna and forest elephants at the species level (e.g., Roca et al., 2001, 2005; Murata et al., 2009; see discussion below), and suggest that these species diverged sometime between 4.91 ± 1.22 Ma (Murata et al., 2009) or ~4.0 Ma (Rohland et al., 2007) and 2.6 Ma (Roca et al., 2005). The meager fossil record of *L. africana* hinders paleontological efforts to test these genomic-based hypotheses about divergence times of loxodont elephant species.

In Neolithic times, African savanna elephants ranged across nearly the full length of the continent, including what is now the Sahara, up to the Mediterranean coast, and this distribution may have continued up to the time of the first arrival of non-indigenous peoples (Laws, 1970; Poole et al., 2013). Because of habitat loss and poaching, however, as well as climate and environmental change, savanna elephants now have a far more fragmented and reduced distribution, primarily in southern, eastern, and Central Africa as far north as 12° N, in a broad range of habitats from highland forest to grasslands, from sea level to altitudes in exceptional circumstances up to 4,570 m (Stewart and Stewart, 1963); a population of relict, desert-living elephants also persists in Mali (Bouché et al., 2009). Their geographic extent for most of the Pleistocene remains a mystery, as they are rarely documented to have co-occurred on the continent with the dominant elephants of the "*Elephas recki* complex," especially in eastern Africa.

African savanna elephants are characterized postcranially by their long anterior thoracic vertebral spines and tall pelvic girdles, which anchor a vertebral suspensory "bridge" that in lateral view has a distinctive "swayback" look (Sikes, 1971). They are considerably larger than forest elephants. Laursen and Bekoff (1978) reported adult female savanna elephant body masses between 2,160 and 3,232 kg and their stature as 2.2–2.6 m at the shoulder, and adult male body masses between 4,700 and 6,048 kg and stature from 3.2 to 4.0 m. Ontogenetic increase in shoulder height slows or ceases in females around 20 years of age, whereas it may slowly continue in males well into senescence (Lee and Moss, 1995). In a small sample of adult African savanna elephant limb bones, lengths of femora were 767–1,260 mm, tibial L=488–771 mm, humeral L=631–1,059 mm, and radial L=506–813 mm, producing crural indices of 59–66, brachial indices of 77–80, and more disparate intermembral indices of 76–96 (calculated from Christiansen, 2007). Crural indices of extant African elephants overlap with those of stegotetrabelodonts (see above, Chapter 5), indicating that limb proportions of loxodont elephants may have remained conservative since the origin of Elephantidae in the late Miocene.

Loxodonta africana crania are distinguished by the biconvexity of the frontoparietal region, superoanterior angulation of the nuchal planum, broader supraoccipitals than exoccipitals, moderately posteriorly projecting occipital condyles, laterally downturned edges of the nasal aperture, tight malar notch, lack of a midline parietal depression and occipitoparietal bossing, distal divergence of tusk alveoli separated by a distally broad, tear-drop-shaped external incisive fossa, anteroposteriorly short frontoparietal, and rugose occipitoparietal crest (Figure 7.2E–G; Osborn, 1942; Sikes, 1971; Maglio, 1973; Todd, 2010b). Mandibles of *L. africana* typically have relatively elongate, uninflated corpora, pronounced symphyseal "spouts," rami anteroposteriorly narrower at their bases than superiorly, coronoid processes angled outward, articular processes facing upward, modest masseter attachment areas on the rami, and dentaries articulated into a V-shape, viewed superiorly (Figure 7.2F; Todd, 2010b).

Skull elements of *L. africana* are generally substantially larger than those of *L. cyclotis*. Palombo et al. (2021) reported on the dimensions of several savanna elephant crania, MCZR.Os.1302 and 1303, respectively: bizygomatic breadth, 951 and 505 mm; width across the postorbital processes, 798 and 408 mm; occipital width, 864 and 459 mm; occipital height, 636 and 340 mm; maximum premaxillary width, 748 and 310 mm; maximum skull length 1,100 and 643 mm; and width of the nasal aperture, 453 and 251 mm. Dimensions of mandibles in the same study, MCZR.Os.1302b, 1303b, and 1313, respectively, are: greatest length, 570, 375, and 539 mm; symphyseal length, 190, 105, and 140 mm; and maximum width across the corpora, 490, 304, and 392 mm.

The most striking feature of the African savanna elephant, shared with its forest cousins, is the incorporation of anterior and posterior accessory conules into molar plate wear-figures to form diamond- or lozenge-shaped "loxodont sinuses" (Figure 7.2H; "<>"). These wear-figures are most pronounced with moderate attrition of the crown, becoming transversely rectilinear with heavy crown attrition. Enamel may be finely folded or absent, but usually does not achieve the amplitudes seen in some molars of the "*Elephas recki* complex." The loxodont sinuses of *L. africana* are often not symmetrical, instead exhibiting offset of anterior and posterior accessory conules about the midline; in these instances, with intermediate attrition, median projections of sinuses of adjacent plates do not touch at their tips, instead abutting along their sides. Loxodont sinuses or pseudoloxodonty may be observed to some degree in the molars of other, non-*Loxodonta* elephants (e.g., occasionally in the "*Elephas recki* complex" [see Beden, 1983, 1987a] or in some *Palaeoloxodon* individuals [see Matsumoto, 1929]), but not as regularly through the crown or as saliently. Development of loxodont sinuses in extant African elephant molars is more pronounced than in *L. cookei*, *L. adaurora*, and most molars of *L. exoptata*, but may be equaled in molars of *L. atlantica*.

Plate formulae for *L. africana* are modest compared to numbers reached by crown species of other recent elephant genera: dP2/dp2, x3x-5/3–5; dP3/dp3, 6–7/6–8; dP4/dp4, 8–10/8–10; M1/m1, 8–9/7–10; M2/m2, 11–12/9–12; M3/m3, 11–13/9–14 (Morrison-Scott, 1947; Laws, 1966; Sikes, 1966; Hanks, 1972; Roth, 1992; Sanders et al., 2010a; Craig and Peake, 2011; Lee et al., 2011; Stansfield, 2016; Palombo et al., 2021). Despite substantial differences in skull size between savanna and forest elephants, their cheek tooth dimensions are much more similar (see below). Published lengths and widths of *L. africana* cheek teeth are as follows: dP2, 21–25 mm×18–20mm; dp2, 22–24 mm×15–17mm; dP3, 55–66 mm×30–36mm; dp3, 58–62 mm×25–37mm; dP4, 105–111 mm×39–49mm; dp4, 100–119 mm×40–50mm; M1, 157–160 mm×52–65mm; m1, 132–158 mm×42–60mm; M2, 172–224 mm×63–74mm; m2, 177–235 mm×59–74mm; M3, 214–313 mm×76–91 mm; m3, 205–290 mm×68–81 mm (Morrison-Scott, 1947; Stansfield, 2016; Palombo et al., 2021). Crown height of molars in *L. africana* is hypsodont, to a greater or much greater extent than in molars of fossil loxodont elephants except *L. atlantica*. Hypsodonty indices of loxodonta elephant third molars collated by Lister (2013) overlaps completely between *L. atlantica*, *L. africana*, and *L. cyclotis*, ranging approximately between HI=170 to 220.

Molar size variation is largely driven by sexual dimorphism in *L. africana*. Dental variation is even more exaggerated for tusks of the species, which can achieve considerably greater girth and length in males. A recent study on savanna elephants in Kruger National Park in South Africa carefully recorded growth parameters and dimensions ontogenetically between males and females, revealing that tusks continue to grow and accelerate growth with age in males, whereas in females tusk growth ceases around 40 years of age (Whyte and Hall-Martin, 2017–2018). As a result, mature males have significantly larger tusks than females. In the study sample, mean tusk length for females was 1,137 mm, tusk circumference averaged 247 mm, and greatest length was 1,440 mm in a 45-year-old individual; for males, greatest tusk length was 2,640 mm in a 55-year-old individual and greatest circumference was 480 mm. Mass differences between tusks of right and left sides indicate that these elephants are preferentially "right-tusked." There was no indication of genetic-based incidences of tusklessness in the sample. However, at other sites, because of artificial selection driven by slaughter of elephants for ivory, there is evidence of genetic failure to produce tusks (Jachmann et al., 1995; Whitehouse, 2006). Elephants use their tusks for manipulating their environment, social interaction, and protection from predators; in addition, males employ their tusks to fight other males and for display. The trunk, which is often coordinated with the tusks and has a wider range of social, feeding, and water intake roles, is prehensile and capable of astonishing dexterity, driven by 150,000 muscle fascicles (Shoshani, 1997).

Savanna elephants require a large amount of forage to sustain themselves and are obligate drinkers of standing water. A large bull elephant might require 300 kg and an adult female about 170 kg of food per day (Laws, 1970). Adult elephants require about 160 liters of water daily (Poole et al., 2013). Large-scale studies of savanna elephant diet via investigation of dental isotopes revealed that they are browsers or mixed feeders with primary reliance on C_3 plants (Cerling et al., 1999, 2015). However, seasonally and depending on local conditions, savanna elephants may periodically take in considerable percentages of C_4 vegetation in their diets (Buss, 1961; Laws, 1970; Laws et al., 1975; Williamson, 1975; Cerling et al., 2004, 2009; Codron et al., 2006). The strong isotopic signal for browsing in extant African elephants is sometimes at odds with field observations of feeding or composition of dung (e.g., Wing and Buss, 1970). It has been suggested that the mismatch of these observations with isotope data is due to the inability of elephants to digest C_4 grasses as efficiently as C_3 dicots (van der Merwe et al., 1988). Nonetheless, there is no question that the isotope record of extant African elephants does not reflect nearly as dedicated a commitment to grazing as the records for earlier African fossil elephants, including earlier species of the genus (e.g., Manthi et al., 2017), and particularly representatives of the "*Elephas recki* complex" (Cerling et al., 1999).

Laws (1970) referred to African savanna elephants as "bush" elephants, because of their habitat preferences for woodland, forest edge, bushland, and wooded or bushed grassland. There is no question that these elephants have a profound ability to physically transform, fragment, and open up these habitats by destruction of the understory, eating seedlings, and ringbarking trees. Although by themselves elephants appear unable to transform woodlands into savannas, this can be achieved with the added effects of natural fires, and once habitats are altered this way elephants can maintain them as grasslands (Dublin et al., 1990). In Central African forests, forest elephants also have a major impact on their ecosystems, leading to a loss of diversity of smaller tree species and an increase in numbers of larger trees with higher wood density and aboveground biomass by 26–60 t/ha (Berzhagi et al., 2019). In addition, African elephants play a major role as seed dispersers and in cycling nutrients throughout their habitats (Hawthorne and Parren, 2000; Blake et al., 2009; Beaune et al., 2013). Both savanna and forest elephants may be thought of as keystone taxa vital for creating ecological opportunities for smaller ungulates (Laws, 1970; Western, 1989; Kerley et al., 2008).

Loxodonta africana is a highly social species with a long period of pre-adult ontogeny and learning, centered around matrilineal herds (older adolescent/subadult and adult males live in "bachelor" groups or in isolation except for intervals of mating) (Laursen and Bekoff, 1978). Herd organization and social behavior of African elephants may have very ancient roots; there is evidence from ichnofossils of some of the oldest known elephants associating in herds of these kinds, excluding bulls (Bibi et al., 2012;

Sanders, 2022). The rich pattern of collective interaction and seasonal migratory memory bespeaks an elevated level of intelligence that is evidenced by investigations of brain size and structure in African savanna elephants. Average brain weights for males and females of the species are 5,004 g and 4,677 g, respectively, yielding encephalization quotients (EQ) of 1.3–2.3 (average 2.0), relatively larger than predicted for their body mass but far lower than the human EQ of 7.5 (Shoshani et al., 2006; Hart et al., 2008). Similar to neural structures in humans, in African savanna elephants the brain has a markedly large cerebellum with salient convolutions of gyri and sulci, though the greater dominance of olfaction in their sensory array is expressed by the immensity of their temporal lobes (Shoshani et al., 2006). In comparison, the size of the frontal lobe is unimpressive and the occipital and parietal lobes are ill-defined. In addition, neurons comprising the African elephant brain are far less densely packed than in humans or other higher primates (Hart et al., 2008).

Nonetheless, African elephants exhibit cognitive and behavioral traits that tend to be associated with humans, such as emotional intelligence, empathy, concern for distressed or deceased individuals, and rendering compassionate assistance to the ailing (Douglas-Hamilton et al., 2006; Hawley et al., 2017–2018). Strikingly, both savanna and forest elephants exhibit intense interest in the carcasses and skeletal remains of conspecifics (but not of other species), including manipulation and olfactory investigation of elements of dead individuals (McComb et al., 2005; Douglas-Hamilton et al., 2006; Merte et al., 2008–2009; Hawley et al., 2017–2018). Visiting the remains of dead elephants is not limited to individuals with shared kinship; this behavior is highly unusual among non-human animals. It is not known if these rituals signify suffering or compassion, or curiosity about death, among surviving elephants (Douglas-Hamilton et al., 2006), but they appear to indicate a higher intellectual recognition of affinity.

As important as olfaction is for African elephants, a derived set of adaptations for vocalization and auditory reception of complex communications constitutes an equally valuable sensory system for these animals. Along with high-frequency "trumpeting" produced in the trunk, elephants also generate a range of low-frequency rumbles that can reach into the infrasonic range around 15–25 Hz (Venter and Hanekom, 2010), using a myoelastic-aerodynamic, flow-driven mode of communication involving vocal folds in the larynx and modulated across the vocal tract (Herbst et al., 2012). These sounds are imperceptible or nearly so to the human ear, but can be perceived by elephants as far away as 2.5–4.0 km (more typically 1.0–1.5 km, but farther than airborne sounds [Günther et al., 2004; Mortimer et al., 2018]), can be differentiated among the calls of at least 100 different individuals, are useful for coordinating movements of herd members, reuniting individuals, and carry information about mate attraction, individual identity, and the emotional and reproductive states of animals (McComb et al., 2003; Soltis, 2010). The intense low-frequency seismic waves used to communicate over distance are not all in the infrasonic range; they are more complex, and vocal tract resonance spans several harmonics centering around 115 Hz (McComb et al., 2003). Anatomical features correlated with low-frequency communication in elephants include the great size of the external ears, or pinna, up to 1.8 × 1.1 m in adults, cochlea sensitive in the 10.5–17.0 Hz range, large vocal folds of about 7.0–7.5 cm (1.0 cm in humans, by comparison), a long vocal tract of about 75 cm from larynx to mouth and 2.5 m from larynx to tip of the trunk, massive middle ear ossicles, a capacious tympanic membrane of about 855 mm^2, and hyoid apparatus with flexible upper and lower articulations to facilitate flexible laryngeal movements (Reuter et al., 1998; Soltis, 2010; Venter and Hanekom, 2010). The size of the ear ossicles accords well with an auditory system in which acoustic information is received via body vibrations (Reuter et al., 1998). The presence of hyoid bones in the American mastodon that are morphologically very similar to those of modern elephants suggests that infrasonic communication constitutes a shared feature of all elephantimorph proboscideans (Shoshani and Marchant, 2001; Shoshani, 2003).

LOXODONTA CYCLOTIS (Matschie, 1900)
(Tables 7.1 and 7.2; Figures 7.1 and 7.3)

Partial Synonymy—*Elephas cyclotis* Matschie, 1900; *E. africanus pumilio* Noack, 1906; *E. africanus albertensis* Lydekker, 1907; *E. africanus cyclotis* Lydekker, 1907; *Loxodonta cyclotis* Morrison-Scott, 1947; *L. cyclotis* Grubb et al., 2000; *L. africana cyclotis* Debruyne et al., 2003a; *L. cyclotis* Roca et al., 2001, 2005; *L. cyclotis* Murata et al., 2009; *L. africana cyclotis* Todd, 2010b

Classified as either a subspecies of *Loxodonta africana* or elevated to full species status (*Loxodonta cyclotis*), the forest African elephant has no fossil record, and therefore discussions about its phylogeny are limited to information garnered from genetic studies or comparative morphology of specimens from extant populations. Distribution of the species is limited to the western half of sub-Saharan Africa up to the lake region of the Western Rift Valley (Sanders et al., 2010a).

These elephants are smaller than African savanna elephants. Shoulder heights of adult males are between 2.4 and 3.0 m and those of females are 1.8–2.4 m, and body weights of both sexes combined range from 2,000 to 4,000 kg (Grubb et al., 2000; Turkalo and Barnes, 2013). Comparative morphology of the forest African elephant has been studied by Lydekker (1907), Allen (1936), Morrison-Scott (1947), and Grubb et al. (2000), who collated distinctions of the forest elephant compared to its savanna cousin, including smaller body size, smaller and more rounded ears, tusks straighter, more slender, and more downward-directed, forelimbs lower than hindquarters, cranium relatively broader with more widely separated temporal ridges, less cranial pneumatization, forehead sloped back more sharply, shorter, distally less flared rostrum, more vertical occipital planum, relatively wider nasal aperture, lower, longer mandible with a more prominent spoutlike symphysis. Also, forest elephants tend to exhibit five nails in the forefoot and four in the hindfoot (as opposed to four and three, respectively, in the front and hind feet of savanna elephants) (Turkalo and Barnes, 2013).

Todd (2010b) contributed detailed comparison of forest and savanna African elephant skulls, adding more distinctions of forest elephant crania, including: External incisive fossa more rectangular rather than tear-drop-shaped, widest point of the cranium across the zygomatic process of the lacrimal, nuchal ligament fossa broader, supraoccipitals located directly above exoccipitals, and occipital condyles oval-shaped and more posteriorly projecting. However, forest elephants do not appear to have fewer molar plates per tooth, as was suggested by Todd (2010b). Published sources that have collated these formulae (e.g., Morrison-Scott, 1947; Petter, 1958; Palombo et al., 2021) counted dP3/dp3 = 6/7, dP4/dp4 = 7–9/8–10, M1/m1 = ?8–10/7–11, M2/m2 = 9–10/9–?11, and M3/m3 = 10–14/10–14. Average numbers of plates in molars of *L. cyclotis* fall within the ranges of plate counts for *L. africana* molars (see Morisson-Scott, 1947; Roth, 1992). Nonetheless, molars of forest elephants are distinguished by their diminutive size. Recorded lengths and widths for deciduous and permanent molars of forest elephants are dP2, 15 × 12 mm; dP3, 55 × 31 mm; dp3, 56 × 26 mm; dP4, 101–130 × 36–48 mm; dp4, 102–140 × 35–44 mm; M1, 130–160 × 45–72 mm; m1, 130–165 × 47–65 mm; M2, 150–192 × 54–69 mm; m2, 140–200 × 49–66 mm; M3, 210–265 × 60–75 mm; m3, 230–250 × 55–80 mm (Morrison-Scott, 1947; Petter, 1958; Palombo et al., 2021). Interindividual variation of these dimensions may be attributed primarily to sexual dimorphism.

Palombo et al. (2021) reported that skull elements of *L. cyclotis* are substantially smaller than those of *L. africana*. Forest elephant mandible MCZR Os.1307b has a greatest length of 360 mm, maximum width across the corpora of 326 mm, and symphyseal length of 113 mm. A second mandible of the species, MCZR Os.1308b, is 280 mm in length, 222 mm wide, and has a symphyseal length of 96 mm. Crania associated with these mandibles have a bizygomatic width of 415 mm (MCZR Os.1308), breadths across the zygomatic processes of the orbits of 438 mm (MCZR Os.1307) and 303.5 mm, width between the temporal lines of 284 and 228.7 mm, occipital breadths of 488 and 355 mm, occipital height of 369 mm (MCZR Os.1307), maximum premaxillary breadths of 332 and 255 mm, and maximum skull length of 588 mm (MCZR Os.1307).

Additionally, forest elephants are more frugivorous in feeding adaptations than are savanna elephants, with seasonal adjustments away from consuming fruit to eating bark and leaves (Short, 1981; White et al., 1993; White, 1994; Maurois et al., 1997; Blake, 2002). However, there is no evidence that they are more committed browsers than savanna elephants (Tchamba and Seme, 1993; White et al., 1993). Despite these differences in size, morphology, and ecological adaptations, forest and savanna elephants share numerous anatomical resemblances, including the striking lozenge- or loxodont-shaped occlusal enamel wear patterns of their plates ("<>") (Todd, 2010b), and neither exhibit the salient cranial apomorphies of *Elephas*, *Mammuthus*, or *Palaeoloxodon*. Morphologically, there is no compelling reason to separate them at genus level.

Claims of extant "pygmy" Central and West African elephants (e.g., *Loxodonta "pumilio"* and *L. "fransseni"*; Noack, 1906; Schouteden, 1911a, b) have been demonstrated to lack evidence for adequate genetic separation from forest elephants (Debruyne et al., 2003b), and have been dismissed as confusing subadult or juvenile examples of forest elephants with true dwarfed individuals (Petter, 1958; Groves and Grubb, 2000a; Turkalo and Barnes, 2013), although Debruyne and colleagues (2003b) treated "*cyclotis*" and "*pumilio*" as two main subclades within *L. africana cyclotis* and did not rule out cases of dwarfing.

Studies of mitochondrial genes (mtDNA) showed evidence for gene flow between forest and savanna elephant populations and incomplete speciation (Eggert et al., 2002; Debruyne, 2005). Debruyne (2004:335) argued for maintenance of subspecific differentiation of forest and savanna African elephants because of a high level of mitochondrial introgression between them, a general "continuum in the morphology of the skulls of [their] morphotypes," and evidence for interbreeding across intersecting geographic zones. Indeed, morphometric analysis of 295 African elephant crania established the presence of anatomically intermediate individuals in habitats marginal to the distribution of forest elephants (Groves and Grubb, 2000b). However, these intermediates and records of genetic hybrids are uncommon (Roca et al., 2005). In contrast, a study of the cytochrome *b* mitochondrial gene in living and recent elephants with the American mastodon as the outgroup by Barriel et al. (1999) revealed that for sequences of this gene, the forest elephant is as divergent from the savanna elephant as either is from Asian elephants, although the authors continued to maintain the forest elephant as a subspecies of *L. africana*.

Additional genetic investigation of nuclear genes revealed strong genetic distinctiveness (58% of the genetic distance between Asian and African elephant species) and isolation between savanna and forest elephants, with a limited history of hybridization between them (Roca et al., 2001). Moreover, using an African-Asian elephant divergence time of 5.0 Ma, the results of the study indicated the timing of the forest-savanna elephant split at 2.63 ± 0.94 Ma. This may be an underestimate as current understanding of the fossil record (Table 5.1) shows an earlier time for divergence between *Loxodonta* and *Elephas+Mammuthus* (genetically determined coalescence time of 7.77 ± 0.60 Ma), yielding an age of 6.89 ± 0.60 Ma for the divergence of *Elephas* from *Mammuthus* and age of 4.91 ± 1.22 Ma for separation of forest and savanna elephants, based on mtDNA sequencing (Murata et al., 2009). These studies and the mtDNA analysis of Rohland et al. (2010) indicate that mammoths and

Asian elephants form a clade and confirm a deep separation of forest from savanna elephants, while establishing that gene flow between them must have occurred more recently, based on the mtDNA haplogroup shared among some forest and savanna elephants. Thus, according to genomic studies, forest and savanna elephant lineages diverged either at the start of the Pliocene or the beginning of the Pleistocene. Using results from both nuclear and mtDNA, Brandt et al. (2012) favored the younger divergence date and associated the event with fragmentation of forest habitats by climate and ecological change at the start of the Pleistocene.

Perhaps the most important genetic study concerning the relationship of forest and savanna African elephants involved assessment of mtDNA and paternally (Y-chromosome segments) and biparentally (X-chromosome segments) inherited gene sequences (Roca et al., 2005). This study sought to explain the dissociation of results from analyses of nuclear DNA that showed the near absence of forest elephant sequences in savanna elephant populations (and see Roca et al., 2015) and mtDNA that showed the presence of the forest elephant mitochondrial haplotype in populations of savanna elephants, sometimes at high frequency. The explanation provided to account for this incongruency involves multiple generations of unidirectional hybridization of forest or hybrid female elephants to savanna elephant males; the repeated backcrosses would "dilute the proportion of forest nuclear alleles … until the populations had overwhelmingly savanna-like nuclear genes but retained the maternally-inherited forest … mtDNA haplotype" (Roca et al., 2005:98; Roca, 2019). Roca et al. (2005) concluded that the absence of nuclear gene interchange between forest and savanna African elephants warrants their division at the species level.

More recent genomic investigation of broader relationships of African loxodont elephants produced a surprising result using mtDNA and nuclear DNA sequencing, suggesting that forest African elephants are more closely related to *Palaeoloxodon antiquus* than to savanna African elephants (Meyer et al., 2017). There is no morphological or paleontological support for this hypothesis. Subsequent, more comprehensive analysis of elephant genomes (Palkopoulou et al., 2018) revealed a more nuanced, complex phylogenetic history for *Palaeoloxodon* and different interpretation of the relationship of African elephants to *Palaeoloxodon*. In the Palkopoulou et al. (2018) study, the *P. antiquus* lineage is hypothesized to have derived from the ancestral stock that produced *L. africana*+*L. cyclotis*. Moreover, the nuclear DNA results of the study suggest that palaeoloxodonts form a clade separate from that of savanna+forest African elephants. Shared mtDNA haplotypes of *L. cyclotis* and *P. antiquus* were attributed to repeated episodes of hybridization. Thus, in this hypothesis, palaeoloxodonts cluster with African loxodont elephants but do not share a special relationship with forest elephants (Figure 7.3). Furthermore, the results of the study suggest gene flow separately between palaeoloxodonts and the *Elephas* and *Mammuthus* lineages, which the authors use to explain the greater morphological similarity of palaeoloxodonts to these elephants than to African loxodont species. There is, however, no African paleontological evidence to support this hypothesis. Finally, results of the Palkopoulou et al. (2018) study strongly support the closer relationship of *Mammuthus* to *Elephas* than to *Loxodonta*, in contrast to findings that some *Mammuthus* individuals genetically sampled are closest to *Loxodonta africana* (Barriel et al., 1999), and maintain the species-level separation of *L. cyclotis* from *L. africana*.

These distinctions are important for conservation efforts. Tallying forest elephants with savanna elephants as a single species weakens the perception of impending extinction for the far less numerous forest elephant populations. On its own, the prospect for the survival of *L. cyclotis* is dire and calls for an extreme response to protect individuals and their habitats. The IUCN Red List for threatened species lists it as "critically endangered," the category proceeding extinction, and shows its current range as severely reduced from historic records (Gobush et al., 2021).

MAMMUTHUS Brookes, 1828
MAMMUTHUS AFRICANAVUS (Arambourg, 1952)
(Tables 7.1 and 7.2; Figures 7.1 and 7.4)

Partial Synonymy—*Elephas meridionalis* Pomel, 1895; *E. planifrons* Deperet and Mayet, 1923; *E. africanavus* Arambourg, 1952, 1970; *Loxodonta africanava* Cooke, 1960; *L. africanava* Coppens, 1965; *L. africanava* Aguirre 1969b; *Mammuthus africanavus* Maglio 1969b, 1973; *Archidiskodon africanavus* Garutt et al., 1977; *M. africanavus* Coppens et al. 1978; *L. africanavus* Mackaye, 2001; *M. africanavus* Sanders et al., 2010a; *M. africanavus* Zhang, 2020

These mammoths comprise a primarily mid-to-late Pliocene North and Central African species, with a latest occurrence documented at Aïn Boucherit, Algeria in the early Pleistocene (Tables 6.1 and 7.1; Sahnouni et al., 2011). Historically, *Mammuthus africanavus* has been taxonomically problematic, having been placed in a variety of genera and also possibly comprising a mélange of taxa (Mackaye, 2001; Sanders et al., 2010a; Markov, 2012). North African fossils assigned to the species were described by Arambourg (1952, 1970) under the nomen "*Elephas africanavus*," and it is entirely possible that some molars of the sample might belong to *Elephas*, as cheek teeth of primitive *Elephas* and *Mammuthus* are difficult to distinguish without accompanying crania. Others have championed the idea that the fossils of this species should be placed in *Loxodonta* (Cooke, 1960; Coppens, 1965; Aguirre, 1969b; Mackaye, 2001). The original diagnosis (Arambourg, 1952), based on specimens from Lac Ichkeul, Tunisia, exaggerated some traits (for example, "low" lamellar frequency, "extreme" brachyodonty, "massively" thick enamel), and despite describing enamel wear-figures as comprising "loxodont sinuses more or less pronounced," identified the fossils from the site as *Elephas*.

Nonetheless, as recognized by Maglio (1973) and others (Coppens et al., 1978; Sanders et al., 2010a; Zhang, 2020),

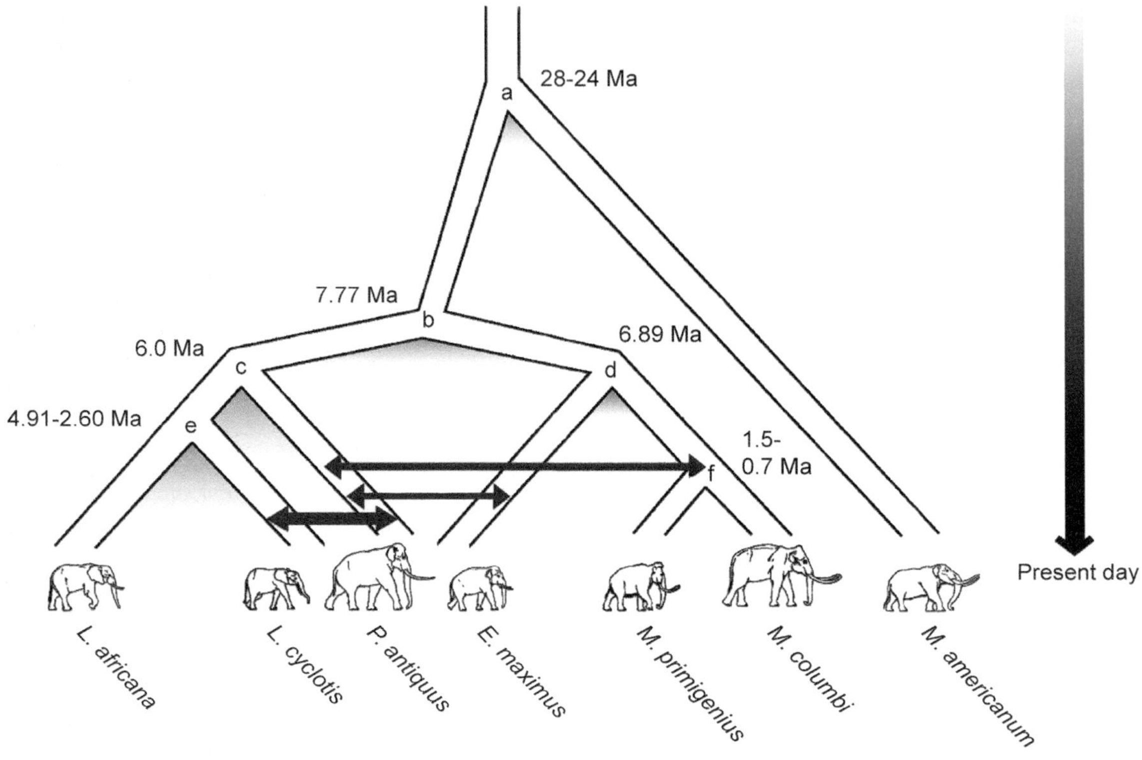

FIGURE 7.3 Elephant phylogeny based on genomic analysis of fossil and extant elephants. Species include the savanna and forest African elephants (*Loxodonta africana*, *L. cyclotis*), straight-tusked elephant (*Palaeoloxodon antiquus*), woolly and Columbian mammoths (*Mammuthus primigenius*, *Mammuthus columbi*), and a mammutid outgroup, the American mastodon (*Mammut americanum*). (Phylogeny adopted from Palkopoulou et al. [2018:fig. 3].) Phylogenetic bifurcations are represented by letters a–f: a, split of Elephantida and Mammutida; b, initial separation of *Loxodonta* from *Elephas*+*Mammuthus*; c, separation of *Loxodonta* and *Palaeoloxodon* lineages; d, separation of *Elephas* and *Mammuthus* lineages; e, split of savanna and forest elephant lineages; f, split of mammoth lineages. Red arrows indicate gene flow between *Palaeoloxodon* and other elephant lineages, with the thickness of the arrow corresponding to the extent of gene flow. Phylogenetic bifurcation dates are based on genetic analysis results from Roca et al. (2005), Rohland et al. (2007), Murata et al. (2009), and Palkopoulou et al. (2018). (Image use with permission of the Proceedings of the National Academy of Sciences, USA.)

a cranium of a large, old male individual from Garet et Tir, Algeria provides evidence for the presence of mammoths in North Africa during the second half of the Pliocene (Figure 7.4A and B). The tusks are massive (W=136mm, H=140mm; L=+2310mm on the left side) and recurved upward and inward distally (Arambourg, 1970; Maglio, 1973). Maglio (1973) felt that the cranium resembled that of European *M. meridionalis*, based on examination of field photographs taken prior to transport and damage of the skull. Notable features of the cranium include its great height (1,452mm vertically from the level of the distal tusk alveoli to level of the vertex), narrow palate, occipital planum much wider (800mm) than high (490mm) and set at a right angle to the palate, great length of the premaxillary tusk sheaths (450mm) and face overall, and relatively narrow width between the tusk alveoli distally (270mm), which are even more narrowly spaced at their bases (160mm), although the tusks sharply diverge exiting their alveoli. The basicranial region appears very short and the temporal foramen is rectangular in shape and not flared laterally. Unfortunately, the dorsum of the specimen, particularly the fronto-nasal region, is damaged and distorted. The occipital planum does not appear to have been inflated (Arambourg, 1970: pl. II, fig. 1).

An accompanying mandible has a tall, downward-inclined interalveolar crest that is broken, and is missing its rami. It has eroded m3s that may have had nine plates, of an approximate size of L=210mm, W=97mm (Arambourg, 1970). M3s in the cranium have a plate formula of 9x and measure L=220mm, W=89mm. Because they are worn, hypsodonty indices cannot be accurately calculated for these molars. Unfortunately, the original specimen is apparently missing (see Zhang, 2020).

Third molars of the species sample from North Africa have a low number of plates compared to the same teeth in more derived mammoths, moderately thick (ET=2.6–4.3mm), occasionally undulated, unfolded enamel, well-spaced plates (LF=3.0–5.2), molars tapered laterally strongly toward the apex of the crown, and accessory central conules restricted to the anterior half of the crown (that do not generally contribute to the formation of loxodont sinuses, contra Mackaye, 2001) (Figure 7.4C

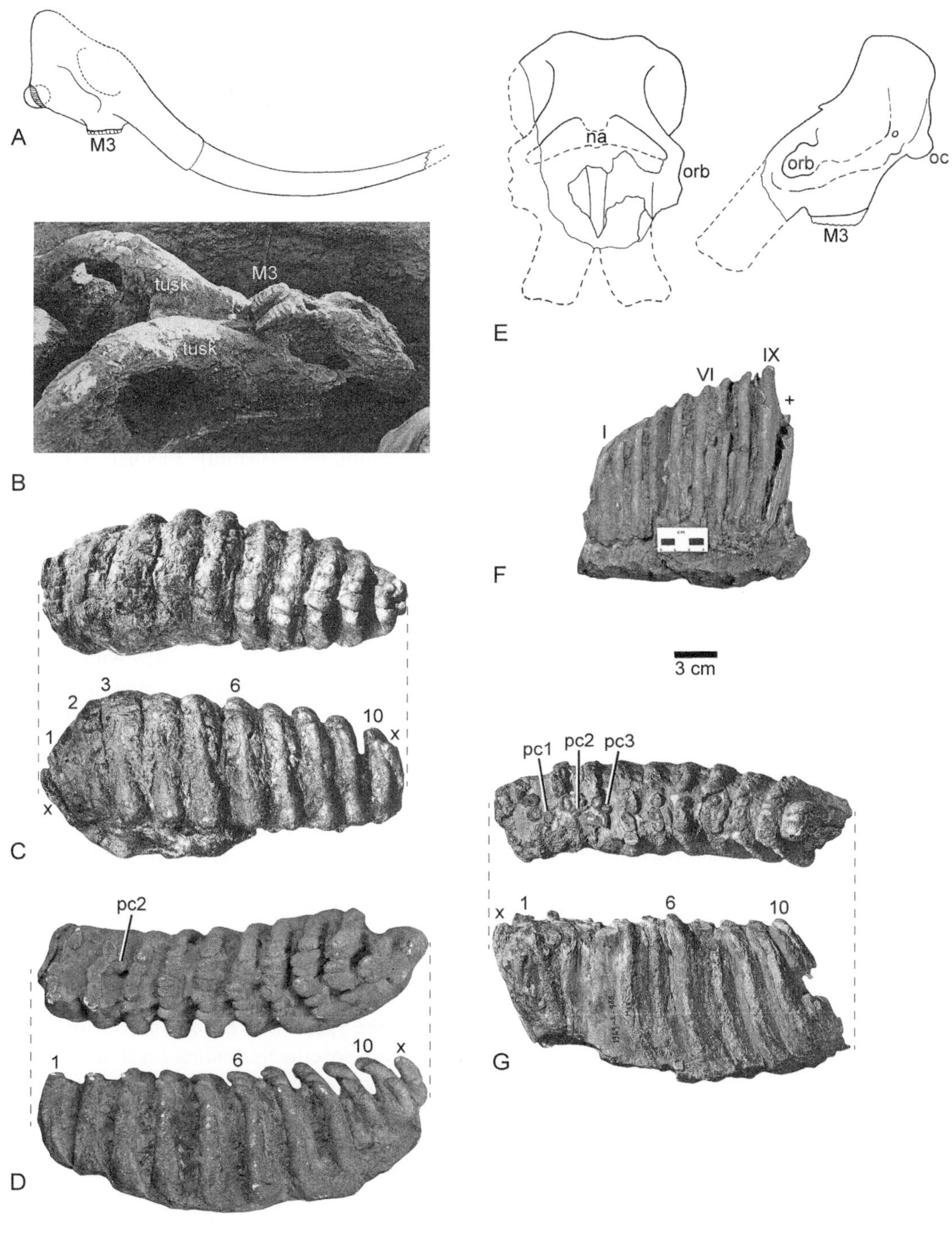

FIGURE 7.4 Mid-Pliocene to early Pleistocene fossil remains of *Mammuthus* spp. from North Africa. Abbreviations: M, upper molar; na, nasal aperture; oc, occipital condyle; oc, occipital condyle; orb, orbit; pc, posterior pretrite accessory conule; x, pre- or postcingulum(id); 1, 2, 3, …, plate number counted from the anterior of the crown; I, II, III, …, plate number counted from the posterior of the crown; +, indicates missing morphology. (C, D, G) anterior to the left. (C, D, F, G) to the same scale. (A) Reconstruction of cranium from Garet et Tir, Algeria, lateral view, *Mammuthus africanavus*. Anterior to the right. (From Arambourg [1970:fig. 25].) (B) Cranium in situ at Garet et Tir, ventral view, *Mammuthus africanavus*. Anterior to the left. Length of the right tusk is 1,870+ mm and of the left tusk is 2,310+ mm. (From Arambourg [1970:fig. 24].) (C) Left M3, 1958-14: 213 from Ain Brimba, Tunisia, occlusal and lateral views, *Mammuthus africanavus*. (From Arambourg [1970:pl. VI, fig. 3].) (D) Right m3, MNHN 1950-1-12 (type) from Lac Ichkeul, Tunisia, occlusal and lateral views, *Mammuthus africanavus*. (E) Cranium, reconstruction based on European cranial fossils of the Geological Institute of Florence, Nos. 1049, 1051, and 1054, anterior and lateral views, *Mammuthus meridionalis*. (From Maglio [1970b:fig. 3A], with permission of the Museum of Comparative Zoology, Harvard University.) (F) Partial M3, 1948-1-127 from Ain Hanech, Algeria, lateral view, *Mammuthus meridionalis*. (G) Left m3, 1949-2-968 from Ain Hanech, Algeria, occlusal and lateral views, *Mammuthus meridionaolis*. (Images from Arambourg [1970] ©Publications Scientifiques du Museum national d'Histoire naturelle, Paris.)

and D). Plate faces are parallel to one another and consequently are separated by U-shaped transverse valleys; plates are formed of between five and seven conelets. The widest point of molars is located one-third to one-quarter of the crown height from the base. Cementum may be abundant in transverse valleys. According to Maglio (1973), hypsodonty may reach HI = 120. Although Arambourg (1970) labeled some specimens as premolars, the sample only includes deciduous premolars and molars. Plate formulae are as follows: dP3/dp3 = 5–6/x6x; dP4/dp4 = 6/6–x7x; M1/m1 = 6–7/7–x8x; M2/m2 = 6–8/8–9; M3/m3 = 9/9–13.

The holotype is MNHN 1950–1–12, a right m3 from Lac Ichkeul, Tunisia (Figure 7.4D). It has a plate formula of x10x and dimensions of L=283 mm, W=95 mm, and H=92 mm, yielding a hypsodonty index of 97 (subhypsodont rather than "extremely brachyodont"). Enamel is moderately thick (ET=4.0 mm) and plate spacing is also moderately wide (LF=4.0). Molar breadth is greatest at one-quarter the height of the plates, and plates angle inward on their lateral margins to the apex of the crown. Occlusal wear is modest, particularly posteriorly. This site and nearby Aïn Brima, Tunisia have produced the majority of North African specimens assigned to *M. africanavus*. A premaxillary segment of a cranium from Lac Ichkeul has a length of 570 mm from the nasal aperture to the opening of the tusk alveolus, which measures W = 111 mm, H=136 mm. An M1 (N° 1950–1–96) from Lac Ichkeul has a length of 139 mm, greatest width of 64 mm, and HI = 110. Lamellar frequency is 5.0 and enamel thickness is 2.0–2.5 mm. It has six plates, accessory conules associated with the anterior plates, and exhibits abundant cementum in the transverse valleys. Plate shape is as in other molars of the species. An M2 (N° 1950–1–8d) from Lac Ichkeul has dimensions of L=205 mm, W=92 mm, H=94 mm (producing HI = 111), and a lamellar frequency of 4.0. It exhibits extensive cementum filling the transverse valleys and accessory conules associated with the first three plates.

These molars are slightly more advanced than those of *M. subplanifrons* in plate number, hypsodonty, deposition of cementum, and in having more parallel-faced plates with U-shaped transverse valleys. Some *M. africanavus* occurrences are probably closer in age to those of the first European mammoth, *M. rumanus*, dated from ca. 3.5 Ma (Lister and van Essen, 2003; Markov and Spassov, 2003; Radulescu et al., 2003; Lister et al., 2005; Markov, 2012). A mammoth species of late Pliocene age from the Sidi Hakoma and Denen Dora Mbs. of the Hadar Fm., Ethiopia (*Mammuthus* sp. "Hadar-type") was identified from field reports of molars, tusks, skulls, and partial skeletons and reported to be morphometrically intermediate between *M. subplanifrons* and *M. africanavus* (Beden, 1979, 1985; White et al., 1984; Kalb and Mebrate, 1993; Kalb and Froelich, 1995), but many specimens remain in situ and the entire sample needs to be adequately described.

M3s from Mesgid Dora in the Woranso-Mille area, Ethiopia (MSD-VP-3/039) (Table 6.1, Figure 6.6), dated to the early Pliocene (Deino et al., 2010), may belong to the slightly more archaic *Mammuthus* sp. "Hadar-type" and are morphometrically similar to molars of *M. rumanus* (Sanders and Haile-Selassie, 2012). In addition, two crania from Omo, Ethiopia, including a juvenile specimen and an adult specimen with M3s, are currently under study but nonetheless are said to display clear mammoth features; the M3s resemble those of *M. africanavus* but are more hypsodont (Zhang, 2020).

Anomalously older specimens attributed to the species derive from the early Pliocene site of Kollé, Chad (Table 6.1; Mackaye, 2001). The most impressive of these include a large mandible (greatest width=670 mm) with m3s, KL5.99.018. Its corpus is elongate, low, and bulging (L=495 mm; anterior H = 171 mm, anterior W = 155 mm; posterior H = 155 mm, W=200 mm). It has a brevirostrine symphysis (L=+195 mm) and impressive-sized rami (L=317 mm, H=370 mm). The m3s in this mandible have a plate formula of 11x and are large (L=305 mm, W=114 mm). Enamel is coarsely undulated and thick (ET=5.0 mm), and lamellar frequency (LF=4.1) suggests the plates are moderately well-spaced. Transverse valleys are U-shaped. Each plate has five to six conelets, and median pillars exhibit anterior and posterior "bombements," or rounded expansions that are interpreted as being accessory conules incorporated into plates 1–4.

An associated palate has M3s with estimated plate formulae of 9x and morphology similar to the m3s. Dimensions are L=270 mm and W=117 mm; the crown is described as brachyodont (Mackaye, 2001), but it is worn. An m3 of a different individual from the site (KL9.96.017) has at least nine plates and dimensions indicating that it is subhypsodont rather than brachyodont (W=90 mm, H=85 mm; HI=94).

Specimens from the younger, mid-to-late Pliocene site of Koro-Toro (Table 7.1) attributed to the species are more abundant and include an m2 (KT8.94.006b) that appears to have too few plates (plate formula=x5x) to belong to *M. africanavus*; an M2 from the site (KT1.94.023) has six plates, at the low end of range for the species. Although Mackaye (2001) placed the species in *Loxodonta*, including the Kollé and Koro-Toro specimens, finding a resemblance to *L. adaurora*, accessory conules are restricted to the anterior moiety of molar crowns, and anterior and posterior "bombements" are not equally prominent. Moreover, with the exception of midline expansion in plates 2 and 3 of the left M3 of KL5.99.018, enamel wear figures of the molars of the Chadian sample do not typically exhibit loxodont sinuses.

Associated fauna from the early Pleistocene site of Aïn Boucherit, Algeria features a mix of modern and extinct mammalian genera, and includes felids, canids, the rhino *Ceratotherium*, the equids *Hipparion* and *Equus*, the suid *Kolpochoerus*, the giraffid *Sivatherium*, and a small diversity of at least six bovids (Sahnouni et al., 2011). The horses, rhino, suid, and some of the bovids support the paleoecological inference of an open environmental conditions for mammoths at the site during the early Pleistocene (Sahnouni et al., 2011).

MAMMUTHUS MERIDIONALIS (Nesti, 1825)
(Tables 7.1 and 7.2; Figures 7.1 and 7.4)

Partial Synonymy—*Elephas meridionalis* Nesti, 1825; *E. antiquus* (in part) Falconer and Cautley, 1845–1849; *E. lyrodon* Weithofer, 1889; *E. planifrons* Depéret and Mayet, 1923; *E. planifrons rumanus* Stefanescu, 1924; *E. planifrons* Doumergue, 1928; *Elephas* aff. *meridionalis* Arambourg, 1952; *E. moghrebiensis* Arambourg, 1970; *Mammuthus meridionalis* Maglio, 1973; *E. recki ileretensis* Geraads and Metz-Muller, 1999; *M. meridionalis* Sanders et al., 2010a

Maglio (1973) speculated that *Mammuthus meridionalis* was present at the North African sites of Aïn Hanech (and Djebel Bel Hacel), Algeria, based on isolated molars identified as "*Elephas moghrebiensis*" in Arambourg's (1970) monograph on North African vertebrates. Aïn Hanech is dated by biochronological correlation and paleomagnetic sequence stratigraphy to between 1.95 and 1.78 Ma (Sahnouni et al., 2002, 2004, 2011), and has an archeological context its site locality of El Kherba of Oldowan tools and large mammal bones scored by cut marks linked with hominin subsistence activities (Sahnouni et al., 2013). Possibly the geologically youngest African anancine gomphotheres are also present at the site (Table 7.1; Sahnouni et al., 2011). The fauna also includes diverse bovids, giraffes, canids, hyaenas, equids, rhinos, hippos, and suids (Sahnouni et al., 2011, 2013); the suid *Kolpochoerus heseloni* and *Equus* species are particularly important taxa for biochronological correlation of the site.

Mammuthus meridionalis is primarily a European mammoth, ranging in age from latest Pliocene to early Pleistocene, and has been subdivided into a number of stages in several different schemes (Maglio, 1973; Zhang, 2020). Maglio (1973) placed the Aïn Hanech sample in his "Bacton Stage," defined by M3s with up to 14 plates, relatively thin enamel, and hypsodonty indices of 140–165, and depicted the species as having evolved from *M. africanavus*. Zhang (2020) did not recognize the presence of this mammoth species in Africa. It is difficult to confirm Maglio's (1973) hypothesis, because of the absence of cranial remains of the species in Africa. European crania of the species are of typical mammoth morphology (Figure 7.4E), exhibiting expanded occipitoparietals, anteroposteriorly strongly concave frontoparietal surfaces that are transversely convex, large nasal apertures that are laterally elongated and downturned at the edges, and tusk alveoli closely spaced proximally that diverge at the distal end to accommodate massive, spiraled, and curved tusks (Maglio, 1973).

Molars of African examples of this species are large and high crowned (Figure 7.4F), with modest enamel thickness. Plates are transversely widest at approximately their mid-heights. Cementum between plates is abundant and fills the transverse valleys. In lateral view, the plates are parallel-sided and the transverse valleys are U-shaped and anteroposteriorly narrow. Worn plates show no signs of incorporating prominent accessory conules, and have irregularly folded enamel. Arambourg (1970) listed the occurrence of three M3s from Aïn Hanech as part of his "*E. moghrebiensis*," including the complete specimens N° 1948-1:126 and N° 1949-2:205, with dimensions/indices of L = 300–306 mm, W = 89–90 mm, ET = 3.5, LF = 4.0–4.5, and H = 122–134 mm (yielding HI = 154–157). An incomplete m3 from the site, N° 1949-2:968 (Figure 7.4G), is over 280 mm in length, preserves 11 plates (with more originally), has a narrow width of 75 mm, and a height of 132 mm, yielding a hypsodonty index of 176. Enamel thickness is 3 mm and the lamellar frequency is between 4.0 and 4.5. Plate formula for these molars is listed as 15. In aggregate, Arambourg (1970) collated plate formulae for "*E. moghrebiensis*" as: dP4/dp4, 9/?12; m2, 12+; M3/m3, 14–15/16. In his 1973 monograph, Maglio listed two other M3s from Aïn Hanech, MNHN 48-1-127 and MNHN 48-22-205, with dimensions/indices of L = 252–290 mm, W = 89.9–90.0 mm, H = 127.6–128.0 mm (yielding HI = 142), ET = 2.8-3.2 mm, and LF = 4.1–4.7. However, for all Aïn Hanech M3s, Maglio (1973:table 29) records plate formulae of 12x-13x.

The general plate outline and metrics of these specimens are similar enough to those of "*E. recki ileretensis*" for Geraads and Metz-Muller (1999) to have placed the Aïn Hanech specimens in that taxon. These North African elephants either represent evidence for the pan-African distribution of the "*Elephas recki* complex" or the last known mammoth populations on the continent. If these elephants are mammoths, it is possible that the European deme of the species evolved from *M. rumanus* (see Lister and van Essen, 2003; Lister et al., 2005) and subsequently migrated into Africa. This interpretation is supported by the more progressive metrics of molars in the African sample. Conversely, Maglio (1973) hypothesized the origin of *M. meridionalis* via evolution from late Pliocene-early Pleistocene *M. africanavus* in Africa.

ELEPHAS Linnaeus, 1758
"*ELEPHAS RECKI*" Dietrich, 1915
(alt. "*ELEPHAS RECKI* COMPLEX")

(Tables 6.1, 7.1, and 7.2; Figures 7.1, 7.5–7.9)

Partial Synonymy—*Elephas* (*Loxodon*) *zulu* Scott, 1907; *E.* cf. *meridionalis* Haug, 1911; *E. meridionalis* Dietrich, 1913; *Elephas* aff. *antiquus* Reck, 1914; *E. antiquus recki* Dietrich, 1915; *E.* cf. *meridionalis* Dietrich, 1915; *E. antiquus* var. *recki* Hopwood, 1926; *E. zulu* Hopwood, 1926; *E.* (*Loxodonta*) *antiquus recki* Dart, 1927; *E.* cf. *planifrons* Joleaud, 1928; *Loxodonta antiquus recki* Haughton, 1929; *E. meridionalis* Joleaud, 1931; *E.* (*antiquus*) *recki*? Leakey et al., 1931; *Pilgrimia* (*antiquus*) *recki* Haughton, 1932; *Elephas* Arambourg, 1934; *Elephas* cf. *recki* Arambourg, 1934; *Palaeoloxodon antiquus recki* Hopwood, 1936; *E. recki* Arambourg, 1938; *E. recki* Arambourg, 1942; *Palaeoloxodon antiquus* Dietrich, 1942; *Palaeoloxodon antiquus recki* MacInnes, 1942:42, plate 8, figs. 4–5; *Loxodonta zulu* Osborn, 1942; *Palaeoloxodon recki* Osborn, 1942; *E.* (*Archidiskodon*) *recki* Arambourg, 1947; *E.* (*Archidiskodon*) *recki*

***atavus* Arambourg, 1947; *E.* (*Archidiskodon*) *recki recki* Arambourg, 1947; *Archidiskodon* cf. *planifrons* Arambourg, 1947; *Archidiskodon recki* Arambourg, 1947; *Palaeoloxodon recki* Cooke, 1947; *Elephas* du groupe *planifrons* Arambourg, 1951; *E. recki* Arambourg, 1951; *Archidiskodon* cf. *planifrons* Arambourg, 1951; *Archidiskodon recki* Arambourg, 1951; *Metarchidiskodon zulu* Dietrich, 1951; *Metarchidiskodon recki* Dietrich, 1951; *Archidiskodon exoptatus* (in part) Dietrich, 1951; *Palaeoloxodon antiquus* Leakey, 1954; *E. hysudricus recki* Vaufrey, 1955; *E. recki* Vaufrey, 1955; *Omoloxodon arambourgi* Deraniyagala, 1955; *Omoloxodon recki* Deraniyagala, 1955; *Loxodonta* (*Palaeoloxodon*) *antiquus recki* Singer and Crawford, 1958; "*Elephas*" *africanavus* Arambourg, 1960; *E. recki* advanced form Leakey, 1967; *E. recki* primitive form Leakey, 1967; *E.* cf. *africanavus* Leakey, 1967; *E. africanavus* Leakey, 1967; *E.* (*Archidiskodon*) cf. *planifrons* Howell, 1968; *E.* (*Archidiskodon*) cf. *exoptatus* Howell, 1968; *Palaeoloxodon recki* Aguirre, 1969b; *E. recki* forme primitive, forme normale, forme progressive Arambourg 1969b; *E. africanavus* Arambourg et al., 1969; *E. recki* stage 1, stage 2, stage 3 Maglio, 1969b; *E. recki* stage 4 Maglio, 1970a; *Mammuthus* (*Archidiskodon*) *transvaalensis* Cooke and Coryndon, 1970; *E. iolensis recki* Arambourg, 1970 *Loxodonta atlantica zulu* Maglio, 1973; *E. recki* Beden, 1980; *E.* (*Palaeoloxodon*) *recki* Beden, 1983; *Omoloxodon recki* Garutt, 1986; *E. recki* Sanders et al., 2010a; *Phanagoroloxodon planifrons* (in part) Zhang, 2020; *Elephas* sp. (in part) Zhang, 2020; *Elephas ekorensis* (in part) Zhang, 2020; *Elephas atavus* Zhang, 2020; *Palaeoloxodon recki* Zhang, 2020; *Palaeoloxodon recki* Larramendi et al., 2020; *Palaeoloxodon recki brumpti*, *P. recki shungurensis*, *P. recki atavus*, *P. recki ileretensis*, *P. recki recki* Chaix et al., 2020**

Elephas recki, or the "*Elephas recki* complex" (sensu Zhang, 2020), was the dominant elephant group in Africa for most of the late Pliocene and Pleistocene, in fossil abundance and commonality of site occurrences (Maglio, 1973; Beden, 1980). The species is named for paleontologist and geologist Hans Reck, who undertook initial field survey of Olduvai Gorge, Tanzania in 1913 and was responsible for recovering a large collection of vertebrate fossils, including specimens of his namesake species (Dietrich, 1915). Presciently, Reck (1914) thought that these Olduvai elephants showed affinity with European *Elephas antiquus* (now *Palaeoloxodon antiquus* [Larramendi et al., 2020; Zhang, 2020]). Although *E. recki* is best documented from eastern Africa, there is evidence that distribution of the group was pan-African and extended to the Arabian Peninsula and the Near East (Table 7.1).

These elephants reached immense body sizes, larger than extant African elephants, and are among the most massive fossil elephants documented (see Larramendi, 2016; Larramendi et al., 2017). Based on a nearly complete skeleton from Area 123 at Koobi Fora, Kenya, dated to between 1.87 and 1.53 Ma and attributed to *E. recki atavus* (Gathogo and Brown, 2006), Larramendi (2016) reconstructed shoulder height of this adult male individual as 427 cm and body mass as equivalent to 11,158 kg. Femoral and humeral lengths of this individual are 1,580 and 1,400 mm, respectively. As impressive as those dimensions are, a femur recovered from the Upper Burgi Mb. of the Koobi Fora Fm. at Area 105, with a geological age of about 1.9 Ma and presumed to belong to *E. recki atavus*, KNM-ER 63870, has a length of 1,700 mm (Figure 7.5A). Regression analysis of this length (following Christiansen, 2004) yields a body mass estimate of approximately 17,300 kg. The individual to which this femur belonged was almost certainly taller than the Area 123 elephant.

Elephants as massive as *E. recki* must have had tremendous effects on local ecosystems, possibly competitively excluding other elephants (Cerling et al., 2015). This might explain the coincident rarity of *Loxodonta* and *Mammuthus* in eastern Africa during the greater than three Myr of *E. recki*'s existence. Consistent with its increasingly hypsodont molars, isotopic studies of dentition of individuals in the "*E. recki* complex" reveal that these elephants became committed grazers, particularly in the Pleistocene (Cerling et al., 1999, 2011a, 2015; Lister, 2013). Even early members of the "*E. recki* complex," such as "*E. recki brumpti*" at Pliocene-aged Allia Bay, Kenya, are documented as having consumed C_4 grasses as a significant percentage of their diets (e.g., Schoeninger et al., 2003). Dental isotopic and microwear analyses suggest that Eurasian palaeoloxodont elephants, derived from the "*Elephas recki* complex," may have been graze-dominated mixed feeders in the middle Pleistocene, showing retention of ancestral dietary habits, but by the latter part of the middle Pleistocene and late Pleistocene, they had embraced a more browse-dominated diet (Palombo et al., 2005; Rivals et al., 2012).

Elephas recki is an extremely contentious taxon beset by a number of critical problems relating to its composition, taxonomic legitimacy, and mode of evolution (Maglio, 1973; Beden, 1980; Ferretti et al., 2003; Sanders et al., 2010a; Todd, 2010a; Sanders and Haile-Selassie, 2012; Zhang, 2020). First, there is the question of whether it is a legitimate species, or alternatively represents a miscellany of different genera and species cobbled together from different lineages. Second, does its fossil record preserve evidence of repeated cladogenetic episodes, or alternatively document anagenetic evolution of a single lineage? Third, if it was an anagenetically evolving lineage, can it be meaningfully subdivided arbitrarily into time-successive species, stages or subspecies, and what should the criteria be for separating these lineage units? Finally, what is the relationship of the fossil elephants in the "*Elephas recki* complex" to other elephant taxa, including Eurasian species?

In his sweeping revision of the Elephantidae, Maglio (1973) implied that *Elephas recki* was a phylogenetically unbroken Plio-Pleistocene lineage that had evolved via

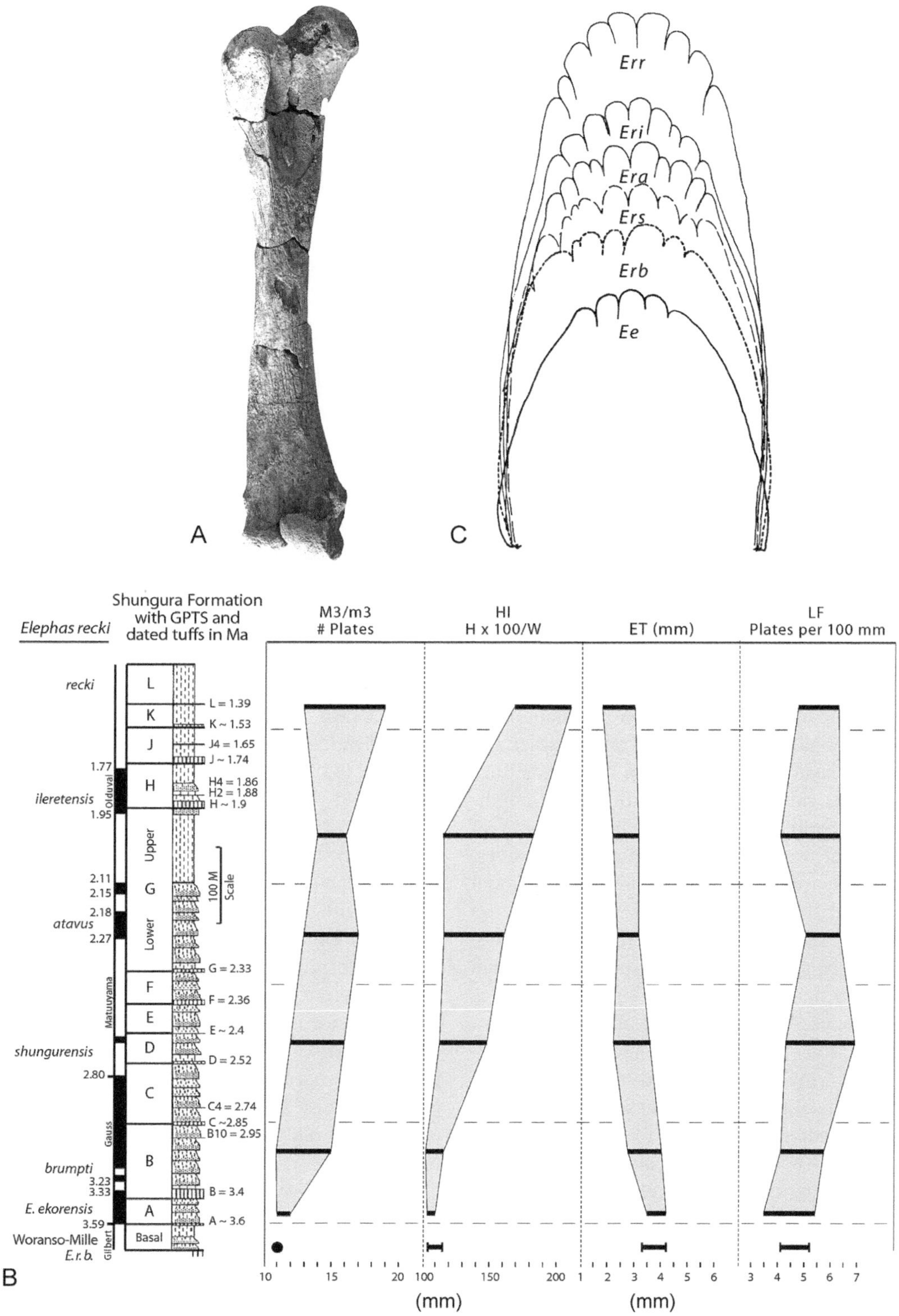

FIGURE 7.5 Morphological and chronostratigraphic aspects of the Plio-Pleistocene "*Elephas recki* complex." Abbreviations: *Ee*, *Elephas ekorensis*; *Era*, "*Elephas recki atavus*"; *Erb*, "*Elephas recki brumpti*"; *Eri*, "*Elephas recki ileretensis*"; *Err*, "*Elephas recki recki*"; *Ers*, "*Elephas recki shungurensis*"; ET, enamel thickness, HI, hypsodonty index (molar height × 100/width); LF, lamellar frequency (number of plates per 100 mm); M/m, upper/lower molar; mm, millimeter(s). (A) Left femur, KNM-ER 63870 from Upper Burgi Mb., Area 105, Ileret, Kenya, posterior view, *Elephas atavus*. Femoral length is 1,700 mm. (B) Chronostratigraphic scheme of subdivision of the "*Elephas recki*" lineage into subspecies, showing gradual, directional change of morphometric features of third molars over time. (From Sanders and Haile-Selassie [2012:fig. 8].) (C) Outlines of transverse cross-sections of third molar plates of *Elephas ekorensis* and subspecies of the "*Elephas recki* complex," representative of changes over time in hypsodonty and plate shape in the lineage. (From Sanders et al. [2010a:fig. 15.7].)

anagenesis. In a common paleontological device to work with such lineages, Maglio adopted a time-successive arbitrary partition scheme, that subdivided the species into four stages. This stage-based scenario relied heavily on identification of evolved morphometric features of molars. Stage 1 was characterized by fossil material from the lowest levels of the Shungura Fm. in the Omo, Ethiopia; stage 2 was typified by fossils from above Mb. C of the Shungura Fm.; stage 3 included more progressive specimens from Olduvai Bed I and lower Bed II, Tanzania; and molars from Olduvai Bed IV (the original type horizon) were used to characterize stage 4. In his conceptualization of *E. recki*, Maglio (1973) hypothesized that it was a descendant of "*E.*" *ekorensis* and direct ancestor of *E.* "*iolensis*" (correctly, "*jolensis*" [Manthi et al., 2019]). As Maglio (1973) placed all non-mammoth elephants from Eurasia in *Elephas*, in his view presumably *Elephas recki* was ultimately ancestral to all of them.

The stage-grade schema was adopted by Beden (1976), who subdivided *E. recki* into stage 1 from the Usno Fm. and Shungura Fm. Mb. B, stage 2A from Mbs. B-E of the Shungura Fm., stage 2B from Mbs. F-G of the Shungura Fm., and stage 3 from the uppermost levels of the Shungura Fm. Beden's (1976) stage 2 is approximately equivalent to Maglio's (1973) stage 2.

The decision to subdivide *E. recki* into stages was justified on the grounds that time-successive populations of the species morphologically graded into one another but had evolved progressively; metrically, this meant continuous (and directional) shifting over time of modal conditions of dimensions and their ratios, such as number of plates or lamellae, crown height and hypsodonty indices, lamellar frequencies, frequency and amplitude of enamel folding, and enamel thickness, with *substantial overlap between succeeding stages in the range of these metrics*. Temporally, crown height increased substantially, enamel became thinner and more finely folded, and molar plates were more closely spaced and became greater in number.

This gradation was likened by Sanders and Haile-Selassie (2012) to a "ring species" rolled out through time instead of geographically. They emphasized the highly arbitrary nature of subdividing anagenetically, directionally evolving lineages. Each succeeding population or stage was similar enough to the preceding population to place them in the same species, but the "end members" of the lineage were substantially different enough from one another to justify recognition of them as separate species. In fact, the proposed African end member of the lineage was placed in *E.* "*iolensis*" (more correctly, "*jolensis*") rather than being considered as another stage of *E. recki* (Maglio, 1973). As with the stage schema of Maglio (1973) and Beden (1976), the organization of the lineage later utilized by Sanders and Haile-Selassie (2012) was substantially based on morphometric aspects of molars.

The succession of *E. recki* stages was further refined and codified by Beden (1980), who subdivided the species into five subspecies, from oldest to youngest *E. recki* "*brumpti*," "*shungurensis*," "*atavus*," "*ileretensis*," and "*recki*." As he had invested considerable work into studying the proboscideans of the Omo Shungura Fm., it is not surprising that the geological framework of the formation, cleanly divisible into discrete members by tuffaceous horizons, provided the basis of Beden's (1980) arbitrary division of the lineage into chronologically successive subspecies (Figure 7.5B). Thus, these subspecies were circumscribed by boundaries of geological units defined by tephra and tuffaceous sediments (e.g., McDougall et al., 2012), with morphological and metric aspects of molars collated for subspecies from these geological members rather than dictating biological limits between them (Table 7.1), a device probably influenced by the difficulty of clearly differentiating temporally adjacent subspecies morphometrically. Hence, "*brumpti*" was *Elephas recki* from Upper Mb. A and Mb. B of the Shungura Fm., "*shungurensis*" from Mb. C to Lower Mb. F, "*atavus*" from Upper Mb. F and G, "*ileretensis*" from Upper Mb. G and H (proboscidean samples are very poor from these units), and "*recki*" is from Mbs. K and L (Beden, 1980). In Beden's (1980:table 1) subspecific scheme, there is considerable overlap of molar dimensions and indices between each subspecies, but substantial differences between end taxa *E. recki brumpti* and *E. r. recki* in m3 plate numbers (11–14 versus 14–18), enamel thicknesses (2.8–4.0 mm versus 2.3–3.0 mm), lamellar frequencies (4.2–5.5 versus 5.5–6.0), and hypsodonty indices (102–116 versus 161–190). The pattern of evolution of these features in *E. recki* over time in the Omo Shungura sequence is summarized in Sanders and Haile-Selassie (2012:fig. 8). A caveat of these research results is that the dimensional range and variance of these features is dependent on fossil molar sample size from each geological unit (see Ferretti et al., 2003b; Sanders and Haile-Selassie, 2012); nonetheless, despite significant contrasts in sizes of samples from different members of the Shungura Formation, the picture that emerges is one of directionally shifting metrics over time, associated with molars becoming composed of increasing numbers of ever-higher-crowned plates (Figure 7.5C), consequent greater anteroposterior crowding of plates, and thinner, more regularly and tightly folded enamel (Sanders and Haile-Selassie, 2012).

Beden's (1980, 1983, 1987a) taxonomic and phylogenetic schema for organizing the constituents of the "*Elephas recki* complex" over its approximately three Myr of existence have become the orthodox view of the group (Sanders et al., 2010a), confirming the evident utility of *E. recki* molar metrics for biochronological correlation of (primarily eastern) African sites, originally suggested by Cooke and Maglio (1972) and Maglio (1972b), and have become an essential part of the narrative about mammalian adaptive response to ecological change in Plio-Pleistocene Africa (e.g., Levin et al., 2011; Uno et al., 2011; Cerling et al., 2015).

However, validity of subspecies are more readily assessed in living taxa where biogeography and genetics are well delineated, rather than in paleospecies. Subspecies are generally defined as a group of populations with a distinct breeding range that are morphologically distinct from other such groups but are not reproductively isolated (Patten, 2015). A "75% rule" is sometimes used to diagnose extant subspecies, where ≥75% of individuals in one subspecies should exceed 99% of the range of variation in a character or measure of another subspecies (Patten, 2015 and references therein). Moreover, the formal definition of a subspecies necessarily indicates that time-successive subspecies of a paleospecies evolved anagenetically (no reproductive isolation).

It can be argued that most molar metrics of successive *E. recki* subspecies, as arbitrarily erected by Beden (1980) based on geological subdivisions, fail to meet the strict diagnostic conditions of the 75% rule (for example, plate number, the range of which shifts directionally upward over time but overlaps considerably between succeeding subspecies). The most important molar features used to chronicle evolution of the *E. recki* lineage have been plate number and crown height (absolute and relative, or hypsodonty index). To a lesser extent, lamellar frequency, enamel thickness, and frequency and amplitude of enamel folding have also been examined to document change over time (e.g., Maglio, 1973; Beden, 1980; Sanders and Haile-Selassie, 2012).

A critical challenge to Beden's (1980) molar- and geological-unit–based subspecific subdivisions and the hypothesis of anagenetic evolution of *E. recki* was posed by Todd (2001, 2005). She set the conditions of her critique of the species as follows: "If … [paleontological] subspecies *overlap in time* and the *variation* within each taxonomic unit *is high*, then the evolution of the lineage can no longer be considered as anagenesis" (Todd, 2001:p. 694; italics inserted). Her study concluded that (a) variance is high in molar dimensions of each *E. recki* subspecies; (b) there is substantial morphological similarity, between successive subspecies; (c) the amount of morphometric variation within the entirety of *E. recki* is unacceptably too great to be accommodated in a single species; and (d) there is substantial overlap of the temporal ranges of subspecies. Thus, according to her results, *E. recki* is not a single species and did not evolve anagenetically. Additionally, her cladistic analysis based on molar characters indicated that Beden's (1980) subspecies units are not monophyletic, with *E. recki brumpti*, *shungurensis*, and *atavus* comprising one clade (albeit also including *L. exoptata*), and *E. recki ileretensis* and *recki* forming a different and widely separated clade (with unlikely clade members *L. atlantica*, *E. hysudricus*, and "*E.*" (=*Palaeoloxodon*) *mnaidriensis* [Todd, 2005:fig. 2]).

Analysis of elephant molars from the Dioli and Aalad Areas at Buia (Table 7.1), in the northern Afar Depression, Eritrea (the northernmost incidence of the "*E. recki* complex" in eastern Africa), indicated that the sample is morphometrically intermediate between *E. recki ileretensis* and *E. recki recki* (Ferretti et al., 2003b), as are molars from intervals 4 to 6 of the Konso Fm. at Konso, Ethiopia (Suwa et al., 2003), supporting Todd's (2001, 2005) hypothesis that these subspecies are part of the same clade, but obscuring recognition of *ileretensis* and *recki* as subspecies with distinct morphological boundaries. Ferretti and colleagues (2003b) considered that the intermediate state of the Buia elephant molars establishes *E. recki ileretensis-E. recki recki* as a single phyletic lineage, but also believed that problems determining the morphometric and chronological boundaries of Beden's (1980) subspecies (see also Todd and Roth, 1996) necessitated a thorough revision of the classification of *E. recki*.

Temporal ranges of *E. recki* subspecies depicted in Todd (2001:fig. 1, 2005:fig. 1) exhibit substantial overlap, particularly if sites outside of the Omo are incorporated. However, advances in absolute dating, chronostratigraphy, correlation of tephra between sites, and magnetostratigraphy (for example, Brown and McDougall, 2011; Levin et al., 2011; McDougall et al., 2012; Kidane et al., 2014) drastically reduce Todd's (2001) intervals of overlying subspecies' temporal ranges (Table 7.1)—even despite the circularity of examining subspecific chronological boundaries when chronostratigraphy is the primary criterion for arbitrarily subdividing the lineage (!). The problem is exacerbated by (a) some first or last occurrences of subspecies are assigned to intervals within geological units and may not actually be present throughout the entire interval; (b) assignment of subspecies to geological units have been altered between publications (e.g., *E. r. ileretensis* was placed in?Mb. J, Mb. K, and Lower Mb. L and *E. r. recki* in Upper Mb. L of the Shungura Fm. by Beden (1980) but in?Upper Mb. G,?Mb. H, and Mbs. K-L, respectively, by Beden (1985), shifting both subspecies in the formation to older temporal ranges); and (c) precise stratigraphic provenience data are lacking for many specimens. As currently constituted, the succession between *E. recki brumpti* and *shungurensis* is approximately at 3.2 Ma, between *shungurensis* and *atavus* at ~2.3 Ma, between *atavus* and *ileretensis* at approximately 1.8 Ma, and the succession between *ileretensis* and *recki* is sometime in the interval 1.6–1.4 Ma (Table 7.1). Nonetheless, chronological boundaries between subspecies are still not sharp, requiring more precise re-examination of the chronostratigraphic controls on locality/horizon samples. Also, assessing variation in lineage segments ("subspecies" or "stages") that approach or exceed a million years of existence is an improper exercise if morphological change was continual over time. It would be more useful to evaluate variation in lineage segments within shorter—perhaps 100 kyr—increments of time. This degree of chronological resolution is only possible at a few sites, such as the Omo, Woranso-Mille, and Hadar, Ethiopia, and Koobi Fora, Kenya, and not throughout the entirety of the stratigraphic sections at those sites.

A more substantial challenge to the orthodoxy of *E. recki* stems from the recent ambitious taxonomic revision of Elephantidae comprising the doctoral work of Zhang (2020). In contrast to Maglio (1973) and Beden's (1980, 1983, 1985, 1987a) emphasis on molars, Zhang's (2020) review of *E. recki* concentrated on the fertile but underexplored utility of elephant skull morphology for phylogenetic systematics. As a result of this focus, Zhang (2020) and Larramendi et al. (2020) placed *E. recki recki* in *Palaeoloxodon* (a genus with broad Eurasian distribution), redesignating it as *Palaeoloxodon recki*. Moreover, *E. recki brumpti* was subsumed into a new taxonomic combination, *Phanagoroloxodon planifrons* (primarily with South Asian distribution), and *E. recki atavus* was elevated to *Elephas atavus* (Zhang, 2020). This revolutionary perspective of the "*E. recki* complex" has merit, particularly for carefully documentating stark contrasts in cranial morphology within the group and generating discussion about the meaning of these contrasts in a phylogenetic-systematic-evolutionary framework.

The idea that *E. recki* is phylogenetically connected with the Eurasian *Palaeoloxodon* group, including *P. antiquus*, *P. namadicus*, and *P. naumanni*, is not new. For example, MacInnes (1942) and Osborn (1942) both incorporated *E. recki* into *Palaeoloxodon*. Aguirre (1969a, b) notably considered *Palaeoloxodon* to be a subgenus of *E.* (*Palaeoloxodon*) *recki* and the species to be ancestral to the Eurasian forms of the subgenus. Beden (1983, 1987a) also hypothesized that *E. recki* was the most primitive species of the subgenus *Palaeoloxodon*, and that "*E.*" *ekorensis* was ancestral to the other major subgenus of *Elephas*, in which he included the extant Asian elephant and species such as *E.* (*Elephas*) *planifrons*, *E.* (*E.*) *hysudricus*, and *E.* (*E.*) *hysudrindicus*. However, in Beden's (1983, 1987a) view, in contrast to Zhang's (2020) phylogeny, "*E.* (*P.*) *recki atavus*" (and not "*E.* (*P.*) *recki recki*") was the ancestor of Eurasian species of *Elephas* (*Palaeoloxodon*).

Saegusa and Gilbert (2008) also considered subspecies of *E. recki* to belong in the subgenus *Palaeoloxodon*, except for *E. recki brumpti*, for which they identified no synapomorphies linking the subspecies with the *E.* (*Palaeoloxodon*) clade. Furthermore, they presented evidence documenting that the first Eurasian members of this clade occurred around 1.0 Ma (Saegusa and Gilbert, 2008 and references therein), suggesting derivation from "*E.* (*P.*) *recki recki*." This is concordant with Zhang (2020) and Larramendi et al.'s (2020) hypothesis that Eurasian species of *Palaeoloxodon* descended from *Palaeoloxodon recki*.

In a more fanciful classification, Deraniyagala (1955) placed Arambourg's (1947) elephants from the middle beds of the Omo Shungura Fm. (now included in "*E. recki atavus*" or "*E. atavus*") in a new genus, *Omoloxodon* (*O. recki*), the type series of *E. recki* from Olduvai in "*Loxodonta africana zulu*," and bundled both taxa into the elephant subfamily Loxodontinae. This taxonomy is wildly imaginative and unsustainable as the cranial and dental fossil remains do not exhibit loxodontine synapomorphies.

"*ELEPHAS RECKI BRUMPTI*" Beden, 1980 (alt. *ELEPHAS BRUMPTI* [Beden, 1980]) (alt. *PHANAGOROLOXODON PLANIFRONS* [Falconer and Cautley, 1845]) (alt. *PHANAGOROLOXODON BRUMPTI* [Beden, 1980])

(Tables 6.1, 7.1, and 7.2; Figures 7.1, 7.5, and 7.6)

This subspecies is best documented from localities in the Afar region and greater Turkana Basin in Kenya and Ethiopia, dating from the early to late Pliocene (Tables 6.1 and 7.1). Designated as the earliest and most archaic stage of the species (Beden, 1980), *E. recki brumpti* has been taxonomically problematic for unity of the species because its dentition (and cranial remains) differ substantially from those of the species type from Olduvai Bed IV and similar-aged *E. recki recki* fossils from elsewhere. Moreover, molars of *E. recki brumpti* have been described as very similar to those of "*E.*" *ekorensis* (Beden, 1987a). Although fossils now attributable to *E. recki brumpti* were first recovered at the onset of the 1900s, they were thought to be too primitive to include in *E. recki* as characterized by molars from the type site/horizon (e.g., Dietrich, 1915, 1921), and therefore were not subsumed into the species until much later (Maglio, 1973).

The subspecies has been defined by molars composed of a modest number of plates, mostly parallel anterior and posterior faces of plates separated by U-shaped transverse valleys that are anteroposteriorly thicker than the plates, accessory conules apically free on the first few plates, simple and rounded median plate sinuses, moderately thick enamel, lamellar frequencies higher than in penecontemporaneous *L. adaurora*, and slight molar hypsodonty (Figure 7.6A and B; Beden, 1976, 1980, 1983, 1987a; Sanders et al., 2010). Beden (1976) described *E. recki brumpti* as having M3s with plate formula 9–12, LF=4.0–4.5, HI=107–115, and ET=3.2–4.5 mm. This sample probably includes specimens from the Usno Fm. that later were included in "*E.*" *ekorensis*. Subsequent collation of data of the subspecies from Koobi Fora and Kubi Algi, Kenya, and the Omo and Woranso-Mille, Ethiopia, indicate plate formulae of dP2/dp2, 4x/x3x-x4; dP3/dp3, 6x-7/6x-7x; dP4/dp4, x7x/-x7x-9; M3/m3, 11-?/x11x-14x, and third molar metrics of LF=4.0–6.0, HI=101–124, and ET=2.8–4.5 mm (Beden, 1983, 1987a; Sanders and Haile-Selassie, 2012). Cross-sectional profile of molar plates shows the widest transverse points to be about a quarter of the height from the cervix of the crown, with lateral plate sides parallel for the first third of their extent and then gently curving toward the apex, with unworn central conelets higher than lateral conelets (Figure 7.6C). Molar plates are formed of between five and eight conelets. In lateral view, cementum coats the unworn plates and they are parallel-sided, appearing characteristically like narrow silos. Enamel folding is highly variable, may be irregular along a plate, and is strongest toward the median area of the crown. Accessory conules may be present anteriorly and posteriorly in association with plates, but do not contribute to loxodont sinuses with wear, and often are "unbalanced," with either anterior or posterior conules larger and more projecting.

Included in elephant dental specimens from Koro-Toro, Chad, dated to the mid-to-late Pliocene (Table 7.1), is a complete M3 attributed to *E. recki*, KT8.94.009b, preserving a plate formula of 12x (Mackaye, 2001). Dimensions of this specimen are L=285 mm, W=101 mm, H=110 mm (yielding HI of 122), ET=3.5 mm, and LF=4.5. It is reminiscent of molars from the late Pliocene Sidi Hakoma Mb. of the Hadar Fm. Cçementum covers the plates and infills the U-shaped transverse valleys, which are anteroposteriorly

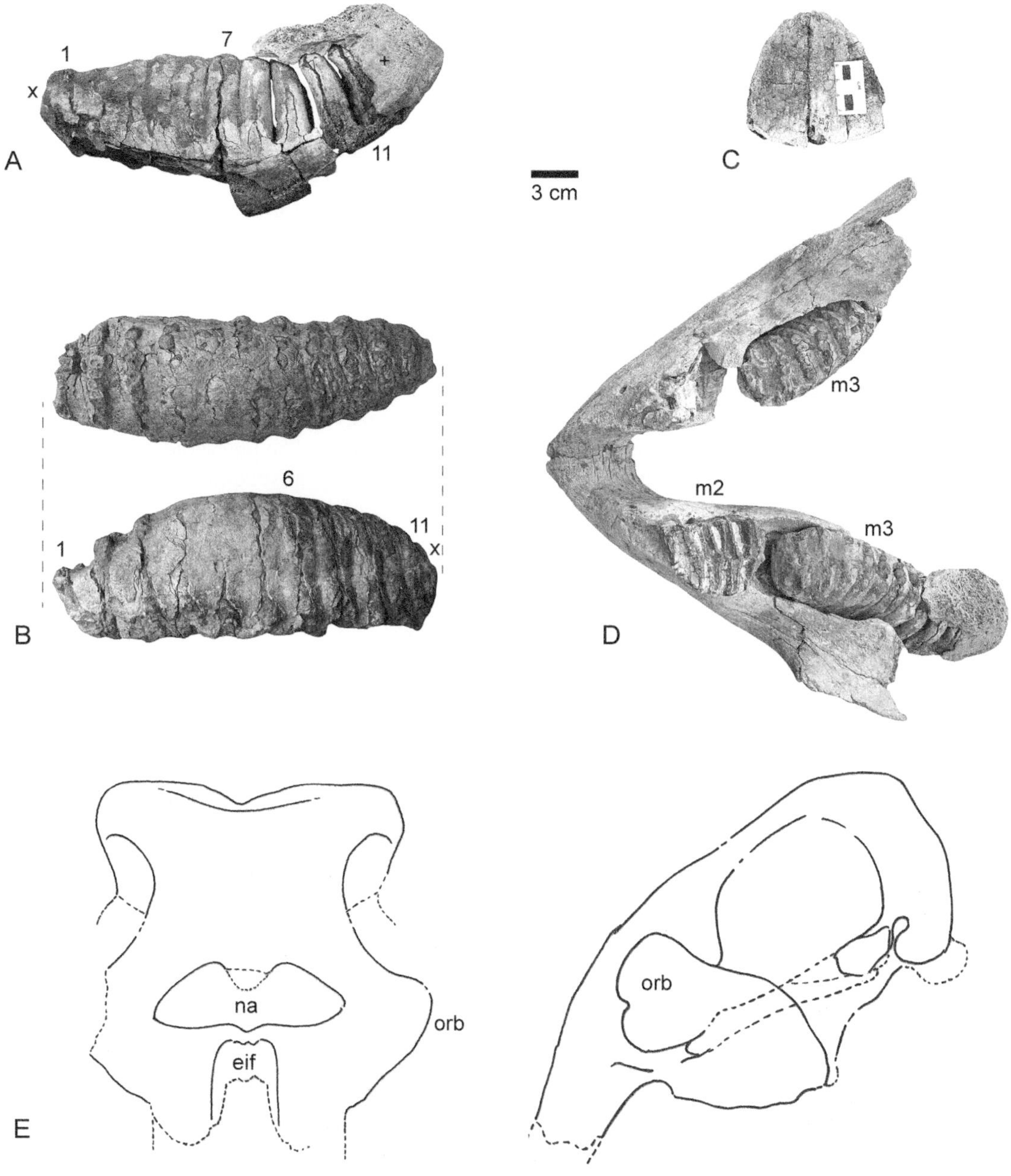

FIGURE 7.6 Skull and molar fossils of Pliocene *Elephas brumpti*. Abbreviations: eif, external incisive fossa; m, lower molar; na, nasal aperture; orb, orbit; x, pre- or postcingulum(id); 1, 2, 3, …, plates counted from the anterior of the crown. (A, B, D) anterior to the left. (A–C) to the same scale. (D, E) not to the same scale. (A) Emergent left m3, MSD-VP-5/048 from early Pliocene Mesgid Dora, Woranso-Mille, Ethiopia, lateral view. (B) Right M3, A.L. 126-3a from the late Pliocene Sidi Hakoma Mb. of the Hadar Fm., Hadar, Ethiopia, occlusal and lateral views. (C) Molar plate, KNM-ER 5810 from the late Pliocene Tulu Bor Mb. of the Koobi Fora Fm., Area 117, Koobi Fora, Kenya. Note the modest hypsodonty and that the greatest width is about one-quarter the distance from the base of the crown. (D) Partial mandible with right m3 and left m2–3, MSD-VP-5/033 and 048 from early Pliocene Mesgid Dora, Woranso-Mille, Ethiopia, ventral view. (E) Partial cranium, L 1.70 from Upper Mb. B, Shungura Fm., Omo, Ethiopia, anterior and lateral views. Note the lack of significant bossing, broad flare of the orbit(s), and rounded temporal fossa. (Adapted from Beden [1987a:fig. 12A, B], with permission of the Museum National d'Histoire naturelle, Paris.)

thicker than the plates. The lateral edges of plates curve gently toward their apices in their upper halves; plates are transversely widest above the base. As a broken, worn M3 from Koro-Toro (KT9.94.005) shows, enamel is coarsely folded or undulated, accessory conules are incorporated into enamel wear figures without formation of loxodont sinuses, and plates are formed of six or seven apical digitations.

Geologically older, early Pliocene gnathodental specimens attributed to *E. recki brumpti* from Woranso-Mille, Ethiopia

(Table 6.1; Sanders and Haile-Selassie, 2012), include a mandible with a worn left m2 and m3s emergent from the crypt, MSD-VP-5/033, and 5/048 (Figure 7.6D). The most complete m3 has 11 plates plus a half-plate wedged between plates 6 and 7, and dimensions/indices of L=278.0 mm, W=80.4 mm, H=93.0 mm (yielding HI=116), and LF=4.75 (Figure 7.6A). Plates are variably widest between the base and one-quarter of their height from the base. The mandible is missing its rami, posterior corpora, and the symphysis is broken anteriorly. Height and width of the corpus at the anterior of m2 are 140.5 mm×94.8 mm, respectively, and there are three mental foramina in each dentary. A more robust dentary from the site, MSD-VP-5/041, has height and width dimensions at m3 of +162.0 mm×175 mm, respectively. An emergent m3, MSD-VP-8/010, has an incipient 12th plate, moderate hypsodonty (HI=e.104), plates formed of seven conelets with greatest width one-quarter the height above the base, and exhibits strong posterior central accessory conules at least to the fifth plate that are nearly as tall as the plates themselves.

M3s from the site have 11 plates, and dimensions/indices of LF=4.0–5.0 and ET=3.4–4.2 mm. Partial M3s in a very fragmentary cranium from the site, MKM-VP-2/008, exhibit fine enamel folding along the breadth of its plates, which is more advanced-looking than the enamel undulations in the Koro-Toro worn M3, but has single, prominent central conelets, which appears more primitive. M3s may exhibit a greater degree of enamel folding than m3s in the same sample. Alternatively, Zhang (2020) placed MKM-VP-2/008 in "*Elephas* sp." and MSD-VP-5/033 and 5/048 in "*E. ekorensis*," but this introduces unnecessary complications as these specimens morphometrically fit comfortably in samples of *E. recki brumpti* from other sites (e.g., Beden, 1983, 1987a).

Cranium L.1.70 from unit Mb. B-11 of the Shungura Fm., dated to approximately 2.9 Ma (see Levin et al., 2011), preserves much of its morphology, except for the occipital condyles, zygomatic arches, and lower part of the rostrum, as well as the edge of the right orbit and terminus of the nasals (Figure 7.6E). The specimen is of modest overall size and is relatively low and wide: incomplete length=670 mm, width across the orbits is estimated to be 640 mm, least transverse width between the temporal lines=280 mm (narrow relative to overall cranial width), and width across the base of the rostrum=360 mm.

In lateral view, the profile of the cranium is gently rounded from the nuchal crest to the broken end of the rostrum. Although the reconstructed position of the missing occipital condyles is well above the level of the palate, the top of the orbits are not lower than the level of the condyles. The occipital planum is marked by a sagittal depression, but although the nuchal crest and supraoccipital bones are robust, development of occipitoparietal bosses is only incipient, and the midline depression in the nuchal crest is nascent. The fossa for the nuchal ligament is placed high in the occipital planum sagittal depression, and is deep, dorsoventrally long, and narrow. The occipital planum is much wider (550 mm) than high (380 mm) and ovoid. The rostrum is strongly downturned. The orbits are very anteriorly positioned, project strongly laterally, and are at the level of the lower aspect of the nasal aperture. Within the orbits, the lacrimal processes are well developed. The external nasal opening forms a laterally narrowed, elongated oval, of relatively large size ([contra Beden, 1987a; Zhang, 2020] anteroposterior diameter=140 mm, breadth=300 mm). Tusk alveoli are parallel to one another and are separated by a deep external incisive fossa. The temporal fossae are deeply excavated, indicating the presence of powerful temporal muscles.

A second cranium attributed to the subspecies, KNM-ER 5710, derives from below the Hasuma Tuff in Area 204 at Allia Bay, Kenya and is far less complete. It is missing the superior aspect of its occipital, zygomatic arches, orbits, most of its dorsal surface, and rostrum (Beden, 1983). An anteriorly broken M3 has all plates in wear, indicating that the specimen is from a full adult. The form of the deep external incisive fossa indicates that it was bordered by parallel tusk alveoli. The pterygoid region differs from that of *L. adaurora*; the alisphenoid bone is long and extends up to the level of the palate, obscuring the pterygoid process of the palatine bone, in posterior view (Beden, 1983:67). Lateral swelling of the nuchal margin bordering the temporal fossa is equivalent to the condition in specimen L.1.70 from the Omo. Maximum width of the cranium is across the occipital and measures 770 mm, suggesting that KNM-ER 5710 is from a larger individual than L.1.70.

Based primarily on the morphology of cranium L.1.70, Zhang (2020) erected an alternative hypothesis placing *E. recki brumpti* into *Phanagoroloxodon planifrons* (=*E. planifrons* of Maglio, 1973). As detailed above, it is unlikely that the cranium of *L. adaurora kararae* also belongs in this species or that it is conspecific with *E. recki brumpti* (contra Zhang, 2020). *Phanagoroloxodon* and the tribe "Phanagoroloxodontini" was erected in 1957 by Garutt to accommodate elephant remains from northwestern Caucaus region of Russia of late Pliocene or early Pleistocene age ("*Ph. mammontoides*"), and was thought to represent the ancestral taxon for both Asian elephants and mammoths (Garutt, 1986, 1992, 1995). According to Garutt (1986), its primitive sister taxon is *Protelephas planifrons*. Zhang (2020) incorporated *Pr. planifrons* of Garutt into *Phanagoroloxodon* based on shared cranial features that distinguish the genus from *Elephas*: rudimentary development of occipitoparietal bosses and supraoccipital swellings, lesser constriction of the cranial vault across the temporal lines, no strong concavity of the parietofrontal part of the vault, and a relatively narrower occipital planum that may not be as wide as the transverse diameter across the supraorbitals.

Cranial remains of South Asian *Elephas planifrons* resemble specimen L.1.70 in having low, rounded lateral profiles, laterally-flared orbits, orbits at about the same level as the occipital condyles, a mostly flat frontoparietal surface, deep temporal fossae, and diminutive occipitoparietal bosses that variably may be separated by a shallow sagittal nuchal furrow (Maglio, 1973; Nanda et al., 2000). Projection of the rostrum in South Asian *E. planifrons* is not as downturned as it is in L.1.70, and the nasal apertures in these elephants are relative smaller, as well. *Elephas planifrons* is best known from the late Pliocene-early Pleistocene Tatrot and Pinjor formations in India and Pakistan, and appears to have entered the subcontinent at approximately 3.0 Ma (Barry et al., 1982; Patnaik et al., 2019). This is important for establishing the timing of emigration of the ancestral source of the lineage out of Africa. Isotopic analysis of *E. planifrons* molars reveals that they were predominantly C_4 grazers (Patnaik et al., 2019), with similar feeding preferences to those of constituents of the *E. recki* complex.

Garutt's (1957, 1986, 1992, 1995) taxonomic and phylogenetic ideas about *Phanagoroloxodon* are needlessly byzantine. The tribe erected by Garutt for this genus, Phanagoroloxodontini, is irrelevant given what is known about the early fossil record of *Mammuthus* (see Chapter 5) and genetic-based hypotheses about the timing of divergence of these groups (Rohland et al., 2007; Murata et al., 2009). Conversely, it is reasonable to view *E. recki brumpti* and *E. planifrons* as sister taxa or even as congenerics (Zhang, 2020), although they may not be conspecific. Possibly, they are primitive members of the *Elephas* clade prior to the origin of *Palaeoloxodon*, and their dearth of derived *Elephas* cranial features indicates that they are stem, basal, species of the clade. One indication of this is that there is evidence for the retention of permanent premolars in *E. planifrons* (Hooijer, 1982; Sanders, 2017). They do not possess synapomorphies definitive of the *Palaeoloxodon* clade or many features of crown members of *Elephas*, but it seems a needless overreach to place them in *Phanagoroloxodon*. Nonetheless, if *E. recki recki* belongs in *Palaeoloxodon*, as appears probable at either the subgeneric or generic level (Saegusa and Gilbert, 2008; Larramendi et al., 2020; Zhang, 2020), then a new species name is needed for *E. recki brumpti* (perhaps "*E. brumpti*").

Also, because there are no undeniable adult crania of "*E.*" *ekorensis* (see Chapter 6) or *E. recki shungurensis*, and molars of both may be difficult to distinguish morphometrically from those of *E. recki brumpti*, it is only a working hypothesis to consider them as part of the same clade of elephants. Alternatively, this clade is (a) phylogenetically unconnected to *Elephas* and *Palaeoloxodon*, (b) may be the phylogenetic source for *Elephas* (including *E. recki atavus* and extant Asian elephants), and (c) may ultimately have also been ancestral to palaeoloxodonts, as well. Hypothesis (a) implies that no Pliocene or earliest Pleistocene evidence for the African ancestry of the *Elephas* and *Palaeoloxodon* lineages has been recovered.

Elephas recki brumpti belongs to very diverse mammalian faunas at its site occurrences, characterized by a mix of modern and extinct genera, early hominins, such as *Australopithecus anamensis*, *A. deyiremeda*, *A. afarensis*, *Paranthropus boisei*, and *Kenyanthropus platyops*, and was sympatric with a number of other proboscideans, such as "*E.*" *ekorensis*, *L. adaurora*, *L. exoptata*, *Mammuthus* sp. "Hadar type," *Anancus ultimus*, *Stegodon kaisensis*, and *Deinotherium bozasi* (Tables 6.1 and 7.1; Beden, 1976; Kalb et al., 1982; White et al., 1984; Harris et al., 1988b; Bobe, 1997; Leakey et al., 2001; Bobe et al., 2002; Alemseged, 2003; Haile-Selassie, 2010; Sanders and Haile-Selassie, 2012; Haile-Selassie et al., 2015; Su and Haile-Selassie, 2022). For example, at West Turkana, Kenya, *E. recki brumpti* co-occurs with *D. bozasi*, *L. adaurora*, *L. exoptata*, unspecified hominins, *Paranthropus boisei*, a number of extinct species or genera of cercopithecoids including *Theropithecus brumpti*, *Parapapio* spp., and *Paracolobus*, modern genera of hyaenas alongside saber-toothed cats (*Homotherium*, *Dinofelis*), hipparionine horses, modern genera of rhinos, extinct suid genera *Metridiochoerus*, *Nyanzachoerus*, *Kolpochoerus* and *Notochoerus*, modern and extinct genera of hippos (*Hippopotamus*, *Hexaprotodon*), modern and extinct genera of giraffes (*Giraffa*, *Sivatherium*), and a wealth of bovids mostly of modern generic affiliation but comprising extinct species (Harris et al., 1988a, b). Soil carbonates and faunal evidence from Koobi Fora, West Turkana, the Omo region, and in the Awash around the time of *E. recki brumpti* (~3.7–3.0 Ma) indicate close to 40% or more woody canopy cover contributing to mosaic habitats that fostered this rich diversity of mammalian taxa (Levin et al., 2011; Su and Haile-Selassie, 2022). Within these heterogeneous ecologies, elephants were competing for graze with gomphotheres, stegodonts, rhinos, hippos, suids, horses, possibly some cercopithecoids, and bovids. By the end of this interval, anancine gomphotheres, stegodonts, and non-hypsodont elephants had vanished, and the "*E. recki* complex" had evolved molars with more plates and greater hypsodonty.

"*ELEPHAS RECKI SHUNGURENSIS*" Beden, 1980 (alt. *ELEPHAS SHUNGURENSIS* [Beden, 1980])

(Tables 7.1 and 7.2; Figures 7.1, 7.5 and 7.7)

Elephas recki shungurensis is the subspecies of Beden (1980) that temporally bridged the late Pliocene-early Pleistocene (Table 7.1). Well-documented in the Turkana basin region, it is also the elephant of the Lower beds at Olduvai Gorge, Tanzania and appears to have had broad distribution into North Africa, the Western Rift, and southern Africa (Table 7.1). Metrically, its molars are transitional between *E. recki brumpti* and later *E. recki atavus* that exhibits distinctive cranial synapomorphies linking it with other *Elephas* species, and overlaps with both. Geologically older molar specimens assigned to *E. recki shungurensis*, such as isolated specimens from the Sidi Hakoma Mb. of the Hadar Fm. (pers. obs.), are nearly indistinguishable from those of *E. recki brumpti* (Figure 7.7A). Plate formulae for *E. recki shungurensis* are: dP2/dp2, 3x–5/3x–x4; dP3/dp3, x5x–x7x/x6x–7x; dP4/dp4, x9–?10x/9x–10x; M1/m1, 8x–10x/10x; M2/m2, 11x/10–11x; M3/m3, x12x–15/x12x–15 (Beden, 1983, 1987a). Third molar hypsodonty indices were collated as 113–138; ET ranges from 2.2 to 3.6 mm, and LF = 4.2–6.5, indicating moderate to tight spacing of plates (Beden, 1980). A typical m3 of this taxon, KNM-ER 4564 from the Lower Burgi Mb. of the Koobi Fora Fm. at Area 100 of Koobi Fora, has a plate formula of x14x and dimensions/indices of L = 285 mm, W = 76 mm, H = 93 mm (yielding HI = 122), ET = 2.8–3.2 mm, and LF = 5.1–5.4 (Beden, 1983). Molars of "*E. r. shungurensis*" are broadest about half-way up from the cervix of the crown (Figure 7.7B)

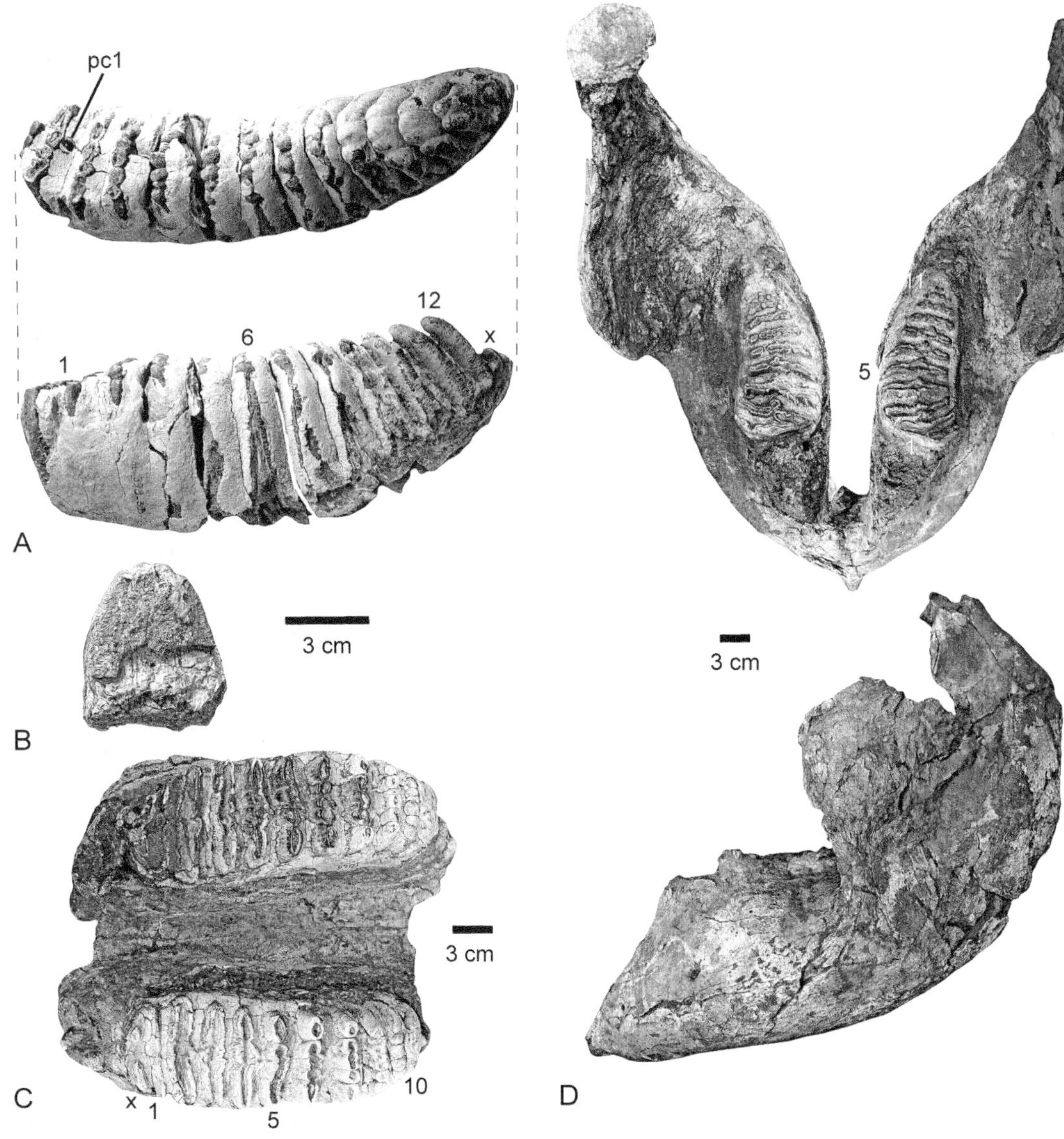

FIGURE 7.7 Mandibular and molar fossils of late Pliocene-early Pleistocene "*Elephas recki shungurensis*." Abbreviations: pc, posterior accessory conule; x, pre- or post-cingulum(id); 1, 2, 3, …, plates counted from the anterior of the crown. (A, C) anterior to the left. (A, B) to the same scale. (A) Right m3, A.L. 398-3b from the Sidi Hakoma Mb. of the Hadar Fm., Hadar Ethiopia, occlusal and lateral views. (B) Molar plate, A.L. 1367-2 from the late Pliocene Kada Hadar Mb. of the Hadar Fm., Hadar, Ethiopia, anterior view. Note that the greatest width is between one-quarter and one-half the distance from the base of the crown. Contrast this plate profile with that of *Elephas brumpti* in Figure 7.6C. (C) Palate with right and left M2, KNM-ER 719 from the early Pleistocene Upper Burgi Mb. of the Koobi Fora Fm., Area 131, Koobi Fora, Kenya, occlusal view. (D) Mandible with right and left m3, KNM-ER 342 from the early Pleistocene Upper Burgi Mb. of the Koobi Fora Fm., Area 105, Koobi Fora, Kenya, dorsal and lateral views. Note the relatively short corpus and anteroposteriorly elongated ramus. Length of the mandible is greater than 700 mm. (Images C and D courtesy of J. Saarinen and P. Mbatha.)

with convex lateral plates sides, are a little more expanded laterally near their apices than in "*E. r. brumpti*," and are composed of five to seven conelets have irregular enamel folding of occlusal wear figures that is usually restricted to the medial part of the crown, may exhibit anterior and/or posterior accessory conules (the anterior conules are always more reduced than the posterior accessory conules [Beden, 1980]) and pseudo-loxodont sinuses in their occlusal wear figures (or just simple enamel loops), and have abundant cementum covering plates and infilling U-shaped transverse valleys (Figure 7.7A and C).

Molars from Kikagati and Kaiso Village in the Western Rift, Uganda (Table 7.1), some of which were previously assigned to "*E.*" *ekorensis* or stage 1 of *Elephas recki* (Maglio, 1973), may be attributable to either *E. recki brumpti* (Sanders, 2020) or *E. recki shungurensis*. Specimen M15211 from Kikagati, an m3, has a plate formula of +11+ with evidence of having had 14 plates when complete.

Incorporated into its worn anterior plates are both anterior and posterior accessory conules. Maglio (1973) provided dimensions/indices for this specimen of L=+258+ mm, W=84.5 mm, H=112.2 mm (yielding HI=133), LF=5.0, and ET=2.8–3.4 mm. A pair of very narrow m3s from Kaiso Village, M12639 (Hopwood, 1926), have either 12 or 13 plates, and when sectioned horizontally revealed very prominent posterior accessory conules to at least plate 6. Complete length of the left m3 is 273 mm, and because of the unusual narrowness of the crown (71.8 mm), the relatively modest height of 105 mm yields a hypsodonty index of 146, which is technically outside of the hypsodonty range of "*E.*" *ekorensis* or the early stages of *E. recki* calculated by Beden (1980, 1983, 1987a). Nonetheless, its overall morphology is consistent with assignment to either *E. recki brumpti* or *shungurensis*.

Another specimen from Kaiso Village, partial m3 M12641, preserves the precingulid and first seven plates, and also exhibits prominent posterior accessory conules throughout. Some plates of this molar exhibit irregular enamel folding in their wear figures. An m3 from Kikagati, M15212 and 15213, has 13 or 14 plates, dimensions of L=+253 mm, W=85.1 mm, H=107.5 mm (yielding HI=126), ET=2.9–3.3 mm, and LF=5.5. Vertical sectioning of the crown reveals that the accessory conules are considerably lower than the main conelets in unworn plates, with heights of 63–82 mm. This same strong expression of posterior accessory conules throughout the crown is observable in Kaiso Village molar M12641, which also displays irregular enamel folding across the breadth of its enamel loops. This Western Rift sample is distinctive and requires more comparative study; however, enamel folding and prominence of accessory conules suggest closest affinity to *E. recki shungurensis*, of known elephant taxa.

No crania of fully adult individuals have been collected for this subspecies. However, a juvenile cranium with right and left dP3–4s, KNM-ER 5712, is mostly complete, missing the majority of its rostrum, dorsal aspect of the left side, and zygomatic arches (Beden, 1983). It is better preserved on the right than on the left side. It derives from the northwest edge of the Kokoi Horst at Koobi Fora, Kenya, of probably the same geological age as the southwest flank of Kokoi at Area 117, late Pliocene equivalent to Mb. C of the Shungura Fm. and younger than specimens attributed to *E. recki brumpti* from nearby (Beden, 1983; Harris, 1983a). Ontogenetic age based on dP3 in full wear and wear on the anterior plates of dP4, compared with modern African elephants, is equivalent to Laws' (1966) age grades IV-V, or two to three years old, a very young juvenile.

Because of the youth of this individual, it is difficult to make meaningful taxonomic comparisons of KNM-ER 5712 with crania of other elephants in the "*E. recki* complex." Tusk alveoli are parallel at their preserved bases and are separated by a narrow, deep external incisive fossa. The base of the rostrum is strongly downturned at a more severe angle than the vault and upper face. There is no development of occipitoparietal bosses, which is the condition expected of such a young individual. The occipital condyle is situated much higher than the palate and projects posteriorly beyond the superoanteriorly angled occipital planum. Overall, the cranium is rounded and the vault is convex. The orbits are very anteriorly advanced and located lower than the level of the condyle(s). The nasal aperture is relatively high on the cranium and located above the level of the orbits. As noted by Beden (1983), for the young age of the individual, KNM-ER 5712 is already very short anteroposteriorly. When complete, the zygomatic arch would have been long and plunged downward anteriorly. The bullae are proportionally massive in both anteroposterior and lateral directions, as is expected for the ontogenetic stage of KNM-ER 5712.

A beautifully-preserved adult mandible, KNM-ER 342, derives from Area 105 at Koobi Fora, Kenya, below the KBS Tuff, probably in the lower part of the Upper Burgi Mb. of the Koobi Fora Fm (Beden, 1983; Harris, 1983a). This mandible is brevirostrine and broad (520 mm), with short, massive corpora rounded in section near the rami (Figure 7.7D). The rami are immense anteroposteriorly, with deep, expansive masseteric fossae, and are tall (435 mm). The coronoid processes are rounded, anteroposteriorly large, thin, not everted, and at a height only a little lower than that of the condyles. The condyles are more prominent toward the inside of the mandible and are oval in shape. Molars are strongly shifted to the lingual side of the alveolar planes of each dentary, a configuration that is accentuated by the breadth of the corpora. The symphyseal trough is U-shaped. Features of KNM-ER 342 are very similar to the morphology of *Elephas maximus* mandibles (see Todd, 2010b), particularly proportions of the rami and corpora, angulation of the condyles, extent of the masseteric fossa, and overall massiveness of the dentaries.

Ecosystems in the greater Turkana Basin and in the Awash appear to have continued to reduce canopy cover and an increase in abundance of C_4 grasses through the late Pliocene (this was more notable at Koobi Fora than in the Omo Shungura area) (Levin et al., 2011), which in *E. recki shungurensis* may have favored selection for evolution of features associated with grazing such as greater molar hypsodonty and plate number. In Shungura Mbs. C-lower F, the rodents indicate the presence of dry and mesic savanna woodlands and bovids shifted from dominance of reducines to greater presence of alcelaphines, bovines, and the genus *Aepyceros*, compared with Mbs. A-B; alcelaphines increased even more in importance from Mbs. C to F; suids shifted from strong representation by *Notochoerus* in Mbs. A-B to decline of this genus and concomitant increased dominance of *Metridiochoerus* through Mbs. C-F (Alemseged, 2003), which is consistent with relative increase in grasslands. Hominins in the genera *Australopithecus* and *Paranthropus* occupied the same landscapes as "*E. recki shungurensis*" during its late Pliocene-early Pleistocene existence (Harris, 1978b; White et al., 1984; Alemseged, 2003; MacLatchy et al., 2010).

"*ELEPHAS RECKI ATAVUS*" Arambourg, 1947 (alt. *ELEPHAS ATAVUS* [Arambourg, 1947]) (alt. *ELEPHAS* [*PALAEOLOXODON*] *RECKI ATAVUS* Arambourg, 1947)

(Tables 7.1 and 7.2; Figures 7.1, 7.5, and 7.8)

Elephas recki atavus is an early Pleistocene phase of the "*E. recki* complex" (Table 7.1) that contrasts starkly with *E. recki brumpti* in cranial morphology. Plate formulae for the subspecies, as summarized by Beden (1980, 1983, 1987a), are dP2/dp2, 5-x5/4-4x; dP3/dp3, 6x-7x/6x-x8; dP4/dp4, 9-9x/x8x-10; M1/m1, 9-x11x/10–11; M2/m2, 9–11x/10–12x; M3/m3, 14x-x18x/13x-x18x. Dimensions and indices of third molars may vary wildly in the *E. recki atavus* sample, and the challenge for understanding this variation is to sort out the contribution to this disparity by time, geography, ontogenetic age and wear of molars, and interindividual variation. Molar lengths (M3/m3 ranging from 265 to 360 mm) and postcranial remains (see Figure 7.5A) indicate large to immense body size. Third molar crown height is generally augmented relative to earlier subspecies but varies from 85 to 147 mm (yielding HI = 110–151), LF = 4.8–6.0, and ET = 2.4–3.2 mm.

Variation of molar occlusal morphology is so great in the subspecies that Beden (1983, 1987a) designated three sub-types of *E. recki atavus* molars, including (a) those with greater number of plates and size than *E. recki shungurensis* molars but with similar "primitive" folding of enamel; (b) those with little increase in size or number of plates and enamel folding very irregular, tending to exhibit deep median folds, and with biconvex enamel loops that are nearly loxodont; and (c) those with size and number of plates as in sub-type 2 but considerably more hypsodont and with thin enamel powerfully folded along wear figures. More generally, lateral surfaces of molar plates curve lightly and gently from the cervix to the point of greatest transverse width across them, well above the base of the crown, and then curve inward toward the apex; anterior and posterior plate faces are parallel, contributing to "silo"-shaped lamella; accessory conules may be well developed at least in the anterior moiety of the crown and may be apically free; enamel is best folded in the median segment of enamel loops at mid-height, but may be folded throughout the breadth of enamel loops with greater wear (Figure 7.8A); and cementum covers plates and infills U-shaped transverse valleys.

The subspecies is well represented in the fossil record, including a number of crania and mandibles. A nearly complete cranium, N° 300, with all plates of M3 in wear (Figure 7.8B), from uncertain stratigraphic provenience in the Shungura Fm., recovered during fieldwork in 1932–1933 and described subsequently by Arambourg (1947), probably derives from a middle horizon of the formation and belongs in *E. recki atavus*. It is the lectotype of the subspecies. Its molar emergence and wear fit with an age >40 years for extant *Elephas maximus* (Roth and Shoshani, 1988) or age grades XV-XVI for extant *Loxodonta africana*, about 47–49 ± 2 years at death (Laws, 1966). The cranium measures 1,030 mm in greatest length and its morphology is striking: It is very tall (672 mm), extremely anteroposteriorly compressed, concomitantly has a very deep and anteriorly-shifted temporal fossa in nearly rectangular form, an anteroinferiorly-plunging zygomatic arch, large, projecting occipital condyles, a very broad (470 mm), laterally downturned nasal aperture, and has tremendous swelling of the nuchal region such that it is very swollen laterally and substantially encroaches over the parietofrontal. The rostrum is relatively short, downturned, and the tusk alveoli are parallel throughout their length and separated by a narrow, deep external incisive fossa. The posterodorsal process of the premaxillae has two superior projections into the nasal aperture that are separated by a small notch. In anterior view, the occipitoparietals are bossed (792 mm transversely across, the widest chord of the cranium), the nuchal crest is "en bombée," there is a broad, shallow midline trough separating the bosses, the forehead/supranasal region is concave, and the minimum distance across the temporal lines is relatively narrow (370 mm). The occipital planum is relatively low because of its breadth, with a height of 334 mm. In lateral view, the orbits are very anteriorly placed, below the level of the occipital condyles, and are relatively small. They do not project laterally as prominently as those of *E. recki brumpti* ("*Phanagoroloxodon planifrons*" of Zhang [2020]), and are located well below the level of the nasal aperture. Because of the hypertrophy of the superior occipital and nuchal regions, the occipital planum appears to be posteriorly inclined at its superior margin, and the posterior extension of this hypertrophied bone diminishes the effect of posterior projection of the occipital condyles. The supraoccipitals are also swollen. The nuchal ligament fossa is extremely vertically elongate and deep. A second cranium from the sample, N° 775, is from a younger individual, with M2 and only the anterior plates of M3 in wear and the rest of the molar still emerging from its crypt. This fits with an age 30–40 years for extant *E. maximus* (Roth and Shoshani, 1988), or age-grade XX of Laws (1966), equivalent to 34 ± 2 years of age at death for extant *L. africana*. Arambourg (1947) briefly mentioned it as similar morphologically to N° 300.

Cranium N° 300 has M3s preserving +11 plates, with dimensions of L = 264 mm, W = 80 mm, LF = 5.5, and ET = 2.5–3.5 mm. Other M3s described from the same sample are similar in dimensions, with 13 plates when complete, and their heights range up to 124 mm, yielding hypsodonty indices between 110 and 150 (Arambourg, 1947).

A mostly complete cranium, KNM-ER 5711, from slightly below the KBS at Koobi Fora, Kenya, closely matches cranium N° 300 in morphology (Figure 7.8C and D). When first collected, the cranium was missing part of the lateral aspect of the left nasal aperture and orbit, as well as the left zygomatic arch; it has suffered considerable

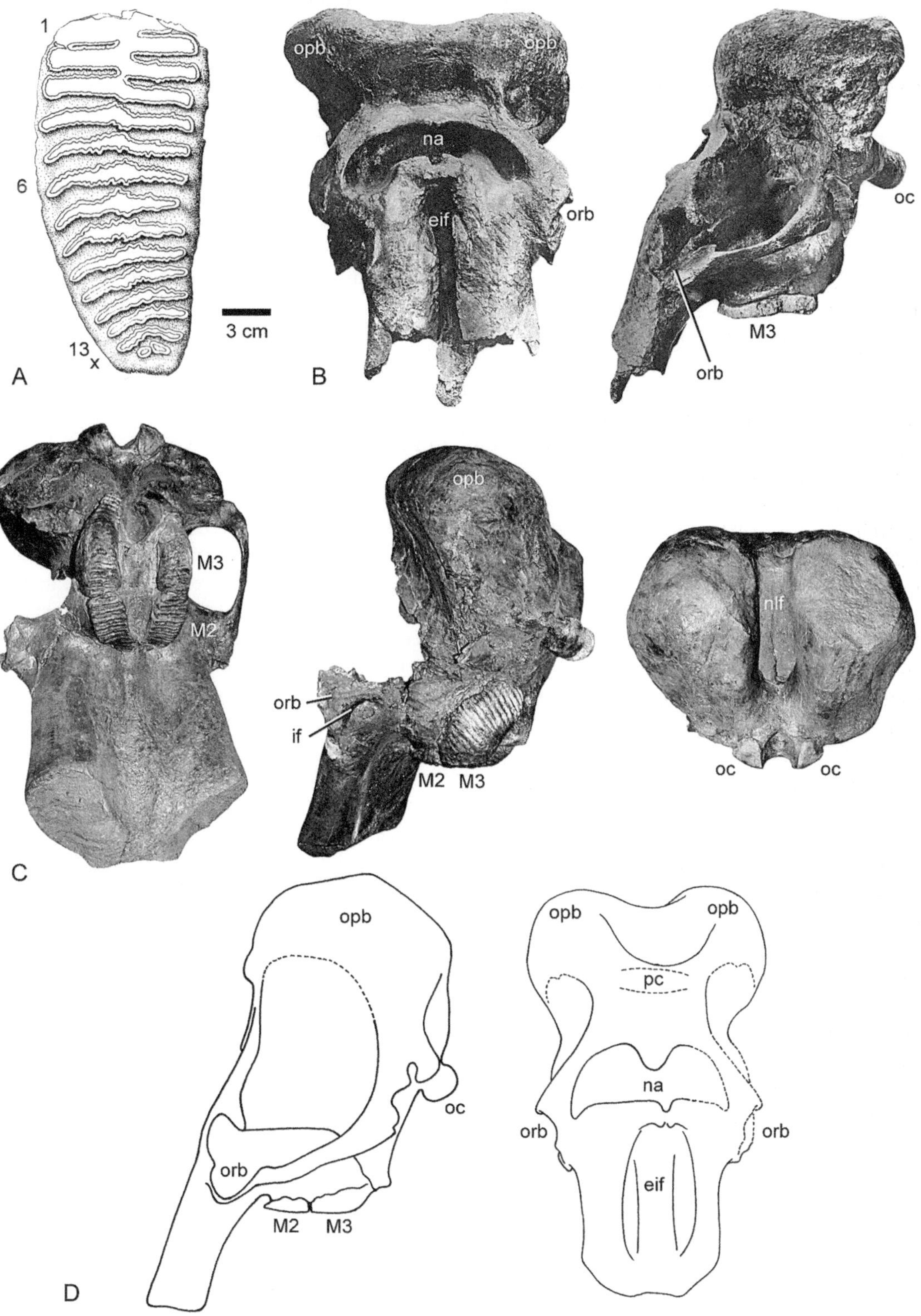

FIGURE 7.8 Cranial and dental fossils of early Pleistocene *Elephas atavus*. Abbreviations: eif, external incisive fossa; if, infraorbital foramen; M, upper molar; na, nasal aperture; nlf, nuchal ligament fossa; oc, occipital condyle; opb, occipitoparietal boss; orb, orbit; pc, parietal crest; x, postcingulid; 1, 2, 3, …, plates counted from the anterior of the crown. (A) Left m3, KNM-ER 4305 from the Upper Burgi Mb., Koobi Fora Fm., Area 100, Koobi Fora, Kenya, occlusal view. Anterior to the top. (From Beden [1983:fig.3.18B], with permission of Oxford Publishing Limited.) (B) Cranium, No. 300 from unknown horizon, Shungura Fm., Omo, Ethiopia, anterior and lateral views. (From Arambourg [1947:fasc. 3, pl. II and fasc. 3, pl. III], with permission of the Museum National d'Histoire naturelle, Paris.) Length = 1,030 mm. Note the very low, anterior position of the orbits. (C) Partial cranium, KNM-ER 5711 from Area 102, Upper Burgi Mb., Koobi Fora Fm., Koobi Fora, Kenya, ventral, posterodorsal, and posterior views, partial cranium. Length = 1,350 mm. Note the salient inflation of the occipital planum and bossing of the occipitoparietal corners of the vault. (D) Reconstructe partial cranium, KNM-ER 5711, lateral and anterior views. (From Beden [1983:figs 3.17A, B], with permission of Oxford Publishing Limited.)

damage since then, especially on its dorsal and right lateral aspect. It exhibits salient development of occipitoparietal bosses, midline nuchal depression, swelling of the nuchal crest over the parietofrontal region, very swollen supraoccipitals, concavity of the frontonasal region, relatively narrow transverse diameter across the temporal lines (330 mm), low and very anteriorly-located orbits set below the position of the nasal aperture and only modestly projecting laterally, temporal fossae anteriorly shifted, nearly rectangular in shape, a deep, very broad, (530 mm) laterally-downturned nasal aperture, a thin zygomatic arch that angles downward anteriorly, large occipital condyles overshadowed by posterior swelling of the occipital region, and parallel tusk alveoli separated by an external incisive fossa that is deepest superiorly and broader than in the N° 300 cranium (Beden, 1983). A transverse swelling across the anterior extent of the occipitoparietal bossing comprises a parietal crest. Maximum cranial length is estimated at 1,350 mm and greatest width of the cranial vault is 890 mm; maximum width across the zygomatic arches is estimated at 820 mm, equivalent to width across the postorbital processes (Beden, 1983). The third molar of this cranium reaches the maximum plate number for the subspecies, x18x, and is very large, with dimensions/indices of L=335 mm, W=~100 mm, H=146 mm (yielding HI=146), LF=5.6, and ET=3.1–3.5 mm (Beden, 1983). Age-grade status of this individual is similar but not identical to that of Omo cranium N° 775, with M2–3 in place.

Other crania attributed to *E. recki atavus* include Omo 75.69.3190, from the lower part of Mb. G of the Shungura Fm., with a heavily damaged vault and occipital and missing its occipital condyles (Beden, 1987a). Its geological age is close to that of KNM-ER 5711, in the interval ~2.19–2.0 Ma (see Levin et al., 2011:fig. 2). Greatest length of the cranium is estimated at 1,260 mm, greatest width across the occipital planum is >800 mm, biorbital width is 800 mm, and bizygomatic width is 720 mm (Beden, 1987a). Despite the postdepositional damage, Omo 75.69.3190 preserves evidence of a narrow vault across the temporal lines, a broad nasal aperture (W=480 mm) that is narrowed and slightly downturned laterally, and bordered by thin frontal struts, parallel tusk alveoli separated by a deep external incisive fossa, deep and anteriorly-shifted temporal fossae, low and anteriorly-located orbits, and an anteriorly thin, downwardly oriented zygomatic arch. Position of the tusks in the cranium show that they diverged strongly after exiting their alveoli (Beden, 1987a:fig.18). The M3 of this cranium has a plate formula of x15, and dimensions/indices of L=300 mm, W=102 mm, H=125 mm (yielding HI=122), LF=4.8–5.5, and ET=3.0 mm. It was probably close to 40 years old at death (Roth and Shoshani, 1988).

Omo cranium 75.69.3217 is damaged dorsally and missing its zygomatic arches, part of the rostrum, and part of the occipital on the right side (Beden, 1987a). It also derives from Mb. G of the Shungura Fm. Although less complete, specimen 75.69.3217 exhibits features that reinforce the bauplan of *E. recki atavus* crania: The orbits are very anteriorly positioned and low, with the occipital condyles mostly superior to the vertical position of the orbits; the occipital condyle is raised well above the level of the palate; the supraoccipital is distinctly swollen and expanded on both sides (width of the occipital is 540 mm), although less massive than in 75.69.3190 or KNM-ER 5711, and the occipital angles posteriorly at its superior end; the rostrum is more steeply downturned than the dorsum of the vault; the temporal foramen is relatively small (L=200 mm, W=150 mm); the nasal aperture is broad (W=390 mm). M2–3 are present in Omo 75.69.3217, suggesting a similar age-grade status to Omo 75.69.3190.

Omo specimen 75.70.826 is also from Mb. G of the Shungura Fm.. It is the most poorly preserved cranium of the subspecies. M2 is nearly fully in wear but M3 is not emergent, suggesting it was closer to 20–30 years old at death, with reference to extant *E. maximus* ontogeny (Roth and Shoshani, 1988), or close to 30 years old compared with extant *L. africana* (Laws, 1966). It has large occipital condyles that are superolaterally dominated by the hyperinflation of the occipital. The occipital appears low because of its great width relative to its height, and is marked by a strong midline vertical trough that encloses the narrow and deep nuchal fossa.

Considered together, the cranial assemblage of *E. recki atavus* (or *E. atavus* sensu Zhang [2020]) from the Omo and Koobi Fora shares a number of features in common with extant *E. maximus*, including a concave forehead, narrow vault between the temporal lines, parallel tusk alveoli, medially concave nuchal crest, large, swollen bosses, maximum width across the occipitoparietals, nasals directed forward, and rostra more downturned than cranial vaults in lateral view (see Todd, 2010b). Conversely, crania of *E. recki atavus* differ from those of Eurasian and/or African *Palaeoloxodon* species by not forming a distinct parietofrontal crest, lower position of the orbits, longer, more slender zygomatic arches, presence of non-divergent and generally shorter tusk alveoli, more rounded posterior profile of the occipital, possession of more slender frontral struts framing the nasal aperture, and nasal aperture more downturned and narrowed laterally (see Saegusa and Gilbert, 2008). Although Saegusa and Gilbert (2008) dispute Beden's (1983) claim that the parietal crest in KNM-ER 5711 (Figure 7.8D) is homologous with the parietofrontal crest of Eurasian *Palaeoloxodon* spp., they hypothesize that presence of a post-temporal crest is a synapomorphy of *Palaeoloxodon* spp. and *E. recki atavus* (their *E.* (*Palaeoloxodon*) *recki atavus*).

Mandibles of *E. recki atavus* (Arambourg, 1947; Beden, 1983, 1987a) exhibit massive corpora that are short and robust, rounded posteriorly, and anteriorly tall. Symphyses are brevirostrine, except for an anterior "beak" of 160 mm in the mandible associated with cranium KNM-ER 5711, which Beden considered to be an "individual variant without taxonomic importance" (1983:81). The symphyseal fossa is U-shaped. Although there are no complete rami for these mandibles, they are tall with ovoid articular condyles that are inclined forward and slightly internally, and have thin coronoid processes that are either anteroposteriorly straight or lean slightly outward, as seen in Omo Shungura mandible N° 774 that is associated with cranium N° 775. Masseteric fossae are extensive. As in mandibles of extant Asian elephants, the molars are strongly shifted to the internal side of the alveolar plane of the corpora. Dimensions of specimen N° 774 are overall length=520mm, ramus H=550mm, bicondylar breadth=560mm, and corpora dimensions below m3 are W=225mm and H=200mm (Arambourg, 1947). Mandibles of *E. maximus* appear to have relatively shorter corpora, more anteroposteriorly extensive rami, and greater medial angulation of their coronoid processes (see Todd, 2010b).

The Omo proboscidean collections of 1932–33 also included a small number of postcranial remains (Arambourg, 1947), which tentatively can be included in the *E. recki atavus* sample. Among these are an elongate (1,280mm), gracile humerus (Arambourg, 1947: pl. VII, figs. 1, 1a), nearly complete humerus (1,110+ mm), and similarly elongate ulna (1,150mm) and radius (1,030mm). The humerus is much more lightly constructed than the humeri of stegotetrabelodonts from Sahabi, but its length approximates that of elephants reaching 390cm in shoulder height (see Larramendi, 2016). The olecranon process of the ulna is low with a transversely enlarged tuberosity but is not reflected much posteriorly, and the bone is generally robustly constructed. It is apparent that the radius was fixed in pronation on the ulna, although not fused (see Arambourg, 1947:pl. VIII, figs. 5, 5a). Other elements in the sample are an unciform, metacarpals (MC) II, III, and V, and metatarsals (MT) III, IV, varying in robustness and size as they derive from different individuals. Lengths of these metapodials are: MCs II, 181 and 192mm; MC III, 231 mm; MCs V, 141.5 and 173mm; MTs III, 141 and 191 mm; MTs IV, 114.4 and 214mm (Arambourg, 1947). The postcranial remains in Arambourg's collection are consistent with evidence from Koobi Fora (see above) that "*E. recki atavus*" individuals reached very large sizes, probably typically taller and heavier than extant adult elephants.

Faunas associated with *E. atavus* (or *E.* [*Palaeoloxodon*] *atavus*) are characterized by a mix of extinct and extant genera of living mammalian families, with few extant species. In the ~2.35 Ma occurrence of *E. atavus* at Maka'amitalu, Ethiopia, for example, *E. atavus* is accompanied by a primitive representative of *Homo*, the extinct gelada baboon *Theropithecus oswaldi oswaldi*, extinct suids in the genera *Kolpochoerus* and *Metridiochoerus*, the extant giraffe genus *Giraffa* and extinct giraffe *Sivatherium maurusium*, the extinct hippo *Hippopotamus gorgops*, the modern cheetah genus *Acinonyx* alongside the extinct machairodontine *Dinofelis*, and a large mix of extant and extinct genera and species of bovids (Rowan et al., 2022). The dietary preferences of the Maka'amitalu herbivores was dominated by grazers, with browsers approximately equivalent in proportion of the fauna to mixed feeders (Rowan et al., 2022). Paleoecological inferences about this sort of fauna suggests that *E. atavus* was living proximate to or in settings with abundant open-country, grassy habitats, which is consistent with its hypsodont molars and the slender isotopic evidence of its diet (Cerling et al., 1999).

PALAEOLOXODON Matsumoto, 1924b

Palaeoloxodon comprises a significant group of elephants that reached its apex in diversity and most progressive set of morphological features in middle-late Pleistocene Eurasia (Zhang, 2020). The geographic reach of its impressive adaptive radiation is marked by the occurrence of its type species, *P. naumanni*, in Japan (Makiyama, 1924; Matsumoto, 1924b). It originated in Africa as part of the "*Elephas recki* complex," with distinctive generic features nascent in "*E. recki recki*" (Larramendi et al., 2020), and migrated out into Eurasia in late early or early middle Pleistocene, probably through a Levantine corridor (Saegusa and Gilbert, 2008; Lister and Sher, 2015; Larramendi et al., 2020). The close similarity of molars of "*E. r. ileretensis*" and "*E. r. recki*" suggests that the former taxon also belongs in *Palaeoloxodon*. Cranial features appear paramount for recognition of the genus, and thus early forms of the "*E. recki* complex," lacking definitive apomorphies of the genus, such as "*E. r. brumpti*," have been taxonomically separated from *Palaeoloxodon* (e.g., Saegusa and Gilbert, 2008; Zhang, 2020). There is no consensus about whether early Pleistocene "*E. r. atavus*" exhibits synapomorphies of *Palaeoloxodon*, but if it does, an argument can be made on paleontological-morphological grounds for *Palaeoloxodon* to have derived from and to be a subgenus of *Elephas* (Saegusa and Gilbert, 2008; following Beden, 1983, 1987a); if not, "*E. r. atavus*" can be elevated to specific status, *E. atavus* (Zhang, 2020). However, despite the absence of supporting paleontological evidence, genomic studies of ancient DNA indicate that *Palaeoloxodon* shares a deep ancestry with *Loxodonta*, and that genetic material and morphology shared with *Elephas* are the result of more recent cross-specific introgression (see discussions elsewhere in this chapter; Figure 7.3; Meyer et al., 2017; Palkopoulou et al., 2018).

Cranial apomorphies of *Palaeoloxodon*, which are not always present in their entirety in different individuals and species of the genus, include: Elongate, laterally divergent and distally flared premaxillaries (reaching wider than 1,000mm), separated by a distally broad, V-shaped external incisive fossa; premaxillaries in line with a steeply slanted facial plane; prominently wide, flat frontals; forwardly-bent occipital surface; marked development and anterior placement of the parieto-occipital crest (presumably for anchoring powerful splenius and rhomboideus cervicis muscles to support the head) with consequent foreshortening of the visible frontal surface; high frontal; nasal aperture modest in size, laterally rounded, and not downturned; orbits low, below the level of the nasal aperture and occipital condyles; occipital planum broad (reaching 1,100mm), low, with low placement of nuchal ligament fossa; temporal fossa rectangular in shape; and salient swelling of occipital and parietals (Larramendi et al., 2017, 2020). These were massively large elephants, with skulls greater than 1,500 mm in length, reaching 400 cm in shoulder height and ~12,000 kg in body mass, with extreme expression of sexual dimorphism in body size (Larramendi et al., 2017; Konidaris et al., 2018). Reconstruction of their distinctive body profile indicates their backs sloped irregularly downward posteriorly, in lateral view (Larramendi et al., 2017). African extinction of such magnificent elephants occurred early in the late Pleistocene (see below), while their demise in Eurasia took place toward the end of the late Pleistocene (Zhang, 2020).

"*ELEPHAS RECKI ILERETENSIS*" Beden, 1980 (*alt. ELEPHAS [PALAEOLOXODON] RECKI ILERETENSIS* Beden, 1980) (alt. *PALAEOLOXODON RECKI ILERETENSIS* [Beden, 1980])

(Tables 7.1 and 7.2; Figures 7.1, 7.5, and 7.9)

Elephas recki ileretensis is the segment of the "*E. recki* complex" recognized from Mbs. K and L (and possibly Mb. J) of the Shungura Fm., Upper Bed II at Olduvai, and the Okote Mb. of the Koobi Fora Fm. (Table 7.1). There are also important occurrences of the taxon at several sites in Djibouti (Faure and Guérin, 1997; Berthelet, 2001; Berthelet and Chavaillon, 2001). It is a problematic "subspecies" in that its molars can be difficult to differentiate morphometrically from those of *E. recki recki* (=*Palaeoloxodon recki recki*) (e.g., Ferretti et al., 2003b; Suwa et al., 2003). Third molars of *E. recki ileretensis* have enamel thicknesses of 2.2–3.2 mm, lamellar frequencies of 4–6, and crown heights ranging from 95 to 138 mm, yielding hypsodonty indices of 116–172 (Beden, 1980). Plate formulae for the taxon are: dp3, 7x; dP4/dp4, 9/8x-9; M1/m1, 10/x10x; M2/m2, 12/11-x12x; and M3/m3, x15x-x16/x11-x14 (Beden, 1987a).

Molars of *E. recki ileretensis* are generally very tall, with lateral sides of plates parallel to about mid-height of unworn specimens, where they are widest (Figure 7.9A). Enamel is thin and often irregularly folded, particularly at mid-height in worn molars (Figure 7.9A). Lower molars may exhibit tighter, more regular folding across the width of enamel loops (Beden, 1983). Third molar plates typically are composed of between five and seven conelets. Enamel folds may be prominent and multiple, occasionally producing the effect of loxodont sinuses ("<>") in worn enamel loops.

No adult crania of *Elephas recki ileretensis* have been recovered. A small (L=390+ mm) juvenile cranium with dP4 in wear, KNM-ER 799 (Figure 7.9B), was collected at Area 103 at Koobi Fora, high in the section (Beden, 1983). Ontogenetic age of this individual is equivalent to about five years, in comparison with extant Asian elephants (Roth and Shoshani, 1988). KNM-ER 799 is missing much of its rostrum and some of its left side, as well as the right zygomatic arch and condyles (Figure 7.9B). Because of the breakage and loss of bone of the premaxillaries, it is difficult to ascertain whether the tusk alveoli were parallel or flared, but they appear to be divergent proximally (Figure 7.9B) and the external incisive fossa appears to have been deep proximally, as well. In lateral view, the occipital plane and vault are rounded, the forehead is anteroposteriorly concave, and the rostrum projects forward and downward in equal measure. In anterior view, the forehead is broad and transversely convex, the orbits flare moderately laterally, and the nasal aperture is bean-shaped and rounded but not downturned laterally. Maximum width of the vault is 390 mm and minimal width across the temporal lines is 225 mm. The anterior tip of the nasals is small and overhangs the nasal aperture, but does not project forward. The nasal aperture is at the level of the upper part of the orbits, and has a width of 195 mm. At this young age, the level of the orbit is not different from the elevation of the condyles, which in turn are not much higher than the palate. There is a transverse ridge across the parietals that continues posterolaterally over the temporal fossae, becoming confluent with strong lateral occipitoparietal bosses. It is probable that the zygomatic arch was elongate, gracile, and anteriorly angled downward. Beden (1983) felt that the overall similarity of the features of KNM-ER 799 is with primitive *Elephas* crania, although there are major differences between this juvenile cranium and that of *E. recki shungurensis* (KNM-ER 5712) that are probably not explicable solely by their difference in ontogenetic age.

There are several mandibles or dentaries of the subspecies known from Koobi Fora and Ileret, Kenya, and Omo, Ethiopia. The most complete of these is KNM-ER 927, with m3s in place. Measurements provided by Beden (1983) show that this mandible is tall (490 mm), has a relatively short corpus (L=250 mm between base of ramus and anterior alveolar border) and anteroposteriorly broad ramus (230 mm), and is brevirostrine (symphyseal L=90 mm). Posteriorly, the corpus is almost round, with dimensions of H=155 mm and W=145 mm. Another mandible from Ileret, KNM-ER 787, exhibits a protuberant symphyseal beak, although the symphysis itself is quite short (70 mm) (Beden, 1983). A short, rotund corpus, expansive, tall ramus, and brevirostrine condition of the symphysis are all features shared with extant *Elephas* (see Todd, 2010b).

Difficulties in differentiating some elephant molars between "*E. recki ileretensis*" and "*E. recki recki*" (e.g., Ferretti et al., 2003b; Suwa et al., 2003), along with cogent arguments that "*E. recki recki*" should be reclassified as *Palaeoloxodon recki* (Larramendi, 2020; Zhang, 2020) or *Elephas* (*Palaeoloxodon*) *recki* (Saegusa and Gilbert, 2012) support suggestions that "*E. recki ileretensis*" should be included in the same species and designated as "*Palaeoloxodon recki ileretensis*" (see Faure and Guérin, 1997). Although this may be a taxonomic improvement for generic affiliation and recognizing clades within the "*E. recki* complex," it does not resolve the problem of morphometric indistinguishability or inappropriate application of the subspecies concept between putative lineage successors.

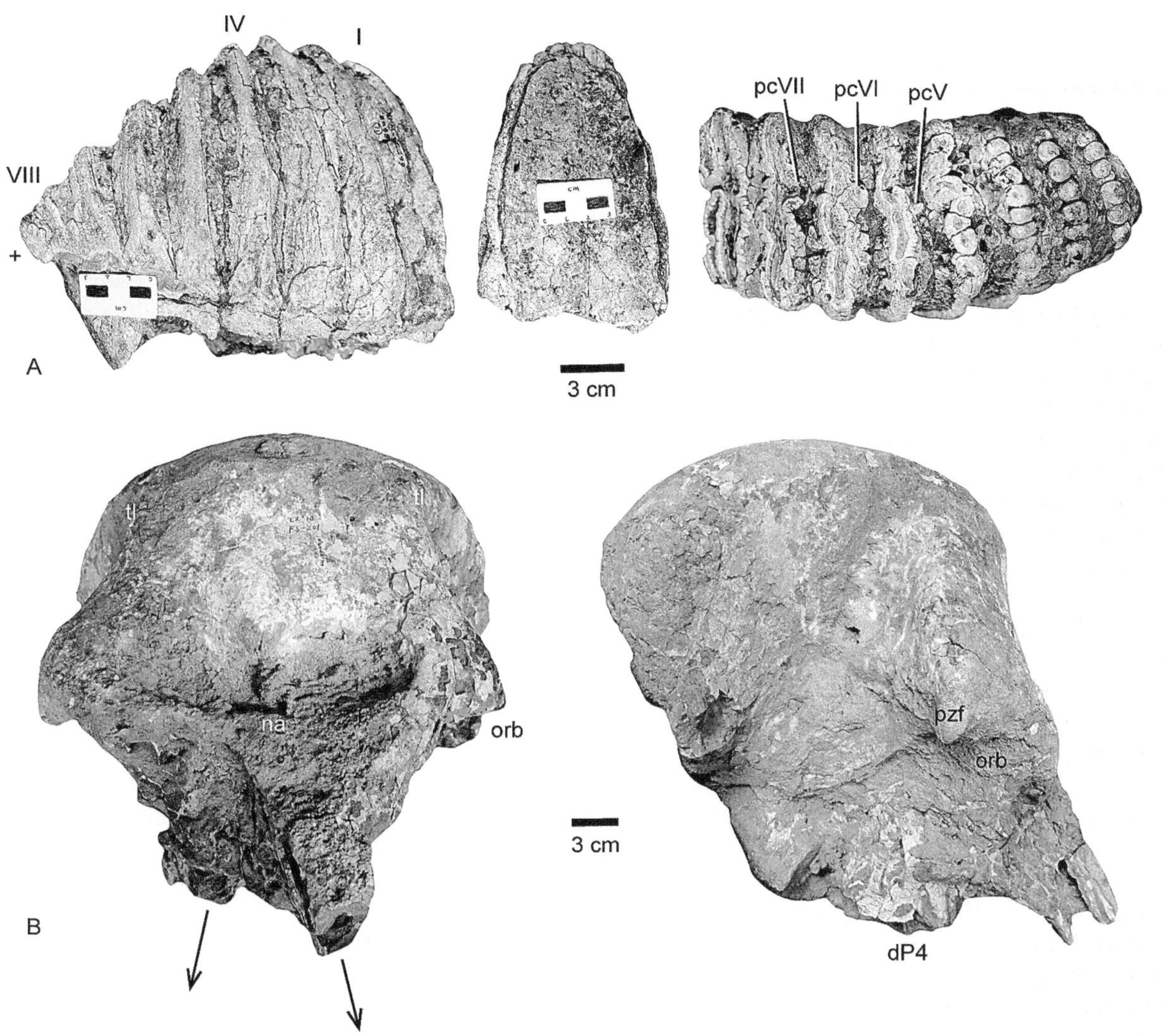

FIGURE 7.9 Cranial and dental fossils of *Palaeoloxodon recki ileretensis*. Abbreviations: dP, upper deciduous premolar; na, nasal aperture; orb, orbit; pc, posterior accessory conule; pzf, zygomatic process, frontal; tl, temporal line; I, II, III, ..., plates counted from the posterior of the crown; +, indicates missing morphology. (A) Partial right M3, KNM-ER 299 from Koobi Fora Fm., Area 103, Koobi Fora, Kenya, lateral, anterior, and occlusal views. Anterior to the left in lateral and occlusal views. All views to the same scale. (B) Juvenile partial cranium, KNM-ER 799 from KBS Mb., Koobi Fora Fm., Area 103, Koobi Fora, Kenya. Length = 390 + mm. Note arrows indicating probable orientation of the long axes of the tusk alveoli. Images to the same approximate scale. (Images courtesy of J. Saarinen and P. Mbatha.)

Several cranial specimens with associated skeletons from Djibouti might be valuable for deciphering the phylogenetic position of *E. recki ileretensis*, when they are available to be studied more intensively. A schematic of a partially scattered skeleton lacking limbs, from the Gobaad region near Gafalo, Djibouti shows a specimen in reasonably good condition (Faure and Guérin, 1997:fig. p. 1018), but little can be gleaned about the morphology of the skull from the illustration in the publication. Molars in the Gafalo cranium are probably M2s; the left M2 appears to have a total of 12 plates estimated to be about 200 mm in length, with a greatest width of 73 mm; lamellar frequency of this molar is 7.0, enamel thickness reaches 4.0 mm but is well-folded, and the crown has a hypsodonty index of 177 (Faure and Guérin, 1997), a little above the high end of variation recorded by Beden (1980, 1983, 1987a) for this subspecies. The elephant remains were accompanied by an immense mandible of *Hippopotamus amphibius*.

Associated skeletal remains of an elephant individual from a butchery site at Barogali, Djibouti, are about the same early Pleistocene age as those from Gafalo (Table 7.1). It is clear that the bones of this individual had been hacked and scattered about (Berthelet and Chavaillon, 2001). The tusks are still inserted in the rostrum, but it is impossible to gauge from the schematic of the excavation whether the tusk alveoli were divergent or parallel. The vault was disarticulated from the rest of the cranium, probably to access the brain. Technologically archaic, the butchery tools at the site are typical of a Developed Oldowan assemblage, numerically dominated by hammer stones, flaked pebbles, and choppers, with little retouch evident, many of them broken (Berthelet, 2001). For taxonomic reasons, it would be especially valuable if the rostrum and calotte could be reconstructed.

African proboscidean butchery occurrences, such as at Barogali, are rare, whereas sites with proboscidean remains in association with artifacts are less scarce but not common. Haynes (2020) provided a useful summary of some of these occurrences. Among these, the complete skeleton of an elephant, probably *E. recki atavus*, was found with a few bones of a second individual, over 100 lithics, and remains of about three dozen other animals at site FLK North ("Zinj" site) in Upper Bed I at Olduvai Gorge, Tanzania. Leakey (1971) interpreted the occurrence as a butchery site. Although there has been debate about whether the primary agency of damage to the bones at the site was due to early hominins or carnivores (see Bunn et al., 1986), Domínguez-Rodrigo and Barba (2006) supported Leakey's hypothesis that carnivores had secondary access to hominin butchery of fleshed carcasses and marrow extraction. Later, however, Domínguez-Rodrigo et al. (2007) reinterpreted damage to the bones as the result of animal trampling.

Also at Olduvai, an astragalus of an unidentified proboscidean with cut marks on it was recovered from site HWK EE in Bed II in association with a large Oldowan lithic assemblage (de la Torres et al., 2018; Pante et al., 2018). Remains of deinotheres and elephants (possibly *E. recki ileretensis*) were also found at the site. A partial skeleton of a deinothere is associated with lithics at site FLK North in Lower Bed II at Olduvai, and elephant bones (possibly *E. recki atavus*) with evidence of flaking, battering, and abrasion were also observed at the site. Leakey (1971) thought that the deinothere had been butchered with Oldowan tools. This interpretation was questioned by Domínguez-Rodrigo and colleagues (2007), as well.

Bell's Korongo site in Upper Bed II at Olduvai produced a large number of Developed Oldowan tools and lithic debitage, along with cut-marked and "green-broken" bones of a diversity of animals from different size classes, including those of an elephant (Leakey, 1971). Given the age of the deposits (Table 7.1), the elephant could be *E. recki ileretensis*. This appears to be a valid case of elephant butchery, with evidence of deliberate long-bone breakage (Domínguez-Rodrigo et al., 2014; Yravedra et al., 2017).

Excavation of Site 15 in Upper Mb. 1 at Olorgesailie, Kenya, dated to 990 kyr, revealed most of the skeleton of an *E. recki recki* individual, spatially associated with a large number of sharp stone flakes, including Acheulean bifaces, some of which that had been utilized (Potts 1989; Potts et al., 1999). The occurrence is interpreted as a butchery area of an elephant that had died at the site.

Not listed by Haynes (2020) is the northern Sudan site of Kaddanarti, dated broadly to between 1.6 and 0.5 Ma (Chaix et al., 2000). Mammalian remains of *Hippopotamus amphibius*, ?*Pelorovis* sp., medium-sized bovids, and ?cf. *Equus africanus* were found in association with two partial molars attributed to "*Palaeoloxodon recki*" and stone choppers, scrapers, and flakes, which may represent a Developed Oldowan assemblage. The site is referred to by the authors of the study as "lower Pleistocene," but the mammalian assemblage and metrics of the elephant suggest it is at the younger end of its estimated age range. Enamel thickness of one molar is 3.0 mm and hypsodonty index of the tooth is calculated to be 170, which suggests potential identity with *E. recki recki*. The association of the mammalian remains and stone tools is interesting, but as pointed out by Haynes (2020), it does not automatically imply hunting or scavenging.

The site of Nadung'a 4, at Lake Turkana, Kenya, is dated to ~700 Ka and contains bones of a young adult *E. recki recki* individual, a bird, crocodile, suid, rare fish and bovid remains, and a large number of artifacts (Delagnes et al., 2006). The artifacts include worked pebbles, stone cores, hammerstones, flakes, and debitage. Classification of the artifacts resists identification with classic Acheulean tool kits typical of the time period. The heavy-duty tools at the site were interpreted by Delagnes and colleagues (2006) as likely used for butchery and carcass processing, though they cautiously hypothesize that the elephant individual could have died of natural causes rather than from hunting.

The site of Gesher Benot Ya'aqov, Israel, dated to near the beginning of the middle Pleistocene (Table 7.1), features an association of a butchered young male or female *E. recki recki* individual and thousands of lithics, including Acheulean handaxes and cleavers (Goren-Inbar et al., 1994). Although not in Africa, the site is located in the northern extension of the African Rift Valley, in a corridor connecting Africa with Eurasia (Haynes, 2020). Evidently, the site was occupied multiple times by hominins over a span of many tens of thousands of years. Archeologists interpret the elephant remains as having been killed by a small group of hominins and breakage of the cranium as consistent with extracting the brain (Goren-Inbar et al., 1994).

Revadim Quarry, Israel is of similar age and also features numerous Acheulean artifacts, alongside thousands of animal bones, including remains of either *E. recki recki* or *Palaeoloxodon antiquus* (Rabinovich et al., 2012). These elephant species, from this time period, might be difficult to differentiate. Some of the bones exhibit cut marks, suggesting deliberate defleshing. Also in the Africa-Eurasia corridor is the Saudi Arabian site of Ti's Al Gadah, with a rich fauna comprising birds, reptiles, and a mix of extinct and extant mammalian taxa befitting its middle Pleistocene age (Table 7.1; Stimpson et al., 2016). The mammals include a large extinct Eurasian panther (*Panthera* cf. *gombaszogensis*), the big extinct African buffalo *Pelorovis*, extinct hippo *Hexaprotodon*, and representatives of extant mammal genera *Equus, Hippopotamus, Vulpes, Canis, Crocuta, Bos, Bubalus, Oryx*, and *Gazella*, along with hystricids, leporids, mustelids, camels, and a number of individuals attributed to *Palaeoloxodon recki recki*. Artifacts from the site include Levallois flakes and retouched flakes of Middle Paleolithic attribution (Stewart et al., 2019). Sedimentological study and taphonomic interpretation of the site accumulation indicates "a predation hotspot where large felids and probably hominins ambushed mostly medium-sized ungulates …," with evidence for some hominin butchery, including of elephants (Stewart et al., 2019:250–251).

Younger sites with associations of artifacts and elephant remains include Mwanganda's Village, Malawi, dated to ?~400 Ka, with part of an unidentified elephant along with hippo, giraffe, turtle, and equid fossils, and about 300 late Middle Stone Age tools (Clark and Haynes, 1970; Kaufulu, 1990, Wright et al., 2014). Whether the association of artifacts and elephant bones is casual or associated with butchery has been debated by the researchers who have studied the site. The site of Namib IV, Namibia, dated to ~350 Ka, has scattered bones of several antelopes and an elephant identified as *E. recki recki*, scattered around handaxes, small flake tools, and cleavers, an association interpreted as a butchery site reused multiple times by hominins (Shackley, 1980; Shackley et al., 1985). At the site of Shishan Marsh 1, Jordan, late middle Pleistocene artifacts are associated with a fauna that includes *E. recki recki* and a second elephant species, *E.* cf. *hysudricus*, along with the remains of lions, gazelles, buffalo, camels, horses, and the extinct Eurasian/North African two-horned rhino *Stephanorhinus*, thought to have comprised hominin dietary resources (Pokines et al., 2018). The youngest site of this kind listed by Haynes (2020) is Zoo Park, Namibia, radiocarbon-dated to about 5,200 y BP (MacCalman, 1967; Clark and Haynes, 1970). Bones of two disarticulated elephant skeletons, unidentified to genus and species, were scattered in a marsh deposit with choppers, hammerstones, and unretouched flakes, interpreted by MacCalman (1967) as a kill site where elephants had been driven into the marsh.

As Haynes (2020) indicated in his synopsis, there are additional African Pleistocene sites where elephant bones either show cut marks, are associated with lithics, or both. Although the stone tool industries at many of these sites are Acheulean, Klein (1988) felt that Acheulean hominins rarely hunted or killed elephants, and Middle and Late Stone Age sites in Africa rarely show evidence of elephant hunting or scavenging, perhaps reflecting the danger and/or energetic costs of such activities (Haynes, 2020). Nonetheless, it is evident that hominins such as *Homo habilis* and *Homo erectus* had the capacity and at least occasional determination to exploit elephants for meat and marrow, as early as the beginning of the Pleistocene.

Little has been written collectively about Omo Shungura faunas of Mbs. J-L, but the study of Plummer et al. (2015) provides valuable information about the constitution of bovid tribes and paleobotany of those members, reflecting paleoecological conditions between 1.74 and <1.39 Ma. Fossil woods indicate very dry habitats during that time; plant biomarkers and paleosol carbonate data indicate a significant increase in C_4 plant mass and grassland habitats in the lower Omo Valley just before the start of this interval (Bobe, 2011). Open-country antilopines, alcelaphines, and hippotragines comprise 22.2% of the bovid assemblages, clearly associated with a greater amount of open-country habitats than in Mbs. B-lower G (Plummer et al., 2015), and a number of grazing bovids made their first appearances in eastern Africa in this interval (Bobe, 2011). As well, the start of this interval witnessed the origin of the highly terrestrial, large-bodied monkey *Theropithecus oswaldi*, and a cohort of bipedal (and presumably predominantly terrestrial) hominins, including *Paranthropus boisei*, *Homo habilis*, *H. rudolfensis*, and *H. erectus/ergaster*, were also contemporaneous in this time period (Bobe, 2011). Similarly, there was a shift to cooler temperatures and greater percentage of open-country habitats, as well as a jump in the percentage of grazing herbivores, in the Turkana Basin around 2.0 Ma that continued into this 1.74 to <1.39 Ma interval, although shifts in habitat and feeding structures were apparently not necessarily tied to trends in paleoaridity (Blumenthal et al., 2017). At Ileret, Karari, and the Koobi Fora Ridge, Kenya, as in the earlier Upper Burgi and KBS Mbs. of the Koobi Fora Fm., in the Okote Mb. grazers overwhelming dominated mixed feeders and especially browsers among herbivores (O'Brien et al., 2020).

"*ELEPHAS RECKI RECKI*" (Dietrich, 1915) (alt. *ELEPHAS [PALAEOLOXODON] RECKI RECKI* [Dietrich, 1915]) (alt. ELEPHAS [PALAEOLOXODON] RECKI Dietrich, 1915) (alt. PALAEOLOXODON RECKI [Dietrich, 1915])

(Tables 7.1 and 7.2; Figures 7.1, 7.5, 7.10, and 7.11)

Elephas recki recki or its equivalent stage has been widely considered to be the most progressive phase of the species (Maglio, 1973; Beden, 1980, 1983, 1987a; Sanders et al., 2010a; Sanders and Haile-Selassie, 2012). However, if *Elephas jolensis* (=*Palaeoloxodon jolensis*) is considered a lineal descendant of *E. recki recki*, then it comprised the terminal and most progressive segment of the *E. recki* lineage in Africa (Manthi et al., 2019). Chronostratigraphically, *E. recki recki* spanned the later part of the early Pleistocene through most of the middle Pleistocene, with well-documented occurrences in the upper levels of the Shungura, Koobi Fora, and Nachukui Fms., as well as from the top horizons of the Olduvai sequence (Table 7.1; Maglio, 1973; Beden, 1983, 1987a; Harris et al., 1988a, b). Although Beden (1980) noted the presence of *E. recki recki* at Garba IV at Melke Kunture, Ethiopia, the provenience of these fossils needs to be confirmed, as recent geochronological work (Morgan et al., 2012) established a new geological age for the locality horizon that is anomalously old for this subspecies (Table 7.1).

Cheek tooth formulae of *E. recki recki* are collated as: dP2/dp2, --/3x-4x; dP3/dp3, --/7x-8; dP4/dp4, --/--; M1/m1, x9x/--; M2/m2, x10–12/x10-x12; M3/m3, x13x-x19/14x-18x (Beden, 1980, 1983, 1987a; Saegusa and Gilbert, 2008). Third molar dimensions range from L=285–362 mm; W=68–104 mm; H=123–190 mm (yielding hypsodonty indices of 161–211); ET=1.8–3.0 mm; and LF=4.6–6.0 (Beden, 1980, 1987a). The metrics of the molars indicate that these were large elephants with relatively narrow, very high-crowned molars. Enamel is thin and finely and compactly folded toward the apex of the crown, at high frequency per cm; it is more irregularly folded and may have higher amplitude at mid-crown height, and enamel folding is more regular and retains a high amplitude near the base of the crown. Plate faces may be anteriorly and posteriorly convex, producing pseudoloxodont enamel wear figures. Accessory conules tend to be reduced in prominence.

Mandibles of *E. recki recki* are similar to *Elephas* and *Palaeoloxodon* mandibles (see Saegusa and Gilbert, 2008; Larramendi et al., 2017) in having a very brevirostrine symphysis, relatively short, thick corpus, and anteriorposteriorly very extensive ramus that dominates the lateral view of the element (Figure 7.10A and B). They contrast with mandibular specimens of *Loxodonta*, which have relatively longer corpora and symphyses and a less dominant ramus. In instances where specimens of the "*Elephas recki* complex" exhibit a protuberant symphyseal spout (e.g., *E. atavus* mandible KNM-ER 5711 and *P. recki recki* mandible BOU-VP-3/131), the projection of the "bec" or spout is very slender and atavistic, but the symphysis itself is very short.

Crania recovered from the Daka Mb. of the Bouri Fm. at Bouri, Ethiopia, dated to ~1.0 Ma, were referred along with abundant other fossils to *Elephas* (*Palaeoloxodon*) *recki recki* (Saegusa and Gilbert, 2008). They are extremely valuable for assessing phylogenetic and systematic hypotheses about this part of the "*E. recki* complex," as they are the only well described crania of *E. recki recki.* A partial skeleton with a skull from Haïdalo, Djibouti (Figure 7.11), of similar geological age (Chavaillon et al., 1990), could provide even more useful information, but still awaits proper description and analysis. The crania from Bouri include a specimen that is weathered away ventrally but preserves good details of the dorsum of the vault and the face (BOU-VP-3/131; Figure 7.10C), a specimen that has suffered damage to the facial region but retains the occiput, rostrum, tusks in place, and its ventral aspect (BOU-VP-25/50; Figure 7.10D), and a juvenile specimen heavily damaged and very incomplete (BOU-VP-4/144). There were described and studied in exquisite detail by Saegusa and Gilbert (2008), together yielding a wealth of information that distinguishes them from other crania of the "*E. recki* complex" and crania of Eurasian elephants.

Notably, the tusk alveoli of these Bouri crania have distally divergent tusk alveoli, separated by an external incisive fossa that is proximally narrow and deep but distally shallow and broad (dissimilar to the condition in *E. recki atavus*). Even the juvenile specimen shows signs of strong distal flare of the premaxillae, which distinguishes it slightly from the lesser extent of premaxillary flare in *E. recki ileretensis* juvenile cranium KNM-ER 799. Although there is no adequate description of the Haidalo skull, a schematic of the skeleton shows it to have distally flaring tusk alveoli (Figure 7.11; Chavaillon et al., 1990). This condition resembles the rostra of palaeoloxodonts (Larramendi et al., 2020).

In the Bouri crania, the nasal aperture is anteroposteriorly high, very wide, and primitively its lateral margins are gently downcurved. In contrast, the nasal aperture of *E. recki atavus* cranium KNM-ER 5711 has an even larger nasal aperture with lateral margins more sharply downturned. In the Bouri specimens, the anterior nasal protuberance of the nasals is large but not outwardly pointed. In addition, the zygomatic processes of the frontal protrude laterally and the lacrimal processes are prominent and project strongly laterally from the anterior rim of the orbit.

The Bouri crania are distinguished from those of Eurasian palaeoloxodonts by a higher frontoparietal surface and narrower minimum width across the temporal lines ("isthmus frontalis"). Rather than displaying a parietofrontal crest as might be found in crania of Eurasian palaeoloxodonts, there is a central swelling in the frontoparietal surface that appears as a small prominence in lateral view (Figure 7.10D), similar to what is sometimes observed in extant *E. maximus* crania (Saegusa and Gilbert, 2008). In *E. recki atavus* cranium KNM-ER 5711, in the upper half of the frontoparietal surface there is a transversely extended elevation that Beden (1983, 1987a) felt was homologous with the parietofrontal crest of palaeoloxodonts (the "parietooccipital" crest of Larramendi et al. [2020]), but Saegusa and Gilbert (2008) believed that this structure to more resemble the transverse elevation occasionally observed in crania of *Elephas maximus.*

Other features of note of the Bouri crania include: The orbit is higher than in *E. recki atavus* and accompanied by a zygomatic arch which is notably shorter and thicker; the auditory bulla is positioned high and is wide, as in *Elephas* species; the basicranium is similar to that of *E. recki atavus*; the temporal fossa is subrectangular in shape; the fossa for the nuchal ligament is deep and surrounded by exoccipitals that protrude strongly posteriorly; and the occipitoparietal bosses are markedly developed, but there is no sign of the extreme forward expansion of the nuchal plane observed in crania of *Palaeoloxodon namadicus* males (Saegusa and Gilbert, 2020).

In the phylogenetic systematic analysis of Saegusa and Gilbert (2020), the occipital bossing in the Bouri crania is a synapomorphy of *Elephas* and *Palaeoloxodon.* Synapomorphies shared with *Palaeoloxodon* include: Absence of strong angulation between the vault and rostrum; rectangular shape of the temporal fossa; the strong distal flare of the premaxillae; shallow, wide aspect of the external incisive fossa; short, robust zygomatic arch; and high position of the orbit. These features lend support to the hypotheses of Beden (1983, 1987a), Saegusa and Gilbert (2008), Zhang (2020), and Larremendi et al. (2020) that *E. recki recki* belongs in a clade with Eurasian palaeoloxodonts, with far more certainty than hypotheses that include *E. recki atavus* in the clade (e.g., Beden, 1983, 1987a; Saegusa and Gilbert, 2008).

Occurrences of *E. recki recki* in Middle Eastern and Arabian Peninsula sites (Table 7.1) suggest that the route of dispersal of the taxon to "seed" the radiation of palaeoloxodonts in Eurasia was through these regions. Dating of the initial occurrences of palaeoloxodonts in Eurasia around or a little younger than 1.0 Ma (e.g., Bonifay, 1996; Aouadi, 2001) indicates that their immediate phylogenetic source was "*E. recki recki*," not "*E. recki atavus.*" A more contentious question involves deciphering the ancestry of

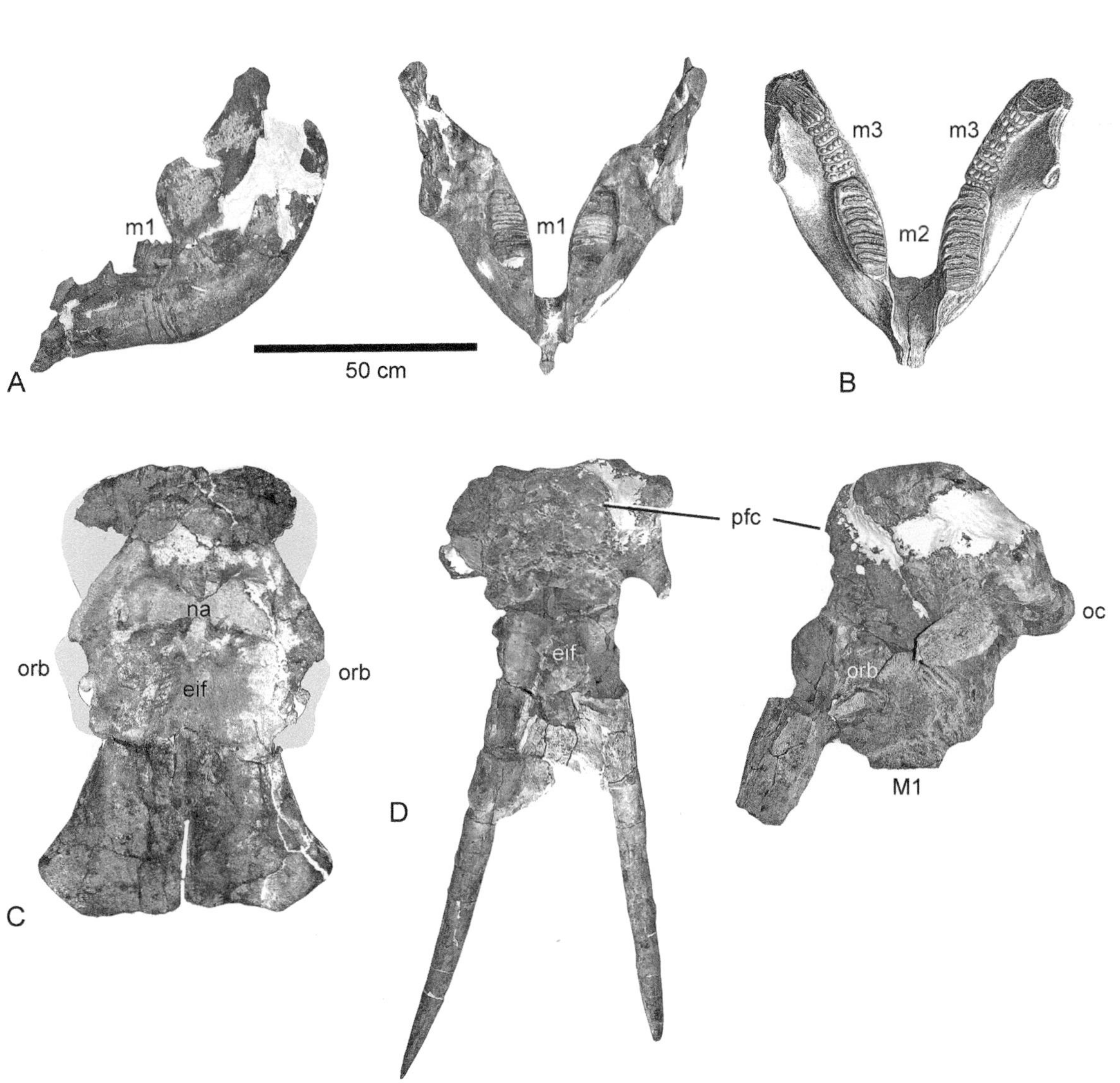

FIGURE 7.10 Skull fossils of late early Pleistocene *Palaeoloxodon recki recki*. Abbreviations: eif, external incisive fossa; M/m, upper/lower molar; na, nasal aperture; oc, occipital condyle; orb, orbit; pfc, parietofrontal crest. (A, B) to the same approximate scale. (A) Mandible, BOU-VP-3/131 from the Daka Mb., Bouri Fm., Middle Awash, Ethiopia, lateral and anterodorsal views. (B) Mandible, specimen XVIII 1439–1440 from indeterminate level of the Upper Beds at Olduvai, Tanzania, anterodorsal view. (From Dietrich [1916:Tafel II, fig. 2].) (C) Partial cranium, BOU-VP-3/131 (associated with mandible) from the Daka Mb., Bouri Fm., Middle Awash, Ethiopia, anterior view. Length = 1,350 mm. (D) Partial cranium with tusks and M1s, BOU-VP-25/50 from the Daka Mb., Bouri Fm., Middle Awash, Ethiopia, anterior and lateral views. Length = 846 mm. (Images in A, C, D with permission of W. H. Gilbert [Saegusa and Gilbert, 2008:figs.9.3a, b and 9.4b–d].)

palaeoloxodonts, including "*E. recki recki*" (or *P. recki*) and their relationship to other African elephants, fossil and extant.

Palaeogenomic investigation of this question by Meyer and colleagues (2017) produced unexpected results implying that *Palaeoloxodon antiquus* is a close sister taxon to forest African elephants (*L. cyclotis*), more closely related than *L. cyclotis* is to the savanna African elephant (*L. africana*). The authors note the disparity of their genomic results with morphological evidence showing *Elephas* features in *Palaeoloxodon* and a basic lack of *Loxodonta* anatomy in palaeoloxodonts, but hint at introgression by *Elephas* on *Palaeoloxodon* as the explanation. Their results are in conflict with an earlier study on mitochondrial DNA in dwarfed palaeoloxodont elephants from Charkadio Cave on the island of Tilos, Greece, which indicated a palaeoloxodont-*Elephas* clade and was thus consistent with morphological evidence supporting this clade (Poulakakis et al., 2002). More recently, a more thorough genomic study by Palkopoulou et al. (2018:p. 3) found that palaeoloxodonts do not comprise a simple clade with forest African elephants, and instead indicated that the largest major genetic component of palaeoloxodonts reveals derivation from a "lineage basal to the common ancestor of forest and savanna elephants." Moreover, the results of their study show significant genetic introgressions to the genome

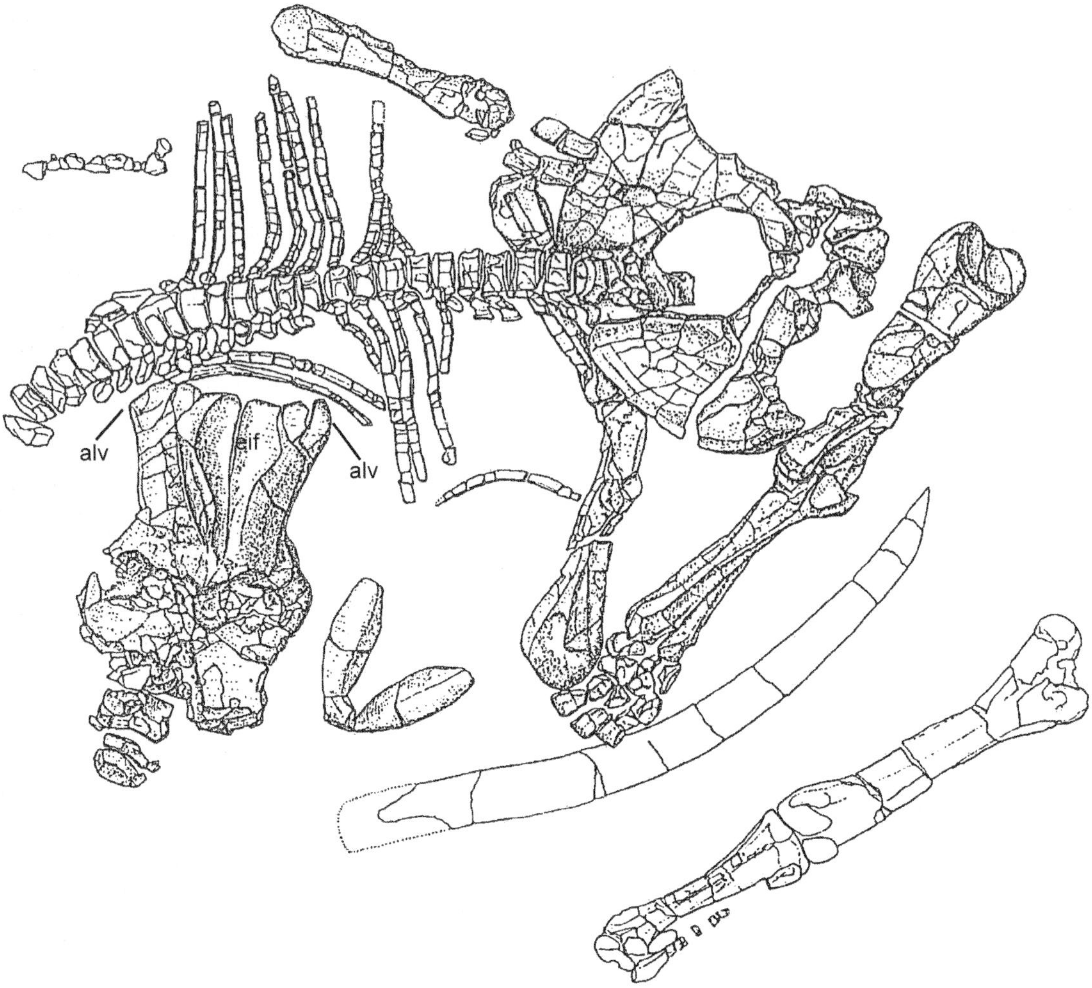

FIGURE 7.11 Schematic drawing of partial skeleton of *Palaeoloxodon recki recki* in semi-articulation from the late early Pleistocene of Haïdalo, Djibouti. Abbreviations: alv, tusk alveolus; eif, external incisive fossa. The skeleton occupies an area of about 16 m^2. Note the divergent distal tusk alveoli and broad distal aspect of the external incisive fossa. (Drawing by J. L. Bosaubert, from Chavaillon et al. [1990:fig. p. 444]. Permission from the Académie des Sciences, France under the Creative Commons CC BY 4.0 license.)

of palaeoloxodonts by woolly mammoths (*Mammuthus primigenius*) and forest African elephants; in addition, woolly mammoths and Asian elephants share significantly more derived alleles with palaeoloxodonts than expected. Palkopoulou et al. (2018:3 of 9) explain the morphological similarity between Asian elephants and palaeoloxodonts by "hybridization from an ancestral population that split off from the mammoth lineage early in its history, close in time to the common ancestor of Asian elephants and mammoths." In a tortuous ad hoc argument, they continued that the morphological synapomorphies of Asian elephants must have been present in the common ancestor of Asian elephants and mammoths, and were subsequently lost in the mammoth lineage.

These genomic scenarios display a profound lack of understanding of, or perhaps disinterest in, the paleontological record. No amount of speculation can presently align paleobiogeography, chronology, and morphology of fossil African elephants with the genomic hypotheses that place African early palaeoloxodonts phylogenetically closer to *Loxodonta* than to *Elephas*. For example, at the time of emergence of palaeoloxodont features in African elephants, initially in eastern Africa, co-occurrences of loxodont and palaeoloxodont elephants were scant (Table 7.1), and presumably forest elephants (if they had even separated from savanna African elephants by that time) were restricted to western and Central African locales, far from the "*Elephas recki* complex." The most ancient documented mammoths are from the late Miocene Lukeino Fm. in the Tugen Hills, Kenya (Table 5.1), and presumably this means that the common ancestor of *Elephas* and *Mammuthus* was at least this old; hybridization between a population of early mammoths close in time to this common ancestor and palaeoloxodonts would have been impossible as there is no credible evidence for palaeoloxodonts as ancient as the late Miocene or early Pliocene.

What the fossil record does clearly indicate is the origin of palaeoloxodonts within the "*Elephas recki* complex" during the early Pleistocene. Thus, it is proper to place *E. recki recki* in the *Palaeoloxodon* clade (Larramendi, 2020; Zhang, 2020). If *E. recki atavus* also belongs in this clade, it might be taxonomically more accurate to refer the clade to "*Elephas* (*Palaeoloxodon*)" (Beden, 1983, 1987a; Saegusa and Gilbert, 2008). However, if the genomic hypotheses of Meyer et al. (2017) and Palkopoulou et al. (2018) are correct, *Palaeoloxodon* has no direct phylogenetic connection to *Elephas*, and *Palaeoloxodon* cannot be a subgenus of *Elephas*.

Palaeoloxodon recki was part of mammalian faunas that exhibited some archaic taxa alongside more modern genera and/or species. A diverse assemblage that demonstrates are the mammals that were collected from the Daka Mb. of the Bouri Fm. at Bouri, Ethiopia (Gilbert, 2012a). This assemblage includes the hominin *Homo erectus* (replete with a good sample of Acheulean lithics), extinct species of the cercopithecoid monkeys *Cercopithecoides* and *Theropithecus*, the modern leopards and lions *Panthera pardus* and *Panthera leo*, advanced and primitive giraffids (*Giraffa* sp. and *Sivatherium* sp.), modern and extinct genera of horses (*Equus* and *Eurygnathohippus*), modern black and white rhinos (*Diceros* sp. and *Ceratotherium simum*), a cohort of extinct suid species in the genera *Kolpochoerus* and *Metridiochoerus* living next to the extant warthog genus *Phacochoerus*, and a large number of bovids, mostly of extant genera (but not all, notably, for example, *Pelorovus antiquus* and *Pelorovus oldowayensis*) in the tribes Aepycerotini, Alcelaphini, Antilopini, Bovini, Caprini, Hippotragini, Reduncini, and Tragelaphini. The overall ecological reconstruction indicated by these mammals is dominantly grassland or savanna interfingered by wooded areas, gallery forests, and floodplains, as well as xeric habitats (Gilbert, 2012b). Presumably, *Palaeoloxodon recki* was competing with suids, horses, bovids, and gelada baboons for graze in the more open-country settings at Bouri.

"*ELEPHAS JOLENSIS*" Pomel, 1895 (alt. *PALAEOLOXODON JOLENSIS* [Pomel, 1895])

(Tables 7.1 and 7.2; Figures 7.1 and 7.12)

Partial Synonymy—*Elephas jolensis* Pomel, 1895; *E. iolensis* Boule, 1900; *E. iolensis* de Lamothe, 1904; *E. iolensis* Depéret and Mayet, 1923; *Archidiskodon transvaalensis* Dart, 1927, 1929; *A. sheppardi* Dart 1927, 1929; *A. broomi* Osborn, 1928; *A. hanekomi* Dart, 1929; *Palaeoloxodon kuhni* Dart, 1929; *Pilgrimia wilmani* Dart, 1929; *A. yorki* Dart, 1929; *Pilgrimia yorki* Dart, 1929; *Pilgrimia archidiskodontoides* Haughton, 1932; *Pilgrimia subantiqua* Haughton, 1932; *Palaeoloxodon transvaalensis* Osborn 1934, 1942; *E. iolensis* Arambourg, 1938; *Loxodonta* (*Palaeoloxodon*) *darti* Cooke, 1939; *Palaeoloxodon darti* Cooke and Clark, 1939; *A. yorki* Osborn, 1942; *Palaeoloxodon hanekomi* Osborn, 1942; *Palaeoloxodon jolensis* Osborn, 1942; *L.* (*Palaeoloxodon*) *transvaalensis* Cooke, 1947, 1949; *L.* (*Palaeoloxodon*) *darti* Cooke, 1947; *L.* (*Palaeoloxodon*) *hanekomi* Cooke 1947, 1949; *E. pomeli* (in part) Arambourg, 1952; *E. iolensis* (in part) Arambourg, 1952; *E. recki* (in part) Biberson and Ennouchi, 1952; *L. jolensis* Hopwood and Hollyfield, 1954; *E. iolensis* Arambourg, 1960; *E. jolensis* Cooke, 1960; *A. transvaalensis* Cooke, 1960; *A.* cf. *transvaalensis* Hendey, 1967; *E. iolensis iolensis* Arambourg, 1970; cf. *Mammuthus* (*Archidiskodon*) *tranvaalensis* Cooke and Coryndon, 1970; *E. iolensis* Maglio, 1973; *E. iolensis* Coppens and Gaudant, 1976; *E. iolensis* Coppens et al., 1978; *E. iolensis* Geraads, 1980; *E. iolensis* Sanders et al., 2010a; *E. iolensis* Marinheiro et al., 2014; *E. jolensis* Manthi et al., 2019

"*Elephas jolensis*" is a dental species, represented only by isolated molars and a few tusk fragments, but the morphometric similarity of its cheek teeth to those of *Palaeoloxodon recki* indicates that it was the final African successor of the "*Elephas recki* complex," and therefore should be referred to *Palaeoloxodon jolensis* (see Zhang, 2020). Originally, it was diagnosed as a dwarfed species because of misidentification of the type specimen as a diminutive third molar (Pomel, 1895). Later, the type specimen was recognized as a more anterior molar and the species was determined to be of medium-to-large size for an elephant (Arambourg, 1960; Maglio, 1973; Coppens and Gaudant, 1976; Coppens et al., 1978; Geraads, 1980). If *Palaeoloxodon jolensis* is an advanced stage of *Palaeoloxodon recki* rather than a separate species, technically the lineage should be renamed "*P. jolensis*" as it has taxonomic priority. Although Pomel (1895) clearly used the specific nomen "*jolensis*," for unknown reasons other researchers soon began using "*iolensis*" instead, and this name became embedded in the literature until it was corrected by Manthi et al. (2019), hopefully for the last time (see also Osborn, 1942; Hopwood and Hollyfield, 1954; Cooke, 1960).

Palaeoloxodon jolensis has its most substantial representation in North African sites located in Algeria, Morocco, and Tunisia, of reported Tyrrhenian Stage antiquity, spanning the latest middle Pleistocene to the end of the late Pleistocene (Table 7.1; Arambourg, 1952, 1960; Biberson 1952a, b; Biberson and Ennouchi, 1952; Coppens and Gaudant, 1976; Geraads, 1980; Marinheiro et al., 2014). Nonetheless, these ages are approximations rather than calculations as they are not based on radiometric methods. The species appears to have had a pan-African distribution, as molars from the Vaal River, South Africa (Dart, 1927, 1929; Osborn, 1928; Haughton, 1932), near the Victoria Falls, Zimbabwe (Cooke and Clark, 1939), Behanga I and Kaiso Village, Uganda (Cooke and Coryndon, 1970; Maglio, 1973), and from the Kibish Fm. at Natodomeri in Kenyan-administered territory (Manthi et al., 2019) closely resemble the North African fossils morphometrically. The eastern and southern African occurrences are very imprecisely dated from the late middle to late Pleistocene (Table 7.1), except for the recovery of molars from Mbs. I and II of the Kibish Fm., which were radiometrically constrained to late middle Pleistocene dates of 205–195 Ka and >130 Ka, respectively (McDougall et al., 2005; Brown and Fuller, 2008; Brown et al., 2012).

Palaeoloxodon jolensis was evidently replaced in northern Kenya-southern Ethiopia by *Loxodonta africana* by the time of deposition of Mbs. III and IV of the Kibish Fm., dated to 130–74 Ka and <10 Ka, respectively (Butzer et al., 1972; Assefa et al., 2008; Brown et al., 2012). It did not typically co-occur with loxodont elephants except for its sympatry with *L. atlantica* at Gué de Constantine (Arambourg, 1952) and early middle Pleistocene Aïn Maarouf, Morocco (Geraads and Amani, 1997), and possibly with *L. africana* in the middle Pleistocene Apoko Fm. at Kanjera, Kenya, around 500 Ka (Plummer and Potts, 1989; Behrensmeyer et al., 1995). Its presence and chronological framework in eastern Africa is bolstered by unpublished *P. jolensis* molars from Herto, Ethiopia, dated to about 160 Ka (Clark et al., 2003). There is additional information from the Afar of *P. jolensis* from sediments twice that age (pers. comm., Yontan Sahle Chemere).

The archeological context for this elephant is generally Middle Stone Age (MSA) or associated with lithic phases that are transitional between the Acheulean and MSA, such as the Fauresmith, characterized by prepared cores, blades, Levallois points, and bifaces (Beaumont and Vogel, 2006). This industrial transition is marked earliest at Olorgesailie, Kenya, between 320 and 305 Ka (Potts et al., 2018); *Palaeoloxodon recki* at the site is only documented from older, pre-MSA horizons (O'Regan et al., 2005). This is true in southern Africa, as well, where *P. recki* is not found in levels above those yielding Acheulean stone tools (Klein, 1988, 2000; Porat et al., 2010). The stone tools from the Vaal River horizons that produced *P. jolensis* molars recall the transitional phase of the Fauresmith, featuring Levallois blades and bifaces (Cooke and Clark, 1939). In addition, *P. jolensis* coincides with early anatomically modern *Homo sapiens* (Fleagle et al., 2008) in outcrops of the Mb. I of the Kibish Fm. in Ethiopia and in the Herto Fm., Herto, Ethiopia (White et al., 2003).

Molars of *P. jolensis* are impressively high crowned, routinely reaching hypsodonty indices well above 200, with crown heights of third molars attaining over 200 mm (Manthi et al., 2019). In cross-sectional view, plates are nearly parallel-sided or widest well above the cervix of the crown, tapering gently toward the apex (Figure 7.12A). In lateral view, plates are typically thicker than the spacing of the transverse valleys between them. Cementum is abundant and in unworn molar crowns fills the transverse valleys and covers the plates. In moderately worn crowns, enamel loops exhibit strong enamel folding of varying amplitude, occasionally with small anterior or posterior enamel projections. In crowns worn more deeply, enamel folding is more regular and tighter, with lower amplitude but higher frequency (Figure 7.12B and C). There is no development of loxodont sinuses in these enamel loops. Enamel is moderate in thickness, ranging from about 2.2 to 4.0 mm in third molars. There are few complete molars known for the species, and plate number of third molars is variable, reaching an extreme of 21 plates in M3 specimen KNM-WT 59676 from Natodomeri (Figure 7.12C and D). Plate formulae are incompletely and poorly established as dP4/dp4, 8/x10x; M1/m1, --/8; M2/m2, --/x11x–12 or 13; M3/m3, 13x–21/13 (Dart, 1927, 1929; Arambourg, 1960; Maglio, 1973; Coppens and Gaudant, 1976; present study). Variation in molar plate numbers may be interpopulational (across the vastness of Africa) or indicative of rapid evolution in the species over a relative condensed interval of a few hundred thousand years.

The tremendous hypsodonty of molars in *P. jolensis* suggests adaptation to resisting enamel abrasion from grazing and consequent acquisition of plants growing low to the ground, which might have incorporated grit into the diet. The results of enamel carbon isotope analysis of molars from Natodomeri shows that *P. jolensis* from the site had predominantly grazing diets, with $\delta^{13}C$ values averaging $+0.3 \pm 0.5‰$ (Manthi et al., 2019). Depending on which isotope enrichment value is applied to these results (Cerling and Harris, 1999; Passy et al., 2005), and using the mixing lines of Cerling et al. (2015), the resultant diet values of −12‰ to −14‰ for the Natodomeri elephants indicate between 75% and 100% C_4 biomass contributing to their diets (Manthi et al., 2019). Although extant Asian and African elephants are mixed feeders and primarily browsers (Sukumar et al., 1987; Sukumar and Ramesh, 1992; Cerling et al., 1999, 2007, 2015), because extinct fossil representatives of *Loxodonta* and the "*Elephas recki* complex" were primarily grazers (Cerling et al., 1999, 2015), evidence that *P. jolensis* was a committed grazer is hardly surprising.

However, given the long dominance of elephants in the "*Elephas recki* complex," particularly in eastern Africa, and the impressive molar adaptations of *P. jolensis*, the sudden evolutionary demise of this group in Africa and its replacement on the landscape by loxodont elephants (which had not been ascendant since the early Pliocene) in retrospect appears astonishing. Although results of molecular analyses indicate a significant expansion in numbers of *L. africana* between 500 and 100 Ka, nearly synchronous with the disappearance of *P. recki* (Murata et al., 2009; Brandt et al., 2012), *P. jolensis* filled the intervening gap until the late Pleistocene re-emergence of *L. africana*. The work on the Kibish Fm. elephants in the northern Turkana Basin suggests that while there is no plausible evidence for human hunting as the agency for the disappearance of the "*E. recki* complex" and *P. jolensis* (Klein, 1988), the timing of the replacement is correlative with intensity and increased rate of fluctuation of climate change. While isotopic and mesowear evidence shows that *P. recki* and *P. jolensis* were living as dedicated grazers (Cerling et al., 1999; Saarinen et al., 2015) in warm, moist conditions during the middle Pleistocene (Klein, 2000), a harsh

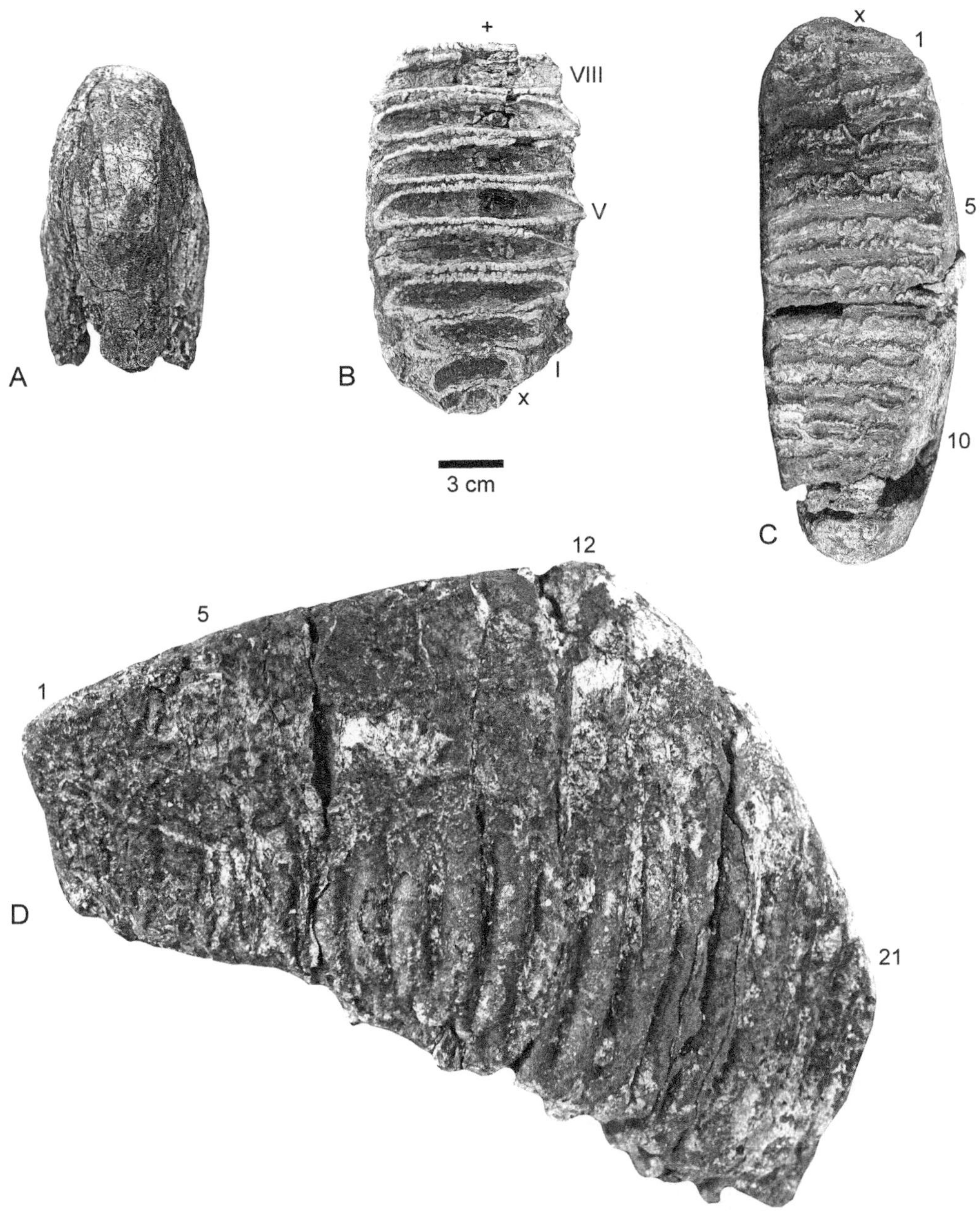

FIGURE 7.12 Molar fossils of *Palaeoloxodon jolensis* from late middle Pleistocene Natodomeri, Kenya. Abbreviations: x, pre- or postcingulum; 1, 2, 3, …, plates counted from the anterior of the crown; I, II, III, …, plates counted from the posterior of the crown; +, indicates missing morphology. (A–D) to the same scale. (B, C) anterior to the top. (A) ?M3, KNM-ND 68505, posterior view. (B) Partial left M3, KNM-ND 68136a, occlusal view. Note the fine folding of enamel. (C) Left M3, KNM-WT 59676, occlusal view of wear surface. Note the fine folding of enamel and its irregular amplitude along the loops. (D) Left M3, KNM-WT 59676, lateral view. Anterior to the left. Occlusal surface to the top. This specimen has the most plates of any elephant molar found in the African fossil record. (Unpublished images in C, D courtesy of J. Saarinen.)

dry, cold climate interval marked by intense episodes of drought initiated around the start of the late Pleistocene (Cohen et al., 2007; Scholz et al., 2007; Tierney et al., 2017). These conditions may have stressed food supplies for specialized grazing elephants and favored the dietary plasticity of generalized mixed feeders such as *L. africana* (see Cerling et al., 1999; Saarinen et al., 2015). With that, one of the most successful phases in proboscidean evolution came to a close, leaving the savanna and forest loxodont elephants as the survivors of the rich, diverse phylogenetic journey of proboscideans on the continent and to inherit the super-megaherbivore niche in Africa. Whether they will long fill that role, given the rapacity and environmental impact of human activity, is questionable.

SUMMARY

1. The youngest episodes of African proboscidean evolution documented in the fossil record occurred during the mid-Pliocene to Holocene, distinguished by substantial changes in global climate, regional paleoecology, and landscape configuration. These changes involved continuation of large-scale rifting, particularly along the length of eastern Africa, resulting in uplift of rift shoulders, downwarping of basins, formation of lakes, and significant deposits of sediment (which accounts for the accumulation of a superb fossil record in this part of Africa during the interval).
2. Additionally, extreme climate transitions involved greater amplitude of global temperature change, accumulation of ice at the poles, and overall lower world-wide temperatures. In Africa, this was associated with increasing aridification, widespread arid and mesic conditions, and greater rainfall seasonality. In response, in local ecological settings, C_4 vegetation, including grasses, expanded to contribute to more widespread development of wooded grasslands, and by the middle Pleistocene, to the formation of pure grasslands.
3. Between the mid-Pliocene and 2.35 Ma, African herbivores comprised a greater number of mixed C_3/C_4 feeders than in modern faunas, probably taking advantage of resources in more open conditions; by 2.35–1.0 Ma, perissodactyls, artiodactyls, suids, and proboscideans featured a greater percentage of grazers than in modern faunas on the continent.
4. The last non-elephant proboscideans went extinct in Africa during this interval. Stegodonts disappeared during the late Pliocene and anancine gomphotheres during the early Pleistocene, perhaps due to increasing competition with herbivores better adapted for grazing efficiency, such as more progressive species of loxodont elephants and mammoths, as well as a new cohort of hypsodont elephants in the "*Elephas recki* complex." The last deinotheres persisted to the end of the Pleistocene, avoiding grazing competition by remaining as committed browsers, and most likely suffered the effects of loss or degradation of habitat.
5. In eastern Africa, particularly in the greater Turkana Basin, elephants of the early Pliocene "relay" interval, such as *Loxodonta adaurora*, were replaced by elephants of the "*Elephas recki* complex," which dominated African open woodland and grassland habitats until the close of the middle Pleistocene. Primitive mammoths evolved into more hypsodont species during this interval as well, although they appear to have been largely relegated to more peripheral continental areas in northern Africa; there are no published records of mammoths on the continent after the early Pleistocene. Although the first good evidence of *L. africana* dates to the early Pleistocene, its occurrences and biogeography during this interval are exceedingly rare and mostly unknown. A hypsodont loxodont elephant, *L. atlantica*, is recorded in late early and early middle Pleistocene northern African locales, but it may have eventually been outcompeted by the expansion of massive-sized, hyper-grazing elephants in the *Palaeoloxodon* lineage.
6. Morphological transformations of the advanced elephants of this interval involved evolution of progressively more hypsodont molars that achieved greater crown height by vertical elongation of occlusal surfaces via extension of Phase I development, greater numbers of plates (plate formulae of elephant third molars increased from 6 to 21 between the late Miocene origin of the Family and middle-late Pleistocene *Palaeoloxodon jolensis*), thicker coatings of cementum, thinner enamel with greater amplitudes and frequencies of folding, and closer spacing of plates (higher lamellar frequencies). Skulls of some African species of elephants from this interval appear highly apomorphic with occipitoparietal bosses and swollen nuchal planum surfaces, and some exhibit signs of greater encroachment of nuchal musculature toward the frontoparietal vertex that became a salient feature of Eurasian *Palaeoloxodon*.
7. The classic approach to the "*Elephas recki* complex" was to consider these Plio-Pleistocene elephants as members of an anagenetically-evolved lineage, and to subdivide the lineage arbitrarily (based primarily on successive stratigraphic units) either as stages or subspecies of *Elephas recki*. This approach relied heavily on morphometric, directional changes in molars over time. Lineage members adjacent in time were seen as analogous to neighboring populations of a ring species, rolled out in time rather than geographically, with the end members (most basal and most recent) different enough to be considered as equivalent to separate species. The subspecies include Pliocene "*E. r. brumpti*," late Pliocene-early Pleistocene "*E. r. shungurensis*," early Pleistocene "*E. r. atavus*," late early Pleistocene "*E. r. ileretensis*," late early-middle Pleistocene "*E. r. recki*," along with middle-late Pleistocene "*E. jolensis*." Recent studies focused on cranial morphology suggested episodes of cladogenetic evolution within this group, and have reorganized these stages into separate species, as either "*E. brumpti*," "*E.* (*Palaeoloxodon*) *atavus*," "*E.* (*Palaeoloxodon*) *ileretensis*," and "*E.* (*Paleoloxodon*) *recki*," or, alternatively, "*Phanagoroloxodon planifrons*," "*E.*

atavus," and "*Palaeoloxodon recki*." With cranial evidence at hand, there is no question that Eurasian species of *Palaeoloxodon*, which underwent a very successful and geographically extensive radiation beginning in the late Pleistocene, derived from African *P. recki*. Moreover, it appears highly likely that Asian *Elephas*, including the contemporary Asian elephant, derived from some earlier ancestor in the "*E. recki* complex."

8. These new phylogenetic scenarios of elephant evolution, however, are complicated by the results of recent ancient DNA studies that indicate a common ancestry of African loxodont elephants and *Palaeoloxodon*, and point to substantial genetic exchange between early African palaeoloxodonts and forest African elephants. This work poses a conundrum because there is no morphological or paleontological support for these evolutionary events. According to some geneticists, the divergence of African savanna and forest elephants occurred as early as the beginning of the Pliocene, or no later than the end of the epoch. However, the oldest evidence of *L. africana* is early Pleistocene in age and there is virtually no fossil record for *L. cyclotis*, so these hypotheses cannot be tested paleontologically. Similarly, it has not yet been established that when African elephants re-emerged in the late Pleistocene, coincident with the demise of the last vestiges of the "*Elephas recki* complex" in the guise of *Palaeoloxdodon jolensis*, they were already mixed feeders with substantial intake of browse as in extant elephants on the continent. However, this is a viable hypothesis because the disappearance of *Palaeoloxodon* in Africa has been associated with late middle Pleistocene-late Pleistocene episodes of repeated intense droughts and the broader extinction of open savanna grazing specialists in multiple herbivore taxa. Genetic studies indicate results consistent with a population expansion of the *L. africana* lineage during the interval 500–100 Kyr; the paleontological evidence points to that expansion occurring in the latter part of that interval.

Whether hominins contributed substantially to the extinction of the "*Elephas recki* complex" is debatable. Although it is evident that hominins had the capacity to kill and butcher elephants as long ago as the beginning of the early Pleistocene, African proboscidean kill sites with Acheulean, MSA, or Late Stone Age stone tools are rare occurrences, and in some cases, the association of elephant carcasses and stone tools are coincidental or the result of scavenging. With body masses and heights impressively exceeding those of extant African elephants, *Palaeoloxodon recki* and *Palaeoloxodon jolensis* would have been dangerous and imposing quarries.

References

Abbas, S. G., M. A. Khan, M. A. Babar, M. Hanif, and M. Akhtar. 2018. New materials of *Choerolophodon* (Proboscidea) from Dhok Pathan Formation of Siwaliks, Pakistan. *Vertebrata Palasiatica* 56:294–305.

Abbas, S. G., M. A. Khan, M. A. Babar, R. Shahid, M. K. Nawaz, and M. Akhtar. 2016. Some new fossils of *Protanancus chinjiensis* (Proboscidea, Mammalia) from middle Miocene of Pakistan. *Biologia (Pakistan)* 62:345–349.

Abbate, E., A. Albianelli, A. Awad, P. Billi, P. Bruni, M. Delfino, M. P. Ferretti, O. Filippi, G. Gallai, M. Ghinassi, S.-E. Lauritzen, D. Lo Vetro, B. Martínez-Navarro, F. Martini, G. Napoleone, O. Bedri, M. Papini, L. Rook, and M. Sagri. 2010. Pleistocene environments and human presence in the middle Atbara valley (Khashm el Girba, eastern Sudan). *Palaeogeography, Palaeoclimatology, Palaeoecology* 292:12–34.

Abbate, E., P. Bruni, A. Coppa, D. Aria, M. P. Ferretti, Y. Libsekal, L. Rook, and M. Sagri. 2012. A new Oligo-Miocene mammal-bearing site from a sedimentary intercalation in the Trap basalts of central Eritrea. *Rivista Italiana di Paleontologia e Stratigrafia* 118:545–550.

Abbate, E., P. Bruni, M. P. Ferretti, C. Delmer, M. A. Laurenzi, M. Hagos, O. Bedri, L. Rook, M. Sagri, and Y. Libsekal. 2014. The East African Oligocene intertrappean beds: Regional distribution, depositional environments and Afro/Arabian mammal dispersals. *Journal of African Earth Sciences* 99:463–489.

Abel, O. 1914. *Die Vorzeitlichen Säugetiere*. Gustav Fischer, Jena, 309 pp.

—. 1919. *Die Stämme der Wirbeltiere*. Vereinigung Wissenschaftlicher Verleger Walter de Gruyter, Berlin and Leipzig, 914 pp.

—. 1920. *Lehrbuch der Paläozoologie*. Gustav Fischer, Jena, 500 pp.

Abell, P. I., S. M. Awramik, R. H. Osborne, and S. Tomellini. 1982. Plio-Pleistocene lacustrine stromatolites from Lake Turkana, Kenya: Morphology, stratigraphy and stable isotopes. *Sedimentary Geology* 32:1–26.

Abouessa, A., J. Pelletier, P. Duringer, S. Mathieu, P. Schaeffer, E. Métais, M. Benammi, M. J. Salem, O. Hlal, M. Brunet, J.-J. Jaeger, and J.-L. Rubino. 2012. New insight into the sedimentology and stratigraphy of the Dur at Talah tidal-fluvial transition sequence (Eocene-Oligocene, Sirt Basin, Libya). *Journal of African Earth Sciences* 65:72–90.

Adnet, S., H. Cappetta, and R. Tabuce. 2010. A middle-late Eocene vertebrate fauna (marine fish and mammals) from southwestern Morocco; preliminary report: Age and palaeobiogeographical implications. *Geological Magazine* 147:860–870.

Adrian, B., L. Werdelin, and A. Grossman. 2018. New Miocene Carnivora (Mammalia) from Moruorot and Kalodirr, Kenya. *Palaeontologia Electronica* 21.1. 10A:1–19. https://doi.org/10.26879/778

Agrasar, E. L. 2004a. Crocodilian remains from the Upper Eocene of Dor-El-Talha, Libya. *Annales de Paléontologie* 90:209–222.

Agrasar, E. L. 2004b. Crocodile remains from the Burdigalian (lower Miocene) of Gebel Zelten (Libya). *Geodiversitas* 26:309–321.

Aguirre, E. E. 1969a. Evolutionary history of the elephant. *Science* 164:1366–1376.

—. 1969b. Revisión sistematica de los Elephantidae por su morfologiá y morfometria dentaria. *Estudios Geológicos* 25:123–177; 317–367.

Aguirre, E. E., and M. T. Alberdi. 1974. *Hipparion* remains from the northern part of the Rift Valley (Kenya). *Proceedings Koninklijke Nederlandse Akademie van Wetenschappen* 77:146–156.

Aguirre, E. E., and P. H. Leakey. 1974. Nakali: Nueva fauna de *Hipparion* del Rift Valley de Kenya. *Estudios Geológicos* 30:219–227.

Alberdi, M. T. 1971. Primer ejemplar complete de un *Tetralophodon longirostris* Kaup, 1835, encontrado en España. *Estudios Geologicos* 27:181–196.

Alemseged, Z. 2003. An integrated approach to taphonomy and faunal change in the Shungura Formation (Ethiopia) and its implication for hominid evolution. *Journal of Human Evolution* 44:451–478.

Alemseged, Z., and D. Geraads. 2000. A new middle Pleistocene fauna from the Busidima-Telalak region of the Afar, Ethiopia. *Comptes rendus de l'Académie des Sciences, Paris: Sciences de la Terre et des Planètes* 331:549–556.

Al-Kindi, M. 2018. The Cenozoic rocks and life in Oman; pp. 141–169 in M. H. Al-Kindi (Ed.), *Evolution of Land and Life in Oman: An 800 Million Year Story*. Springer, Cham, Switzerland.

Al-Kindi, M., M. Pickford, Y. Al-Sinani, I. Al-Ismaili, A. Hartman, and A. Heward. 2017. Large mammals from the Rupelian of Oman—recent finds. *Fossil Imprint* 73:300–321.

Allen, G. M. 1936. Zoological results of the George Vanderbilt African expedition of 1934: Part II, The forest elephant of Africa. *Proceedings of the Academy of Natural Sciences, Philadelphia* 88:15–44.

Al-Sayigh, A. R., S. Nasir, A. S. Schulp, and N. J. Stevens. 2008. The first described *Arsinoitherium* from the Upper Eocene Aydim Formation of Oman: Biogeographic implications. *Palaeoworld* 17:41–46.

Ambrose, S. H., L. J. Hlusko, D. Kyule, A. Deino, and M. Williams. 2003. Lemundong'o: A new 6 Ma paleontological site near Narok, Kenya Rift Valley. *Journal of Human Evolution* 44:737–742.

Andrews, C. W. 1901a. Fossil vertebrates from Egypt. *Zoologist* 4(5):318–319.

—. 1901b. A new name for an ungulate. *Nature* 64:577.

—. 1901c. Preliminary note on some recently discovered extinct vertebrates from Egypt. Part I. *Geological Magazine, Series IV* 8:400–409.

—. 1901d. Über das Vorkommen von Proboscidiern in untertertiären Ablagerungen Aegyptens. *Tagesblatt des V Internationalen Zoologischen-Kongresses, Berlin* 4:4–5.

—. 1902. Über das Vorkommen von Proboscidiern in untertertiären Ablagerungen Aegyptens. In *Verhandlungen des V Internationalen Zoologischen-Kongresses, Berlin*, 528 pp.

—. 1903a. On the evolution of the Proboscidea. *Philosophical Transactions of the Royal Society of London* 196:99–118.

—. 1903b. Notes on an expedition to the Fayum, Egypt, with descriptions of some new mammals. *Geological Magazine*, Decade IV, n.s. 10:338–343.

—. 1904a. Further notes on the mammals of the Eocene of Egypt. *Geological Magazine*, Decade V, n.s. 1:109–115.

—. 1904b. Note on the Barypoda, a new order of ungulate mammals. *Geological Magazine*, Decade V, n.s. 1:481–482.

—. 1905. Note on the species of *Palaeomastodon. Geological Magazine, Series V* 2:512–563.

—. 1906. *A Descriptive Catalogue of the Tertiary Vertebrates of the Fayum, Egypt.* Natural History Museum, London, 319 pp.

—. 1907. The recently discovered Tertiary Vertebrata of Egypt. *Science Progress in the Twentieth Century (1906–1916)* 1:669–682.

—. 1908. On the skull, mandible, and milk dentition of *Palaeomastodon*, with some remarks on the tooth change in Proboscidea in general. *Philosophical Transactions of the Royal Society B* 199:393–407.

—. 1909a. The systematic position of *Moeritherium. Nature* 81:305.

—. 1909b. The feeding habits of *Moeritherium* and *Palaeomastodon. Nature* 81:135–136.

—. 1911. On a new species of *Dinotherium* from British East Africa. *Proceedings of the Zoological Society of London* 1911:943–945.

Andrews, C. W., and H. J. L. Beadnell. 1902. *A Preliminary Note on Some New Mammals from the Upper Eocene of Egypt.* Survey Department, Public Works Ministry, Cairo, pp. 1–9.

Andrews, P. J. 1974. *Report on the Karungu 1973 Expedition.* Kenya National Museum, Nairobi.

Andrews, P. J., and L. Martin. 1987. The phyletic position of the Ad Dabtiyah hominoid. *Bulletin of the British Museum (Natural History) (Geology)* 41:383–393.

Andrews, P. J., and J. A. H. Van Couvering. 1975. Palaeoenvironments in the East African Miocene; pp. 62–103 in F. S. Szalay (Ed.), *Approaches to Primate Paleobiology: Contributions to Primatology*, Vol. 5. Karger, Basel.

Andrews, P. J., and A. Walker. 1976. The primate and other faunas from Fort Ternan, Kenya; pp. 279–304 in G. Isaac, and E. R. McCown (Eds.), *Human Origins.* Benjamin, Menlo Park.

Andrews, P. J., W. R. Hamilton, and P. J. Whybrow. 1978. Dryopithecines from the Miocene of Saudi Arabia. *Nature* 274:249–251.

Andrews, P. J., J. M. Lord, and E. A. Nesbit Evans. 1979. Patterns of ecological diversity in fossil and modern mammalian faunas. *Biological Journal of the Linnean Society* 11:177–205.

Andrews, P. J., G. E. Meyer, D. R. Pilbeam, J. A. Van Couvering, and J. A. H. Van Couvering. 1981. The Miocene fossil beds of Maboko Island, Kenya: Geology, age, taphonomy, and palaeontology. *Journal of Human Evolution* 10:35–48.

Anonymous. 1827. Analytical Notices of Books. Histoire Naturelle des Mammifères, avec des Figures originale, dessinées d'après des Animaux vivans; &c. Par MM. Geoffroy-Saint-Hilaire, et F. Cuvier. Livraison 52 et 53. *The Zoological Journal* 3:140–143.

Anonymous. 1976. Miocene vertebrates from South West Africa. *South African Journal of Science* 72:355.

Antoine, P.-O., L. Marivaux, D. A. Croft, G. Billet, M. Ganerød, C. Jaramillo, T. Martin, M. J. Orliac, J. Tejada, A. J. Altamirano, F. Duranthon, G. Fanjat, S. Rousse, and R. S. Gismondi. 2012. Middle Eocene rodents from Peruvian Amazonia reveal the pattern and timing of caviomorph origins and biogeography. *Proceedings of the Royal Society B* 279:1319–1326.

Antoine, P.-O., G. Métais, M. J. Orliac, J.-Y. Crochet, L. J. Flynn, L. Marivaux, A. R. Rajpar, G. Roohi, and J.-L. Welcomme. 2013. Mammalian Neogene biostratigraphy of the Sulaiman Province, Pakistan; pp. 400–422 in X. Wang, L. J. Flynn, and M. Fortelius (Eds.), *Fossil Mammals of Asia: Neogene Biostratigraphy and Chronology.* Columbia University Press, New York.

Antoine, P.-O., J.-L. Welcomme, L. Marivaux, I. Baloch, M. Benammi, and P. Tassy. 2003. First record of Paleogene Elephantoidea (Mammalia, Proboscidea) from the Bugti Hills of Pakistan. *Journal of Vertebrate Paleontology* 23:977–980.

Anyonge, W. 1991. Fauna from a new Lower Miocene locality west of Lake Turkana, Kenya. *Journal of Vertebrate Paleontology* 11:378–390.

Aouadi, N. 2001. New data on the diversity of elephants (Mammalia, Proboscidea) in the early and early middle Pleistocene of France; pp. 81–84 in G. Cavarretta, P. Gioia, M. Mussi, and M. R. Palombo (Eds.), *The World of Elephants: Proceedings of the First International Congress (October 16–21, 2001, Rome).* Consiglio Nazionale delle Richerche, Roma.

Arambourg, C. 1934a. Le *Dinotherium* des Gisements de l'Omo. *Comptes rendus de la Société Géologique de France* 1932:86–87.

—. 1934b. Le *Dinotherium* des Gisements de l'Omo (Abyssinie). *Bulletin de la Société Géologique de France* 4:305–309.

—. 1934c. Les resultats geologiques de la mission de l'Omo (1932–1933). *Comptes rendus de la Société Géologique de France* 5:63–64.

—. 1938. Mammiferes fossils du Maroc. *Mémoires de la Société des Sciences Naturelles du Maroc* 46:1–74.

—. 1942. L'*Elephas recki* Dietrich. Exposition systématique et ses affinités. *Bulletin de la Société Géologique de France* 12:73–87.

—. 1943. Contribution a l'étude géologique et paléontologique du Bassin du Lac Rodolphe et de la Basse Vallee de l'Omo. *Mission Scientifique de l'Omo 1932–1933. Tome 1. Géologie—Anthropologie.* Muséum National d'Historie Naturelle, Paris, pp. 231–559.

—. 1945. *Anancus osiris*, un mastodonte nouveau du Pliocène inférieur d'Egypte. *Bulletin de la Société Géologique de France* 15:479–495.

—. 1947. *Mission Scientifique de l'Omo. 1932–1933. Tome I. Géologie-Anthropologie. Fascicule III.* Muséum National d'Histoire Naturelle, Paris, pp. 1–562.

—. 1951. La succession des faunes mammalogiques en Afrique du Nord, au cours du Tertiare et du Quaternaire. *Comptes Rendus Sommaire des Seances de la Société Biogéographique de France* 241:49–56.

—. 1952. Note préliminaire sur quelques Eléphants fossils de Berberie. *Bulletin du Muséum national d'Histoire naturelle de Paris* 24:407–418.

—. 1959. Vertébrés continentaux du Miocène Supérieur de l'Afrique du Nord. *Service de la Carte Géologique de l'Algérie, Paléontologie Mémoire* 4:1–159.

—. 1960. Au sujet de *Elephas iolensis* Pomel. *Bulletin d'Archéologie Marocaine* 3:93–105.

—. 1961. Note préliminaire sur quelques Vertébrés nouveaux du Burdigalien de Libye. *Comptes Rendus Sommaire des Seances de la Société Géologique de France* 1961: 107–109.

—. 1963. Continental vertebrate faunas of the Tertiary of North Africa; pp. 55–60 in F. C. Howell and F. Bourliere (Eds.), *African Ecology and Human Evolution*. Aldine, Chicago.

—. 1969a. La nouvelle expedition scientifique de l'Omo. *Rivista di Scienze Preistoriche* 24:3–13.

—. 1969b. Les corrélations paléontologiques et chronologiques entre le Pléistocène inférieur de l'Europe et celui de l'Afrique. *Bulletin Société Géologique de France* 11:106–115.

—. 1970. Les Vertébrés du Pléistocène de l'Afrique du Nord. *Archives du Muséum National d'Histoire Naturelle, Paris, 7e Série* 10:1–126.

Arambourg, C., and P. F. Burollet. 1962. Restes de Vertèbrés oligocènes en Tunisie centrale. *Compte Rendu Sommaire des Séances de la Société Géologique de France* 2:42–43.

Arambourg, C., and P. Magnier. 1961. Gisements de Vertébrés dans le basin tertiaire de Syrte (Libye). *Comptes Rendus Hebdomadaires des Séances de l'Académie des Sciences* 252:1181–1183.

Arambourg, C., J. Chavaillon, and Y. Coppens. 1969. Résultats de la Nouvelle Mission de l'Omo (2e Campagne 1968). *Comptes rendus des séances de l'Académie des Sciences* 268:759–762.

Archibald, J. D., and D. H. Deutschman. 2001. Quantitative analysis of the timing of origin and diversification of extant placental orders. *Journal of Mammalian Evolution* 8:107–124.

Arney, I. 2019. *East African Miocene Catarrhine Evolutionary Ecology*. Unpublished PhD dissertation, University of Michigan, 203 pp.

Arnould-Saget, S., and P. Magnier. 1961. Decouverte de dents de palaeomastodontes dan la region de Zella (Tripolitaine). *Bulletin de la Société Géologique de France* S7–111(3):283–287.

Asher, R. J. 2001. Cranial anatomy in tenrecid insectivorans: Character evolution across competing phylogenies. *American Museum Novitates* 3352:1–54.

Asher, R. J., and T. Lehmann. 2008. Dental eruption in afrotherian mammals. *BMC Biology* 6:14. doi:10.1186/1741-7007-6-14.

Asher, R. J., and E. R. Seiffert. 2010. Systematics of endemic African mammals; pp. 903–920, in L. Werdelin and W. J. Sanders (Eds.), *Cenozoic Mammals of Africa*. University of California Press, Berkeley.

Asher, R. J., M. J. Novacek, and J. H. Geisler. 2003. Relationships of endemic African mammals and their fossil relatives based on morphological and molecular evidence. *Journal of Mammalian Evolution* 10:131–194.

Assefa, Z., S. Yirga, and K. E. Reed. 2008. The large-mammal fauna from the Kibish Formation. *Journal of Human Evolution* 55:501–512.

Augé, M., and J.-C. Rage. 2006. Herpétofaunes du Paléocène supérieur et de l'Eocène inférieur du Maroc. *Annales de Paléontologie* 92:235–253.

Averianov, A.O., and A.V. Lopatin. 2014. High-level systematics of placental mammals: Current status of the problem. *Biology Bulletin* 41:801–816.

Aymard, A. 1855. *Anancus* Aymard, *Anancus macroplus* Aymard; p. 505 in M. J. Dorlhac (Ed.), *Notice géologique sur le cratère de Coupet et sur son gisement de gemmes et d'ossements fossiles. Annales de la Societé d'Agriculture, Sciences, Arts et Commerce du Puy* 19:497–517.

Bachmann, I. 1875. Beschreibung eines Unterkiefers von *Dinotherium bavaricum* H. v. Meyer. *Abhandlungen der Schweizerischen Paläontologischen Gesellschaft* 2:1–20.

Barbour, E. H. 1927. Preliminary notice of a new proboscidean *Amebelodon fricki*, gen. et sp. nov. *Bulletin of the University of Nebraska State Museum* 13:131–134.

—. 1930. *Amebelodon sinclairi* sp. nov. *Nebraska State Museum Bulletin* 17:155–158.

Barriel, V., E. Thuet, and P. Tassy. 1999. Molecular phylogeny of Elephantidae. Extreme divergence of the extant forest African elephant. *Comptes rendus de l'Académie des Sciences, Paris, Sciences de la vie* 322:447–454.

Barry, J. C. 1999. Late Miocene carnivore from the Emirate of Abu Dhabi, United Arab Emirates; pp. 203–208 in P. J. Whybrow and A. Hill (Eds.), *Fossil Vertebrates of Arabia. With Emphasis on the Late Miocene Faunas, Geology, and Palaeoenvironments of the Emirate of Abu Dhabi, United Arab Emirates*. Yale University Press, New Haven.

Barry, J. C., and L. J. Flynn. 1989. Key biostratigraphic events in the Siwalik sequence; pp. 557–571 in E. H. Lindsay, V. Fahlbusch, and P. Mein (Eds.), *European Neogene Mammal Chronology*. Plenum Press, New York.

Barry, J. C., E. H. Lindsay, and L. L. Jacobs. 1982. A biostratigraphic zonation of the Middle and Upper Siwaliks of the Potwar Plateau of northern Pakistan. *Palaeogeography, Palaeoclimatology, Palaeoecology* 37:95–130.

Beaune, D., B. Fruth, L. Bollache, G. Hohmann, and F. Bretagnolle. 2013. Doom of the elephant-dependent trees in a Congo tropical forest. *Forest Ecology Management* 295:109–117.

Beden, M. 1975. A propos des Proboscidiens Plio-Quaternaires des gisements de l'Omo (Éthiopie); pp. 693–705 in *Colloque International C.N.R.S. n° 218 (Paris 4–9 juin 1973): Problèmes Actuels de Paléontologie-Évolution des Vertébrés*.

—. 1976. Proboscidea; pp. 193–208 in Y. Coppens, F. C. Howell, G. L. Isaac, and R. F. Leakey (Eds.), *Earliest Man and Environments in the Lake Rudolf Basin*. University of Chicago Press, Chicago.

—. 1979. *Les Eléphants (Loxodonta et Elephas) d'Afrique Orientale: Systématique, phylogénie, intérêt biochronologique*. Thèse d'Etat, Faculté de Sciences, Université de Poitiers, 567 pp.

—. 1980. *Elephas recki* Dietrich, 1915 (Proboscidea, Elephantidae). Évolution au cours du Plio-Pléistocène en Afrique Orientale. *Geobios* 13:891–901.

—. 1983. Family Elephantidae; pp. 40–129 in J. M. Harris (Ed.), *Koobi Fora Research Project. Volume 2. The Fossil Ungulates: Proboscidea, Perissodactyla, and Suidae*. Clarendon Press, Oxford.

—. 1985. Les Proboscidiens des Grands Gisements a Hominidés Plio-Pléistocène d'Afrique Orientale; pp. 21–44 in *L'Environnnement des Hominidés au Plio-Pléistocène*. Fondation Singer-Polignac, Masson, Paris.

—. 1987a. *Les Faunes Plio-Pléistocènes de la Vallée de l'Omo. Tome 2. Les Eléphantidés (Mammalia, Proboscidea)*. Éditions du Centre National de la Recherche Scientifique, Paris, pp. 162.

—. 1987b. Fossil Elephantidae from Laetoli; pp. 259–294 in M. D. Leakey and J. M. Harris (Eds.), *Laetoli: A Pliocene Site in Northern Tanzania*. Clarendon Press, Oxford.

Behrensmeyer, A. K. 1976. Lothagam Hill, Kanapoi, and Ekora: A general summary of stratigraphy and faunas; pp. 163–170 in Y. Coppens, F. C. Howell, G. L. Isaac, and R. E. F. Leakey (Eds.), *Earliest Man and Environments in the Lake Rudolf Basin*. University of Chicago Press, Chicago.

Behrensmeyer, A. K., A. L. Deino, A. Hill, J. D. Kingston, and J. J. Saunders. 2002. Geology and geochronology of the middle Miocene Kipsaramon site complex, Muruyur Beds, Tugen Hills, Kenya. *Journal of Human Evolution* 42:11–38.

Behrensmeyer, A. K., R. Potts, T. Plummer, L. Tauxe, N. Opdyke, and T. Jorstad. 1995. The Pleistocene locality of Kanjera, western Kenya: Stratigraphy, chronology and paleoenvironments. *Journal of Human Evolution* 29:247–274.

Benoit, J. 2015. A new method of estimating brain mass through cranial capacity in extinct proboscideans to account for the non-tissues surrounding their brain. *Journal of Vertebrate Paleontology* 35:e991021.

Benoit, J., L. J. Legendre, R. Tabuce, T. Obada, V. Mararescul, and P. Manger. 2019. Brain evolution in Proboscidea (Mammalia, Afrotheria) across the Cenozoic. *Scientific Reports* 9:9323. https://doi.org/10.1038/s41598-019-45888-4

Benoit, J., S. Merigeaud, and R. Tabuce. 2013. Homoplasy in the ear region of Tethytheria and the systematic position of Embrithopoda (Mammalia, Afrotheria). *Geobios* 46:357–370.

Bergounioux, F. M., and F. Crouzel. 1956. Présence de *Tetralophodon longirostris* dans le Vindobonien inférieur de Tunisie. *Bulletin de la Société Géologique de France* 6:547–557.

—. 1959. Nouvelles observations sur un petit Mastodonte du Cartennien de Kabylie. *Comptes Rendus Sommaire des Seances de la Société Géologique de France* 1959:101–102.

Bernor, R. L. 1983. Geochronology and zoogeographic relationships of Miocene Hominoidea; pp. 21–64 in R. L. Ciochon and R. S. Corruccini (Eds.), *New Interpretations of Ape and Human Ancestry*. Springer, New York.

Bernor, R. L., and P. Pavlakis. 1987. Zoogeographic relationships of the Sahabi large mammal fauna (early Pliocene, Libya); pp. 349–383 in N. T. Boaz, A. El-Ernauti, A. W. Gaziry, J. de Heinzelin, and D. D. Boaz (Eds.), *Neogene Paleontology and Geology of Sahabi*. Liss, New York.

Bernor, R. L., M. Brunet, L. Ginsburg, P. Mein, M. Pickford, F. Rogl, S. Sen, F. Steininger, and H. Thomas. 1987. A consideration of some major topics concerning Old World Miocene mammalian chronology, migrations and paleogeography. *Geobios* 20:431–439.

Berthelet, A. 2001. L'outillage lithique du site de depecage a *Elephas recki ileretensis* de Barogali (republique de Djibouti). *Comptes rendus de l'Académie des Sciences de la Terre et des planètes* 332:411–416.

Berthelet, A., and J. Chavaillon. 2001. The Early Palaeolithic butchery site of Barogali (Republic of Djibouti); pp. 176–179 in G. Cavarretta, P. Gioia, M. Mussi, and M. R. Palombo (Eds.), *Proceedings of the First International Congress of la Terra degli Elefanti: The World of Elephants*. Consiglio Nazionale delle Richerche, Rome.

Berzhagi, F., M. Longo, P. Ciais, S. Blake, F. Bretagnolle, S. Vieira, M. Scaranello, G. Scarascia-Mugnozza, and C. E. Doughty. Carbon stocks in central African forests enhanced by elephant disturbance. *Nature Geoscience*. https://doi.org/10.1038/s41561-019-0395-6

Biberson, P. 1952a. Découverte d'une molaire d'Éléphant a la carrière de Sidi-Abder-Rahmane. *Bulletin Société des sciences naturelles du Maroc* 2:34–35.

—. 1952b. Présentation d'un fragment de mandibule d'Éléphant découverte a la carrière de la S.T. I.C. près de Casablanca. *Bulletin Société des sciences naturelles du Maroc* 4:81–83.

Biberson, P., and E. Ennouchi. 1952. Présence de *Elephas recki* Dietrich dans la carriere de Sidi-Abder-Rahmane, á Casablanca. *Comptes Rendus de la Société Géologique de France* 6:90–92.

Bibi, F., A. Hill, M. Beech, and W. Yasin. 2013. Late Miocene fossils from the Baynunah Formation, United Arab Emirates; pp. 583–594 in X. Wang, L. J. Flynn, and M. Fortelius (Eds.), *Fossil Mammals of Asia. Neogene Biostratigraphy and Chronology*. Columbia University Press, New York.

Bibi, F., F. Kaya, and S. Varela. 2022. Paleoecology and paleobiogeography of the Baynunah fauna; pp. 333–350 in F. Bibi, B. Kratz, M. Beech, and A. Hill (Eds.), *Sands of Time: Late Miocene Fossils from the Baynunah Formation, U.A.E.* Springer, Cham, Switzerland.

Bibi, F., B. Kraatz, N. Craig, M. Beech, M. Schuster, and A. Hill. 2012. Early evidence for complex social structure in Proboscidea from a late Miocene trackway site in the United Arab Emirates. *Biology Letters* 8:670–673.

Bininda-Emonds, O. R. P., M. Cardillo, K. E. Jones, R. D. E. MacPhee, R. M. D. Beck, R. Grenyer, S. A. Price, R. A. Vos, J. L. Gittleman, and A. Purvis. 2007. The delayed rise of present-day mammals. *Nature* 446:507–512.

Bishop, W. W. 1958. Miocene Mammalia from the Napak Volcanics, Karamoja, Uganda. *Nature* 182:1480–1482.

—. 1964. More fossil primates and other Miocene mammals from north-east Uganda. *Nature* 203:1327–1331.

—. 1967. The later Tertiary in East Africa: Volcanics, sediments and faunal inventory; pp. 31–56 in W. W. Bishop and J. D. Clark (Eds.), *Background to Evolution in Africa*. University of Chicago Press, Chicago.

—. 1971. The late Cenozoic history of East Africa in relation to hominoid evolution; pp. 493–527 in K. K. Turekian (Ed.), *The Late Cenozoic Glacial Ages*. Yale University Press, New Haven.

—. 1972. Stratigraphic succession "versus" calibration in East Africa; pp. 219–246 in W. W. Bishop and J. A. Miller (Eds.), *Calibration of Hominoid Evolution*. Scottish Academic Press, Edinburgh.

Bishop, W. W., and A. Hill. 1999. Fossil Suidae from the Baynunah Formation, Emirate of Abu Dhabi, United Arab Emirates; pp. 254–270 in P. J. Whybrow and A. Hill (Eds.), *Fossil Vertebrates of Arabia. With Emphasis on the Late Miocene Faunas, Geology, and Palaeoenvironments of the Emirate of Abu Dhabi, United Arab Emirates*. Yale University Press, New Haven.

Bishop, W. W., and M. H. L. Pickford. 1975. Geology, faunas and palaeoenvironments of the Ngorora Formation, Kenya Rift Valley. *Nature* 254:185–192.

Bishop, W. W., and F. Whyte. 1962. Tertiary mammalian faunas and sediments in Karamoja and Kavirondo, East Africa. *Nature* 196:1283–1287.

Bishop, W. W., G. R. Chapman, A. Hill, and J. A. Miller. 1971. Succession of Cainozoic vertebrate assemblages from the northern Kenya Rift Valley. *Nature* 233:389–394.

Bishop, W. W., J. A. Miller, and F. J. Fitch. 1969. New potassium-argon date determinations relevant to the Miocene fossil mammal sequence in East Africa. *American Journal of Science* 267:669–699.

Blake, S. 2002. *The Ecology of Forest Elephant Distribution, Ranging, and Habitat Use in the Ndoki Forest, Central Africa*. Unpublished Doctoral Thesis, University of Edinburgh, UK.

Blake, S., and S. Hedges. 2004. Sinking the flagship: The case of forest elephants in Asia and Africa. *Conservation Biology* 18:1191–1202.

Blake, S., S. L. Deem, E. Mossimbo, F. Maisels, and P. Walsh. 2009. Forest elephants: Tree planters of the Congo. *Biotropica* 41:459–468.

Blanckenhorn, M. 1901. Neues zur Geologie und paläontologie Aegyptens. III: Das Miozan. *Zeitschrift der Deutschen Geologischen Gesellschaft* 53:52–132.

Blumenbach, J. F. 1797. *Handbuch der Naturgeschichte* (5th edition). Dieterich, Göttingen. 714 pp.

Blumenthal, S. A., N. E. Levin, F. H. Brown, J.-P. Brugal, K. L. Chritz, J. M. Harris, G. E. Jehle, and T. E. Cerling. 2017. Aridity and hominin environments. *Proceedings of the National Academy of Sciences, USA* 114:7331–7336.

Boaz, D. D. 1982. Preliminary assessment of taphonomy and paleoecology at Sahabi. *Garyounis Scientific Bulletin* 1982:109–121.

Boaz, N. T. 1994. Significance of the Western Rift for hominid evolution; pp. 321–343 in R. S. Corruccini and R. L. Ciochon (Eds.), *Integrative Paths to the Past, Paleoanthropological Advances in Honor of F. Clark Howell.* Prentice Hall, Englewood Cliffs, N.J.

Boaz, N. T., A. El-Arnauti, J. Agusti, R. L. Bernor, P. Pavlakis, and L. Rook. 2008. Temporal, lithostratigraphic, and biochronologic setting of the Sahabi Formation, North-Central Libya. *Geology of East Libya* 3:959–972.

Boaz, N. T., J. de Heinzelin, A. W. Gaziry, and A. El-Arnauti. 1982. Results from the International Sahabi Research Project. *Garyounis Scientific Bulletin* Special Issue No. 4:142 pp. University of Garyounis, Benghazi.

Bobe, R. 1997. *Hominid Environments in the Pliocene: An Analysis of Fossil Mammals from the Omo Valley, Ethiopia.* Unpublished doctoral dissertation, University of Washington, 252 pp.

—. 2006. The evolution of arid ecosystems in eastern Africa. *Journal of Arid Environments* 66:564–584.

—. 2011. Fossil mammals and paleoenvironments in the Omo-Turkana Basin. *Evolutionary Anthropology* 20:254–263.

Bobe, R., and A. K. Behrensmeyer. 2004. The expansion of grassland ecosystems in Africa in relation to mammalian evolution and the origin of the genus *Homo. Palaeogeography, Palaeoclimatology, Palaeoecology* 207:399–420.

Bobe, R., A. K. Behrensmeyer, and R. E. Chapman. 2002. Faunal change, environmental variability and late Pliocene hominin evolution. *Journal of Human Evolution* 42:475–497.

Bocherens, H., P. L. Koch, A. Mariotti, D. Geraads, and J.-J. Jaeger. 1996. Isotopic biogeochemistry ^{13}C, ^{18}O) of mammalian enamel from African Pleistocene hominid sites. *Palaios* 11:306–318.

Boeuf, O. 1992. *Anancus arvernensis chilhiacensis* nov. subsp. (Proboscidea, Mammalia), un Mastodonte du Plio-Pléistocène de Haute-Loire, France. *Geobios Mémoire Special* 14:179–188.

Böhme, M. 2003. The Miocene Climatic Optimum: Evidence from ectothermic vertebrates of Central Europe. *Palaeogeography, Palaeoclimatology, Palaeoecology* 195:389–401.

Boisserie, J.-R. 2005. The phylogeny and taxonomy of Hippopotamidae (Mammalia: Artiodactyla): A review based on morphology and cladistics analysis. *Zoological Journal of the Linnean Society* 143:1–26.

Boisserie, J.-R., M. Schuster, M. J. Beech, A. Hill, and F. Bibi. 2017. A new species of hippopitamine (Cetartiodactyla, Hippopotamidae) from the late Miocene Baynunah Formation, Abu Dhabi, United Arab Emirates. *Palaeovertebrata* 41:1–16.

Bonaparte, C. L. 1845. *Catalogo Metodico dei Mammiferi Europei.* Valenciennes, Milan, 36 pp.

Bonifay, M. F. 1996. The importance of mammalian faunas from the early middle Pleistocene of France; pp. 255–262 in A. Turner (Ed.), *The Early Middle Pleistocene in Europe.* Proceedings of the SEQS Cromer Symposium, Balkema, Rotterdam.

Bonis, L. de, J.-J. Jaeger, B. Coiffait, and P.-E. Coiffait. 1988. Découverte du plus ancient primate Catarrhinien connu dans l'Éocène supérieur d'Afrique du Nord. *Comptes Rendus de l'Académie des Sciences, Paris, Série II* 206:929–934.

Bonnefille, R., R. Potts, F. Chalie, D. Jolly, and O. Peyron. 2004. High resolution vegetation and climate change associated with Pliocene *Australopithecus afarensis. Proceedings of the National Academy of Sciences, USA* 101:12125–12129.

Borissiak, A. A. 1929. On a new mastodon from the Chokrak Beds (middle Miocene) of the Kuban region. *Platybelodon danovi*, n. gen., n. sp. *Annual of the Paleontological Society Russie* 7:105–120.

Borths, M. R., and N. J. Stevens. 2019. *Simbakubwa kutokaafrika* gen. et sp. nov. (Hyainailourinae, Hyaeondonta, 'Creodonta,' Mammalia), a gigantic carnivore from the earliest Miocene of Kenya. *Journal of Vertebrate Paleontology* 39:1, e1570222. https://doi.org/10.1080/02724634.2019.1570222

Boschetto, H. B., F. H. Brown, and I. McDougall. 1992. Stratigraphy of the Lothidok Range, northern Kenya, and K/Ar ages of its Miocene primates. *Journal of Human Evolution* 22:47–71.

Bouché, P., B. Doamba, B. Sissoko, and S. Bouju. 2009. Les éléphants du Gourma, Mali: Statut et menaces pour leur conservation. *Pachyderm* 45:47–56.

Bown, T. M. 1982. Ichnofossils and rhizoliths of the nearshore fluvial Jebel Qatrani Formation (Oligocene), Fayum Province, Egypt. *Palaeogeography, Palaeoclimatology, Palaeoecology* 40:255–309.

Bown, T. M., and M. J. Kraus. 1988. Geology and paleoenvironment of the Oligocene Jebel Qatrani Formation and adjacent rocks, Fayum Depression, Egypt. *United States Geological Survey Professional Paper* 1452, 60 pp.

Bown, T. M., J. J. Kraus, S. L. Wing, J. G. Fleagle, B. H. Tiffney, E. L. Simons, and C. F. Vondra. 1982. The Fayum primate forest revisited. *Journal of Human Evolution* 11:603–632.

Brandt, A. L., Y. Ishida, N. J. Georgiadis, and A. L. Roca. 2012. Forest elephant mitochondrial genomes reveal that elephantid diversification in Africa tracked climate transitions. *Molecular Ecology* 21:1175–1189.

Braun, D. R., N. E. Levin, D. D. Stynder, A. I. R. Herries, W. Archer, F. Forest, D. L. Roberts, L. C. Bishop, T. Matthews, S. B. Lehmann, R. Pickering, and K. E. Fitzsimmons. 2013. Mid-Pleistocene hominin occupation at Elandsfontein, Western Cape Province, South Africa. *Quaternary Science Reviews* 82:145–166.

Brives, A. 1919. Sur la découverte d'une dent de *Dinotherium* dans la sablière du Djebel Knoif, près Tébessa. *Bulletin de la Société d'Histoire Naturelle d"Afrique du Nord* 10:90–93.

Brock, P. W. G., and R. Macdonald. 1969. Geological environment of the Bukwa mammalian fossil locality, eastern Uganda. *Nature* 223:593–596.

Bromage, T. G., F. Schrenk, and Y. M. Juwayeyi. 1995. Paleobiogeography of the Malawi Rift: Age and vertebrate paleontology of the Chiwondo Beds, northern Malawi. *Journal of Human Evolution* 28:37–57.

Brook, G. A., A. Cherkinsky, E. Marais, and N. Todd. 2014. Rare elephant molar (*Loxodonta africana zulu*) from the Windhoek Spring Deposit, Namibia. *Transactions of the Royal Society of South Africa* 69:145–150.

Brookes, J. 1828. *A Catalogue of the Anatomical and Zoological Museum of Joshua Brookes Esq.* George Robins, London.

Brown, F. H. 1994. Development of Pliocene and Pleistocene chronology of the Turkana Basin, East Africa, and its relation to other sites; pp. 285–312 in R. S. Corruccini and R. L. Ciochon (Eds.), *Integrative Paths to the Past. Paleoanthropological Advances in Honor of F. Clark Howell*. Prentice Hall, Englewood Cliffs, N.J.

Brown, F. H., and C. S. Feibel. 1986. Revision of lithostratigraphic nomenclature in the Koobi Fora region, Kenya. *Journal of the Geological Society, London* 143:297–310.

Brown, F. H., and C. R. Fuller. 2008. Stratigraphy and tephra of the Kibish Formation, southwestern Ethiopia. *Journal of Human Evolution* 55:366–403.

Brown, F. H., and I. McDougall. 2011. Geochronology of the Turkana Depression of northern Kenya and southern Ethiopia. *Evolutionary Anthropology* 20:217–227.

Brown, F. H., B. R. Jicha, and R. E. Leakey. 2016. An age for Kajong, a Miocene fossil site east of Lake Turkana, Kenya. *Journal of African Earth Sciences* 114:74–77.

Brown, F. H., I. McDougall, and J. G. Fleagle. 2012. Correlation of the KHS tuff of the Kibish Formation to volcanic ash layers at other sites, and the age of early *Homo sapiens* (Omo I and Omo II). *Journal of Human Evolution* 63:577–585.

Brugal, J.-P., H. Roche, and M. Kibunjia. 2003. Faunes et paléoenvironements des principaux sites archéologiques plio-pléistocènes de la formation de Nachukui (Ouest-Turkana, Kenya). *Comptes Rendus Palevol* 2:675–684.

Brunet, M. 2001. Chadian australopithecines: Biochronology and environmental context; pp. 103–106 in P. V. Tobias, M. A. Raath, J. Moggi-Cecchi, and G. A. Doyle (Eds.), *Humanity from African Naissance to Coming Millenia*. Florence University Press, Florence.

Brunet, M., A. Beauvilain, D. Billiou, H. Bocherons, J.-R. Boisserie, L. de Bonis, P. Branger, A. Brunet, Y. Coppens, R. Daams, J. Dejax, C. Denys, P. Duringer, V. Eisenmann, F. Fanoné, P. Fronty, M. Gayet, D. Geraads, F. Guy, M. Kasser, G. Koufos, A. Likius, N. Lopez-Martinez, A. Louchart, L. MacLatchy, H. T. Mackaye, B. Marandat, G. Mouchelin, C. Mourer-Chauviré, O. Otero, S. Peigné, P. P. Campomanes, D. Pilbeam, J. C. Rage, D. de Ruitter, M. Schuster, J. Sudre, P. Tassy, P. Vignaud, L. Viriot, and A. Zazzo. 2000. Chad: Discovery of a vertebrate fauna close to the Miocene-Pliocene boundary. *Journal of Vertebrate Paleontology* 20:205–209.

Brunet, M., A. Beauvilain, D. Geraads, F. Guy, M. Kasser, H. T. Mackaye, L. M. MacLatchy, G. Mouchelin, J. Sudre, and P. Vignaud. 1998. Tchad: Découverte d'une faune de mammifères du Pliocène inferieur. *Comptes Rendus de l'Académie des Sciences, Paris, Sciences de la Terre et des Planètes* 326:153–158.

Brunet, M., F. Guy, D. Pilbeam, H. T. Mackaye, A. Likius, D. Ahounta, A. Beauvilain, C. Blondel, H. Bocherens, J.-R. Boisserie, L. De Bonis, Y. Coppens, J. Dejax, C. Denys, P. Duringer, V. Eisenmann, G. Fanone, P. Fronty, D. Geraads, T. Lehmann, F. Lihoreau, A. Louchart, A. Mahamat, G. Merceron, G. Mouchelin, O. Otero, P. P. Campomanes, M. P. De Leon, J.-C. Rage, M. Sapanet, M. Schuster, J. Sudre, P. Tassy, X. Valentin, P. Vignaud, L. Viriot, A. Zazzo, C. Zollikofer. 2002. A new hominid from the Upper Miocene of Chad, Central Africa. *Nature* 418:141–151.

Buffetaut, E. 1984. On the occurrence of *Crocodylus pigotti* in the Miocene of Saudi Arabia, with remarks on the origin of the Nile crocodile. *Neues Jahrbuch für Geologie und Paläontologie Monatshefte* 9:513–520.

Bulot, C., and L. Ginsburg. 1993. Gisements à mammifères miocènes du Haut-Armagnac et âge des plus anciens Proboscidiens d'Europe occidentale. *Comptes Rendus de l'Académie des Sciences, Paris* 316, série II:1011–1016.

Bunter, M. A. G., T. Debretsion, and L. Weldegiorgis. 1998. New developments in the pre-rift prospectivity of the Eritrean Red Sea. *Journal of Petroleum Geology* 21:373–400.

Burmeister, H. 1837. *Handbuch der Naturgeschichte zum Gebrauch bei Vorlesungen*. T. C. F. Enslin, Berlin. 858 pp.

Buss, I. O. 1961. Some observations on food habits and behaviour of the African elephant. *Journal of Wildlife Management* 25:131–148.

Butts, C., S. Cote, and J. Kingston. 2018. Reconstructing the paleoenvironments of Kalodirr and Moruorot, Kenya using stable carbon isotopes. *The University of Alberta Palaeontological Society's 7th Biennial Symposium, Vertebrate Anatomy Morphology Palaeontology Abstracts* 5:12–13.

Butzer, K. W., G. L. Isaac, J. L. Richardson, and C. Washbourn-Kamau. 1972. Radiocarbon dating of East African lake levels. *Science* 175:1069–1076.

Bye, B. A., F. H. Brown, T. E. Cerling, and I. McDougall. 1987. Increased age estimate for the Lower Paleolithic hominid site at Olorgesailie, Kenya. *Nature* 329:237–239.

Capelli, C., R. D. E. MacPhee, A. L. Roca, F. Brisighelli, N. Georgiadis, S. J. O'Brien, and A. D. Greenwood. 2006. A nuclear DNA phylogeny of the woolly mammoth (*Mammuthus primigenius*). *Molecular Phylogenetics and Evolution* 40:620–627.

Cappellini, E., A. Gentry, E. Palkopoulou, Y. Ishida, D. Cram, A.-M. Roos, M. Watson, U. F. Johansson, B. Fernholm, P. Agnelli, F. Barbagli, D. T. J. Littlewood, C. D. Kelstrup, J. V. Olsen, A. M. Lister, A. L. Roca, L. Dalén, and M. T. P. Gilbert. 2014. Resolution of the type material of the Asian elephant, *Elephas maximus* Linnaeus, 1758 (Proboscidea, Elephantidae). *Zoological Journal of the Linnean Society* 170:222–232.

Cappetta, H., J.-J. Jaeger, M. Sabatier, B. Sigé, J. Sudre, and M. Vianey-Liaud. 1978. Decouverte dans le Paléocène du Maroc des plus anciens mammifères eutheriens d'Afrique. *Geobios* 11:257–263.

Cappetta, H., J.-J. Jaeger, B. Sige, J. Sudre, and M. Vianey-Liaud. 1987. Compléments et Précisions Biostratigraphiques sur la Faune Paléocène a Mammifères et Sélaciens du Bassin d'Ouarzazate (Maroc). *Tertiary Research* 8:147–157.

Carter, A. M., and A. Mess. 2007. Evolution of the placenta in eutherian mammals. *Placenta* 28:259–262.

Cerling, T. E. 1992. Development of grasslands and savannas in East Africa during the Neogene. *Palaeogeography, Palaeoclimatology, Palaeoecology* 97:241–247.

Cerling, T. E., and J. M. Harris. 1999. Carbon isotope fractionation between diet and bioapatite in ungulate mammals and implications for ecological and paleoecological studies. *Oecologia* 120:347–363.

Cerling, T. E., S. A. Andanje, S. A. Blumenthal, F. H. Brown, K. L. Chritz, J. M. Harris, J. A. Hart, F. M. Kirera, P. Kaleme, L. N. Leakey, M. G. Leakey, N. E. Levin, F. K. Manthi, B. H. Passey, and K. T. Uno. 2015. Dietary changes of large herbivores in the Turkana Basin, Kenya from 4 to 1 Ma. *Proceedings of the National Academy of Sciences, USA*. www.pnas.org/cgi/doi/10.1073/pnas.1513075112

Cerling, T. E., J. M. Harris, S. H. Ambrose, M. G. Leakey, and N. Solounias. 1997a. Dietary and environmental reconstruction with stable isotope analyses of herbivore tooth enamel from the Miocene locality of Fort Ternan, Kenya. *Journal of Human Evolution* 33:635–650.

Cerling, T. E., J. M. Harris, and M. G. Leakey. 1999. Browsing and grazing in elephants: The isotope record of modern and fossil proboscideans. *Oecologia* 120:364–374.

—. 2003. Isotope paleoecology of the Nawata and Nachukui Formations at Lothagam, Turkana Basin, Kenya; pp. 605–624 in M. G. Leakey and J. M. Harris (eds.), *Lothagam: The Dawn of Humanity in Eastern Africa*. Columbia University Press, New York.

—. 2005. Environmentally driven dietary adaptations in African mammals; pp. 258–272 in J. R. Ehleringer, M. D. Dearing, and T. E. Cerling (Eds.), *A History of Atmospheric CO_2 and Its Effects on Plants, Animals, and Ecosystems. Ecological Studies*, Volume 177, part 2. Springer, New York, N.Y.

Cerling, T. E., J. M. Harris, B. J. MacFadden, M. G. Leakey, J. Quade, V. Eisenmann, and J. R. Ehleringer. 1997b. Global change through the Miocene/Pliocene boundary. *Nature* 389:153–158.

Cerling, T. E., N. E. Levin, and B. H. Passey. 2011a. Stable isotope ecology in the Omo-Turkana Basin. *Evolutionary Anthropology* 20:228–237.

Cerling, T. E., F. K. Manthi, E. N. Mbua, L. N. Leakey, M. G. Leakey, R. E. Leakey, F. H. Brown, F. E. Grine, J. A. Hart, P. Kaleme, H. Roche, K. T. Uno, and B. A. Wood. 2013. Stable isotope-based diet reconstructions of Turkana Basin hominins. *Proceedings of the National Academy of Sciences, USA* 110:10501–10506.

Cerling, T. E., P. Omondi, and A. N. Macharia. 2007. Diets of Kenyan elephants from stable isotopes and the origin of confiscated ivory in Kenya. *African Journal of Ecology* 45:614–623.

Cerling, T. E., B. H. Passey, L. K. Ayliffe, C. S. Cook, J. R. Ehleringer, J. M. Harris, M. B. Dhidha, and S. M. Kasiki. 2004. Orphans' tales: Seasonal dietary changes in elephants from Tsavo National Park, Kenya. *Palaeogeography, Palaeoclimatology, Palaeoecology* 206:367–276.

Cerling, T. E., Y. Wang, and J. Quade. 1993. Expansion of C_4 ecosystems as an indicator of global ecological change in the late Miocene. *Nature* 361:344–345.

Cerling, T. E., G. Wittemyer, J. R. Ehleringer, C. H. Remien, and I. Douglas-Hamilton. 2009. History of animals using isotope records (HAIR): A 6-year dietary history of one family of African elephants. *Proceedings of the National Academy of Sciences, USA*. https://doi.org/10.1073/pnas.0902192106

Cerling, T. E., J. G. Wynn, S. A. Andanje, M. I. Bird, D. K. Korir, N. E. Levin, W. Mace, A. N. Macharia, J. Quade, and C. H. Remien. 2011b. Woody cover and hominin environments in the past 6 million years. *Nature* 476:51–56.

Chaid-Saoudi, Y., D. Geraads, and J.-P. Raynal. 2006. The fauna and associated artefacts from the Lower Pleistocene site of Mansourah (Constantine, Algeria). *Comptes Rendus Palevol* 5:963–971.

Chaix, L., M. Faure, C. Guerin, and M. Honegger. 2000. Kaddanarti, a Lower Pleistocene assemblage from northern Sudan. *Recent Research into the Stone Age of Northeastern Africa*. Studies in African Archaeology 7, Poznan Archaeological Museum.

Chavaillon, J., A. Berthelet, J. L. Boisaubert, M. Faure, C. Guérin, and S. A. Warsama. 1990. Un *Elephas recki* decouvert en connexion anatomique dans le site de Haidalo, pres de As Eyla (republique de Djibouti). *Comptes rendus Académie des Sciences, Paris* Série II 310:441–446.

Chazan, M., H. Monchot, N. Porat, A. Lister, P. Davies, and L. K. Horwitz. 2001. Le site acheuléen de plein air d'Holon (Isräel): Premiers résultats. *Comptes rendus de l'Academie des Sciences, Paris. Sciences de la Terre et des planètes* 332:201–207.

Chorowicz, J. 2005. The East African rift system. *Journal of African Earth Sciences* 43:379–410.

Christiansen, P. 2004. Body size in proboscideans, with notes on elephant metabolism. *Zoological Journal of the Linnean Society* 140:523–549.

—. 2007. Long-bone geometry in columnar-limbed animals: Allometry of the proboscidean appendicular skeleton. *Zoological Journal of the Linnean Society* 149:423–436.

Cirilli, O., and S. Zouhri. 2018. Preliminary report on the taxonomic revision of fossil Equidae from Tizi'n Tadderht (Ouarzazate, Morocco). *Fossilia* 2018:11–12.

Clark, J. D., and C. V. Haynes, Jr. 1970. An elephant butchery site at Mwanganda's Village, Karonga, Malawi, and its relevance for Palaeolithic archaeology. *World Archaeology* 1:390–411.

Clark, J. D., Y. Beyene, G. WoldeGabriel, W. K. Hart, P. R. Renne, H. Gilbert, A. Defleur, G. Suwa, S. Katoh, K. R. Ludwig, J.-R. Boisserie, B. Asfaw, and T. D. White. 2003. Stratigraphic, chronological and behavioural contexts of Pleistocene *Homo sapiens* from Middle Awash, Ethiopia. *Nature* 423:727–732.

Clark, J. D., J. D. de Heinzelin, K. D. Schick, W. K. Hart, T. D. White, G. WoldeGabriel, R. C. Walter, G. Suwa, B. Asfaw, E. Vrba, and Y. Haile-Selassie. 1994. African *Homo erectus* old radiometric ages and young Oldowan assemblages in the Middle Awash valley, Ethiopia. *Science* 264:1907–1910.

Clementz, M. T., P. A. Holroyd, and P. L. Koch. 2008. Identifying aquatic habits of herbivorous mammals through stable isotope analysis. *Palaios* 23:574–585.

Clos, L. M. 1995. A new species of *Varanus* (Reptilia: Sauria) from the Miocene of Kenya. *Journal of Vertebrate Paleontology* 15:254–267.

Codron, J., J. A. Lee-Thorp, M. Sponheimer, D. Codron, R. C. Grant, and D. J. de Ruiter. 2006. Elephant (*Loxodonta africana*) diets in Kruger National Park, South Africa: Spatial and landscape differences. *Journal of Mammalogy* 87:27–34.

Coffing, K., C. Feibel, M. Leakey, and A. Walker. 1994. Four-million-year-old hominids from East Lake Turkana, Kenya. *American Journal of Physical Anthropology* 93:55–65.

Cohen, A. S., J. R. Stone, K. R. M. Beuning, L. E. Park, P. N. Reinthal, D. Dettman, C. A. Scholz, T. C. Johnson, J. W. King, M. R. Talbot, E. T. Brown, and S. J. Ivory. 2007. Ecological consequences of early late Pleistocene megadroughts in tropical Africa. *Proceedings of the National Academy of Sciences, USA* 104:16422–16427.

Cooke, H. B. S. 1939. Discussion on H. B. Maufe's paper, "new sections in the Kalahari beds at the Victoria Falls, Rhodesia." *Proceedings of the Geological Society of South Africa* 42:51–52.

—. 1947. Variation in the molars of the living African elephant and a critical revision of the fossil Proboscidea of southern Africa. *American Journal of Science* 245:434–457.

—. 1949. Fossil mammals of the Vaal River deposits. *Geological Survey (South Africa) Memoir* 35 (III):1–169.

—. 1960. Further revision of the fossil Elephantidae of southern Africa. *Palaeontologia Africana* 7:46–58.

—. 1978. Africa: The physical setting; pp. 17–45 in V. J. Maglio and H. B. S. Cooke (Eds.), *Evolution of African Mammals*. Harvard University Press, Cambridge.

—. 1993. Fossil proboscidean remains from Bolt's Farm and other Transvaal cave breccias. *Palaeontologia Africana* 30:25–34.

Cooke, H. B. S., and J. D. Clark. 1939. New fossil elephant remains from the Victoria Falls, northern Rhodesia, and a preliminary note on the geology and archaeology of the deposit. *Transactions of the Royal Society of South Africa* 27:287–319.

Cooke, H. B. S., and S. Coryndon. 1970. Fossil mammals from the Kaiso Formation and other related deposits in Uganda; pp. 107–224 in L. S. B. Leakey and R. G. J. Savage (Eds.), *Fossil Vertebrates of Africa, vol. 2*. Academic Press, Edinburgh.

Cooke, H. B. S., and V. J. Maglio. 1972. Plio-Pleistocene stratigraphy in East Africa in relation to proboscidean and suid evolution; pp. 303–329 in W. W. Bishop and J. A. Miller (Eds.), *Calibration of Hominoid Evolution: Recent Advances in Isotopic and Other Dating Methods Applicable to the Origin of Man*. Wenner-Gren Foundation for Anthropological Research, Scottish Academic Press, New York.

Coppens, Y. 1961. Découverte d'un Australopithéciné dans le Villafranchien du Tchad. *Comptes Rendus Hebdomadaires des Séances de l'Académie des Sciences* 252:3851–3852.

—. 1965. Les Proboscidiens du Tchad. *Actes du Ve Congrés Panafricain de Préhistoire et de l'Étude du Quaternaire (Santa Cruz de Tenerife)*, I 5:331–387.

—. 1967. Les Faunes de Vertebres Quaternaires du Tchad; pp. 89–97 in W. W. Bishop and J. D. Clark (Eds.), *Background to Evolution in Africa*. University of Chicago Press, Chicago.

—. 1972. Un nouveau Proboscidien du Pliocène du Tchad, *Stegodibelodon schneideri* nov. gen. nov. sp., et le phylum des Stegotetrabelodontinae. *Comptes Rendus de l'Academie des Sciences*, Paris 274:2962–2965.

Coppens, Y., and M. Beden. 1978. Moeritherioidea; pp. 333–335 in V. J. Maglio and H. B. S. Cooke (Eds.), *Evolution of African Mammals*. Harvard University Press, Cambridge.

Coppens, Y., and M. Gaudant. 1976. Découverte d'*Elephas iolensis* Pomel dans le Tyrrhénien de Tunisie. *Bulletin de la Société Géologique de France* 18:171–177.

Coppens, Y., and F. C. Howell. 1974. Les faunes de mammifères des formations plio-pléistocènes de l'Omo en Éthiopie (Proboscidea, Perissodactyla, Artiodactyla). *Comptes Rendus de l'Academie des Sciences, Paris* 278:2275–2278.

Coppens, Y., V. J. Maglio, C. T. Madden, and M. Beden. 1978. Proboscidea; pp. 336–367 in V. J. Maglio and H. B. S. Cooke (Eds.), *Evolution of African Mammals*. Harvard University Press, Cambridge.

Corvinus, G., and Q. B. Hendey. 1978. A new Miocene vertebrate locality at Arrisdrift in South West Africa (Namibia). *Neues Jahrbuch für Geologie und Paläontologie, Monatshefte* 1978(4):193–205.

Coster, P. M. C., K. C. Beard, M. J. Salem, Y. Chaimanee, M. Brunet, and J.-J. Jaeger. 2015. A new early Oligocene mammal fauna from the Sirt Basin, central Libya: Biostratigraphic and paleobiogeographic implications. *Journal of African Earth Sciences* 104:43–55.

Coster, P., M. Benammi, M. Mahboubi, R. Tabuce, M. Adaci, L. Marivaux, M. Bensalah, S. Mahboubi, A. Mahboubi, F. Mebrouk, C. Maameri, and J.-J. Jaeger. 2012. Chronology of the Eocene continental deposits of Africa: Magnetostratigraphy and biostratigraphy of the El Kohol and Glib Zegdou Formations, Algeria. *Geological Society of American Bulletin* 124:1590–1606.

Cote, S., J. Kingston, A. Deino, A. Winkler, R. Kityo, and L. MacLatchy. 2018. Evidence for rapid faunal change in the early Miocene of East Africa based on revised biostratigraphic and radiometric dating of Bukwa, Uganda. *Journal of Human Evolution* 116:95–107.

Court, N. 1992. Cochlea anatomy of *Numidotherium koholense*: Auditory acuity in the oldest known proboscidean. *Lethaia* 25:211–215.

—. 1993. A dental feeding peculiarity in *Numidotherium koholense*: Evidence of feeding behaviour in a primitive proboscidean. *Zeitschrift für Saugetierkunde* 58:194–196.

—. 1994a. Limb posture and gait in *Numidotherium koholense*, a primitive proboscidean from the Eocene of Algeria. *Zoological Journal of the Linnean Society* 111:297–338.

—. 1994b. The periotic of *Moeritherium* (Mammalia, Proboscidea): Homology or homoplasy in the ear region of Tethytheria McKenna, 1975? *Zoological Journal of the Linnean Society* 112:13–28.

—. 1995. A new species of *Numidotherium* (Mammalia: Proboscidea) from the Eocene of Libya and the early phylogeny of the Proboscidea. *Journal of Vertebrate Paleontology* 15:650–671.

Court, N., and M. Mahboubi. 1993. Reassessment of Lower Eocene *Seggeurius amourensis*: Aspects of primitive dental morphology in the mammalian order Hyracoidea. *Journal of Paleontology* 67:883–888.

Craig, G. C., and D. Peake. 2011. *The Elephants of Northern Botswana, Age Determinations, Age and Size of Hunting Trophies*. Conservation Trust Fund of the Ministry of Environment, Wildlife and Tourism, Gabarone, pp. 27.

Cuvier, F. 1825. Pp. 117–118 in É. Geoffroy-Saint-Hilaire and F. Cuvier (Eds.), *Histoire Naturelle des Mammifères, avec des figures originales, coloriées, dessinées d'après des Animaux vivans. Tome. 6*. A Belain, Paris.

Cuvier, G. 1798. *Tableau élémentaire de l'histoire naturelle des animaux*. Baudoin, Paris, 710 pp.

Dart, R. 1927. Mammoths and man in the Transvaal. *Supplement to Nature* 3032:41–48.

—. 1929. Mammoths and other fossil elephants of the Vaal and Limpopo watersheds. *South African Journal of Science* 26:698–731.

Davies, P., and A. M. Lister. 2007. *Palaeoloxodon*; pp. 123–131 in M. Chazan and L. K. Horwitz (Eds.), *Holon: A Lower Paleolithic Site in Israel*. American School of Prehistoric Research Bulletin, vol. 50, Peabody Museum of Archaeology and Ethnology, Harvard University, Cambridge.

Davies, T. W., M. A. Bell, A. Goswami, and T. J. D. Halliday. 2017. Completeness of the eutherian mammal fossil record and implications for reconstructing mammal evolution through the Cretaceous/Paleogene mass extinction. *Paleobiology* 43:521–536.

Death, C., and G. Coulson. 2016. A method for age estimation in the swamp wallaby (*Wallabia bicolor*). *Australian Mammalogy* 38:246–248.

De Bruijn, H. 1999. A late Miocene insectivore and rodent fauna from the Baynunah Formation, Emirate of Abu Dhabi, United Arab Emirates; pp. 186–197 in P. J. Whybrow and A. Hill (Eds.), *Fossil Vertebrates of Arabia. With Emphasis on the Late Miocene Faunas, Geology, and Palaeoenvironments of the Emirate of Abu Dhabi, United Arab Emirates*. Yale University Press, New Haven.

De Bruijn, H., and P. J. Whybrow. 1994. A late Miocene rodent fauna from the Baynunah Formation, Emirate of Abu Dhabi, United Arab Emirates. *Proceedings Koninklijke Nederlandse Akademie van Wetenschappen* 97:407–422.

Debruyne, R. 2004. Apports de la phylogénie moléculaire et de la morphométrie à la systématique des éléphants d'Afrique. *Journal de la Société de Biologie* 198:335–342.

—. 2005. A case study of apparent conflict between molecular phylogenies: The interrelationships of African elephants. *Cladistics* 21:31–50.

Debruyne, R., V. Barriel, and P. Tassy. 2003a. Mitochondrial cytochrome *b* of the Lyakhov mammoth (Proboscidea, Mammalia): New data and phylogenetic analyses of Elephantidae. *Molecular Phylogenetics and Evolution* 26:421–434.

Debruyne, R., A. Van Holt, V. Barriel, and P. Tassy. 2003b. Status of the so-called African pygmy elephant (*Loxodonta pumilio* (Noack 1906)): Phylogeny of cytochrome b and mitochondrial control region sequences. *Comptes Rendus Biologies* 126:687–697.

DeConto, R. M., and D. Pollard. 2003. Rapid Cenozoic glaciation of Antarctica induced by declining atmospheric CO2. *Nature* 421:245–249.

Deino, A. L. 2011. $^{40}Ar/^{39}Ar$ dating of Laetoli, Tanzania; pp. 77–98 in T. Harrison (Ed.), *Paleontology and Geology of Laetoli: Human Evolution in Context. Volume 1: Geology, Geochronology, Paleoecology and Paleoenvironment*. Springer, New York.

—. 2012. $^{40}Ar/^{39}Ar$ dating of Bed I, Olduvai Gorge, Tanzania, and the chronology of early Pleistocene climate change. *Journal of Human Evolution* 63:251–273.

Deino, A., and R. Potts. 1990. Single-crystal $^{40}AR/^{39}Ar$ dating of the Olorgesailie Formation, southern Kenya Rift. *Journal of Geophysical Research* 95:8453–8470.

Deino, A. L., C. Heil Jr., J. King, L. J. McHenry, I. G. Stanistreet, H. Stollhofen, J. K. Njau, J. Mwankunda, K. D. Schick, and N. Toth. 2021. Chronostratigraphy and age modeling of Pleistocene drill cores from the Olduvai Basin, Tanzania (Olduvai Gorge Coring Project). *Palaeogeography, Palaeoclimatology, Palaeoecology* 571:109990.

Deino, A. L., G. Scott, B. Saylor, M. Alene, J. D. Angelini, and Y. Haile-Selassie. 2010. $^{40}Ar/^{39}Ar$ dating, paleomagnetism, and tephrochemistry of Pliocene strata of the hominid-bearing Woranso-Mille area, west-central Afar Rift, Ethiopia. *Journal of Human Evolution* 58:111–126.

Deino, A. L., L. Tauxe, M. Monaghan, and A. Hill. 2002. $^{40}Ar/^{39}Ar$ geochronology and paleomagnetic stratigraphy of the Lukeino and lower Chemeron Formations at Tabarin and Kapcheberek, Tugen Hills, Kenya. *Journal of Human Evolution* 42:117–140.

Delagnes, A., A. Lenoble, S. Harmand, J.-P. Brugal, S. Prat, J.-J. Tiercelin, and H. Roche. 2006. Interpreting pachyderm single carcass sites in the African Lower and Early Middle Pleistocene record: A multidisciplinary approach to the site of Nadung'a 4 (Kenya). *Journal of Anthropological Archaeology* 25:448–465.

Delfino, M., F. Candilio, G. Carnevale, A. Coppa, T. Medin, M. Pavia, L. Rook, A. Urciuoli, and A. Villa. 2018. The early Pleistocene vertebrate fauna of Mulhuli-Amo (Buia area, Danakil Depression, Eritrea). *Bollettino della Società Palaeontologica Italiana* 57:27–44.

Delmer, C. 2005. *Les premières phases de différenciation des proboscidiens (Tethytheria, Mammalia): Le rôle du* Barytherium grave *de Libye*. Unpublished PhD dissertation, Muséum National d'Histoire Naturelle, Paris, 470 pp.

—. 2009. Reassessment of the generic attribution of *Numidotherium savagei* and the homologies of lower incisors in proboscideans. *Acta Palaeontologica Polonica* 54:561–580.

Delmer, C., M. Mahboubi, R. Tabuce, and P. Tassy. 2006. A new species of *Moeritherium* (Proboscidea, Mammalia) from the early late Eocene of Algeria: New perspectives on the ancestral morphotype of the genus. *Palaeontology* 49:421–434.

Denton, G. H. 1999. Cenozoic climate change; pp. 94–114 in T. G. Bromage and F. Schrenk (Eds.), *African Biogeography, Climate Change, and Human Evolution*. Oxford University Press, New York.

Depéret, C. 1897. Découverte du *Mastodon angustidens* dans l'étage Cartennien de Kabylie. *Bulletin de la Société Géologique de France* Series 3, 25:518–521.

Depéret, C., and L. Mayet. 1923. Monographies des Éléphants pliocènes de l'Europe et de l'Afrique du Nord; pp. 91–224 in C. Depéret, L. Mayet, and F. Roman, *Les Éléphants pliocènes, Ile part, Annales de l'Université de Lyon* (n.s.) I:43.

Depéret, C., L. Lavauden, and M. Solignag. 1925. Sur la déecouverte du *Mastodon arvernensis* dans le Pliocène de Ferryville (Tunisie). *Comptes Rendus Sommaire des Séeances, Sociéé Géologique de France* 1–2:518–521.

Deraniyagala, P. E. P. 1948. Some scientific results of two visits to Africa. *Spolia Zeylan* 25:1–42.

—. 1955. Some extinct elephants, their relatives and the two living species. *Ceylon National Museum Publication, Colombo* 1955:1–161.

De Ruiter, D. J. 2003. Revised faunal lists for Members 1–3 of Swartkrans, South Africa. *Annals of the Transvaal Museum* 40:29–41.

De Vries, D., S. Heritage, M. R. Borths, H. S. Sallam, and E. R. Seiffert. 2021. Widespread loss of mammalian lineage and dietary diversity in the early Oligocene of Afro-Arabia. *Communications Biology* 4:1172. https://doi.org/10.1038/s42003-021-02707-9

Dietrich, W. O. 1913. Zur stammesgeschichte des afrikanischen Elefanten. *Zeitschrift fur inductive Abstammungs- und Vererbungslehre, Berlin* 10:49–73.

—. 1915. *Elephas antiquus recki* n. f. aus dem Diluvium Deutsch-Ostafrikas, nebst Bemerkungen über die stammesgeschichtlichen Veränderungen des Extremitätenskeletts der Proboscidier. *Wissenschaftliche Ergebnisse Oldoway Expedition* Hans Reck edit., Leipzig, pp. 1–80.

—. 1921. Über den "horizontalen Zahnwechsel" bei *Mastodon* und *Elephas*. *Zentralblatt für Mineralogie, Geologie und Paläontologie, Stuttgart*, Hans Reck ed.:1–80.

—. 1941. Die Säugetierpaläontologischen Ergebnisse der Kohl-Larsen'shen Expedition 1937–1939 in nördlichen Deutsch-Ostafrika. *Zentralblatt für Mineralogie, Geologie und Paläontologie, Stuttgart* 1941(B):217–223.

—. 1942. Altestquartare Saugetiere aus der sudlichen Serengeti, Deutsch-Ostafrika. *Palaeontographica* 94(A):43–133.

—. 1943. Ueber innerafrikanische Mastodonten. *Zoological Deutsch Geologische Gesellschaft* 95:46–48.

—. 1951. Daten zu den fossilen Elefanten Afrikas und Ursprung der Gattung *Loxodonta*. *Neues Jahrbuch für Mineralogie, Geologie und Paläontologie, Stuttgart* 93:325–378.

Diez-Martín, F., P. Sánchez-Yustos, D. Uribelarrea, M. Domínguez-Rodrigo, C. Fraile-Márquez, R.-A. Obregón, I. Díaz-Muñoz, A. Mabulla, E. Baquedano, A. Pérez-González, H. T. Bunn. 2014. New archaeological and geological research at SHK main site (Bed II, Olduvai Gorge, Tanzania). *Quaternary International* 322–323:107–128.

Ditchfield, P., J. Hicks, T. Plummer, L. C. Bishop, and R. Potts. 1999. Current research on the late Pliocene and Pleistocene deposits north of Homa Mountain, southwestern Kenya. *Journal of Human Evolution* 36:123–150.

Doman, J. H. 2017. *Paleontology and Paleoecology of the Late Miocene Mpesida Beds and Lukeino Formation, Tugen Hills Succession, Baringo, Kenya*. Unpublished doctoral dissertation, Yale University, 205 pp.

Domning, D. P., and L.-A. C. Hayek. 1984. Horizontal tooth replacement in the Amazonian manatee. *Mammalia* 48:105–127.

Domning, D. P., C. E. Ray, and M. C. McKenna. 1986. Two new Oligocene desmostylians and a discussion of tethytherian systematics. *Smithsonian Contributions to Paleobiology* 59:1–56.

Domning, D. P., I. S. Zalmout, and P. D. Gingerich. 2010. Sirenia; pp. 147–160 in L. Werdelin and W. J. Sanders (Eds.), *Cenozoic Mammals of Africa*. University of California Press, Berkeley.

Douglas-Hamilton, I., S. Bhalla, G. Wittemyer, and F. Vollrath. 2006. Behavioural reactions of elephants towards a dying and deceased matriarch. *Applied Animal Behaviour Science* 100:87–102.

Doumergue, F. 1928. Decouverte de l'*Elephas planifron* Falconer a Rachgoun (Dept. d'Oran). *Bulletin de la Société de Géographie et d'Archéologie d'Oran* 1928:114–132.

Drake, R., and G. H. Curtis. 1987. K-Ar geochronology of the Laetoli fossil localities; pp. 48–61 in M. D. Leakey and J. M. Harris (Eds.), *Laetoli: A Pliocene Site in Northern Tanzania*. Clarendon Press, Oxford.

Drake, R. E., J. A. Van Couvering, M. Pickford, G. H. Curtis, and J. A. Harris. 1988. New chronology for the early Miocene faunas of Kisingiri, western Kenya. *Journal of the Geological Society, London* 145:479–491.

Drapeau, M. S. M., R. Bobe, J. G. Wynn, C. J. Campisano, L. Dumouchel, and D. Geraads. 2014. The Omo Mursi Formation: A window into the East African Pliocene. *Journal of Human Evolution* 75:64–79.

Drury, S. A., S. P. Kelley, S. M. Berhe, R. E. L. I. Collier, and M. Abraha. 1994. Structures related to Red Sea evolution in northern Eritrea. *Tectonics* 13:1371–1380.

Dublin, H. T., A. R. E. Sinclair, and J. McGlade. 1990. Elephants and fire as causes of multiple stable states in the Serengeti-Mara woodlands. *Journal of Animal Ecology* 59:1147–1164.

Dubrovo, I. A. 1974. Some new data on mastodonts from western Mongolia; pp. 64–73 in N. M. Kramarenko (Ed.), *Mesozoic and Cenozoic Faunas and Biostratigraphy of Mongolia*. Izdatelstro "Nauka," Moscow.

Ducrocq, S., J.-R. Boisserie, J.-J. Tiercelin, C. Delmer, G. Garcia, F. K. Manthi, M. G. Leakey, L. Marivaux, O. Otero, S. Peigné, P. Tassy, and F. Lihoreau. 2010. New Oligocene vertebrate localities from northern Kenya (Turkana Basin). *Journal of Vertebrate Paleontology* 30:293–299.

Dudley, J. P. 1996. Mammoths, gomphotheres, and the Great American Faunal Interchange; pp. 289–295 in J. Shoshani and P. Tassy (Eds.), *The Proboscidea: Evolution and Palaeoecology of Elephants and Their Relatives*. Oxford University Press, Oxford.

Dudley, J. P., G. C. Craig, D. St. C. Gibson, G. Haynes, and J. Klimowicz. 2001. Drought mortality of bush elephants in Hwange National Park, Zimbabwe. *African Journal of Ecology* 39:187–194.

Dupont, L. M., F. Rommerskirchen, G. Mollenhauer, and E. Schefuss. 2013. Miocene to Pliocene changes in South African hydrology and vegetation in relation to the expansion of C_4 plants. *Earth and Planetary Science Letters* 375:408–417.

Edwards, E. J., C. P. Osborne, C. A. E. Strömberg, S. A. Smith, W. J. Bond, P.-A. Christin, A. B. Cousins, M. R. Duvall, D. L. Fox, R. P. Freckleton, O. Ghannoum, J. Hartwell, Y. Huang, C. M. Janis, J. E. Keeley, E. A. Kellogg, A. K. Knapp, A. D. B. Leakey, D. M. Nelson, J. M. Saarela, R. F. Sage, O. E. Sala, N. Salamin, C. J. Still, and B. Tipple. 2010. The origins of C_4 grasslands: Integrating evolutionary and ecosystem science. *Science* 328:587–591.

Eggert, L. S., C.A. Rasner, and D. S. Woodruff. 2002. The evolution and phylogeography of the African elephant inferred from mitochondrial DNA sequence and nuclear microsatellite markers. *Proceedings of the Royal Society of London, B* 269:1993–2006.

Éhik, J. 1930. *Prodinotherium* [sic] *hungaricum* n.g., n.sp. *Geologica Hungarica Series Palaeontologica* 6:1–24.

Eisenmann, V., and P. J. Whybrow. 1999. Hipparions from the late Miocene Baynunah Formation, Emirate of Abu Dhabi, United Arab Emirates; pp. 234–253 in P. J. Whybrow and A. Hill (Eds.), *Fossil Vertebrates of Arabia. With Emphasis on the Late Miocene Faunas, Geology, and Palaeoenvironments of the Emirate of Abu Dhabi, United Arab Emirates*. Yale University Press, New Haven.

Eizirik, E., W. J. Murphy, and S. J. O'Brien. 2001. Molecular dating and biogeography of the early placental mammalian radiation. *The American Genetic Association* 92:212–219.

El-Araby, A., and A. N. El-Barkooky. 2000. Stratigraphic architecture and facies evolution of the Oligocene Gebel Qatrani, Formation, North Fayum, Egypt. *5th International Conference on the Geology of the Arab World*, Cairo University, pp. 1175–1208.

El-Khashab, B. 1979. A brief account on Egyptian Paleogene Proboscidea. *Annals of the Geological Survey of Egypt* 9:245–260.

El-Shahat, A., S. N. Ayyad, and M. A. Abdalla. 1997. Pliocene facies and fossil contents of Qaret El-Mulul Formation at Wadi El-Natrun Depression, Western Desert, Egypt. *Facies* 37:211–224.

El-Shawaihdi, M. H., P. S. Mozley, N. T. Boaz, F. Salloum, P. Pavlakis, A. Muftah, and M. Triantaphyllou. 2016. Stratigraphy of the Neogene Sahabi units in the Sirt Basin, northeast Libya. *Journal of African Earth Sciences* 118:87–106.

Emerling, C. A., F. Delsuc, and M. W. Nachman. 2018. Chitinase genes (*CHIAs*) provide genomic footprints of a post-Cretaceous dietary radiation in placental mammals. *Science Advances* 4:eaar6478.

Evans, E. M. N., J. A. H. Van Couvering, and P. Andrews. 1981. Palaeoecology of Miocene sites in western Kenya. *Journal of Human Evolution* 10:99–116.

Faith, J. T. 2014. Late Pleistocene and Holocene mammal extinctions on continental Africa. *Earth-Science Reviews* 128:105–121.

Falconer, H. 1857. On the species of Mastodon and Elephant occurring in the fossil state in Great Britain. Part I. Mastodon. *Quarterly Journal of the Geological Society* 13:307–360.

—. 1868. Notes on fossil remains found in the valley of the Indus below Attock, and at Jubbulpoor; pp. 414–417 in C. Murchison (Ed.), *Palaeontological Memoirs and Notes of the Late Hugh Falconer, A.M., M.D., Volume 1: Fauna Antiqua Sivalensis*. Robert Hardwicke, London.

Falconer, H., and P. T. Cautley. 1845–1849. *Fauna Antiqua Sivalensis, Being the Fossil Zoology of the Sewalik Hills in the North of India*. Smith, Elder and Co., London. Part I, 1846; Part II, 1847; Parts III-VIII, 1849.

Fara, E., A. Likius, H. T. Mackaye, P. Vignaud, and M. Brunet. 2005. Pliocene large-mammal assemblages from northern Chad: Sampling and ecological structure. *Naturwissenschaften* 92:537–541.

Faure, M., and C. Guérin. 1997. Gafalo, nouveau site a *Palaeoloxodon recki* et *Hippopotamus amphibius* du Pleistocene ancien du Gobaad (Republique de Djibouti). *Comptes rendus de l'Académie des Sciences, Paris* série IIa 324:1017–1021.

Feibel, C. S. 2003. Stratigraphy and depositional setting of the Pliocene Kanapoi Formation, Lower Kerio Valley, Kenya; pp. 9–20 in J. M. Harris and M. G. Leakey (Eds.), *Geology and Vertebrate Paleontology of the Early Pliocene Site of Kanapoi, Northern Kenya*. Contributions in Science, Natural History Museum of Los Angeles County, 498.

Feibel, C. S., and F. H. Brown. 1991. Age of the primate-bearing deposits on Maboko Island, Kenya. *Journal of Human Evolution* 21:221–225.

Feibel, C. S., F. H. Brown, and I. McDougall. 1989. Stratigraphic context of fossil hominids from the Omo Group Deposits: Northern Turkana Basin, Kenya and Ethiopia. *American Journal of Physical Anthropology* 78:595–622.

Fernando, P., M. E. Pfrender, S. E. Encalada, and R. Lande. 2000. Mitochondrial DNA variation, phylogeography and population structure of the Asian elephant. *Heredity* 84:362–372.

Ferretti, M. P. 2016. A species-level phylogeny of elephants (Elephantinae, Proboscidea) based on cranial characters. Abstract, *26º Giornate di Paleontologia 2016, Faenza*.

Ferretti, M. P., and R. Debruyne. 2011. Anatomy and phylogenetic value of the mandibular and coronoid canals and their associated foramina in proboscideans (Mammalia). *Zoological Journal of the Linnean Society* 161:391–413.

Ferretti, M. P., G. Ficcarelli, Y. Libsekal, T. M. Tecle, and L. Rook. 2003b. Fossil elephants from Buia (northern Afar Depression, Eritrea) with remarks on the systematics of *Elephas recki* (Proboscidea, Elephantidae). *Journal of Vertebrate Paleontology* 23:244–257.

Ferretti, M. P., L. Rook, and D. Torre. 2003a. *Stegotetrabelodon* (Proboscidea, Elephantidae) from the late Miocene of southern Italy. *Journal of Vertebrate Paleontology* 23:659–666.

Fisher, D. C. 1996. Extinction of proboscideans in North America; pp. 296–315 in J. Shoshani and P. Tassy (Eds.), *The Proboscidea: Evolution and Palaeoecology of Elephants and Their Relatives*. Oxford University Press, Oxford.

Fitch, F. J., and J. A. Miller. 1976. Conventional potassium-argon and argon-40/argon-39 dating of volcanic rocks from East Rudolf; pp. 123–147 in Y. Coppens, F. C. Howell, G. L. Isaac, and R. E. F. Leakey (Eds.), *Earliest Man and Environments in the Lake Rudolf Basin*. University of Chicago Press, Chicago.

Fleagle, J. G., Z. Assefa, F. H. Brown, and J. J. Shea. 2008. Paleoanthropology of the Kibish Formation, southern Ethiopia: Introduction. *Journal of Human Evolution* 55:360–365.

Fleischer, R. C., E. A. Perry, K. Muralidharan, E. S. Stevens, and C. M. Wemmer. 2001. Phylogeography of the Asian elephant (*Elephas maximus*) based on mitochondrial DNA. *Evolution* 55:1882–1892.

Flower, B. P., and J. P. Kennett. 1994. The middle Miocene climatic transition: East Antarctic ice sheet development, deep ocean circulation and global carbon cycling. *Palaeogeography, Palaeoclimatology, Palaeoecology* 108:537–555.

Foley, C., N. Pettorelli, and L. Foley. 2008. Severe drought and calf survival in elephants. *Biology Letters* 4:541–544.

Foote, M., J. P. Hunter, C. M. Janis, and J. J. Sepkoski. 1999. Evolutionary and preservational constraints on origins of biologic groups: Divergence times of eutherian mammals. *Science* 283:1310–1314.

Forster Cooper, C. 1922. Miocene Proboscidia from Baluchistan. *Proceedings of the Zoological Society of London* 1922:606–626.

Fournet, A. 1971. Les Gisements à Faune Villafranchienne de Tunisie. *Notes du Service Géologique Tunisie* 34:53–69.

Fourtau, R. 1918. *Contribution a l'Étude Vertébrés Miocènes de l'Égypte*. Survey Department, Ministry of Finance, Egypt, Government Press, Cairo, 121 pp.

—. 1920. *Contribution a l'Étude Vertébrés Miocènes de l'Égypte*. Survey Department, Ministry of Finance, Egypt, Government Press, Cairo, 121 pp. [2nd printing]

Fox, D. L., D. J. Peppe, K. P. McNulty, W. Harcourt-Smith, H. M. Dunsworth, A. L. Deino, T. Lehman, and A. L. Andrews. 2009. Geochronology, stratigraphy, and paleoenvironments of early Miocene faunas from Mfangano and Rusinga Islands, eastern Lake Victoria, Kenya. *Geological Society of America Abstracts with Programs* 41:362.

Franz-Odendall, T. A., Lee-Thorp, J. A., and Chinsamay, A. 2002. New evidence for the lack of C_4 grassland expansions during the early Pliocene at Langebaanweg, South Africa. *Paleobiology* 28:378–388.

Frick, C. 1933. New remains of trilophodont-tetrabelodont mastodons. *Bulletin of the American Museum of Natural History* 59:505–652.

Friscia, A. R., M. Macharwas, S. Muteti, F. Ndiritu, and D. T. Rasmussen. 2020. A transitional mammalian carnivore community from the Paleogene-Neogene boundary in northern Kenya. *Journal of Vertebrate Paleontology*. https://doi.org/10.1080/02724634.2020.1833895

Froehlich, D. J., and J. E. Kalb. 1995. Internal reconstruction of elephantid molars: Applications for functional anatomy and systematics. *Paleobiology* 21:379–392.

Fuchs, V. E. 1934. The geological work of the Cambridge expedition to the East African lakes. *Geological Magazine* 71:97–116, 145–166.

Fukuchi, A., B. Ratanasthien, S. Tanaka, S. Nagaoka, and S. Suzuki. 2007. Stratigraphy and sedimentary environment of late Middle-early Late Miocene Chiang Muan Formation, Phayao Province, Thailand. *Nature and Human Activities* 11:1–15.

Gaeth, A. P., R. V. Short, and M. B. Renfree. 1999. The developing renal, reproductive, and respiratory systems of the African elephant suggest an aquatic ancestry. *Proceedings of the National Academy of Sciences, USA* 96:5555–5558.

Gagnon, M. 1997. Ecological diversity and community ecology in the Fayum sequence (Egypt). *Journal of Human Evolution* 32:133–160.

Garevski, R., and G. N. Markov. 2011. A *Deinotherium gigantissimum* (Mammalia, Peoboscidea) palate with deciduous dentition from the area of Veles, Republic of Macedonia. *Paläontolische Zeitschrift* 85:33–36.

Garrido, G., and A. Arribas. 2014. The last Iberian gomphothere (Mammalia, Proboscidea): *Anancus arvernenesis mencalensis* nov. ssp. from the earliest Pleistocene of the Guadix Basin (Granada, Spain). *Palaeontologia Electronica* 17.1:4A.

Garutt, W. E. 1957. On a new fossil elephant *Phanagoroloxodon mammontoides* gen. et sp. nov. from the Caucasus. *Reports of the Academy of Sciences of the USSR* 112:333–335.

—. 1986. Origin of elephants Elephantidae and their phylogeny. *Proceedings of the Zoological Institute, Leningrad* 149:15–32.

—. 1992. Phanagorian elephant *Phanagoroloxodon mammontoides* Garutt, 1957, and the evolutionary pathways of the subfamily Elephantinae. *Proceedings of the Zoological Institute of the Russian Academy of Sciences* 246:29–40.

—. 1995. The phanagorian elephant *Phanagoroloxodon mammontoides* Garutt, 1957 from the Pliocene of the north-western Caucasus. *Cranium* 12:87–92.

Garutt, W. E., L. I. Alexejeva, and V. S. Baigusheva. 1977. On the oldest *Archidiskodon* elephant from Anthropogene of the USSR. *Journal of the Paleontological Society of India* 20:4–9.

Gasparik, M., and G. N. Markov. 2009. *Gomphotherium 'annectens* group' (Proboscidea) in Hungary. *Fragmenta Palaeontologica Hungarica* 27:73–79.

Gathogo, P. N., and F. H. Brown. 2006. Revised stratigraphy of Area 123, Koobi Fora, Kenya, and new age estimates of its fossil mammals, including hominins. *Journal of Human Evolution* 51:471–479.

Gaudry, A. 1891. Quelques remarques sur les Mastodontes à propos de l'animal du Cherichera. *Mémoires de la Société Géologique de France, Paléontologie* 8:1–6.

Gaziry, A. W. 1976. Jungtiertiäre Mastodonten aus Anatolien (Türkei). *Geologisches Jahrbuch* 22:3–143.

—. 1982. Proboscidea from the Sahabi Formation. *Garyounis Scientific Bulletin* 1982:101–108.

—. 1987a. New mammals from the Jabal Zaltan site, Libya. *Senckenbergiana Lethaea* 68:69–89.

—. 1987b. Remains of Proboscidea from the early Pliocene of Sahabi, Libya; pp. 183–203 in N. T. Boaz, A. El-Arnauti, A.W. Gaziry, J. de Heinzelin, and D. D. Boaz (Eds.), *Neogene Paleontology and Geology of Sahabi*. Liss, New York.

Gebo, D. L., L. MacLatchy, R. Kityo, A. Deino, J. Kingston, and D. Pilbeam. 1997. A hominid genus from the early Miocene of Uganda. *Science* 276:401–404.

Gentry, A. W. 1987a. Mastodons from the Miocene of Saudi Arabia. *Bulletin of the British Museum (Natural History) (Geology)* 41:409–432.

—. 1987b. Rhinoceroses from the Miocene of Saudi Arabia. *Bulletin of the British Museum (Natural History) (Geology)* 41:409–432.

—. 1987c. Ruminants from the Miocene of Saudi Arabia. *Bulletin of the British Museum (Natural History) (Geology)* 41:433–439.

—. 1999a. A fossil hippopotamus from the Emirate of Abu Dhabi, United Arab Emirates; pp. 271–289 in P. J. Whybrow and A. Hill (Eds.), *Fossil Vertebrates of Arabia. With Emphasis on the Late Miocene Faunas, Geology, and Palaeoenvironments of the Emirate of Abu Dhabi, United Arab Emirates*. Yale University Press, New Haven.

—. 1999b. Fossil pecorans from the Baynunah Formation, Emirate of Abu Dhabi, United Arab Emirates; pp. 290–316 in P. J. Whybrow and A. Hill (Eds.), *Fossil Vertebrates of Arabia. With Emphasis on the Late Miocene Faunas, Geology, and Palaeoenvironments of the Emirate of Abu Dhabi, United Arab Emirates*. Yale University Press, New Haven.

Gentry, A. W., and A. Gentry. 1969. Fossil camels in Kenya and Tanzania. *Nature* 222:898.

Geraads, D. 1980. Le faune des sites à *Homo erectus* des carrières Thomas (Casablanca, Maroc). *Quaternaria* 22:65–94.

Geraads, D. 1982. Paléobiogéographie de l'Afrique du Nord depuis de Miocène terminal, d'apres des grand mammiféres. *Geobios Mémoire Spécial* 6:473–481.

—. 1983. Paléobiogeographie de l'Afrique du Nord depuis de Miocène terminal, d'aprés des grand mammiféres. *Geobios Mémoire Spécial* 6:473–481.

—. 1989. Vertébrés fossils du Miocène supérieur du Djebel Krechem el Artsouma (Tunisie Centrale): Comparisons biostratigraphiques. *Geobios* 22:777–801.

—. 2002. Plio-Pleistocene mammalian biostratigraphy of Atlantic Morocco. *Quaternaire* 13:43–53.

—. 2010. Rhinocerotidae; pp. 669–683 in L. Werdelin and W. J. Sanders (Eds.), *Cenozoic Mammals of Africa*. University of California Press, Berkeley.

—. 2016. Pleistocene Carnivora (Mammalia) from Tighennif (Ternifine), Algeria. *Geobios* 49:445–458.

Geraads, D., and F. Amani. 1997. La faune du gisement a *Homo erectus* de l'Aïn Maarouf, près de El Hajeb (Maroc). *L'Anthropologie (Paris)* 101:522–530.

—. 1998. Bovidae (Mammalia) du Pliocene final d'Ahl al Oughlam, Casablanca, Maroc. *Paläontologisches Zeitschrift* 72:191–205.

Geraads, D., and F. Metz-Muller. 1999. Proboscidea (Mammalia) du Pliocène final d'Ahl al Oughlam (Casablanca, Maroc). *Neues Jahrbuch für Geologie und Paläontologie, Monatshefte* 1999(1):52–64.

Geraads, D., and E. Miller. 2013. *Brachypotherium minor* n. sp., and other Rhinocerotidae from the early Miocene of Buluk, northern Kenya. *Geodiversitas* 35:359–375.

Geraads, D., Z. Alemseged, and H. Bellon. 2002. The late Miocene mammalian fauna of Chorora, Awash Basin, Ethiopia: Systematics, biochronology and the ^{40}K-^{40}Ar ages of the associated volcanics. *Tertiary Research* 21:113–122.

Geraads, D., F. Amani, and J.-J. Hublin. 1992. The middle Pleistocene site of Ain Maarouf near El Hajeb, Morocco: Discovery of a fossil hominid. Comptes rendus de Académie des sciences. *Série 2, Mecanique, Physique, Chimie, Sciences de l'univers, Sciences de la Terre* 314:319–323.

Geraads, D., S. El Boughabi, and S. Zouhri. 2012. A new caprin bovid (Mammalia) from the late Miocene of Morocco. *Palaeontologia Africana* 47:19–24.

Geraads, D., J. Hublin, J.-J. Jaeger, H. Tong, S. Sen, and P. Toubeau. 1986. The Pleistocene hominid site of Ternifine, Algeria: New results on the environment, age and human industries. *Quaternary Research* 25:380–386.

Geraads, D., J.-P. Raynal, and V. Eisenmann. 2004. The earliest human occupation of North Africa: A reply to Sahnouni et al. (2002). *Journal of Human Evolution* 46:751–761.

Geraads, D., S. Zouhri, and G. N. Markov. 2019. The first *Tetralophodon* (Mammalia, Proboscidea) cranium from Africa. *Journal of Vertebrate Paleontology* 29:3, e1632321. https://doi.org/10.1080/0272634.2019.1632321

Gheerbrant, E. 1990. On the early biogeographic history of the African placentals. *Historical Biology* 4:107–116.

—. 1998. The oldest known proboscidean and the role of Africa in the radiation of modern orders of placentals. *Bulletin of the Geological Society of Denmark* 44:181–185.

—. 2009. Paleocene emergence of elephant relatives and the rapid radiation of African ungulates. *Proceedings of the National Academy of Sciences, USA* 106:10717–10721.

Gheerbrant, E., and J.-C. Rage. 2006. Paleobiogeography of Africa: How distinct from Gondwana and Laurasia? *Palaeogeography, Palaeoclimatology, Palaeoecology* 241:224–246.

Gheerbrant, E., and P. Tassy. 2009. L'origine et lévolution des éléphants. *Comtes Rendus Palevol* 8:281–294.

Gheerbrant, E., M. Amaghzaz, B. Bouya, F. Goussard, and C. Letenneur. 2014. *Ocepeia* (middle Paleocene of Morocco): The oldest skull of an afrotherian mammal. *Plos One* 9:1–30.

Gheerbrant, E., B. Bouya, and M. Amaghzaz. 2012. Dental and cranial anatomy of *Eritherium azzouzorum* from the Paleocene of Morocco, earliest known proboscidean mammal. *Palaeontographica* 297:151–183.

Gheerbrant, E., H. Cappetta, M. Feist, J.-J. Jaeger, J. Sudre, M. Vianey-Liaud, and B. Sigé. 1992. La succession des faunes de vertébrés d'âge paléocène supérieur et éocène inferieur dans le bassin d'Ouarzazate, Maroc. Contexte geologique, bilans biostratigraphique et paléogéographique. *Newsletter on Stratigraphy* 28:33–58.

Gheerbrant, E., H. Cappetta, F. de Lapparent de Broin, J.-C. Rage, and R. Tabuce. 2017. Les faunes de vertébrés marins et terrestres du Paléogène du Bassin d'Ouarzazate, Maroc. *Mémoires de la Société Géologique de France* 180:485–525.

Gheerbrant, E., A. Filippo, and A. Schmitt. 2016. Convergence of afrotherian and laurasiatherian ungulate-like mammals: First morphological evidence from the Paleocene of Morocco. *Plos One* 11:1–35.

Gheerbrant, E., A. Schmitt, and L. Kocsis. 2018. Early African fossils elucidate the origin of embrithopod mammals. *Current Biology* 28:1–7.

Gheerbrant, E., J. Sudre, and H. Cappetta. 1996. A Paleocene proboscidean from Morocco. *Nature* 383:68–70.

Gheerbrant, E., J. Sudre, H. Cappetta, and G. Bignot. 1998a. *Phosphatherium escuilliei* du Thanétien du Bassin des Ouled Abdoun (Maroc), plus ancien proboscidien (Mammalia) d'Afrique. *Geobios* 30:247–269.

Gheerbrant, E., J. Sudre, H. Cappetta, M. Iarochène, M. Amaghzaz, and B. Bouya. 2002. A new large mammal from the Ypresian of Morocco: Evidence of surprising diversity of early proboscideans. *Acta Palaeontologica Polonica* 47:493–506.

Gheerbrant, E., J. Sudre, H. Cappetta, C. Mourer-Chauviré, E. Bourdon, M. Iarochene, M. Amaghzaz, and B. Bouya. 2003. Les localities à mammifères des carrièrs de Grand Daoui, bassin des Ouled Abdoun, Maroc, Yprésien: premier état des lieux. *Bulletin de la Société Géologique de France* 174:279–293.

Gheerbrant, E., J. Sudre, S. Sen, C. Abrial, B. Marandat, B. Sigé, and M. Vianey-Liaud. 1998b. Nouvelles donnees sur les mammiferes du Thanetien et de l'Ypresien du Bassin d'Ouarzazate (Maroc) et leur context stratigraphique. *Palaeovertebrata* 27:155–202.

Gheerbrant, E., J. Sudre, P. Tassy, M. Amaghzaz, B. Bouya, and M. Iarochène. 2005. Nouvelles données sur *Phosphatherium escuilliei* (Mammalia, Proboscidea) de l'Éocène inférieur du Maroc, apports à la phylogénie des Proboscidea et des ongulés lophodontes. *Geodiversitas* 27:239–333.

Gilbert, W H. 2012a. Introduction; pp. 1–12 in W. H. Gilbert and B. Asfaw (Eds.), *Homo Erectus. Pleistocene Evidence from the Middle Awash, Ethiopia*. University of California Press, Berkeley.

—. 2012b. Ecological and biogeographic context of the Daka Member; pp. 397–411 in W. H. Gilbert and B. Asfaw (Eds.), *Homo Erectus. Pleistocene Evidence from the Middle Awash, Ethiopia*. University of California Press, Berkeley.

Gilbert, C. C., F. Bibi, A. Hill, and M. J. Beech. 2014. Early guenon from the late Miocene Baynunah Formation, Abu Dhabi, with implications for cercopithecoid biogeography and evolution. *Proceedings of the National Academy of Sciences, USA* 111:10119–10124.

Gingerich, P. D. 1993. Oligocene age of the Gebel Qatrani Formation, Fayum, Egypt. *Journal of Human Evolution* 24:207–218.

—. 2006. Environment and evolution through the Paleocene-Eocene thermal maximum. *Trends in Ecology and Evolution* 21:246–253.

—. 2010. Cetacea; pp. 873–899 in L. Werdelin and W. J. Sanders (Eds.), *Cenozoic Mammals of Africa*. University of California Press, Berkeley.

Gingerich, P. D., D. E. Russell, and N. A. Wells. 1990. Astragalus of *Anthracobune* (Mammalia, Proboscidea) from the early-middle Eocene of Kashmir. *Contributions from the Museum of Paleontology, The University of Michigan* 28:71–77.

Gobush, K. S., C. T. T. Edwards, F. Maisels, G. Wittemyer, D. Balfour, and R. D. Taylor. 2021. *Loxodonta cyclotis. The IUCN Red List of Threatened Species* 2021:e. T181007989A181019888.

Godinot, M. 2010. Paleogene prosimians; pp. 319–332 in L. Werdelin and W. J. Sanders (Eds.), *Cenozoic Mammals of Africa*. University of California Press, Berkeley.

Göhlich, U. B. 1998. Elephantoidea (Proboscidea, Mammalia) aus dem Mittel- und Obermiozän der Oberen Sußwassermolasse Suddeutschlands: Odontologie und Osteologie. *Münchner Geowissenschafliche Abhandlungen* (A) 36:1–245.

—. 1999. Order Proboscidea; pp. 157–168 in G. Rossner and K. Heissig (Eds.), *The Miocene Land Mammals of Europe*. Friedrich Pfeil Verlag, Munich.

—. 2010. The Proboscidea (Mammalia) from the Miocene of Sandelzhausen (southern Germany). *Paläontologische Zeitschrift* 84:163–204.

—. 2020. The proboscidean fauna (Mammalia) from the middle Miocene lignites of Gracanica near Bugojno (Bosnia-Herzegovina). *Palaeobiodiversity and Palaeoenvironments*. https://doi.org/10.1007/s12549-020-00436-1

Göhlich, U. B., and K. Huttunen. 2009. Proboscidea. *Annalen des Naturhistorischen Museums in Wien. Serie A für Mineralogie und Petrographie, Geologie und Paläontologie, Anthropologie und Prähistorie* 111:635–645.

Goldfuss, G. A. 1821. Osteologische Beiträge zur Kenntniss verschiedener Saügetiere der Vorwelt. *Nova Acta Academa Ceasar Leopoldino-Carolinae Germanicae Natura Curiosorum* 10:453–502 and 723–724.

Gomez, F., W. Beauchamp, and M. Barazangi. 2000. Role of the Atlas Mountains (northwest Africa) within the African-Eurasian plate-boundary zone. *Geology* 28:775–778.

Goren-Inbar, N., C. S. Feibel, K. L. Verosub, Y. Melamed, M. E. Kislev, E. Tchernov, and I. Saragusti. 2000. Pleistocene milestones on the out-of-Africa corridor at Gesher Benot Ya'aqov, Israel. *Science* 289:944–947.

Goren-Inbar, N., A. Lister, E. Werker, and M. Chech. 1994. A butchered elephant skull and associated artifacts from the Acheulian site of Gesher Benot Ya'aqov. *Paléorient* 20:99–112.

Gorodiski, A., and R. Lavocat. 1953. Première découverte de Mammifères dans le Tertiare (Lutétien) du Sénégal. *Comptes Rendus Sommaires de la Société Géologique de France* 15:314–316.

Gräf, I. E. 1957. Die prinzipien der Arbestimmung bei *Deinotherium*. *Palaeontographica Abt. A* 180:131–185.

Gray, J. E. 1821. On the natural arrangement of vertebrate animals. *London Medical Repository* 15:296–310.

—. 1843. *List of the Specimens of Mammalia in the Collection of the British Museum*. George Woodfall and Son, London. 216 pp.

Green, J. L. 2002. *Mammut americanum* (Kerr, 1792). *Fossil Species of Florida* 1:1–11.

Green, J. L., and R. C Hulbert, Jr. 2005. The deciduous premolars of *Mammut americanum* (Mammalia, Proboscidea). *Journal of Vertebrate Paleontology* 25:702–715.

Greenwood, P. H. 1987. Early Miocene fish from eastern Saudi Arabia. *Bulletin of the British Museum (Natural History) (Geology)* 41:451–453.

Groenewald, P. A., J. Sealy, D. Stynder, and K. M. Smith. 2020. Dietary resource partitioning among three coeval proboscidean taxa (*Anancus capensis*, *Mammuthus suplanifrons*, *Loxodonta cookei*) from the South African early Pliocene locality of Langebaanweg E Quarry. *Palaeogeography, Palaeoclimatology, Palaeoecology* 543:109606.

Grohe, C., M. Morlo, Y. Chaimanee, C. Blondel, P. Coster, X. Valentin, M. Salem, A. A. Bilal, J.-J. Jaeger, and M. Brunet. 2012. New Apterodontinae (Hyaenodontida) from the Eocene locality of Dur at-Talah (Libya): Systematic, paleoecological and phylogenetical implications. *PLoS ONE* 7:e49054.

Grossman, A., C. Liutkus-Pierce, B. Kyongo, and F. M'Kirera. 2014. New fauna from Loperot contributes to the understanding of early Miocene catarrhine communities. *International Journal of Primatology* 35:1253–1274.

Groves, C. P., and P. Grubb. 2000a. Are there pygmy elephants? *Elephant* 2:8–10.

Groves, C. P., and P. Grubb. 2000b. Do *Loxodonta cyclotis* and *L. africana* interbreed? *Elephant* 2:4–7.

Grubb, P., C. P. Groves, J. P. Dudley, and J. Shoshani. 2000. Living African elephants belong to two species: *Loxodonta africana* (Blumenbach, 1797) and *Loxodonta cyclotis* (Matschie, 1900). *Elephant* 2:1–4

Grundling, T., and A. Hill. 2000. Geological context of fossil Cercopithecoidea from eastern Africa; pp. 180–211 in P. F. Whitehead and C. J. Jolly (Eds.), *Old World Monkeys*. Cambridge University Press, Cambridge.

Guan, J. 1996. On the shovel-tusked elephantoids from China; pp. 124–135 in J. Shoshani and P. Tassy (Eds.), *The Proboscidea: Evolution and Palaeoecology of Elephants and Their Relatives*. Oxford University Press, Oxford.

Guerin, C., and M. Pickford. 2003. *Ougandatherium napakense* nov. gen. nov. sp., le plus ancien Rhinocerotidae Iranotheriinae d'Afrique. *Annales de Paléontologie* 89:1–35.

Gunnell, G. F. 2010. Chiroptera; pp. 581–598 in L. Werdelin and W. J. Sanders (Eds.), *Cenozoic Mammals of Africa*. University of California Press, Berkeley.

Gunnell, G. F., D. M. Boyer, A. R. Friscia, S. Heritage, F. K. Manthi, E. R. Miller, H. M. Sallam, N. B. Simmons, N. J. Stevens, and E. R. Seiffert. 2018. Fossil lemurs from Egypt and Kenya suggest an African origin for Madgascar's aye-aye. *Nature Communications* 9:3193. https://doi.org/10.1038/s41467-108-05648-w/www.nature.com/naturecommunications

Gunnell, G. F., P. D. Gingerich, and P. A. Holroyd. 2010. Ptolemaiidae; pp. 83–88 in L. Werdelin and W. J. Sanders (Eds.), *Cenozoic Mammals of Africa*. University of California Press, Berkeley.

Gunnell, G. F., E. L. Simons, and E. R. Seiffert. 2008. New bats (Mammalia: Chiroptera) from the late Eocene and early Oligocene, Fayum Depression, Egypt. *Journal of Vertebrate Paleontology* 28:1–11.

Günther, R. H., C. E. O'Connell-Rodwell, and S. L. Klemperer. 2004. Seismic waves from elephant vocalizations: A possible communication mode? *Geophysical Research Letters* 31:L11602. https://doi.org/10.1029/2004GL019671

Gussone, N., A. Eisenhauer, R. Tiedemann, G. H. Haug, A. Heusser, B. Bock, Th.F. Nagler, and A. Muller. 2004. Reconstruction of Caribbean sea surface temperature and salinity fluctuations in response to the Pliocene closure of the Central American Gateway and radiative forcing, using $\delta^{44/40}$Ca, δ^{18}O and Mg/Ca ratios. *Earth and Planetary Sciences Letters* 227:201–214.

Gutiérrez, M. 2011. *Taxonomic and Ecological Characterization of a Late Oligocene mammalian fauna from Kenya*. Unpublished PhD dissertation, Washington University, St. Louis.

Gutiérrez, M., and D. Rasmussen. 2007. Late Oligocene mammals from northern Kenya. *Journal of Vertebrate Paleontology* 27 (suppl. to no. 3):85A.

Hagelberg, E., M. G. Thomas, C. E. Cook, Jr., A. V. Sher, G. F. Barishnikov, and A. M. Lister. 1994. DNA from ancient mammoth bones. *Nature* 370:333–334.

Haile-Selassie, Y. 2000. A newly discovered early Pliocene hominid-bearing paleontological site in the Mulu Basin, Ethiopia. *American Journal of Physical Anthropology* supplement 20:170.

—. 2001. *Late Miocene mammalian fauna from the Middle Awash Valley, Ethiopia*. Unpublished PhD dissertation, University of California, Berkeley, 425 pp.

—. 2010. Phylogeny of early *Australopithecus*: New fossil evidence from the Woranso-Mille (central Afar, Ethiopia). *Philosophical Transactions of the Royal Society B* 365:3323–3331.

Haile-Selassie, Y., and G. WoldeGabriel. 2009. Introduction; pp. 1–25 in Y. Haile-Selassie and G. WoldeGabriel (Eds.), *Ardipithecus Kadabba. Late Miocene Evidence from the Middle Awash, Ethiopia*. University of California Press, Berkeley.

Haile-Selassie, Y., L. Gibert, S. M. Melillo, T. M. Ryan, M. Alene, A. Deino, N. E. Levin, G. Scott, and B. Z. Saylor. 2015. New species from Ethiopia further expands Middle Pliocene hominin diversity. *Nature* 521:483–488.

Haile-Selassie, Y., S. M. Melillo, A. Vazzana, S. Benazzi, and T. M. Ryan. 2019. A 3.8-million-year-old hominin cranium from Woranso-Mille, Ethiopia. *Nature* 573:214–219.

Haile-Selassie, Y., B. Saylor, A. Deino, M. Alene, and B. M. Latimer. 2010. New hominid fossils from Woranso-Mille (Central Afar, Ethiopia) and taxonomy of early *Australopithecus*. *American Journal of Physical Anthropology* 141:406–417.

Haile-Selassie, Y., G. WoldeGabriel, T. D. White, R. L. Bernor, D. Degusta, P. R. Renne, W. K. Hart, E. Vrba, S. Ambrose, and F. C. Howell. 2004. Mio-Pliocene mammals from the Middle Awash, Ethiopia. *Geobios* 37:536–552.

Halliday, T. J. D., P. Upchurch, and A. Goswami. 2015. Resolving the relationships of Paleocene placental mammals. *Biological Reviews*. https://doi.org/10.1111/brv.12242, 30 pp.

—. 2016. Eutherians experienced elevated evolutionary rates in the immediate aftermath of the Cretaceous-Palaeogene mass extinction. *Proceedings of the Royal Society B* 283:20153026.

Hamilton, W. R. 1973. A lower Miocene mammalian fauna from Siwa, Egypt. *Palaeontology* 16:275–281.

Hamilton, W. R., P. J. Whybrow, and H. A. McClure. 1978. Fauna of fossil mammals from the Miocene of Saudi Arabia. *Nature* 274:248–249.

Hanks, J. 1972. Aspects of dentition of the African elephant. *Arnoldia Rhodesia* 5:1–8.

Harris, J. M. 1969. *Prodeinotherium from Gebel Zelten*, Libya. Unpublished PhD Thesis, University of Bristol, Bristol, 303 pp.

—. 1973. *Prodeinotherium* from Gebel Zelten, Libya. *Bulletin of the British Museum (Natural History, Geology)* 23:285–348.

—. 1975. Evolution of feeding mechanisms in the Family Deinotheriidae. *Zoological Journal of the Linnean Society* 56:332–362.

—. 1976. Cranial and dental remains of *Deinotherium bozasi* (Mammalia: Proboscidea) from East Rudolf, Kenya. *Journal of Zoology, London* 178:57–75.

—. 1977a. Deinotheres from southern Africa. *South African Journal of Science* 73:281–282.

—. 1977b. Mammalian faunas from East African hominid-bearing localities; pp. 21–48 in T. H. Wilson (Ed.), *A Survey of the Prehistory of Eastern Africa*. Pan-African Congress for Prehistory and Quaternary Studies, Nairobi.

—. 1978a. Deinotherioidea and Barytherioidea; pp. 315–322 in V. J. Maglio and H. B. S. Cooke (Eds.), *Evolution of African Mammals*. Harvard University Press, Cambridge.

—. 1978b. Palaeontology; pp. 32–61 in M. G. Leakey and R. E. Leakey (Eds.), *The Fossil Hominids and an Introduction to their Context 1968–1974. Koobi Fora Research Project, Volume 1*. Clarendon Press, Oxford.

—. 1983a. Background to the study of the Koobi Fora fossil faunas; pp. 1–21 in J. M. Harris (Ed.), *Koobi Fora Research Project: Volume 2. The Fossil Ungulates: Proboscidea, Perissodactyla, and Suidae*. Clarendon Press, Oxford.

—. 1983b. Family Deinotheriidae; pp. 22–39 in J. M. Harris (Ed.), *Koobi Fora Research Project: Volume 2. The Fossil Ungulates: Proboscidea, Perissodactyla, and Suidae*. Clarendon Press, Oxford.

—. 1987a. Fossil Deinotheriidae from Laetoli; pp. 294–297 in M. D. Leakey and J. M. Harris (Eds.), *Laetoli: A Pliocene Site in Northern Tanzania*. Clarendon Press, Oxford.

—. 1987b. Summary; pp. 524–531 in M. D. Leakey and J. M. Harris (Eds.), *Laetoli: A Pliocene Site in Northern Tanzania*. Clarendon Press, Oxford.

—. 2003. Deinotheres from the Lothagam succession; pp. 359–362 in M. G. Leakey and J. M. Harris (Eds.), *Lothagam: The Dawn of Humanity in Eastern Africa*. Columbia University Press, New York.

Harris, J. M., and M. G. Leakey. 2003. Introduction; pp. 1–8 in J. M. Harris and M. G. Leakey (Eds.), *Geology and Vertebrate Paleontology of the Early Pliocene Site of Kanapoi, Northern Kenya*. Contributions in Science, Natural History Museum of Los Angeles County, 498.

Harris, J. M., and R. Watkins. 1974. New early Miocene vertebrate locality near Lake Rudolf, Kenya. *Nature* 252: 576–577.

Harris, J. M., F. H. Brown, and M. G. Leakey. 1988a. Stratigraphy and paleontology of Pliocene and Pleistocene localities west of Lake Turkana, Kenya. *Contributions in Science, Natural History Museum of Los Angeles County* 399:1–128.

Harris, J. M., F. H. Brown, M. G. Leakey, A. C. Walker, and R. E. Leakey. 1988b. Pliocene and Pleistocene hominid-bearing sites from West of Lake Turkana, Kenya. *Science* 239:27–33.

Harris, J. M., M. G. Leakey, and T. E. Cerling. 2003. Early Pliocene tetrapod remains from Kanapoi, Lake Turkana Basin, Kenya; pp. 39–113 in J. M. Harris and M. G. Leakey (Eds.), *Geology and Vertebrate Paleontology of the Early Pliocene Site of Kanapoi, Northern Kenya*. Contributions in Science, Natural History Museum of Los Angeles County, 498.

Harrison, C. J. O. 1980. Fossil birds from afrotropical Africa in the collection of the British Museum (Natural History). *Ostrich* 51:92–98.

Harrison, T. 2002. The first record of fossil hominins from the Ndolanya Beds, Laetoli, Tanzania. *American Journal of Physical Anthropology* 119:83.

—. 2011. Introduction: The Laetoli hominins and associated fauna; pp. 1–14 in T. Harrison (Ed.), *Paleontology and Geology of Laetoli: Human Evolution in Context. Volume 2: Fossil Hominins and the Associated Fauna*. Springer, New York.

Harrison, T., and P. Andrews. 2009. The anatomy and systematic position of the early Miocene proconsulid from Meswa Bridge, Kenya. *Journal of Human Evolution* 56:479–496.

Harrison, T., and E. Baker. 1997. Paleontology and biochronology of fossil localities in the Manonga Valley, Tanzania; pp. 361–393 in T. Harrison (Ed.), *Neogene Paleontology of the Manonga Valley, Tanzania*. Plenum Press, New York.

Hart, B. L., L. A. Hart, and N. Pinter-Wollman. 2008. Large brains and cognition: Where do elephants fit in? *Neuroscience and Biobehavioral Reviews* 32:86–98.

Hart, W. K., G. WoldeGabriel, Y. Haile-Selassie, and P. R. Renne. 2009. Volcanic record of the Adu-Asa Formation; pp. 63–91 in Y. Haile-Selassie and G. WoldeGabriel (Eds.), *Ardipithecus kadabba. Late Miocene Evidence from the Middle Awash, Ethiopia*. University of California Press, Berkeley.

Harzhauser, M., A. Kroh, O. Mandic, W. E. Piller, U. Göhlich, M. Reuter, and B. Berning. 2007. Biogeographic responses to geodynamics: A key study all around the Oligo-Miocene Tethyan Seaway. *Zoologischer Anzeiger—A Journal of Comparative Zoology* 246:241–256.

Haug, E. 1911. *Traité de Géologie. Les Périodes Géologiques*, Fascicule 3:1397–2024.

Haughton, S. H. 1929. A review of the fossil mammals of Central Africa. *American Journal of Science* 17:101–118.

—. 1932. On some South African Proboscidea. *Transactions of the Royal Society of South Africa* 21:1–18.

Hautier, L., H. T. Mackaye, F. Lihoreau. P. Tassy, P. Vignaud, and M. Brunet. 2009. New material of *Anancus kenyensis* (Proboscidea, Mammalia) from Toros-Menalla (late Miocene, Chad): Contribution to the systematics of African anancines. *Journal of African Earth Sciences* 53:171–176.

Hautier, L., R. Tabuce, M. J. Mourlam, K. E. Kassegne, Y. Z. Amoudji, M. Orliac, F. Quillévére, A.-L. Charruault, A. L. C. Johnson, G. Guinot. 2021. New middle Eocene proboscidean from Togo illuminates the early evolution of the elephant-like dental pattern. *Proceedings of the Royal Society B* 288:20211439.

Hawley, C. R., C. Beirne, A. Meier, and J. R. Poulson. 2017–2018. Conspecific investigation of a deceased forest elephant (*Loxodonta cyclotis*). *Pachyderm* 59:97–100.

Hawthorne, W. D., and M. P. E. Parren. 2000. How important are forest elephants to the survival of woody plant species in Upper Guinean forests? *Journal of Tropical Ecology* 16:133–150.

Hay, O. P. 1922. Further observations on some extinct elephants. *Proceedings of the Biological Society of Washington* 35:97–101.

Hay, R. L. 1976. *Geology of the Olduvai Gorge: A Study of Sedimentation in a Semiarid Basin*. University of California Press, Berkeley, 203 pp.

—. 1986. Role of tephra in the preservation of fossils in Cenozoic deposits of East Africa; pp. 339–344 in L. E. Frostick, R. W. Renaut, I. Reid, and J. J. Tiercelin (Eds.), *Sedimentation in the African Rifts*. Geological Society Special Publication No. 25, London.

—. 1987. Geology of the Laetoli Area; pp. 23–47 in M. D. Leakey and J. M. Harris (Eds.), *Laetoli: A Pliocene Site in Northern Tanzania*. Clarendon Press, Oxford.

Heinzelin, J. de. 1957. Les fouilles d'Ishango. I.P.N.C.B., *Exploration Parc Nationale Albert, Mission J. de Heinzelin de Braucourt (1950)*. Fasc. 2:128 pp.

—. 1983. *The Omo Group: Vol. I. Text. Archives of the International Omo Research Expedition*. Musée Royal de l'Afrique Centrale, Tervuren. *Annalen. Reeks in-8o. Geologische Wetenschappen* nr. 85:1–365.

Heinzelin, J. de, and A. El-Arnauti. 1982. Stratigraphy and geological history of the Sahabi and related formations. *Garyounis Scientific Bulletin* 1982:5–12.

Heinzelin, J. de, J. D. Clark, T. White, W. Hart, P. Renne, G. WoldeGabriel, Y. Beyene, and E. Vrba. 1999. Environment and behavior of 2.5-million-year-old Bouri hominids. *Science* 284:625–629.

Heller, E., and T. Roosevelt. 1914. *Life-histories of African Game Animals*. Charles Scribner's Sons, New York. 420 pp.

Hendey, Q. B. 1967. A specimen of *'Archidiskodon'* cf. *transvaalensis* from the south-western Cape Province. *South African Archaeological Bulletin* 22:53–56.

—. 1970. The age of the fossiliferous deposits at Langebaanweg, Cape Province. *Annals of the South African Museum* 56:119–131.

—. 1976. The Pliocene fossil occurrences in 'E' Quarry, Langebaanweg, South Africa. *Annals of the South African Museum* 69:215–247.

—. 1978. Preliminary report on the Miocene vertebrates from Arrisdrift, South West Africa. *Annals of the South African Museum* 76:1–41.

—. 1981. Palaeoecology of the late Tertiary fossil occurrences in 'E' Quarry, Langebaanweg, South Africa, and a reinterpretation of their geological context. *Annals of the South African Museum* 84:1–104.

Herbst, C. T., A. S. Stoeger, R. Frey, J. Lohscheller, I. R. Titze, M. Gumpenberger, and W. T. Fitch. 2012. How low can you go? Physical production mechanism of elephant infrasonic vocalizations. *Science* 337:595–599.

Hill, A. 1994. Late Miocene and early Pliocene hominoids from Africa; pp. 123–145 in R. S. Corruccini and R. L. Ciochon (Eds.), *Integrative Paths to the Past*. Prentice Hall, Englewood Cliffs, NJ.

—. 1999. The Baringo Basin, Kenya: From Bill Bishop to BPRP; pp. 85–97 in P. Andrews and P. Banham (Eds.), *Late Cenozoic Environments and Hominid Evolution: A Tribute to Bill Bishop*. Geological Society, London.

—. 2002. Paleoanthropological research in the Tugen Hills, Kenya. *Journal of Human Evolution* 42:1–10.

Hill, A., and T. Grundling. 1999. A monkey (Primates: Cercopithecidae) from the late Miocene of Abu Dhabi, United Arab Emirates; pp. 198–202 in P. J. Whybrow and A. Hill (Eds.), *Fossil Vertebrates of Arabia. With Emphasis on the Late Miocene Faunas, Geology, and Palaeoenvironments of the Emirate of Abu Dhabi, United Arab Emirates*. Yale University Press, New Haven.

Hill, A., G. Curtis, and R. Drake. 1986. Sedimentary stratigraphy of the Tugen Hills, Baringo, Kenya; pp. 285–295 in L. E. Frostick, R. W. Renaut, I. Reid, and J.-J. Tiercelin (Eds.), *Sedimentation in the African Rifts*. Blackwell and Geological Society of London Special Publication 25, Oxford.

Hill, A., R. Drake, L. Tauxe, M. Monaghan, J. C. Barry, A. K. Behrensmeyer, G. Curtis, B. Fine Jacobs, L. Jacobs, N. Johnson, and D. Pilbeam. 1985. Neogene palaeontology and geochronology of the Baringo Basin, Kenya. *Journal of Human Evolution* 14:759–773.

Holroyd, P. A., Lihoreau, F., Gunnell, G. F., and E. R. Miller. 2010. Anthracotheriidae; pp. 843–851 in L. Werdelin and W. J. Sanders (Eds.), *Cenozoic Mammals of Africa*. University of California Press, Berkeley.

Holroyd, P. A., E. L. Simons, T. M. Bown, P. D. Polly, and M. J. Kraus. 1996. New records of terrestrial mammals from the Upper Eocene Qasr el Sagha Formation, Fayum Depression, Egypt. *Palaeovertebrata* 25:175–192.

Hooijer, D. A. 1954. Pleistocene vertebrates from Celebes. XI. Milk molars and a tusked mandible of *Archidiskodon celebensis* Hooijer. *Zoologische Mededelingen, Leiden* 32:221–231.

—. 1963. Miocene Mammalia of Congo. *Koninklijk Museum voor Midden-Afrika, Tervuren, Belgie, Annalen, Series in 8°, Geologische Wetenschappen* 46:1–77.

—. 1971. A new rhinoceros from the late Miocene of Loperot, Turkana District, Kenya. *Bulletin of the Museum of Comparative Zoology* 142:339–392.

—. 1982. Premolars of *Elephas planifrons* Falconer & Cautley from the Pleistocene of Java. *Palaeontology Proceedings B* 85:265–272.

Hooker, P. J., and J. A. Miller. 1979. K-Ar dating of the Pleistocene fossil hominid site of Chesowanja, North Kenya. *Nature* 282:710–712.

Hopwood, A. T. 1926. Fossil Mammalia; pp. 13–36 in E. J. Wayland (Ed.), *The Geology and Palaeontology of the Kaiso Bone Beds*. Occasional Papers of the Geological Survey of Uganda, Uganda.

—. 1936. New and little-known fossil mammals from the Pleistocene of Kenya colony and Tanganyika Territory. *The Annals and Magazine of Natural History, London* 17:636–641.

—. 1939. The mammalian fossils; pp. 308–316 in T. P. O'Brien (Ed.), *The Prehistory of Uganda Protectorate*. Cambridge University Press, Cambridge.

Hopwood, A. T., and J. P. Hollyfield. 1954. *An Annotated Bibliography of the Fossil Mammals of Africa (1742–1950). Fossil Mammals of Africa No. 8*. British Museum (Natural History), London.

Hormann, K. 1963. Note on a mastodontid from Libya. *Zeitschrift für Säugetierkunde* 28:88–93.

Howell, F. C. 1968. Omo research expedition. *Nature* 219:567–572.

Howell, F. C., Fichter, L. S., and G. Eck. 1969. Vertebrate assemblages from the Usno Formation, white Sands and brown Sands localities, lower Omo Basin, Ethiopia. *Quaternaria* 11:65–88.

Hublin, J.-J. 1992. The middle Pleistocene human femur from Aïn Maarouf (El Hajeb, Morocco). *Comptes rendus de l'Académie des sciences. Série 2, Mecanique, Physique, Chimie, Sciences de l'univers, Sciences de la Terre* 314:975–980.

Hublin, J.-J. 2001. Northwestern African Middle Pleistocene hominids and their bearing on the emergence of *Homo sapiens*; pp. 99–121 in L. Barham and K. Robson-Brown (Eds.), *Human Roots. Africa and Asia in the Middle Pleistocene*. Western Academic and Specialist Press, Bristol.

Hujer, W., K. Kuiper, T. B. Viola, M. Wagreich, and P. Faupl. 2015. Lithostratigraphy of the late Miocene to early Pleistocene, hominid-bearing Galili Fm., southern Afar Depression, Ethiopia. *Austrian Journal of Earth Sciences* 108:105–127.

Hurford, A. J., and R. T. Watkins. 1987. Fission-track age of the tuffs of the Buluk Member, Bakate Formation, northern Kenya: A suitable fission-track age standard. *Chemical Geology (Isotope Geoscience Section)* 66:209–216.

Huttunen, K. J. 2000. *Deinotheriidae (Proboscidea, Mammalia) of the Miocene of Lower Austria, Burgenland and Franzenbad, Czech Republic: Systematics, Odontology and Osteology*. Unpublished Thesis, University of Vienna, 76 pp.

—. 2002. Systematics and taxonomy of European Deinotheriidae (Proboscidea, Mammalia). *Annalen des Naturhistorischen Museums in Wien Serie A* 103A:237–250.

Illiger, C. D. 1811. *Prodromus systematis mammalium et avium additis terminis zoographicis attriusque classis*. Sumptibus C. Salfield, Berolini, 300 pp.

Ishida, H., and S. Ishida. 1982. *Study of the Tertiary Hominoids and Their Paleoenvironments: Vol. 1. Report of the Field Survey in Kirimun, Kenya, 1980*. Osaka University Press, Osaka, 181 pp.

Jachmann, H. 1988. Estimating age in African elephants: A revision of Laws' molar evaluation technique. *African Journal of Ecology* 26:51–56.

Jachmann, H., P. S. M. Berry, and H. Imae. 1995. Tusklessness in African elephants: A future trend. *African Journal of Ecology* 33:230–235.

Jacobs, B. F. 2004. Palaeobotanical studies from tropical Africa: Relevance to the evolution of forest, woodland and savannah biomes. *Philosophical Transactions of the Royal Society of London, B* 359:1573–1583.

Jacobs, B. F., and C. H. S. Kabuye. 1987. A middle Miocene (12.2 my old) forest in the East African Rift Valley, Kenya. *Journal of Human Evolution* 16:147–155.

Jacobs, L. L., W. Anyonge, and J. C. Barry. 1987. A new tenrecid from Kenya. *Journal of Mammalogy* 68:10–16.

Jacobs, B. F., J. D. Kingston, and L. L. Jacobs. 1999. The origin of grass-dominated ecosystems. *Annals of the Missouri Botanical Garden* 86:590–643.

Jacobs, B. F., N. Tabor, M. Feseha, A. Pan, J. Kappelman, D. T. Rasmussen, W. J. Sanders, M. Wiemann, J. Crabaugh, and J. L. G. Massini. 2005. Oligocene terrestrial strata of northwestern Ethiopia: A preliminary report on paleoenvironments and paleontology. *Palaeontologia Electronica* 8:19.

Jaeger, J.-J., C. K. Beard, Y. Chaimanee, M. J. Salem, M. Benammi, O. Hlal, P. Coster, A. A. Bilal, P. Duringer, M. Schuster, X. Valentin, B. Marandat, L. Marivaux, G. Métais, O. Hammuda, and M. Brunet, M. 2010a. Late middle Eocene epoch of Libya yields earliest known radiation of African anthropoids. *Nature* 467:1095–1099.

Jaeger, J.-J., L. Marivaux, M. Salem, A. A. Bilal, M. Benammi, Y. Chaimanee, P. Duringer, B. Marandat, E. Metais, M. Schuster, X. Valentin, and M. Brunet. 2010b. New rodent assemblages from the Eocene Dur at-Talah escarpment (Sahara of central Libya): Systematic, biochronological, and palaeobiogeographical implications. *Zoological Journal of the Linnean Society* 160:195–213.

Jaeger, J.-J., M. J. Salem, A. A. Bilal, M. Benammi, Y. Chaimanee, P. Duringer, B. Marandat, L. Marivaux, E. Metais, M. Schuster, X. Valentin, L. Viriot, O. Hammuda, and M. Brunet. 2012. Dur at Talah vertebrate locality revisited: New data on its stratigraphy, age, sedimentology and mammalian fossils; pp. 241–258 in M. J. Salem and M. T. Busrewil (Eds.), *Geology of Southern Libya, Vol. 1*. Academic Press, London.

Jaeger, J.-J., M. J. Salem, A. A. Bilal, M. Benammi, L. Viriot, M. Schuster, Y. Chaimanee, L. Marivaux, B. Marandat, and M. Brunet. 2008. Dur at Talhah locality revisited, but with new fossil mammal discoveries! Abstract, *Sedimentary Basins of Libya Fourth Symposium: Geology of Southern Libya, Tripoli, Libya*, pp. 56.

Janis, C. M. 1993. Tertiary mammal evolution in the context of changing climates, vegetation, and tectonic events. *Annual Review of Ecology and Systematics* 24:467–500.

Johanson, D. C., T. D. White, and Y. Coppens. 1978. A new species of the genus *Australopithecus* (Primates: Hominidae) from the Pliocene of eastern Africa. *Kirtlandia* 28:12–14.

Joleaud, L. 1928. Eléphants et Dinothériums Pliocène de l'Ethiopie: Contribution à l'étude paléogéographique des Proboscidiens africains. *International Geological Congress* 14:1001–1007.

—. 1931. Etudes de géographie zoologique sur la Berbérie. Les Proboscidiens. I—L'Elephant d'Afrique. *Bulletin de la Société Zoologique de France* 56:683–699.

Joleaud, L., and J. Lombard. 1933a. Conditions de fossilization et de gisement des mammifères quaternaires d'Ounianga Kebir. *Bulletin de la Société Géologique de France* 3:239–243.

—. 1933b. Mammifères quaternaires d'Ounianga Kebir (Tibesti sud-oriental). *Comptes Rendus Hebdomadaires des Séances de l'Académie des Sciences* 196:497–499.

Kalb, J. E. 1995. Fossil elephantoids, Awash paleolake basins, and the Afar triple junction, Ethiopia. *Palaeogeography, Palaeoclimatology, Palaeoecology* 114:357–368.

Kalb, J. E., and D. J. Froehlich. 1995. Interrelationships of late Neogene elephantoids: New evidence from the Middle Awash Valley, Ethiopia. *Geobios* 28:727–736.

Kalb, J. E., and A. Mebrate. 1993. Fossil elephantoids from the hominid-bearing Awash Group, Middle Awash Valley, Afar Depression, Ethiopia. *Transactions of the American Philosophical Society* 83:1–114.

Kalb, J. E., D. J. Froehlich, and G. L. Bell. 1996a. Phylogeny of African and Eurasian Elephantoidea of the late Neogene; pp. 101–116 in J. Shoshani and P. Tassy (Eds.), *The Proboscidea: Evolution and Palaeoecology of Elephants and Their Relatives*. Oxford University Press, Oxford.

—. 1996b. Palaeobiogeography of late Neogene African and Eurasian Elephantoidea, pp. 117–123 in J. Shoshani and P. Tassy (Eds.), *The Proboscidea: Evolution and Palaeoecology of Elephants and Their Relatives*. Oxford University Press, Oxford.

Kalb, J. E., C. J. Jolly, A. Mebrate, S. Tebedge, C. Smart, E. B. Oswald, D. L. Cramer, P. F. Whitehead, C. B. Wood, G. C. Conroy, T. Adefris, L. Sperling, and B. Kana. 1982a. Fossil mammals and artefacts from the Middle Awash Valley, Ethiopia. *Nature* 298:25–29.

Kalb, J. E., C. J. Jolly, S. Tebedge, A. Mebrate, C. Smart, E. B. Oswald, P. F. Whitehead, C. B. Wood, T. Adefris, and V. Rawn-Shatzinger. 1982b. Vertebrate faunas from the Awash Group, Middle Awash Valley, Afar, Ethiopia. *Journal of Vertebrate Paleontology* 2:237–258.

Kappelman, J. 1991. The paleoenvironment of *Kenyapithecus* at Fort Ternan. *Journal of Human Evolution* 20:95–129.

—. 1992. The age of the Fayum primates as determined by paleomagnetic reversal stratigraphy. *Journal of Human Evolution* 22:495–503.

Kappelman, J., D. T. Rasmussen, W. J. Sanders, M. Feseha, T. Bown, P. Copeland, J. Crabaugh, J. Fleagle, M. Glantz, A. Gordon, B. Jacobs, M. Maga, K. Muldoon, A. Pan, L. Pyne, B. Richmond, T. Ryan, E. Seiffert, S. Sen, L. Todd, M. C. Wiemann, and A. Winkler. 2003. Oligocene mammals from Ethiopia and faunal exchange between Afro-Arabia and Eurasia. *Nature* 426:549–552.

Kappelman, J., E. L. Simons, and C. C. Swisher III. 1992. New age determinations for the Eocene-Oligocene boundary sediments in the Fayum Depression, northern Egypt. *The Journal of Geology* 100:647–667.

Kaufulu, Z. M. 1990. Sedimentary environments at the Mwanganda site, Malawi. *Geoarchaeology* 5:15–27.

Kaup, J. 1829. Neues Saugethiere, *Deinotherium*: *Deinotherium giganteum*. *Isis* 22:401–404.

Kender, S., V. L. Peck, R. W. Jones, and M. A. Kaminski. 2009. Middle Miocene oxygen minimum zone expansion offshore West Africa: Evidence for global cooling precursor events. *Geology* 37:699–702.

Kennett, J. P. 1995. A review of polar climatic evolution during the Neogene, based on the marine sediment record; pp. 49–64 in E. S. Vrba, G. H. Denton, T. C. Partridge, and L. H. Burckle (Eds.), *Paleoclimate and Evolution, with Emphasis on Human Origins*. Yale University Press, New Haven.

Kent, P. E. 1941. The recent history and Pleistocene deposits of the plateau north of Lake Eyasi, Tanganyika. *Geological Magazine* 78:173–184.

—. 1942. The Pleistocene beds of Kanam and Kanjera, Kavirondo, Kenya. *Geological Magazine* 79:72–77.

Kerley, G. I. H., M. Landman, L. Kruger, N. Owen-Smith, D. Balfour, W. F. de Boer, A. Gaylord, K. Lindsay, and R. Slotow. 2008. Effects of elephants on ecosystems and biodiversity; pp. 146–205 in R. J. Scholes and K. G. Mennell (Eds.), *Elephant Management. A Scientific Assessment for South Africa*. Wits University Press, Johannesburg.

Khalaf-Sakerfalke von Jaffa, N. A. B. A. T. 2010. *Stegotetrabelodon syrticus emiratus* Khalaf, 2010: A new fossil four-tusked elephant subspecies from the Emirate of Abu Dahbi, United Arab Emirates. *Gazelle: The Palestinian Biological Bulletin* 98:1–60.

Kidane, T., F. H. Brown, and C. Kidney. 2014. Magnetostratigraphy of the fossil-rich Shungura Formation, southwest Ethiopia. *Journal of African Earth Sciences* 97:207–223.

Kielan-Jaworowska, Z., R. L. Cifelli, and Z. X. Luo. 2004. *Mammals from the Age of Dinosaurs*. Columbia University Press, New York, 700 pp.

Kingdon, J., E. R. Seiffert, B. Hedges, and G. Rathbun. 2013. Supercohort Afrotheria; pp. 143–146 in J. Kingdon, D. Happold, M. Hoffmann, T. Butynski, M. Happold, and J. Kalina (Eds.), *Mammals of Africa. Volume I. Introductory Chapters and Afrotheria*. Bloomsbury, London.

Kingston, J. 1999. Isotopes and environments of the Baynunah Formation, Emirate of Abu Dhabi, United Arab Emirates; pp. 354–372 in P. J. Whybrow and A. Hill (Eds.), *Fossil Vertebrates of Arabia. With Emphasis on the Late Miocene Faunas, Geology, and Palaeoenvironments of the Emirate of Abu Dhabi, United Arab Emirates*. Yale University Press, New Haven.

—. 2011. Stable isotopic analyses of Laetoli fossil herbivores; pp. 293–328 in T. Harrison (Ed.), *Paleontology and Geology of Laetoli: Human Evolution in Context. Volume 1: Geology, Geochronology, Paleoecology and Palaeoenvironment*. Springer, New York.

Kingston, J., and T. Harrison. 2007. Isotopic dietary reconstructions of Pliocene herbivores at Laetoli: Implications for early hominin paleoecology. *Palaeogeography, Palaeoclimatology, Palaeoecology* 243:272–306.

Kingston, J., B. F. Jacobs, A. Hill, and A. Deino. 2002. Stratigraphy, age and environments of the late Miocene Mpesida Beds, Tugen Hills, Kenya. *Journal of Human Evolution* 42:95–116.

Kingston, J., L. MacLatchy, S. Cote, R. Kityo, and W. J. Sanders. 2011. Isotopic evidence of paleoenvironments and niche partitioning of early Miocene fossil fauna from Napak and Moroto, Uganda. *Journal of Vertebrate Paleontology, Program and Abstracts* 2011:136.

Kingston, J., B. D. Marino, and A. Hill. 1994. Isotopic evidence for Neogene hominid paleoenvironments in the Kenya Rift Valley. *Science* 264:955–959.

Klein, R. G. 1976. A preliminary report on the 'Middle Stone Age' open-air site of Duinefontein 2 (Melkbosstrand, South-Western Cape Province, South Africa). *The South African Archaeological Bulletin* 31:12–20.

—. 1984. The large mammals of southern Africa: Late Pliocene to Recent; pp. 107–146 in R. G. Klein (Ed.), *Southern African Prehistory and Paleoenvironments*. Balkema, Rotterdam.

—. 1985. The archaeological significance of animal bones from Acheulean sites in southern Africa. *African Archaeological Review* 6:3–25.

—. 1988. The archaeological significance of animal bones from Acheulean sites in southern Africa. *The African Archaeological Review* 6:3–25.

—. 2000. The earlier Stone Age of southern Africa. *South African Archaeological Bulletin* 55:107–122.

Klein, R. G., and K. Cruz-Uribe. 1991. The bovids from Elandsfontein, South Africa, and their implications for the age, palaeoenvironment, and origins of the site. *African Archaeological Review* 9:21–79.

Klein, R. G., G. Avery, K. Cruz-Uribe, and T. E. Steele. 2007. The mammalian fauna associated with an archaic hominin skullcap and later Acheulean artifacts at Elandsfontein, Western Cape Province, South Africa. *Journal of Human Evolution* 52:164–186.

Koch, P. L. 1991. The isotopic ecology of Pleistocene proboscideans. *Journal of Vertebrate Paleontology* 11 (Supplement to no. 3):40A.

Koch, P. L., J. Heisinger, C. Moss, R. W. Carlson, M. L. Fogel, and A. K. Behrensmeyer. 1995. Isotopic tracking of change in diet and habitat use in African elephants. *Science* 267:1340–1343.

Kocsis, L., E. Gheerbrant, M. Mouflih, H. Cappetta, J. Yans, and M. Amaghzaz. 2014. Comprehensive stable isotope investigation of marine biogenic apatite from the late Cretaceous-early Eocene phosphate series of Morocco. *Palaeogeography, Palaeoclimatology, Palaeoecology* 394:74–88.

Koenigswald, W. von. 2011. Diversity of hypsodont teeth in mammalian dentitions—Construction and classification. *Palaeontographica, Abt. A: Palaeozoology-Stratigraphy* 294:63–94.

—. 2014. Functional diversity in the masticatory patterns of Proboscidea. *Scientific Annals, School of Geology, Aristotle University of Thessaloniki, Greece* Special Volume 102:88–89.

—. 2016a. The diversity of mastication processes in Neogene and Quaternary proboscideans. *Palaeontographica* Abt. A 307:1–41.

—. 2016b. Specialized wear facets and late ontogeny in mammalian dentitions. *Historical Biology*. https://doi.org/10.1080/08912963.2016.1256399

Konidaris, G. E., and S. J. Roussiakis. 2019. The first record of *Anancus* (Mammalia, Proboscidea) in the late Miocene of Greece and reappraisal of the primitive anancines from Europe. *Journal of Vertebrate Paleontology*. https://doi.org/10.1080/02724634.2018.1534118

Konidaris, G. E., A. Athanassiou, V. Tourloukis, N. Thompson, D. Giusti, E. Panagopoulou, and K. Harvati. 2018. The skeleton of a straight-tusked elephant (*Palaeoloxodon antiquus*) and other large mammals from the Middle Pleistocene butchering locality Marathousa 1 (Megalopolis Basin, Greece): Preliminary results. *Quaternary International* 497:65–84.

Konidaris, G. E., G. D. Koufos, D. S. Kostopoulos, and G. Merceron. 2014a. Taxonomy, biostratigraphy and palaeoecology of the genus *Choerolophodon* (Proboscidea, Mammalia) during the Miocene of the peri-Mediterranean region. *Scientific Annals, School of Geology, Aristotle University of Thessaloniki, Greece* Special Volume 102:92–94.

—. 2014b. Taxonomy, biostratigraphy and palaeoecology of *Choerolophodon* (Proboscidea, Mammalia) in the Miocene of SE Europe-SW Asia: Implications for phylogeny and biogeography. *Journal of Systematic Palaeontology* 14:1–27.

Konidaris, G. E., S. J. Roussiakis, A. Athanassiou, and G. E. Theodorou. 2017. The huge-sized deinothere *Deinotherium proavum* (Proboscidea, Mammalia) from the late Miocene localities Pikermi and Halmyropotamos (Greece). *Quaternary International* 445:5–22.

Konidaris, G. E., S. J. Roussiakis, G. E. Theodorou, and G. D. Koufos. 2014c. The Eurasian occurrence of the shovel-tusker *Konobelodon* (Mammalia, Proboscidea) as illuminated by its presence in the late Miocene of Pikermi (Greece). *Journal of Vertebrate Paleontology* 34:1437–1453.

Kowalski, K., and B. Rzebik-Kowalska. 1991. *Mammals of Algeria*. Polish Academy of Sciences, Warsaw, 370 pp.

Kraatz, B., F. Bibi, and A. Hill. 2009. New rodents from the late Miocene of the United Arab Emirates. *Journal of Vertebrate Paleontology* 29:129A.

Krause, J., P. H. Dear, J. L. Pollack, M. Slatkin, H. Spriggs, I. Barnes, A. M. Lister, I. Ebersberger, S. Pääbo, and M. Hofreiter. 2006. Multiplex amplification of the mammoth mitochondrial genome and the evolution of Elephantidae. *Nature* 439:724–727.

Kullmer, O., L. Sandrock, T. B. Viola, W. Hujer, H. Said, and H. Seidler. 2008. Suids, elephantoids, paleochronology, and paleoecology of the Pliocene hominid site Galili, Somali Region, Ethiopia. *Palaios* 23:452–464.

Kunimatsu, Y., Y. Sawada, T. Sakai, M. Saneyoshi, H. Nakaya, A. Yamamoto, and M. Nakatsukasa. 2017. The latest occurrence of the nyanzapithecines from the early late Miocene Nakali Formation in Kenya, East Africa. *Anthropological Science*. https://doi.org/10.1537/ase.170126

Kusky, T. M., T. M. Ramadan, M. M. Hassan, and S. Gabr. 2011. Structural and tectonic evolution of El-Faiyum Depression, north western Desert, Egypt based on analysis of Landsat ETM+, and SRTM data. *Journal of Earth Science* 22:75–100.

Lambert, W. D. 1990. Rediagnosis of the genus *Amebelodon* (Mammalia, Proboscidea, Gomphotheriidae). *Journal of Paleontology* 64:1032–1040.

—. 1992. The feeding habits of the shovel-tusked gomphotheres: Evidence from tusk wear patterns. *Paleobiology* 18:132–147.

—. 1996. The biogeography of the gomphotheriid proboscideans of North America; pp. 143–148 in J. Shoshani and P. Tassy (Eds.), *The Proboscidea: Evolution and Palaeoecology of Elephants and Their Relatives*. Oxford University Press, Oxford.

—. 2007. Tetralophodont gomphothere material from Nebraska and its implications for the status of North American *Tetralophodon*. *Journal of Vertebrate Paleontology* 27:676–682.

—. 2016. The phylogenetic affinities of the shovel-tusked gomphothere *Konobelodon* (Mammalia, Proboscidea): Is its sister group *Amebelodon* or *Platybelodon*? *Journal of Vertebrate Paleontology* Program and Abstracts 2016:171–172.

Lambert, W. D., and J. Shoshani. 1998. Proboscidea; pp. 606–621 in C. M. Janis, K. M. Scott, and L. L. Jacobs (Eds.), *Evolution of Tertiary Mammals of North America: Volume 1. Terrestrial Carnivores, Ungulates, and Ungulatelike Mammals*. Cambridge University Press, Cambridge.

Larramendi, A. 2016. Shoulder height, body mass and shape of proboscideans. *Acta Palaeontographica Polonica* 61:537–574.

Larramendi, A., M. R. Palombo, and F. Marano. 2017. Reconstructing the life appearance of a Pleistocene giant: Size, shape, sexual dimorphism and ontogeny of *Palaeoloxodon antiquus* (Proboscidea: Elephantidae) from Neumark-Nord 1 (Germany). *Bollettino della Società Paleontologica Italiana* 56:299–317.

Larramendi, A., H. Zhang, M. R. Palombo, and M. P. Ferretti. 2020. The evolution of *Palaeoloxodon* skull structure: Disentangling phylogenetic, sexually dimorphic, ontogenetic, and allometric morphological signals. *Quaternary Science Reviews* 229:106090.

Laursen, L., and M. Bekoff. 1978. *Loxodonta africana*. *Mammalian Species* 92:1–8.

Lavocat, R. 1961. Le gisement de Vertébrés miocènes de Beni Mellal (Maroc): Étude systematique de la faune de Mammifères. *Notes et Mémoires du Service Géologique du Maroc, Rabat* 155:29–94.

Laws, R. M. 1966. Age criteria for the African elephant, *Loxodonta a. africana*. *East African Wildlife Journal* 4:1–37.

—. 1970. Elephants as agents of habitat and landscape change in East Africa. *Oikos* 21:1–15.

Laws, R. M., I. S. C. Parker, and R. C. Johnstone. 1975. *Elephants and Their Habitats: The Ecology of Elephants in North Bunyoro, Uganda*. Clarendon Press, Oxford. 376 pp.

Leakey, L. S. B. 1965. *Olduvai Gorge 1951–1961: A Preliminary Report on the Geology and Fauna. Volume 1.* Cambridge University Press, Cambridge, 118 pp.

—. 1967. Notes on the mammalian faunas from the Miocene and Pleistocene of East Africa; pp. 7–28 in W. W. Bishop and J. D. Clark (Eds.), *Background to Evolution in Africa*. University of Chicago Press, Chicago.

Leakey, L. S. B., A. T. Hopwood, and H. Reck. 1931. New yields from the Oldoway Bone Beds, Tanganyika Territory. *Nature* 128:1075

Leakey, M. D. 1987. Introduction; pp. 490–523 in M. D. Leakey and J. M. Harris (Eds.), *Laetoli: A Pliocene Site in Northern Tanzania*. Clarendon Press, Oxford.

Leakey, M. G. 1985. Early Miocene cercopithecids from Buluk, northern Kenya. *Folia Primatologia* 44:1–14.

Leakey, M. G., and J. M. Harris. 2003. Lothagam: Its significance and contributions; pp. 625–655 in M. G. Leakey and J. M. Harris (Eds.), *Lothagam: The Dawn of Humanity in Eastern Africa*. Columbia University Press, New York.

Leakey, M. G., C. S. Feibel, R. L. Bernor, J. M. Harris, T. E. Cerling, K. M. Stewart, G. W. Storrs, A. Walker, L. Werdelin, and A. J. Winkler. 1996. Lothagam: A record of faunal change in the Late Miocene of East Africa. *Journal of Vertebrate Paleontology* 16:556–570.

Leakey, M. G., C. S. Feibel, I. McDougall, and A. Walker. 1995. New four-million-year-old hominid species from Kanapoi and Allia Bay, Kenya. *Nature* 376:565–571.

Leakey, M. G., A. Grossman, M. Gutíerrez, and J.G. Fleagle. 2011. Faunal change in the Turkana Basin during the late Oligocene and Miocene. *Evolutionary Anthropology* 20:238–253.

Leakey, M. G., F. Spoor, F. H. Brown, P. N. Gathogo, C. Kiarie, L. N. Leakey, and I. McDougall. 2001. New hominin genus from eastern Africa shows diverse middle Pliocene lineages. *Nature* 410:433–440.

Leakey, R. E., and M. G. Leakey. 1986. A new Miocene hominoid from Kenya. *Nature* 324:143–146.

Leakey, R. E. F., and A. Walker. 1985. New higher primates from the early Miocene of Buluk, Kenya. *Nature* 318:173–175.

LeBlanc, J. 2000. A guide to macrofossil localities of Libya: With notes on mineral sites. http://sites.google.come/site/leblancjacques/fossilhome

Lebtard, A.-E., D. L. Bourlés, P. Duringer, M. Jolivet, R. Braucher, J. Carcaillet, M. Schuster, N. Arnaud, P. Monié, F. Lihoreau, A. Likius, H. T. Mackaye, P. Vignaud, and M. Brunet. 2008. Cosmogenic nuclide dating of *Sahelanthropus tchadensis* and *Australopithecus bahrelghazali*: Mio-Pliocene hominids from Chad. *Proceedings of the National Academy of Sciences, USA* 105:3226–3231.

Lee, P. C., and C. J. Moss. 1995. Statural growth in known-age African elephants (*Loxodonta africana*). *Journal of Zoology, London* 236:29–41.

Lee, P. C., S. Sayialel, W. K. Lindsay, and C. J. Moss. 2011. African elephant age determination from teeth: Validation from known individuals. *African Journal of Ecology* 50:9–20.

Le Fur, S., E. Fara, H. T. Mackaye, P. Vignaud, and M. Brunet. 2009. The mammal assemblage of the hominid site TM266 (late Miocene, Chad Basin): Ecological structure and paleoenvironmental implications. *Naturwissenschaften* 96:565–574.

—. 2014. Toros-Menalla (Chad, 7 Ma), the earliest hominin-bearing area: How many mammal paleocommunities? *Journal of Human Evolution* 69:79–90.

Legler, B., H. D. Johnson, G. J. Hampson, B. Y. G. Massart, C. A.-L. Jackson, M. D. Jackson, A. El-Barkooky, and R. Ravnas. 2013. Facies model of a fine-grained, tide-dominated delta: Lower Dir Abu Lifa Mb. (Eocene), Western Desert, Egypt. *Sedimentology* 60:1313–1356.

Lehmann, S. B., D. R. Braun, K. J. Dennis, D. B. Patterson, D. D. Stynder, L. C. Bishop, F. Forrest, and N. E. Levin. 2016. Stable isotopic composition of fossil mammal teeth and environmental change in southwestern South Africa during the Pliocene and Pleistocene. *Palaeogeography, Palaeoclimatology, Palaeoecology* 457:396–408.

Lehmann, U. 1950. Über Mastodontenreste in der Bayerischen Staatssammlung in München. *Palaeontographica, Abt. A* 99:121–232.

Lepersonne, J. 1959. Données géologiques et stratigraphiques et comparaison entre le lac Albert et le lac Edouard; in W. Adam, Mollusques Pléistocènes de la Region du Lac Albert et de la Semliki. *Annales, Musee Royal du Congo Belge, series in 8°, Sciences Géologiques* 25:78–119.

Levin, N. E. 2015. Environment and climate of early human evolution. *Annual Review of Earth and Planetary Sciences* 43:405–429.

Levin, N. E., F. H. Brown, A. K. Behrensmeyer, R. Bobe, and T. E. Cerling. 2011. Paleosol carbonates from the Omo Group: Isotopic records of local and regional environmental change in East Africa. *Palaeogeography, Palaeoclimatology, Palaeoecology* 307:75–89.

Levin, N. E., S. W. Simpson, J. Quade, T. E. Cerling, and S. R. Frost. 2008. Herbivore enamel carbon isotopic composition and the environmental context of *Ardipithecus* at Gona, Ethiopia; pp. 215–234 in J. Quade and J. G. Wynn (Eds.), *The Geology of Early Humans in the Horn of Africa*. Geological Society of America Special Paper 446, Boulder.

Lewis, M. E., and M. Morlo. 2010. Creodonta; pp. 543–560 in L. Werdelin and W. J. Sanders (Eds.), *Cenozoic Mammals of Africa*. University of California Press, Berkeley.

Lihoreau, F., J.-R. Boisserie, L. Viriot, Y. Coppens, A. Likius, H. T. Mackaye, P. Tafforeau, P. Vignaud, and M. Brunet. 2006. Anthracothere dental anatomy reveals a late Miocene Chado-Libyan bioprovince. *Proceedings of the National Academy of Sciences, USA* 23:8763–8767.

Lihoreau, F., L. Hautier, and M. Mahboubi. 2015. The new Algerian locality of Bir el Ater 3: Validity of *Libycosaurus algeriensis* (Mammalia, Hippopotamoidea) and the age of the Nementcha Formation. *Palaeovertebrata* 39: 1–11.

Lihoreau, F., R. Sarr, D. Chardon, J.-R. Boisserie, R. Lebrun, S. Adnet, J. E. Martin, L. Pallas, B. Sambou, R. Tabuce, M. M. Thiam, and L. Hautier. 2021. A fossil terrestrial fauna from Tobène (Senegal) provides a unique early Pliocene window in western Africa. *Gondwana Research* 99:21–35.

Linder, H. P. 2017. East African Cenozoic vegetation history. *Evolutionary Anthropology* 26:300–312.

Lindsay, E. H., N. D. Opdyke, and N. M. Johnson. 1980. Pliocene dispersal of the horse *Equus* and late Cenozoic mammal dispersal events. *Nature* 287:135–138.

Lister, A. M. 1996. Evolution and taxonomy of Eurasian mammoths; pp. 203–213, 295 in J. Shoshani and P. Tassy (Eds.), *The Proboscidea: Evolution and Palaeoecology of Elephants and Their Relatives*. Oxford University Press, Oxford.

—. 2001. "Gradual" evolution and molar scaling in the evolution of the mammoth; pp. 648–651 in G. Cavarretta, P. Gioia, M. Mussi, and M. R. Palombo (Eds.), *Proceedings of the First International Congress of La Terra degli Elefanti: The World of Elephants*. Consiglio Nazionale delle Richerche, Rome.

—. 2004. Ecological interactions of elephantids in Pleistocene Eurasia: *Palaeoloxodon* and *Mammuthus*; pp. 53–60 in N. Goren-Inbar and J. D. Speth (Eds.), *Human Paleoecology in the Levantine Corridor*. Oxbow Books, Oxford.

—. 2013. The role of behaviour in adaptive morphological evolution of African proboscideans. *Nature* 500:331–334.

—. 2017. On the type material and evolution of North American mammoths. *Quaternary International*. https://doi.org/10.1016/j.quaint.2017.02.027

Lister, A. M., and P. Bahn. 2007. *Mammoths: Giants of the Ice Age*. 3rd Edition. University of California Press, 168 pp.

Lister, A. M., and A. V. Sher. 2001. The origin and evolution of the woolly mammoth. *Science* 294:1094–1097.

—. 2015. Evolution and dispersal of mammoths across the Northern Hemisphere. *Science* 350:805–809.

Lister, A. M., and H. van Essen. 2003. *Mammuthus rumanus* (Stefanescu), the earliest mammoth in Europe; pp. 47–52 in A. Petculecu and E. Stiuca (Eds.), *Advances in Vertebrate Paleontology "Hen to Panta."* Romanian Academy "Emile Recovitza" Institute of Speleology, Bucharest.

Lister, A. M., A. V. Sher, H. van Essen, and G. Wei. 2005. The pattern and process of mammoth evolution in Eurasia. *Quaternary International* 126–128:49–64.

Liu, A. G. S. C., E. R. Seiffert, and E. L. Simons. 2008. Stable isotope evidence for an amphibious phase in early proboscidean evolution. *Proceedings of the National Academy of Sciences, USA* 105:5786–5791.

Liu, L., J. Zhang, F. E. Rheindt, F. Lei, Y. Qu, Y. Wang, Y. Zhang, C. Sullivan, W. Nie, J. Wang, F. Yang, J. Chen, S. V. Edwards, J. Meng, and S. Wu. 2017. Genomic evidence reveals a radiation of placental mammals uninterrupted by the KPg boundary. *Proceedings of the National Academy of Sciences, USA* 114:E7282–E7290.

Liutkus-Pierce, C. M., K. Takashita-Bynum, L. A. Beane, C. T. Edwards, O. E. Burns, S. Mana et al. 2017. Data from: Ar-Ar results for Loperot, Kenya. *Mendeley Data* https://doi.org/10.17632/ff7yz7jzsy.1

Liutkus-Pierce, C. M., K. K. Takashita-Bynum, L. A. Beane, C. T. Edwards, O. E. Burns, S. Mana, S. Hemming, A. Grossman, J. D. Wright, and F. M. Kirera. 2019. Reconstruction of the early Miocene critical zone at Loperot, southwestern Turkana, Kenya. *Frontiers in Ecology and Evolution* 7:44. https://doi.org/10.3389/fevo.2019.00044

Lohrke, N. 2017. *African Deinothere Isotopic Paleoecology: Developing a Diet and Habitat Proxy for an Extinct Proboscidean*. Unpublished Honors Thesis, University of Michigan Department of Anthropology, 28 pp.

Louchart, A., P. Vignaud, Y. Haile-Selassie, A. Likius, and M. Brunet. 2008. Fossil birds from the late Miocene of Chad and Ethiopia and zoogeographical implications. *Oryctos* 7:147–167.

Loveridge, A. J., J. E. Hunt, F. Murindagomo, and D. W. Macdonald. 2006. Influence of drought on predation of elephant calves by lions in an African wooded savannah. *Journal of Zoology* 270:523–530.

Ludwig, K. R., and P. R. Renne. 2000. Geochronology on the paleoanthropological time scale. *Evolutionary Anthropology* 9:101–110.

Lukens, W. E., T. Lehmann, D. J. Peppe, D. L. Fox, S. G. Driese, and K. P. McNulty. 2017. The early Miocene critical zone at Karungu, western Kenya: An equatorial, open habitat with few primate remains. *Frontiers in Earth Science* 5:87. https://doi.org/10.3389/feart.2017.00087

Luo, Z. X., R. L. Cifelli, and Z. Kielan-Jaworowska. 2001. Dual origin of tribosphenic mammals. *Nature* 409:53–57.

Lydekker, R. 1907. The ears as a race-character in the African elephant. *Proceedings of the Zoological Society, London* 1907:380–403.

Ma, A., and H. Tang. 1992. On discovery and significance of a Holocene *Ailuropoda-Stegodon* fauna from Jinhua, Zhejiang. *Vertebrata PalAsiatica* 30:295–312.

MacCalman, H. R. 1967. The Zoo Park elephant site, Windhoek. *Palaeoecology of Africa* 2:102–103.

Macgregor, D. 2015. History of the development of the East African Rift System: A series of interpreted maps through time. *Journal of African Earth Sciences* 101:232–252.

MacInnes, D. G. 1942. Miocene and post-Miocene Proboscidia [sic] from East Africa. *Transactions of the Zoological Society, London* 25:33–106.

Mackaye, H. T. 2001. *Les Proboscidiens du Mio-Pliocène du Tchad: Biodiversité, Biochronologie, Paléoécologie et Paléobiogeographie*. Unpublished doctoral dissertation, Ingénierie Chimique Biologique et Géologique, Université de Poitiers, 214 pp.

Mackaye, H. T., M. Brunet, and P. Tassy. 2005. *Selenetherium kolleensis* nov. gen. nov. sp.: un nouveau Proboscidea (Mammalia) dans le Pliocene tchadien. *Geobios* 38:765–777.

Mackaye, H. T., Y. Coppens, P. Vignaud, F. Lihoreau, and M. Brunet. 2008. De nouveaux restes de *Primelephas* dans le Mio-Pliocène du Nord du Tchad et révision du genre *Primelephas*. *Compte Rendus Palevol* 7:227–236.

MacLatchy, L. 2004. The oldest ape. *Evolutionary Anthropology* 13:90–103.

MacLatchy, L., A. Deino, and J. Kingston. 2006. An updated chronology for the early Miocene of NE Uganda. *Journal of Vertebrate Paleontology* 26 (Supplement to Number 3):93A.

MacLatchy, L., J. Desilva, W. J. Sanders, and B. Wood. 2010. Hominini; pp. 471–540 in L. Werdelin and W. J. Sanders (Eds.), *Cenozoic Mammals of Africa*. University of California Press, Berkeley.

MacLatchy, L., D. Gebo, R. Kityo, and D. Pilbeam. 2000. Postcranial functional morphology of *Morotopithecus bishopi*, with implications for the evolution of modern ape locomotion. *Journal of Human Evolution* 39:159–183.

MacLatchy, L., J. Rossie, A. Houssaye, A. J. Olejniczak, and T. M. Smith. 2019. New hominoid fossils from Moroto II, Uganda and their bearing on the taxonomic and adaptive status of *Morotopithecus bishopi*. *Journal of Human Evolution* 132:227–246.

Madden, C. T. 1972. Miocene mammals, stratigraphy and environment of Muruarot Hill, Kenya. *Paleobios* 14:1–12.

—. 1977. *Tetralophodon* (Proboscidea, Gomphotheriidae) from Subsaharan Africa. *Revue de Zoologie Africaine* 91:153–160.

—. 1980. *Zygolophodon* from Subsaharan Africa, with observations on the systematics of palaeomastodontid proboscideans. *Journal of Paleontology* 54:57–64.

—. 1982. Primitive *Stegotetrabelodon* from latest Miocene of Subsaharan Africa. *Revue de Zoologie Africaine* 96:782–796.

Madden, C. T., K. W. Glennie, R. Dehm, F. C. Whitmore, R. J. Schmidt, R. J. Ferfoglia, and P. J. Whybrow. 1982. *Stegotetrabelodon (Proboscidea, Gomphotheriidae from the Miocene of Abu Dhabi*. United States Geological Survey, Jiddah.

Maglio, V. J. 1969a. A shovel-tusked gomphothere from the Miocene of Kenya. *Breviora* 310:1–10.

—. 1969b. The status of the East African elephant '*Archidiskodon exoptatus*' Dietrich 1942. *Breviora* 336:1–25.

—. 1970a. Early Elephantidae of Africa and a tentative correlation of African Plio-Pleistocene deposits. *Nature* 225:328–332.

—. 1970b. Four new species of Elephantidae from the Plio-Pleistocene of northwestern Kenya. *Breviora* 341:1–43.

—. 1971. Vertebrate faunas from the Kubi Algi, Koobi Fora and Ileret areas, East Rudolf, Kenya. *Nature* 231:248–249.

—. 1972a. Evolution of mastication in the Elephantidae. *Evolution* 26:638–658.

—. 1972b. Vertebrate faunas and chronology of hominid-bearing sediments east of Lake Rudolf, Kenya. *Nature* 239:379–385.

—. 1973. Origin and evolution of the Elephantidae. *Transactions of the American Philosophical Society* 63:1–149.

—. 1974. A new proboscidean from the late Miocene of Kenya. *Palaeontology* 17:699–705.

Maglio, V. J., and H. B. S. Cooke. 1978. *Evolution of African Mammals*. Harvard University Press, Cambridge, 641 pp.

Maglio, V. J., and Q. B. Hendey. 1970. New evidence relating to the supposed stegolophodont ancestry of the Elephantidae. *South African Archaeological Bulletin* 25:85–87.

Maglio, V. J., and A. B. Ricca. 1977. Dental and skeletal morphology of the earliest elephants. *Verhandelingen der Koninklijke Nederlandse Akademie van Wetenschappen, Afd. Natuurkunde Eerste Reeks, Deel* 29:1–51.

Magnier, P. 1962. Etude geologique du gisement de Vertebres du Gebel Zelten (Libye). *Comptes Rendus Sommaire des Seances de la Société Géologique de France* 2:55–57.

Mahboubi, M., R. Ameur, J. Y. Crochet, and J.-J. Jaeger. 1984. Earliest known proboscidean from early Eocene of northwest Africa. *Nature* 308:543–544.

—. 1986. El Kohol (Saharan Atlas, Algeria): A new Eocene mammal locality in northwestern Africa. *Palaeontographica Abt. A* 192:15–49.

Mahboubi, S., H. Bocherens, M. Scheffler, M. Benammi, and J.-J. Jaeger. 2014. Was the early Eocene proboscidean *Numidotherium koholense* semi-aquatic of terrestrial? Evidence from stable isotopes and bone histology. *Comptes Rendus Palevol* 13:501–509.

Mahboubi, M., R. Tabuce, F. Mebrouck, B. Coiffait, P. E. Coiffait, and J. J. Jaeger. 2003. L'Eocène continental à vertébrés de la bordure Sud des Monts des Nementcha (Atlas saharien oriental, Algérie)—Précisions stratigraphiques et implications paléobiogéographiques. *Bulletin du Service Géologique de l'Algérie* 14:27–35.

Makinouchi, T., S. Ishida Y. Sawada, N. Kuga, N. Kimura, Y. Orihashi, B. Bajope, M. wa Yemba, and H. Ishida. 1992. Geology of the Sinda-Mohari Region, Haut-Zaire Province, eastern Zaire. *African Study Monographs* suppl. 17:3–18.

Manthi, F. K., T. E. Cerling, and K. L. Chritz. 2017. Diets of mammalian fossil fauna from Kanapoi, northwestern Kenya. *Journal of Human Evolution*. https://doi.org/10.1016/j.jhevol.2017.05.005

Manthi, F. K., J. M. Plavcan, and C. V. Ward. 2012. New hominin fossils from Kanapoi, Kenya, and the mosaic evolution of canine teeth in early hominins. *South African Journal of Science* 108:1–9.

Manthi, F. K., W. J. Sanders, J. M. Plavcan, T. E. Cerling, and F. H. Brown. 2019. Late middle Pleistocene elephants from Natodomeri, Kenya and the disappearance of *Elephas* (Proboscidea, Mammalia) in Africa. *Journal of Mammalian Evolution*. https://doi.org/10.1007/s10914-019-90474-9

Marinheiro, J. 2015. *Proboscideans and other vertebrates from Anchrif, Morocco*. Unpublished doctoral dissertation, Universidade Nova, Lisbon, 71 pp.

Marinheiro, J., O. Mateus, A. Alaoui, F. Amani, M. Nami, and C. Ribeiro. 2014. New Quaternary fossil sites from the Middle Atlas of Morocco. *Comunicações Geológicas* 101, Especial I:485–488.

Markov, G. N. 2008. The Turolian proboscideans (Mammalia) of Europe: Preliminary observations. *Historia naturalis bulgarica* 19:153–178.

—. 2012. *Mammuthus rumanus*, early mammoths, and migration out of Africa: Some interrelated problems. *Quaternary International* 276–277:23–26.

Markov, G. N., and H. Saegusa. 2008. On the validity of *Stegoloxodon* Kretzoi, 1950 (Mammalia: Proboscidea). *Zootaxa* 1861:55–56.

Markov, G. N., and N. Spassov. 2003. Primitive mammoths from northeast Bulgaria in the context of the earliest mammoth migrations in Europe; pp. 53–58 in A. Petculescu and E. Stuica (Eds.), *Advances in Vertebrate Paleontology*. Hen to Panta, Bucharest.

Markov, G. N., and S. Vergiev. 2010. First report of cf. *Protanancus* (Mammalia, Proboscidea, Ambelodontidae) from Europe. *Geodiversitas* 32:493–500.

Markov, G. N., N. Spassov, and V. Simeonovski. 2001. A reconstruction of the facial morphology and feeding behaviour of the deinotheres. *The World of Elephants-International Congress, Rome* 2001:652–655.

Matschie, P. 1900. Ueber die geographische Abarten des afrikanischen Elefanten. *Sitzungsberichte des Gesellschaft der Naturforschender Freunde, Berlin* 1900:189–197.

Matsumoto, H. 1922. Revision of *Palaeomastodon* and *Moeritherium*. *Palaeomastodon intermedius*, and *Phiomia osborni*, new species. *American Museum Novitates* 51:1–6.

—. 1923. A contribution to the knowledge of *Moeritherium*. *Bulletin of the American Museum of Natural History* 48:97–140.

—. 1924a. A revision of *Palaeomastodon* dividing it into two genera, and with descriptions of two new species. *Bulletin of the American Museum of Natural History* 50:1–58.

—. 1924b. Preliminary note on fossil elephants in Japan. *Journal of the Geological Society of Tokyo* 31:255–272.

—. 1929. On *Loxodonta* (*Palaeoloxodon*) *namadica* (Falconer and Cautley) in Japan. *Science Reports of the Tohoku Imperial University 2nd Series, Geology* 13:1–5.

Mauffret, A. 2007. The northwestern (Maghreb) boundary of the Nubia (Africa) Plate. *Tectonophysics* 429:21–44.

Maurois, C., C. Chamberlan, and C. Marechal. 1997. An outline of the diet of forest elephants, *Loxodonta africana cyclotis*, in the Parc National d'Odzala, Republic of the Congo. *Mammalia* 61:127–130.

Mawby, J. E. 1970. Fossil vertebrates from northern Malawi: Preliminary report. *Quaternaria* 13:319–324.

Mazo, A. V. 1996. Gomphotheres and mammutids from the Iberian Peninsula, pp. 136–142 in J. Shoshani and P. Tassy (Eds.), *The Proboscidea: Evolution and Palaeoecology of Elephants and Their Relatives*. Oxford University Press, Oxford.

Mazo, A. V., and P. Montoya. 2003. Proboscidea (Mammalia) from the Upper Miocene of Crevillente (Alicante, Spain). *Scripta Geologica* 126:79–109.

Mazzini, I., and M. Kovacova. 2022. Ostracods, charophytes, and pollen: Late Miocene paleoenvironments of the Baynunah Formation, pp. 65–77 in F. Bibi, B. Krarz, M. Beech, and A. Hill (Eds.), *Sands of Time: Late Miocene Fossils from the Baynunah Formation, U.A.E.* Springer, Cham, Switzerland.

Mazzini, I., F. Bibi, M. Schuster, M. Beech, and A. Hill. 2013. The "elephants" and the ostracods: A 7 My old tale from the United Arab Emirates. *Il Naturalista Siciliano* 37:209–211.

McComb, K., L. Baker, and C. J. Moss. 2005. African elephants show high levels of interest in the skulls and ivory of their own species. *Biology Letters*. https://doi.org/10.1098/rsbl.2005.0400

McComb, K., D. Reby, L. Baker, and C. Moss. 2003. Long-distance communication of acoustic cues to social identity in African elephants. *Animal Behaviour* 65:317–329.

McDougall, I., and F. H. Brown, 2008. Geochronology of the pre-KBS Tuff sequence, Omo Group, Turkana Basin. *Journal of Geological Science* 165:549–562.

McDougall, I., and C. S. Feibel. 2003. Numerical age control for the Miocene-Pliocene succession at Lothagam, a hominoid-bearing sequence in the northern Kenya Rift; pp. 43–64 in M. G. Leakey and J. M. Harris (Eds.), *Lothagam: The Dawn of Humanity in Eastern Africa*. Columbia University Press, New York.

McDougall, I., and R. T. Watkins. 1985. Age of hominoid-bearing sequence at Buluk, northern Kenya. *Nature* 318:175–178.

—. 1988. Potassium-argon ages of volcanic rocks from northeast of Lake Turkana, northern Kenya. *Geological Magazine* 125:15–23.

McDougall, I., F. H. Brown, and J. G. Fleagle. 2005. Stratigraphic placement and age of modern humans from Kibish, Ethiopia. *Nature* 433:733–736.

McDougall, I., F. H. Brown, P. M. Vasconcelos, B. E. Cohen, D. S. Thiede, and M. J. Buchanan. 2012. New single crystal $^{40}Ar/^{39}Ar$ ages improve time scale for deposition of the Omo Group, Omo-Turkana Basin, East Africa. *Journal of the Geological Society, London* 169:213–226.

McHenry, L. J., and I. G. Stanistreet. 2018. Tephrochronology of Bed II, Olduvai Gorge, Tanzania, and placement of the Oldowan-Acheulean transition. *Journal of Human Evolution* 120:7–18.

McInerney, F. A., and S. L. Wing. 2011. The Paleocene-Eocene thermal maximum: A perturbation of carbon cycle, climate, and biosphere with implications for the future. *Annual Review of Earth and Planetary Sciences* 39:489–516.

McKenna, M. C. 1975. Toward a phylogenetic classification of mammals; pp. 21–46 in W. P. Luckett and F. S. Szalay (Eds.), *Phylogeny of the Primates: A Multidisciplinary Approach*. Plenum Press, New York.

Mead, J. G. 1975. A fossil beaked whale (Cetacea: Ziphiidae) from the Miocene of Kenya. *Journal of Paleontology* 49:745–751.

Mebrate, A. 1983. *Late Miocene-Middle Pleistocene Proboscidean Fossil Remains from the Middle Awash Valley, Afar Depression, Ethiopia*. Unpublished Masters Thesis, Department of Systematics and Ecology, University of Kansas, 119 pp.

Mebrate, A., and J. E. Kalb. 1985. Anancinae (Proboscidea: Gomphotheriidae) from the Middle Awash Valley, Afar, Ethiopia. *Journal of Vertebrate Paleontology* 5:93–102.

Meiring, A. J. D. 1955. Fossil proboscidean teeth and ulna from Virginia, O.F.S. *Navorsinge van die Nasionale Museum, Bloemfontein* 1(8):187–201.

Merte, C. E., K. F. Gough, and B. A. Schulte. 2008–2009. Investigation of a fresh African elephant carcass by conspecifics. *Pachyderm* 45:124–126.

van der Merwe, N. J., J. A. Lee-Thorp, and R. H. V. Bell. 1988. Carbon isotopes as indicators of elephant diets and African environments. *African Journal of Ecology* 26:163–172.

Metz-Muller, F. 1995. Mise en évidence d'une variation intraspécifique des caractères dentaires chez *Anancus arvernensis* (Proboscidea, Mammalia) du gisement de Dorkovo (Pliocène ancien de Bulgarie, Biozone MN 14). *Geobios* 28:737–743.

—. 2000. *La population d'Anancus arvernensis (Proboscidea, Mammalia) du Pliocène de Dorkovo (Bulgarie); étude des modalités évolutives d'Anancus arvernensis et phylogénie du genre Anancus*. Unpublished PhD Thesis, Museum National d'Histoire Naturelle de Paris (2 vol.), pp. 478.

von Meyer, H. 1831. Note on *Dinotherium giganteum* and *D. bavaricum*. *Neues Jahrbuch für Mineralogie* 1831:296–297.

Meyer, M., E. Palkopoulou, S. Baleka, M. Stiller, K. E. H. Penkman, K. W. Alt, Y. Ishida, D. Mania, S. Mallick, T. Meijer, H. Meller, S. Nagel, B. Nickel, S. Ostritz, N. Rohland, K. Schauer, T. Schüler, A. L. Roca, D. Reich, B. Shapiro, and M. Hofreiter. 2017. Palaeogenomes of Eurasian straight-tusked elephants challenge the current view of elephant evolution. *eLife*. https://doi.org/10.7554/eLife.25413.001

Meylan, P., and W. Auffenberg. 1986. New land tortoises (Testudines: Testudinidae) from the Miocene of Africa. *Zoological Journal of the Linnean Society* 86:279–307.

Miller, E. R. 1996. *Mammalian Paleontology of an Old World Monkey Locality, Wadi Moghara, Early Miocene, Egypt*. Unpublished PhD dissertation, Washington University, St. Louis, 372 pp.

—. 1999. Faunal correlation of Wadi Moghara, Egypt: Implications for the age of *Prohylobates tandyi*. *Journal of Human Evolution* 36:519–533.

Miller, E. R., B. R. Benefit, M. L. McCrossin, J. M. Plavcan, M. G. Leakey, A. N. El-Barkooky, M. A. Hamdan, M.K. Abdel Gawad, S. M. Hassan, and E. L. Simons. 2009. Systematics of early and middle Miocene Old World monkeys. *Journal of Human Evolution* 57:195–211.

Miller, E. R., G. F. Gunnell, M. A. Gawad, M. Hamdan, A. N. El-Barkooky, M. T. Clementz, and S. M. Hassan. 2014. Anthracotheres from Wadi Moghra, early Miocene, Egypt. *Journal of Paleontology* 88:967–981.

Mlíkovsky, J. 2003. Early Miocene birds of Djebel Zelten, Libya. *Casopis Národního muzea, Rada prírodovedná (Journal of Natural History Museum, Natural History Series)* 172:114–120.

Mol, D. 2006. The Langebaanweg locality: An important site for a better view of the evolutionary history of the mammoths. *African Natural History* 2:190–191.

Moody, R. T. J., and P. J. C. Sutcliffe. 1993. The sedimentology and palaeontology of the Upper Cretaceous-Tertiary deposits of central West Africa. *Modern Geology* 18:539–554.

Morales, J., and M. Pickford. 2008. Creodonts and carnivores from the middle Mioceme Muruyur Formation at Kipsaraman and Cheparawa, Baringo District, Kenya. *Comptes Rendus Palevol* 7:487–497.

Morales, J., P. Brewer, and M. Pickford. 2010. Carnivores (Creodonta and Carnivora) from the basal middle Miocene of Gebel Zelten, Libya, with a note on a large amphicyonid from the middle Miocene of Ngorora, Kenya. *Bulletin of the Tethys Geological Society, Cairo* 5:43–54.

Morales, J., M. Pickford, S. Fraile, M. J. Salesa, and D. Soria. 2003a. Creodonta and Carnivora from Arrisdrift, early middle Miocene of southern Namibia. *Memoir of the Geological Survey of Namibia* 19:177–194.

Morales, J., M. Pickford, D. Soria, and S. Fraile. 2001b. New Viverrinae (Carnivora: Mammalia) from the basal middle Miocene of Arrisdrift, Namibia. *Palaeontologica Africana* 37:99–102.

Morales, J., M. J. Salesa, M. Pickford, and D. Soria. 2001a. A new tribe, new genus and two new species of Barbourofelinae (Felidae, Carnivora, Mammalia) from the early Miocene of East Africa and Spain. *Transactions of the Royal Society of Edinburgh: Earth Sciences* 92:97–102.

Morales, J., D. Soria, and M. Pickford. 1999. New stem giraffoid ruminants from the early and middle Miocene of Namibia. *Geodiversitas* 21:229–253.

Morales, J., D. Soria, I. M. Sánchez, V. Quiralte, and M. Pickford. 2003b. Tragulidae from Arrisdrift, basal middle Miocene, southern Namibia. *Memoir of the Geological Survey of Namibia* 19:359–369.

Morley, C. K., D. K. Ngenoh, and J. K. Ego. 1999. Introduction to the East African rift system; pp. 1–18 in C. K. Morley (Ed.), *Geoscience of Rift Systems—Evolution of East Africa*. AAPG Studies in Geology, vol. 44, Houston, TX.

Morley, R. J. 2005. *Origin and Evolution of Tropical Rain Forests*. Wiley&Sons, New York, 378 pp.

Morlo, M., E. R. Miller, and A. N. El-Barkooky. 2007. Creodonta and Carnivora from Wadi Moghra, Egypt. *Journal of Vertebrate Paleontology* 27:145–159.

Morrison-Scott, T. C. S. 1947. A revision of our knowledge of African elephants' teeth, with notes on forest and "pygmy" elephants. *Proceedings of the Zoological Society of London* 117:505–527.

Mortimer, B., W. L. Rees, P. Koelemeijer, and T. Nissen-Meyer. 2018. Classifying elephant behaviour through seismic vibrations. *Current Biology* 28:R547–548.

Moss, C. 1992. Elephant calves: The story of two sexes; pp. 106–115 in J. Shoshani (Ed.), *Elephants*. Simon and Schuster, New York.

Moullé, P. E., A. Echassoux, Z. Alemseged, and E. Desclaux. 2001. On the presence of *Elephas recki* at the Oldowan prehistoric site of Fejej FJ-1 (Ethiopia); pp. 122–125 in G. Cavarretta, P. Gioia, M. Mussi, and M. R. Palombo (Eds.), *Proceedings of the First International Congress of la Terra degli Elefanti: The World of Elephants*. Consiglio Nazionale delle Richerche, Rome.

Moustafa, Y. S. 1974a. The Oligocene African Proboscidea: Part I. Introduction. *Annals of the Geological Survey of Egypt* 4:385–415.

—. 1974b. The Oligocene African Proboscidea: Part II. The *Phiomia-Palaeomastodon* question. *Annals of the Geological Survey of Egypt* 4:417–432.

Mufarreh, Y. A., D. C. Fisher, A. M. Memesh, S. A. Soubhi, A. J. Abdulshakoor, A. M. Masary, M. A. Haptari, and I. S. Zalmout. 2017. Skeletal characteristics of a Pleistocene elephant from the Nafud Desert, northwestern Saudi Arabia. *Journal of Vertebrate Paleontology, Program and Abstracts* 2017(Supplement to 37):167A.

Muftah, A. M., A. A. El-Mehaghag, and S. Starkie. 2008. Biostratigraphical notes on the As Sahabi stratigraphic boreholes 1 and 2, Sirt Basin, Libya. *Garyounis Scientific Bulletin* 5:47–57.

Mundinger, G. S., and W. J. Sanders. 2001. Taxonomic and systematic re-assessment of the Mio-Pliocene elephant *Primelephas gomphotheroides*. *Scientific Programme and Abstracts, 8th International Theriological Congress* 2001:102.

Murata, Y., T. Yonezawa, I. Kihara, T. Kashiwamura, Y. Sugihara, M. Nikaido, N. Okada, H. Endo, and M. Hasegawa. 2009. Chronology of the extant African species and case study of the species identification of the small African elephant with the molecular phylogenetic method. *Gene* 441:176–186.

Murphy, W. J., E. Eizirik, W. E. Johnson, Y. P. Zhang, O. A. Ryder, and S. J. O'Brien. 2001. Molecular phylogenetics and the origins of placental mammals. *Nature* 409:614–618.

Musalizi, S., B. Senut, M. Pickford, and E. Musiime. 2009. Geological and palaeontological archives relating to early Miocene localities of Uganda, 1957–1969. *Geo-Pal Uganda* 1:2–96.

Nakatsukasa, M., Y. Kunimatsu, H. Nakaya, Y. Sawada, T. Sakai, H. Hyodo, T. Itaya, and E. Mbua. 2007. Late Miocene fossil locality Nakali in Kenya and its paleoenvironment. *American Journal of Physical Anthropology* suppl. 44:177.

Nakaya, H. 1993. Mammalian fauna of late Miocene Samburu Hills, Kenya, East Africa and environments of pre-hominids. *L'Anthropologie* 97:9–16.

Nakaya, H., M. Pickford, Y. Nakano, and H. Ishida. 1984. The late Miocene large mammal fauna from the Namurungule Formation, Samburu Hills, northern Kenya. *African Study Monographs* suppl. 2:87–131.

Nakaya, H., M. Pickford, K. Yasui, and Y. Nakano. 1987. Additional large mammalian fauna from the Namurungule Formation, Samburu Hills, northern Kenya. *African Study Monographs* suppl. 5:79–129.

Nanda, A. C., D. Dun, and G. Corvinus. 2000. Skull characteristics of two proboscideans from the Upper Siwalik Subgroup of Nepal. *Neues Jahrbuch für Geologie und Paläontologie Abh.* 217:89–110.

Nengo, I. O., and T. C. Rae. 1992. New hominoid fossils from the early Miocene site of Songhor, Kenya. *Journal of Human Evolution* 23:423–429.

Nesti, F. 1825. Sulla nuove species de elephante fossile del Valdarno all 'Illustrissimo sig. Dott. Prof. Ottaviano Targioni Tozzetti (Lettre sopra alcune ossa fossili de Valdarno non per anco descritte). *Nuovo Giorno Lettera* 11:195–216.

Noack, T. 1906. Eine Zwergform des afrikanischen Elefanten. *Zoologische Anzeiger Leipzig* 29:631.

Nogués-Bravo, D., J. Rodríguez, J. Hortal, P. Batra, and M. B. Araújo. 2008. Climate change, humans, and the extinction of the Woolly Mammoth. *PLoS Biology* 6:e79.

Noro, M., R. Masuda, I. A. Dubrovo, M. C. Yoshida, and M. Kato. 1998. Molecular phylogenetic inference of the woolly mammoth *Mammuthus primigenius*, based on complete sequences of mitochondrial cytochrome *b* and 12S ribosomal RNA genes. *Journal of Molecular Evolution* 46:314–326.

Noubhani, A., L. Hautier, J.-J. Jaeger, M. Mahboubi, and R. Tabuce. 2008. Dental and cranial variability of *Numidotherium koholense* (Mammalia, Proboscidea) from El Kohol, Eocene, Algeria. *Geobios* 41:515–531.

O'Brien, K., D. B. Patterson, M. D. Biernat, D. R. Braun, T. E. Cerling, A. McGrosky, and J. T. Faith. 2020. Ungulate turnover in the Koobi Fora Formation: Spatial and temporal variation in the Early Pleistocene. *Journal of African Earth Sciences* 161:103658.

Ogunfolakan, B. A., and A. Olayemi. 2008. Preliminary report of elephant remains in an archaeological site within Idanre Township, Ondo State, Nigeria. *Nyame Akuma* 69:31–40.

O'Leary, M. A., J. I. Bloch, J. J. Flynn, T. J. Gaudin, A. Giallombardo, N. P. Giannini, S. L. Goldberg, B. P. Kratz, Z.-X. Luo, J. Meng, X. Ni, M. J. Novacek, F. A. Perini, Z. S. Randall, G. W. Rougier, E. J. Sargis, M. T. Silcox, N. B. Simmons, M. Spaulding, P. M. Velazco, M. Weksler, J. R. Wible, and A. L. Cirranello. 2013. The placental mammal ancestor and the post-K-Pg radiation of placentals. *Science* 339:662–667.

O'Leary, M. A., M. L. Bouaré, K. Heilbronn, R. V. Hill, J. McCartney, J. A. Sessa, F. Sissoko, L. Tapanila, E. Wheeler, and E. M. Roberts. 2019. Stratigraphy and paleobiology of the Upper Cretaceous-Lower Paleogene sediments from the Trans-Saharan Seaway in Mali. *Bulletin of the American Museum of Natural History* 436:1–177.

O'Leary, M. A., E. M. Roberts, M. Bouare, F. Sissoko, and L. Tapanila. 2006. Malian Paenungulata (Mammalia: Placentalia): New African afrotheres from the early Eocene. *Journal of Vertebrate Paleontology* 26:981–988.

Orliac, M., J.-R. Boisserie, L. Maclatchy, and F. Lihoreau. 2010. Early Miocene hippopotamids (Cetartiodactyla) constrain the phylogenetic and spatiotemporal settings of hippopotamid origin. *Proceedings of the National Academy of Sciences, USA*. www.pnas.org/cgi/doi/10.1073/pnas.1001373107

Osborn, H. F. 1909. The feeding habit of *Moeritherium* and *Palaeomastodon*. *Nature* 81:139–140.

—. 1918. A long-jawed mastodon skeleton from South Dakota and phylogeny of the Proboscidea. *Bulletin of the Geological Society of America* 29:133–137.

—. 1919. *Palaeomastodon*, the ancestor of the long-jawed mastodons only. *Proceedings of the National Academy of Sciences, USA* 5:265–266.

—. 1921. The evolution, phylogeny and classification of the Proboscidea. *American Museum Novitates* 1:1–15.

—. 1928. Mammoths and man in the Transvaal. *Nature* 121:672–673.

—. 1932. *Trilophodon cooperi*, sp. nov., of Dera Bugti, Baluchistan. *American Museum Novitates* 585:1–6.

—. 1934. Primitive *Archidiskodon* and *Palaeoloxodon* of South Africa. *American Museum Novitates* 741:1–15.

—. 1936. *Proboscidea: A Monograph of the Discovery, Evolution, Migration and Extinction of the Mastodonts and Elephants of the World: Vol. 1 Moeritherioidea, Deinotherioidea, Mastodontoidea*. American Museum Press, New York, 802 pp.

—. 1942. Proboscidea: *A Monograph of the Discovery, Evolution, Migration and Extinction of the Mastodonts and Elephants of the World: Vol. 2 Stegodontoidea, Elephantoidea*. American Museum Press, New York, 870 pp.

Osborn, H. F., and W. Granger. 1931. The shovel-tuskers, Amebelodontinae, of Central Asia. *American Museum Novitates* 470:1–12.

—. 1932. *Platybelodon grangeri*, three growth stages, and a new serridentine from Mongolia. *American Museum Novitates* 537:1–13.

Ozawa, T., S. Hayashi, and V. M. Mikhelson. 1997. Phylogenetic position of mammoth and Steller's sea cow within Tethytheria demonstrated by mitochondrial DNA sequences. *Journal of Molecular Evolution* 44:406–413.

Pagani, M., K. H. Freeman, and M. A. Arthur. 1999. Late Miocene atmospheric CO2 concentrations and the expansion of C4 grasses. *Science* 288:876–879.

Palkopoulou, E., M. Lipson, S. Mallick, S. Nielsen, N. Rohland, S. Baleka, E. Karpinski, A. M. Ivancevic, T.-H. To, R. D. Kortschak, J. M. Raison, Z. Qu, T.-J. Chin, K. W. Alt, S. Claesson, L. Dalén, R. D. E. MacPhee, H. Meller, A. L. Roca, O. A. Ryder, D. Heiman, S. Young, M. Breen, C. Williams, B. L. Aken, M. Ruffier, E. Karlsson, J. Johnson, F. Di Palma, J. Alfoldi, D. L. Adelson, T. Mailund, K. Munch, K. Lindblad-Toh, M. Hofreiter, H. Poinar, and D. Reich. 2018. A comprehensive genomic history of extinct and living elephants. *Proceedings of the National Academy of Sciences, USA* 115(11). www.pnas.org/cgi/doi/10.1073/pnas.1720554115

Palmer, R. W. 1924. An incomplete skull of *Dinotherium*, with notes on the Indian forms. *Palaeontologia Indica* 7:1–14.

Palombo, M. R., R. Carlini, and S. Gippoliti. 2021. Critical inventory of *Loxodonta* and *Elephas* (Mammalia, Proboscidea) cranial remains in the collections of the Museo Civico di Zoologia of Rome (Italy). *Bollettino del Museo Civico di Storia Naturale di Verona, Botanica Zoologia* 45:17–60.

Palombo, M. R., M. L. Filippi, P. Iacumin, A. Longinelli, M. Barbieri, and A. Maras. 2005. Coupling tooth microwear and stable isotope analyses for paleodiet reconstruction: The case study of late middle Pleistocene *Elephas* (*Palaeoloxodon*) *antiquus* teeth from Central Italy (Rome area). *Quaternary International* 126–128:153–170.

Pan, A. D. 2010. Rutaceae leaf fossils from the late Oligocene (27.23 Ma) Guang River flora of northwestern Ethiopia. *Review of Palaeobotany and Palynology* 159:188–194.

Pan, A. D., B. F. Jacobs, J. Dransfield, and W. J. Baker. 2006. The fossil history of palms (Arecaceae) in Africa and new records from the late Oligocene (28-27 Ma) of northwestern Ethiopia. *Botanical Journal of the Linnean Society* 151:69–81.

Pan, A. D., B. F. Jacobs, and P. S. Herendeen. 2010. Detarieae *sensu lato* (Fabaceae) from the late Oligocene (27.23 Ma) Guang River flora of north-western Ethiopia. *Botanical Journal of the Linnean Society* 163:44–54.

Partridge, T. C. 1997. Late Neogene uplift in eastern and southern Africa and its paleoclimatic implications; pp. 64–86 in W. F. Ruddiman (Ed.), *Tectonic Uplift and Climate Change*. Plenum Press, New York.

—. 2000. Hominid-bearing cave and tufa deposits; pp. 100–125 in T. C. Partridge and R. R. Maud (Eds.), *The Cenozoic of Southern Africa*. Oxford University Press, Oxford.

Partridge, T. C., A. G. Latham, and D. Heslop. 2000. Appendix on magnetostratigraphy of Makapansgat, Sterkfontein, Taung, and Swartkrans; pp. 126–129 in T. C. Partridge and R. R. Maud (Eds.), *The Cenozoic of Southern Africa*. Oxford University Press, Oxford.

Passey, B. H., T. F. Robinson, L. K. Ayliffe, T. E. Cerling, M. Sponheimer, M. D. Dearing, B. L. Roeder, and J. R. Ehleringer. 2005. Carbon isotopic fractionation between diet, breath, and bioapatite in different mammals. *Journal of Archaeological Science* 32:1459–1470.

Patnaik, R., N. P. Singh, D. Paul and R. Sukumar. 2019. Dietary and habitat shifts in relation to climate of Neogene-Quaternary proboscideans and associated mammals of the Indian subcontinent. *Quaternary Science Reviews* 224:105968.

Patten, M. A. 2015. Subspecies and the philosophy of science. *The Auk. Ornithological Advances* 132:481–485.

Patterson, B. 1966. A new locality for early Pleistocene fossils in northwestern Kenya. *Nature* 212:577–581.

Patterson, C., and A. E. Longbottom. 1989. An Eocene Amiid fish from Mali, West Africa. *Copeia* 4:827–836.

Patterson, B., A. K. Behrensmeyer, and W. D. Sill. 1970. Geology and fauna of a new Pliocene locality in northwestern Kenya. *Nature* 226:918–921.

Pearson, P. N., I. K. McMillan, B. S. Wade, T. D. Jones, H. K. Coxall, P. R. Bown, and C. H. Lear. 2008. Extinction and environmental change across the Eocene-Oligocene boundary in Tanzania. *Geology* 36:179–182.

Peppe, D. J., K. P. McNulty, S. M. Cote, W. E. H. Harcourt-Smith, H. M. Dunsworth, and J. A. Van Couvering. 2009. Stratigraphic interpretation of the Kulu Formation (early Miocene, Rusinga Island, Kenya) and its implications for primate evolution. *Journal of Human Evolution* 56:447–461.

Peters, J. 1990. Late Pleistocene hunter-gatherers at Ishango (eastern Zaire): The faunal evidence. *Revue de Paléobiologie* 9:73–112.

Petrocchi, C. 1934. I ritrovamenti faunistici di es-Sahabi. *Rivista delle Colonie Italiane* 7:733–742.

—. 1941. Il giacimento fossilifero di Sahabi. *Bollettino della Società Geologica Italiana* 60:107–114.

—. 1943. Il giacimento fossilifero di Sahabi. *Collezione Scientifica e Documentaria a Cura del Ministero dell' Africa Italiana, Verbania* 12:1–169.

—. 1954. Paleontologia di Sahabi: Parte I. Proboscidati di Sahabi. *Rendiconti Accademia nazionale dei XL* 4–5:8–74.

Petronievics, B. 1923. Remarks upon the skull of *Moeritherium* and *Palaeomastodon*. *Annals and Magazine of Natural History, London, Series 9* 12:55–61.

Petter, G. 1958. A propos de quelques petite elephants de foret attribues a *Loxodonta cyclotis* Matschie. *Mammalia* 70:575–590.

Phillips, M. J. 2016. Geomolecular dating and the origin of placental mammals. *Systematic Biology* 65:546–557.

Pickford, M. 1981. Preliminary Miocene mammalian biostratigraphy for western Kenya. *Journal of Human Evolution* 10:73–97.

—. 1982. The tectonics, volcanics and sediments of the Nyanza Rift Valley, Kenya. *Zeitshcrift für Geomorphologie*, Neue Folge Supplement-Band 42:1–33.

—. 1984. A revision of the Sanitheriidae, a new family of Suiformes (Mammalia). *Geobios* 17:133–154.

—. 1985. A new look at *Kenyapithecus* based on recent discoveries in western Kenya. *Journal of Human Evolution* 14:113–144.

—. 1986a. Cainozoic paleontological sites of western Kenya. *Münchner Geowissenschaftliche Abhandlungen A* 8:1–151.

—. 1986b. Recognition of an early Oligocene or late Eocene mammal fauna from Cabinda, Angola. *Musée Royal de l'Afrique Centrale (Belgique), Rapport Annuel du Département de Géologie et de Mineralogie* 1985–1986:89–92.

—. 1987a. Miocene Suidae from Ad Dabtiyah, eastern Saudi Arabia. *Bulletin of the British Museum (Natural History) (Geology)* 41:441–446.

—. 1987b. Recognition of an early Oligocene or late Eocene mammal fauna from Cabinda, Angola. *Musée Royal de l'Afrique Centrale (Belgique), Rapport Annuel du Département de Géologie et de Mineralogie* 1985–1986:89–92.

—. 1990. Uplift of the roof of Africa and its bearing on the evolution of mankind. *Human Evolution* 5:1–20.

—. 1991a. Biostratigraphic correlation of the middle Miocene mammal locality of Jabal Zaltan, Libya; pp. 1483–1490 in M. J. Salem, O. S. Hammuda, and B. A. Eliagoubi (Eds.), *The Geology of Libya, Vol. IV.* Elsevier, Amsterdam.

—. 1991b. Revision of Neogene Anthracotheriidae of Africa; pp. 1491–1525 in M. J. Salem, O. S. Hammuda, and B. A. Eliagoubi (Eds.), *The Geology of Libya, Vol. IV.* Elsevier, Amsterdam.

—. 1994. A new species of *Prohyrax* (Mammalia, Hyracoidea) from the middle Miocene of Arrisdrift, Namibia. *Communications of the Geological Survey of Namibia* 9:43–62.

—. 1995. Suidae (Mammalia, Artiodactyla) from the early middle Miocene of Arrisdrift, Namibia: *Namachoerus* (gen. nov.) *moruoroti*, and *Nguruwe kijivium*. *Comptes Rendus de l'Académie des Sciences Série IIa* 320:319–326.

—. 2000. Neogene and Quaternary vertebrate biochronology of the Sperrgebiet and Otavi Mountainland, Namibia. *Communications of the Geological Survey of Namibia* 12:411–419.

—. 2001. *Afrochoerodon* nov. gen. *kisumuensis* (MacInnes) (Proboscidea, Mammalia) from Cheparawa, middle Miocene, Kenya. *Annales de Paléontologie* 87:99–117.

—. 2002. Early Miocene grassland ecosystem at Bukwa, Mt. Elgon, Uganda. *Comptes Rendus Palevol* 1:213–219.

—. 2003a. New Proboscidea from the Miocene strata in the lower Orange River Valley, Namibia. *Memoir Geological Survey of Namibia* 19:207–256.

—. 2003b. Suidae from the middle Miocene of Arrisdrift, Namibia. *Memoir of the Geological Survey of Namibia* 19:291–303.

—. 2004. Partial dentition and skeleton of *Choerolophodon pygmaeus* (Deperet) from Ngenyin, 13 Ma, Tugen Hills, Kenya: Resolution of a century old enigma. *Zona Arqueologica: Miscelànea en Homenaje a Emiliano Aguirre, Palaeontologia. Madrid Museo Arqueologica Regional* 2:429–463.

—. 2005a. *Choerolophodon pygmaeus* (Proboscidea, Mammalia) from the middle Miocene of southern Africa. *South African Journal of Science* 101:175–177.

—. 2005b. Preliminary report on a proboscidean tooth from Daberas (middle Miocene), Sperrgebeit, Namibia. Unpublished report, 2 pp.

—. 2007a. Suidae and Hippopotamidae from the middle Miocene of Kipsaraman, Kenya and other sites in East Africa. *Paleontological Research* 11:85–105.

—. 2007b. New mammutid proboscidean teeth from the middle Miocene of tropical and southern Africa. *Palaeontologia Africana* 42:29–35.

—. 2008. Proboscidea from the early Miocene of the northern Sperrgebiet, Namibia. *Memoir of the Geological Survey of Namibia* 20:327–329.

—. 2012. *Mocambique Palaeontology Reconnaissance, November 2012.* Field Notes, pp. 1–22.

—. 2015a. Large ungulates from the basal Oligocene of Oman: 1—Embrithopoda. *Spanish Journal of Palaeontology* 30:139–148.

—. 2015b. Large ungulates from the basal Oligocene of Oman: 2—Proboscidea. *Spanish Journal of Palaeontology* 30:209–222.

—. 2015c. Large ungulates from the basal Oligocene of Oman: 3—Anthracotheriidae. *Spanish Journal of Palaeontology* 30:257–264.

—. 2020. Napak Gomphotheriidae (Proboscidea). *Geo-Pal Uganda* 16:1–21.

Pickford, M., and P. Andrews. 1981. The Tinderet Miocene sequence in Kenya. *Journal of Human Evolution* 10:11–33.

Pickford, M., and Y. Kunimatsu. 2005. Catarrhines from the middle Miocene (ca. 14.5 Ma) of Kipsaraman, Tugen Hills, Kenya. *Anthropological Science* 113:189–224.

Pickford, M., and P. Mein. 2006. Early middle Miocene mammals from Moroto II, Uganda. *Beiträge zur Paläontologie* 30:361–386.

—. 2011. New Pedetidae (Rodentia: Mammalia) from the Mio-Pliocene of Africa. *Estudios Geológicos* 67:455–469.

Pickford, M., and Z. Pourabrishami. 2013. Deciphering Dinotheriensande deinotheriid diversity. *Palaeobiodiversity and Palaeoenvironments* 93:121–150.

Pickford, M., and B. Senut. 1997. Cainozoic mammals from coastal Namaqualand, South Africa. *Palaeontologica Africana* 34:199–217.

—. 2000. Geology and palaeobiology of the Namib Desert, southwestern Africa. *Memoir Geological Survey of Namibia* 18:1–155.

—. 2001. 'Millenium ancestor,' a 6-million-year-old bipedal hominid from Kenya. *South African Journal of Science* 97:22.

Pickford, M., and P. Tassy. 1980. A new species of *Zygolophodon* (Mammalia, Proboscidea) from the Miocene hominoid localities of Meswa Bridge and Moroto (East Africa). *Neues Jahrbuch für Geologie und Paläontologie, Monatshefte* 4:235–251.

Pickford, M., H. Ishida, Y. Nakaya, and K. Yasui. 1987. The middle Miocene fauna from the Nachola and Aka Aiteputh Formations, northern Kenya. *African Study Monographs* Supplementary issue 5:141–154.

Pickford, M., E. R. Miller, and A. N. El-Barkooky. 2010. Suidae and Sanitheriidae from Wadi Moghra, early Miocene, Egypt. *Acta Palaeontologica Polonica* 55:1–11.

Pickford, M., J. Morales, and D. Soria. 1995. Fossil camels from the Upper Miocene of Europe: Implications for biogeography and faunal change. *Geobios* 28:641–650.

Pickford, M., Y. Sawada, R. Tayama, Y. Matsuda, T. Itaya, H. Hyodo, and B. Senut. 2006. Refinement of the age of the middle Miocene Fort Ternan Beds, western Kenya, and its implications for Old World biochronology. *Comptes Rendus Geoscience* 338:545–555.

Pickford, M., B. Senut, and K. Cheboi. 2009. The geology and palaeobiology of the Tugen Hills, Kenya. *Geo-Pal Kenya* 1:1–133.

Pickford, M., B. Senut, and D. Hadoto. 1993. *Geology and Palaeobiology of the Albertine Rift Valley, Uganda-Zaire. Volume I: Geology.* Occasional Publication 24, Centre International pour la Formation et les Echanges Géologiques, Orleans (France), pp. 1–190.

Pickford, M., B. Senut, D. Hadoto, J. Muisi, and C. Kariira. 1986. Decouvertes recentes dans le sites miocenes de Moroto (Ouganda Oriental): aspects biostratigraphiques et paleoecologiques. *Comptes Rendus de l'Academie des Sciences, Paris, Serie* II 9:681–686.

Pickford, M., B. Senut, D. Gommery, and E. Musiime. 2003. New catarrhine fossils from Moroto II, early middle Miocene (ca. 17.5 Ma) Uganda. *Comptes Rendus Palevol* 2:649–662.

Pickford, M., B. Senut, P. Mein, J. Morales, D. Soria, M. Nieto, J. Ward, and M. Bamford. 1995. The discovery of lower and middle Miocene vertebrates at Auchas, southern Namibia. *Comptes Rendus de l'Académie des Sciences, Série IIa* 322:901–906.

Pik, R., B. Marty, J. Carignan, G. Yirgu, and T. Ayalew. 2008. Timing of East African Rift development in southern Ethiopia: Implication for mantle plume activity and evolution of topography. *Geology* 36:167–170.

Plummer, T. W., and R. Potts. 1989. Excavations and new findings at Kanjera, Kenya. *Journal of Human Evolution* 18:269–276.

Plummer, T. W., J. V. Ferraro, J. Louys, F. Hertel, Z. Alemseged, R. Bobe, and L. C. Bishop. 2015. Bovid ecomorphology and hominin paleoenvironments of the Shungura Formation, lower Omo River Valley, Ethiopia. *Journal of Human Evolution* 88:108–126.

Pokines, J. T., A. M. Lister, C. J. H. Ames, A. Nowell, and C. E. Cordova. 2018. Faunal remains from recent excavations at Shishan Marsh 1 (SM1), a late Lower Paleolithic open-air site in the Azraq Basin, Jordan. *Quaternary Research.* https://doi.org/10.1017/qua.2018.113

Pomel, A. 1879. Ossements d'Éléphants et d'Hippopotames découvertes dans une station préhistorique de la pleine d'Eghis (Province d'Oran). *Bulletin de la Société Géologique de France ser. 3* 7:44–51.

—. 1895. Les éléphants Quaternaires. *Carte Géologique l'Algérie, Paléontologie Monographies Locales* 6:1–68.

Pontier, G. 1907. Sur une espèce nouvelle de *Palaeomastodon* (*Palaeomastodon barroisi*). *Annales de la Société Géologique du Nord* (*Lyon*) 36:150–154.

Poole, J., P. Kahumbu, and I. Whyte. 2013. *Loxodonta africana.* Savanna elephant (African Bush elephant); pp. 181–194 in J. Kingdon, D. Happold, M. Hoffmann, T. Butynski, M. Happold, and J. Kalina (Eds.), *Mammals of Africa, Volume I.* Bloomsbury, London.

Potter, P. E., and P. Szatmari. 2009. Global Miocene tectonics and the modern world. *Earth-Science Reviews* 96:279–295.

Poulakakis, N., G. E. Theodorou, E. Zouros, and M. Mylonas. 2002. Molecular phylogeny of the extinct Pleistocene dwarf elephant *Palaeoloxodon antiquus falconeri* from Tilos Island, Dodekanisa, Greece. *Journal of Molecular Evolution* 55:364–374.

Prothero, D. R. 1994. The late Eocene-Oligocene extinctions. *Annual Review of Earth and Planetary Sciences* 22:145–165.

Qiu, Z.-X., B.-Y. Wang, H. Li, T. Deng, and Y. Sun. 2007. First discovery of deinotheres in China. *Vertebrata PalAsiatica* 45:261–277.

Qiu, Z., W. Wu, and Z. Qiu. 1999. Miocene mammal faunal sequence of China: Palaeozoogeography and Eurasian relationships; pp. 443–455 in G. Rossner and K. Heissig (Eds.), *The Miocene Land Mammals of Europe.* Verlag, Munich.

Quade, J., N. E. Levin, S. W. Simpson, R. Butler, W. C. McIntosh, S. Semaw, L. Kleinsasser, G. Dupont-Nivet, P. Renne, and N. Dunbar. 2008. The geology of Gona, Afar, Ethiopia. *The Geological Society of America Special Paper* 446:1–33.

Rabinovich, R., O. Ackermann, E. Aladjem, R. Barkai, R. Biton, I. Milevski, N. Solodenko, and O. Marder. 2012. Elephants at the middle Pleistocene Acheulian open-air site of Revadim Quarry, Israel. *Quaternary International* 276–277:183–197.

Rage, J.-C., and E. Gheerbrant. 2020. Island Africa and vertebrate evolution: A review of data and working hypotheses; pp. 251–264 in G. V. R. Prasad and R. Patnaik (Eds.), *Biological Consequences of Plate Tectonics: New Perspectives on Post-Gondwana Break-Up—A Tribute to Ashok Sahni.* Springer, Switzerland.

Rasmussen, D. T., and M. Gutiérrez. 2009. A mammalian fauna from the late Oligocene of northwestern Kenya. *Palaeontographica Abt. A* 288:1–52.

—. 2010. Hyracoidea; pp. 123–145 in L. Werdelin and W. J. Sanders (Eds.), *Cenozoic Mammals of Africa.* University of California Press, Berkeley.

Rasmussen, D. T., A. R. Friscia, M. Gutiérrez, J. Kappelman, E. R. Miller, S. Muteti, D. Reynoso, J. B. Rossie, T. L. Spell, N. J. Tabor, E. Gierlowski-Kordesch, B. F. Jacobs, B. Kyongo, M. Macharwas, and F. Muchemi. 2019. Primitive Old World monkey from the earliest Miocene of Kenya and the evolution of cercopithecoid bilophodonty. *Proceedings of the National Academy of Sciences, USA* 116:6051–6056.

Rasmussen, D. T., M. Gagnon, and E. L. Simons. 1990. Taxeopody in the carpus and tarsus of Oligocene Pliohyracidae (Mammalia: Hyracoidea) and the phyletic position of hyraxes. *Proceedings of the National Academy of Sciences, USA* 87:4688–4691.

Rasmussen, D. T., S. O. Tshakreen, M. M. Abugares, and J. B. Smith. 2008. Return to Dor al-Talha: Paleontological reconnaissance of the early Tertiary of Libya; pp. 181–196 in J. G. Fleagle and C. C. Gilbert (Eds.), *Elwyn Simons: A Search for Origins.* Springer, New York.

Raup, D. M., and L. G. Marshall. 1980. Variation between groups in evolutionary rates: A statistical test of significance. *Paleobiology* 6:9–23.

Raynal, J.-P., F. Z. S. Alaoui, L. Magoga, A. Mohib, and M. Zouak. 2004a. The Lower Paleolithic sequence of Atlantic Morocco revisited after recent excavations at Casablanca. *Bulletin d'Archéologie Marocaine* 20:44–76.

Raynal, J.-P., G. Kieffer, and G. Bardin. 2004b. Garba IV and the Melka Kunture Formation: A preliminary lithostratigraphic approach; pp. 137–166 in J. Chavaillon and M. Piperno (Eds.), *Studies on the Early Paleolithic Site of Melka Kunture, Ethiopia.* Instituto Italiano di Preistoria e Protostoria, Florence.

Raza, S. M., and G. E. Meyer. 1984. Early Miocene geology and paleontology of the Bugti Hills, Pakistan. *Memoirs of the Geological Survey of Pakistan* 11:43–63.

Reck, H. 1914. Zweite vorläufige Mitteilung über fossile Tier- und Menschenfunde aus Oldoway in Zentralafrika. *Sitzungsberichte der Gesellschaft Naturforschender Freunde, Berlin* 7:305–318.

Remy, J.-A. 1976. Presence de *Deinotherium* sp. Kaup (Proboscidea, Mammalia) dans la faune miocène de Beni Mellal (Maroc). *Géologie Mediterranéenne* 3:109–114.

Renne, P. R., G. WoldeGabriel, W. K. Hart, G. Heiken, and T. D. White. 1999. Chronostratigraphy of the Miocene-Pliocene Sagantole Formation, Middle Awash Valley, Afar Rift, Ethiopia. *Geological Society of America Bulletin* 111:869–885.

Retallack, G. J., J. G. Wynn, B. R. Benefit, M. L. McCrossin. 2002. Paleosols and paleoenvironments of the middle Miocene, Maboko Formation, Kenya. *Journal of Human Evolution* 42:659–703.

Reuter, T., S. Nummela, and S. Hemilä. 1998. Elephant hearing. *The Journal of the Acoustical Society of America* 104:1122–1123.

Reynoso, D. E. 2014. *New Early Miocene $^{40}Ar/^{39}Ar$ Ages for Nakwai, Northwest Kenya, Africa, and Paleontological Analysis of* Meroehyrax kyongoi *Dentition*. Unpublished Masters of Science Thesis, Department of Geoscience, University of Nevada, Las Vegas.

Richmond, B. G., J. G. Fleagle, J. Kappelman, and C. C. Swisher III. 1998. First hominoid from the Miocene of Ethiopia and the evolution of the catarrhine elbow. *American Journal of Physical Anthropology* 105:257–277.

Rightmire, G. P. 1996. The human cranium from Bodo, Ethiopia: Evidence for speciation in the Middle Pleistocene? *Journal of Human Evolution* 31:21–39.

Rivals, F., D. Mol, F. Lacombat, A. M. Lister, and G. M. Semprebon. 2014. Resource partitioning and niche separation between mammoths (*Mammuthus rumanus* and *Mammuthus meridionalis*) and gomphotheres (*Anancus arvernensis*) in the early Pleistocene of Europe. *Quaternary International*. http://dx.doi.org/10.1016/j.quaint.2014.12.031

Rivals, F., G. Semprebon, and A. Lister. 2012. An examination of dietary diversity patterns in Pleistocene proboscideans (*Mammuthus*, *Palaeoloxodon*, and *Mammut*) from Europe and North America as revealed by dental microwear. *Quaternary International* 255:188–195.

Robbins, L. M. 1987. Hominid footprints from Site G; pp. 496–502 in M. D. Leakey and J. M. Harris (Eds.), *Laetoli: A Pliocene Site in Northern Tanzania*. Clarendon Press, Oxford.

Roberts, D. L., M. D. Bateman, C. V. Murray-Wallace, A. S. Carr, and P. J. Holmes. 2008. Last interglacial fossil elephant trackways dated by OSL/AAR in costal aeolianites, Still Bay, South Africa. *Palaeogeography, Palaeoclimatology, Palaeoecology* 257:261–279.

Robinson, P. 1974. The Beglia Formation of Tunisia. *Memoires, Bureau de Recherches Géologiques et Minières (France)* 78:235.

—. 2015. *Gomphotherium angustidens* (Mammalia, Proboscidea, Gomphotheriidae) from the El Haouaria Formation, Fortuna Group, Djbel Nara, East-Central Tunisia. *Notes du Service Géologique (Tunisia)* 80:11–21.

Robinson, P., and C. C. Black. 1973. A small Miocene faunule from near Testour, Beja Gouvernorat, Tunisia. *Annales des Mines et de la Géologie* 26:445–449.

—. 1974. Vertebrate fauanas from the Neogene of Tunisia. *Annals of the Geological Survey of Egypt* 4:319–332.

Robinson, P. T. 2013. *Choeropsis liberiensis*. Pygmy hippopotamus; pp. 80–83 in J. Kingdon and M. Hoffman (Eds.), *Mammals of Africa. Volume VI. Pigs, Hippopotamuses, Chevrotain, Giraffes, Deer and Bovids*. Bloomsbury, London.

Robinson, T. J., and E. R. Seiffert. 2004. Afrotherian origins and interrelationships: New views and future prospects. *Current Topics in Developmental Biology* 63:37–60.

Roca, A. L. 2019. African elephant genetics: Enigmas and anomalies. *Journal of Genetics* 98:83. https://doi.org/10.1007/s12041-109-1125-y

Roca, A. L., and S. J. O'Brien. 2005. Genomic inferences from Afrotheria and the evolution of elephants. *Current Opinion in Genetics & Development* 15:652–659.

Roca, A. L., N. Georgiadis, and S. J. O'Brien. 2005. Cytonuclear genomic dissociation in African elephant species. *Nature Genetics* 37:96–100.

Roca, A. L., N. Georgiadis, J. Pecon-Slattery, and S. J. O'Brien. 2001. Genetic evidence for two species of elephant in Africa. *Science* 293:1473–1477.

Roca, A. L., Y. Ishida, A. L. Brandt, N. R. Benjamin, K. Zhao, and N. J. Georgiadis. 2015. Elephant natural history: A genomic perspective. *Annual Review of Animal Biosciences* 3:139–167.

Roche, D., L. Ségalen, B. Senut, and M. Pickford. 2013. Stable isotope analyses of tooth enamel carbonate of large herbivores from the Tugen Hills deposits: Palaeoenvironmental context of the earliest Kenyan hominids. *Earth and Planetary Science Letters* 381:39–51.

Rodrigues, D., P.I. Abell, and S. Kröpelin. 2000. Seasonality in the early Holocene climate of northwest Sudan: Interpretation of *Etheria elliptica* shell isotopic data. *Global and Planetary Change* 26:181–187.

Rogaev, E. I., Y. K. Moliaka, B. A. Malyarchuk, F. A. Kondrashov, M. V. Derenko, I. Chumakov, and A. P. Grigorenko. 2006. Complete mitochondrial genome and phylogeny of Pleistocene mammoth *Mammuthus primigenius*. *PLoS Biology* 4(3). https://doi.org/10.1371/journal.pbio.0040073

Roger, J., M. Pickford, H. Thomas, F. de Lapparent de Broin, P. Tassy, W. Van Neer, C. Bourdillon-de-Gissac, and S. Al-Busaidi. 1994. Découverte de Vertébrés Fossiles dan le Miocène de la Région du Huqf au Sultanat d'Oman. *Annales de Paléontologie (Vert.-Invert.)* 80:253–273.

Roger, O. 1886. Ueber *Dinotherium bavaricum*. *Palaeontographica* 32:215–226.

Rögl, F. 1999. Oligocene and Miocene palaeogeography and stratigraphy of the circum-Mediterranean Region; pp. 485–500 in P. J. Whybrow and A. Hill (Eds.), *Fossil Vertebrates of Arabia*. Yale University Press, New Haven.

Rohland, N., A. Malaspinas, J. L. Pollack, M. Slatkin, P. Matheus, and M. Hofreiter. 2007. Proboscidean mitogenomics: Chronology and mode of elephant evolution using mastodon as outgroup. *PLoS Biology* 5:1663–1671.

Rohland, N., D. Reich, S. Mallick, M. Meyer, R. E. Green, N. J. Georgiadis, A. L. Roca, and M. Hofreiter. 2010. Genomic DNA sequences from mastodon and woolly mammoth reveal deep speciation of forest and savanna elephants. *PLoS Biology* 8:e10000564.

Rook, L. 2008. The discovery of the Sahabi Site: Ardito Desio or Carlo Petrocchi? *Garyounis Scientific Bulletin* Special Issue 2008:13–21.

Rook, L., R. L. Bernor, L. S. Avilla, O. Cirilli, L. Flynn, A. M. Jukar, W. Sanders, E. Scott, and X. Wang. 2019. Mammal biochronology (land mammal ages) around the world from late Miocene to middle Pleistocene and major events in horse evolutionary history. *Frontiers in Ecology and Evolution*. https://doi.org/10.3389/fevo.2019.00278

Rose, K. D., L. T. Holbrook R. S. Rana, K. Kumar, K. E. Jones, H. E. Ahrens, P. Missiaen, A. Sahni, and T. Smith. 2014. Early Eocene fossils suggest that the mammalian order Perissodactyla originated in India. *Nature Communications* 5:5570. https://doi.org/10.1038/ncomms6570 (2014)

Roth, V. L. 1989. Fabricational noise in elephant dentitions. *Paleobiology* 15:165–179.

—. 1992. Quantitative variation in elephant dentitions: Implications for the delimitation of fossil species. *Paleobiology* 18:184–202.

Rowan, J., I. A. Lazagabaster, C. J. Campisano, F. Bibi, R. Bobe, J.-R. Boisserie, S. R. Frost, T. Getachew, C. C. Gilbert, M. E. Lewis, S. Melaku, E. Scott, A. Souron, L. Werdelin, W. J. Kimbel, and K. E. Reed. 2022. Early Pleistocene large mammals from Maka'amaitalu, Hadar, lower Awash Valley, Ethiopia. *PeerJ* 10:e13210. https://doi.org/10.7717/peerj.13210

Saarinen, J. 2019. The palaeontology of browsing and grazing; pp. 5–59 in I. J. Gordon and H. H. T. Prins (Eds.), *The Ecology of Browsing and Grazing II*. Ecological Studies 239, Springer Nature, Switzerland.

Saarinen, J., A. Karme, T. Cerling, K. Uno, L. Säilä, S. Kasiki, S. Ngene, T. Obari, E. Mbua, F. K. Manthi, and M. Fortelius. 2015. A new tooth wear-based dietary analysis method for Proboscidea (Mammalia). *Journal of Vertebrate Paleontology*. https://doi.org/10.1080/02724634.2014.918546

Sach, V. J., and E. P. J. Heizmann. 2001. Stratigraphie and Säugetierfaunen der Brackwassermolasse in der Umgebung von Ulm (Südwestdeutschland). *Stuttgarter Beiträge zur Naturkunde, Serie B* 310:1–95.

Saegusa, H. 1987. Cranial morphology and phylogeny of the stegodonts. *The Compass* 64:221–243.

—. 1996a. Sister group relationships within Elephantoidea as suggested by the occlusal direction of their teeth; pp. 133–145 in B. Ratanasthien and S. L. Rieb (Eds.), *Proceedings International Symposium on Geology and Environment 31 January-2 February 1996*.

—. 1996b. Stegodontidae: Evoloutionary relationships; pp. 178–190 in J. Shoshani and P. Tassy (Eds.), *The Proboscidea: Evolution and Palaeoecology of Elephants and Their Relatives*. Oxford University Press, Oxford.

—. 2018. New proboscidean remains from the late Cenozoic of Central Myanmar. *Fossils. The Palaeontological Society of Japan* 104:35–49.

—. 2020. Stegodontidae and *Anancus*: Keys to understanding dental evolution in Elephantidae. *Quaternary Science Reviews* 231:106176.

Saegusa, H., and W. H. Gilbert. 2008. Elephantidae; pp. 193–226 in W. H. Gilbert and B. Asfaw (Eds.), *Homo Erectus. Pleistocene Evidence from the Middle Awash, Ethiopia*. University of California Press, Berkeley.

Saegusa, H., and Y. Haile-Selassie. 2009. Proboscidea; pp. 469–516 in Y. Haile-Selassie and G. WoldeGabriel (Eds.), *Ardipithecus Kadabba. Late Miocene Evidence from the Middle Awash, Ethiopia*. University of California Press, Berkeley.

Saegusa, H., and L. J. Hlusko. 2007. New late Miocene elephantoid (Mammalia: Proboscidea) fossils from Lemudong'o, Kenya. *Kirtlandia* 56:140–147.

Saegusa, H., H. Nakaya, Y. Kumimatsu, M. Nakatsukasa, H. Tsujikawa, Y. Sawada, M. Saneyoshi, and T. Sakai. 2014. Earliest elephantid remains from the late Miocene locality, Nakali, Kenya. *Scientific Annals, School of Geology, Aristotle University of Thessaloniki, Greece. VIth International Conference on Mammoths and their Relatives, Greneva-Siatista* Special Volume 102:175.

Saegusa, H., Y. Thasod, and B. Ratanasthien. 2005. Notes on Asian stegodontids. *Quaternary International* 126–128:31–48.

Sage, R. F. 2004. The evolution of C4 photosynthesis. *New Phytologist* 161:341–370.

Sagri, M., Abbate, E., Azzaroli, A., Balestrieri, M. L., Benvenuti, M., Bruni, P., Fazzuoli, M., Ficcarelli, G., Marcucci, M., Papini, M., Pavia, G., Reale, V., Rook, L., and T. M. Tecle. 1998. New data on the Jurassic and Neogene sedimentation in the Danakil Horst and northern Afar Depression, Eritrea; pp. 193–214 in S. Crasquin-Soleau and E. Barrier (Eds.), *Peri-Tethys Memoir 3: Stratigraphy and Evolution of Peri-Tethyan Platforms. Mémoires du Muséum National d'Histoire Naturelle, Paris*, vol. 177. Éditions du Muséum, Paris.

Sahle, Y., D. Giusti, and V. Tourkloukis. 2019. Newly discovered Middle Pleistocene hominid-bearing deposits from the Lower Awash basin, Ethiopia. *Anthropological Science*. https://doi.org/10.1537/ase.190603

Sahni, M. R., and C. Tripathi. 1957. A new classification of the Indian deinotheres and description of *D. orlovii* sp. nov. *Memoirs of the Geological Society of India, Paleontologia India* 33:1–33.

Sahnouni, M., and J. de Heinzelin. 1998. The site of Aïn Hanech revisited: New investigations at this Lower Pleistocene site in northern Algeria. *Journal of Archaeological Science* 25:1083–1101.

Sahnouni, M., D. Hadjouis, J. van der Made, A. Derradji, A. Canals, M. Medig, H. Belarech, Z. Harichane, and M. Rabhi. 2002. Further research at the Oldowan site of Aïn Hanech, north-eastern Algeria. *Journal of Human Evolution* 43:925–937.

—. 2004. On the earliest human occupation in North Africa: A response to Geraads et al. *Journal of Human Evolution* 46:763–775.

Sahnouni, M., J. van der Made, and M. Everett. 2011. Ecological background to Plio-Pleistocene hominin occupation in North Africa: The vertebrate faunas from Aïn Boucherit, Aïn Hanech and El-Kherba, and paleosol stable-carbon-isotope studies from El-Kherba, Algeria. *Quaternary Science Reviews* 30:1307–1317.

Sahnouni, M., J. Rosell, J. van der Made, J. M. Vergès, A. Ollé, N. Kandi, Z. Harichane, A. Derradji, and M. Medig. 2013. The first evidence of cut marks and usewear traces from the Plio-Pleistocene locality of El-Kherba (Ain Hanech), Algeria: Implications for early hominin subsistence activities circa 1.8 Ma. *Journal of Human Evolution* 64:137–150.

Said, R. 1962a. *The Geology of Egypt*. Elsevier, New York, 377 p.

—. 1962b. Über das Miozän in der westlichen wüste Ägyptens. *Geologisches Jahrbuch* 80:349–366.

Sallam, H. M., and E. R. Seiffert. 2016. New phiomorph rodents from the Eocene of Egypt, and the impact of Bayesian "clock"-based phylogenetic methods on estimates of basal hystricognath relationships and biochronology. *PeerJ* 4:e1717. https://doi.org/10.7717/peerj.1717

Sallam, H. M., E. R. Seiffert, and E. L. Simons. 2011. Craniodental morphology and systematics of a new family of hystricognathous rodents (Gaudeamuridae) from the late Eocene and early Oligocene of Egypt. *PLoS One* 6(2):e16525. https://doi.org/10.1371/journal.pone.001625

Sánchez, I. M., J. Morales, J. L. Cantalapiedra, V. Quiralte, and M. Pickford. 2018. Preliminary phylogenetic analysis of the Tragulidae (Mammalia, Cetartiodactyla, Ruminantia) from Arrisdrift: Implications for the African Miocene tragulids. *Communications of the Geological Survey of Namibia* 19:110–122.

Sánchez, I. M., V. Quiralte, J. Morales, B. Azanza, and M. Pickford. 2010. Sexual dimorphism of the frontal appendages of the early Miocene African pecoran *Prolibytherium* Arambourg, 1961 (Mammalia, Ruminantia). *Journal of Vertebrate Paleontology* 30:1306–1310.

Sánchez-Villagra, M. R., Y. Narita, and S. Kuratani. 2007. Thoracolumbar vertebral number: The first skeletal synapomophy for afrotherian mammals. *Systematics and Biodiversity* 5:1–7.

Sanders, W. J. 1990. Fossil Proboscidea from the Pliocene Lusso Beds of the Western Rift, Zaire; pp. 171–187 in N. T. Boaz (Ed.), *Evolution of Environments and Hominidae in the African Western Rift Valley*. Virginia Museum of Natural History Memoir No. 1, Martinsville, Virginia.

—. 1996. Fossil proboscideans of the Manonga Valley, Tanzania. *Journal of Vertebrate Paleontology* 16:62A–63A.

—. 1997. Fossil Proboscidea from the Wembere-Manonga Formation, Manonga Valley, Tanzania; pp. 265–310 in T. Harrison (Ed.), *Neogene Paleontology of the Manonga Valley, Tanzania*. Plenum Press, New York.

—. 1999. Oldest record of *Stegodon* (Mammalia: Proboscidea). *Journal of Vertebrate Paleontology* 19:793–797.

—. 2003. Proboscidea; pp. 209–219 in M. Fortelius, J. Kappelman, S. Sen, and R. L. Bernor (Eds.), *Geology and Paleontology of the Miocene Sinap Formation, Turkey*. Columbia University Press, New York.

—. 2004. Taxonomic and systematic review of Elephantidae based on late Miocene-early Pliocene fossil evidence from Afro-Arabia. *Journal of Vertebrate Paleontology* 24(suppl. to no. 3):109A.

—. 2005. New Pliocene fossil proboscidean specimens from Laetoli, Tanzania. *Journal of Vertebrate Paleontology* 25(-suppl. to no. 3):109A.

—. 2006. Comparative description and taxonomy of proboscidean fossils from Langebaanweg, South Africa. *African Natural History* 2:196–197.

—. 2007. Taxonomic review of fossil Proboscidea (Mammalia) from Langebaanweg, South Africa. *Transactions of the Royal Society of South Africa* 62:1–16.

—. 2008a. Review of fossil Proboscidea from the early-middle Miocene site of Jabal Zaltan, Libya. *Garyounis Scientific Bulletin* Special Issue No. 5:217–239.

—. 2008b. Review of fossil Proboscidea from the late Miocene-early Pliocene site of As Sahabi, Libya. *Garyounis Scientific Bulletin* Special Issue No. 5:217–239.

—. 2011. Proboscidea; pp. 233–262 in T. Harrison (Ed.), *Paleontology and Geology of Laetoli: Human Evolution in Context. Volume 2: Fossil Hominins and the Associated Fauna*. Springer, New York.

—. 2017. Horizontal tooth displacement and premolar occurrence in elephants and other elephantiform proboscideans. *Historical Biology*. https://doi.org/10.1080/08912963.2017.1297436

—. 2020. Proboscidea from Kanapoi, Kenya. *Journal of Human Evolution* 140:102547. https://doi.org/10.1016/j.jhevol.2018.10.013

—. 2022. Proboscidea; pp. 141–177 in F. Bibi, B. Kratz, M. Beech, and A. Hill (Eds.), *Sands of Time: Late Miocene Fossils from the Baynunah Formation, U.A.E.* Springer, Cham, Switzerland.

Sanders, W. J., and B. Bodenbender. 1994. Morphometric analysis of lumbar vertebra UMP 67-28: Implications for spinal function and phylogeny of the Miocene Moroto hominoid. *Journal of Human Evolution* 26:203–237.

Sanders, W. J., and Y. Haile-Selassie. 2012. A new assemblage of mid-Pliocene proboscideans from the Woranso-Mille Area, Afar Region, Ethiopia: Taxonomic, evolutionary, and paleoecological considerations. *Journal of Mammalian Evolution* 19:105–128.

Sanders, W. J., and E.R. Miller. 2002. New proboscideans from the early Miocene of Wadi Moghara, Egypt. *Journal of Vertebrate Paleontology* 22:388–404.

Sanders, W. J., and E. Seiffert. 2014. Proboscideans from the late Eocene Birket Qarun Formation of northern Egypt, and their biochronological implications. *Journal of Vertebrate Paleontology, Program and Abstracts* 2014:220–221.

Sanders, W. J., E. Gheerbrant, J. M. Harris, H. Saegusa, and C. Delmer. 2010a. Proboscidea; pp. 161–251 in L. Werdelin and W. J. Sanders (Eds.), *Cenozoic Mammals of Africa*. University of California Press, Berkeley.

Sanders, W. J., J. Kappelman, and D. T. Rasmussen. 2004. New large-bodied mammals from the late Oligocene site of Chilga, Ethiopia. *Acta Palaeontologica Polonica* 49:365–392.

Sanders, W. J., M. G. Leakey, L. N. Leakey, C. S. Feibel, T. G. Ibui, C. Nyete., M. P. Mbete, and F. H. Brown. 2021. Morphological description and identification of an extraordinary new fossil elephant cranium from the early Pliocene of Ileret, Kenya. *Palaeovertebrata* 44(2):e3. https://doi.org/10.18563/pv.44.2.e3

Sanders, W. J., E. R. Miller, I. Nengo, G. Semprebon, and I. Arney. 2020. Taxonomy and paleoecology of a remarkably diverse new proboscidean assemblage from the terminal early Miocene site of Buluk, Kenya. *Journal of Vertebrate Paleontology, Program and Abstracts* 2020:292.

Sanders, W. J., D. T. Rasmussen, and J. Kappelman. 2010b. Embrithopoda; pp. 115–122 in L. Werdelin and W. J. Sanders (Eds.), *Cenozoic Mammals of Africa*. University of California Press, Berkeley.

Sandrock, O., O. Kullmer, F. Schrenk, Y. M. Juwayeyi, and T. G. Bromage. 2007. Fauna, taphonomy, and ecology of the Plio-Pleistocene Chiwondo Beds, northern Malawi; pp. 315–332 in R. Bobe, Z. Alemseged, and A. K. Behrensmeyer (Eds.), *Hominin Environments in the East African Pliocene: An Assessment of the Faunal Evidence*. Springer, Dordrecht.

Sankhyan, A. R., and S. L. Sharma. 2014. *In situ* dental remains of *Deinotherium* from northwest Indian Siwaliks. *Himalyan Geology* 35:75–81.

Sarwar, M. 1977. Taxonomy and distribution of the Siwalik Proboscidea. *Bulletin of the Department of Zoology, University of the Punjab* 10:1–172.

Saunders, J. J. 1996. North American Mammutidae; pp. 271–279 in J. Shoshani and P. Tassy (Eds.), *The Proboscidea: Evolution and Palaeoecology of Elephants and Their Relatives*. Oxford University Press, Oxford.

Savage, R. J. G. 1967. Early Miocene mammal faunas of the Tethyan region. *Systematics Association Publication* 7:247–282.

—. 1969. Early Tertiary mammal locality in southern Libya. *Proceedings of the Geological Society of London* 1657:167–171.

—. 1971. Review of the fossil mammals of Libya. *Symposium on the Geology of Libya, Faculty of Science, University of Libya*, pp. 217–225.

—. 1976. Review of early Sirenia. *Systematic Zoology* 25:344–351.

—. 1989. The African dimension in European early Miocene mammal faunas; pp. 587–599 in E. H. Lindsay, V. Fahlbusch, and P. Mein (Eds.), *European Neogene Mammal Chronology*. Plenum Press, New York.

Savage, R. J. G., and W. R. Hamilton. 1973. Introduction to the Miocene mammalian faunas of Gebel Zelten, Libya. *Bulletin of the British Museum (Natural History), Geology* 22:515–527.

Savage, R. J. G., and M. E. White. 1965. Two mammal faunas from the early Tertiary of central Libya. *Proceedings of the Geological Society, London* 1623:89–91.

Savage, R. J. G., and P. G. Williamson. 1978. The early history of the Turkana Depression; pp. 375–394 in W. W. Bishop (Ed.), *Geological Background to Fossil Man*. Scottish Academic Press, Edinburgh.

Savage, R. J. G., D. P. Domning, and J. G. Thewissen. 1994. Fossil Sirenia of the West Atlantic and Caribbean region. V. The most primitive known sirenian, *Prorastomus sirenoides* Owen, 1855. *Journal of Vertebrate Paleontology* 14:427–449.

Sawada, Y., M. Pickford, T. Itaya, T. Makinouchi, M. Tateishi, K. Kabeto, S. Ishida, and H. Ishida. 1998. K-Ar ages of Miocene Hominoidea (*Kenyapithecus* and *Samburupithecus*) from Samburu Hills, northern Kenya. *Comptes Rendus de l'Académie des Sciences, Paris, Sciences de la Terre et des Planetès* 326:445–451.

Scally, M., O. Madsen, C. J. Douady, W. W. de Jong, M. J. Stanhope, and M. S. Springer. 2001. Molecular evidence for the major clades of placental mammals. *Journal of Mammalian Evolution* 8:239–277.

Schanz, T., Y. Lins, H. Viefhas, T. Barciaga, S. Läbe, H. Preuschoft, U. Witzel, and P. M. Sander. 2013. Quantitative interpretation of tracks for determination of body mass. *PLoS ONE* 8(10):e77606.

Schinz, R. H. 1824. *Naturgeschichte und Abbildungen der Säutethiere*. Brodtmanns Lithographische Kunstantalt, Zürich. 348 pp.

Schlesinger, G. 1917. Die mastodonten des K. K. naturhistorischen Hofmuseums. *Denkschriften des K. K. Naturhistorischen Hofmuseums, Geologischl-Paläontologische Reihe* 1:181–222.

Schlosser, M. 1911. Beiträge zur Kenntnis der Oligozänen Landsäugetiere aus dem Fayum: Ägypten. *Beiträge zur Paläontologie und Geologie Österreich-Ungarns* 24:51–167.

Schmitt, A., and E. Gheerbrant. 2016. The ear region of earliest known elephant relatives: New light on the ancestral morphotype of proboscideans and afrotherians. *Journal of Anatomy* 228:137–152.

Schoeninger, M. J., H. Reeser, and K. Hallin. 2003. Paleoenvironment of *Australopithecus anamensis* at Allia Bay, East Turkana, Kenya: Evidence from mammalian herbivore enamel stable isotopes. *Journal of Anthropological Archaeology* 22:200–207.

Scholz, C. A., T. C. Johnson, A. S. Cohen, J. W. King, J. A. Peck, J. T. Overpeck, M. R. Talbot, E. T. Brown, L. Kalindekafe, P. Y. O. Amoako, R. P. Lyons, T. M. Shanahan, I. S. Castañeda, C. W. Heil, S. L. Forman, L. R. McHargue, K. R. Beuning, J. Gomez, and J. Pierson. 2007. East African megadroughts between 135 and 75 thousand years ago and bearing on early-modern human origins. *Proceedings of the National Academy of Sciences, USA* 104:16416–16421.

Schouteden, H. 1911a. L'éléphant nain du Congo. *Revue de Zoologique africaine* 1:222–229.

Schouteden, H. 1911b. L'éléphant nain du Lac Leopold II (Congo). *Revue de Zoologique africaine* 1:391–397.

Schreiter, S., S. I. Higgins, C. P. Osborne, C. Bradshaw, D. Lunt, B. S. Ripley, L. L. Taylor, and D. J. Beerling. 2012. Fire and fire-adapted vegetation promoted C_4 expansion in the late Miocene. *New Phytologist*. https://doi.org/10.1111/j/1469-8137.2012.04202.x

Schrenk, F., T. G. Bromage, C. G. Betzler, U. Ring, and Y. M. Juwayeyi. 1993. Oldest *Homo* and Pliocene biogeography of the Malawi Rift. *Nature* 365:833–836.

Schuster, M., P. Duringer, J.-F. Ghienne, C. Roquin, P. Sepulchre, A. Moussa, A.-E. Lebatard, H. T. Mackaye, A. Likius, P. Vignaud, and M. Brunet. 2009. Chad Basin: Paleoenvironments of the Sahara since the late Miocene. *Compte Rendus Geoscience* 341:603–611.

Scott, W. B. 1907. A collection of fossil mammals from the coast of Zululand; pp. 253–262 in W. Anderson (Ed.), *Third and Final Report of the Geological Survey of Natal and Zululand*. West, Newman and Co., London.

Ségalen, L. J. A. Lee-Thorp, and T. Cerling. 2007. Timing of C_4 grass expansion across sub-Saharan Africa. *Journal of Human Evolution* 53:549–559.

Seiffert, E. R. 2006. Revised age estimates for the later Paleogene mammal faunas of Egypt and Oman. *Proceedings of the National Academy of Sciences, USA* 103:5000–5005.

—. 2007a. Evolution and extinction of Afro-Arabian primates near the Eocene-Oligocene boundary. *Folia Primatologica* 78:314–327.

—. 2007b. A new estimate of afrotherian phylogeny based on simultaneous analysis of genomic, morphological, and fossil evidence. *BMC Evolutionary Biology* 7:224. https://doi.org/10.1186/1471-2148-7-224.

—. 2013a. Cohort Paenungulata; p. 147 in J. Kingdon, D. Happold, M. Hoffmann, T. Butynski, M. Happold, and J. Kalina (Eds.), *Mammals of Africa, Volume I*. Bloomsbury, London.

—. 2013b. Superorder Tethytheria; p. 172 in J. Kingdon, D. Happold, M. Hoffmann, T. Butynski, M. Happold, and J. Kalina (Eds.), *Mammals of Africa, Volume I*. Bloomsbury, London.

Seiffert, E. R., T. M. Bown, W. C. Clyde, and E. Simons. 2008. Geology, paleonenvironment, and age of Birket Qarun Locality 2 (BQ-2), Fayum Depression, Egypt; pp. 71–86 in J. G. Fleagle and C. C. Gilbert (Eds.), *Elwyn Simons: A Search for Origins*. Springer, New York.

Seiffert, E. R., S. Nasir, A. Al-Harthy, J. R. Groenke, B. P. Kraatz, N. J. Stevens, and A. R. Al-Sayigh. 2012. Diversity in the later Paleogene proboscidean radiation: A small barytheriid from the Oligocene of Dhofar Governorate, Sultanate of Oman. *Naturwissenschaften* 99:133–141.

Seiffert, E. R., Simons, E. L., Clyde, W. C., Rossie, J. B., Attia, Y., Bown, T. M., Chatrath, P., and M. Mathison. 2005b. Basal anthropoids from Egypt and the antiquity of Africa's higher primate radiation. *Science* 310:300–304.

Seiffert, E. R., E. L. Simons, J. G. Fleagle, and M. Godinot. 2010. Paleogene anthropoids; pp. 369–392 in L. Werdelin and W. J. Sanders (Eds.), *Cenozoic Mammals of Africa*. University of California Press, Berkeley.

Seiffert, E. R., E. L. Simons, T. M. Ryan, and Y. Attia. 2005a. Additional remains of *Wadilemur elegans*, a primitive stem galagid from the late Eocene of Egypt. *Proceedings of the National Academy of Sciences, USA* 102:11396–11401.

Selley, R. C. 1967. Palaeocurrents and sediment transport in nearshore sediments of the Sirte basin, Libya. *Journal of Geology* 75:215–223.

—. 1969. Near-shore marine and continental sediments of the Sirte basin, Libya. *Quarterly Journal of the Geological Society, London* 124:419–460.

Semaw, S., S. W. Simpson, J. Quade, P. R. Renne, R. F. Butler, W. C. McIntosh, N. Levin, M. Dominguez-Rodrigo, and M. J. Rogers. 2005. Early Pliocene hominids from Gona, Ethiopia. *Nature* 433:301–304.

Semprebon, G. M., W. J. Sanders, A. Lister, M. E. Morgan, T. E. Cerling, G. Rivals, U. Gohlich, and J. M. Fahlke. 2015. Dietary reconstruction of fossil proboscideans from the

Siwalik Series of Pakistan using enamel and carbon isotopes. *Journal of Vertebrate Paleontology, Program and Abstracts* 2015:211–212.

Semprebon, G. M., L. M. Uttrecht, and W. J. Sanders. 2018. What can enamel microwear reveal about the dietary behavior of *Moeritherium* and *Phiomia* from the Fayum region of Egypt? *Journal of Vertebrate Paleontology, Program and Abstracts* 2018:213–214.

Semprebon, G. M., L. Utrecht, W. J. Sanders, J. M. Fahlke, and F. Rivals. 2017. Dietary reconstruction of *Moeritherium* and *Phiomia* from the Fayum Sequence (Egypt) using enamel microwear. *VII International Conference of Mammoths and Their Relatives, Taiwan, Abstracts Book*, P–04.

Sen, S. 2013. Dispersal of African mammals in Eurasia during the Cenozoic: Ways and whys. *Geobios* 46:159–172.

Sénégas, F., and D. M. Avery. 1998. New evidence for the murine origins of the Otomyinae (Mammalia, Rodentia) and the age of Bolt's Farm (South Africa). *South African Journal of Science* 94:503–507.

Senut, B., M. Pickford, D. Gommery, P. Mein, K. Cheboi, and Y. Coppens. 2001. First hominid from the Miocene (Lukeino Formation, Kenya). *Comptes Rendus de l'Academie des Sciences, Series IIA—Earth and Planetary Science* 332:137–144.

Senut, B., M. Pickford, M. de Wit, J. Ward, R. Spaggiari, and J. Morales. 1996. Biochronology of sediments at Bosluis Pan, Northern Cape Province, South Africa. *South African Journal of Science* 92:249–251.

Sepulchre, P., G. Ramstein, F. Fluteau, M. Schuster, M. Tiercelin, and J.-J. Brunet. 2006. Tectonic uplift and eastern African aridification. *Science* 313:1419–1423.

Servant-Vildary, S. 1973. Stratigraphie et néotectonique du Plio-Pléistocène ancien du Tchad d'après l'étude des Diatomées. *Comptes Rendus de l'Academie des Sciences, Série D* 276:2633–2636.

Shackley, M. 1980. An Acheulean industry with *Elephas recki* fauna from Namib IV South West Africa (Namibia). *Nature* 284:340–341.

Shackley, M., K. Komura, T. Hayashi, M. Ikeya, S. Matsu'ura, and N. Ueta. 1985. Chronometric dating of bone from Namib IV Acheulean site, South West Africa/Namibia. *Bulletin of the National Science Museum, Tokyo (Series D, Anthropology)* 11:7–12.

Shipman, P. 1986. Paleoecology of Fort Ternan reconsidered. *Journal of Human Evolution* 15:193–204.

Shipman, P., A. Walker, J. A. Van Couvering, P. J. Hooker, and J. A. Miller. 1981. The Fort Ternan hominoid site, Kenya: Geology, age, taphonomy and paleoecology. *Journal of Human Evolution* 10:49–72.

Short, J. 1981. Diet and feeding behaviour of the forest elephant. *Mammalia* 45:177–186.

Shoshani, J. 1992. Anatomy and Physiology; pp. 66–81 in J. Shoshani (Ed.), *Elephants*. Weldon Owen Pty Limited, Sydney.

—. 1996. Para- or monophyly of the gomphotheres and their position within Proboscidea; pp. 149–177 in J. Shoshani and P. Tassy (Eds.), *The Proboscidea: Evolution and Palaeoecology of Elephants and Their Relatives*. Oxford University Press, Oxford.

—. 1997. It's a nose! It's a hand! It's an elephant's trunk! *Natural History* 106:37–43.

—. 1998. Understanding proboscidean evolution: A formidable task. *Tree* 13:480–487.

—. 2002. Proboscidea (Elephants); pp. 1–16. *Encyclopedia of Life Sciences*. Wiley, Hoboken, NJ.

—. 2003. *Mammut* hyoid elements from the Hiscock site; description and implications. *Bulletin of the Buffalo Society of Natural Sciences* 37:114–120.

—. 2004. Paleozoogeography and neozoogeography of mammals in Eritrea. *Bulletin of Carnegie Museum of Natural History* 36:267–276.

Shoshani, J., and J. F. Eisenberg. 1982. *Elephas maximus*. *Mammalian Species* 182:1–8.

Shoshani, J., and G. Marchant. 2001. Hyoid apparatus: A little known complex of bones and its "contribution" to proboscidean evolution; pp. 668–675 in G. Cavarretta, P. Gioia, M. Mussi, and M. R. Palombo (Eds.), *Proceedings of the First International Congress of la Terra degli Elefanti: The World of Elephants*. Consiglio Nazionale delle Richerche, Rome.

Shoshani, J., and P. Tassy. 1992. Classifying elephants; pp. 22–23 in J. Shoshani (Ed.), *Elephants*. Weldon Owen Pty Limited, Sydney.

—. 1996. Summary, conclusions, and a glimpse into the future; pp. 335–348 in J. Shoshani and P. Tassy (Eds.), *The Proboscidea: Evolution and Palaeoecology of Elephants and Their Relatives*. Oxford University Press, Oxford.

—. 2005. Advances in proboscidean taxonomy & classification, anatomy & physiology, and ecology & behavior. *Quaternary International* 126–128:5–20.

—. 2013. Order Proboscidea–Elephants; pp. 173–176 in J. Kingdon, D. Happold, M. Hoffmann, T. Butynski, M. Happold, and J. Kalina (Eds.), *Mammals of Africa, Volume I*. Bloomsbury, London.

Shoshani, J., P. Bloomer, and E. R. Seiffert. 2013. Order Hyracoidea—hyraxes; pp. 148–149 in J. Kingdon, D. Happold, M. Hoffmann, T. Butynski, M. Happold, and J. Kalina (Eds.), *Mammals of Africa, Volume I*. Bloomsbury, London.

Shoshani, J., M. P. Ferretti, A. M. Lister, L. D. Agenbroad, H. Saegusa, D. Mol, and K. Takahashi. 2007. Relationships within the Elephantinae using hyoid characters. *Quaternary International* 169–170:174–185.

Shoshani, J., E. M. Golenberg, and H. Yang. 1998. Elephantidae phylogeny: Morphological versus molecular results. *Acta Theriologica* Suppl. 5:89–122.

Shoshani, J., W. J. Kupsky, and G. H. Marchant. 2006. Elephant brain. Part I: Gross morphology, functions, comparative anatomy, and evolution. *Brain Research Bulletin* 70:124–157.

Shoshani, J., W. J. Sanders, and P. Tassy. 2001a. Elephants and other proboscideans: A summary of recent findings and new taxonomic suggestions; pp. 676–679 in G. Cavarretta, P. Gioia, M. Mussi, and M. R. Palombo (Eds.), *Proceedings of the First International Congress of la Terra degli Elefanti: The World of Elephants*. Consiglio Nazionale delle Richerche, Rome.

Shoshani, J., R. J. G. Savage, and R. M. West. 1989. A new species of *Barytherium* (Mammalia, Proboscidea) from Africa, and a discussion of early proboscidean systematics and paleoecology. In *Abstracts of Papers and Posters, Fifth International Theriological Congress, Rome*, p. 160.

Shoshani, J., R. C. Walter, M. Abraha, S. Berhe, R. T. Buffler, and B. Negassi. 2001c. *Elephas recki* from Dandero, northern Danakil Depression, Eritrea; pp. 143–147 in G. Cavarretta, P. Gioia, M. Mussi, and M. R. Palombo (Eds.), *Proceedings of the First International Congress of La Terra degli Elefanti: The World of Elephants*. Consiglio Nazionale dell Richerche, Rome.

Shoshani, J., R. C. Walter, M. Abraha, S. Berhe, P. Tassy, W. J. Sanders, G. H. Marchant, Y. Libsekal, T. Ghirmai, and D. Zinner. 2004. A proboscidean from the late Oligocene of Eritrea, a "missing link" between early Elephantiformes and Elephantimorpha, and biogeographic implications. *Proceedings of the National Academy of Sciences, USA* 103:17296–17301.

Shoshani, J., R. C. Walter, Y. Libeskal, M. Abraha, and S. Berhe. 2001b. The oldest gomphothere in Africa: A latest Oligocene proboscidean mandible from the Dogali Formation, eastern Eritrea. *Abstracts of Papers and Posters, Eighth International Theriological Congress, Sun City, South Africa, 12–17 August 2001*, Abstract No. 337:128–129.

Shoshani, J., R. M. West, N. Court, R. J. G. Savage, and J. M. Harris. 1996. The earliest proboscideans: General plan, taxonomy and palaeoecology; pp. 57–75 in J. Shoshani and P. Tassy (Eds.), *The Proboscidea: Evolution and Palaeoecology of Elephants and Their Relatives*. Oxford University Press, Oxford.

Sigogneau-Russell, D., J. J. Hooker, and P. C. Ensom. 2001. The oldest tribosphenic mammal from Laurasia (Purbeck Limestone Group, Berriasian, Cretaceous, UK) and its bearing on the 'dual origin' of Tribosphenida. *Comptes rendus de l'Académie des Sciences* 333:141–147.

Sikes, S. K. 1967. The African elephant, *Loxodonta africana*: a field method for the estimation of age. *Journal of Zoology, London* 154:235–248.

—. 1971. *The Natural History of the African Elephant*. Weidenfeld and Nicolson, London, 397 pp.

Simmons, N. B., E. R. Seiffert, and G. F. Gunnell. 2016. A new family of large omnivorous bats (Mammalia, Chiroptera) from the late Eocene of the Fayum Depression, Egypt, with comments on use of the name "Eochiroptera." *American Museum Novitates* 3857:1–43.

Simons, E. L. 1968. Early Cenozoic mammalian faunas. Fayum Province, Egypt. Part I. *Bulletin of the Peabody Museum of Natural History* 28:1–21.

Simpson, G. G. 1945. Principles of classification and a classification of mammals. *Bulletin of the American Museum of Natural History* 85:1–350.

Singer, R., and J. R. Crawford. 1958. The significance of the archaeological discoveries at Hopefield, South Africa. *The Journal of the Royal Anthropological Institute of Great Britain and Ireland* 88:11–19.

Singer, R., and D. A. Hooijer. 1958. A *Stegolophodon* from South Africa. *Nature* 182:101–102.

Singh, N. P., A. M. Jukar, R. Patnaik, K. M. Sharma, N. A. Singh, and Y. P. Singh. 2020. The first specimen of *Deinotherium indicum* (Mammalia, Proboscidea, Deinotheriidae) from the late Miocene of Kutch, India. *Journal of Paleontology*. https://doi.org/10/1017/jpa.2020.3

Smart, C. 1976. The Lothagam 1 fauna: Its phylogenetic, ecological, and biogeographic significance; pp. 361–369 in Y. Coppens, F. C. Howell, G. L. Isaac, and R. E. F. Leakey (Eds.), *Earliest Man and Environments in the Lake Rudolph Basin: Stratigraphy, Paleoecology, and Evolution*. University of Chicago Press, Chicago.

Smith, K. M., and D. D. Stynder. 2015. Biogeography and molar morphology of Pleistocene African elephants: New evidence from Elandsfontein, Western Cape Province, South Africa. *Quaternary Science Reviews* 115:101–111.

Smith, N. A. 2013. Avian fossils from the early Miocene Moghra Formation of Egypt. *Ostrich* 84:181–189.

Smuts, M. M. S., and A. J. Bezuidenhout. 1994. Osteology of the pelvic limb of the African elephant (*Loxodonta africana*). *Oderstepoort Journal of Veterinary Research* 61:51–66.

Solé, F., E. M. Essid, W. Marzougui, R. Temani, H. K. Ammar, M. Mahboubi, L. Marivaux, M. Vianey-Liaud, and R. Tabuce. 2016. New fossils of Hyaenodonta (Mammalia) from the Eocene localities of Chambi (Tunisia) and Bir el Ater (Algeria), and the evolution of the earliest African hyaenodontids. *Palaeontolgia Electronica* 19(3):41A.

Solé, F., E. Gheerbrant, M. Amaghzaz, and B. Bouya. 2009. Further evidence of the African antiquity of hyaenodontid ('Creodonta,' Mammalia) evolution. *Zoological Journal of the Linnean Society* 156:827–846.

Soltis, J. 2010. Vocal communication in African elephants (*Loxodonta africana*). *Zoo Biology* 29:192–209.

Spiegel, C., B. P. Kohn, D. X. Belton, and A. J. W. Gleadow. 2007. Morphotectonic evolution of the central Kenya rift flanks: Implications for late Cenozoic environmental change in East Africa. *Geology* 35:427–430.

Springer, M. S., G. C. Cleven, O. Madsen, W. W. de Jong, V. G. Waddell, H. M. Amrine, and M. J. Stanhope. 1997. Endemic African mammals shake the phylogenetic tree. *Nature* 388:61–64.

Springer, M. S., N. M. Foley, P. L. Brady, J. Gatesy, and W. J. Murphy. 2019. Evolutionary models for the diversification of placental mammals across the KPg boundary. *Frontiers in Genetics* 10:1–22. https://doi.org/10.3389/fgene.2019.01241

Springer, M. S., R. W. Meredith, J. E. Janecka, and W. J. Murphy. 2011. The historical biogeography of Mammalia. *Philosophical Transactions of the Royal Society B* 366:2478–2502.

Springer, M. S., W. J. Murphy, E. Eizirik, and S. J. O'Brien. 2003. Placental mammal diversification and the Cretaceous Tertiary boundary. *Proceedings of the National Academy of Sciences, USA* 100:1056–1061.

Stanhope, M. J., V. G. Waddell, O. Madsen, W. de Jong, S. B. Hedges, G. C. Cleven, D. Kao, and M. S. Springer. 1998. Molecular evidence for multiple origins of Insectivora and for a new order of endemic African insectivore mammals. *Proceedings of the National Academy of Sciences, USA* 95:9967–9972.

Stansfield, F. J. 2016. Assessment of African elephant (*Loxodonta africana*) molars and mandibles and their association with estimated age and sex. *Pachyderm* 57:68–77.

Stefanescu, S. 1924. Sur la présence de l'*Eléphas planifrons* et de trois mutations de l'*Eléphas antiquus* dans les couches géologiques de Roumanie. *Comptes Rendus de l'Academie des Sciences, Paris* 179:1418–1419.

Stevens, N. J., E. R. Seiffert, P. M. O'Connor, E. M. Roberts, M. D. Schmitz, C. Krause, E. Gorscak, S. Ngasala, T. L. Hieronymus, and J. Temu. 2013. Palaeontological evidence for an Oligocene divergence between Old World monkeys and apes. *Nature* 497:611–614.

Stewart, D. R. M., and J. Stewart. 1963. The distribution of some large mammals in Kenya. *Journal of the East African Natural History Society* 24:1–52.

Stewart, J. R. 2005. Miocene geology and fossils of Abu Dhabi; pp. 17–42 in M. Beech and P. Hellyer (Eds.), *Abu Dhabi–8 million Years Ago. Late Miocene Fossils from the Western Region*. Abu Dhabi Islands Archaeological Survey, Abu Dhabi, United Arab Emirates.

Stewart, K. 2003. Fossil fish remains from the Pliocene Kanapoi Site, Kenya; pp. 21–38 in J. M. Harris and M. G. Leakey (Eds.), *Geology and Vertebrate Paleontology of the Early Pliocene*

Stewart, M., J. Louys, H. S. Groucutt, I. Candy, R. Clark-Wilson, P. S. Breeze, N. A. Drake, G. J. Price, Y. S. A. Al-Mufarreh, A. al Omari, and M. D. Petraglia. 2019. Taphonomic and zooarchaeological investigations at the middle Pleistocene site of Ti's al Ghadah, western Nefud Desert, Saudi Arabia. *Quaternary Science Reviews* 218:228–253.

Stimpson, C. M., A. Lister, A. Parton, L. Clark-Balzan, P. S. Breeze, N. A. Drake, H. S. Groucutt, R. Jennings, E. M. L. Scerri, T. S. White, M. Zahir, M. Duval, R. Grun, A. A-Omari, K. S. M. Al Murayyi, A. S. Zalmout, Y. A. Mufarreh, A. M. Memesh, and M. D. Petraglia. 2016. Middle Pleistocene vertebrate fossils from the Nefud Desert, Saudi Arabia: Implications for biogeography and palaeoecology. *Quaternary Science Reviews* 143:13–36.

Strömberg, C. A. E. 2011. Evolution of grasses and grassland ecosystems. *Annual Review Earth and Planetary Sciences* 39:517–544.

Su, D., and Y. Haile-Selassie. 2022. Mosaic habitats at Woranso-Mille (Ethiopia) during the Pliocene and implications for *Australopithecus* paleoecology and taxonomic diversity. *Journal of Human Evolution* 163:103076.

Su, D., and T. Harrison. 2015. The paleoecology of the Upper Laetolil Beds, Laetoli Tanzania: A review and synthesis. *Journal of African Earth Sciences* 101:405–419.

Sudre, J., J.-J. Jaeger, B. Sigé, and M. Vianey-Liaud. 1993. Nouvelles données sur les condylarthres du Thanétien et de l'Yprésien du Bassin d'Ouarzazate (Maroc). *Geobios* 26:609–615.

Sukumar, R., and R. Ramesh. 1992. Stable carbon isotope ratios in Asian elephant collagen: Implications for dietary studies. *Oecologia* 91:536–539.

Sukumar, R., and C. Santiapillai. 1996. *Elephas maximus*: Status and distribution; pp. 327–331 in J. Shoshani and P. Tassy (Eds.), *The Proboscidea: Evolution and Palaeoecology of Elephants and Their Relatives*. Oxford University Press, Oxford.

Sukumar, R., S. K. Bhattacharaya, and R. V. Krishnamurthy. 1987. Carbon isotope evidence for different feeding patterns in an Asian elephant population. *Current Science* 56:11–14.

Suwa, G., H. Nakaya, B. Asfaw, H. Saegusa, A. Amzaye, R. T. Kono, Y. Beyene, and S. Katoh. 2003. Plio-Pleistocene terrestrial mammal assemblage from Konso, southern Ethiopia. *Journal of Vertebrate Paleontology* 23: 901–916.

Suwa, G., T. White, B. Asfaw, G. WoldeGabriel, and T. Yemane. 1991. Miocene faunal remains from the Burji-Soyama area, Amaro Horst, southern sector of the main Ethiopian rift. *Palaeontologia Africana* 28:23–28.

Sweyden, S., G. Merzeraud, E. M. Essid, W. Marzougui, R. Temani, H. K. Ammar, L. Marivaux, M. Vianey-Liaud, R. Tabuce, and S. Adnet. 2019. A reworked elasmobranch fauna from Tunisia providing a snapshot of Eocene-Oligocene Tethyan faunas. *Journal of African Earth Sciences* 149:194–206.

Swezey, C. S. 2009. Cenozoic stratigraphy of the Sahara, northern Africa. *Journal of African Earth Science* 53:89–121.

Symeonidis, N. K. 1970. Ein *Dinotherium*-Fund in Zentralmakedonien (Griechenland). *Annales Géologiques des Pays Helléniques* 3:1153–1165.

Tabuce, R. 2002. *Mammifères du Paléogène ancien d'Afrique du Nord-Occidentale: Systématique, paléobiogéographie et apport à la phylogénie des "orders endemiques africains."* Unpublished PhD Thesis, University Montpellier II, Montpellier, 226 pp.

Tabuce, R., B. Coffait, P. E. Coffait, M. Mahboubi, and J. J. Jaeger. 2000. A new species of *Bunohyrax* (Hyracoidea, Mammalia) from the Eocene of Bir El Ater (Algeria). *Comptes Rendus de l'Académie des Sciences de Paris, Série II, Sciences de la Terre et des Planètes* 331:61–66.

—. 2001. A new genus of macroscelidea (Mammalia) from the Eocene of Algeria: A possible origin for elephant-shrews. *Journal of Vertebrate Paleontology* 21:535–546.

Tabuce, R., C. Delmer, and E. Gheerbrant. 2007b. Evolution of the tooth enamel microstructure in the earliest proboscideans (Mammalia). *Zoological Journal of the Linnean Society* 149:611–628.

Tabuce, R., J.-J. Jaeger, L. Marivaux, M. Salem, A. A. Bilal, M. Benammi, Y. Chaimanee, P. Coster, B. Marandat, X. Valentin, and M. Brunet. 2012. New stem elephant-shrews (Mammalia, Macroscelidea) from the Eocene of Dur at-Talah, Libya. *Palaeontology* 55:945–955.

Tabuce, R., F. Lihoreau, F. Mees, M. J. Orliac, T. De Putter, and T. Smith. 2021. A reassessment of the Oligocene hyracoid mammals from Malembo, Cabinda, Angola. *Geobios* 66–67:207–215.

Tabuce, R., L. Marivaux, M. Adaci, M. Bensalah, J.-L. Hartenberger, M. Mahboubi, F. Mebrouk, P. Tafforeau, and J.-J. Jaeger. 2007a. Early Tertiary mammals from North Africa reinforce the molecular Afrotheria clade. *Proceedings of the Royal Society of London B* 274:1159–1166.

Tabuce, R., L. Marivaux, R. Lebrun, M. Adaci, M. Bensalah, P.-H. Fabre, H. Gomes Rodrigues, L. Hautier, J.-J. Jaeger, V. Lazzari, F. Mebrouk, S. Peigne, J. Sudre, P. Tafforeau, X. Valentin, and M. Mahboubi. 2009. Anthropoid versus strepsirrhine status of the African Eocene primates *Algeripithecus* and *Azibius*: Craniodental evidence. *Proceedings of the Royal Society B: Biological Sciences* 276:4087–4094.

Tabuce, R., R. Sarr, S. Adnet, R. Lebrun, F. Lihoreau, J. E. Martin, B. Sambou, M. Thiam, and L. Hautier. 2019. Filling a gap in the proboscidean fossil record: A new genus from the Lutetian of Senegal. *Journal of Paleontology*. https://doi.org/10.1017/jpa.2019.98

Taieb, M., D. C. Johanson, Y. Coppens, and J. L. Aronson. 1976. Geological and paleontological background of Hadar hominid site, Afar, Ethiopia. *Nature* 260:289–293.

Takai, F. 1954. An addition to the mammalian fauna of the Japanese Miocene. *Journal of the Faculty of Sciences, University of Tokyo*, section II, 9:331–335.

Takai, F., H. Saegusa, Thaung-Htike, and Zin-Maung-Maung-Thein. 2006. Neogene mammalian fauna in Myanmar. *Asian Paleoprimatology* 4:143–172.

Tamrat, E., N. Thouveny, M. Taieb, and N. D. Opdyke. 1995. Revised magnetostratigraphy of the Plio-Pleistocene sedimentary sequence of the Olduvai Formation (Tanzania). *Palaeogeography, Palaeoclimatology, Palaeoecology* 114:273–283.

Tanner, L. G. 1978. Embrithropoda; pp. 277–283 in V. J. Maglio and H. B. S. Cooke (Eds.), *Evolution of African Mammals*. Harvard University Press, Cambridge.

Tassy, P. 1977a. Découverte de *Zygolophodon turicensis* (Schinz) (Proboscidea, Mammalia) au lieu-dit Malartic a Simorre, Gers (Vidobonien moyen); implications paléoécologiques et biostratigraphiques. *Geobios* 10:655–669.

—. 1977b. Présence du genre *Choerolophodon* Schlesinger (Proboscidea, Mammalia) dans le Miocène est-africain. *Comptes Rendus de l'Académie des Sciences, Paris, Série D* 248:2487–2490.

—. 1979a. Les Proboscidiens (Mammalia) du Miocène d'Afrique orientale: résultats préliminaires. *Bulletin de la Société Géologique de France* 21:265–269.
—. 1979b. Relations phylogénétiques du genre *Moeritherium* Andrews, 1901 (Mammalia). *Comptes Rendus de l'Académie des Sciences de Paris, Série D* 189:85–88.
—. 1981. Le crâne de *Moeritherium* (Proboscidea, Mammalia) de l'Eocène de Dor el Talha (Libye) et le problème de la classification phylogénétique du genre les Tethytheria McKenna, 1975. *Bulletin du Muséum National d'Histoire Naturelle de Paris, 4e Série, 3, Section C* 1:87–147.
—. 1982. Les principales dichotomies dans l'histoire des Proboscidea (Mammalia): Une approche phylogénétique. *Geobios* Special Edition 6:225–245.
—. 1983a. Les Elephantoidea miocènes du Plateau du Potwar, Groupe de Siwalik, Pakistan. Ire partie: introduction, cadre chronologique et géographique, mammutidés, amébélodontidés. *Annales de Paléontologie* 69:99–136.
—. 1983b. Les Elephantoidea Miocènes du Plateau du Potwar, Groupe de Siwalik, Pakistan. IIe Partie: choerolophodonts et gomphothères. *Annales de Paléontologie* 69:235–298.
—. 1983c. Les Elephantoidea Miocènes du Plateau du Potwar, Groupe de Siwalik, Pakistan. IIIe partie: stégodontidés, éléphantoïdes indéterminés. Restes postcrâniens. Conclusions. *Annales de Paléontologie* 69:317–354.
—. 1984. Le mastodonte à dents étroites, le grade trilophodonte et la radiation initiale des Amebelodontidae; pp. 459–473 in E. Buffetaut, J.-M. Mazin, and E. Salmon (Eds.), *Actes du Symposium paléontologique Georges Cuvier.* Montbeliard.
—. 1985. *La place des mastodontes miocènes de l'ancien monde dans la phylogénie des Proboscidea (Mammalia): Hypothèses et conjectures.* Unpublished PhD dissertation, Université Pierre et Marie Curie, Paris., vols. 1–3, 861 pp.
—. 1986. *Nouveaux Elephantoidea (Mammalia) dans le Miocène du Kenya.* Cahiers de Paléontologie, Éditions du Centre de la Recherche Scientifique, Paris, 135 pp.
—. 1988. The classification of Proboscidea: How many cladistic classifications? *Cladistics* 4:43–57.
—. 1990a. The "Proboscidean Datum Event": How many proboscideans and how many events?; pp. 237–252 in E.H. Lindsay, V. Fahlbusch, and P. Mein (Eds.), *European Neogene Mammal Chronology.* Plenum Press, New York.
—. 1990b. Phylogénie et classification des Proboscidea (Mammalia): Historique et actualité. *Annales de Paléontologie* 76:159–224.
—. 1994a. Gaps, parsimony, and early Miocene elephantoids (Mammalia), with a re-evaluation of *Gomphotherium annectens* (Matsumoto, 1925). *Zoological Journal of the Linnean Society* 112:101–117.
—. 1994b. Origin and differentiation of the Elephantiformes (Mammalia, Proboscidea). *Verhandlungen des Naturwissenschaftlichen Vereins in Hamburg* 34:73–94.
—. 1995. Les Proboscidiens (Mammalia) Fossiles du Rift Occidental, Ouganda; pp. 217–257 in M. Pickford and B. Senut (Eds.), *Geology and Palaeobiology of the Albertine Rift Valley, Uganda-Zaire. Vol. II: Palaeobiology.* CIFEG Occasional Publications, 1994/29, Orléans.
—. 1996a. Dental homologies and nomenclature in the Proboscidea; pp. 21–25 in J. Shoshani and P. Tassy (Eds.), *The Proboscidea: Evolution and Palaeoecology of Elephants and Their Relatives.* Oxford University Press, Oxford.
—. 1996b. Who is who among the Proboscidea?; pp. 39–48 in J. Shoshani and P. Tassy (Eds.), *The Proboscidea: Evolution and Palaeoecology of Elephants and Their Relatives.* Oxford University Press, Oxford.
—. 1996c. Growth and sexual dimorphism among Miocene elephantoids: The example of *Gomphotherium angustidens*; pp. 92–100 in J. Shoshani and P. Tassy (Eds.), *The Proboscidea: Evolution and Palaeoecology of Elephants and Their Relatives.* Oxford University Press, Oxford.
—. 1996c. The earliest gomphotheres; pp. 89–91 in J. Shoshani and P. Tassy (Eds.), *The Proboscidea: Evolution and Palaeoecology of Elephants and Their Relatives.* Oxford University Press, Oxford.
—. 1999. Miocene elephantids (Mammalia) from the Emirate of Abu Dhabi, United Arab Emirates: Palaeobiogeographic implications; pp. 209–233 in P. J. Whybrow and A. Hill (Eds.), *Fossil Vertebrates of Arabia. With Emphasis on the late Miocene Faunas, Geology, and Palaeoenvironments of the Emirate of Abu Dhabi, United Arab Emirates.* Yale University Press, New Haven.
—. 2003. Elephantoidea from Lothagam; pp. 331–358 in M. G. Leakey and J. M. Harris (Eds.), *Lothagam: The Dawn of Humanity in Eastern Africa.* Columbia University Press, New York.
—. 2013. L'anatomie cranio-mandibulaire de *Gomphotherium angustidens* (Cuvier, 1817) (Proboscidea, Mammalia): données issues du gisement d'En Péjouan (Miocène moyen du Gers, France). *Geodiversitas* 35:377–445.
—. 2014. L'odontologie de *Gomphotherium angustidens* (Cuvier, 1817) (Proboscidea, Mammalia): données issues du gisement d'En Péjouan (Miocène moyen du Gers, France). *Geodiversitas* 36:35–115.
—. 2018. Remarks on the cranium of *Eozygodon morotoensis* (Proboscidea, Mammalia) from the early Miocene of Africa, and the question of the monophyly of Elephantimorpha. *Revue de Paléobiologie, Genève* 37:592–607.
Tassy, P., and P. Darlu. 1986. Les Elephantidae: Nouveau regard sur les analyses de parcimonie. *Geobios* 20:487–494.
Tassy, P., and R. Debruyne. 2001. The timing of early Elephantinae differentiation: The palaeontological record, with a short comment on molecular data; pp. 685–687 in G. Cavarretta, P. Gioia, M. Mussi, and M.R. Palombo (Eds.), *Proceedings of the First International Congress of La Terra degli Elefanti: The World of Elephants.* Consiglio Nazionale delle Richerche, Rome.
Tassy, P., and M. Pickford. 1983. Un nouveau Mastodonte zygolophodonte (Proboscidea, Mammalia) dans le Miocène Inférieur d'Afrique Orientale: Systématique et Paléonenvironnement. *Geobios* 16:53–77.
Tassy, P., and J. Shoshani. 1988. The Tethytheria: Elephants and their relatives; pp. 283–315 in M. J. Benton (Ed.), *The Phylogeny and Classification of the Tetrapods, Vol. 2: Mammals.* Clarendon, Oxford.
Tassy, P., P. Anupandhanant, L. Ginsburg, P. Mein, B. Ratanasthien, and V. Suteethorn. 1992. A new *Stegolophodon* (Proboscidea, Mammalia) from the early Miocene of northern Thailand. *Geobios* 25:511–523.
Tassy, P., F. Goussard, and M. G. Sanz. 2013. The status of *Mastodon angustidens pygmaeus* Deperet, 1897 (Proboscidea, Mammalia): The contribution of X-ray tomography. *Geobios* 46:329–334.

Tchamba, M. N., and P. M. Seme. 1993. Diet and feeding behaviour of the forest elephant in the Santchou Reserve, Cameroon. *African Journal of Ecology* 31:165–171.

Thasod, Y., P. Jintasakul, and B. Ratanasthien. 2012. Proboscidean fossil from the Tha Chang Sand Pits, Nakhon Ratchasima Province, Thailand. *Journal of Science and Technology Mahasarakham University* 31:33–39.

Thomas, H., and G. Petter. 1986. Révision de la faune de mammifères du Miocène supérieur de Manacer (ex-Marceau), Algérie: Discussion sur l'âge du gisement. *Geobios* 19:357–373.

Thomas, H., D. Geraads, D. Janjou, D. Vaslet, A. Memesh, D. Billiou, H. Bocherons, G. Dobigny, V. Eisenmann, M. Gayet, F. de Lapparent de Broin; G. Petter, and M. Halawani. 1998. First Pleistocene faunas from the Arabian Peninsula, Nafud Desert, Saudi Arabia. *Comptes rendus de l'Académie des Sciences, Paris* 326:145–152.

Thomas, H., J. Roger, S. Sen, C. Bourdillon-de-Grissac, and Z. Al-Sulaimani. 1989. Découverte de vertébrés fossiles dans l'Oligocène inférieur du Dhofar (Sultanat d'Oman). *Geobios* 22:101–120.

Thomas, H., J. Roger, S. Sen, M. Pickford, E. Gheerbrant, Z. Al-Sulaimani, and S. Al-Busaidi. 1999. Oligocene and Miocene terrestrial vertebrates in the southern Arabian Peninsula (Sultanate of Oman) and their geodynamic and palaeogeographic settings; pp. 430–442 in P. J. Whybrow and A. Hill (Eds.), *Fossil Vertebrates of Arabia*. Yale University Press, New Haven.

Thomas, H., S. Sen, M. Khan, B. Battail, and G. Ligabue. 1982. The Lower Miocene fauna of Al-Sarrar (Eastern province, Saudi Arabia). *Atlal* 5:109–136.

Thomas, M. G., and A. M. Lister. 2001. A statistical appraisal of molecular and morphological evidence for mammoth-elephant relationships; pp. 688–692 in G. Cavarretta, P. Gioia, M. Mussi, and M. R. Palombo (Eds.), *Proceedings of the First International Congress of La Terra degli Elefanti: The World of Elephants*. Consiglio Nazionale delle Richerche, Rome.

Thomas, M. G., E. Hagelberg, H. B. Jones, Z. Yang, and A. M. Lister. 2000. Molecular and morphological evidence on the phylogeny of the Elephantidae. *Proceedings of the Royal Society of London, B* 267:2493–2500.

Thouless, C. R., H. T. Dublin, J. J. Blanc, D. P. Skinner, T. E. Daniel, R. D. Taylor, F. Maisels, H. L. Frederick, and P. Bouché. 2016. IUCN African elephant status report 2016. An update from the African elephant database. *Occasional Paper of the IUCN Species Survival Commission* no. 60:1–317.

Tibuleac, P. 2014. Presence of the genus *Choerolophodon* (Proboscidea: Mammalia) within the Moldavian Platform framework (Eastern Carpathians Foreland, Romania). *Analele Stiintifice ale Universitatii "Al. I. Cuza" din lasi Seria Geologie* 60:55–67.

Tibuleac, P., T. Obada, and I. Cojocaru. 2015. *Tetralophodon longirostris* (Mammalia: Proboscidea) in the eastern Carapathians Foreland (Romania and the Republic of Moldova). *North-Western Journal of Zoology* 11:138–150.

Tiercelin, J.-J., J. Michaux, and Y. Bandet. 1979. Le Miocène supérieur du Sud de la Dépression de l'Afar, Éthiopie: Sédiments, faunes, âges isotopiques. *Bulletin de la Société Géologique de France* 21:255–258.

Tierney, J. E., P. B. deMenocal, and P. D. Zander. 2017. A climatic context for the out-of-Africa migration. *Geology*. https://doi.org/10.1130/G39457.1

Tiffney, B. H. 1991. Paleoenvironment of the Oligocene Jebel Qatrani Formation, Fayum Depression, northern Egypt, based on floral remains. *Geological Society of America Abstracts* 23:456.

Tiffney, B. H., J. G. Fleagle, and T. M. Bown. 1994. Early to middle Miocene angiosperm fruits and seeds from Fejej, Ethiopia. *Tertiary Research* 15:25–42.

Tillet, T. 1985. The Palaeolithic and its environment in the northern part of the Chad basin. *The African Archaeological Review* 3:163–177.

Tipple, B. J., and M. Pagani. 2007. The early origins of terrestrial C_4 photosynthesis. *Annual Review Earth and Planetary Sciences* 35:435–461.

Tiwari, B. N., B. C. Verma, and A. Bhandari. 2006. Record of *Prodeinotherium* (Proboscidea: Mammalia) from the mid-Tertiary Dharmsala Group of the Kangra Valley, NW Himalaya, India: Biochronologic and palaeobiogeographic implications. *Journal of the Palaeontological Society of India* 51:93–100.

Tobien, H. 1971. *Moeritherium, Palaeomastodon, Phiomia* aus dem Paläogen Nordafrikas und die Abstammung der Mastodonten (Proboscidea, Mammalia). *Mitteilungen aus dem Geologischen Institut der Technischen Universität Hannover* 10:141–163.

—. 1973a. On the evolution of mastodonts (Proboscidea, Mammalia). Part 1: The bunodont trilophodont group. *Notizblatt des Hessischen Landesamtes für Bodenfurschung zu Wiesbaden* 101:202–276.

—. 1973b. The structure of the mastodont molar (Proboscidea, Mammalia): Part I. The bunodont pattern. *Mainzer Schafliche Mitteilungen* 2:115–147.

—. 1975. The structure of the mastodont molar (Proboscidea, Mammalia). Part 2: The zygodont and zygobunodont patterns. *Mainzer Geowissenschaften Mitteilungsblatt* 4:195–233.

—. 1978a. The structure of the mastodont molar (Proboscidea, Mammalia). Part 3: The Oligocene mastodont genera *Palaeomastodon, Phiomia* and the Eo/Oligocene Paenungulate *Moeritherium*. *Mainzer Geowissenschaftliche Mitteilungen* 6:177–208.

—. 1978b. On the evolution of mastodonts (Proboscidea, Mammalia). Part 2: The bunodont tetralophodont groups. *Geologisches Jahrbuch Hessen* 106:159–208.

—. 1980a. A note on the skull and mandible of a new choerolophodont mastodont (Proboscidea, Mammalia) from the middle Miocene of Chios (Aegean Sea, Greece); pp. 299–307 in L. L. Jacobs (Ed.), *Aspects of Vertebrate History: Essays in Honor of Edwin Harris Colbert*. Museum of Northern Arizona Press, Flagstaff.

—. 1980b. A note on the mastodont taxa (Proboscidea, Mammalia) of the "Dinotheriensande" (Upper Miocene, Rheinhessen, Federal Republic of Germany). *Mainzer Geowissenschaften Mitteilungen* 9:187–201.

—. 1988. Les Proboscidiens Deinotheriidae. *Palaeovertebrata*, Mémoire Extraordinaire 1988:135–175.

—. 1996. Evolution of zygodons with emphasis on dentitions; pp. 76–85 in J. Shoshani and P. Tassy (Eds.), *The Proboscidea: Evolution and Palaeocology of Elephants and Their Relatives*. Oxford University Press, Oxford.

Tobien, H., G. Chen, and Y. Li. 1986. Mastodonts (Proboscidea, Mammalia) from the late Neogene and early Pleistocene of the People's Republic of China Part 1: Historical account: The genera *Gomphotherium, Choerolophodon, Synconolophus, Amebelodon, Platybelodon, Sinomastodon*. *Mainzer Geowissenschaftliche Mitteilungen* 15:119–181.

—. 1988. Mastodonts (Proboscidea, Mammalia) from the late Neogene and early Pleistocene of the People's Republic of China. Part 2: the genera *Tetralophodon, Anancus, Stegotetrabelodon, Zygolophodon, Mammut, Stegolophodon*; some generalities on the Chinese mastodonts. *Mainzer Geowissenschaften Mitteilungen* 17:95–220.

Todd, N. E. 2001. African *Elephas recki*: Time, space and taxonomy; pp. 693–697 in G. Cavarretta, P. Gioia, M. Mussi, and M. R. Palombo (Eds.), *Proceedings of the First International Congress of la Terra degli Elefanti: The World of Elephants*. Consiglio Nazionale delle Richerche, Rome.

—. 2005. Reanalysis of African *Elephas recki*: Implications for time, space and taxonomy. *Quaternary International* 126–128:65–72.

—. 2010a. New phylogenetic analysis of the Family Elephantidae based on cranio-dental morphology. *Anatomical Record* 293:74–90.

—. 2010b. Qualitative comparison of the craniodental osteology of the extant elephants, *Elephas maximus* (Asian elephant) and *Loxodonta africana* (African elephant). *The Anatomical Record* 293:62–73.

Todd, N. E., and V. L. Roth. 1996. Origin and radiation of the Elephantidae; pp. 193–202 in J. Shoshani and P. Tassy (Eds.), *The Proboscidea: Evolution and Palaeoecology of Elephants and Their Relatives*. Oxford University Press, Oxford.

Tsoukala, E. S., and J. K. Melentis. 1994. *Deinotherium giganteum* Kaup (Proboscidea) from Kassandra Peninsula (Chalkidiki), Macedonia, Greece. *Geobios* 27: 633–640.

Tsujikawa, H. 2005a. The palaeoenvironment of *Samburupithecus kiptalami* based on its associated fauna. *African Study Monographs* suppl. 32:51–62.

—. 2005b. The updated late Miocene large mammal fauna from Samburu Hills, northern Kenya. *African Study Monographs* suppl. 32:1–50.

Tsujikawa, H., and M. Pickford. 2006. Additional specimens of Hyracoidea (Mammalia) from the early and midde Miocene of Kenya. *Annales de Paléontologie* 92:1–12.

Turkalo, A., and R. Barnes. 2013. *Loxodonta cyclotis*. Forest elephant; pp. 195–200 in J. Kingdon, D. Happold, M. Hoffmann, T. Butynski, M. Happold, and J. Kalina (Eds.), *Mammals of Africa, Volume I*. Bloomsbury, London.

Underwood. 2017. A new Oligocene site with terrestrial mammals and a selachian fauna from Minqar Tabaghbagh, the Western Desert of Egypt. *Acta Palaeontologica Polonica* 62:509–525.

Ungar, P. S. 2010. *Mammal Teeth. Origins, Evolution, and Diversity*. The Johns Hopkins University Press, Baltimore. 304 pp.

Uno, K., and F. Bibi. 2022. Stable isotope paleoecology of the Baynunah Formation, pp. 299–331 in F. Bibi, B. Krarz, M. Beech, and A. Hill (Eds.), *Sands of Time: Late Miocene Fossils from the Baynunah Formation, U.A.E.* Springer, Cham, Switzerland.

Uno, K. T., T. E. Cerling, J. M. Harris, Y. Kunimatsu, M. G. Leakey, M. Nakatsukasa, and H. Nakaya. 2011. Late Miocene to Pliocene carbon isotope record of differential diet change among East African herbivores. *Proceedings of the National Academy of Sciences, USA* 108:6509–6514.

Vacek, M. 1877. Uber osterreichische Mastodonten und Ihre Beziehungen zu den Mastodon-Arten Europas. *Abhandlungen der Kaiserliche-Königlichen geologischen Reichsantalt* 7:1–45.

Van Couvering, J. A. H., and J. A. Van Couvering. 1976. Early Miocene mammal fossils from East Africa: Aspects of geology, faunistics and palaeoecology; pp. 155–207 in G. L. Isaac and E. R. McCown (Eds.), *Human Origins*. W. A. Benjamin, Menlo Park.

Van der Made, J. 2010. The evolution of the elephants and their relatives in the context of a changing climate and geography; pp. 340–360 in D. Hohne and W. Schwarz (Eds.), *Elefantenreich—Eine Fossilwelt in Europa*. Landesamt für Denkmalplfege und Archälogie Sachsen-Anhalt & Landesmuseum für Vorgeschichte, Halle.

Van der Made, J., and A. V. Mazo. 2003. Proboscidean dispersals from Africa towards western Europe; in J. W. F. Reumer, J. De Vos, and D. Mol (Eds.), *Advances in Mammoth Research (Proceedings of the Second International Mammoth Conference, Rotterdam, May 16–20 1999), DEINSEA* 9:437–452.

Van der Maarel, F. H. 1932. Contributions to the knowledge of the fossil mammalian fauna of Java. *Wetenschappelijke Mededelingen Dienst van den Mijnbouw in Nederlandsch-Indië* 15:1–208.

Vanoverstraeten, M., J. Van Gysel, P. Tassy, B. Senut, and M. Pickford. 1990. Découverte d'une molaire éléphantine dans le Pliocène de la région d'Ishasha, Parc national des Virunga, sud du lac Édouard, Province du Kivu, Zaire. *Comptes Rendus de l'Académie des Sciences, Paris, Série II* 311:887–892.

Van Vliet, H. J., A. S. Schulp, G. A. M. M. Abu El-Kheir, T. M. Paijmans, M. Bosselaers, and C. J. Underwood. 2017. A new Oligocene site with terrestrial mammals and a selachian fauna from Minqar Tibaghbagh, the Western Desert of Egypt. *Acta Palaeontologica Polonica* 62:509–525.

Van Zinderen Bakker, E. M., and J. H. Mercer. 1986. Major late Cainozoic climatic events and palaeoenvironmental changes in Africa viewed in a world wide context. *Palaeogeography, Palaeoclimatology, Palaeoecology* 56:217–235.

Vaufrey, R. 1955. Proboscidiens fossils; pp. 784–767 in P. P. Grasse (Ed.), *Traité de Zoologie, vol. 17*. Masson, Paris.

Vaufrey, R. 1958. Proboscidea, étude systématique; pp. 203–295 in J. Piveteau (Ed.), *Traité de Paléontologie, Tome 6, Volume 2*. Masson & Cie, Paris.

Venter, P. J., and J. J. Hanekom. 2010. Automatic detection of African elephant (*Loxodonta africana*) infrasonic vocalisations from recordings. *Biosystems Engineering* 106:286–294.

Vialli, V. 1966. Sul rinvenimento de dinoterio (*Deinotherium* cf. *hobleyi* Andrews) nelle lignite di Adi Ugri (Eritrea). *Giornale di Geologia (Bologna)*, ser. 2, 33:447–458.

Vidya, T. N. C., P. Fernando, D. J. Melnick, and R. Sukumar. 2005. Population differentiation within and among Asian elephant (*Elephas maximus*) populations in southern India. *Heredity* 94:71–80.

Vignaud, P., P. Duringer, H. T. Mackaye, A. Likius, C. Blondel, J.-B. Boisserie, L. de Bonis, V. Eisenmann, M.-E. Etienne, D. Geraads, F. Guy, T. Lehmann, F. Lihoureau, N. Lopez-Martinez, C. Mourer-Chauviré, O. Otero, J.-C. Rage, M. Schuster, L. Viriot, A. Zazzo, and M. Brunet. 2002. Geology and palaeontology of the Upper Miocene Toros-Menalla hominid locality, Chad. *Nature* 418:152–155.

Walker, A. 1969. Fossil mammal locality on Mount Elgon, eastern Uganda. *Nature* 223:591–593.

Walker, C. A., and G. J. Dyke. 2006. New records of fossil birds of prey from the Miocene of Kenya. *Historical Biology* 18:95–98.

Wang, S.-Q. 2014. *Gomphotherium inopinatum*, a basal *Gomphotherium* species from the Linxia Basin, China, and other Chinese members of the genus. *Vertebrata PalAsiatica* 52:183–200.

Wang, S.-Q., and T. Deng. 2011. The first *Choerolophodon* (Proboscidea, Gomphotheriidae) skull from China. *Science China Earth Sciences* 54:1326–1337.

Wang, S.-Q., T. Deng, T. Tang, G. Xie, Y. Zhang, and D. Wang. 2015. Evolution of *Protanancus* (Proboscidea, Mammalia) in East Asia. *Journal of Vertebrate Paleontology* 35:e881830.

Wang, S.-Q., T. Deng, J. Ye, W. He, and S.-Q. Chen. 2016. Morphological and ecological diversity of Amebelodontidae (Proboscidea, Mammalia) revealed by a Miocene fossil accumulation of an upper-tuskless proboscidean. *Journal of Systematic Palaeontology*. http://dx.doi.org/10.1080/14772019.2016.1208687

Wang, S.-Q., Y. Li, J. Duangkrayom, X.-W. Yang, W. He, and S.-Q. Chen. 2017a. A new species of *Gomphotherium* (Proboscidea, Mammalia) from China and the evolution of *Gomphotherium* in Eurasia. *Journal of Vertebrate Paleontology* 37:e1318284. https://doi.org/10.1080/02724634.2017.1318284

Wang, S.-Q., H. Saegusa, J. Duangkrayom, W. He, and S.-Q. Chen. 2017b. A new species of *Tetralophodon* from the Linxia Basin and the biostratigraphic significance of tetralophodont gomphotheres from the Upper Miocene of northern China. *Palaeoworld* 26:703–717.

Ward, C. 2002. Interpreting the posture and locomotion of *Australopithecus afarensis*: Where do we stand? *Yearbook of Physical Anthropology* 45:185–215.

Ward, C., M. Leakey, and A. Walker. 1999. The new hominid species *Australopithecus anamensis*. *Evolutionary Anthropology* 7:197–205.

—. 2001. Morphology of *Australopithecus anamensis* from Kanapoi and Allia Bay, Kenya. *Journal of Human Evolution* 41:255–368.

Ward, S., B. Brown, A. Hill, J. Kelley, and W. Downs. 1999. *Equatorius*: A new hominoid genus from the middle Miocene of Kenya. *Science* 285:1382–1386.

Warny, S. A., P. J. Bart, and J.-P. Suc. 2003. Timing and progression of climatic, tectonic and glacioeustatic influences on the Messinian Salinity Crisis. *Palaeogeography, Palaeoclimatology, Palaeoecology* 202:59–66.

Watkins, R. T. 1989. The Buluk Member, a fossil hominoid-bearing sedimentary sequence of Miocene age from northern Kenya. *Journal of African Earth Sciences* 8:107–112.

Weinsheimer, O. 1883. Über *Dinotherium giganteum* Kaup. *Palaeontologische Abhandlungen* 1:205–282.

Weithofer, K. A. 1889. Ueber die Tertiären Landsäugetiere Italiens. *Jahrbuch der Kaiserlich-Köninglichen Geologischen Reichsanstalt* 39:55–82.

Welcomme, J. L. 1993. Le proboscidien du Mazet et les faunes de vertébrés continentaux du Burdigalien de la Basse Vallée de l'Hérault (Languedoc, Sud de la France). Diplôme de l'Ecole Pratique des Hautes Études, Montpellier (unpublished).

—. 1994. Le plus ancient crâne de proboscidien d'Europe, *Gomphotherium hannibali* nov. sp. (Proboscidea, Mammalia) du Miocène inférieur du Languedoc (France). *Comptes Rendus de l'Académie des Sciences. Série 2. Sciences de la Terre et des Planètes* 319:135–140.

Werdelin, L. 2010. Chronology of Neogene mammal localities; pp. 27–44 in L. Werdelin and W. J. Sanders (Eds.), *Cenozoic Mammals of Africa*. University of California Press, Berkeley.

Werdelin, L., and W. J. Sanders. 2010. *Cenozoic Mammals of Africa*. University of California Press, Berkeley, 986 pp.

Wessels, W., O. Fejfar, P. Peláez-Campomanes, A. van der Meulen, and H. de Bruijn. 2003. Miocene small mammals from Jebel Zelten, Libya. *Coloquios de Palaeontología* Vol. Ext. 1:699–715.

Western, D. 1989. The ecological role of elephants in Africa. *Pachyderm* 12:45–55.

Western, D., C. Moss, and N. Georgiadis. 1983. Age estimation and population age structure of elephants from footprint dimensions. *The Journal of Wildlife Management* 47:1192–1197.

White, L. J. T. 1994. *Sacoglottis gabonensis* fruiting and the seasonal movements of elephants in the Lope Reserve, Gabon. *Journal of Tropical Ecology* 10:121–125.

White, L. J. T., C. Tutin, and M. Fernandez. 1993. Group composition and diet of forest elephants, *Loxodonta africana cyclotis* Matschie, 1900, in the Lope Reserve, Gabon. *African Journal of Ecology* 31:181–199.

White, R., and McKenzie, D. 1989. Magmatism at rift zones: The generation of volcanic continental margins and flood basalts. *Journal of Geological Research Solid Earth* 94:7685–7729.

White, T. D., and G. Suwa. 1987. Hominid footprints at Laetoli: Facts and interpretations. *American Journal of Physical Anthropology* 72:485–514.

White, T. D., B. Asfaw, D. DeGusta, H. Gilbert, G. D. Richards, G. Suwa, and F. C. Howell. 2003. Pleistocene *Homo sapiens* from Middle Awash, Ethiopia. *Nature* 423:742–747.

White, T. D., R. V. Moore, and G. Suwa. 1984. Hadar biostratigraphy and hominid evolution. *Journal of Vertebrate Paleontology* 4:575–583.

White, T. D., G. WoldeGabriel, B. Asfaw, S. Ambrose, Y. Beyene, R. L. Bernor, J.-R. Boisserie, B. Currie, H. Gilbert, Y. Haile-Selassie, W. K. Hart, L. J. Hlusko, F. C. Howell, R. T. Kono, T. Lehmann, A. Louchart, C. O. Lovejoy, P. R. Renne, H. Saegusa, E. S. Vrba, H. Wesselman, and G. Suwa. 2006. Asa Issie, Aramis and the origin of *Australopithecus*. *Nature* 440:883–889.

Whitehouse, A. M. 2006. Tusklessness in the elephant population of Addo Elephant National Park, South Africa. *Journal of Zoology* 257:249–254.

Whybrow, P. J., and D. Clements. 1999. Arabian Tertiary fauna, flora, and localities; pp. 460–473 in P. J. Whybrow and A. Hill (Eds.), *Fossil Vertebrates of Arabia*. Yale University Press, New Haven.

Whybrow, P. J., and A. Hill. 1999. Introduction to Fossil Vertebrates of Arabia; pp. 3-6 in P. J. Whybrow and A. Hill (Eds.), *Fossil Vertebrates of Arabia*. Yale University Press, New Haven.

Whybrow, P. J., A. Hill, W. Y. al-Tikriti, and E. A. Hailwood. 1990. Late Miocene primate fauna, flora and initial paleomagnetic data from the Emirate of Abu Dhabi, United Arab Emirates. *Journal of Human Evolution* 19:583–588.

Whybrow, P. J., H. A. McClure, and G. F. Elliot. 1987. Miocene stratigraphy, geology and flora (Algae) of eastern Saudi Arabia and the Ad Dabtiyah vertebrate locality. *Bulletin of the British Museum (Natural History)* 41:371–382.

Whyte, I. J., and A. Hall-Martin. 2017–2018. Growth characteristics of tusks of elephants in Kruger National Park. *Pachyderm* 59:31–40.

Wichura, H., R. Bousquet, R. Oberhänsli, M. R. Strecker, and M. H. Trauth. 2010. Evidence for middle Miocene uplift of the East African Plateau. *Geology* 38:543–546.

Wight, A. W. R. 1971. Palaeogene Vertebrates from Libya. Unpublished Ph.D. dissertation, University of Bristol.

—. 1980. Palaeogene vertebrate fauna and regressive sediments of Dur at Talhah, southern Sirt Basin, Libya; pp. 309–325 in M. J. Salem and M. T. Busrewil (Eds.), *The Geology of Libya, Vol. 1*. Academic Press, London.

Williams, B. A., R. F. Kay, and E. C. Kirk. 2010. New perspectives on anthropoid origins. *Proceedings of the National Academy of Sciences, USA* 107:4797–4804.

Williams, C., S. K. Tiwari, V. R. Goswami, S. de Silva, A. Kumar, N. Baskaran, K. Yoganand, and V. Menon. 2020. *Elephas maximus. IUCN Red List of Threatened Species*. https://doi.org/10.2305/IUCN.UK.2020-3.RLTS.T7140A45818198.en

Williamson, B. R. 1975. The condition and nutrition of elephant in Wankie National Park. *Arnoldia Rhodesia* 7:1–20.

Williamson, P. G., and R. J. G. Savage. 1986. Early rift sedimentation in the Turkana basin, northern Kenya; pp. 267–283 in L. E. Frostick, R. W. Renaut, I. Reid, and J. J. Tiercelin (Eds.), *Sedimentation in the African Rifts*. Geological Society Special Publication No. 25, London.

Wing, L. D., and I. O. Buss. 1970. Elephants and forests. *Wildlife Monographs* 19:3–92.

Winkler, A. J. 1992. Systematics and biogeography of middle Miocene rodents from the Muruyur Beds, Baringo District, Kenya. *Journal of Vertebrate Paleontology* 12:236–249.

Winkler, A. J., C. Denys, and D. M. Avery. 2010. Rodentia; pp. 263–304 in L. Werdelin and W. J. Sanders (Eds.), *Cenozoic Mammals of Africa*. University of California Press, Berkeley.

WoldeGabriel, G., W. H. Gilbert, W. K. Hart, P. R. Renne, and S. H. Ambrose. 2008. Geology and geochronology; pp. 13–43 in W. H. Gilbert and B. Asfaw (Eds.), *Homo Erectus. Pleistocene Evidence from the Middle Awash, Ethiopia*. University of California Press, Berkeley.

WoldeGabriel, G., Y. Haile-Selassie, P. R. Renne, W. K. Hart, S. H. Ambrose, B. Asfaw, G. Helken, and T. White. 2001. Geology and palaeontology of the late Miocene Middle Awash Valley, Afar Rift, Ethiopia. *Nature* 412:175–178.

WoldeGabriel, G., W. K. Hart, S. Katoh, Y. Beyene, and G. Suwa. 2005. Correlation of Plio-Pleistocene tephra in Ethiopian and Kenyan rift basins: Temporal calibration of geological features and hominid fossil records. *Journal of Volcanology and Geothermal Research* 147:81–108.

WoldeGabriel, G., T. D. White, G. Suwa, P. Renne, J. de Heinzelin, W. K. Hart, and G. Helken. 1994. Ecological and temporal placement of early Pliocene hominids at Aramis, Ethiopia. *Nature* 371:330–333.

WoldeGabriel, G., T. Yemane, G. Suwa, T. White, and B. Asfaw. 1991. Age of volcanism and rifting in the Burji-Soyoma area, Amaro Horst, southern Main Ethiopian Rift: Geo- and biochronologic data. *Journal of African Earth Sciences* 13:437–447.

Wood, B. 2002. Hominid revelations from Chad. *Nature* 418:133–135.

Woolley, L.-A., R. L. Mackey, B. R. Page, and R. Slotow. 2008. Modelling the effect of age-specific mortality on elephant *Loxodonta africana* populations: Can natural mortality provide regulation? *Oryx* 42:49–57.

Wright, D. K., J. Thompson, A. Mackay, M. Welling, S. L. Forman, G. Price, J.-X. Zhao, A. S. Cohen, O. Malijani, and E. Gomani-Chindebvu. 2014. Renewed geoarchaeological investigations of Mwanganda's Village (Elephant Butchery Site), Karonga, Malawi. *Geoarchaeology: An International Journal* 29:98–120.

Wynn, J. G. 2000. Paleosols, stable carbon isotopes, and paleoenvironmental interpretation of Kanapoi, northern Kenya. *Journal of Human Evolution* 39:411–432.

Wynn, J. G., and G. J. Retallack. 2001. Paleoenvironmental reconstruction of middle Miocene paleosols bearing *Kenyapithecus* and *Victoriapithecus*, Nyakach Formation, southwestern Kenya. *Journal of Human Evolution* 40:263–288.

Wynn, J. G., Z. Alemseged, R. Bobe, D. Geraads, D. Reed, and D. C. Roman. 2006. Geological and palaeontological context of a Pliocene juvenile hominin at Dikika, Ethiopia. *Nature* 443:332–336.

Wynn, J. G., K. E. Reed, M. Sponheimer, W. H. Kimbel, Z. Alemseged, Z. K. Bedaso, and C. J. Campisano. 2016. Dietary flexibility of *Australopithecus afarensis* in the face of paleoecological change during the middle Pliocene: Faunal evidence from Hadar, Ethiopia. *Journal of Human Evolution* 99:93–106.

Wynn, J. G., M. Sponheimer, W. H. Kimbel, and J. N. Wilson. 2013. Diet of *Australopithecus afarensis* from the Pliocene Hadar Formation, Ethiopia. *Proceedings of the National Academy of Sciences, USA* 110:10495–10500.

Yanagisawa, Y. 2011. Diatom biostratigraphy of the lower Miocene Yunagaya Group, Joban Area, Japan. *Diatom* 27:33–45.

Yang, H. E. M. Golenberg, and J. Shoshani. 1996. Phylogenetic resolution within the Elephantidae using fossil DNA sequence from the American mastodon (*Mammut americanum*) as an outgroup. *Proceedings of the National Academy of Sciences, USA* 93:1190–1194.

Yans, J., M. Amaghzaz, B. Bouya, H. Cappetta, P. Iacumin, L. Kocsis, M. Mouflih, O. Selloum, S. Sen, J.-Y. Storme, and E. Gheerbrant. 2014. First carbon isotope chemostratigraphy of the Ouled Abdoun phosphate Basin, Morocco; implications for dating and evolution of earliest African placental mammals. *Gondwana Research* 25:257–269.

Yasui, K., Y. Kunimatsu, N. Kuga, B. Bajope, and H. Ishida. 1992. Fossil mammals from the Neogene strata in the Sinda Basin, Eastern Zaire. *African Study Monographs* supplement 17:87–107.

Yoder, A. D. 2013. Fossils versus clocks. *Science* 339:656–658.

You, Y., M. Huber, R. D. Müller, C. J. Poulsen, and J. Ribbe. 2009. Simulation of the middle Miocene climate optimum. *Geophysical Research Letters* 36:L04702. https://doi.org/10.1029/2008GL036571

Youcef, S. 2020. African origins of modern asses as seen from paleontology and DNA: What about the Atlas wild ass? *Geobios* 58:73–84.

Zachos, J., M. Pagani, L. Sloan, E. Thomas, and K. Billups. 2001. Trends, rhythms, and abberations in global climate 65 Ma to present. *Science* 292:686–693.

Zalmout, I., W. J. Sanders, L. M. MacLatchy, G. F. Gunnell, Y. A. Al-Mufarreh, A. Mohammad, M. A. Ali, A.-A. Nasser, A. M. Al-Masari, S. A. Al-Sobhi, A. O. Nadhra, A. J. Matari, J. A. Wilson, and P. D. Gingerich. 2010. New Oligocene primate from Saudi Arabia and the divergence of apes and Old World monkeys. *Nature* 466:360–364.

Zanettin, B., G. Bellieni, and E. J. Visentin. 2006. New radiometric ages of volcanic rocks in the central Eritrean plateau (from Asmara to Adi Quala): Considerations on stratigraphy and correlations. *Journal of African Earth Sciences* 45:156–161.

Zazzo, A., H. Bocherens, M. Brunet, A. Beauvilain, D. Billiou, H. T. Mackaye, P. Vignaud, and A. Mariotti. 2000. Herbivore paleodiet and paleoenvironmental changes in Chad during the Pliocene using stable isotope ratios of tooth enamel carbonate. *Paleobiology* 26:294–309.

Zhang, H. 2020. *Evolution and Systematics of the Elephantidae (Mammalia, Proboscidea) from the Late Miocene to Recent*. Unpublished doctoral dissertation, University of Bristol, 388 pp.

Zhang, H., and A. M. Lister. 2017. A review of *Loxodonta atlantica* (Pomel, 1879) based on newly identified materials from the Museum national d'Histoire naturelle (Paris, France). *Programme and Abstracts, Annual Meeting, Palaeontological Association* 2017:118.

Zhang, R., V. A. Kravchinsky, and L. Yue. 2012. Link between global cooling and mammalian transformation across the Eocene-Oligocene boundary in the continental interior of Asia. *International Journal of Earth Science* 101:2193–2200.

Zhang, X., and S. Wang. 2020. First report of *Eozygodon* (Mammutidae, Proboscidea) in Eurasia. *Historical Biology*. https://doi.org/10.1080/08912963.2020.1723579

Zhou, M., and Y. Zhang. 1983. Occurrence of the proboscidean genus *Stegotetrabelodon* in China. *Vertebrata PalAsiatica* 21:52–58.

Ziegler, S., S. Merker, B. Streit, M. Boner, and D. E. Jacob. 2016. Towards understanding isotope variability in elephant ivory to establish isotopic profiling and source area determination. *Biological Conservation* 197:154–163.

Zouhri, S., D. Geraads, S. El Boughabi, and A. El Harfi. 2012. Discovery of an Upper Miocene vertebrate fauna near Tizi N'Tadderht, Skoura, Ourzazate Basin (Central High Atlas, Morocco). *Comptes Rendus Palevol* 11:455–461.

Index

Note: **Bold** page numbers refer to tables and *italic* page numbers refer to figures.

For Product Safety Concerns and Information please contact our EU representative GPSR@taylorandfrancis.com Taylor & Francis Verlag GmbH, Kaufingerstraße 24, 80331 München, Germany

Batch number: 10731517

Printed by Printforce, the Netherlands